ADSORPTION ANALYSIS: EQUILIBRIA AND KINETICS

SERIES ON CHEMICAL ENGINEERING

Series Editor: Ralph T. Yang *(Univ. of Michigan)*
Advisory Board: Robert S. Langer *(Massachusetts Inst. of Tech.)*
Donald R. Paul *(Univ. of Texas)*
John M. Prausnitz *(Univ. of California, Berkeley)*
Eli Ruckenstein *(State Univ. of New York)*
James Wei *(Princeton Univ.)*

Vol. 1 Gas Separation by Adsorption Processes
Ralph T. Yang (Univ. of Michigan)

Forthcoming

Bulk Solids Mixing
János Gyenis (Hungary Acad. Sci.) and L T Fan (Kansas State Univ.)

ADSORPTION ANALYSIS: EQUILIBRIA AND KINETICS

Duong D. Do
Department of Chemical Engineering
University of Queensland
Queensland, Australia

Imperial College Press

Published by

Imperial College Press
203 Electrical Engineering Building
Imperial College
London SW7 2BT

Distributed by

World Scientific Publishing Co. Pte. Ltd.
P O Box 128, Farrer Road, Singapore 912805
USA office: Suite 1B, 1060 Main Street, River Edge, NJ 07661
UK office: 57 Shelton Street, Covent Garden, London WC2H 9HE

British Library Cataloguing-in-Publication Data
A catalogue record for this book is available from the British Library.

ADSORPTION ANALYSIS: EQUILIBRIA AND KINETICS

Copyright © 1998 by Imperial College Press

All rights reserved. This book, or parts thereof, may not be reproduced in any form or by any means, electronic or mechanical, including photocopying, recording or any information storage and retrieval system now known or to be invented, without written permission from the Publisher.

For photocopying of material in this volume, please pay a copying fee through the Copyright Clearance Center, Inc., 222 Rosewood Drive, Danvers, MA 01923, USA. In this case permission to photocopy is not required from the publisher.

ISBN 1-86094-130-3
ISBN 1-86094-137-0 (pbk)

This book is printed on acid-free paper.

Printed in Singapore by Uto-Print

Dedication

I dedicate this book to my parents.

Preface

The significant research in adsorption in the 70s through the 90s could be attributed to the discovery of many new porous materials, such as carbon molecular sieve, and the invention of many new clever processes, notably Pressure Swing Adsorption (PSA) processes. This evolution in adsorption research is reflected in many books on adsorption, such as the ones by Ruthven (1984), Yang (1987, 1997), Jaroniec and Madey (1988), Suzuki (1990), Karger and Ruthven (1992) and Rudzinski and Everett (1992). Conferences on adsorption are organized more often than before, such as the Fundamentals of Adsorption, the conference on Characterization of Porous Solids, the Gas Separation Technology symposium, the Symposium in Surface Heterogeneity, and the Pacific Rim workshop in Adsorption Science and Technology. The common denominator of these books and proceedings is the research on porous media since it is the heart for the understanding of diffusion and adsorption. Since porous media are very complex, the understanding of many practical solids is still far from complete, except solids exhibiting well defined structure such as synthetic zeolites. It is the complex interplay between the solid structure, diffusion and adsorption that makes the analysis of adsorption more complicated than any other traditional unit operations process such as distillation, etc.

Engineers dealing with adsorption processes, therefore, need to deal with model equations usually in the form of partial differential equation, because adsorption processes are inherently transient. To account for the details of the system, phenomena such as film diffusion, interparticle diffusion, intragrain diffusion, surface barrier and adsorption in addition to the complexities of solid structure must be allowed for. The books of Ruthven, Yang, and Suzuki provide excellent sources for engineers to fulfill this task. However, missing in these books are many recent results in studying heterogeneous solids, the mathematics in dealing with differential equations, the wider tabulation of adsorption solutions, and the many methods of measuring diffusivity. This present book will attempt to fill this gap. It starts with five chapters covering adsorption equilibria, from fundamental to practical approaches. Multicomponent equilibria of homogeneous as well as heterogeneous solids are also dealt with, since they are the cornerstone in designing separation systems.

After the few chapters on equilibria, we deal with kinetics of the various mass transport processes inside a porous particle. Conventional approaches as well as the new approach using Maxwell-Stefan equations are presented. Then the analysis of adsorption in a single particle is considered with emphasis on the role of solid structure. Next we cover the various methods to measure diffusivity, such as the Differential Adsorption Bed (DAB), the time lag, the diffusion cell, chromatography, and the batch adsorber methods.

It is our hope that this book will be used as a teaching book as well as a book for engineers who wish to carry out research in the adsorption area. To fulfill this niche, we have provided with the book many programming codes written in MatLab language so that readers can use them directly to understand the behaviour of single and multicomponent adsorption systems

Duong D. Do
University of Queensland
January 1998

Table of Contents

Preface vii
Table of contents ix

PART I: EQUILIBRIA

Chapter 1	Introduction	1
Chapter 2	Fundamentals of Pure Component Adsorption Equilibria	11
Chapter 3	Practical approaches of Pure Component Adsorption Equilibria	49
Chapter 4	Pure Component Adsorption in Microporous Solids	149
Chapter 5	Multicomponent Adsorption Equilibria	191
Chapter 6	Heterogeneous Adsorption Equilibria	249

PART II: KINETICS

Chapter 7	Fundamentals of Diffusion and Adsorption in Porous Media	337
Chapter 8	Diffusion in Porous Media: Maxwell-Stefan Approach	415
Chapter 9	Analysis of Adsorption Kinetics in a Single Homogeneous Particle	519
Chapter 10	Analysis of Adsorption Kinetics in a Zeolite Particle	603
Chapter 11	Analysis of Adsorption Kinetics in a Heterogeneous Particle	679

PART III: MEASUREMENT TECHNIQUES

Chapter 12	Time Lag in Diffusion and Adsorption in Porous Media	701
Chapter 13	Analysis of Steady State and Transient Diffusion Cells	755
Chapter 14	Adsorption and Diffusivity Measurement by a Chromatography Method	775
Chapter 15	Analysis of a Batch Adsorber	795

Table of Computer MatLab Programs 811
Nomenclature 815
Constants and Units Conversion 821
Appendices 825
References 879
Index 889

Detailed Table of Contents

PART I: EQUILIBRIA

CHAPTER 1 Introduction

1.1	Introduction	1
1.2	Basis of separation	1
1.3	Adsorbents	2
1.3.1	Alumina	3
1.3.2	Silica gel	3
1.3.3	Activated carbon	4
1.3.4	Zeolite	6
1.4	Adsorption processes	7
1.5	The structure of the book	7

CHAPTER 2 Fundamentals of Pure Component Adsorption Equilibria

2.1	Introduction	11
2.2	Langmuir equation	13
2.2.1	Basic theory	13
2.2.2	Isosteric heat of adsorption	17
2.3	Isotherms based on the Gibbs approach	18
2.3.1	Basic theory	19
2.3.2	Linear isotherm	22
2.3.3	Volmer isotherm	22
2.3.4	Hill-de Boer isotherm	24
2.3.5	Fowler-Guggenheim equation	26
2.3.6	Harkins-Jura isotherm	31
2.3.7	Other isotherms from Gibbs equation	34
2.4	Multisite occupancy model of Nitta	35
2.4.1	Estimation of the adsorbate-adsorbate interaction energy	37
2.4.2	Special case	38
2.4.3	Extension to multicomponent systems	39

2.5		Mobile adsorption model of Nitta et al.	39
2.6		Lattice vacancy theory	42
2.7		Vacancy solution theory (VSM)	43
	2.7.1	VSM-Wilson model	43
	2.7.2	VSM-Flory-Huggin model	44
	2.7.3	Isosteric heat of adsorption	45
2.8		2-D Equation of state (2D-EOS) adsorption isotherm	46
2.9		Concluding remarks	48
CHAPTER 3		**Practical Approaches of Pure Component Adsorption Equilibria**	
3.1		Introduction	49
3.2		Empirical isotherm equations	49
	3.2.1	Freundlich equation	50
	3.2.2	Sips equation (Langmuir-Freundlich)	57
	3.2.3	Toth equation	64
	3.2.4	Unilan equation	70
	3.2.5	Keller, Staudt and Toth's equation	76
	3.2.6	Dubinin-Radushkevich equation	77
	3.2.7	Jovanovich equation	82
	3.2.8	Temkin equation	82
	3.2.9	Summary of empirical equations	83
3.3		BET isotherm and modified BET isotherm	84
	3.3.1	BET equation	84
	3.3.2	Differential heat	94
	3.3.3	BDDT classification	94
	3.3.4	Comparison between van der Waals and the capillary condensation	99
	3.3.5	Other modified versions of the BET equation	99
	3.3.6	Aranovich's modified BET equations	101
3.4		Harkins-Jura, Halsey isotherms	103
3.5		Further discussion on the BET theory	104
	3.5.1	Critical of the BET theory	104
	3.5.2	Surface with adsorption energy higher than heat of liquefaction	107
3.6		FHH multilayer equation	107
3.7		Redhead's empirical isotherm	108
3.8		Summary of multilayer adsorption equations	110
3.9		Pore volume and pore size distribution	112
	3.9.1	Basic theory	112
3.10		Approaches for the pore size distribution determination	130
	3.10.1	Wheeler and Schull's method	130
	3.10.2	Cranston and Inkley's method	136

	3.10.3	De Boer method	140
3.11		Assessment of pore shape	142
	3.11.1	Hysteresis loop	142
	3.11.2	The t-method	143
	3.11.3	The α_S method	147
3.12		Conclusion	148

Chapter 4   Pure Component Adsorption in Microporous Solids

4.1		Introduction	149
	4.1.1	Experimental evidence of volume filling	150
	4.1.2	Dispersive forces	151
	4.1.3	Micropore filling theory	154
4.2		Dubinin equations	156
	4.2.1	Dubinin-Radushkevich equation	156
	4.2.2	Dubinin-Astakhov equation	159
	4.2.3	Isosteric heat of adsorption and heat of immersion	168
4.3		Theoretical basis of the potential adsorption isotherms	171
4.4		Modified Dubinin equations for inhomogeneous microporous solids	172
	4.4.1	Ideal inhomogeneous microporous solids	172
	4.4.2	Solids with distribution in characteristic energy E_0	173
4.5		Solids with micropore size distribution	183
	4.5.1	DR local isotherm and Gaussian distribution	185
	4.5.2	DA local isotherm and Gamma micropore size distribution	187
4.6		Other approaches	188
	4.6.1	Yang's approach	188
	4.6.2	Schlunder's approach	189
	4.6.3	Modified Antoine equation	189
4.7		Concluding remarks	190

CHAPTER 5   Multicomponent Adsorption Equilibria

5.1		Introduction	191
5.2		Langmurian multicomponent theory	191
	5.2.1	Kinetic approach	191
	5.2.2	Equilibrium approach	195
	5.2.3	Empirical approaches based on the Langmuir equation	197
5.3		Ideal adsorption solution theory	198
	5.3.1	The basic thermodynamic theory	198
	5.3.2	Myers and Prausnitz theory	201
	5.3.3	Practical considerations of the Myers-Prausnitz IAS equations	203
	5.3.4	The Lewis relationship	205
	5.3.5	General IAS algorithm: Specification of P and y	206

	5.3.6	Thermodynamic justification of the extended Langmuir equation	213
	5.3.7	Inverse IAS algorithm: Specification of $C_{\mu T}$ and x_i	216
	5.3.8	Numerical example of the IAS theory	217
5.4		Fast IAS theory	222
	5.4.1	Original fast IAS procedure	223
	5.4.2	Modified fast IAS procedure	227
	5.4.3	The initial guess for the hypothetical pure component pressure	230
	5.4.4	The amount adsorbed	231
	5.4.5	The FastIAS algorithm	231
	5.4.6	Other cases	233
	5.4.7	Summary	233
5.5		LeVan and Vermeulen approach for binary systems	234
	5.5.1	Pure component Langmuir isotherm	235
	5.5.2	Pure component Freundlich isotherm	239
5.6		Real adsorption solution theory	240
5.7		Multisite occupancy model of Nitta et al.	243
5.8		Mobile adsorption model of Nitta et al.	245
5.9		Potential theory	246
5.10		Other approaches	248
5.11		Conclusions	248

CHAPTER 6   Heterogeneous Adsorption Equilibria

6.1		Introduction	249
6.2		Langmuir approach	252
	6.2.1	Isosteric heat of adsorption	253
6.3		Energy distribution approach	257
	6.3.1	Random topography	257
	6.3.2	Patchwise topography	257
	6.3.3	The maximum adsorption capacity	258
	6.3.4	Other local adsorption isotherm & energy distribution	262
6.4		Isosteric heat	265
6.5		Brunauer, Love and Keenan approach	268
	6.5.1	BLK equation versus the Unilan equation	269
6.6		Hobson approach	270
6.7		DR/DA as local isotherm	273
6.8		Distribution of Henry constant	273
	6.8.1	The energy distribution	275
6.9		Distribution of free energy approach	276
	6.9.1	Water adsorption in activated carbon	277
	6.9.2	Hydrocarbon adsorption in activated carbon	280
6.10		Relationship between slit shape micropore and adsorption energy	282

	6.10.1	Two atoms or molecules interaction	282
	6.10.2	An atom or molecule and a lattice plane	284
	6.10.3	An atom or molecule and a slab	287
	6.10.4	A species and two parallel lattice planes	290
	6.10.5	A species and two parallel slabs	296
	6.10.6	Adsorption isotherm for slit shape pore	299
	6.10.7	An atom or molecule and two parallel lattice planes with sublattice layers	308
6.11		Horvarth and Kawazoe's approach on micropore size distribution	315
	6.11.1	The basic theory	315
	6.11.2	Differential heat	318
	6.11.3	Model parameters	318
	6.11.4	Applications	320
6.12		Cylindrical pores	322
	6.12.1	A molecule and a cylindrical surface	322
	6.12.2	A molecule and a cylindrical slab	326
	6.12.3	Adsorption in a cylindrical pore	328
6.13		Adsorption-condensation theory of Sircar	331
	6.13.1	Mesoporous solid	331
	6.13.2	Micropore-mesoporous solids	335
6.14		Conclusion	336

PART II KINETICS

CHAPTER 7 Fundamentals of Diffusion and Adsorption in Porous Media

7.1		Introduction	337
	7.1.1	Historical development	338
7.2		Devices used to measure diffusion in porous solids	339
	7.2.1	Graham's system	340
	7.2.2	Hoogschagen's system	341
	7.2.3	Graham and Loschmidt's systems	342
	7.2.4	Stefan tube	343
	7.2.5	Diffusion cell	344
7.3		Modes of transport	344
7.4		Knudsen diffusion	348
	7.4.1	Thin orifice	350
	7.4.2	Cylindrical capillary	352
	7.4.3	Converging or diverging capillary	359
	7.4.4	Porous solids	362
	7.4.5	Graham's law of effusion	367

7.5		Viscous flow	369
	7.5.1	Viscous flux in a capillary	369
	7.5.2	Porous media: Parallel capillaries model	372
	7.5.3	Porous media: Unconsolidated packed bed model	376
7.6		Transition between the viscous flow and Knudsen flow	380
	7.6.1	Extension from viscous flow analysis	381
	7.6.2	Steady state flow when viscous and slip mechanisms are operating	383
	7.6.3	Semi-empirical relation by Knudsen	384
	7.6.4	Porous media	386
7.7		Continuum diffusion	387
	7.7.1	Binary diffusivity	389
	7.7.2	Constitutive flux equation for a binary mixture in a capillary	391
	7.7.3	Porous medium	393
7.8		Combined bulk and Knudsen diffusion	394
	7.8.1	Uniform cylindrical capillary	394
	7.8.2	Porous solids	396
	7.8.3	Model for tortuosity	397
7.9		Surface diffusion	399
	7.9.1	Characteristics of surface diffusion	399
	7.9.2	Flux equation	401
	7.9.3	Temperature dependence of surface diffusivity	404
	7.9.4	Surface diffusion variation with pore size	405
	7.9.5	Surface diffusivity models	406
7.10		Concluding remarks	414
CHAPTER 8		**Diffusion in Porous Media: Maxwell-Stefan Approach**	
8.1		Introduction	415
8.2		Diffusion in ideal gaseous mixture	415
	8.2.1	Stefan-Maxwell equation for binary systems	416
	8.2.2	Stefan-Maxwell equation for ternary systems	421
	8.2.3	Stefan-Maxwell equation for the N-multicomponent system	422
	8.2.4	Stefan tube with binary system	431
	8.2.5	Stefan tube for ternary system	438
	8.2.6	Stefan tube with n component mixtures	442
8.3		Transient diffusion of ideal gaseous mixtures in Loschmidt's tube	449
	8.3.1	The mass balance equations	449
	8.3.2	The overall mass balance	452
	8.3.3	Numerical analysis	452
8.4		Transient diffusion of ideal gaseous mixtures in two bulb method	457
	8.4.1	The overall mass balance equation	458
	8.4.2	Nondimensionalization of the mass balance equations	459

8.5		Diffusion in nonideal fluids	462
	8.5.1	The driving force for diffusion	462
	8.5.2	The Maxwell-Stefan equation for nonideal fluids	463
	8.5.3	Special case: Ideal fluids	465
	8.5.4	Table of formula of constitutive relations	465
8.6		Maxwell-Stefan formulation for bulk-Knudsen diffusion in capillary	470
	8.6.1	Non-ideal systems	472
	8.6.2	Formulas for bulk-Knudsen diffusion case	474
	8.6.3	Steady state multicomponent system at constant pressure conditions	482
8.7		Stefan-Maxwell approach for bulk-Knudsen diffusion in complex ..	487
	8.7.1	Bundle of parallel capillaries	487
	8.7.2	Capillaries in series	490
	8.7.3	A simple pore network	493
8.8		Stefan-Maxwell approach for bulk-Knudsen-viscous flow	495
	8.8.1	The basic equation written in terms of fluxes N	496
	8.8.2	The basic equations written in terms of diffusive fluxes J	499
	8.8.3	Another form of basic equations in terms of N	502
	8.8.4	Limiting cases	502
8.9		Transient analysis of bulk-Knudsen-viscous flow in a capillary	510
	8.9.1	Nondimensional equations	511
8.10		Maxwell-Stefan for surface diffusion	515
	8.10.1	Surface diffusivity of single species	517
8.11		Conclusion	517

CHAPTER 9		**Analysis of Adsorption Kinetics in a Single Homogeneous Particle**	
9.1		Introduction	519
9.2		Adsorption models for isothermal single component systems	521
	9.2.1	Linear isotherms	521
	9.2.2	Nonlinear models	545
9.3		Adsorption model for nonisothermal single component systems	562
	9.3.1	Problem formulation	562
9.4		Finite kinetics adsorption model for single component systems	580
9.5		Multicomponent adsorption models for a porous solid: Isothermal	584
	9.5.1	Pore volume flux vector $\underline{N}_p$	585
	9.5.2	Flux vector in the adsorbed phase	586
	9.5.3	The working mass balance equation	589
	9.5.4	Nondimensionalization	590
9.6		Nonisothermal model for multicomponent systems	596
	9.6.1	The working mass and heat balance equations	599
	9.6.2	The working nondimensional mass and heat balance equations	600
	9.6.3	Extended Langmuir isotherm	601

9.7		Conclusion	602
CHAPTER 10		**Analysis of Adsorption Kinetics in a Zeolite Particle**	
10.1		Introduction	603
10.2		Single component micropore diffusion (Isothermal)	604
	10.2.1	The necessary flux equation	605
	10.2.2	The mass balance equation	608
10.3		Nonisothermal single component adsorption in a crystal	623
	10.3.1	Governing equations	624
	10.3.2	Nondimensional equations	625
	10.3.3	Langmuir isotherm	629
10.4		Bimodal diffusion models	634
	10.4.1	The length scale and the time scale of diffusion	635
	10.4.2	The mass balance equations	637
	10.4.3	Linear isotherm	639
	10.4.4	Irreversible isotherm	644
	10.4.5	Nonlinear isotherm and nonisothermal conditions	650
10.5		Multicomponent adsorption in an isothermal crystal	656
	10.5.1	Diffusion flux expression in a crystal	656
	10.5.2	The mass balance equation in a zeolite crystal	661
10.6		Multicomponent adsorption in a crystal: Nonisothermal	667
	10.6.1	Flux expression in a crystal	667
	10.6.2	The coupled mass and heat balance equations	670
10.7		Multicomponent adsorption in a zeolite pellet: Non isothermal	675
10.8		Conclusion	677
CHAPTER 11		**Analysis of Adsorption Kinetics in a Heterogeneous Particle**	
11.1		Introduction	679
11.2		Heterogeneous diffusion & sorption models	679
	11.2.1	Adsorption isotherm	679
	11.2.2	Constitutive flux equation	680
11.3		Formulation of the model for single component systems	683
	11.3.1	Simulations	686
11.4		Experimental section	689
	11.4.1	Adsorbent and gases	689
	11.4.2	Differential adsorption bed apparatus (DAB)	689
	11.4.3	Differential Adsorption Bed procedure	690
11.5		Results & Discussion	691
11.6		Formulation of sorption kinetics in multicomponent systems	694
	11.6.1	Adsorption isotherm	694
	11.6.2	Local flux of species k	696

	11.6.3	Mass balance equations	697
11.7		Micropore size distribution induced heterogeneity	698
11.8		Conclusions	699

PART III: MEASUREMENT TECHNIQUES

CHAPTER 12 — Time Lag in Diffusion and Adsorption in Porous Media

12.1		Introduction	701
12.2		Nonadsorbing gas with Knudsen flow	702
	12.2.1	Adsorption: Medium is initially free from adsorbate	705
	12.2.2	Medium initially contains diffusing molecules	716
12.3		Frisch's analysis (1957-1959) on time lag	718
	12.3.1	Adsorption	719
	12.3.2	General boundary conditions	723
12.4		Nonadsorbing gas with viscous flow	728
12.5		Time lag in porous media with adsorption	732
	12.5.1	Linear isotherm	732
	12.5.2	Finite adsorption	735
	12.5.3	Nonlinear isotherm	739
12.6		Further consideration of the time lag method	746
	12.6.1	Steady state concentration	747
	12.6.2	Functional dependence of the diffusion coefficient	748
	12.6.3	Further about time lag	750
12.7		Other considerations	753
12.8		Conclusion	754

CHAPTER 13 — Analysis of Steady State and Transient Diffusion Cells

13.1		Introduction	755
13.2		Wicke-Kallanbach diffusion cell	758
13.3		Transient diffusion cell	762
	13.3.1	Mass balance around the two chambers	763
	13.3.2	The type of perturbation	764
	13.3.3	Mass balance in the particle	765
	13.3.4	The moment analysis	769
	13.3.5	Moment analysis of non-adsorbing gas	770
	13.3.6	Moment analysis of adsorbing gas	773
13.4		Conclusion	774

CHAPTER 14 — Adsorption & Diffusivity Measurement by Chromatography Method

14.1		Introduction	775

14.2		The methodology	776
	14.2.1	The general formulation of mass balance equation	778
	14.2.2	The initial condition	779
	14.2.3	The moment method	780
14.3		Pore diffusion model with local equilibrium	781
	14.3.1	Parameter determination	782
	14.3.2	Quality of the chromatographic response	784
14.4		Parallel diffusion model with local equilibrium	786
14.5		Pore diffusion model with linear adsorption kinetics	786
14.6		Bi-dispersed solid with local equilibrium	787
	14.6.1	Uniform grain size	787
	14.6.2	Distribution of grain size	790
14.7		Bi-dispersed solid (alumina type) chromatography	791
14.8		Perturbation chromatography	793
14.9		Concluding remarks	794

CHAPTER 15   Analysis of Batch Adsorber

15.1		Introduction	795
15.2		The general formulation of mass balance equation	796
	15.2.1	The initial condition	797
	15.2.2	The overall mass balance equation	797
15.3		Pore diffusion model with local equilibrium	798
	15.3.1	Linear isotherm	804
	15.3.2	Irreversible adsorption isotherm	806
	15.3.3	Nonlinear adsorption isotherm	809
15.4		Concluding remarks	809

Table of Computer MatLab Programs — 811

Nomenclature — 815

Constants and Units Conversion — 821

Appendices — 825

Appendix 3.1: Isosteric heat of the Sips equation (3.2-18) — 825
Appendix 3.2: Isosteric heat of the Toth equation (3.2-19) — 826
Appendix 3.3: Isosteric heat of the Unilan equation (3.2-23) — 827
Appendix 6.1: Energy potential between a species and surface atoms — 828
Appendix 8.1: The momentum transfer of molecular collision — 829
Appendix 8.2: Solving the Stefan-Maxwell equations (8.2-97 and 8.2-98) — 831
Appendix 8.3: Collocation analysis of eqs. (8.3-16) and (8.3-17) — 833
Appendix 8.4: Collocation analysis of eqs. (8.4-13) to (8.4-15) — 838
Appendix 8.5: The correct form of the Stefan-Maxwell equation — 840

Appendix 8.6: Equivalence of two matrix functions — 842
Appendix 8.7: Alternative derivation of the basic equation for bulk-Knudsen-vis... — 843
Appendix 8.8: Derivation of eq.(8.8-19a) — 844
Appendix 8.9: Collocation analysis of model equations (8.9-10) — 846
Appendix 9.1: Collocation analysis of a diffusion equation (9.2-3) — 850
Appendix 9.2: The first ten eigenvalues for the three shapes of particle — 853
Appendix 9.3: Collocation analysis of eq. (9.2-47) — 854
Appendix 9.4: Collocation analysis of eqs. (9.3-19) — 856
Appendix 9.5: Mass exchange kinetics expressions — 858
Appendix 9.6: Collocation analysis of model equations (9.5-26) — 858
Appendix 9.7: Collocation analysis of eqs. (9.6-24) — 860
Appendix 10.1: Orthogonal collocation analysis of eqs. (10.2-38) to (10.2-40) — 863
Appendix 10.2: Orthogonal collocation analysis eqs. (10.3-8) to (10.3-10) — 864
Appendix 10.3: Order of magnitude of heat transfer parameters — 866
Appendix 10.4: Collocation analysis eqs. (10.4-45) — 868
Appendix 10.5: Orthogonal collocation analysis of eq. (10.5-22) — 870
Appendix 10.6: Orthogonal collocation analysis of (eqs. 10.6-25) — 873
Appendix 12.1: Laplace transform for the finite kinetic case — 875

References — 879

Index — 889

1
Introduction

1.1 Introduction

This book deals with the analysis of equilibria and kinetics of adsorption in a porous medium. Although gas phase systems are particularly considered in the book, the principles and concepts are applicable to liquid phase systems as well.

Adsorption phenomena have been known to mankind for a very long time, and they are increasingly utilised to perform desired bulk separation or purification purposes. The heart of an adsorption process is usually a porous solid medium. The use of porous solid is simply that it provides a very high surface area or high micropore volume and it is this high surface area or micropore volume that high adsorptive capacity can be achieved. But the porous medium is usually associated with very small pores and adsorbate molecules have to find their way to the interior surface area or micropore volume. This "finding the way" does give rise to the so-called diffusional resistance towards molecular flow. Understanding of the adsorptive capacity is within the domain of equilibria, and understanding of the diffusional resistance is within the domain of kinetics. To properly understand an adsorption process, we must understand these two basic ingredients: equilibria and kinetics, the analysis of which is the main theme of this book.

1.2 Basis of Separation

The adsorption separation is based on three distinct mechanisms: steric, equilibrium, and kinetic mechanisms. In the steric separation mechanism, the porous solid has pores having dimension such that it allows small molecules to enter while excluding large molecules from entry. The equilibrium mechanism is based on the solid having different abilities to accommodate different species, that is the stronger adsorbing species is preferentially removed by the solid. The kinetic mechanism is based on the different rates of diffusion of different species into the pore; thus by controlling the time of exposure the faster diffusing species is preferentially removed by the solid.

1.3 Adsorbents

The porous solid of a given adsorption process is a critical variable. The success or failure of the process depends on how the solid performs in both equilibria and kinetics. A solid with good capacity but slow kinetics is not a good choice as it takes adsorbate molecules too long a time to reach the particle interior. This means long gas residence time in a column, hence a low throughput. On the other hand, a solid with fast kinetics but low capacity is not good either as a large amount of solid is required for a given throughput. Thus, a good solid is the one that provides good adsorptive capacity as well as good kinetics. To satisfy these two requirements, the following aspects must be satisfied:
(a) the solid must have reasonably high surface area or micropore volume
(b) the solid must have relatively large pore network for the transport of molecules to the interior

To satisfy the first requirement, the porous solid must have small pore size with a reasonable porosity. This suggests that a good solid must have a combination of two pore ranges: the micropore range and the macropore range. The classification of pore size as recommended by IUPAC (Sing et al., 1985) is often used to delineate the range of pore size

Micropores	$d < 2$ nm
Mesopores	$2 < d < 50$ nm
Macropores	$d > 50$ nm

This classification is arbitrary and was developed based on the adsorption of nitrogen at its normal boiling point on a wide range of porous solids. Most practical solids commonly used in industries do satisfy these two criteria, with solids such as activated carbon, zeolite, alumina and silica gel. The industries using these solids are diversified, with industries such as chemical, petrochemical, biochemical, biological, and biomedical industries.

What to follow in this section are the brief description and characterisation of some important adsorbents commonly used in various industries.

1.3.1 Alumina

Alumina adsorbent is normally used in industries requiring the removal of water from gas stream. This is due to the high functional group density on the surface, and it is those functional groups that provide active sites for polar molecules (such as water) adsorption. There are a variety of alumina available, but the common solid used in drying is γ-alumina. The characteristic of a typical γ-alumina is given below (Biswas et al., 1987).

Table 1.2-1: Typical characteristics of γ-alumina

True density	2.9 - 3.3 g/cc
Particle density	0.65 - 1.0 g/cc
Total porosity	0.7 - 0.77
Macropore porosity	0.15 - 0.35
Micropore porosity	0.4 - 0.5
Macropore volume	0.4 - 0.55 cc/g
Micropore volume	0.5 - 0.6 cc/g
Specific surface area	200 - 300 m^2/g
Mean macropore radius	100 - 300 nm
Mean micropore radius	1.8 - 3 nm

As seen in the above table, γ-alumina has a good surface area for adsorption and a good macropore volume and mean pore size for fast transport of molecules from the surrounding to the interior.

1.3.2 Silica gel

Silica gel is made from the coagulation of a colloidal solution of silicic acid. The term gel simply reflects the conditions of the material during the preparation step, not the nature of the final product. Silica gel is a hard glassy substance and is milky white in colour. This adsorbent is used in most industries for water removal due to its strong hydrophilicity of the silica gel surface towards water. Some of the applications of silica gel are
(a) water removal from air
(b) drying of non-reactive gases
(c) drying of reactive gases
(d) adsorption of hydrogen sulfide
(e) oil vapour adsorption
(f) adsorption of alcohols

The following table shows the typical characteristics of silica gel.

Table 1.2-2: Typical characteristics of silica gel

Particle density	0.7 - 1.0 g/cc
Total porosity	0.5 - 0.65
Pore volume	0.45 - 1.0 cc/g
Specific surface area	250 - 900 m^2/g
Range of pore radii	1 to 12 nm

Depending on the conditions of preparation, silica gel can have a range of surface area ranging from about 200 m^2/g to as high as 900 m^2/g. The high end of surface area is achievable but the pore size is very small. For example, the silica gel used by Cerro and Smith (1970) is a high surface area Davison silica gel having a specific surface area of 832 m^2/g and a mean pore radius of 11 Angstrom.

1.3.3 Activated Carbon

Among the practical solids used in industries, activated carbon is one of the most complex solids but it is the most versatile because of its extremely high surface area and micropore volume. Moreover, its bimodal (sometimes trimodal) pore size distribution provides good access of sorbate molecules to the interior. The structure of activated carbon is complex and it is basically composed of an amorphous structure and a graphite-like microcrystalline structure. Of the two, the graphitic structure is important from the capacity point of view as it provides "space" in the form of slit-shaped channel to accommodate molecules. Because of the slit shape the micropore size for activated carbon is reported as the micropore half-width rather than radius as in the case of alumina or silica gel. The arrangement of carbon atoms in the graphitic structure is similar to that of pure graphite. The layers are composed of condensed regular hexagonal rings and two adjacent layers are separated with a spacing of 0.335nm. The distance between two adjacent carbon atoms on a layer is 0.142nm. Although the basic configuration of the graphitic layer in activated carbon is similar to that of pure graphite, there are some deviations, for example the interlayer spacing ranges from 0.34nm to 0.35nm. The orientation of the layers in activated carbon is such that the turbostratic structure is resulted. Furthermore, there are crystal lattice defect and the presence of built-in hetero-atoms.

The graphitic unit in activated carbon usually is composed of about 6-7 layers and the average diameter of each unit is about 10nm. The size of the unit can increase under the action of graphitization and this is usually done at very high temperature (>1000°C) and in an inert atmosphere.

The linkage between graphite units is possible with strong cross linking. The interspace between those graphite units will form pore network and its size is usually in the range of mesopore and macropore.

Typical characteristics of activated carbon are listed below.

Table 1.2-3: Typical characteristics of activated carbon

True density	2.2 g/cc
Particle density	0.73 g/cc
Total porosity	0.71
Macropore porosity	0.31
Micropore porosity	0.40
Macropore volume	0.47 cc/g
Micropore volume	0.44 cc/g
Specific surface area	1200 m^2/g
Mean macropore radius	800 nm
Mean micropore half width	1 - 2 nm

Macropore having a size range of greater than 100 nm is normally not filled with adsorbate by capillary condensation (except when the reduced pressure is approaching unity). The volume of macropore is usually in the order of 0.2-0.5 cc/g and the area contributed by the macropore is usually very small, of the order of 0.5 m^2/g, which is negligible compared to the area contributed by the micropore. Macropores, therefore, are of no significance in terms of adsorption capacity but they act as transport pores to allow adsorbate molecules to diffuse from the bulk into the particle interior.

Mesopore has a size range from 2 nm to 100 nm, and it is readily filled during the region of capillary condensation ($P/P_o \gtrsim 0.3$). The volume of mesopore is usually in the range of 0.1 to 0.4 cc/g and the surface area is in the range of 10-100 m^2/g. Mesopore contributes marginally to the capacity at low pressure and significantly in the region of capillary condensation. Like macropores, mesopores act as transport pore when capillary condensation is absent and they act as conduit for condensate flow in the capillary condensation region.

Micropores are pores having size less than 2 nm. These pores are slit-shaped and because of their high dispersive force acting on adsorbate molecule they provide space for storing most of adsorbed molecules and the mechanism of adsorption is via the process of volume filling.

Chemical nature of the surface of activated carbon is more complex than the pore network. This property depends on many factors, for example the source of carbon as well as the way how the carbon is activated. Activated carbon is made from raw materials which are usually rich in oxygen and therefore many functional groups on activated carbon have oxygen atom. Moreover, oxygen also is introduced during the course of preparation, for example coal activation by air or gasified by water vapor. Oxygen carrying functional groups can be classified into two main types: acidic group and basic group. The functional groups of an activated carbon can be increased by treating it with some oxidizing agents or decreased by exposing it to a vacuum environment at very high temperature.

Commercial activated carbon has a very wide range of properties depending on the application. If the application is for gas phase separation, then the characteristics given in Table 1.2-3 is typical. For liquid phase applications, however, due to the large molecular size of adsorbate activated carbon used in such applications will possess larger mesopore volume and larger average pore radius for the ease of diffusion of molecules to the interior.

1.3.4 Zeolite

Another important class of solid used as widely as activated carbon is zeolite. Zeolite can be found naturally or made synthetically. Application of natural zeolite is not as widely as that of synthetic zeolite because of the more specificity of the synthetic zeolite. There are many types of synthetic zeolite, such as type A, X, Y, mordenite, ZSM, etc. The book by Ruthven (1984) provides a good overview of these zeolites. The typical characteristics of the zeolite A are listed below.

Table 1.2-3: Typical characteristics of zeolite 5A

Crystal density*	1.57 g/cc
Particle density	1.1 g/cc
Macropore porosity	0.31
Macropore volume	0.28 cc/g
Micropore volume	0.3 cc/g
Exterior surface area	1 -20 m^2/g
Mean macropore radius	30-1000 nm
Mean micropore radius	0.5 nm

* mass/volume of crystal

1.4 Adsorption Processes

With many solids available to the industries, there are many important processes which currently enjoy their applications. In general, we can classify adsorption processes into two classes. The first is the bulk separation and the other is the purification. Some important processes are listed in the following table.

Table 1.3-1: Typical processes using adsorption technology

Normal paraffins, iso-paraffins	Zeolite 5A
Nitrogen/ Oxygen	Zeolite 5A
Oxygen/ Nitrogen	Carbon molecular sieve
Carbon oxides/Methane	Zeolite, activated carbon
Ethylene/vent stream	Activated carbon
VOCs removal from air	Activated carbon
Carbon dioxide, ethylene from natural gas	Zeolite
Sulfur compound from natural gas	Zeolite
Drying of reactive gases	Zeolite 3A, silica gel, alumina
Solvent removal from air	Activated carbon
Ordors from air	Activated carbon
NO_x, SO_2 from flue gas	Zeolite, activated carbon

1.5 The Structure of the Book

This book will address the various fundamental aspects of adsorption equilibria and dynamics in microporous solids such as activated carbon and zeolite. The treatment of equilibria and kinetics, when properly applied, can be used for solids other than microporous solid, such as alumina, silica gel, etc. Recognizing that practical solids are far from homogeneous, this book will also cover many recent results in dealing with heterogeneous media.

We start this book with a chapter (Chapter 2) on the fundamentals of pure component equilibria. Results of this chapter are mainly applicable to ideal solids or surfaces, and rarely applied to real solids. Langmuir equation is the most celebrated equation, and therefore is the cornerstone of all theories of adsorption and is dealt with first. To generalise the fundamental theory for ideal solids, the Gibbs approach is introduced, and from which many fundamental isotherm equations, such as Volmer, Fowler-Guggenheim, Hill-de Boer, Jura-Harkins can be derived. A recent equation introduced by Nitta and co-workers is presented to allow for the multi-site adsorption. We finally close this chapter by presenting the vacancy solution theory of Danner and co-workers. The results of Chapter 2 are used as a basis for the development of equilibria theory in dealing with practical solids, and we do this in Chapter 3 by presenting a number of useful empirical as well as semi-empirical equations for describing adsorption equilibria. Some equations are useful to describe adsorption of gases and vapors below the capillary condensation region, equations such as Freundlich, Langmuir-Freundlich (Sips), Toth, Unilan, and Dubinin-Radushkevich. To describe equilibrium data in the region of multilayering adsorption, the classical equation BET is presented in Chapter 3. Various modifications of the BET equation are also presented to account for the various features inherent with real solids. Other semi-empirical equations, such as Harkins-Jura, FHH are also discussed. Finally, we close this chapter with a section on pore volume and pore size distribution.

Chapter 4 particularly deals with microporous solids, and for these solids the most celebrated equation for adsorption equilibrium is the Dubinin-Radushkevich equation. Since its publication, there are many versions of such equation to deal with a variety of cases, equations such as Dubinin-Astakhov equation to allow for solid heterogeneity, and Dubinin-Stoeckli equation to account for the structure heterogeneity. Although the Dubinin equations are popular in describing adsorption isotherm for activated carbon as well as zeolite, they have a serious limitation which is the zero slope of the isotherm equation at zero loading. To remedy this, various approaches have been attempted, and we have presented those approaches in this chapter. We finally close this chapter by discussing micropore size distribution and the various versions of the Dubinin equations in dealing with heterogeneous microporous solids.

Chapters 2 to 4 deal with pure component adsorption equilibria. Chapter 5 will deal with multicomponent adsorption equilibria. Like Chapter 2 for pure component systems, we start this chapter with the now classical theory of Langmuir for multicomponent systems. This extended Langmuir equation applies only to ideal solids, and therefore in general fails to describe experimental data. To account for this deficiency, the Ideal Adsorption Solution Theory (IAST) put forward by Myers and Prausnitz is one of the practical approaches, and is presented in some details in Chapter 5. Because of the reasonable success of the IAS, various versions have been proposed, such as the FastIAS theory and the Real Adsorption Solution Theory (RAST), the latter of which accounts for the non-ideality of the adsorbed phase. Application of the RAST is still very limited because of the uncertainty in the calculation of activity coefficients of the adsorbed phase. There are other factors such as the geometrical heterogeneity other than the adsorbed phase nonideality that cause the deviation of the IAS theory from experimental data. This is the area which requires more research.

Practical solids are generally heterogeneous, and this subject of heterogeneity is the topic of Chapter 6, where the concept of distribution of the interaction energy between adsorbate molecules and solid atoms is discussed. For systems, such as non-polar hydrocarbons on activated carbon, where the adsorption force is dispersive by nature, the role of micropore size distribution is important in the description of solid heterogeneity. The concept of distribution is not restricted to the interaction energy between adsorbate molecules and solid atoms, it can be applied to the Henry constant, the approach of which has been used by Sircar, and it can be applied to free energy, which was put forward by Aharoni and Evans.

The rest of the book is dedicated to adsorption kinetics. We start with the detailed description of diffusion and adsorption in porous solids, and this is done in Chapter 7. Various simple devices used to measure diffusivity are presented, and the various modes of transport of molecules in porous media are described. The simplest transport is the Knudsen flow, where the transport is dictated by the collision between molecules and surfaces of the pore wall. Other transports are viscous flow, continuum diffusion and surface diffusion. The combination of these transports is possible for a given system, and this chapter will address this in some detail.

The same set of transport mechanisms learnt in Chapter 7 is again considered in Chapter 8, but is dealt with in the framework of Maxwell-Stefan. This is the cornerstone in dealing with multicomponent diffusion in homogeneous media as well as heterogeneous media. We first address this framework to a homogeneous medium so that readers can grasp the concept of friction put forwards by Maxwell and Stefan in dealing with multicomponent systems. Next, we deal with diffusion of a multicomponent mixture in a capillary and a porous medium where continuum diffusion, Knudsen diffusion as well as viscous flow can all play an important role in the transport of molecules.

Adsorption kinetics of a single particle (activated carbon type) is dealt with in Chapter 9, where we show a number of adsorption / desorption problems for a single particle. Mathematical models are presented, and their parameters are carefully identified and explained. We first start with simple examples such as adsorption of one component in a single particle under isothermal conditions. This simple example will bring out many important features that an adsorption engineer will need to know, such as the dependence of adsorption kinetics behaviour on many important parameters such as particle size, bulk concentration, temperature, pressure, pore size and adsorption affinity. We then discuss the complexity in the dealing with multicomponent systems whereby governing equations are usually coupled nonlinear differential equations. The only tool to solve these equations is the numerical method. Although there are a number of numerical methods available to effectively solve these equations (Rice and Do, 1995), I would prefer to use the orthogonal collocation method to solve these equations. Although the choice is purely of personal taste, it is a very convenient method and very stable in solving most adsorption kinetics problems. Isothermal as well as nonisothermal conditions are dealt with in this chapter.

Chapter 10 deals with zeolite type particle, where the particle is usually in bidisperse form, that is small pores (channels inside zeolite crystal) are grouped together within a crystal, and the intercrystal void would form a network of larger pores. In other words, there are two diffusion processes in the particle, namely micropore diffusion and macropore diffusion. In the micropore network, only one phase is possible: the adsorbed phase. Depending on the relative time scales between these two diffusion processes, a system can be either controlled by the macropore diffusion, or by micropore diffusion, or by a combination of both. Isothermal as well as nonisothermal conditions will be addressed in this chapter.

Chapter 11 will deal with heterogeneous particle. Like Chapter 6 for equilibria, the area of heterogeneity is a topic of current research in adsorption, especially in kinetics, and much is needed before a full understanding of the effects of heterogeneity can be realized. This chapter, however, will provide some results in this area, and students are encouraged to develop their own thoughts in such a fruitful area.

The remainder of the book deals with various methods commonly used in the literature for the measurement of diffusivity. We start with Chapter 12 with a time lag method, which belongs to the class of permeation method, of which another method employing a diffusion cell is presented in Chapter 13. The time lag method was pioneered by Barrer in the early 50's, and is a very useful tool to study diffusion through porous media as well as polymeric membranes. Chromatography method is presented in Chapter 14, and finally we conclude with a chapter (Chapter 15) on the analysis of batch adsorber.

About the notations used in all chapters, I have attempted to use the same notations throughout the text to ensure the uniformity in nomenclature. A table of nomenclature is provided at the end of the text. The numbering of equations, tables and figures is done with the section used as the prefix. For example, the first equation in section 7.4 is numbered as eq. (7.4-1). Similarly, the second figure in section 6.1 is labelled as Figure (6.1-2). Finally, the book is provided with numerous computer codes written in MatLab language for solving many adsorption equilibria and kinetics problems. Students are encouraged to use them for effectively learning the various concepts of adsorption.

2
Fundamentals of Pure Component Adsorption Equilibria

2.1 Introduction

Adsorption equilibria information is the most important piece of information in understanding an adsorption process. No matter how many components are present in the system, the adsorption equilibria of pure components are the essential ingredient for the understanding of how much those components can be accommodated by a solid adsorbent. With this information, it can be used in the study of adsorption kinetics of a single component, adsorption equilibria of multicomponent systems, and then adsorption kinetics of multicomponent systems.

In this chapter, we present the fundamentals of pure component equilibria. Various fundamental equations are shown, and to start with the proceeding we will present the most basic theory in adsorption: the Langmuir theory (1918). This theory allows us to understand the monolayer surface adsorption on an ideal surface. By an ideal surface here, we mean that the energy fluctuation on this surface is periodic (Figure 2.1-1) and the magnitude of this fluctuation is larger than the thermal energy of a molecule (kT), and hence the troughs of the energy fluctuation are acting as the adsorption sites. If the distance between the two neighboring troughs is much larger than the diameter of the adsorbate molecule, the adsorption process is called localised and each adsorbate molecule will occupy one site. Also, the depth of all troughs of the ideal surface are the same, that is the adsorption heat released upon adsorption on each site is the same no matter what the loading is.

After the Langmuir theory, we will present the Gibbs thermodynamics approach. This approach treats the adsorbed phase as a single entity, and Gibbs adapted the classical thermodynamics of the bulk phase and applied it to the adsorbed phase. In doing this the concept of volume in the bulk phase is replaced by the area, and the pressure is replaced by the so-called spreading pressure. By assuming some forms of thermal equation of state relating the number of mole of adsorbate, the area and the spreading pressure (analogue of equations of state in the gas phase) and using them in the Gibbs equation, a number of fundamental equations can be derived, such as the linear isotherm, the Volmer isotherm, etc.

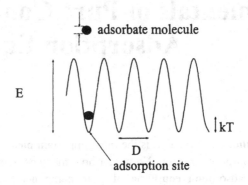

Figure 2.1-1: Surface energy fluctuations

Following the Gibbs approach, we will show the vacancy solution theory developed by Suwanayuen and Danner in 1980. Basically in this approach the system is assumed to consist of two solutions. One is the gas phase and the other is the adsorbed phase. The difference between these two phases is the density. One is denser than the other. In the context of this theory, the vacancy solution is composed of adsorbates and vacancies. The latter is an imaginary entity defined as a vacuum space which can be regarded as the solvent of the system.

Next, we will discuss one of the recent equations introduced by Nitta and his co-workers. This theory based on statistical thermodynamics has some features similar to the Langmuir theory, and it encompasses the Langmuir equation as a special case. Basically it assumes a localised monolayer adsorption with the allowance that one adsorbate molecule can occupy more than one adsorption site. Interaction among adsorbed molecules is also allowed for in their theory. As a special case, when the number of adsorption sites occupied by one adsorbate molecule is one, their theory is reduced to the Fowler-Guggenheim equation, and further if there is no adsorbate-adsorbate interaction this will reduce to the Langmuir equation. Another model of Nitta and co-workers allowing for the mobility of adsorbed molecules is also presented in this chapter.

The fundamental equations, Langmuir, Volmer, Fowler-Guggenheim and Hill de Boer, will form a basis for the study of heterogeneous adsorbents as we shall discuss briefly in Chapter 3 and in further detail in Chapter 6.

Finally, we will discuss briefly the lattice vacancy theory of Honig and Mueller (1962) who adapted the Flory-Huggin polymer-monomer solution theory. The form of their equation is identical to that derived by Nitta using the statistical thermodynamics approach.

2.2 Langmuir Equation

2.2.1 Basic Theory

Langmuir (1918) was the first to propose a coherent theory of adsorption onto a *flat* surface based on a kinetic viewpoint, that is there is a continual process of bombardment of molecules onto the surface and a corresponding evaporation (desorption) of molecules from the surface to maintain zero rate of accumulation at the surface at equilibrium.

The assumptions of the Langmuir model are:
1. Surface is homogeneous, that is adsorption energy is constant over all sites (we will discuss heterogeneous surfaces in Chapter 6)
2. Adsorption on surface is localised, that is adsorbed atoms or molecules are adsorbed at definite, localised sites (mobile adsorption will be dealt with in Sections 2.3.3 and 2.5)
3. Each site can accommodate only one molecule or atom

The Langmuir theory is based on a kinetic principle, that is the rate of adsorption (which is the striking rate at the surface multiplied by a sticking coefficient, sometimes called the accommodation coefficient) is equal to the rate of desorption from the surface.

The rate of striking the surface, in mole per unit time and unit area, obtained from the kinetic theory of gas is:

$$R_s = \frac{P}{\sqrt{2\pi M R_g T}} \qquad (2.2\text{-}1)$$

To give the reader a feel about the magnitude of this bombardment rate of molecule, we tabulate below this rate at three pressures

P (Torr)	R_s (molecules/cm²/sec)
760	3×10^{23}
1	4×10^{20}
10^{-3}	4×10^{17}

This shows a massive amount of collision between gaseous molecules and the surface even at a pressure of 10^{-3} Torr.

A fraction of gas molecules striking the surface will condense and is held by the surface force until these adsorbed molecules evaporate again (see Figure 2.2-1). Langmuir (1918) quoted that there is good experimental evidence that this fraction is unity, but for a real surface which is usually far from ideal this fraction could be much less than unity. Allowing for the sticking coefficient α (which accounts for non perfect sticking), the rate of adsorption in mole adsorbed per unit *bare* surface area per unit time is:

$$R_a = \frac{\alpha P}{\sqrt{2\pi M R_g T}} \tag{2.2-2}$$

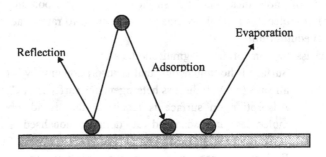

Figure 2.2-1: Schematic diagram of Langmuir adsorption mechanism on a flat surface

This is the rate of adsorption on a bare surface. On an occupied surface, when a molecule strikes the portion already occupied with adsorbed species, it will evaporate very quickly, just like a reflection from a mirror. Therefore, the rate of adsorption on an occupied surface is equal to the rate given by eq. (2.2-2) multiplied by the fraction of empty sites, that is:

$$R_a = \frac{\alpha P}{\sqrt{2\pi M R_g T}}(1-\theta) \tag{2.2-3}$$

where θ is the fractional coverage. Here R_a is the number of moles adsorbed per unit area (including covered and uncovered areas) per unit time.

The rate of desorption from the surface is equal to the rate, which corresponds to fully covered surface (k_d), multiplied by the fractional coverage, that is:

$$R_d = k_d \theta = k_{d\infty} \exp\left(-\frac{E_d}{R_g T}\right) \theta \qquad (2.2\text{-}4)$$

where E_d is the activation energy for desorption, which is equal to the heat of adsorption for physically sorbed species since there is no energy barrier for physical adsorption. The parameter $k_{d\infty}$ is the rate constant for desorption at infinite temperature. The inverse of this parameter is denoted as

$$\tau_{d\infty} = \frac{1}{k_{d\infty}}$$

The average residence time of adsorption is defined as:

$$\tau_a = \tau_{d\infty} e^{E_d/RT}$$

This means that the deeper is the potential energy well (higher E_d) the longer is the average residence time for adsorption. For physical adsorption, this surface residence time is typically ranging between 10^{-13} to 10^{-9} sec, while for chemisorption this residence time has a very wide range, ranging from 10^{-6} (for weak chemisorption) to about 10^9 for systems such as CO chemisorbed on Ni. Due to the Arrhenius dependence on temperature this average surface residence time changes rapidly with temperature, for example a residence time of 10^9 at 300K is reduced to only 2 sec at 500K for a system having a desorption energy of 120 kJoule/mole.

Equating the rates of adsorption and desorption (eqs. 2.2-3 and 2.2-4), we obtain the following famous Langmuir isotherm written in terms of fractional loading:

$$\theta = \frac{bP}{1+bP} \qquad (2.2\text{-}5)$$

where

$$b = \frac{\alpha \, \exp(Q/R_g T)}{k_{d\infty} \sqrt{2\pi M R_g T}} = b_\infty \exp(Q/R_g T) \qquad (2.2\text{-}6)$$

Here Q is the heat of adsorption and is equal to the activation energy for desorption, E_d. The parameter b is called the affinity constant or Langmuir constant. It is a measure of how strong an adsorbate molecule is attracted onto a surface. The pre-exponential factor b_∞ of the affinity constant is:

$$b_\infty = \frac{\alpha}{k_{d\infty} \sqrt{2\pi M R_g T}} \qquad (2.2\text{-}7)$$

which is inversely proportional to the square root of the molecular weight. When P is in Torr, the magnitude of b_∞ for nitrogen is given by Hobson (1965) as:

$$b_\infty = 5.682 \times 10^{-5} (MT)^{-1/2} \quad \text{Torr}^{-1} \qquad (2.2\text{-}8)$$

The isotherm equation (2.2-5) reduces to the Henry law isotherm when the pressure is very low (bP << 1), that is the amount adsorbed increases linearly with pressure, a constraint demanded by statistical thermodynamics. When pressure is sufficiently high, the amount adsorbed reaches the saturation capacity, corresponding to a complete coverage of all adsorption sites with adsorbate molecules (this is called monolayer coverage, $\theta \rightarrow 1$). The behaviour of the Langmuir isotherm (θ versus P) is shown in Figure 2.2-2.

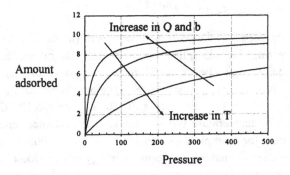

Figure 2.2-2: Behaviour of the Langmuir equation

When the affinity constant b is larger, the surface is covered more with adsorbate molecule as a result of the stronger affinity of adsorbate molecule towards the surface. Similarly, when the heat of adsorption Q increases, the adsorbed amount increases due to the higher energy barrier that adsorbed molecules have to overcome to evaporate back to the gas phase. Increase in the temperature will decrease the amount adsorbed at a given pressure. This is due to the greater energy acquired by the adsorbed molecule to evaporate.

The isotherm equation (2.2-5) written in the form of fractional loading is not useful for the data correlation as isotherm data are usually collated in the form of amount adsorbed versus pressure. We now let C_μ be the amount adsorbed in mole per unit mass or volume[1], and $C_{\mu s}$ be the maximum adsorbed concentration

---

[1] This volume is taken as the particle volume minus the void volume where molecules are present in free form.

corresponding to a complete monolayer coverage, then the Langmuir equation written in terms of the amount adsorbed useful for data correlation is:

$$C_\mu = C_{\mu s} \frac{b(T)P}{1 + b(T)P} \qquad (2.2\text{-}9a)$$

where

$$b(T) = b_\infty \exp\left(\frac{Q}{R_g T}\right) \qquad (2.2\text{-}9b)$$

Here we use the subscript μ to denote the adsorbed phase, and this will be applied throughout this text. For example, C_μ is the concentration of the adsorbed phase, and D_μ is the diffusion coefficient of the adsorbed phase, V_μ is the volume of the adsorbed phase, etc.

The temperature dependence of the affinity constant (e.g. 2.2-6) is $T^{-1/2} \exp(Q/R_g T)$. This affinity constant decreases with temperature because the heat of adsorption is positive, that is adsorption is an exothermic process. Since the free energy must decrease for the adsorption to occur and the entropy change is negative because of the decrease in the degree of freedom, therefore

$$\Delta H = \Delta G + T \Delta S < 0 \qquad (2.2\text{-}10)$$

The negativity of the enthalpy change means that heat is released from the adsorption process.

The Langmuir equation can also be derived from the statistical thermodynamics, based on the lattice statistics. Readers interested in this approach are referred to the book by Rudzinski and Everett (1992) for more detail.

2.2.2 Isosteric Heat of Adsorption

One of the basic quantities in adsorption studies is the isosteric heat, which is the ratio of the infinitesimal change in the adsorbate enthalpy to the infinitesimal change in the amount adsorbed. The information of heat released is important in the kinetic studies because when heat is released due to adsorption the released energy is partly absorbed by the solid adsorbent and partly dissipated to the surrounding. The portion absorbed by the solid increases the particle temperature and it is this rise in temperature that slows down the adsorption kinetics because the mass uptake is controlled by the rate of cooling of the particle in the later course of adsorption. Hence the knowledge of this isosteric heat is essential in the study of adsorption kinetics.

The isoteric heat may or may not vary with loading. It is calculated from the following thermodynamic van't Hoff equation:

$$\frac{\Delta H}{R_g T^2} = -\left(\frac{\partial \ln P}{\partial T}\right)_{C_\mu} \tag{2.2-11}$$

For Langmuir isotherm of the form given in eq. (2.2-9), we take the total differentiation of that equation and substitute the result into the above van't Hoff equation to get:

$$\frac{\Delta H}{R_g T^2} = \frac{Q}{R_g T^2} + \delta(1 + bP) \tag{2.2-12}$$

in which we have allowed for the maximum adsorbed concentration ($C_{\mu s}$) to vary with temperature and that dependence is assumed to take the form:

$$\frac{1}{C_{\mu s}}\frac{dC_{\mu s}}{dT} = -\delta \tag{2.2-13}$$

Since $(1+bP) = 1/(1-\theta)$, eq. (2.2-12) will become:

$$-\Delta H = Q + \frac{\delta R_g T^2}{1 - \theta} \tag{2.2-14}$$

The negativity of the enthalpy change indicates that the adsorption process is an exothermic process. If the maximum adsorbed concentration, $C_{\mu s}$, is a function of temperature and it decreases with temperature, the isosteric heat will increase with the loading due to the second term in the RHS of eq. (2.2-14). For the isosteric heat to take a finite value at high coverage (that is $\theta \to 1$) the parameter δ (thermal expansion coefficient of the saturation concentration) must be zero. This is to say that the saturation capacity is independent of temperature, and as a result the heat of adsorption is a constant, independent of loading.

2.3 Isotherms based on the Gibbs Approach

The last section dealt with the basic Langmuir theory, one of the earliest theories in the literature to describe adsorption equilibria. One should note that the Langmuir approach is kinetic by nature. Adsorption equilibria can be described quite readily by the thermodynamic approach. What to follow in this section is the approach due to Gibbs. More details can be found in Yang (1987) and Rudzinski and Everett (1992).

2.3.1 Basic Theory

In the bulk α-phase containing N components (Figure 2.3-1), the following variables are specified: the temperature T^α, the volume V^α and the numbers of moles of all species n_i^α (for i = 1, 2, ..., N). The upperscript is used to denote the phase. With these variables, the total differential Helmholtz free energy is:

$$dF = -S^\alpha dT^\alpha - P^\alpha dV^\alpha + \sum_{i=1}^{N} \mu_i^\alpha dn_i^\alpha \qquad (2.3\text{-}1)$$

where S^α is the entropy of the α phase, P^α is the pressure of that phase, n_i^α is the number of molecule of the species i, and μ_i^α is its chemical potential.

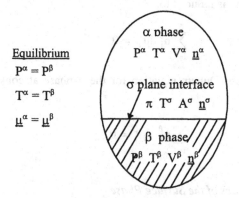

Figure 2.3-1: Equilibrium between the phases α and β separated by a plane interface σ

Similarly, for the β-phase, we can write a similar equation for the differential Helmholtz free energy:

$$dF = -S^\beta dT^\beta - P^\beta dV^\beta + \sum_{i=1}^{N} \mu_i^\beta dn_i^\beta \qquad (2.3\text{-}2)$$

If equilibrium exists between the two phases with a plane interface (Figure 2.3-1), we have:

$$T^\alpha = T^\beta \quad ; \quad P^\alpha = P^\beta \quad ; \quad \mu_i^\alpha = \mu_i^\beta \qquad (2.3\text{-}3)$$

that is equality in temperature, pressure and chemical potential is necessary and sufficient for equilibrium for a plane interface.

To obtain the Helmholtz free energy at a given temperature, pressure and chemical potential, we integrate eq. (2.3-1) keeping T, P and μ_i constant and obtain:

$$F^\alpha = -PV^\alpha + \sum_{i=1}^{N} \mu_i^\alpha n_i^\alpha \qquad (2.3\text{-}4)$$

Differentiating eq. (2.3-4) and subtracting the result from eq. (2.3-1) will give the following Gibbs-Duhem equation:

$$-V^\alpha dP - S^\alpha dT + \sum_{i=1}^{N} n_i^\alpha d\mu_i^\alpha = 0 \qquad (2.3\text{-}5a)$$

for the bulk α phase. As a special case of constant temperature and pressure, the Gibbs-Duhem's relation is reduced to:

$$\sum_{i=1}^{N} n_i^\alpha d\mu_i^\alpha = 0 \qquad (2.3\text{-}5b)$$

Similarly, the Gibbs-Duhem equation for the β-phase at constant temperature and pressure is:

$$\sum_{i=1}^{N} n_i^\beta d\mu_i^\beta = 0 \qquad (2.3\text{-}5c)$$

2.3.1.1 Thermodynamics of the Surface Phase

We now can develop a similar thermodynamic treatment for the surface phase σ, which is the interface between the phases α and β, and is in equilibrium with these two phases. When the adsorbed phase is treated as a two dimensional surface, fundamental equations in classical thermodynamics can still be applied. Applying the same procedure to surface free energy, we will obtain the Gibbs adsorption equation. This is done as follows. The total differentiation of the surface free energy takes the form similar to eq. (2.3-1) with $P^\alpha dV^\alpha$ being replaced by πdA:

$$dF^\sigma = -S^\sigma dT - \pi dA + \sum_{i=1}^{N} \mu_i dn_i^\sigma \qquad (2.3\text{-}6)$$

where the surface chemical potentials μ_i have the same values as those of the two joining phases, π is the spreading pressure, playing the same role as pressure in the bulk phase.

Integrating eq. (2.3-6) with constant T, π and μ_i yields:

$$F^\sigma = -\pi A + \sum_{i=1}^{N} \mu_i n_i^\sigma \qquad (2.3\text{-}7)$$

which is an analogue of eq. (2.3-4). Differentiation of this equation yields:

$$dF^\sigma = -Ad\pi - \pi dA + \sum_{i=1}^{N} \mu_i dn_i^\sigma + \sum_{i=1}^{N} n_i^\sigma d\mu_i \qquad (2.3\text{-}8)$$

Subtracting eq. (2.3-8) from eq. (2.3-6), we have the Gibbs equation for a planar surface:

$$-Ad\pi - S^\sigma dT + \sum_{i=1}^{N} n_i^\sigma d\mu_i = 0 \qquad (2.3\text{-}9)$$

Adsorption equilibria experiments are usually carried out at constant temperature, therefore the Gibbs adsorption isotherm equation is:

$$-Ad\pi + \sum_{i=1}^{N} n_i^\sigma d\mu_i = 0 \qquad (2.3\text{-}10)$$

For pure component systems ($N = 1$), we have:

$$-Ad\pi + nd\mu = 0 \qquad (2.3\text{-}11)$$

where we have dropped the superscript σ for clarity.

At equilibrium, the chemical potential of the adsorbed phase is equal to that of the gas phase, which is assumed to be ideal, i.e.

$$\mu = \mu_g = \mu_{g0} + R_g T \ln P \qquad (2.3\text{-}12)$$

Substituting eq.(2.3-12) into eq.(2.3-11), the following Gibbs isotherm equation is derived:

$$\left(\frac{d\pi}{d\ln P}\right)_T = \frac{n}{A} R_g T \qquad (2.3\text{-}13)$$

This equation is the fundamental equation relating gas pressure, spreading pressure and amount adsorbed. It is very useful in that if the equation of state relating the spreading pressure and the number of mole on the adsorbed phase is provided, the isotherm expressed as the number of mole adsorbed in terms of pressure can be obtained. We shall illustrate this with the following two examples: linear isotherm and Volmer isotherm.

2.3.2 Linear Isotherm

For an ideal surface at infinite dilution, the equation of state relating the spreading pressure and the number of mole on the surface has the following form:

$$\pi A = nR_g T \qquad (2.3\text{-}14)$$

an analogue of the ideal gas law (i.e. diluted systems), that is the spreading pressure is linear with the number of molecules on a surface of area A. Substituting this equation of state into the Gibbs equation (eq. 2.3-13), we get:

$$\left(\frac{d\pi}{d\ln P}\right)_T = \pi \qquad (2.3\text{-}15)$$

Integrating this equation at constant T, we obtain $\pi = C(T)P$, where $C(T)$ is some function of temperature. This equation means that at equilibrium the spreading pressure in the adsorbed phase is linearly proportional to the pressure in the gas phase. The spreading pressure is not, however, useful in the correlation of adsorption equilibrium data. To relate the amount adsorbed in the adsorbed phase in terms of the gas phase pressure, we use the equation of state (eq. 2.3-14) to finally get:

$$\frac{n}{A} = K(T)P \qquad (2.3\text{-}16a)$$

where

$$K(T) = \frac{C(T)}{R_g T} \qquad (2.3\text{-}16b)$$

The parameter $K(T)$ is called the Henry constant. The isotherm obtained for the diluted system is a linear isotherm, as one would anticipate from such condition of infinite dilution.

2.3.3 Volmer Isotherm

We have seen in the last section that when the system is dilute (that is the equation of state follows eq. 2.3-14), the isotherm is linear because each adsorbed molecule acts independently from other adsorbed molecules. Now let us consider the case where we allow for the finite size of adsorbed molecules. The equation of state for a surface takes the following form:

$$\pi(A - A_0) = nR_g T \qquad (2.3\text{-}17)$$

where A_0 is the minimum area occupied by n molecules.

The Gibbs equation (2.3-13) can be written in terms of the area per unit molecule as follows:

$$\left(\frac{\partial \pi}{\partial \ln P}\right)_T = \frac{R_g T}{\sigma} \qquad (2.3\text{-}18a)$$

where the variable σ is the area per unit molecule of adsorbate

$$\sigma = \frac{A}{n} \qquad (2.3\text{-}18b)$$

Integrating equation (2.3-18a) at constant temperature, we have:

$$\ln P = \frac{1}{R_g T}\int \sigma\, d\pi \qquad (2.3\text{-}19)$$

We rewrite the equation of state in terms of the new variable σ and get:

$$\pi(\sigma - \sigma_0) = R_g T \qquad (2.3\text{-}20)$$

Substituting the spreading pressure from the equation of state into the integral form of the Gibbs equation (2.3-19), we get:

$$\ln P = -\int \frac{\sigma\, d\sigma}{(\sigma - \sigma_0)^2} \qquad (2.3\text{-}21)$$

But the fractional loading is simply the minimum area occupied by n molecules divided by the area occupied by the same number of molecules, that is

$$\theta = \frac{A_0}{A} = \frac{(A_0/n)}{(A/n)} = \frac{\sigma_0}{\sigma} \qquad (2.3\text{-}22)$$

Written in terms of the fractional loading, θ, eq. (2.3-21) becomes:

$$\ln P = \int \frac{d\theta}{\theta(1-\theta)^2} \qquad (2.3\text{-}23)$$

Carrying out the integration, we finally get the following equation:

$$b(T)P = \frac{\theta}{1-\theta}\exp\left(\frac{\theta}{1-\theta}\right) \qquad (2.3\text{-}24a)$$

where the affinity constant $b(T)$ is a function of temperature, which can take the following form:

$$b(T) = b_\infty \exp\left(\frac{Q}{R_g T}\right) \qquad (2.3\text{-}24b)$$

Eq. (2.3-24) is known as the Volmer equation, a fundamental equation to describe the adsorption on surfaces where the mobility of adsorbed molecules is allowed, but no interaction is allowed among the adsorbed molecules.

The factor $\exp(\theta/(1-\theta))$ in eq. (2.3-24a) accounts for the mobility of the adsorbate molecules. If we arrange eq. (2.3-24a) as follows:

$$\frac{\theta}{1-\theta} = b \exp\left(-\frac{\theta}{1-\theta}\right) P \qquad (2.3\text{-}25a)$$

the Volmer equation is similar to the Langmuir isotherm equation (2.2-5) with the <u>apparent</u> affinity as

$$b_{app} = b \exp\left(-\frac{\theta}{1-\theta}\right) \qquad (2.3\text{-}25b)$$

The difference between the Volmer equation and the Langmuir equation is that while the affinity constant remains constant in the case of Langmuir mechanism, the "apparent" affinity constant in the case of Volmer mechanism decreases with loading. This means that the rate of increase in loading with pressure is much lower in the case of Volmer compared to that in the case of Langmuir.

2.3.3.1 Isosteric Heat

If the saturation capacity is independent of temperature, the isosteric heat of adsorption can be obtained for the Volmer equation (using Van't Hoff equation 2.2-11) as:

$$(-\Delta H) = R_g T^2 \left(\frac{\partial \ln P}{\partial T}\right)_{C_\mu} = Q$$

Thus, the isosteric heat is a constant, the same conclusion we obtained earlier for the Langmuir isotherm. This means that the mobility of the adsorbed molecules does not affect the way solid atoms and adsorbate molecule interact vertically with each other.

2.3.4 Hill-deBoer Isotherm

It is now seen that the Gibbs isotherm equation (2.3-13) is very general, and with any proper choice of the equation of state describing the surface phase an isotherm equation relating the amount on the surface and the gas phase pressure can be obtained as we have shown in the last two examples. The next logical choice for the equation of state of the adsorbate is an equation which allows for the co-volume term and the attractive force term. In this theme the following van der Waals equation can be used:

$$\left(\pi + \frac{a}{\sigma^2}\right)(\sigma - \sigma_0) = R_g T \qquad (2.3\text{-}26)$$

With this equation of state, the isotherm equation obtained is:

$$bP = \frac{\theta}{1-\theta} \exp\left(\frac{\theta}{1-\theta}\right) \exp(-c\theta) \qquad (2.3\text{-}27)$$

where

$$b = b_\infty \exp\left(\frac{Q}{R_g T}\right), \quad c = \frac{2a}{R_g T \sigma_0} = \frac{zw}{R_g T} \qquad (2.3\text{-}28)$$

where z is the coordination number (usually taken as 4 or 6 depending on the packing of molecules), and w is the interaction energy between adsorbed molecules. A positive w means attraction between adsorbed species and a negative value means repulsion, that is the apparent affinity is increased with loading when there is attraction between adsorbed species, and it is decreased with loading when there is repulsion among the adsorbed species.

The equation as given in eq. (2.3-27) is known as the <u>Hill-de Boer</u> equation, which describes the case where we have mobile adsorption and lateral interaction among adsorbed molecules. When there is no interaction between adsorbed molecules (that is w = 0), this Hill-de Boer equation will reduce to the Volmer equation obtained in Section 2.3.3.

The first exponential term in the RHS of eq. (2.3-27) describes the mobility of adsorbed molecules, and when this term is removed we will have the case of localised adsorption with lateral interaction among adsorbed molecules, that is:

$$bP = \frac{\theta}{1-\theta} \exp(-c\theta) \qquad (2.3\text{-}29)$$

This equation is known in the literature as the Fowler-Guggenheim equation, or the quasi approximation isotherm. This equation can also be derived from the statistical thermodynamics (Rudzinski and Everett, 1992). Due to the lateral interaction term $\exp(-c\theta)$, the Fowler-Guggenheim equation and the Hill-de Boer equation exhibit a very interesting behaviour. This behaviour is the two dimensional condensation when the lateral interaction between adsorbed molecules is sufficiently strong. We shall illustrate this phenomenon below for the case of Fowler-Guggenheim equation.

2.3.5 Fowler-Guggenheim Equation

Fowler-Guggenheim equation (2.3-29) is one of the simplest equations allowing for the lateral interaction. Before discussing the two dimensional condensation phenomenon, we first investigate the isosteric heat behaviour.

2.3.5.1 Isosteric heat

Using the van't Hoff equation (2.2-11), we can obtain the following heat of adsorption for the Fowler-Guggenheim adsorption isotherm:

$$(-\Delta H) = Q + zw\theta \qquad (2.3\text{-}30)$$

Thus, the heat of adsorption varies linearly with loading. If the interaction between the adsorbed molecules is attractive (that is w is positive), the heat of adsorption will increase with loading and this is due to the increased interaction between adsorbed molecules as the loading increases. This means that if the measured heat of adsorption shows an increase with respect to loading, it indicates the positive lateral interaction between adsorbed molecules. However, if the interaction among adsorbed molecules is repulsive (that is w is negative), the heat of adsorption shows a decrease with loading. If such a decrease of the heat of adsorption with loading is observed experimentally, it does not necessarily mean that the interaction among adsorbed molecules is negative as a decrease in the heat of adsorption could also mean that the surface is heterogeneous, that is the surface is composed of sites having different energy of adsorption. Molecules prefer to adsorb onto sites having the highest energy of adsorption, and as adsorption proceeds molecules then adsorb onto sites of progressively lower energy of adsorption, resulting in a decreased heat of adsorption with loading.

Similarly for the Hill-de Boer equation, we obtain the same isosteric heat of adsorption as that for the case of Fowler-Guggenheim equation. This is so as we have discussed in the section 2.3.3 for the case of Volmer equation that the mobility of adsorbed molecule does not influence the way in which solid interacts with adsorbate.

2.3.5.2 Isotherm Behaviour

Since the Fowler-Guggenheim equation (2.3-29) has the adsorbate-adsorbate interaction term ($e^{-c\theta}$), it will exhibit interesting behaviour when there is attraction among the adsorbed molecules. But first, let us discuss the situation when there is repulsion among the adsorbed molecules (c < 0). When the pressure in the gas phase is very low (hence the surface coverage is also low), the behaviour is identical to that of an ideal surface, that is the amount adsorbed is linearly proportional to the gas phase pressure. When the gas phase pressure increases, more molecules adsorb onto the surface but they have a tendency to stay apart due to the repulsion, and when the pressure is increased sufficiently high the surface will eventually be saturated due to sufficiently high chemical potential in the gas phase. The behaviour is similar to that exhibited by the Langmuir equation, but in this case of c < 0, it takes higher pressure to fill the surface due to the repulsion of molecules.

Now back to the situation where there are strong attractions between adsorbed molecules. If this attraction force is strong enough and when the pressure in the gas phase reaches a certain point there will exist a phenomenon called the two-dimensional condensation, that is the density of the adsorbed phase will change abruptly from a low density to a high density. We will show this analysis of two-dimensional condensation below. To simplify the mathematical notation, we let

$$y = bP \qquad (2.3\text{-}31)$$

Then the Fowler-Guggenheim equation will take the form:

$$y = \frac{\theta}{1-\theta} e^{-c\theta} \qquad (2.3\text{-}32)$$

The behaviour of this equation with respect to the fractional loading can be investigated by studying the first derivative of y with respect to θ:

$$y' = \left[\frac{1 - c\,\theta(1-\theta)}{(1-\theta)^2}\right] e^{-c\theta} \qquad (2.3\text{-}33)$$

The existence of stationary points occurs when $1 - c\theta(1-\theta) = 0$, that is

$$\theta_1 = \frac{1}{2}\left[1 - \left(1 - \frac{4}{c}\right)^{1/2}\right]; \quad \theta_2 = \frac{1}{2}\left[1 + \left(1 - \frac{4}{c}\right)^{1/2}\right] \qquad (2.3\text{-}34)$$

The two solutions for fractional loading are real when <u>c is greater than 4</u>. When this happens, it is not difficult to prove that these two solutions are between 0 and 1 (a physical requirement for the fractional loading). One lies between 0 and 0.5 and the other is between 0.5 and 1. The values of y = bP at these two stationary points are:

$$y_{1,2} = \frac{\theta_{1,2}}{1-\theta_{1,2}} \exp(-c\,\theta_{1,2}) \tag{2.3-35}$$

The following algorithm describes the behaviour of the Fowler-Guggenheim equation.

Algorithm:
For a given y (= bP), that is for a given pressure
1. If y = 0, the only solution for θ of eq. (2.3-32) is 0, irrespective of c
2. If c < 4, there will be only one solution (θ) between 0 and 1.
3. If c > 4, there are three possibilities (see Figure 2.3-3):
   3.1 If $y > y_1$, there is one root which lies between θ_2 and 1.
   3.2 If $y < y_2$, there is one root which lies between 0 and θ_1.
   3.3 If $y_2 < y < y_1$, there will be three solutions mathematically. One will be between 0 and θ_1, one is between θ_1 and θ_2, and the other is between θ_2 and 1.

The problem of multiplicity does not actually happen in the adsorption system, that is the adsorbed phase does not exhibit three regions of different density at the same time. What is occurring in the case of the value of c > 4 is that there exists a threshold pressure y_m such that

$$\int_{\theta_*(y_m)}^{\theta^*(y_m)} \left[\ln\left(\frac{\theta}{1-\theta}\right) - c\theta\right] d\theta = \left[\theta^*(y_m) - \theta_*(y_m)\right] \ln y_m \tag{2.3-36}$$

where θ_* and θ^* are lower and upper solutions for the fractional loading corresponding to the pressure y_m. The other solution for the fractional loading at this threshold pressure is 0.5. Details of the derivation of eq. (2.3-36) can be found in Rudzinski and Everett (1992). The following figure (2.3-2) shows the plot of y (= bP) versus θ (eq. 2.3-32) for the case of c = 7. The figure also shows the line y_m such that eq. (2.3-36) is satisfied.

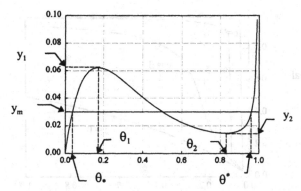

Figure 2.3-2: Plot of y = bP versus the fractional loading for the Fowler-Guggenheim isotherm (c = 7)

This threshold pressure $y_m = (bP_m)$ is called the phase transition pressure. The physical implication of this phase transition pressure is as follows. When the gaseous phase pressure is less than this phase transition pressure ($y < y_m$), the fractional loading will be in the range $(0, \theta_*)$. What we have here is the low density adsorption. At $y = y_m$, the fractional loading is

$$\theta = \frac{\theta_* + \theta^*}{2} = \frac{1}{2} \qquad (2.3\text{-}37a)$$

that is the fractional loading is one half at the phase transition pressure. This phase transition pressure is obtained from (for c > 4):

$$y_m = bP_m = \left.\frac{\theta}{1-\theta} e^{-c\theta}\right|_{\theta=\frac{1}{2}} = \exp\left(-\frac{c}{2}\right) \qquad (2.3\text{-}37b)$$

When the gaseous pressure is greater than the phase transition pressure ($y > y_m$), the fractional loading is in the range $(\theta^*, 1)$. This is what we call the high density adsorption. A computer code Fowler.m is provided with this book, and it calculates the fractional loading for a given value of pressure.

Figure (2.3-3) shows typical plots of the fractional loading versus the nondimensional pressure bP for the case of attraction between adsorbed molecules. Various values of c are used in the generation of these plots.

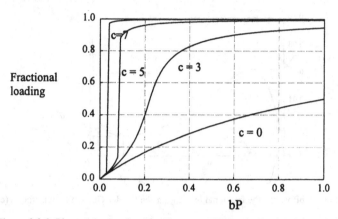

Figure 2.3-3: Plots of the fractional loading versus bP for the Fowler-Guggenheim equation

When the value of c is greater than the critical value of 4, we see the two dimensional condensation when the pressure reaches the phase transition pressure. Take the case of c = 7, there is a two dimensional condensation, and this occurs at the fractional loading of one half and the nondimensional phase transition pressure of

$$bP_m = \exp\left(-\frac{7}{2}\right) \approx 0.03$$

An increase in the interaction (increase in c) will shift the phase transition pressure to the left, that is the phase condensation occurs at a lower pressure, which is attributed to the stronger attraction among adsorbed molecules.

Similar analysis of the Hill-deBoer equation (2.3-27) shows that the two dimensional condensation occurs when the attraction between adsorbed molecules is strong and this critical value of c is 27/4. The fractional loading at the phase transition point is 1/3, compared to 1/2 in the case of Fowler-Guggenheim equation. A computer code Hill.m is provided with this book for the calculation of the fractional loading versus pressure for the case of Hill-de Boer equation. Figure 2.3-4 shows plots of the fractional loading versus nondimensional pressure (bP) for various values of c= {5, 7, 10, 15}.

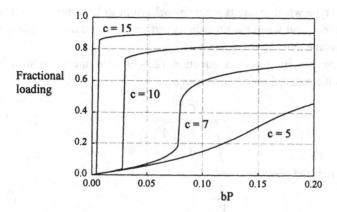

Figure 2.3-4: Plots of the fractional loading versus bP for the Hill-de Boer equation

Here we see the two dimensional condensation when $c > 27/4 \cong 6.75$. For this case, the phase transition pressure is calculated from:

$$bP_m = \frac{\theta}{1-\theta}\exp\left(\frac{\theta}{1-\theta}\right)\exp(-c\theta)\bigg|_{\theta=\frac{1}{3}} = \frac{1}{2}\exp\left(\frac{1}{2}\right)\exp\left(-\frac{c}{3}\right)$$

2.3.6 Harkins-Jura Isotherm

We have addressed the various adsorption isotherm equations derived from the Gibbs fundamental equation. Those equations (Volmer, Fowler-Guggenheim and Hill de Boer) are for monolayer coverage situation. The Gibbs equation, however, can be used to derive equations which are applicable in multilayer adsorption as well. Here we show such application to derive the Harkins-Jura equation for multilayer adsorption. Analogous to monolayer films on liquids, Harkins and Jura (1943) proposed the following equation of state:

$$\pi = b - a\sigma \qquad (2.3\text{-}38)$$

where a and b are constants. Substituting this equation of state into the Gibbs equation (2.3-13) yields the following adsorption equation:

$$\ln\left(\frac{P}{P_0}\right) = B - \frac{C}{v^2} \qquad (2.3\text{-}39)$$

which involves only measurable quantities. Here P_0 is the vapor pressure. This equation can describe isotherm of type II shown in Figure 2.3-5. The classification of types of isotherm will be discussed in detail in Chapter 3. But for the purpose of discussion of the Harkins-Jura equation, we explain type II briefly here. Type II isotherm is the type which exhibits a similar behaviour to Langmuir isotherm when the pressure is low, and when the pressure is further increased the amount adsorbed will increase in an exponential fashion.

Rearranging the Harkins-Jura equation (2.3-39) into the form of adsorbed amount versus the reduced pressure, we have:

$$v = \frac{\sqrt{C/B}}{\sqrt{1 + \frac{1}{B}\ln\left(\frac{1}{x}\right)}} \tag{2.3-40a}$$

where x is the reduced pressure

$$x = \frac{P}{P_o} \tag{2.3-40b}$$

We see that when the pressure approaches the vapor pressure, the adsorbed amount reaches a maximum concentration given below:

$$\lim_{x \to 1} v = v_{max} = \sqrt{\frac{C}{B}} \tag{2.3-41}$$

Thus, the Harkins-Jura isotherm equation can be written as

$$\frac{v}{v_{max}} = \frac{1}{\sqrt{1 + \frac{1}{B}\ln\left(\frac{1}{x}\right)}} \tag{2.3-42}$$

from which we can see that the only parameter which controls the degree of curvature of the isotherm is the parameter B.

2.3.6.1 Characteristics

To investigate the degree of curvature of the Harkins-Jura equation (2.3-42), we study its second derivative:

$$\frac{d^2}{dx^2}\left(\frac{v}{v_{max}}\right) = \frac{1}{2Bx^2} \frac{\left\{\frac{3}{2B} - \left[1 + \frac{1}{B}\ln\left(\frac{1}{x}\right)\right]\right\}}{\left[1 + \frac{1}{B}\ln\left(\frac{1}{x}\right)\right]^{5/2}} \tag{2.3-43}$$

To find the inflexion point, we set the second derivative to zero and obtain the reduced pressure at which the isotherm curve has an inflexion point

$$x_{inf.} = \exp\left[-\left(\frac{3}{2} - B\right)\right] \tag{2.3-44}$$

For the Harkins-Jura equation to describe the Type II isotherm, it must have an inflexion point occurring at the reduced pressure between 0 and 1, that is the restriction on the parameter B is:

$$0 < B < \frac{3}{2} \tag{2.3-45}$$

The restriction of positive B is due to the fact that if B is negative, eq. (2.3-40) does not always give a real solution. With the restriction on B as shown in eq. (2.3-45), the minimum reduced pressure at which the inflexion point occurs is (by putting B to zero in eq. 2.3-44):

$$x_{inf} = \exp\left(-\frac{3}{2}\right) \cong 0.22 \tag{2.3-46}$$

Figure 2.3-5 shows typical plots of the Harkins-Jura equation.

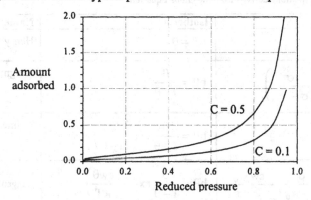

Figure 2.3-5: Plots of the Harkins-Jura equation versus the reduced pressure with B = 0.01

Jura and Harkins claimed that this is the simplest equation found so far for describing adsorption from sub-monolayer to multilayer regions, and it is valid over more than twice the pressure range of any two-constant adsorption isotherms (More about multilayer adsorption will be presented in Chapter 3). They showed that for TiO_2 in the form of anatase, their isotherm agrees with the data at both lower and higher values of pressure than the commonly used BET equation (Section 3.3).

Harkins and Jura (1943) have shown that a plot of $\ln(P/P_0)$ versus $1/v^2$ would yield a straight line with a slope of - C. The square root of this constant is proportional to the surface area of the solid. They gave the following formula:

$$S_g = 4.06\sqrt{C} \tag{2.3-47}$$

where v is the gas volume at STP adsorbed per unit g, and S_g has the unit of m²/g. They also suggested that if the plot of $\ln(P/P_0)$ versus $1/v^2$ exhibits two straight lines, the one at lower pressure range should be chosen for the area calculation as this is the one in which there exists a transition from a monolayer to a polylayer.

2.3.7 Other Isotherms from Gibbs Equation:

We see that many isotherm equations (linear, Volmer, Hill-deBoer, Harkins-Jura) can be derived from the generic Gibbs equation (2.3-13). Other equations of state relating the spreading pressure to the surface concentration can also be used, and thence isotherm equations can be obtained. The following table (Table 2.3-1) lists some of the fundamental isotherm equations from a number of equations of state (Ross and Olivier, 1964; Adamson, 1984).

Table 2.3.1: Isotherm Equations derived from the Gibbs Equation

Equation of state	Isotherm	Name
$\pi\sigma = R_g T$	$bP = \theta$	Henry law
$\pi\sigma = R_g T \dfrac{\sigma}{\sigma_0} \ln\left(\dfrac{\sigma}{\sigma-\sigma_0}\right)$	$bP = \dfrac{\theta}{1-\theta}$	Langmuir
$\pi(\sigma - \sigma_0) = R_g T$	$bP = \dfrac{\theta}{1-\theta}\exp\left(\dfrac{\theta}{1-\theta}\right)$	Volmer
$\pi\sigma = R_g T \dfrac{\sigma}{\sigma_0}\ln\left(\dfrac{\sigma}{\sigma-\sigma_0}\right) - \dfrac{cw}{2}\dfrac{\sigma_0}{\sigma}$	$bP = \dfrac{\theta}{1-\theta}\exp\left(-\dfrac{cw\theta}{RT}\right)$	Fowler-Guggenheim
$\left(\pi + \dfrac{a}{\sigma^2}\right)(\sigma - \sigma_0) = R_g T$	$bP = \dfrac{\theta}{1-\theta}\exp\left(\dfrac{\theta}{1-\theta}\right)\exp(-c\theta)$	Hill-deBoer
$\left(\pi + \dfrac{a}{\sigma^3}\right)(\sigma - \sigma_0) = R_g T$	$bP = \dfrac{\theta}{1-\theta}\exp\left(\dfrac{\theta}{1-\theta}\right)\exp(-c\theta^2)$	
$\left(\pi + \dfrac{a}{\sigma^3}\right)\left(\sigma - \dfrac{\sigma_0}{\sigma}\right) = R_g T$	$bP = \sqrt{\dfrac{\theta}{1-\theta}}\exp\left(\dfrac{1}{1-\theta}\right)\exp(-c\theta)$	

Since there are many fundamental equations which can be derived from various equations of state, we will limit ourselves to a few basic equations such as the Henry law equation, the Volmer, the Fowler-Guggenheim, and the Hill-de Boer equation. Usage of more complex fundamental equations other than those just mentioned needs justification for doing so.

2.4 Multisite Occupancy Model of Nitta

The theory of Nitta et al. (1984) assumes a localised monolayer adsorption on surface with an allowance for multi-site adsorption. This is an extension of the Langmuir isotherm for localised monolayer adsorption. Nitta et al.'s theory is based on the statistical thermodynamics, and its derivation is given briefly below.

Let M be the number of independent active sites and N be the number of adsorbed molecules. When each adsorbed molecule occupies n sites, the <u>partition function of the adsorbed phase</u> Q(M,N,T) is written by:

$$Q(M,N,T) = \left(j_s^N\right) g(N,M) \exp\left[\frac{N\varepsilon}{kT} + \frac{(nN)^2 u}{2MkT}\right] \qquad (2.4\text{-}1)$$

where j_s is the internal and vibrational partition function of an adsorbed molecule, ε is the adsorption energy per molecule for adsorbate-adsorbent interaction and u is the molecular interaction parameter for adsorbate-adsorbate interaction. The factor g(N,M) is the combinatorial factor which describes the number of distinguishable ways of distributing N adsorbed molecules over M sites. This function may be expressed as:

$$g(N,M) = \frac{M!}{N!(M-nN)!} \cdot \frac{\zeta^N}{M^{(n-1)N}} \qquad (2.4\text{-}2)$$

where ζ is a parameter relating to the flexibility and the symmetry of a molecule.

Knowing the partition function Q, the chemical potential of an adsorbate on the surface is equal to the partial derivative of $-kT\ln(Q)$ with respect to N, that is

$$\frac{\mu^s}{kT} = -\ln(j_s\zeta) + \ln\left(\frac{N}{M}\right) - n\ln\left(\frac{M-nN}{M}\right) - \frac{\varepsilon}{kT} - \frac{n^2 Nu}{MkT} \qquad (2.4\text{-}3)$$

At equilibrium, this chemical potential is equal to the chemical potential of the gas, which is expressed in terms of pressure and the internal partition of gas phase:

$$\frac{\mu^g}{kT} = \ln\left(\frac{P\Lambda^3}{j_g kT}\right) \tag{2.4-4}$$

where Λ is the thermal de Broglie wavelength, and j_g is the internal partition of gas phase.

Equating the two chemical potentials at equilibrium, and noting that the fractional loading is

$$\theta = \frac{nN}{M} \tag{2.4-5}$$

the following adsorption isotherm equation is obtained:

$$\ln(nbP) = \ln\theta - n\ln(1-\theta) - \frac{u}{kT}n\theta \tag{2.4-6a}$$

or

$$nbP = \frac{\theta}{(1-\theta)^n} \exp\left(-\frac{nu}{kT}\theta\right) \tag{2.4-6b}$$

where b is the adsorption affinity, defined as:

$$b = \frac{j_s \zeta \Lambda^3}{j_g \, kT} \exp\left(\frac{\varepsilon}{kT}\right) \tag{2.4-7}$$

The behaviour of the Nitta equation is that the slope of the surface coverage versus bP decreases with an increase in n, while it increases with an increase in the interaction parameter u.

The adsorbed amount is related to the fractional loading as follows:

$$C_\mu = C_{\mu s}\theta \tag{2.4-8}$$

where $C_{\mu s}$ is the maximum adsorbed concentration, which is related to the site concentration S_0 as follows.

$$C_{\mu s} = \frac{S_0}{n} \tag{2.4-9}$$

The Nitta et al.'s equation contains four parameters, S_0, n, b, and u. As a first approximation, we can set u = 0 to reduce the number of parameter by one. This is reasonable in systems where the adsorbate-adsorbate interaction is not as strong as the adsorbate-adsorbent interaction. If the fit of the three parameter model with the data is not acceptable, then the four parameter model is used. In the attempt to reduce the number of fitting parameters in the Nitta's isotherm equation, the following section shows how the adsorbate-adsorbate interaction energy can be estimated.

2.4.1 Estimation of the Adsorbate-Adsorbate Interaction Energy

The following analysis shows how the interaction energy can be estimated from the Lennard-Jones 12-6 potential, taken as a model for molecular interaction. The pairwise interaction between two molecules separating by a distance r is given by

$$\phi(r) = 4\varepsilon^* \left[\left(\frac{\sigma}{r}\right)^{12} - \left(\frac{\sigma}{r}\right)^{6} \right] \qquad (2.4\text{-}10)$$

where ϕ is the potential energy, ε^* is the well depth of the potential, and σ is the collision diameter, which is defined as the distance at which the potential energy is zero.

If the adsorbed molecules are randomly distributed, the total energy of interaction between a molecule and all the surrounding molecules is:

$$U_1 = \int_{\sigma}^{\infty} \phi(r) \frac{N}{A} 2\pi r \, dr \qquad (2.4\text{-}11)$$

where N is the number of adsorbed atoms on the surface and A is the surface area.

Substituting eq. (2.4-10) into eq. (2.4-11), we obtain the total energy of interaction between one molecule and all the surrounding molecules as given below:

$$U_1 = -\frac{6\pi}{5}\varepsilon^* \sigma^2 \frac{N}{A} \qquad (2.4\text{-}12)$$

There are N adsorbed molecules on the surface, the total energy due to the molecular interaction is then simply

$$U_T = \frac{N}{2}U_1 = -\frac{3\pi}{5}\varepsilon^* \sigma^2 \frac{N^2}{A} \qquad (2.4\text{-}13)$$

The factor 2 in the above equation is to avoid counting the pairwise interaction twice. This total energy of the adsorbed molecule interaction can also be calculated from

$$U_T = -\frac{1}{2} u \, n \, N\theta \qquad (2.4\text{-}14)$$

where u is the molecular potential energy associated to a site. Thus by equating the above two equations yields the following expression for u

$$u = \frac{6\pi}{5} \varepsilon^* \sigma^2 \frac{N}{n\theta A} \tag{2.4-15}$$

But the fractional loading is given by eq. (2.4-5); hence the above equation becomes:

$$u = \frac{6\pi}{5} \varepsilon^* \sigma^2 \frac{M}{n^2 A} \tag{2.4-16}$$

where M is the number of active sites. When the spherical particles are packed closely together, the fraction occupied by those particles is 0.907, and we write

$$\frac{\pi}{4} \frac{Ma^2}{nA} = 0.907 \tag{2.4-17}$$

where

$$a = (2)^{1/6} \sigma$$

Combining eqs. (2.4-16) and (2.4-17) gives the following estimate for the adsorbate-adsorbate interaction energy:

$$u = 3.44 \left(\frac{\varepsilon^*}{n}\right) \tag{2.4-18}$$

This estimated adsorbate-adsorbate interaction energy may be used in the Nitta et al.'s equation (2.4-6) to reduce the number of parameters of that equation.

2.4.2 Special Case

When n = 1, the Nitta's equation (2.4-6) is reduced to the Fowler-Guggenheim equation (2.3-29). In the case of no adsorbate-adsorbate interaction, that is u = 0, we have the following isotherm:

$$nbP = \frac{\theta}{(1-\theta)^n} \tag{2.4-19}$$

This equation reduces to the famous Langmuir isotherm when n = 1. The adsorption equation (2.4-19) without adsorbate - adsorbate interactions works well with adsorption of hydrocarbons and carbon dioxide on activated carbon and carbon molecular sieve with n ranging from 2 to 6.

2.4.3 Extension to Multicomponent Systems

The Nitta et al.'s equation (2.4-6) is readily extended to multicomponent systems. The relevant equation is:

$$\ln(n_j b_j P_j) = \ln \theta_j - n_j \ln\left(1 - \sum_{i=1}^{N} \theta_i\right) - n_j \sum_{i=1}^{N} \frac{u_{ji}\theta_i}{kT} \qquad (2.4\text{-}20)$$

When all the interaction energies between adsorbed molecules are zero, the multicomponent equations will be:

$$n_i b_i P_i = \frac{\theta_i}{\left(1 - \sum_{j=1}^{N} \theta_j\right)^{n_i}} \qquad (2.4\text{-}21)$$

Further, if $n_i = 1$ for all i, eq. (2.4-12) reduces to the extended Langmuir equation. We shall further discuss multicomponent systems in details in Chapter 5.

2.5 Mobile Adsorption Model of Nitta et al.

The last section deals with the multi-site adsorption model of Nitta and his co-workers. In such model each adsorbate molecule is adsorbed onto n active sites and the adsorption is localised. For surfaces where mobile adsorption is possible, the approach using the scaled particle theory can be used in the statistical thermodynamics to obtain the required adsorption isotherm equation. This has been addressed by Nitta and co-workers and what to follow is the brief account of their theory (Nitta et al., 1991).

At a given temperature T, the system containing N molecules on a surface area A has the following partition function

$$Q = \frac{1}{N!}\left(\frac{j^s A_f}{\Lambda^2}\right)^N \exp\left(-\frac{E}{R_g T}\right) \qquad (2.5\text{-}1)$$

where E is the potential energy, A_f is the free area available to each molecule, Λ is the de Broglie wave length and j^s is the molecular partition function of an adsorbed molecule. The free area for a molecule modelled as a hard disk of diameter d is governed by the following equation

$$\ln\left(\frac{A_f}{A}\right) = \ln(1 - \eta) - \frac{\eta}{1 - \eta} \qquad (2.5\text{-}2)$$

which is developed from the scaled particle theory (Helfand et al., 1961). In the above equation, η is the dimensionless surface density

$$\eta = \left(\frac{\pi d^2}{4}\right)\frac{N}{A} = \beta\frac{N}{A} \qquad (2.5\text{-}3)$$

Figure 2.5-1 shows a plot of the reduced free surface area versus the dimensionless density. We see that the surface free area diminishes to zero when the dimensionless surface density is about 0.8.

The potential energy E in eq. (2.5-1) is contributed by two interactions. One is the vertical interaction: adsorbate-adsorbent interaction, and the other is the horizontal interaction: adsorbate-adsorbate interaction. The vertical interaction energy is the negative of the well depth of the Lennard-Jones potential energy between a molecule and all the atoms on the surface. The horizontal interaction between two adsorbed molecules is (eq. 2.4-13)

$$-N^2\frac{\alpha}{A} \qquad (2.5\text{-}4a)$$

where α is related to the Lennard-Jones potential parameters. It was derived in the last section (eq. 2.4-13) and takes the form

$$\alpha = \frac{3\pi}{5}\sigma^2\varepsilon^* \qquad (2.5\text{-}4b)$$

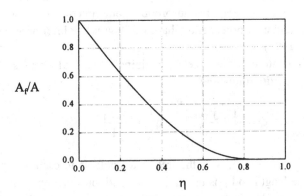

Figure 2.5-1: Plot of the reduced free area versus the dimensionless surface density

Thus the potential energy E is given by the sum of the vertical interaction and the horizontal interaction:

Fundamentals of Pure Component Adsorption Equilibria   41

$$E = -N\varepsilon - \frac{N^2\alpha}{A} \tag{2.5-5}$$

Substitution of eqs. (2.5-2) and (2.5-5) into the partition function (eq. 2.5-1) yields the explicit expression for the partition function:

$$Q = \frac{1}{N!}\left(\frac{j^s A}{\Lambda^2}\right)^N (1-\eta)^N \exp\left(-\frac{N\eta}{1-\eta} + \frac{N\varepsilon}{kT} + \frac{N^2\alpha}{AkT}\right) \tag{2.5-6}$$

Knowing the partition function of a system containing N molecules occupying an area of A, the chemical potential of a molecule is:

$$\frac{\mu^s}{kT} = -\left(\frac{\partial \ln Q}{\partial N}\right)_{T,A} \tag{2.5-7}$$

Hence

$$\frac{\mu^s}{kT} = \ln\left(\frac{N\Lambda^2}{j^s A}\right) - \ln(1-\eta) + \frac{(3-2\eta)\eta}{(1-\eta)^2} - \frac{\varepsilon}{kT} - \frac{2N\alpha}{AkT} \tag{2.5-8}$$

At equilibrium, the chemical potential of an adsorbed molecule is the same as that of a gas molecule

$$\frac{\mu^g}{kT} = \ln\left(\frac{\Lambda^3 \phi P}{j^g kT}\right) \tag{2.5-9}$$

where j^g is the internal molecular partition function and ϕ is the fugacity coefficient.

Equating eqs. (2.5-8) and (2.5-9) yields the following adsorption isotherm for the case of mobile adsorption on a surface

$$\ln(\phi b P) = \ln\left(\frac{\eta}{1-\eta}\right) + \frac{(3-2\eta)\eta}{(1-\eta)^2} - \frac{u\eta}{kT} \tag{2.5-10}$$

where b is the adsorption affinity, defined as

$$b = \frac{j^s \beta \Lambda}{j^g kT} \exp\left(\frac{\varepsilon}{kT}\right) \tag{2.5-11}$$

and u is the horizontal interaction parameter

$$u = \frac{2\alpha}{\beta} \tag{2.5-12}$$

Eq. (2.5-10) has a very similar form to the Hill de Boer equation (2.3-27).

The above equation of Nitta et al. deals with mobility of adsorbed molecules and their lateral interaction. When the lateral interaction is zero, eq. (2.5-10) becomes:

$$\phi bP = \frac{\eta}{1-\eta} \exp\left[\frac{(3-2\eta)\eta}{(1-\eta)^2}\right] \quad (2.5\text{-}13)$$

This form is similar to that of the Volmer equation, which also accounts for the mobility of the adsorbed molecules (but note the difference in the mobility term of the two equations):

$$\phi bP = \frac{\theta}{1-\theta} \exp\left(\frac{\theta}{1-\theta}\right) \quad (2.5\text{-}14)$$

2.6 Lattice Vacancy Theory

Honig and Mueller (1962) adapted the Flory-Huggin polymer-monomer solution theory to obtain the lattice vacancy theory for gas phase adsorption. Here the concept of fractional hole size is introduced. The isotherm equation obtained is:

$$\frac{P}{P_0} = \frac{\theta}{(1-\theta)^r} C^{-z\theta} \quad (2.6\text{-}1a)$$

or

$$\ln(P/P_0) = \ln\theta - r\cdot\ln(1-\theta) - z\theta \ln C \quad (2.6\text{-}1b)$$

where

$$P_0 = P^* \exp(-\varepsilon/R_g T) \frac{e^{(1-r)}}{j_s} \quad (2.6\text{-}2a)$$

$$C = \exp\left(-\frac{W}{R_g T}\right) \quad (2.6\text{-}2b)$$

where ε is the adsorption energy, j_s is the partition function, w is the lateral energy of interaction, z is the coordination number, and r is the number of monomer to form an r-mer. The form obtained by Honig and Mueller is very similar to the form of the equation obtained by Nitta et al. (1984).

2.7 Vacancy Solution Models (VSM)

2.7.1 VSM-Wilson Model

The vacancy solution theory was first developed by Suwanayuen and Danner (1980). Basically it assumes that the system consists of two solutions. One is the gas phase and the other is the adsorbed phase. They differ in their difference in composition (one is denser than the other). The vacancy solution is composed of adsorbate molecules and vacancies. The latter is an imaginary entity defined as a vacuum space which acts as the solvent of the system, and it has the same size as the adsorbate.

The vacancy is denoted as the species v. The chemical potential of the species v in the adsorbed phase, according to Suwanayuen and Danner, is:

$$\mu = \mu_0 + R_g T \ln(\gamma x) + \pi \sigma \tag{2.7-1}$$

where γ is the activity coefficient, x is the mole fraction, π is the surface pressure and σ is the partial molar surface area.

For the gas phase, a similar equation can be written

$$\mu = \mu_0^G + R_g T \ln(\gamma_G x_G) \tag{2.7-2}$$

Equating the chemical potential for the species v between the two phases gives the following equation of state: (after assuming the contribution of the logarithm term in eq. 2.7-2 is negligible)

$$\pi = -\frac{R_g T}{\sigma_v} \ln(\gamma_v x_v) \tag{2.7-3}$$

with the activity coefficient calculated from the Wilson equation as a function of the mole fractions x_1, x_v and two constants Λ_{v1} and Λ_{1v}:

$$\ln \gamma_v = -\ln(x_v + \Lambda_{v1} x_1) - x_1 \left[\frac{\Lambda_{1v}}{x_1 + \Lambda_{1v} x_v} - \frac{\Lambda_{v1}}{x_v + \Lambda_{v1} x_1} \right] \tag{2.7-4}$$

Using the equation of state (2.7-3) into the Gibbs isotherm equation (eq. 2.3-13), they obtained:

$$P = \left[\frac{C_{\mu s}}{K} \frac{\theta}{1-\theta} \right] \left[\Lambda_{1v} \frac{1-(1-\Lambda_{v1})\theta}{\Lambda_{1v}+(1-\Lambda_{1v})\theta} \right] \exp\left[-\frac{\Lambda_{v1}(1-\Lambda_{v1})\theta}{1-(1-\Lambda_{v1})\theta} - \frac{(1-\Lambda_{1v})\theta}{\Lambda_{1v}+(1-\Lambda_{1v})\theta} \right] \tag{2.7-5a}$$

where K is the Henry constant, and

$$\theta = \frac{C_\mu}{C_{\mu s}} \qquad (2.7\text{-}5b)$$

Eq. (2.7-5a) is called the VSM-W as it uses the Wilson equation for the activity coefficient. This equation has been used successfully to describe adsorption of many hydrocarbons on activated carbon.

2.7.2 VSM-Flory-Huggin Model

The VSM-W isotherm equation is a four parameters model (Λ_{1v}, Λ_{v1}, K and $C_{\mu s}$). The pairwise interaction constants Λ_{1v} and Λ_{v1} have been found to be highly correlated. To avoid this problem, Cochran et al. (1985) used the Flory-Huggin equation for the activity coefficient instead of the Wilson equation:

$$\ln \gamma_v = \frac{\alpha_{1v}\theta}{1+\alpha_{1v}\theta} - \ln(1+\alpha_{1v}\theta) \qquad (2.7\text{-}6a)$$

where

$$\alpha_{1v} = \frac{a_1}{a_v} - 1 \qquad (2.7\text{-}6b)$$

with a_1 and a_v being molar areas of the adsorbed species and the vacancy, respectively. Using this equation for activity in the equation of state (2.7-3) and then substituting it into the Gibbs isotherm equation (2.3-13), the adsorption isotherm equation is:

$$P = \left[\frac{C_{\mu s}}{K}\left(\frac{\theta}{1-\theta}\right)\right]\exp\left(\frac{\alpha_{1v}^2 \theta}{1+\alpha_{1v}\theta}\right) \quad ; \quad \theta = \frac{C_\mu}{C_{\mu s}} \qquad (2.7\text{-}7)$$

This equation has one less parameter than eq. (2.7-5) which uses the Wilson equation for activity coefficient.

Since the molar areas of the adsorbate (a_1) and the vacancy (a_v) are positive, the parameter α_{1v} must be greater than -1.

Rearranging eq. (2.7-7) as

$$y = \frac{KP}{C_{\mu s}} = \frac{\theta}{(1-\theta)}\exp\left(\frac{\alpha_{1v}^2 \theta}{1+\alpha_{1v}\theta}\right) \qquad (2.7\text{-}8)$$

of which the first derivative is

$$\frac{dy}{d\theta} = \frac{1}{(1-\theta)} \exp\left(\frac{\alpha_{1v}^2 \theta}{1+\alpha_{1v}\theta}\right) \left[\frac{1}{1-\theta} + \frac{\alpha_{1v}^2 \theta}{(1+\alpha_{1v}\theta)^2}\right] \qquad (2.7\text{-}9)$$

is always positive for $\alpha_{1v} > -1$, suggesting the monotonic increase of the adsorbed amount versus pressure.

The vacancy solution model is applied with good success in many systems, and it is readily extended to multicomponent systems because the inherent feature of this model is the interaction between gas molecules and the vacancies.

The two VSM isotherm equations are given in eqs. (2.7-5) and (2.7-7) depending on whether the Wilson equation or the Flory-Huggin equation is used to calculate the activity coefficient. Observing the form of these equations, the vacancy solution model equation can be written in general form as follows:

$$P = \frac{C_{\mu s}}{K} \frac{\theta}{1-\theta} f(\theta) \qquad (2.7\text{-}10)$$

The function $f(\theta)$ describes the nonideality of the mixture, and is calculated from equation:

$$f(\theta) = \left[\exp\left(-\int \frac{d\ln \gamma_v}{\theta}\right)\right] \left(\lim_{\theta \to 0} \exp \int \frac{d\ln \gamma_v}{\theta}\right) \qquad (2.7\text{-}11)$$

By nonideality here, we mean the deviation from the Langmuir behaviour. When $f(\theta) = 1$, eq. (2.7-10) reduces to the famous Langmuir equation.

2.7.3 Isosteric Heat of Adsorption

The isosteric heat of the FH-VSM (eq. 2.7-7) is obtained by Talu and Kabel (1987) as:

$$\left(-\frac{\Delta H}{R_g T}\right) = -\left(\frac{d\ln K}{dT}\right)_{C_\mu} + \left[\frac{2\alpha\theta + \alpha^2 \theta^2}{(1+\alpha\theta)^2}\right]\left(\frac{\partial \alpha}{\partial T}\right)_{C_\mu} - \left[\frac{C_\mu}{C_{\mu s}(C_{\mu s}-C_\mu)} + \frac{\alpha^2 \theta}{C_{\mu s}(1+\alpha\theta)^2}\right]\left(\frac{\partial C_{\mu s}}{\partial T}\right) \qquad (2.7\text{-}12)$$

The last term is due to the change in the saturation capacity with respect to temperature. This term blows up when the fractional loading approaches unity. The heat of adsorption should take a finite value at high coverage, thus the saturation capacity, according to Talu and Kabel, must be independent of temperature (that is the third term in the RHS of eq. 2.7-12 must be zero).

Cochran et al. (1985) related the parameter α in the FH equation to temperature through the saturation capacity as

46   Equilibria

$$\alpha = mC_{\mu s} - 1 \qquad (2.7\text{-}13)$$

where m is a constant. If the saturation capacity is assumed constant, then α is also a constant. This implies that the second term in the RHS of eq.(2.7-10) is also zero. Thus,

$$\left(-\frac{\Delta H}{R_g T}\right) = -\left(\frac{d \ln K}{dT}\right) \qquad (2.7\text{-}14)$$

is independent of loading, similar to the Langmuir equation.

For the W-VSM equation (2.7-5), the isosteric calculated for this model is:

$$\left(-\frac{\Delta H}{R_g T}\right) = -\left(\frac{d \ln K}{dT}\right) + \frac{2\theta \Lambda_{lv} + (1-\Lambda_{lv})\theta^2}{\Lambda_{lv}[\Lambda_{lv} + (1-\Lambda_{lv})\theta]^2} \frac{\partial \Lambda_{lv}}{\partial T} + \frac{\theta \Lambda_{vl}[2-(1-\Lambda_{vl})\theta]}{[1-(1-\Lambda_{vl})\theta]^2} \frac{\partial \Lambda_{vl}}{\partial T} \qquad (2.7\text{-}15)$$

Among the two models of VSM, the W-VSM describes experimental isosteric heat much closer than the FH-VSM model, although the isosteric heat calculated from FH-VSM does not differ more than 20% from experimental data.

2.8 2-D Equation of State (2-D EOS) Adsorption Isotherm

A number of fundamental approaches have been taken to derive the necessary adsorption isotherm. If the adsorbed fluid is assumed to behave like a two dimensional nonideal fluid, then the Equation of State developed for three dimensional fluids can be applied to two dimensional fluids with a proper change of variables. The 2D-EOS adsorption isotherm equations are not popularly used in the description of data, but they have an advantage of easily extending to multicomponent mixtures by using a proper mixing rule for the adsorption parameters.

For 3D fluids, the following 3 parameter EOS equation is popularly used

$$\left(p + \frac{a}{v^2 + \alpha bv + \beta b^2}\right)(v-b) = R_g T \qquad (2.8\text{-}1)$$

where p is the pressure, v is the volume per unit mole, a and b are parameters of the fluid and α and β represent numerical values. Different values of α and β give different forms of equation of state. For example, when $\alpha = \beta = 0$, we recover the famous van der Waals equation.

Written in terms of molar density ρ (mole/volume), the 3D-EOS will become:

$$\left(p + \frac{a\rho^2}{1+\alpha b\rho + \beta(b\rho)^2}\right)(1-b\rho) = \rho R_g T \qquad (2.8\text{-}2)$$

Adopting the above form, we can write the following equation for the 2D-EOS as follows:

$$\left(\pi + \frac{a_s \sigma^2}{1+\alpha b_s \sigma + \beta(b_s \sigma)^2}\right)(1-b_s \sigma) = \sigma R_g T \qquad (2.8\text{-}3)$$

where π is the spreading pressure, σ is the surface density (mole/area) and the parameters a_s and b_s are the 2D analogs of a and b of the 3D-EOS. Written in terms of the surface concentration (mole/mass), the above equation becomes:

$$\left(A\pi + \frac{a'_s w^2}{1+\alpha b'_s w + \beta(b'_s w)^2}\right)(1-b'_s w) = w R_g T \qquad (2.8\text{-}4)$$

where A is the specific area (m²/g).

To provide an EOS to properly fit the experimental data, Zhou et al. (1994) suggested the following form containing one additional parameter

$$\left(A\pi + \frac{a_s w^2}{1+\alpha b_s w + \beta(b_s w)^2}\right)\left(1-(b_s w)^m\right) = w R_g T \qquad (2.8\text{-}5)$$

This general equation reduces to special equations when the parameters α, β and m take some specific values. The following table shows various special cases deduced from the above equation.

α	β	m	EOS
0	0	1	van der Waals
0	0	½	Eyring
1	0	1	Soave-Redlich-Kwong
2	-1	1	Peng-Robinson

To fit many experimental data, Zhou et al. (1994) have found that m has to be less than ½. They suggested a value of 1/3 for m to reduce the number of parameters in the 2D-EOS equation (2.8-5).

At equilibrium, the chemical potential of the adsorbed phase is the same as that of the gas phase, that is

$$\mu_a = \mu_g = \mu_g^0 + R_g T \ln P \qquad (2.8\text{-}6)$$

The chemical potential of the adsorbed phase is related to the spreading pressure according to the Gibbs thermodynamics equation (eq. 2.3-11) rewritten here for clarity:

$$-Ad\pi + nd\mu = 0 \qquad (2.8\text{-}7)$$

Thus

$$\left(\frac{d\pi}{d \ln P}\right) = \frac{n}{A} R_g T = \sigma R_g T \qquad (2.8\text{-}8)$$

But the spreading pressure is a function of σ as governed by the equation of state (2.8-3). We write

$$d \ln P = \frac{1}{R_g T} \frac{1}{\sigma} \left(\frac{\partial \pi}{\partial \sigma}\right)_T d\sigma \qquad (2.8\text{-}9)$$

Integrating the above equation, we get

$$\int^P d \ln P = \frac{1}{R_g T} \int^\sigma \frac{1}{\sigma} \left(\frac{\partial \pi}{\partial \sigma}\right) d\sigma \qquad (2.8\text{-}10)$$

Eq. (2.8-10) is the adsorption isotherm equation relating the surface density σ (mole/m^2) in terms of the gas phase pressure. The applicability of this isotherm equation rests on the ability of the 2D-EOS (eq. 2.8-5) to describe the state of the adsorbed molecule. Discussions on the usage of the above equation in the fitting of experimental data are discussed in Zhou et al. (1994).

2.9 Concluding Remarks

This chapter has addressed the fundamentals of adsorption equilibria of a pure component. A number of fundamental equations have been discussed. Although they are successful in describing some experimental data, they are unfortunately unable to describe experimental data of practical solids. This is usually attributed to the complexity of the solid, and to some extent the complexity of the adsorbate molecule. There are two approaches adopted to address this problem. One is the empirical approach, which we will address in Chapter 3, and the other involves the concept of heterogeneity of the system whether this heterogeneity is from the solid or from the adsorbate or a combination of both. This second approach is addressed in some detail in Chapter 6.

3

Practical Approaches of Pure Component Adsorption Equilibria

3.1 Introduction

In the last chapter, we discussed the description of pure component adsorption equilibrium from the fundamental point of view, for example Langmuir isotherm equation derived from the kinetic approach, and Volmer equation from the Gibbs thermodynamic equation. Practical solids, due to their complex pore and surface structure, rarely conform to the fundamental description, that is very often than not fundamental adsorption isotherm equations such as the classical Langmuir equation do not describe the data well because the basic assumptions made in the Langmuir theory are not readily satisfied. To this end, many semi-empirical approaches have been proposed and the resulting adsorption equations are used with success in describing equilibrium data. This chapter will particularly deal with these approaches. We first present a number of commonly used empirical equations, and will discuss some of these equations in more detail in Chapter 6.

3.2 Empirical isotherm equations

In this section, we present a number of popularly used isotherm equations. We start first with the earliest empirical equation proposed by Freundlich, and then Sips equation which is an extension of the Freundlich equation, modified such that the amount adsorbed in the Sips equation has a finite limit at sufficiently high pressure (or fluid concentration). We then present the two equations which are commonly used to describe well many data of hydrocarbons, carbon oxides on activated carbon and zeolite: Toth and Unilan equations. A recent proposed equation by Keller et al. (1996), which has a form similar to that of Toth, is also discussed. Next, we describe the Dubinin equation for describing micropore filling, which is popular in fitting data of many microporous solids. Finally we present the relatively less used equations in physical adsorption, Jovanovich and Tempkin, the latter of which is more popular in the description of chemisorption systems.

3.2.1 Freundlich Equation

The Freundlich equation is one of the earliest empirical equations used to describe equilibria data. The name of this isotherm is due to the fact that it was used extensively by Freundlich (1932) although it was used by many other researchers. This equation takes the following form:

$$C_\mu = KP^{1/n} \qquad (3.2\text{-}1)$$

where C_μ is the concentration of the adsorbed species, and K and n are generally temperature dependent. The parameter n is usually greater than unity. The larger is this value, the adsorption isotherm becomes more nonlinear as its behaviour deviates further away from the linear isotherm. To show the behaviour of the amount adsorbed versus pressure (or concentration) we plot ($C_\mu / C_{\mu 0}$) versus (P/P$_0$) as shown in Figure 3.2-1, that is

$$\frac{C_\mu}{C_{\mu 0}} = \left(\frac{P}{P_0}\right)^{1/n}$$

where P_0 is some reference pressure and $C_{\mu 0}$ is the adsorbed concentration at that reference pressure, $C_{\mu 0} = K P_0^{1/n}$.

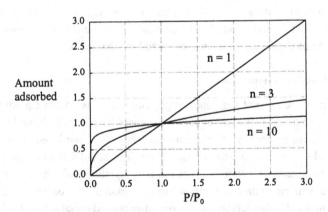

Figure 3.2-1: Plots of the Freundlich isotherm versus P/P$_0$

We see from Figure 3.2-1 that the larger is the value of n, the more nonlinear is the adsorption isotherm, and as n is getting larger than about 10 the adsorption isotherm is approaching a so-called rectangular isotherm (or irreversible isotherm). The term "irreversible isotherm" is normally used because the pressure (or concentration) needs to go down to an extremely low value before adsorbate molecules would desorb from the surface.

The Freundlich equation is very popularly used in the description of adsorption of organics from aqueous streams onto activated carbon. It is also applicable in gas phase systems having heterogeneous surfaces, provided the range of pressure is not too wide as this isotherm equation does not have a proper Henry law behaviour at low pressure, and it does not have a finite limit when pressure is sufficiently high. Therefore, it is generally valid in the narrow range of the adsorption data.

Parameters of the Freundlich equation can be found by plotting $\log_{10}(C_\mu)$ versus $\log_{10}(P)$

$$\log_{10}(C_\mu) = \log_{10}(K) + \frac{1}{n}\log_{10}(P)$$

which yields a straight line with a slope of $(1/n)$ and an intercept of $\log_{10}(K)$.

Example 3.2-1: *Fitting of propane/ activated carbon data*

To illustrate the linearity of this plot, we apply the Freundlich equation to the adsorption data of propane onto a sample of activated carbon at 10, 30 and 60 °C. The data are shown in Table 3.2-1 and Figure 3.2-2. This activated carbon is a typical commercial activated carbon, having a BET surface area of 1100 m²/g, and a porosity of 0.7 (including macropores and micropores)

Table 3.2-1: Adsorption data of propane on AC at 283, 303 and 333 K (C_μ, mmol/g)

283 K		303 K		333 K	
P (kPa)	C_μ	P (kPa)	C_μ	P (kPa)	C_μ
0.21	1.13	0.60	1.12	2.03	1.09
0.64	1.74	1.71	1.71	5.16	1.63
1.39	2.28	3.55	2.23	9.69	2.09
3.03	2.89	7.13	2.79	17.02	2.56
5.67	3.37	12.08	3.22	25.47	2.91
12.66	3.96	22.57	3.72	39.89	3.33
31.99	4.58	45.85	4.26	67.07	3.80
44.79	4.80	59.77	4.48	82.17	4.00
62.45	5.05	78.28	4.71	102.00	4.22
81.41	5.27	98.03	4.92	122.50	4.44
106.10	5.51	123.50	5.15		
126.40	5.68				

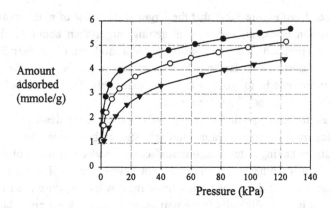

Figure 3.2.-2: Adsorption isotherm of propane on activated carbon (T = 283, 303, 333 K)

Since the Freundlich equation is not applicable over the complete range of pressure, we limit the fitting of this equation to only a few selected data of the three isotherms. The selection of data is arbitrary. Here we choose the first four data points to illustrate the application of the Freundlich equation. The plots of $\log_{10}(C_\mu)$ versus $\log_{10}(P)$ for the three temperatures are shown in Figure 3.2-3, and they all can be fitted with a straight line.

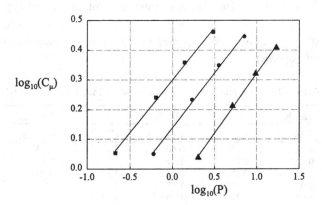

Figure 3.2-3: Plot of $\log_{10}(C_\mu)$ versus $\log_{10}(P)$

Using the linear regression, we find the constants K and n at these three temperatures, and the results are tabulated in Table 3.2-2.

Table 3.2-2: Values for the optimal parameters K and n

T (K)	K	n
283	1.99	2.82
303	1.37	2.70
333	0.83	2.50

We note that the parameter K decreases with temperature, and so does the parameter n. This particular temperature dependence will be discussed next in Section 3.2.1.2.

3.2.1.1 Theoretical Justification of the Freundlich Equation

Although the Freundlich equation was proposed originally as an empirical equation, it can be derived, however, from a sound theoretical footing. It is obtained by assuming that the surface is heterogeneous in the sense that the adsorption energy is distributed and the surface topography is patchwise, that is sites having the same adsorption energy are grouped together into one patch (the adsorption energy here is the energy of interaction between adsorbate and adsorbent). Each patch is independent from each other, that is there is no interaction between patches. Another assumption is that on each patch adsorbate molecule only adsorbs onto one and only one adsorption site; hence, the Langmuir equation is applicable for the description of equilibria of each patch, that is for a patch with an adsorption energy E, the **local** adsorption isotherm equation of that patch is:

$$C_\mu(E) = C_{\mu s} \frac{b(E)P}{1+b(E)P} = C_{\mu s} \frac{b_\infty \exp(E/R_g T)P}{1+b_\infty \exp(E/R_g T)P} \qquad (3.2\text{-}2)$$

where E is the interaction energy between solid and adsorbate molecule.

Zeldowitsch assumed that the energy distribution of all patches follows the exponential decay function. Let the number of sites having adsorption energy between E and E + dE be F(E)dE, where F(E) is given by:

$$F(E) = a \cdot \exp(-E/E_0) \qquad (3.2\text{-}3)$$

where a and E_0 are constants with the product aE_0 being the total number of sites. The overall fractional coverage is simply the average of the local adsorption isotherm over the full energy distribution, that is:

$$\theta = \int_0^\infty \theta(E) F(E) dE \Big/ \int_0^\infty F(E) dE \qquad (3.2\text{-}4a)$$

where

$$\theta(E) = C_\mu(E)/C_{\mu s} \qquad (3.2\text{-}4b)$$

Substitution of the local isotherm equation (3.2-2) and the energy distribution (eq. 3.2-3) into eq. (3.2-4a), and after some approximations, Zeldowitsch derived the Freundlich isotherm. Thus, the Freundlich equation has some theoretical basis at least for heterogeneous solids having an exponential decay energy distribution and Langmuir adsorption mechanism is operative on all patches.

3.2.1.2 Temperature Dependence of K and n

The parameters K and n of the Freundlich equation (3.2-1) are dependent on temperature as seen in Example 3.2-1. Their dependence on temperature is complex, and one should not extrapolate them outside their range of validity. The system of CO adsorption on charcoal (Rudzinski and Everett, 1992) has temperature-dependent n such that its inverse is proportional to temperature. This exponent was found to approach unity as the temperature increases. This, however, is taken as a specific trend rather than a general rule.

To derive the temperature dependence of K and n, we resort to an approach developed by Urano et al. (1981). They assumed that a solid surface is composed of sites having a distribution in surface adsorption potential, which is defined as:

$$A' = R_g T \ln\left(\frac{P_0}{P'}\right) \qquad (3.2\text{-}5)$$

The adsorption potential A' is the work (energy) required to bring molecules in the gas phase of pressure P' to a condensed state of vapor pressure P_0. This means that sites associated with this potential A' will have a potential to condense molecules from the gas phase of pressure P'. If the adsorption potential of the gas

$$A = R_g T \ln\left(\frac{P_0}{P}\right) \qquad (3.2\text{-}6)$$

is less than the adsorption potential A' of a site, then that site will be occupied by an adsorbate molecule. On the other hand, if the gas phase adsorption potential is greater, then the site will be unoccupied (Figure 3.2-4). Therefore, if the surface has a distribution of surface adsorption potential F(A') with F(A')dA' being the amount adsorbed having adsorption potential between A' and A'+dA', the adsorption isotherm equation is simply:

$$C_\mu = \int_A^\infty F(A') dA' \qquad (3.2\text{-}7)$$

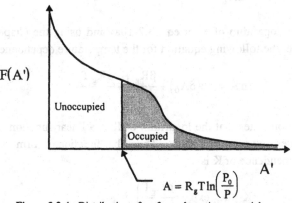

Figure 3.2-4: Distribution of surface adsorption potential

If the density function F(A') takes the form of decaying exponential function

$$F(A) = \delta \cdot \exp(-A/A_0) \qquad (3.2\text{-}8)$$

where A_0 is the characteristic adsorption potential, the above integral can be integrated to give the form of the Freundlich equation:

$$C_\mu = K P^{1/n} \qquad (3.2\text{-}9)$$

where the parameter K and the exponent (1/n) are related to the distribution parameters δ, A_0, and the vapor pressure and temperature as follows:

$$K = (\delta A_0) P_0^{-R_g T/A_0} \qquad (3.2\text{-}10a)$$

$$\frac{1}{n} = \frac{R_g T}{A_0} \qquad (3.2\text{-}10b)$$

The parameter n for most practical systems is greater than unity; thus eq. (3.3-10b) suggests that the characteristic adsorption energy of surface is greater than the molar thermal energy $R_g T$. Provided that the parameters δ and A_0 of the distribution function are constant, the parameter 1/n is a linear function of temperature, that is nRT is a constant, as experimentally observed for adsorption of CO in charcoal for the high temperature range (Rudzinski and Everett, 1992). To find the temperature dependence of the parameter K, we need to know the temperature dependence of the vapor pressure, which is assumed to follow the Clapeyron equation:

$$\ln P_0 = \alpha - \frac{\beta}{T} \qquad (3.2\text{-}11)$$

Taking the logarithm of K in eq. (3.2-10a) and using the Clapeyron equation (3.2-11), we get the following equation for the temperature dependence of lnK:

$$\ln K = \left[\ln(\delta A_0) + \frac{\beta R_g}{A_0}\right] - \frac{\alpha R_g T}{A_0} \qquad (3.2\text{-}12a)$$

This equation states that the logarithm of K is a linear function of temperature, and it decreases with temperature. Thus the functional form to describe the temperature dependence of K is

$$K = K_0 \exp\left(-\frac{\alpha R_g T}{A_0}\right) \qquad (3.2\text{-}12b)$$

and hence the explicit temperature dependence form of the Freundlich equation is:

$$C_\mu = K_0 \exp\left(-\frac{\alpha R_g T}{A_0}\right) \times P^{R_g T/A_0} \qquad (3.2\text{-}12c)$$

Since $\ln(C_\mu)$ and $1/n$ are linear in terms of temperature, we can eliminate the temperature and obtain the following relationship between lnK and n:

$$\ln K = \left[\ln(\delta A_0) + \frac{R_g \beta}{A_0}\right] - \frac{\alpha}{n} \qquad (3.2\text{-}13)$$

suggesting that the two parameters K and n in the Freundlich equation are not independent. Huang and Cho (1989) have collated a number of experimental data and have observed the linear dependence of ln(K) and (1/n) on temperature. We should, however, be careful about using this as a general rule for extrapolation as the temperature is sufficiently high, the isotherm will become linear, that is n = 1, meaning that 1/n no longer follows the linear temperature dependence as suggested by eq. (3.2-10b). Thus, eq. (3.2-13) has its narrow range of validity, and must be used with extreme care. Using the propane data on activated carbon (Table 3.2-1), we show in Figure 3.2-5 that lnK and 1/n (tabulated in Table 3.2-2) are linearly related to each other, as suggested by eq.(3.2-13).

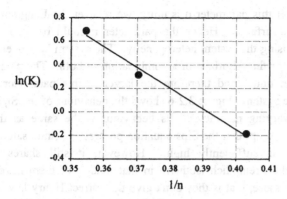

Figure 3.2-5: Plot of ln(K) versus 1/n for propane adsorption on activated carbon

3.2.1.3 Heat of Adsorption

Knowing K and n as a function of temperature, we can use the van't Hoff equation

$$\Delta H = -R_g T^2 \left(\frac{\partial \ln P}{\partial T}\right)_{C_\mu} \quad (3.2\text{-}14)$$

to determine the isosteric heat of adsorption. The result is (Huang and Cho, 1989):

$$\Delta H = -\left[\ln(\delta A_0') + \frac{R_g \beta}{A_0}\right] A_0 + A_0 \ln(C_\mu) \quad (3.2\text{-}15)$$

Thus, the isosteric heat is a linear function of the logarithm of the adsorbed amount.

3.2.2 Sips Equation (Langmuir-Freundlich)

Recognising the problem of the continuing increase in the adsorbed amount with an increase in pressure (concentration) in the Freundlich equation, Sips (1948) proposed an equation similar in form to the Freundlich equation, but it has a finite limit when the pressure is sufficiently high.

$$C_\mu = C_{\mu s} \frac{(bP)^{1/n}}{1 + (bP)^{1/n}} \quad (3.2\text{-}16)$$

In form this equation resembles that of Langmuir equation. The difference between this equation and the Langmuir equation is the additional parameter "n" in the Sips equation. If this parameter n is unity, we recover the Langmuir equation applicable for ideal surfaces. Hence the parameter n could be regarded as the parameter characterising the system heterogeneity. The system heterogeneity could stem from the solid or the adsorbate or a combination of both. The parameter n is usually greater than unity, and therefore the larger is this parameter the more heterogeneous is the system. Figure 3.2-6 shows the behaviour of the Sips equation with n being the varying parameter. Its behaviour is the same as that of the Freundlich equation except that the Sips equation possesses a finite saturation limit when the pressure is sufficiently high. However, it still shares the same disadvantage with the Freundlich isotherm in that neither of them have the right behaviour at low pressure, that is they don't give the correct Henry law limit. The isotherm equation (3.2-16) is sometimes called the Langmuir-Freundlich equation in the literature because it has the combined form of Langmuir and Freundlich equations.

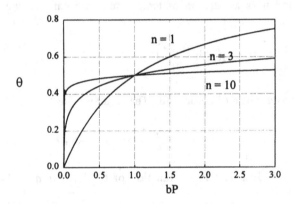

Figure 3.2-6: Plots of the Sips equation versus bP

To show the good utility of this empirical equation in fitting data, we take the same adsorption data of propane onto activated carbon (Table 3.2-1) used earlier in the testing of the Freundlich equation. The following figure (Figure 3.2-7) shows the degree of good fit between the Sips equation and the data. The fit is excellent and it is fairly widely used to describe data of many hydrocarbons on activated carbon with good success. For each temperature, the fitting between the Sips equation and experimental data is carried out with MatLab nonlinear optimization routine, and the optimal parameters from the fit are tabulated in the following table. A code ISO_FIT1 provided with this book is used for this optimisation, and students are encouraged to use this code to exercise on their own adsorption data.

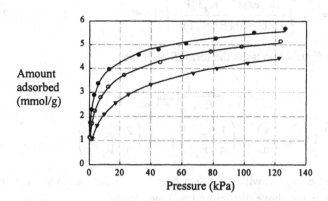

Figure 3.2-7: Fitting of the propane/activated carbon data with the Sips equation (symbol -data; line: fitted equation)

The optimal parameters from the fitting of the Sips equation with the experimental data are tabulated in Table 3.2-3.

Table 3.2-3: Optimal parameters for the Sips equation in fitting propane data on activated carbon

T (K)	$C_{\mu s}$ (mmole/g)	b (kPa^{-1})	n (-)
283	7.339	0.1107	2.306
303	7.232	0.04986	2.117
333	7.583	0.01545	1.956

The parameter n is greater than unity, suggesting some degree of heterogeneity of this propane/ activated carbon system. The larger is this parameter, the higher is the degree of heterogeneity. However, this information does not point to what is the source of the heterogeneity, whether it be the solid structural property, the solid energetical property or the sorbate property. We note from the above table that the parameter n decreases with temperature, suggesting that the system is "apparently" less heterogeneous as temperature increases.

3.2.2.1 The Theoretical Basis of the Sips Equation

Using the energy distribution approach (the approach which we shall discuss in more detail in Chapter 6), Sips derived eq. (3.2-16) from the following integral equation, which is the average of the local Langmuir isotherm equation over an energy distribution as shown below:

$$C_\mu = C_{\mu s} \int_{-\infty}^{\infty} \frac{b_\infty \exp(E/R_gT)P}{1+b_\infty \exp(E/R_gT)P} F(E) \, dE \qquad (3.2\text{-}17a)$$

where the energy distribution takes the form (Sips, 1948)

$$F(E) = \frac{1}{\pi R_g T} \frac{\exp\left(\frac{1}{n}\frac{E_m - E}{R_g T}\right) \sin\left(\frac{\pi}{n}\right)}{1 + 2\cos\left(\frac{\pi}{n}\right) \exp\left(\frac{1}{n}\frac{E_m - E}{R_g T}\right) + \exp\left(\frac{2}{n}\frac{E_m - E}{R_g T}\right)} \qquad (3.2\text{-}17b)$$

Here E_m is the energy at which the distribution is maximum. When the energy is large and positive, the above distribution will reduce to:

$$F(E) \cong \frac{1}{R_g T} \exp\left(\frac{1}{n}\frac{E_m - E}{R_g T}\right) \qquad (3.2\text{-}17c)$$

This distribution is an exponential decay function with respect to the adsorption energy. Figure 3.2-8 shows the plot of the energy distribution (3.2-17b) versus $E_m - E$ at 273 K.

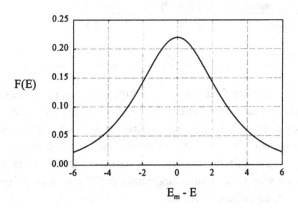

Figure 3.2-8: Plot of the energy distribution versus $E_m - E$ fot T = 273 K

It exhibits a Gaussian distribution shape, and when the energy E is either larger or smaller than E_m the distribution exhibits an exponential decay.

3.2.2.2 The Temperature Dependence of the Sips Equation

For useful description of adsorption equilibrium data at various temperatures, it is important to have the temperature dependence form of an isotherm equation. The temperature dependence of the Sips equation

$$C_\mu = C_{\mu s} \frac{(bP)^{1/n}}{1+(bP)^{1/n}} \qquad (3.2\text{-}18a)$$

for the affinity constant b and the exponent n may take the following form:

$$b = b_\infty \exp\left(\frac{Q}{R_g T}\right) = b_0 \exp\left[\frac{Q}{R_g T_0}\left(\frac{T_0}{T}-1\right)\right] \qquad (3.2\text{-}18b)$$

$$\frac{1}{n} = \frac{1}{n_0} + \alpha\left(1-\frac{T_0}{T}\right) \qquad (3.2\text{-}18c)$$

Here b_∞ is the adsorption affinity constant at infinite temperature, b_0 is that at some reference temperature T_0, n_0 is the parameter n at the same reference temperature and α is a constant parameter. The temperature dependence of the affinity constant b is taken from the that of the Langmuir equation. Unlike Q in the Langmuir equation, where it is the isosteric heat, invariant with the surface loading, the parameter Q in the Sips equation is only the measure of the adsorption heat. We shall discuss its physical meaning in Section 3.2.2.2.1. The temperature-dependent form of the exponent n is empirical and such form in eq. (3.2-18c) is chosen because of its simplicity. The saturation capacity can be either taken as constant or it can take the following temperature dependence:

$$C_{\mu s} = C_{\mu s,0} \exp\left[\chi\left(1-\frac{T}{T_0}\right)\right] \qquad (3.2\text{-}18d)$$

Here $C_{\mu s,0}$ is the saturation capacity at the reference temperature T_0, and χ is a constant parameter. This choice of this temperature-dependent form is arbitrary.

This temperature dependence form of the Sips equation (3.2-18) can be used to fit adsorption equilibrium data of various temperatures simultaneously to yield the parameter b_0, $C_{\mu s,0}$, Q/RT_0, n_0 and α.

Example 3.2-2: *Fitting of propane/AC data with temperature dependent Sips equation*

Using the data of propane at three temperatures 283, 303 and 333 K (Table 3.2-1) simultaneously in the fitting of the Sips equation (3.2-18), we get the following optimal parameters:

Table 3.2-4: Optimal parameters for the temperature dependent Sips equation

$C_{\mu s,0}$	7.348 mmole/g
χ	0
b_0	0.1075 (kPa)$^{-1}$
Q/RT_0	12.22
n_0	2.312
α	0.5559

where the reference temperature used was $T_0 = 283$ K. The Matlab code ISO_FIT2 is provided with this book for this task of fitting of multiple isotherm data at various temperatures.

Knowing the adsorption heat number $Q/R_g T_0$, the parameter Q can be calculated as

$$Q = (12.22) \times \left(8.314 \frac{\text{Joule}}{\text{mole} \cdot \text{K}}\right) \times (283 \text{ K}) = 28{,}750 \frac{\text{Joule}}{\text{mole}}$$

This parameter Q is a measure of the adsorption heat, and the above value is typical for adsorption of low alkanes and alkenes on activated carbon. The parameter n decreases with temperature. The same behaviour is found for the exponent (1/n) in the Freundlich equation. The dependence of n on temperature in eq. (3.2-18c) can be used to extrapolate to temperatures outside the range used in the fitting (in this case [283, 333K]) provided that they are not too far away from the fitted range. For example, the extrapolated value of n at 373K is

$$\frac{1}{n_{373}} = \frac{1}{2.312} + 0.5559\left(1 - \frac{283}{373}\right) = 0.5667$$

Thus

$$n_{373} = 1.765$$

3.2.2.2.1 Isosteric Heat and the Physical Meaning of the Parameter Q

To obtain the isosteric heat for the temperature dependence form of the Sips equation as given in eq. (3.2-18), we use the van't Hoff equation

$$(-\Delta H) = R_g T^2 (\partial \ln P / \partial T)_\theta$$

and obtain the following result for the isosteric heat (Appendix 3.1).

$$(-\Delta H) = Q - (\alpha R_g T_0) \cdot n \ln(bP) \qquad (3.2\text{-}18e)$$

or written in terms of fractional loading, we have:

$$(-\Delta H) = Q - (\alpha R_g T_0) n^2 \ln\left(\frac{\theta}{1-\theta}\right) \qquad (3.2\text{-}18f)$$

or in terms of the amount adsorbed C_μ

$$(-\Delta H) = Q - (\alpha R_g T_0) n^2 \ln\left(\frac{C_\mu}{C_{\mu s} - C_\mu}\right) \qquad (3.2\text{-}18g)$$

in which we have assumed that the temperature variation of $C_{\mu s}$ is negligible. The above equation states that the isosteric heat of adsorption decreases with pressure (i.e. with loading). It takes a value of infinity at zero loading and minus infinity at saturation. Thus, the Sips equation, despite having the correct finite capacity at sufficiently large pressure, has its applicability only in the intermediate range of pressure. Eq. (3.2-18e) also states that beside its dependence on the loading it is also a function of temperature, reflecting through the temperature dependence of the parameters n and b.

The isosteric heat equation (3.2-18f) reveals the physical meaning of the parameter Q in the affinity equation (3.2-18b). It shows that when the fractional loading is equal to one half, the isosteric heat is equal to Q. Thus <u>the parameter Q defined in the affinity constant b is the isosteric heat at the fractional loading of 0.5</u>:

$$Q = (-\Delta H)\big|_{\theta=1/2}$$

Using the parameters obtained earlier for the system of propane on activated carbon into eq. (3.2-18g), we get the following expression for the isosteric heat of adsorption as a function of the adsorbed concentration at 283 K

$$(-\Delta H) = 28750 - 6991 \times \ln\left(\frac{C_\mu}{7.348 - C_\mu}\right) \quad \frac{\text{Joule}}{\text{mole}}$$

The following table shows the variation of the isosteric heat the amount adsorbed.

Table 3.2-5: Isosteric heat as a function of loading using the Sips equation

C_μ (mmole/g)	$(-\Delta H)$ (Joule/mole)
1	41,670
2	35,626
3	31,344
4	27,506
5	23,466
6	18,311

Similarly we can obtain the isosteric heat as a function of loading at the other two temperatures 303 and 333K, and Figure 3.2-9 shows plots of the isosteric heat of propane on activated carbon versus loading for the three temperatures 283, 303 and 333 K.

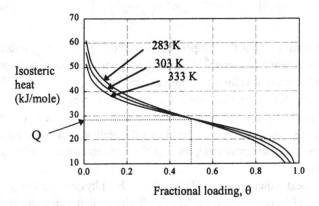

Figure 3.2-9: Plot of the isosteric heat versus fractional loading

Note the pattern of the isosteric heat with respect to the fractional loading, and the three curves intersect at the same point corresponding to the fractional loading of 0.5 and the isosteric heat of Q. This is the characteristics of the Sips equation.

3.2.3 Toth Equation

The previous two equations have their limitations. The Freundlich equation is not valid at low and high end of the pressure range, and the Sips equation is not valid at the low end as they both do not possess the correct Henry law type behaviour. One of the empirical equations that is popularly used and satisfies the two end limits is the Toth equation. This equation describes well many systems with sub-monolayer coverage, and it has the following form:

$$C_\mu = C_{\mu s} \frac{bP}{\left[1+(bP)^t\right]^{1/t}} \quad (3.2\text{-}19a)$$

Here t is a parameter which is usually less than unity. The parameters b and t are specific for adsorbate-adsorbent pairs.

When t = 1, the Toth isotherm reduces to the famous Langmuir equation; hence like the Sips equation the parameter t is said to characterize the system heterogeneity. If it is deviated further away from unity, the system is said to be more heterogeneous. The effect of the Toth parameter t is shown in Figure 3.2-10, where we plot the fractional loading ($C_\mu/C_{\mu s}$) versus bP with t as the varying parameter. Again we note that the more the parameter t deviates from unity, the more heterogeneous is the system. The Toth equation has correct limits when P approaches either zero or infinity.

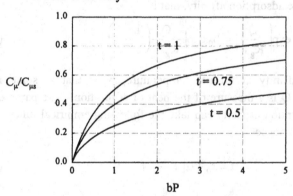

Figure 3.2-10: Plot of the fractional loading versus bP for the Toth equation

Being the three-parameter model, the Toth equation can describe well many adsorption data. We apply this isotherm equation to fit the isotherm data of propane on activated carbon. For example taking the isotherm data at 303 K in Table 3.2-1, the extracted optimal parameters are (using the ISO_FIT1 routine):

$$C_{\mu s} = 33.56 \text{ mmole/g}, \ b = 0.069 \text{ (kPa)}^{-1}, \ t = 0.233$$

The parameter t takes a value of 0.233 (well deviated from unity) indicates a strong degree of heterogeneity of the system.

Several hundred sets of data for hydrocarbons on Nuxit-al charcoal obtained by Szepesy and Illes (Valenzuela and Myers, 1989) can be described well by this equation. Because of its simplicity in form and its correct behaviour at low and high pressures, the Toth equation is recommended as the first choice of isotherm equation for fitting data of many adsorbates such as hydrocarbons, carbon oxides, hydrogen sulfide, alcohols on activated carbon as well as zeolites. Sips equation presented in the last section is also recommended but when the behaviour in the Henry law region is needed, the Toth equation is the better choice.

3.2.3.1 Temperature Dependence of the Toth Equation

Like the other equations described so far, the temperature dependence of equilibrium parameters in the Toth equation is required for the purpose of extrapolation or interpolation of equilibrium at other temperatures as well as the purpose of calculating isosteric heat.

The parameters b and t are temperature dependent, with the parameter b taking the usual form of the adsorption affinity, that is

$$b = b_\infty \exp\left(\frac{Q}{R_g T}\right) = b_0 \exp\left[\frac{Q}{R_g T_0}\left(\frac{T_0}{T} - 1\right)\right] \quad (3.2\text{-}19b)$$

where b_∞ is the affinity at infinite temperature, b_0 is that at some reference temperature T_0 and Q is a measure of the heat of adsorption. The parameter t and the maximum adsorption capacity can take the following empirical functional form of temperature dependence

$$t = t_0 + \alpha\left(1 - \frac{T_0}{T}\right) \quad (3.2\text{-}19c)$$

$$C_{\mu s} = C_{\mu s,0} \exp\left[\chi\left(1 - \frac{T}{T_0}\right)\right] \quad (3.2\text{-}19d)$$

The temperature dependence of the parameter t does not have any sound theoretical footing; however, we would expect that as the temperature increases this parameter will approach unity.

Example 3.2-3: *Fitting of propane/AC data with temperature dependent Toth equation*

The temperature-dependent form of the Toth equation (3.2-19) is used to simultaneously fit the isotherm data at 283, 303 and 333 K of propane onto activated carbon (using the ISO_FIT2 routine), and we extract the following optimally fitted parameters

Table 3.2-6: Optimal parameters for the temperature dependent Toth equation

$C_{\mu s,0}$	8.562 mmole/g
b_0	10.54 (kPa)$^{-1}$
$Q/R_g T_0$	19.59
t_0	0.2842
α	0.284
χ	0

where the reference temperature is $T_0 = 283$ K. Knowing the adsorption heat number Q/RT_0, the parameter Q is calculated as:

$$Q = 46093 \text{ Joule/mole}$$

This measure of heat of adsorption is much higher than that derived from the fitting of the data with the Sips equation earlier, where we have found a value of 28750 J/mol for Q. This large difference should cause no alarm as the parameter Q is only the measure of adsorption heat. For example in the case of the Sips equation, Q is the isosteric heat of adsorption at a fractional loading of 0.5, while the parameter Q in the case of the Toth equation is the isosteric heat of adsorption at zero fractional loading as we shall show in the next section.

3.2.3.1.1 The Isosteric Heat and the Physical Meaning of the Parameter Q

Using the temperature dependence of b and t as given in eq. (3.2-19) and applying the van't Hoff equation

$$\left(-\frac{\Delta H}{R_g T^2}\right) = \left(\frac{\partial \ln P}{\partial T}\right)_\theta$$

we obtain the following isosteric heat equation for the Toth equation (Appendix 3.2).

$$(-\Delta H) = Q - \frac{1}{t}\left(\alpha R_g T_0\right)\left\{\ln(bP) - \left[1+(bP)^t\right]\ln\left[\frac{bP}{\left(1+(bP)^t\right)^{1/t}}\right]\right\} \quad (3.2\text{-}19e)$$

Written in terms of the fractional loading or adsorbed concentration C_μ, the isosteric heat is:

$$(-\Delta H) = Q - \frac{1}{t}(\alpha R_g T_0)\left\{\ln\left[\frac{\theta}{(1-\theta^t)^{1/t}}\right] - \frac{\ln \theta}{(1-\theta^t)}\right\} \qquad (3.2\text{-}19f)$$

or

$$(-\Delta H) = Q - \frac{1}{t}(\alpha R_g T_0)\left\{\ln\left[\frac{C_\mu}{(C_{\mu s}^t - C_\mu^t)^{1/t}}\right] - \frac{\ln(C_\mu/C_{\mu s})}{1-(C_\mu/C_{\mu s})^t}\right\} \qquad (3.2\text{-}19g)$$

Like the Sips equation, the isosteric heat of adsorption is a function of pressure (or loading), and it takes a value of infinity at zero loading and minus infinity at very high loading, which limits the applicability of the Toth equation in its use in the calculation of isosteric heat at two ends of the loading. The meaning of the parameter Q in the Toth equation is now clear in eq. (3.2-19f). It is equal to the isosteric heat when the fractional loading is zero

$$Q = (-\Delta H)\big|_{\theta=0}$$

Since the Sips and Toth empirical equations fit the equilibrium data reasonably well, it may be possible to use the isosteric heat as a function of loading as a criterion in choosing the correct isotherm equation. Using eq. (3.2-19g) and the optimal parameters in Table 3.2-6 for the Toth equation, we obtain the following equation for the isosteric heat at 283 K with values tabulated in Table 3.2-7:

$$(-\Delta H) = 46{,}093 - 2{,}351 \times \left\{\ln\left[\frac{C_\mu}{(1.841 - C_\mu^{0.2842})^{3.52}}\right] - \frac{\ln(C_\mu/8.562)}{1-(C_\mu/8.562)^{0.2842}}\right\} \quad \frac{\text{Joule}}{\text{mole}}$$

Table 3.2-7: Isosteric heat as a function of loading using the Toth equation

C_μ (mmol/g)	$(-\Delta H)$ (Joule/mole)
1	33,608 (-19%)
2	30,451 (-14%)
3	27,775 (-11%)
4	25,137 (-9%)
5	22,274 (-5%)
6	18,854 (-3%)

The isosteric heats calculated by the Toth equation are lower than those calculated by the Sips equation (Table 3.2-5). The above table shows the percentage differences between the values calculated by the Sips and Toth equations. The difference is seen to be significant enough for the isosteric heat to be used as the criterion to better select the isotherm equation.

Take the example of propane on activated carbon, we plot the isosteric heat (eq. 3.2-19f) versus the fractional loading for the Toth equation using the parameters obtained in Table 3.2-6 (Figure 3.2-11). Just like the case of the Sips isotherm, the isosteric heat decreases with loading, and it shows a weak temperature dependence.

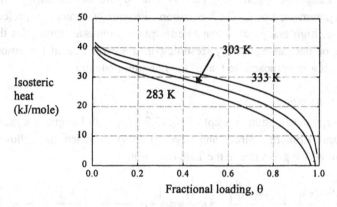

Figure 3.2-11: Plots of the isosteric heat versus fractional loading using the Toth equation

3.2.3.2 Other Properties of the Toth Equation

Although the Toth equation, like Sips and Freundlich, is an empirical equation, it has an advantage over the other two in that it has the following Henry constant at zero loading:

$$C_\mu = (C_{\mu s} b) P \qquad (3.2\text{-}20a)$$

where the Henry constant (H) is $bC_{\mu s}$, which is similar to that of the Langmuir equation. Let us investigate the slope of the Toth adsorption isotherm at finite loadings:

$$\frac{dC_\mu}{dP} = \frac{bC_{\mu s}}{\left[1+(bP)^t\right]^{(1/t+1)}} \qquad (3.2\text{-}20b)$$

70 Equilibria

or written in terms of the fractional loading:

$$\frac{dC_\mu}{dP} = bC_{\mu s}(1-\theta^t)^{1/t+1} \qquad (3.2\text{-}20c)$$

where

$$\theta = C_\mu / C_{\mu s} \qquad (3.2\text{-}20d)$$

The slope of the Toth isotherm (3.2-20c) has a constant limit at zero loading, and at a given loading it decreases with the loading at a rate much faster than that for the case of Langmuir equation. This is due to the heterogeneity effect manifested in the parameter t of the Toth equation. Physically, molecules prefer to adsorb onto sites of high energy and then as adsorption progresses molecules then adsorb onto sites of decreasing energy, resulting in a slower rise in the amount adsorbed versus pressure compared to that of the Langmuir equation.

3.2.3.3 Energy Distribution

Using the energy distribution concept of eq. (3.2-17a) with Langmuir equation describing the local isotherm, Sircar and Myers (1984) provided the following energy distribution which gives rise to the Toth equation:

$$F(E) = \frac{a}{\pi R_g T} \left\{ \frac{[a^t + bx^t \cos(\pi t)]^2 + [bx^t \sin(\pi t)]^2}{[a^{2t} + 2a^t bx^t \cos(\pi t) + b^2 x^{2t}]^2} \right\}^{1/2t} \times \sin\left\{ \frac{1}{t} \arcsin\left[\frac{bx^t \sin(\pi t)}{\left\{ [a^t + bx^t \cos(\pi t)]^2 + [bx^t \sin(\pi t)]^2 \right\}^{1/2}} \right] \right\}$$

where

$$a = 1/b_0; \quad b = (\bar{b})^{-t}; \quad x = \exp\left(\frac{E}{R_g T}\right) - 1 \qquad (3.2\text{-}21)$$

Eq. (3.2-21) is the energy distribution of the Toth equation. Having an energy distribution only means that the system is not homogeneous, and it does not point to the source of heterogeneity as well as the physical meaning of the parameters in the energy distribution.

3.2.4 Unilan equation

Unilan equation is another empirical relation obtained by assuming a patchwise topography on the surface and each patch is ideal such that the local Langmuir isotherm is applicable on each patch. The distribution of energy is assumed uniform. Integrating eq. (3.2-17a) with the following uniform energy distribution

$$F(E) = \begin{cases} \dfrac{1}{E_{max} - E_{min}} & \text{for } E_{min} < E < E_{max} \\ 0 & \text{for } E < E_{min} \text{ or } E > E_{max} \end{cases} \quad (3.2\text{-}22)$$

we obtain the following result, called the Unilan equation (The term Unilan comes from <u>uni</u>form distribution and <u>Lan</u>gmuir local isotherm)

$$C_\mu = \frac{C_{\mu s}}{2s} \ln\left(\frac{1 + \bar{b}e^s P}{1 + \bar{b}e^{-s} P}\right) \quad (3.2\text{-}23a)$$

where

$$\bar{b} = b_\infty \exp\left(\frac{\bar{E}}{R_g T}\right) = b_0 \exp\left[-\frac{\bar{E}}{R_g T_0}\left(1 - \frac{T_0}{T}\right)\right] \quad (3.2\text{-}23b)$$

$$\bar{E} = \frac{E_{max} + E_{min}}{2} \qquad s = \frac{E_{max} - E_{min}}{2R_g T} \quad (3.2\text{-}23c)$$

Here E_{max} and E_{min} are the maximum and minimum energies of the distribution, and b_∞ is the adsorption affinity at infinite temperature. The parameter s characterises the heterogeneity of the system. The larger this parameter is, the more heterogeneous is the system. If s = 0, the Unilan equation (3.2-23) reduces to the classical Langmuir equation as in this limit the range of energy distribution is zero. Figure (3.2-12) shows the behaviour of the Unilan equation with s being the varying parameter.

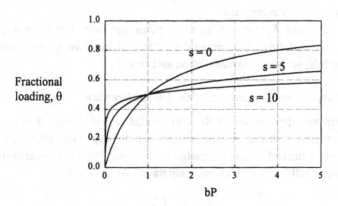

Figure 3.2-12: Plot of the fractional loading versus bP for the Unilan equation

72 Equilibria

There we note that the larger is the value of s (around 10) the closer is the isotherm to the rectangular (irreversible) behaviour.

The Unilan equation has the correct behaviour at low and high pressures. The Henry behaviour at zero loading is:

$$C_\mu = H P \qquad (3.2\text{-}24a)$$

where H is the Henry constant defined as:

$$H = C_{\mu s} \bar{b} \left(\frac{\sinh s}{s} \right) \qquad (3.2\text{-}24b)$$

When $s = 0$ (homogeneous solids), the Henry constant will become $H = C_{\mu s} \bar{b}$.

The slope of the Unilan equation in terms of pressure is:

$$\frac{dC_\mu}{dP} = C_{\mu s} \bar{b} \, \frac{\sinh s}{s} \times \frac{1}{\left(1 + \bar{b} e^{s} P\right)\left(1 + \bar{b} e^{-s} P\right)} \qquad (3.2\text{-}25a)$$

or in terms of fractional loading, it is (using eq. 3.2-23)

$$\frac{dC_\mu}{dP} = C_{\mu s} \bar{b} \, \frac{\sinh s}{s} \times e^{-2s\theta} \left[\frac{1 - e^{-2s(1-\theta)}}{1 - e^{-2s}} \right]^2 \qquad (3.2\text{-}25b)$$

As seen in Figure 3.2-12, the slope of the isotherm is large for large value of s initially and then it decreases rapidly as the loading increases, indicating adsorption at low pressure is favourable at stronger energy sites and then adsorption takes place on progressingly weaker energy sites.

Like the Sips and Toth equation, the Unilan equation is a three-parameter isotherm equation, and they are commonly used to correlate adsorption equilibrium data of many solids, such as activated carbon and zeolite.

3.2.4.1 The Temperature Dependence of the Unilan Equation

The temperature dependence of the Unilan equation is shown in eqs.(3.2-23) assuming the maximum and minimum energies are not dependent on temperature. Like the last two empirical isotherm equations (Sips and Toth), we assume that the saturation capacity follows the following temperature dependence:

$$C_{\mu s} = C_{\mu s,0} \exp\left[\chi \left(1 - \frac{T}{T_0} \right) \right] \qquad (3.2\text{-}26)$$

Example 3.2-4: *Fitting of propane/AC data with temperature dependent Unilan equation*

Using such temperature dependence, the Unilan equation can be used to fit simultaneously the data of propane on activated carbon at 283, 303 and 333 K (Table 3.2-1). We obtain the following extracted fitted parameters by using the ISO_FIT2 routine:

Table 3.2-8: *Optimal parameters of the Unilan equation*

$C_{\mu s,0}$	13.91 mmole/g
b_0	0.001355 (kPa)$^{-1}$
$\bar{E}/RT_0$	8
ΔE	43590
χ	0

The reference temperature T_0 used in the above fitting is 283 K. The degree of fitting between the Unilan equation and the experimental data is comparable to that for the cases of Sips and Toth equations. The discrimination between these three empirical equations now rests on the behaviour of the isosteric heat. This is what we will do next.

3.2.4.1.1 The Isosteric Heat

Using the temperature dependence form of the Unilan equation (3.2-23) into the van't Hoff equation (3.2-14), we derive the following expression for the isosteric heat as a function of loading (Appendix 3.3).

$$(-\Delta H) = \bar{E} + \frac{2(1-\theta)}{\bar{b}P}\left[\frac{(e^s + \bar{b}P)(e^{-s} + \bar{b}P)}{(e^s - e^{-s})}\right]\frac{\Delta E}{2} - \left(\frac{2 + e^s\bar{b}P + e^{-s}\bar{b}P}{e^s - e^{-s}}\right)\frac{\Delta E}{2\bar{b}P} \quad (3.2\text{-}27a)$$

The term $\bar{b}P$ in the above equation is related to the fractional loading through the isotherm equation (3.2-23a), that is:

$$\bar{b}P = \frac{e^{s\theta} - e^{-s\theta}}{e^{s(1-\theta)} - e^{-s(1-\theta)}}$$

The limit of the isosteric heat (eq. 3.2-27a) at zero loading is:

$$\lim_{P\to 0}(-\Delta H) = \bar{E} + \left(\coth s - \frac{1}{s}\right)\frac{\Delta E}{2} \quad (3.2\text{-}27b)$$

Also, the limit at very high loading is:

$$\lim_{P\to\infty}(-\Delta H) = \bar{E} - R_g T\,(s\coth s - 1) \quad (3.2\text{-}27c)$$

Example 3.2-5: *Isosteric heat calculated from the limits of the Unilan equation*

The two limits of the isosteric heat at low and high loading (eqs. 3.2-27b and c) can be obtained by starting from the limits of the isotherm equation (3.2-23). At zero loading, the fractional loading of the Unilan equation is:

$$\lim_{P \to 0} \theta = \left(\overline{b} \cdot \frac{\sinh s}{s}\right) P \qquad (3.2\text{-}28a)$$

from which the isosteric heat can be obtained by substituting it into the van't Hoff equation:

$$\lim_{P \to 0}(-\Delta H) = \overline{E} + \left(\coth s - \frac{1}{s}\right)\frac{\Delta E}{2} \qquad (3.2\text{-}28b)$$

which is exactly the same as eq. (3.2-27b).

Similarly, at very high loading, the Unilan equation (3.2-23) reduces to:

$$\lim_{P \to \infty} \theta = 1 - \left(\frac{\sinh s}{s}\right)\left(\frac{1}{bP}\right) \qquad (3.2\text{-}28c)$$

from which the isosteric heat can be obtained as:

$$\lim_{P \to \infty}(-\Delta H) = \overline{E} - RT(s \coth s - 1) \qquad (3.2\text{-}28d)$$

again confirming eq. (3.2-27c).

The meaning of the mean energy $\overline{E}$ is clear when we consider the isosteric heat equation (3.2-27a). When the isosteric heat of adsorption is equal to $\overline{E}$, it is not difficult to show from eq. (3.2-27a) that $bP = 1$, corresponding to a fractional loading of 0.5. Thus the physical meaning of the mean adsorption energy $\overline{E}$ is that it is the isosteric heat of adsorption at a fractional loading of 0.5. This is similar to the Sips equation where its parameter Q is also the isosteric heat of adsorption at the fractional loading of 0.5.

Using the parameters obtained from the fit of the Unilan equation to the adsorption data of propane on activated carbon at 283, 303 and 333 K, we plot the isosteric heat versus loading at these three temperatures (Figure 3.2-13). Unlike the cases of Sips and Toth where we have seen some moderate temperature dependence of the isosteric heat, this quantity calculated from the Unilan equation is practically independent of temperature.

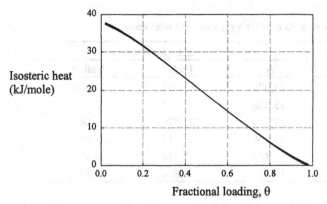

Figure 3.2-13: Plots of the isosteric heat versus fractional loading using the Unilan equation

It is now interesting to compare the isosteric heats calculated from the Sips, Toth and Unilan equations derived from the fitting with propane/activated carbon data at 283, 303 and 333 K. The results are tabulated in Table 3.2-9 and shown graphically in Figure 3.2-14 for T = 283K.

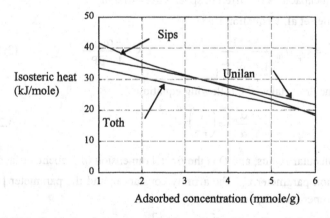

Figure 3.2-14: Comparison between isosteric heats calculated from Sips, Toth and Unilan equations

We see a distinction between the isosteric heat calculated from the Toth equation and those calculated from the Sips and Unilan equations, despite of the fact that the three isotherm equations describe the equilibrium data fairly well. This distinction in the isosteric heat curve could be utilised in the matching between the experimental isosteric heat and the heat equations for the three isotherms (eqs. 3.2-18f, 3.2-19f and 3.2-27a) to decide on the better choice of isotherm equation.

Table 3.2-9: Isosteric heats from the Sips, Toth and Unilan equations

C_μ (mmole/g)	($-\Delta H$) (Joule/mole)		
	Sips	Toth	Unilan
1	41,670	33,608	36,360
2	35,626	30,451	33,881
3	31,344	27,775	31,041
4	27,506	25,137	28,022
5	23,466	22,274	24,930
6	18,311	18,854	21,810

3.2.5 Keller, Staudt and Toth's Equation

Recently, Keller and his co-workers (1996) proposed a new isotherm equation, which is very similar in form to the original Toth equation. The differences between their equation and that of Toth are that:

(a) the exponent α is a function of pressure instead of constant as in the case of Toth
(b) the saturation capacities of different species are different

The form of Keller et al.'s equation is:

$$C_\mu = C_{\mu s} \alpha_m \frac{bP}{\left[1 + (bP)^\alpha\right]^{1/\alpha}} \; ; \qquad \alpha = \frac{1 + \alpha_m \beta P}{1 + \beta P} \qquad (3.2\text{-}29a)$$

where the parameter α_m takes the following equation:

$$\frac{\alpha_m}{\alpha_m^*} = \left(\frac{r}{r^*}\right)^{-D} \qquad (3.2\text{-}29b)$$

Here r is the molecular radius, and D is the fractal dimension of sorbent surface.

The saturation parameter $C_{\mu s}$, the affinity constant b, and the parameter β have the following temperature dependence:

$$C_{\mu s} = C_{\mu s,0} \exp\left[\chi\left(1 - \frac{T}{T_0}\right)\right] \qquad (3.2\text{-}29c)$$

$$b = b_0 \exp\left[\frac{Q_1}{R_g T}\left(\frac{T_0}{T} - 1\right)\right] ; \quad \beta = \beta_0 \exp\left[\frac{Q_2}{R_g T}\left(\frac{T_0}{T} - 1\right)\right] \qquad (3.2\text{-}29d)$$

Here the subscript 0 denotes for properties at some reference temperature T_0. The Keller et al.'s equation contains more parameters than the empirical equations discussed so far.

Fitting the Keller et al's equation with the isotherm data of propane on activated carbon at three temperatures 283, 303 and 333 K, we found the fit is reasonably good, comparable to the good fit observed with Sips and Toth equations. The optimally fitted parameters are:

$C_{\mu s,0}$	16.08 mmole/g
b_0	0.9814 (kPa)$^{-1}$
β_0	3.225 (kPa)$^{-1}$
α_m	0.4438
Q_1/RT_0	10.94
Q_2/RT_0	-0.2863
χ	0.0002476

3.2.6 Dubinin-Radushkevich Equation

The empirical equations dealt with so far, Freundlich, Sips, Toth, Unilan and Keller et al., are applicable to supercritical as well as subcritical vapors. In this section we present briefly a semi-empirical equation which was developed originally by Dubinin and his co-workers for sub critical vapors in microporous solids, where the adsorption process follows a pore filling mechanism. More details about the Dubinin equation will be discussed in Chapter 4.

Hobson and co-workers (1963, 1967, 1969, 1974) and Earnshaw and Hobson (1968) analysed the data of argon on Corning glass in terms of the Polanyi potential theory. They proposed an equation relating the amount adsorbed in equivalent liquid volume (V) to the adsorption potential

$$A = R_g T \ln\left(\frac{P_0}{P}\right) \tag{3.2-30}$$

where P_0 is the vapour pressure. The premise of their derivation is the functional form V(A) which is independent of temperature. They chose the following functional form:

$$\ln V = \ln V_0 - BA^2 \tag{3.2-31}$$

where the logarithm of the amount adsorbed is linearly proportional to the square of the adsorption potential. Eq. (3.2-31) is known as the Dubinin-Radushkevich (DR) equation. Writing this equation explicitly in terms of pressure, we have:

$$V = V_0 \exp\left[-\frac{1}{(\beta E_0)^2}\left(R_g T \ln\frac{P}{P_0}\right)^2\right] \quad (3.2\text{-}32)$$

where E_0 is called the solid characteristic energy towards a reference adsorbate. Benzene has been used widely as the reference adsorbate. The parameter β is a constant which is a function of the adsorptive only. It has been found by Dubinin and Timofeev (1946) that this parameter is proportional to the liquid molar volume. Figure 3.2-15 shows plots of the DR equation versus the reduced pressure with E/RT as the varying parameter.

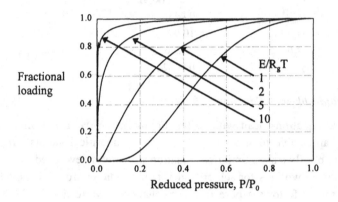

Figure 3.2-15: Plots of the DR equation versus the reduced pressure

We see that as the characteristic energy increases the adsorption is stronger as the solid has stronger energy of interaction with adsorbate. One observation in that equation is that the slope of the adsorption isotherm at zero loading is not finite, a violation of the thermodynamic requirement. This will be discussed in greater detail in Chapter 4, where we will deal with adsorption in microporous solids.

Eq. (3.2-32) when written in terms of amount adsorbed (mole/g) is:

$$C_\mu = C_{\mu s} \exp\left[-\frac{1}{(\beta E_0)^2}\left(R_g T \ln\frac{P}{P_0}\right)^2\right] \quad (3.2\text{-}33a)$$

where the maximum adsorption capacity is:

$$C_{\mu s} = \frac{W_0}{v_M(T)} \quad (3.2\text{-}33b)$$

The parameter W_0 is the micropore volume and v_M is the liquid molar volume. Here we have assumed that the state of adsorbed molecule in micropores behaves like liquid.

Dubinin-Radushkevich equation (3.2-33) is very widely used to describe adsorption isotherm of sub-critical vapors in microporous solids such as activated carbon and zeolite. One debatable point in such equation is the assumption of liquid-like adsorbed phase as one could argue that due to the small confinement of micropore adsorbed molecules experience stronger interaction forces with the micropore walls, the state of adsorbed molecule could be between liquid and solid.

The best utility of the Dubinin-Radushkevich equation lies in the fact that the temperature dependence of such equation is manifested in the adsorption potential A, defined as in eq. (3.2-30), that is if one plots adsorption data of different temperatures as the logarithm of the amount adsorbed versus the square of adsorption potential, all the data should lie on the same curve, which is known as the characteristic curve. The slope of such curve is the inverse of the square of the characteristic energy $E = \beta E_0$.

To show the utility of the DR equation, we fit eq. (3.2-33) to the adsorption data of benzene on activated carbon at three different temperatures, 283, 303 and 333 K. The data are tabulated in Table 3.2-10 and presented graphically in Figure 3.2-16.

Table 3.2-10: Adsorption data of benzene on activated carbon

283 K		303 K		333 K	
P (kPa)	C_μ (mmole/g)	P (kPa)	C_μ (mmole/g)	P (kPa)	C_μ (mmole/g)
0.0133	1.6510	0.0001	0.4231	0.0010	0.4231
0.0933	3.2470	0.0002	0.8462	0.0267	0.8450
0.2932	3.8750	0.0133	1.1110	0.0533	1.1090
0.6798	4.2560	0.0267	1.4060	0.0933	1.4030
1.5590	4.5270	0.0666	1.9540	0.2532	1.9460
2.6520	4.6600	0.0933	2.1660	0.3732	2.1520
4.2920	4.8060	0.1599	2.5090	0.6531	2.4870
6.3580	4.9480	0.3466	2.9730	1.3330	2.9290
8.7440	5.0480	0.6931	3.4310	2.6120	3.3470
10.0200	5.0840	1.2800	3.7610	4.3590	3.6260
		2.8260	4.1490	7.6640	3.9380
		3.9320	4.2770	9.4770	4.0370
		6.6380	4.4410	11.5600	4.2340
		8.5570	4.5370		
		10.410	4.5880		

The vapour pressure and the liquid molar volume of benzene are given in the following table.

Table 3.2-11: Vapour pressure and liquid molar volume of benzene

T (K)	P_0 (kPa)	v_M (cc/mmole)
303	16.3	0.0900
333	52.6	0.0935
363	150	0.0970

By fitting the equilibria data of all three temperatures simultaneously using the ISO_FIT1 program, we obtain the following optimally fitted parameters:

$$W_0 = 0.45 \text{ cc/g}, \quad E = 20,000 \text{ Joule/mole}$$

Even though only one value of the characteristic energy was used in the fitting of the three temperature data, the fit is very good as shown in Figure 3.2-16, demonstrating the good utility of this equation in describing data of sub-critical vapors in microporous solids.

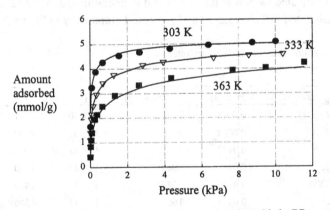

Figure 3.2-16: Fitting the benzene/ activated carbon data with the DR equation

3.2.6.1 Isosteric Heat and the Heat of Immersion

The isosteric heat of adsorption of the DR equation can be calculated from the van't Hoff's equation (3.2-14), and the result is:

$$-\Delta H = \Delta H_{vap} + \beta E_0 \left(\ln \frac{1}{\theta} \right)^{1/2} + \frac{(\beta E_0)\delta T}{2} \left(\ln \frac{1}{\theta} \right)^{-1/2} \quad (3.2\text{-}34a)$$

where ΔH_{vap} is the latent heat of vaporization, δ characterizes the change of the saturation capacity with respect to temperature:

$$\frac{1}{C_{\mu s}}\frac{dC_{\mu s}}{dT} = -\delta \qquad (3.2\text{-}34b)$$

The net heat of adsorption is the isosteric heat minus the heat of vaporization, i.e.:

$$q_{net} = \beta E_0 \left(\ln\frac{1}{\theta}\right)^{1/2} + \frac{(\beta E_0)\delta T}{2}\left(\ln\frac{1}{\theta}\right)^{-1/2} \qquad (3.2\text{-}35)$$

Figure 3.2-17 shows the reduced net heat of adsorption q_{net}/R_gT versus the fractional loading with E/R_gT as the varying parameter. The heat of adsorption is infinite at zero loading, and the higher is the characteristic energy the higher is the isosteric heat of adsorption.

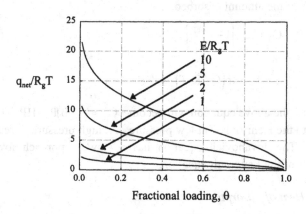

Figure 3.2-17: Plots of the reduced net heat of adsorption versus fractional loading

The enthalpy of immersion is the amount of heat released if adsorption is taken place in a bulk liquid adsorbate. Since adsorption is from the liquid phase there is no phase change associated with the condensation of vapors to liquid phase; hence the net heat of adsorption is used in the calculation of the heat of immersion. This heat of immersion is given by:

$$\Delta h_i = \int_0^1 q_{net}(\theta)d\theta = -(\beta E_0)(1+\delta T)\Gamma\left(\frac{3}{2}\right) \qquad (3.2\text{-}36)$$

which can be readily used to determine the characteristic energy of the system. Stoeckli and his co-wokers (1986, 1989, 1992, 1993) have utilized this dependence to obtain the characteristic energy by simply measuring the heat of immersion of a number of liquids.

3.2.7 Jovanovich Equation

Of lesser use in physical adsorption is the Jovanovich equation. It is applicable to mobile and localised adsorption (Hazlitt et al., 1979). Although it is not as popular as the other empirical equations proposed so far, it is nevertheless a useful empirical equation:

$$1 - \theta = \exp\left[-a\left(\frac{P}{P_0}\right)\right] \tag{3.2-37}$$

or written in terms of the amount adsorbed:

$$C_\mu = C_{\mu s}\left[1 - e^{-bP}\right] \tag{3.2-38a}$$

where

$$b = b_\infty \exp(Q/R_g T) \tag{3.2-38b}$$

At low loading, the above equation will become $C_\mu \approx (C_{\mu s} b)P = HP$. Thus, this equation reduces to the Henry's law at low pressure. At high pressure, it reaches the saturation limit. The Jovanovich equation has a slower approach toward the saturation than that of the Langmuir equation.

3.2.7.1 Isosteric Heat of Adsorption

To calculate the isosteric heat, we use the van't Hoff equation (3.2-14) and obtain:

$$\Delta H = -Q < 0 \tag{3.2-39}$$

Hence the heat of adsorption is constant and is independent of loading, which is the same as in the case of Langmuir isotherm.

3.2.8 Temkin Equation

Another empirical equation is the Temkin equation proposed originally by Slygin and Frumkin (1935) to describe adsorption of hydrogen on platinum electrodes in acidic solutions (chemisorption systems). The equation is (Rudzinski and Everett, 1992):

$$v(P) = C \ln(cP) \qquad (3.2\text{-}40)$$

where C and c are constants specific to the adsorbate-adsorbent pairs. Under some conditions, the Temkin isotherm can be shown to be a special case of the Unilan equation (3.2-23).

3.2.9 Summary of Empirical Equations

The following table summarises the various empirical isotherm equations.

Table 3.2-12 Summary of commonly used empirical isotherm equations

Isotherm	Functional form	Remarks
Frendlich	$C_\mu = KP^{1/n}$	Does not have Henry law limit and no saturation limit
Sips (LF)	$C_\mu = C_{\mu s} \dfrac{(bP)^{1/n}}{1 + (bP)^{1/n}}$	Does not have Henry law limit, but it has finite saturation limit
Toth	$C_\mu = C_{\mu s} \dfrac{bP}{\left[1 + (bP)^t\right]^{1/t}}$	Has Henry law limit and finite saturation limit
Unilan	$C_\mu = \dfrac{C_{\mu s}}{2s} \ln\left(\dfrac{1 + \bar{b}e^s P}{1 + \bar{b}e^{-s} P}\right)$	Has Henry law limit and finite saturation limit
Keller et al.	$C_\mu = C_{\mu s}\alpha_m \dfrac{bP}{\left[1 + (bP)^\alpha\right]^{1/\alpha}}$	Has Henry law limit and finite saturation limit
DR	$V = V_0 \exp\left[-\dfrac{1}{(\beta E_0)^2}\left(R_g T \ln \dfrac{P}{P_0}\right)^2\right]$	Does not have Henry law limit, but reach a finite limit when P approaches P_0
Jovanovich	$C_\mu = C_{\mu s}\left[1 - e^{-bP}\right]$	Has Henry law limit and finite saturation limit
Temkin	$v(P) = C \ln(cP)$	Same as Freundlich. It does not have correct Henry law limit &finite saturation limit

3.3 BET (Brunauer, Emmett and Teller) isotherm and modified BET isotherm

All the empirical equations dealt with in Section 3.2 are for adsorption with "monolayer" coverage, with the exception of the Freundlich isotherm, which does not have a finite saturation capacity and the DR equation, which is applicable for micropore volume filling. In the adsorption of sub-critical adsorbates, molecules first adsorb onto the solid surface as a layering process, and when the pressure is sufficiently high (about 0.1 of the relative pressure) multiple layers are formed. Brunauer, Emmett and Teller are the first to develop a theory to account for this multilayer adsorption, and the range of validity of this theory is approximately between 0.05 and 0.35 times the vapor pressure. In this section we will discuss this important theory and its various versions modified by a number of workers since the publication of the BET theory in 1938. Despite the many versions, the BET equation still remains the most important equation for the characterization of mesoporous solids, mainly due to its simplicity.

3.3.1 BET Equation

The BET theory was first developed by Brunauer et al. (1938) for a *flat* surface (no curvature) and there is *no limit* in the number of layers which can be accommodated on the surface. This theory made use of the same assumptions as those used in the Langmuir theory, that is the surface is energetically homogeneous (adsorption energy does not change with the progress of adsorption in the same layer) and there is no interaction among adsorbed molecules. Let s_0, s_1, s_2 and s_n be the surface areas covered by no layer, one layer, two layers and n layers of adsorbate molecules, respectively (Figure 3.3-1).

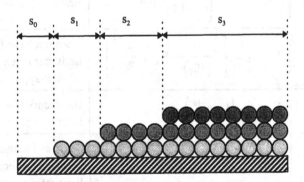

Figure 3.3-1: Mutiple layering in BET theory

The concept of kinetics of adsorption and desorption proposed by Langmuir is applied to this multiple layering process, that is the rate of adsorption on any layer is equal to the rate of desorption from that layer. For the first layer, the rates of adsorption onto the free surface and desorption from the first layer are equal to each other:

$$a_1 P s_0 = b_1 s_1 \exp\left(\frac{-E_1}{R_g T}\right) \quad (3.3\text{-}1)$$

where a_1, b_1, and E_1 are constant, independent of the amount adsorbed. Here E_1 is the interaction energy between the solid and molecule of the first layer, which is expected to be higher than the heat of vaporization.

Similarly, the rate of adsorption onto the first layer must be the same as the rate of evaporation from the second layer, that is:

$$a_2 P s_1 = b_2 s_2 \exp\left(\frac{-E_2}{R_g T}\right) \quad (3.3\text{-}2)$$

The same form of equation then can be applied to the next layer, and in general for the i-th layer, we can write

$$a_i P s_{i-1} = b_i s_i \exp\left(\frac{-E_i}{R_g T}\right) \quad (3.3\text{-}3)$$

The total area of the solid is the sum of all individual areas, that is

$$S = \sum_{i=0}^{\infty} s_i \quad (3.3\text{-}4)$$

Therefore, the volume of gas adsorbed on surface covering by one layer of molecules is the fraction occupied by one layer of molecules multiplied by the monolayer coverage V_m:

$$V_1 = V_m \left(\frac{s_1}{S}\right) \quad (3.3\text{-}5)$$

The volume of gas adsorbed on the section of the surface which has two layers of molecules is:

$$V_2 = V_m \left(\frac{2 s_2}{S}\right) \quad (3.3\text{-}6)$$

The factor of 2 in the above equation is because there are two layers of molecules occupying a surface area of s_2 (Figure 3.3-1). Similarly, the volume of gas adsorbed on the section of the surface having "i" layers is:

$$V_i = V_m \left(\frac{i\, s_i}{S} \right) \tag{3.3-7}$$

Hence, the total volume of gas adsorbed at a given pressure is the sum of all these volumes:

$$V = \frac{V_m}{S} \sum_{i=0}^{\infty} i\, s_i = V_m \frac{\sum_{i=0}^{\infty} i\, s_i}{\sum_{i=0}^{\infty} s_i} \tag{3.3-8}$$

To explicitly obtain the amount of gas adsorbed as a function of pressure, we have to express s_i in terms of the gas pressure. To proceed with this, we need to make a further assumption beside the assumptions made so far about the ideality of layers (so that Langmuir kinetics could be applied). One of the assumptions is that the heat of adsorption of the second and subsequent layers are the same and equal to the heat of liquefaction, E_L:

$$E_2 = E_3 = \cdots = E_i = \cdots = E_L \tag{3.3-9}$$

The other assumption is that the ratio of the rate constants of the second and higher layers is equal to each other, that is:

$$\frac{b_2}{a_2} = \frac{b_3}{a_3} = \cdots = \frac{b_i}{a_i} \stackrel{def}{=} g \tag{3.3-10}$$

where the ratio g is assumed constant. This ratio is related to the vapor pressure of the adsorbate as will be seen later in eq. (3.3-19).

With these two additional assumptions, one can solve the surface coverage that contains one layer of molecule (s_1) in terms of s_0 and pressure as follows:

$$s_1 = \frac{a_1}{b_1} P s_0 \exp(\varepsilon_1) \tag{3.3-11a}$$

where ε_1 is the reduced energy of adsorption of the first layer, defined as

$$\varepsilon_1 = \frac{E_1}{R_g T} \tag{3.3-11b}$$

Similarly the surface coverage of the section containing i layers of molecules is:

$$s_i = \frac{a_1}{b_1} s_0 \, g \exp(\varepsilon_1 - \varepsilon_L) \left[\left(\frac{P}{g}\right) \exp \varepsilon_L\right]^i \qquad (3.3\text{-}12a)$$

for $i = 2, 3, \ldots$, where ε_L is the reduced heat of liquefaction

$$\varepsilon_L = \frac{E_L}{R_g T} \qquad (3.3\text{-}12b)$$

Substituting these surface coverages into the total amount of gas adsorbed (eq. 3.3-8), we obtain:

$$\frac{V}{V_m} = \frac{C \, s_0 \sum_{i=1}^{\infty} i x^i}{s_0 \left(1 + C \sum_{i=1}^{\infty} x^i\right)} \qquad (3.3\text{-}13)$$

where the parameter C and the variable x are defined as follows:

$$y = \frac{a_1}{b_1} P \exp(\varepsilon_1) \qquad (3.3\text{-}14a)$$

$$x = \frac{P}{g} \exp(\varepsilon_L) \qquad (3.3\text{-}14b)$$

$$C = \frac{y}{x} = \frac{a_1 g}{b_1} e^{(\varepsilon_1 - \varepsilon_L)} \qquad (3.3\text{-}14c)$$

By using the following formulas (Abramowitz and Stegun, 1962)

$$\sum_{i=1}^{\infty} x^i = \frac{x}{1-x}; \quad \sum_{i=1}^{\infty} i x^i = \frac{x}{(1-x)^2} \qquad (3.3\text{-}15)$$

eq. (3.3-13) can be simplified to yield the following form written in terms of C and x:

$$\frac{V}{V_m} = \frac{Cx}{(1-x)(1-x+Cx)} \qquad (3.3\text{-}16)$$

Eq. (3.3-16) can only be used if we can relate x in terms of pressure and other known quantities. This is done as follows. Since this model allows for infinite layers on top of a flat surface, the amount adsorbed must be infinity when the gas phase pressure is equal to the vapor pressure, that is $P = P_0$ occurs when $x = 1$; thus the variable x is the ratio of the pressure to the vapour pressure at the adsorption temperature:

$$x = \frac{P}{P_0} \qquad (3.3\text{-}17)$$

With this definition, eq. (3.3-16) will become what is now known as the famous BET equation containing two fitting parameters, C and V_m:

$$\frac{V}{V_m} = \frac{CP}{(P_0 - P)[1 + (C-1)(P/P_0)]} \qquad (3.3\text{-}18)$$

Figure 3.3-2 shows plots of the BET equation (3.3-18) versus the reduced pressure with C being the varying parameter. The larger is the value of C, the sooner will the multilayer form and the convexity of the isotherm increases toward the low pressure range.

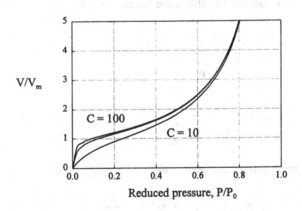

Figure 3.3-2: Plots of the BET equation versus the reduced pressure (C = 10,50, 100)

Equating eqs.(3.3-17) and (3.3-14b), we obtain the following relationship between the vapor pressure, the constant g and the heat of liquefaction:

$$P_0 = g \cdot \exp\left(-\frac{E_L}{R_g T}\right) \qquad (3.3\text{-}19)$$

Within a narrow range of temperature, the vapour pressure follows the Clausius-Clapeyron equation, that is

$$P_0 = \alpha \cdot \exp\left(-\frac{E_L}{R_g T}\right)$$

Comparing this equation with eq.(3.3-19), we see that the parameter g is simply the pre-exponential factor in the Clausius-Clapeyron vapor pressure equation. It is reminded that the parameter g is the ratio of the rate constant for desorption to that for adsorption of the second and subsequent layers, suggesting that these layers condense and evaporate similar to the bulk liquid phase.

The pre-exponential factor of the constant C (eq. 3.3-14c)

$$\frac{a_1 g}{b_1} = \frac{a_1 b_j}{b_1 a_j} \quad \text{for } j > 1$$

can be either greater or smaller than unity (Brunauer et al., 1967), and it is often assumed as unity without any theoretical justification. In setting this factor to be unity, we have assumed that the ratio of the rate constants for adsorption to desorption of the first layer is the same as that for the subsequent layers at infinite temperature. Also by assuming this factor to be unity, we can calculate the interaction energy between the first layer and the solid from the knowledge of C (obtained by fitting of the isotherm equation 3.3-18 with experimental data)

The interaction energy between solid and adsorbate molecule in the first layer is always greater than the heat of adsorption; thus the constant C is a large number (usually greater than 100).

3.3.1.1 Properties of the BET equation

We have seen how the BET equation varies with the reduced pressure as shown in Figure 3.3-2. It is worthwhile to investigate its first and second derivatives to show how the slope as well as the inflexion point vary with the constant C.

The first and second derivatives of the BET equation are

$$\frac{d(V/V_m)}{d(P/P_0)} = C \times \frac{1 + (C-1)(P/P_0)^2}{(1-P/P_0)^2 \left[1 + (C-1)(P/P_0)\right]^2} \qquad (3.3\text{-}20a)$$

$$\frac{d^2(V/V_m)}{d(P/P_0)^2} = 2C \times \frac{(C-1)^2(P/P_0)^3 + 3(C-1)(P/P_0) + 2 - C}{(1-P/P_0)^3[1+(C-1)(P/P_0)]^3} \tag{3.3-20b}$$

Investigation of this slope reveals that the slope decreases with an increase in pressure up to a certain pressure beyond which the slope increases quickly with pressure and becomes infinite when the vapor pressure is reached. This means that there exists an inflexion point, and this point can be found by setting the second derivative to zero. In so doing, we obtain the relative pressure at which the isotherm has an inflexion point (White, 1947):

$$(P/P_0)_I = \frac{(C-1)^{2/3} - 1}{[(C-1) + (C-1)^{2/3}]} \tag{3.3-21a}$$

and the amount adsorbed at this inflexion point is:

$$(V/V_m)_I = \frac{1}{C} \times [(C-1)^{1/3} + 1][(C-1)^{2/3} - 1] \tag{3.3-21b}$$

Figure 3.3-3 shows a plot of the reduced volume adsorbed versus the reduced pressure at the inflexion point with C as the parameter on such plot.

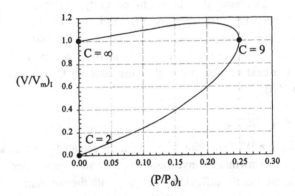

Figure 3.3-3: Plot of V/V_m versus the reduced pressure at the inflexion point

For the inflexion point to exist in the physical range (that is V/V_m must be positive), the lower limit on C will be

$$(C-1)^{2/3} - 1 > 0 \tag{3.3-22a}$$

that is

$$C > 2 \tag{3.3-22b}$$

As the constant C increases, both the amount adsorbed and the pressure at the inflexion point increase up to the reduced pressure of 0.25, at which the constant C is 9. When the constant C is increased beyond 9, the amount adsorbed at the inflexion point only varies slightly between the monolayer coverage V_m and 1.2 V_m, and the reduced pressure at this inflexion point continually decreases with C. As C approaches infinity, the inflexion point moves toward zero pressure and the amount adsorbed approaches the monolayer capacity, V_m. This extreme limit simply means that monolayer is instantaneously formed at minute pressure and multilayering starts immediately at P = 0 according to the following equation:

$$\frac{V}{V_m} = \frac{1}{(1 - P/P_0)}$$

which is the limit of eq. (3.3-18) when $C \to \infty$.

3.3.1.2 Surface area determination

Eq. (3.3-18) is the famous BET equation, and it is used extensively for the area determination because once the monolayer coverage V_m is known and if the area occupied by one molecule is known the surface area of the solid can be calculated.

To conveniently determine V_m, the BET equation can be cast into the form which is amenable for linear plot as follows:

$$\frac{P}{V(P_0 - P)} = \frac{1}{V_m C} + \left(\frac{C-1}{V_m C}\right)\frac{P}{P_0} \qquad (3.3\text{-}23)$$

The pressure range of validity of the BET equation is P/P_0 = 0.05 - 0.3. For relative pressures above 0.3, there exists capillary condensation, which is not amenable to multilayer analysis. The capillary condensation will be dealt with in Section 3.9, and this phenomenon is taken advantage of to determine the pore size distribution as the pressure at which the liquid condenses in a pore depends on the pore size. More about this will be discussed in Section 3.9.

A plot of the LHS of eq.(3.3-23) versus (P/P_0) would yield a straight line with a slope

$$\text{Slope} = \frac{(C-1)}{CV_m}$$

and an intercept

$$\text{Intercept} = \frac{1}{CV_m}$$

Usually the value of C (eq. 3.3-14c) is very large because the adsorption energy of the first layer is larger than the heat of liquefaction, the slope is then simply the inverse of the monolayer coverage, and the intercept is effectively the origin of such plot. Therefore, very often that only one point is sufficient for the first estimate of the surface area.

Once V_m (mole/g) is obtained from the slope, the surface area is calculated from:

$$A = V_m N_A a_m \qquad (3.3\text{-}24)$$

where N_A is the Avogadro number and a_m is the molecular projected area. The following table lists the molecular projected area of a few commonly used adsorbates.

Table 3.3.1: Molecular Projected Area of some common gases

Gas	T (K)	P_0 (Pa)	a_m (Å^2/molecule)
argon	77	2.78×10^4	15
ammonia	209		12.6
carbon dioxide	195	1.013×10^5	20
n-butane	273	1.013×10^5	44
krypton	90	2.74×10^3	20
methane	90	1.08×10^4	16
nitrogen	77	1.013×10^5	16
oxygen	79		14.1
oxygen	90.6		14.6
sulfur dioxide	250		18.7
xenon	90	8.25	23

We illustrate the surface area determination using the BET equation with the following example, where the solid is Carbolac and the adsorbate is CF_2Cl_2 and the adsorption temperature is -33.1 °C.

Example 3.3-1: *Surface area determination of Carbolac*

To illustrate the linear plot of the BET method, we apply it to the data of CF_2Cl_2 adsorption on Carbolac at -33.1 °C (Carman and Raal, 1952). The data are shown in Figure 3.3-4a. The vapour pressure of CF_2Cl_2 at -33.1 °C is 659 Torr. Restricting the range of the reduced pressure below 0.3, we plot $P/V(P_0-P)$ versus P/P_0 as shown in Figure 3.3-4b. A straight line can be drawn through the data points of that plot, and a simple linear regression gives

$$\text{Slope} = \frac{(C-1)}{CV_m} = 0.2493$$

$$\text{Intercept} = \frac{1}{CV_m} = 1.188 \times 10^{-3}$$

from which we can readily calculate the monolayer coverage as $V_m = 4$ mmole/g and the constant $C = 210$.

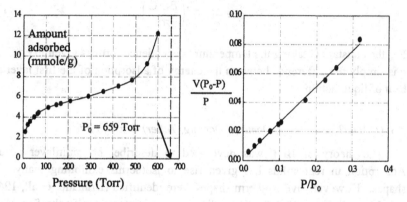

Figure 3.3-4a: Plot of isotherm data of CF_2Cl_2 on Carbolac at -33.1 °C

Figure 3.3-4b: BET plot

Using the projection area of CF_2Cl_2 of 33.6 A^2, we can calculate the surface area as follows:

$$S_g = \frac{0.004 \text{ mole}}{g} \times \frac{6.023 \times 10^{23} \text{ molecules}}{\text{mole}} \times \frac{33.6 \text{ A}^2}{\text{molecule}} \times \frac{1 \text{ m}^2}{10^{20} \text{ A}^2}$$

$$S_g = 809 \frac{m^2}{g}$$

This demonstrates the simplicity of the BET equation in the surface area determination.

3.3.2 Differential heat

The isosteric heat of the BET isotherm is obtained from the following equation (Lopatkin, 1987):

$$-\Delta H = E_L + q\frac{\left(1-\frac{P}{P_0}\right)^2}{1+(C-1)\left(\frac{P}{P_0}\right)^2} \qquad (3.3\text{-}25a)$$

where q is obtained from the temperature dependence of the constant C

$$q = -R_g T^2 \left(\frac{\partial \ln C}{\partial T}\right)_\theta \qquad (3.3\text{-}25b)$$

For the constant C taking the temperature dependence form as given in eq.(3.3-14c), q is simply the difference between the energy of adsorption of the first layer and the heat of liquefaction.

3.3.3 BDDT (Brunauer, Deming, Deming, Teller) Classification

The theory of BET was developed to describe the multilayer adsorption. Adsorption in real solids has given rise to isotherms exhibiting many different shapes. However, five isotherm shapes were identified (Brunauer et al., 1940) and are shown in Figure 3.3-5. The following five systems typify the five classes of isotherm.

1. **Type 1**: Adsorption of oxygen on charcoal at -183°C
2. **Type 2**: Adsorption of nitrogen on iron catalysts at -195°C (many solids fall into this type).
3. **Type 3**: Adsorption of bromine on silica gel at 79°C, water on glass
4. **Type 4**: Adsorption of benzene on ferric oxide gel at 50°C
5. **Type 5**: Adsorption of water on charcoal at 100°C

Type I isotherm is the Langmuir isotherm type (monolayer coverage), typical of adsorption in microporous solids, such as adsorption of oxygen in charcoal. Type II typifies the BET adsorption mechanism. Type III is the type typical of water adsorption on charcoal where the adsorption is not favorable at low pressure because of the nonpolar (hydrophobic) nature of the charcoal surface. At sufficiently high pressures, the adsorption is due to the capillary condensation in mesopores. Type IV and type V are the same as types II and III with the exception that they have finite limit as $P \to P_0$ due to the finite pore volume of porous solids.

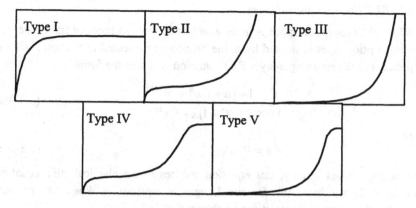

Figure 3.3-5: BDDT classification of five isotherm shapes

The BET equation (eq. 3.3-18) developed originally by Brunauer et al. (1938) is able to describe type I to type III. The type III isotherm can be produced from the BET equation when the forces between adsorbate and adsorbent are smaller than that between adsorbate molecules in the liquid state (i.e. $E_1 < E_L$). Figure 3.3-6 shows such plots for the cases of C = 0.1 and 0.9 to illustrate type III isotherm.

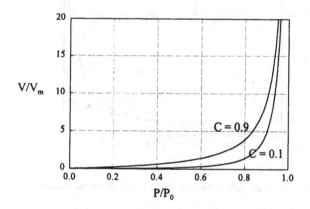

Figure 3.3-6: Plots of the BET equation when C < 1

The BET equation (3.3-18) does not cover the last two types (IV and V) because one of the assumptions of the BET theory is the allowance for infinite layers of molecules to build up on top of the surface. To consider the last two types, we have to limit the number of layers which can be formed above a solid surface. This is dealt with next.

3.3.3.1 BET Equation for n-Layers

When the adsorption space is finite which is the case in pores of finite size, that is the adsorption layer is limited by n, the procedure presented in Section 3.1.1 can be applied and the resulting n-layer BET equation will take the form:

$$\frac{V}{V_m} = \frac{Cx}{1-x} \frac{1-(n+1)x^n + nx^{n+1}}{1+(C-1)x - Cx^{n+1}} \qquad (3.3\text{-}26a)$$

where

$$x = P/P_0 \qquad (3.3\text{-}26b)$$

When n approaches infinity, this equation reduces to the classical BET equation. When n = 1, we have the famous Langmuir equation. When the pressure approaches the vapor pressure, it can be shown that

$$\lim_{x \to 1} \frac{V}{V_m} = \frac{n(n+1)C}{2(nC+1)} \qquad (3.3\text{-}27)$$

Figure 3.3-7 shows plots of eq. (3.3-26) for C = 100 and various values of n.

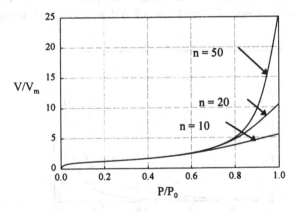

Figure 3.3-7: Plot of the BET-n layer equation (3.3-26) versus the reduced pressure

The parameter C is normally greater than 1 because the heat of adsorption of the first layer is greater than the heat of liquefaction, i.e. the attractive forces between the adsorbed molecule and the adsorbent are greater than the attractive forces between molecules in the liquid state. When this is the case, the amount adsorbed when the pressure reaches the vapor pressure is

$$\lim_{\substack{x \to 1 \\ C \to \infty}} \frac{V}{V_m} = \frac{(n+1)}{2}$$

Although the n-layer BET equation gives a finite limit as $P \to P_0$, this limit is approached in a fashion as shown in Figure 3.3-7, that is it does not exhibit the plateau commonly observed for many solids.

3.3.3.2 Type IV and V isotherms

Although the n-layer BET equation (3.3-26) can deal with types I to III by properly adjusting the parameters n and C in eq.(3.3-26), it does not explain the plateau observed in types IV and V. To account for this plateau, Brunauer et al. (1940) presented a new theory, which is another extension of the BET theory. When the pores are nearly filled, the adsorption energy of binding in some higher layers is greater than the heat of liquefaction. This is because the last adsorbed layers in a capillary is attracted to both sides. We denote this extra energy as Q. Therefore, the rate of evaporation of the last layer is not $k_d\, e^{-E_L/R_gT}$, but rather $k_d\, e^{-(E_L+Q)/R_gT}$, that is molecules of the last layer find it harder to desorb back into the gas phase. The adsorption is assumed to occur between two parallel plates. The maximum number of layer that can fit in the two walls is 2n-1, not 2n because the last layer is common to both walls. Using the following equations of evaporation and condensation equilibria:

$$aPs_0 = bs_1 e^{-E_1/R_gT} \qquad (3.3\text{-}28a)$$

$$aPs_i = bs_{i+1} e^{-E_L/R_gT} \quad \text{for} \quad i = 1, 2, \cdots, n-2 \qquad (3.3\text{-}28b)$$

$$aPs_{n-1} = bs_n e^{-(E_L+Q)/R_gT} \qquad (3.3\text{-}28c)$$

and following the same procedure as done in Section 3.3-1, Brunauer et al.(1940) derived the following equation:

$$\frac{V}{V_m} = \frac{Cx}{1-x} \cdot \frac{1 + (ng/2 - n)x^{n-1} - (ng - n + 1)x^n + (ng/2)x^{n+1}}{1 + (C-1)x + (Cg/2 - C)x^n - (Cg/2)x^{n+1}} \qquad (3.3\text{-}29a)$$

where

$$g = \exp(Q/R_gT) \qquad (3.3\text{-}29b)$$

98 Equilibria

Eq. (3.3-29) was derived for a maximum (2n-1) layers which can be fit into a capillary. If the maximum number is 2n, the isotherm equation takes a slightly different form:

$$\frac{V}{V_m} = \frac{Cx}{1-x} \frac{1+(ng/2-n/2)x^{n-1}-(ng+1)x^n+(ng/2+n/2)x^{n+1}}{1+(C-1)x+(Cg/2-C/2)x^n-(Cg/2+C/2)x^{n+1}} \quad (3.3\text{-}29c)$$

Eqs. (3.3-29a) and (3.3-29c) describe types IV and V, depending on the value of C. If $C \gg 1$, we have type IV isotherm. When $C < 1$, we have type V isotherm. Figure 3.3-8 shows plots generated from these two equations for the case of $C = 100 \gg 1$ (Type IV). The other parameters used are $n = 10$ and $g = \{1, 100, 1000\}$. Figure 3.3-9 shows the isotherm curves for the case of $C = 0.1$ (Type V isotherm).

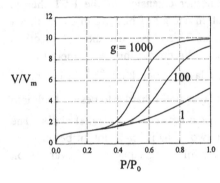

Figure 3.3-8a: Plots of eq. (3.3-29a) versus the reduced pressure for $n = 10$ and $C = 100$

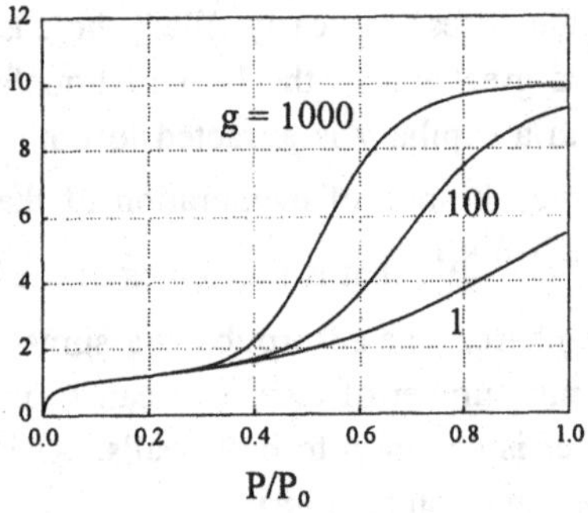

Figure 3.3-8b: Plot of eq. (3.3-29c) versus the reduced pressure for $n = 10$ and $C = 100$

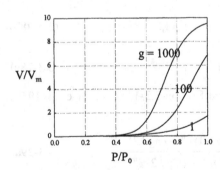 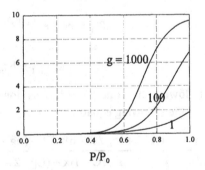

Figure 3.3-9a: Plots of eq. (3.3-29a) versus the reduced pressure for $n = 10$ and $C = 0.05$

Figure 3.3-9b: Plot of eq. (3.3-29c) versus the reduced pressure for $n = 10$ and $C = 0.05$

3.3.4 Comparison between the van der Waals adsorption and the Capillary Condensation

The modified BET equations for n layers and for type IV and V need some discussion. This is done by comparing the theory of van der Waals adsorption and the capillary condensation theory. The porous solid is assumed to consist of capillaries bounded by two parallel planes, and the spacing between the two layers is denoted as D. This simple configuration is sufficient for our discussion.

According to the capillary condensation theory (which will be treated in more details in Section 3.9), the pressure at which the condensation will take place in a pore of width D is:

$$P_k = P_0 \exp\left(-\frac{2\sigma v_M}{R_g T} \cdot \frac{1}{D}\right) \qquad (3.3\text{-}30)$$

where P_0 is the vapor pressure of free liquid, σ is the surface tension, and v_M is the liquid molar volume at temperature T. For gas having a pressure lower than the condensation pressure P_k, there is no capillary condensation except of course a sub-monolayer adsorption, while for gas having pressure equal to or greater than the condensation pressure P_k, the capillary of size D will be filled completely with liquid. This is the basis of the capillary condensation theory.

In the present theory of van der Waals adsorption, the multilayer starts below the pressure P_k and builds up with increasing pressure, and the capillaries are not filled up even for pressures greater than this condensation pressure P_k, which contradicts the capillary condensation. This means that the multilayer theory based on van der Waals adsorption has an upper limit in the pressure range. Despite this limitation, the van der Waals theory puts all five types of adsorption isotherm shape into one framework, that is it can deal with unimolecular adsorption (Langmuir), multilayer adsorption (BET) and enhanced adsorption in capillaries (BDDT).

3.3.5 Other Modified Versions of the BET Equation

Recognising that the BET equation has a narrow range of validity (P/P_0 is between 0.05 to 0.35) and that the amount adsorbed at reduced pressures greater than 0.35 is less than that predicted by the BET equation, Anderson (1946) proposed new theories to extend this range. His reasons are that:
1. the heat of adsorption of the second layer is less than the heat of liquefaction.
2. the structure of solid is such that only finite number of layers is allowed.

Anderson (1946) studied two cases: In one case the modified BET equation with heat of adsorption in the second layer and the next several layers being less than the heat of liquefaction. In the second case the surface available to each subsequent layer is smaller.

In the first case where the heat of adsorption of the second and subsequent layers is less than the heat of liquefaction, he obtained the following equation:

$$\frac{V}{V_m} = \frac{Ckx}{(1-kx)[1+(C-1)kx]} \tag{3.3-31a}$$

where

$$x = \frac{P}{P_0} \tag{3.3-31b}$$

$$k = \exp\left(\frac{d}{R_g T}\right) \tag{3.3-31c}$$

Here the parameter d is the excess of the heat of liquefaction that the second and subsequent layers release.

Bloomquist and Clark (1940) measured the adsorption of nitrogen on microscopic glass sphere at -196 °C, and their data were fitted by the above equation of Anderson, the constant d was found to be -53 cal/mole, or k = 0.715.

It is interesting to note here that the n-layer BET equation (3.3-26), for various values of n, generates a family of curves which are superficially similar to those generated from eq.(3.3-31) with various values of k.

Analysing a number of adsorption systems, Anderson (1946) found that the value of k falling between 0.6 and 0.7, indicating that the excess heat of liquefaction that the second and subsequent layers release is about -60 cal/mole.

In the second case where the surface area available for adsorption is smaller in each subsequent layer, Anderson (1946) obtained the following formula:

$$\frac{V}{V_m} = \frac{Cx}{(1-jx)[1+(C-1)x]} \tag{3.3-32a}$$

where

$$x = \frac{P}{P_0} \tag{3.3-32b}$$

and j is the fraction available in the subsequent layer. This fraction was assumed constant in each layer. When j = 0, this equation reduces to the Langmuir equation, and when j = 1, we recover the famous BET equation.

If we combine the two equations (3.3-31) and (3.3-32), that is when the heat of adsorption of second layer and above is less than the heat of liquefaction and the area of a layer is smaller than the preceding layer, the following formula is obtained:

$$\frac{V}{V_m} = \frac{Ckx}{(1-jkx)[1+(C-1)kx]} \qquad (3.3\text{-}33)$$

where x is the reduced pressure and k is defined in eq.(3.3-31c).

The modified BET equation obtained by Anderson (eq. 3.3-33) can be regarded as replacing the vapor pressure P_0 in the original BET equation by a parameter, P^*. It is reminded that in the Anderson's theory molecules in the second and subsequent layers have a heat of adsorption numerically less than that of liquefaction. It was found that P^*/P_0 can take values up to 1.6. Although there is no convincing argument about the physical meaning of this parameter P^*, the use of the modified BET equation allows the extension of the range of applicability to $P/P_0 = 0.8$, instead of the usual range of 0.05-0.35 of the conventional BET equation. One problem associated with the modified BET equation is that the value P^* greater than the vapor pressure implies that the energy of adsorption in the second and higher layers is lower than the energy of condensation of the liquid, and this is not physically plausible when the number of layers is sufficiently large! The extension of the range of applicability of the Anderson equation is purely by chance as we must remember that any multilayering theory must have a maximum pressure limit beyond which the capillary condensation phenomenon will take over More discussion on the modified BET equation can be found in Everett (1990) and Burgess et al. (1990).

3.3.6 Aranovich's Modified BET Equations

We have addressed the classical BET equation as well as some of its modified versions. Although these modified equations were claimed to add to the original equation some refined features, the classical BET equation is still the one that is used by many workers as the primary tool to study surface area. Before closing this section on BET typed equations, it is worthwhile to point out another equation developed by Aranovich, who proposed a form very similar to that of the BET equation. The difference is in the exponent of the term $(1 - P/P_0)$ in the denominator of the two equations. In the BET case, the exponent is one while in the Aranovich case the exponent is one half.

Aranovich (1988) claimed that his isotherm correctly describes the limiting cases, and the range of validity of the BET equation is narrow (relative pressure of

0.05 to 0.4). Also the BET equation has a factor of $(1-P/P_0)$ in the denominator, giving infinite spreading pressure at saturation pressure.

The assumptions of his model are:
- The adsorbent surface is flat and uniform
- The phase in contact with the adsorbent is a vacancy solution to which a lattice model can be applied.
- The energy change accompanying the evaporation of a molecule depends on the number of layers.
- Only the configurational components of the free energy are considered.

The Aranovich's multimolecular equation is:

$$\frac{V}{V_m} = \frac{C(P/P_0)}{\left[1 + C(P/P_0)\right]\left(1 - P/P_0\right)^{1/2}} \qquad (3.3\text{-}34)$$

For comparison, we compare the BET equation and the Aranovich equation in Figure 3.3-10 with $C = 100$.

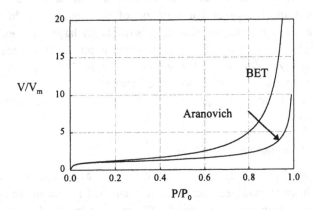

Figure 3.3-10: Comparison between the BET equation and the Aranovich equation ($C = 100$)

Using the Gibbs formula for the spreading pressure:

$$\pi = R_g T \int_0^p V \, d(\ln p) \qquad (3.3\text{-}35)$$

he obtained the following equation of state:

$$\pi = \frac{a_m R_g T}{C_\varphi} \ln\left\{\frac{(C_\varphi + 1)\left[C_\varphi - (1 - P/P_0)^{1/2}\right]}{(C_\varphi - 1)\left[C_\varphi + (1 - P/P_0)^{1/2}\right]}\right\} \tag{3.3-36a}$$

$$C_\varphi = \left(1 + \frac{1}{C}\right)^{1/2} \tag{3.3-36b}$$

This equation relates to the amount adsorbed, V, through the isotherm equation (eq. 3.3-34). Thus, by eliminating P between these two equations an equation of state can be obtained.

Aranovich (1989) used a lattice model to provide a thermodynamics foundation for his model. The differential heat of adsorption at low coverage for his multimolecular isotherm equation as well as other properties are considered in Aranovich (1990, 1992).

3.4 Harkins-Jura, Halsey Isotherms

Although the BET theory is used almost regularly as a convenient tool to evaluate the surface area of a solid, other isotherms such as the Harkins-Jura equation, obtained in Chapter 2 can also be used to determine the surface area. Analogous to a monolayer film on liquids, Harkins and Jura (1943) obtained the following equation:

$$\ln\left(\frac{P}{P_0}\right) = B - \frac{C}{V^2} \tag{3.4-1}$$

where V is the gas volume at STP adsorbed per unit g. Jura and Harkins claimed that this is the simplest equation for describing multilayer adsorption, and it is valid over more than twice the pressure range of any two constant adsorption isotherms. They showed that for TiO_2 in the form of anatase, their isotherm agrees with the experimental data at both lower and higher values of pressure than the commonly used BET equation.

Eq. (3.4-1) suggests that a plot of $\ln(P/P^0)$ versus $1/V^2$ would yield a straight line with a slope of - C. The square root of this constant is proportional to the surface area of the solid. The following formula was provided by Harkins and Jura:

$$S_g = 4.06\sqrt{C} \tag{3.4-2}$$

where S_g has the unit of m^2 per g. They also suggested that if the plot of $\ln(P/P_0)$ versus $1/V^2$ exhibits two straight lines. The one at lower pressure range should be chosen for the area calculation as this is the one in which there exists a transition from a monolayer to multilayers.

3.5 Further Discussion on the BET Theory

3.5.1 Critical of the BET theory

Although the BET theory is used extensively, it still suffers from a number of criticisms. The first is that surfaces of real solids are heterogeneous while the model assumes that all the adsorption sites are energetically identical. The second reason is the assumption of the vertical force between adsorbent and adsorbate molecules. It neglects the horizontal interaction between adsorbed molecules. The third reason was put forward by Halsey (1948) and is detailed below.

The key assumption of the BET theory is that the adsorbate can adsorb a second molecule on top of the first molecule, releasing an amount of heat which is the same as the heat of liquifaction, and this second molecule can allow a third molecule to adsorb on top of it, and so on. Such a picture is very simplistic as it is more reasonable that for hexagonal packing a combination of three adsorbed molecules must form a triangular array before a fourth molecule can adsorb on top of those three triangular adsorbed molecules, and then releasing an amount of heat equal to the liquifaction heat.

With this notion of triangular array of adsorbed molecule, we now consider the situation when the reduced pressure $P/P_0 = 0.5$. At this relative pressure, the fractional coverage of the first layer is 1/2. The fraction of sites for the second layer with filled triangular array underneath is

$$\left(\frac{1}{2}\right)^3 \tag{3.5-1}$$

The exponent 3 is because three molecules in the first layer are required to provide one site for a molecule sitting on top of them. If <u>these sites are filled to the BET amount</u>, that is half of the available sites for the second layer, we have the following fractional coverage for the second layer:

$$\theta_2 = \left(\frac{1}{2}\right)^4 \tag{3.5-2}$$

Using the same argument, the fraction of sites with triangular array underneath for the third layer to form is:

$$\left(\frac{1}{2}\right)^{12} = 0.00024 \tag{3.5-3}$$

which means that only 0.024% of the sites in the third layer available for adsorption, meaning the third layer is practically empty. Now that the third layer is practically empty, the second layer is not protected from evaporation, the fractional loading of the second layer is then controlled by the Langmuirian adsorption, not BET, that is the fractional loading of the second layer is

$$\theta_2 = \left(\frac{1}{2}\right)^3 \frac{(P/P_0)}{1+(P/P_0)} = \left(\frac{1}{2}\right)^3 \times \left(\frac{1}{3}\right) = 0.0417 \tag{3.5-4}$$

instead of that given by eq. (3.5-2). The first factor in the RHS of the above equation is the fraction of active site available for the second layer to form. This low loading of the second layer then renders the first layer from protection, and hence the fractional loading of the first layer is controlled by the Langmuirian adsorption, that is:

$$\theta_1 = \frac{(P/P_0)}{1+(P/P_0)} = \left(\frac{1}{3}\right) \tag{3.5-5}$$

The Langmuirian equation does not allow for the interaction between adsorbed molecules. To allow for this, we need to consider an equation that does so such as the quasi-chemical treatment of Fowler-Guggenheim. The Fowler-Guggenheim equation has the form:

$$\frac{\theta}{(1-\theta)} = bP \times \left[\frac{\beta+1-2\theta}{2-2\theta}\right]^z \tag{3.5-6a}$$

where

$$\beta = \left\{1 - 4\theta(1-\theta)\left[1 - \exp\left(-\frac{2w}{zR_gT}\right)\right]\right\}^{1/2} \tag{3.5-6b}$$

Here w is the interaction energy between adsorbed molecules. A negative value of w means attraction between adsorbed molecules, while a positive value means repulsion. For the case of attraction, two dimensional condensation begins at θ^*, where θ^* is the smallest root of (Section 2.3.5.2).

$$P(\theta^*) = P(1/2) = \frac{1}{b}\exp\left(\frac{w}{R_gT}\right) \qquad (3.5\text{-}7)$$

Here P(1/2) has been obtained from eq. (3.5-6a), and it is the condensation pressure at which the two-dimensional condensation occurs. We denote this condensation pressure as P_c:

$$P_c = P(1/2) = \frac{1}{b}\exp\left(\frac{w}{R_gT}\right) \qquad (3.5\text{-}8)$$

Since this θ^* is small, we can neglect higher order term, and obtain the following approximation for the fractional coverage at condensation:

$$\theta^* = \exp\left(\frac{w}{R_gT}\right) \qquad (3.5\text{-}9)$$

For pressures lower than the condensation pressure P_c, the fractional loading is very low and taking the limit of the Fowler-Guggenheim equation at low pressure we get

$$\theta = bP \qquad (3.5\text{-}10)$$

Combining this with the expression for the condensation pressure (eq. 3.5-8), we get the following expression for the fractional loading below the condensation pressure

$$\theta = \left(\frac{P}{P_c}\right)\exp\left(\frac{w}{R_gT}\right) \qquad (3.5\text{-}11)$$

for $0 < \theta < \theta^*$ and $0 < P < P_c$.

With the assumption of hexagonal packing, there are six neighbors in a two dimensional film and 12 neighbors in the liquid; therefore the interaction energy in the two dimensional film is half of the heat of liquifaction, i.e.

$$w = -\frac{E_L}{2}$$

For the case of nitrogen adsorption at 77K, the heat of liquefaction is 5560 Joule/mole. Therefore, we can calculate the fractional loading at which the two dimensional condensation begins (eq. 3.5-9):

$$\theta^* = \exp\left(-\frac{5560/2}{8.314 \times 77}\right) = 0.013$$

Thus, as long as the pressure is below the condensation pressure P_c, the fractional loading is very low (less than 2%). This is not reflected in the BET equation, and this has led Halsey to conclude that for any value of C the adsorption is confined to the first layer until the condensation pressure has been reached.

3.5.2 Surface with Adsorption Energy Higher than Heat of Liquefaction

By assuming the adsorption energy is greater than the heat of liquefaction by an amount ΔE, the two dimensional condensation is found to occur at the fractional loading as given in eq.(3.5-9), and this occurs at a much lower pressure because of the high adsorption energy of the surface. This new condensation pressure is

$$P^* = P_c \exp\left(-\frac{\Delta E}{R_g T}\right) \tag{3.5-12}$$

If $E_1 > E_2 > E_3 > ... > E_L$, this type of isotherm will lead to stepwise isotherm, with the j-th step occurring at the pressure:

$$P_j^* = P_c \exp\left(-\frac{E_j - E_L}{R_g T}\right) \tag{3.5-13}$$

This type of stepwise isotherm is not observed in practice because discrete values of the adsorption energies are not followed by practical solids. By allowing for the distribution of adsorption energy, which we will show in the next section, Halsey (1948) and others developed a new multilayer isotherm, which is known in the literature as the FHH isotherm.

3.6 FHH Multilayer Equation

Let us assume that the surface is heterogeneous, and such heterogeneity is characterized by the distribution $F(\Delta E)$. For a value of the excess energy ΔE the condensation pressure is given in eq. (3.5-12), hence:

$$\Delta E = R_g T \ln\left(\frac{P_c}{P}\right) \tag{3.6-1}$$

Sites having energies between ΔE and ∞ will be filled with condensed liquid at the pressure P. Hence, the fractional loading is given by:

$$\theta = \int_{R_g T \ln(P_0/P)}^{\infty} F(\Delta E) \times d(\Delta E) \tag{3.6-2}$$

Assuming the energy excess to take the form of an exponential distribution, $F(\Delta E) = c \cdot \exp(-\Delta E/\Delta E_m)$, the above equation becomes:

$$\theta = c' \times \left(\frac{P}{P_c}\right)^{R_g T/\Delta E_m} \tag{3.6-3}$$

which has the form of the Freundlich equation. We have seen in Section 3.2.1 that the Freundlich equation can be obtained from the combination of the Langmuirian local isotherm and an exponential energy distribution.

Instead of assuming a distribution for the excess energy ΔE, we can assume that this excess energy is a function of loading ($\Delta E = a/\theta^r$). With this form, we obtain the following semi-empirical equation (derived from eq. 3.5-12):

$$\frac{P}{P_c} = \exp\left(-\frac{a}{R_g T \theta^r}\right) \tag{3.6-4}$$

which was first derived by Halsey. When $r = 2$, this equation reduces to the Harkins-Jura equation (3.4-1), which means that the Harkin-Jura equation is applicable when the adsorption energy decreases with the second power of the distance from the interface.

Eq. (3.6-4) can be written in terms of the volume of gas adsorbed, v, as follows:

$$\ln\left(\frac{P}{P_0}\right) = -A\,(v)^{-r} \tag{3.6-5}$$

This form of equation has been studied by Frenkel, Halsey and Hill in the study of multilayer adsorption (Frenkel, 1946; Halsey, 1948; Hill, 1949, 1952), and hence it is known in the literature as the FHH equation. The parameter r is regarded as a measure of the rate of decline in the adsorption potential with distance from the surface. For van der Waals forces, r is equal to 3. A value of about 2.7 is commonly observed in practice.

3.7 Redhead's Empirical Isotherm

Recently, Redhead (1995) presented a new empirical equation to cover the multilayer adsorption region, and his purpose is to extend the range of validity of the proposed equation to higher pressure range. His empirical equation takes the following form

$$\frac{V}{V_m} = \left[\frac{(2n-1)(P/P_0)}{1-P/P_0}\right]^{1/n} \tag{3.7-1}$$

where n is the empirical parameter and it was found to be in the range of 2.5 and 4.5 for most cases. Similar to the BET equation, it is a two-parameter equation. Figure 3.7-1 shows plots of this equation with n as the varying parameter. The BET equation with C = 100 is also plotted on the same figure for comparison. Just like the case of Aranovich equation, the Redhead equation lies below the BET equation in the high range of the reduced pressure.

Redhead's equation of the form (3.7-1) was chosen based on the fact that:

(a) the BET equation reduces to

$$\frac{V}{V_m} \approx \frac{1}{1-P/P_0} \qquad (3.7\text{-}2)$$

when C and $C(P/P_0)$ are much larger than unity

(b) the FHH of the form

$$\ln\left(\frac{P}{P_0}\right) = -\frac{k}{(V/V_m)^3} \qquad (3.7\text{-}3)$$

is reduced to

$$\frac{V}{V_m} \approx \left(\frac{kP/P_0}{1-P/P_0}\right)^{1/3} \qquad (3.7\text{-}4)$$

when P/P_0 approaches unity.

The range of validity of the Redhead's equation is such that at the lower limit of the relative pressure, the coverage is a monolayer, that is:

$$\frac{(P/P_0)_m}{1-(P/P_0)_m} = \frac{1}{2n-1} \qquad (3.7\text{-}5a)$$

from which we get

$$(P/P_0)_m = \frac{1}{2n} \qquad (3.7\text{-}5b)$$

Thus, the range of validity of the Redhead's equation is:

$$\frac{1}{2n} < \frac{P}{P_0} < 1 \qquad (3.7\text{-}6)$$

The monolayer coverage can be found by simply plotting $\ln V$ versus

$$\ln\left(\frac{P/P_0}{1-P/P_0}\right)$$

and the slope of such plot is 1/n. The lower applicability limit is then calculated from eq.(3.7-5b). Knowing this lower limit the monolayer coverage can be read directly from the curve V versus the relative pressure. Using experimental data of many systems, the following table shows values of n and the monolayer coverage

Table 3.7-1: The values of n and V_m from the Redhead's method (Redhead, Langmuir, 1996)

Gas	Solid	T (K)	n	V_m (Redhead)	V_m (BET)
H_2O	anatase	298	3.10	3.46	3.67 mL (STP)/g
N_2	anatase	77	3.07	3.39	3.32 mL (STP)/g
N_2	silica	77	4.11	10.66	10.26 μmole/m²
Ar	alumina	77	3.77	37.15	37.05 mg/g
neopentane	silica	273	2.36	0.36	0.43 mmole/g
neopentane	carbon	273	5.81	0.0938	0.102 mmole/g

The monolayer coverages obtained by the BET method are also included in the table, and these values are very comparable to those obtained by the Redhead's method.

3.8 Summary of Multilayer Adsorption Equation

As discussed in previous sections, there are many equations available for the description of multilayer adsorption. Despite the improvement of some equations over the original BET equation, it is still the most popular equation used because of its simplicity. It is in general applied to all solids including zeolite and activated carbon even though its derivation based on surface does not hold for solids such as zeolite and activated carbon. The challenge in this area is the development of an equation which is capable in dealing with adsorption of sub-critical vapors in porous solids having a wide pore size distribution, ranging from micropore to meso and macropore. The subject of microporous solid will be dealt with in Chapter 4 where we will address a different mechanism of adsorption, namely the micropore filling, the various adsorption isotherm equations to deal with this mechanism and the treatment of micropore size distribution.

The following table (Table 3.8-1) summarises all the multilayer equations dealt with in this chapter. The parameter x in some of the equations is the reduced pressure (P/P_0).

Table 3.8-1: Multilayer adsorption isotherm equation

Isotherm	Expression	Parameter	Equation No.
BET	$\dfrac{V}{V_m} = \dfrac{Cx}{(1-x)[1+(C-1)x]}$	V_m, C	(3.8-1)
n-layers BET	$\dfrac{V}{V_m} = \dfrac{Cx}{1-x}\dfrac{1-(n+1)x^n+nx^{n+1}}{1+(C-1)x-Cx^{n+1}}$	V_m, C, n	(3.8-2)
Langmuir case VI	$\dfrac{V}{V_m} = \dfrac{Ckx}{(1-kx)[1+(C-1)kx]}$	V_m, C, k	(3.8-3)
Anderson	$\dfrac{V}{V_m} = \dfrac{Cx}{(1-jx)[1+(C-1)x]}$	V_m, C, j	(3.8-4)
Anderson	$\dfrac{V}{V_m} = \dfrac{Ckx}{(1-jkx)[1+(C-1)kx]}$	V_m, C, j, k	(3.8-5)
Aranovich	$\dfrac{V}{V_m} = \dfrac{Cx}{\sqrt{1-x}(1+Cx)}$	V_m, C	(3.8-6)
Harkins-Jura	$\ln(x) = B - \dfrac{C}{V^2}$	B, C	(3.8-7)
FHH	$\ln(x) = -A(V)^{-B}$	A, B	(3.8-8)
Redhead	$\dfrac{V}{V_m} = \left[\dfrac{(2n-1)x}{1-x}\right]^{1/n}$	V_m, n	(3.8-9)

3.9 Pore volume and pore size distribution

We have studied the various equations in the last sections for the description of multilayer adsorption. Of those equations, the BET equation is the most popular equation for the determination of surface area. The range of validity of this equation is that the relative pressure is between 0.05 and 0.35. Adsorption beyond this range will result in filling of mesopore with sorbate liquid through the action of capillary condensation.

In this section, we will deal with capillary condensation, and investigate how this would vary with the pore size, and from which one could derive useful information about the mesopore size distribution from the data of volume adsorbed versus pressure. The range of validity for the capillary condensation is that the relative pressure is between 0.35 and 0.99. We first start with some basic theories, and then utilize them in the determination of the pore size distribution.

3.9.1 Basic Theory

A typical adsorption-desorption isotherm of a practical porous solid usually exhibits a hysteresis (Figure 3.9-1) over the pressure range where the capillary condensation phenomenon is operating.

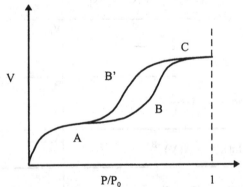

Figure 3.9-1: Adsorption isotherm in mesoporous solids

The onset of the hysteresis loop indicates the start of the capillary condensation mechanism. The desorption curve (AB'C) is always above the adsorption branch (ABC), that is for a given loading adsorbate desorbs from a porous solid at a lower pressure than that required for adsorption. Before proceeding with the analysis of the isotherm, we first start with the basic capillary condensation theory of Lord Kelvin, the former William Thompson.

3.9.1.1 Kelvin Equation for Capillary Condensation

Due to the capillary force in small pores, the vapour pressure of a liquid inside such pores is less than that of a flat surface. The change in the free energy due to evaporation of a differential volume of liquid equals the change in surface area times the surface tension:

$$n\Delta G = -nR_g T \ln\left(\frac{P}{P_0}\right) = (2\pi r dl)\sigma \cos\theta \tag{3.9-1a}$$

Here σ is the surface tension, P_0 is the vapour pressure of the bulk phase, θ is the contact angle, and n is the number of moles in a pore segment having a length of dl and a radius of r:

$$n = \frac{dv}{v_M} = \frac{\pi r^2 dl}{v_M} \tag{3.9-1b}$$

with v_M being the liquid molar volume.

Thus, the pressure at which the liquid will condense in a pore of radius r is obtained by combining eqs. (3.9-1a) and (3.9-1b):

$$\frac{P}{P_0} = \exp\left(-\frac{2\sigma \cos\theta \, v_M}{R_g T} \cdot \frac{1}{r}\right) \tag{3.9-2}$$

which is now known as the famous Kelvin equation. The implicit assumption associated with this equation is that the liquid is incompressible (Gregg and Sing, 1982). Eq. (3.9-2) states that for a capillary of radius r the pressure at which the species condenses or evaporates is less than the free surface vapour pressure, P_0.

When the pore is very large, in the sense that

$$r \gg \frac{2\sigma \cos\theta \, v_M}{R_g T}$$

the pressure at which the species condenses or evaporates is equal to the vapour pressure of the liquid, i.e. $P/P_0 = 1$. On the other hand, for pores of radius r, the liquid will form at pressure $P/P_0 < 1$. This mechanism is valid when the filling and emptying follows a vertical mechanism, that is species is removed out of the pore vertically. This is reasonable for desorption, when the pore is initially filled with liquid and liquid evaporates by the vertical removal (see Figure 3.9-2). We shall discuss the relevant equation for adsorption mechanism but first let us apply eq.(3.9-2) for the case of nitrogen in the following example.

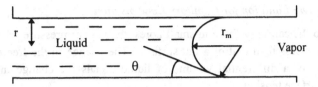

Figure 3.9-2: Evaporation of liquid from a pore of radius r (r_m is the radius of curvature)

Example 3.9-1 *Kelvin equation for nitrogen at 77 K*

To gauge the magnitude of eq. (3.9-2), we take the case of evaporation of liquid nitrogen from a capillary at 77.4 K. The surface tension and liquid molar volume of nitrogen (Gregg and Sing, 1982) are:

$$\sigma = 8.72 \times 10^{-3} \text{ N/m}; \quad v_M = 3.468 \times 10^{-5} \text{ m}^3/\text{mole}$$

Substituting these values into eq. (3.9-2) gives the following equation for the Kelvin radius in terms of the reduced pressure P/P_0

$$r(nm) = \frac{0.9399}{\ln(P_0/P)} = \frac{0.4082}{\log_{10}(P_0/P)}$$

where the contact angle between nitrogen liquid meniscus and the solid surface has been assumed zero. The vapour pressure of nitrogen at 77.4 K is 760 Torr. The following table (Table 3.9-1) shows the capillary pressure at various values of pore radius, and Figure 3.9-3 shows the reduced pressure required for evaporation as a function of pore radius in nm.

Table 3.9-1 Capillary pressure as a function of radius for nitrogen at 77.4 K

r (nm)	P (Torr)	P/P^0
1	297	0.391
2	475	0.625
5	630	0.829
10	691	0.909
20	725	0.954
25	732	0.963

The zero contact angle is the usual assumption made almost automatically. If the correct contact angle is different from zero, say 45 degrees, then the Kelvin radius will differ from the one calculated from the assumption of zero contact angle by a factor of $\cos\theta \cong 0.7$.

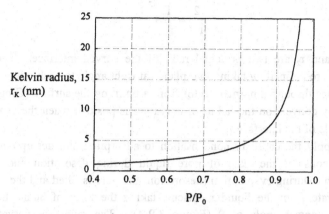

Figure 3.9-3: Plot of the Kelvin radius versus the reduced pressure for N_2 at 77 K

In Table 3.9-1, the range of pore radii is chosen from the lower limit to the upper limit of the mesopore (according to the IUPAC classification, 2 < d < 50 nm). We note that the capillary condensation starts to occur in the smallest mesopore (d = 2 nm = 20 Å) at the relative pressure of 0.39. It is reminded here that the upper limit of the relative pressure for the validity of the multilayer theory is about 0.35. This reduced pressure is usually regarded as the demarcation point between the multilayer adsorption and the capillary condensation mechanism, and it is satisfied by many adsorption systems.

3.9.1.2 Generalised Kelvin Equation

The capillary condensation occurs in the region where the hysteresis exists (Figure 3.9-1). In this region of hysteresis, there are two values of the pressure that give rise to the same amount uptake, that is giving rise to two values of the pore size as calculated from eq. (3.9-2). To properly account for this, one must investigate the way liquid is condensed or evaporated during the adsorption and desorption cycles.

The Kelvin equation written in eq. (3.7.2) is a special case of the following generalized Kelvin equation put forward by Everett (1972, 1975):

$$\frac{P}{P_0} = \exp\left(-\frac{2\sigma v_M}{R_g T} \cdot \frac{1}{r_m}\right) \tag{3.9-3}$$

where σ is the surface tension, v_M is the liquid molar volume and r_m is the mean radius of curvature of the interface defined as (Defay and Prigogine, 1966):

$$\frac{2}{r_m} = \frac{1}{r_1} + \frac{1}{r_2} \tag{3.9-4}$$

Here, r_1 and r_2 are two principal radii of the curved interface. These radii of curvature are defined by taking two planes at right angle to each other, and each of them passes through a normal vector from a point on the surface (Gregg and Sing, 1982). By convention the radius of curvature is positive when the center is in the vapour side of the interface.

To apply the capillary condensation to adsorption and desorption cycles, we need to consider the state of pore liquid during desorption and that during adsorption. During evaporation (desorption) the pore is filled and the sorbate starts to evaporate from the liquid meniscus, taking the form of hemispherical shape having a contact angle of θ (Figure 3.9-4). The radius of curvature of this hemispherical, r_m, is related to the pore radius as follows:

$$r_m = \frac{r}{\cos\theta} \tag{3.9-5}$$

For this radius of curvature, the general equation (eq. 3.9-3) reduces to that of Kelvin (eq. 3.9-2).

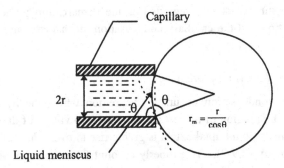

Figure 3.9-4: The schematic diagram of the hemispherical meniscus

For adsorption, liquid is formed via surface layering and at the inception of condensation from a vapour filled pore (radial filling rather than vertical filling), the meniscus takes the cylindrical shape as shown in Figure 3.9-5. For this cylindrical meniscus, two planes are drawn passing through the normal vector from any point on the liquid surface. One plane cuts the pore, and hence the principal radius is $r_1 = r$. The other plane perpendicular to the former will cut the liquid interface along the pore axial direction, and hence its principal pore radius is $r_2 = \infty$. Therefore, the radius of curvature for the case of adsorption is

$$r_m = 2r \qquad (3.9\text{-}6)$$

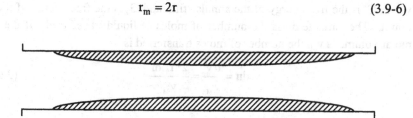

Figure 3.9-5: Filling of a pore during the adsorption cycle

Substituting eq. (3.9-6) into eq. (3.9-3), we obtain the following equation for capillary condensation along the branch of adsorption:

$$\frac{P}{P_0} = \exp\left(-\frac{\sigma v_M}{R_g T} \cdot \frac{1}{r}\right) \qquad (3.9\text{-}7)$$

The capillary condensation equation for adsorption (eq. 3.9-7) can also be obtained from the free energy argument of Cohan discussed in the following section.

3.9.1.3 Cohan's Analysis of Adsorption Branch

Cohan's quantitative analysis in 1938 was based on the suggestion of Foster (1932, 1934) that the hysteresis in adsorption is due to the delay in forming a meniscus in the capillary. For adsorption, this is occurred by radial filling, rather than vertical filling as in the case of desorption. When condensation of the first layer occurs, the effective radius r decreases, causing further condensation at a fixed reduced pressure P/P_0. This means that a pore of radius r, corresponding to the threshold reduced pressure P/P_0, will be filled instantaneously.

If a small volume of liquid $dV = 2\pi r L dr$ is transported from a large body of liquid to the capillary, the decrease in surface area is $dS = 2\pi L dr$ (Figure 3.9-6). Hence the change in surface energy is equal to that decrease in surface area times the surface tension, that is:

$$-2\pi L \sigma dr \qquad (3.9\text{-}8)$$

Equating this change in surface energy to the change in free energy associating with the isothermal transfer of dV of liquid from the bulk to the capillary, we get:

$$\Delta F = dn(G_r - G_0) = -2\pi L \sigma dr \qquad (3.9\text{-}9)$$

where G_r is the free energy of the annular film and G_0 is the free energy of the bulk liquid. The variable dn is the number of moles of liquid transferred. If the liquid molar volume is v_M, the number of moles transferred is:

$$dn = \frac{dV}{v_M} = \frac{2\pi rLdr}{v_M} \qquad (3.9\text{-}10a)$$

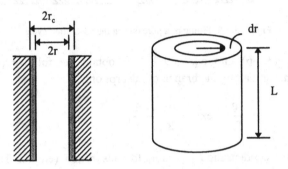

Figure 3.9-6: Radial filling of a cylindrical pore

The free energy change between the annular film and the bulk liquid per unit number of moles is:

$$G_r - G_0 = R_g T \ln \frac{P_r}{P_0} \qquad (3.9\text{-}10b)$$

where P_r is the vapour pressure of the film and P_0 is the vapour pressure of the bulk liquid.

Substituting eqs. (3.9-10) into eq.(3.9-9) we obtain the following equation relating the vapour pressure of the film in terms of the vapour pressure of the bulk liquid and the capillary radius.

$$\frac{P_{ads}}{P_0} = \exp\left(-\frac{\sigma v_M}{rR_g T}\right) \qquad (3.9\text{-}11a)$$

which is the required capillary condensation equation for the adsorption branch.

While vertical emptying of the pores occurs during desorption the vapor pressure of the capillary liquid is given by the classical Kevin equation:

$$\frac{P_{des}}{P_0} = \exp\left(-\frac{2\sigma v_M}{r'R_g T}\right) \qquad (3.9\text{-}11b)$$

where r' is the radius of curvature of the meniscus ($r' = r/\cos\theta$, where θ is the contact angle). For zero wetting angle, the radius of curvature is the same as the capillary radius.

Comparing eq.(3.9-11a) for adsorption to eq.(3.9-11b) for desorption, we obtain the following relationship between the threshold pressure for condensation and that for evaporation for a zero contact angle

$$P_{ads}^2 = P_0 P_{des} \qquad (3.9\text{-}12a)$$

or

$$\left(\frac{P_{ads}}{P_0}\right)^2 = \frac{P_{des}}{P_0} \qquad (3.9\text{-}12b)$$

The reduced pressure required to empty a capillary is equal to the square of that necessary to fill it. For example, a pore of radius r which fills with liquid adsorbate at a reduced pressure P/P_0 of 0.7 would have its filled liquid evaporated at a reduced pressure of about 0.5. The status of filling and emptying at two different relative pressures is the basis of hysteresis in mesoporous solids.

3.9.1.4 Pore Size Distribution

We now know that for each pore of radius r, there exists a threshold pressure for condensation and a threshold pressure for evaporation. This important point now can be used to determine the pore size and its distribution. During the adsorption cycle the filling of pore with adsorbate is in a radial fashion, and hence the rise in the amount adsorbed versus pressure is gradual. After the pore is filled, and when the pressure is reduced the liquid in the pore will remain until the pressure in the gas phase reaches the evaporation pressure governed by eq. (3.9-11b) at which the liquid will instantaneously evaporate, leaving only the adsorbed layer behind. As the pressure is reduced further the amount adsorbed will decrease and the relationship between the amount adsorbed and the pressure is dictated by the equilibrium between the two phases (for example the BET equation).

Knowing this condensation pressure from the desorption branch of point A, the pore radius then can be calculated from the Kelvin equation (3.9-2), assuming that the liquid molar volume, the contact angle and the surface tension are known, i.e.:

$$r = \frac{2\sigma v_M \cos\theta}{R_g T} \left[\ln\left(\frac{P_0}{P_{des}}\right)\right]^{-1} \qquad (3.9\text{-}13)$$

Unfortunately, this ideal situation never occurs in practice as all practical porous solids have a distribution of pore size, so there will be a gradual change in the desorption branch rather than an abrupt change as in the case of ideal solids.

3.9.1.5 Solids Exhibiting no Adsorbed Layer (pure Condensation and Evaporation)

Let us first consider a porous solid having a pore volume distribution $f(r)$, that is $f(r)\, dr$ is the pore volume of pores having radii ranging between r and $r + dr$. For a gas phase of pressure P, the threshold radius is calculated from either eq.(3.9-11a) for condensation mechanism or eq.(3.9-11b) for evaporation mechanism. Let this threshold radius be r_K, then the amount adsorbed at a pressure P is simply the fraction of pores having radii less than this threshold radius r_K, that is:

$$V(P) = \int_0^{r_K(P)} f(r)\, dr \qquad (3.9\text{-}14)$$

where r_K is calculated from the following equation:

$$r_K = \begin{cases} \dfrac{2 v_M \sigma \cos\theta}{R_g T \ln(P_0/P)} & \text{for evaporation} \\[6pt] \dfrac{v_M \sigma}{R_g T \ln(P_0/P)} & \text{for condensation} \end{cases} \qquad (3.9\text{-}15)$$

For simplicity, we have chosen the lower limit in eq. (3.9-14) as zero. Strictly speaking it should be the lower limit of the mesopore range in which the capillary condensation mechanism operates. We now illustrate the capillary condensation in the following examples.

Example 3.9-2: *Maxwellian pore volume distribution*
We take a porous solid having the following Maxwellian pore volume distribution

$$f(r) = \frac{V_s}{r_0}\left(\frac{r}{r_0}\right)\exp\left(-\frac{r}{r_0}\right) \qquad (3.9\text{-}16)$$

where V_s is the total pore volume and r_0 is the characteristic pore radius. Figure 3.9-7a shows a typical plot of this pore volume distribution for $V_s = 0.4$ cc/g, and $r_0 = 3$ nm.

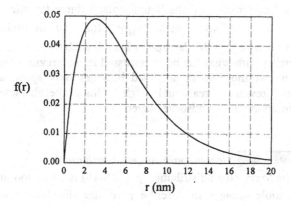

Figure 3.9-7a: Plot of the Maxwellian pore volume distribution with $V_s=0.4$cc/g and $r_0=3$nm

Substitution of eq.(3.9-16) for the pore volume distribution into eq.(3.9-14) yields the following expression for the adsorbed volume as a function of pressure

$$V(P) = V_s \left[1 - \left(1 + \frac{r_K}{r_0}\right) \exp\left(-\frac{r_K}{r_0}\right) \right] \qquad (3.9\text{-}17)$$

where r_K is a function of the reduced pressure, given as in eq.(3.9-15) for adsorption and desorption branches. With the pore volume distribution given in Figure 3.9-7a, Figure 3.9-7b shows the amount adsorbed for the adsorption branch as well as that for the desorption branch for the case of adsorption of nitrogen at 77K.

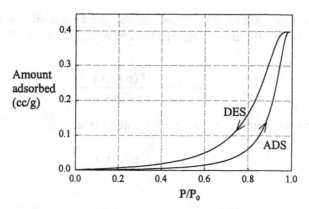

Figure 3.9-7b: Plot of the amount adsorbed versus the reduced pressure

The surface tension and the liquid molar volume for nitrogen at 77K are $\sigma = 8.72 \times 10^{-3}$ Newton/m and $v_M = 3.468 \times 10^{-5}$ m³/mole. From Figure 3.9-7b we note that the desorption branch is above the adsorption branch, concurring with what has been discussed in the previous sections. Also it is seen that the amount adsorbed becomes only appreciably when the reduced pressure is reasonably high. This is the characteristics of pure condensation with no adsorbed layer.

Example 3.9-3: *Gamma pore volume distribution*

The Maxwellian distribution (eq. 3.9-16) is not of too much use as it is not flexible enough to describe pore size distribution of many solids because the mean pore size as well as the variance is controlled by only one parameter, r_0. This example considers a more flexible distribution, the Gamma distribution which has the following form:

$$f(r) = V_s \frac{\alpha^{p+1}}{\Gamma(p+1)} r^p e^{-\alpha r} \qquad (3.9\text{-}18)$$

where $\Gamma(\bullet)$ is the Gamma function, V_s is the total pore volume, and α and p are pore structural parameters. This distribution has the mean and dispersion:

$$\bar{r} = \frac{p+1}{\alpha} \qquad (3.9\text{-}19a)$$

$$\sigma = \frac{\sqrt{p+1}}{\alpha} \qquad (3.9\text{-}19b)$$

With this Gamma distribution for the description of the pore volume distribution, the amount adsorbed can be obtained from eq.(3.9-14) and the result is:

$$V(P) = V_s \frac{\Gamma(p+1, \alpha r_K)}{\Gamma(p+1)} \qquad (3.9\text{-}20)$$

where the function $\Gamma(a,b)$ is the incomplete Gamma function, defined as

$$\Gamma(a,b) = \int_0^b x^{a-1} e^{-x} \, dx \qquad (3.9\text{-}21)$$

To illustrate the amount adsorbed versus pressure, we take the following values for the Gamma pore volume distribution:

$$V_s = 0.5 \text{ cc/g}, p = 3, \alpha = 2 \text{ nm}^{-1}$$

Figure 3.9-8a shows the plot of the pore volume distribution (curve A) and Figure 3.9-8b shows the amounts adsorbed for the adsorption and desorption branches. The mean pore radius for this choice is

$$\bar{r} = \frac{p+1}{\alpha} = \frac{3+1}{2} = 2 \text{ nm}$$

which is the lower limit of the mesopore range according to the IUPAC classification (2 nm < d < 50 nm). This small pore system, therefore, will have the onset of hysteresis in the low pressure range (Figure 3.9-8b)

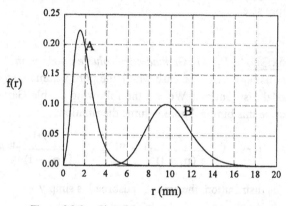

Figure 3.9-8a: Plot of the Gamma pore volume distribution

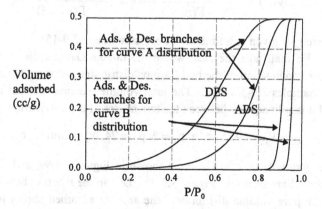

Figure 3.9-8b: Plots of the amount adsorbed versus the reduced pressure

Let us take another set of values

$$V_s = 0.5 \text{ cc/g}, \ p = 24, \ \alpha = 2.5 \text{ nm}^{-1}$$

With these values, the mean pore radius is

$$\bar{r} = \frac{p+1}{\alpha} = \frac{24+1}{2.5} = 10 \text{ nm}$$

which represents a mesoporous solid with large pores. Its pore volume distribution is shown graphically in Figure 3.9-8a (Curve B), and the amounts adsorbed for the adsorption and desorption branches are shown in Figure 3.9-8b. Here we see that with the large mesoporous solid, the onset of the hysteresis does not start until the reduced pressure is relatively high (around $P/P_0 = 0.85$), compared to $P/P_0 = 0.2$ for the solid having small mesopore size.

Example 3.9-4: *Double Gamma pore volume distribution*

Let us now study the case where the pore volume distribution exhibits a bimodal distribution. We use the following double Gamma distribution to describe this bimodal pore volume distribution:

$$f(r) = V_{s,1} \frac{\alpha_1^{p_1+1}}{\Gamma(p_1+1)} r^{p_1} e^{-\alpha_1 r} + V_{s,2} \frac{\alpha_2^{p_2+1}}{\Gamma(p_2+1)} r^{p_2} e^{-\alpha_2 r} \quad (3.9-22)$$

With this distribution, the amount adsorbed is simply

$$V(P) = V_{s,1} \frac{\Gamma(p_1+1, \alpha_1 r_K)}{\Gamma(p_1+1)} + V_{s,2} \frac{\Gamma(p_2+1, \alpha_2 r_K)}{\Gamma(p_2+1)} \quad (3.9-23)$$

where r_K is given as a function of pressure in eq. (3.9-15).

We shall take the case where the double Gamma distribution is the linear combination of the two Gamma distributions that we have dealt with in Examples 3.9-3 above. The total volume remains constant at 0.5 cc/g, and it is split equally between the two modes of the distribution, that is

$$V_{s,1} = V_{s,2} = 0.25 \text{ cc/g}, \ p_1 = 3, \ p_2 = 24, \ \alpha_1 = 2 \text{ nm}^{-1}, \ \alpha_2 = 2.5 \text{ nm}^{-1}$$

The pore volume distribution is shown in Figure 3.9-9a, and the amounts adsorbed are shown in Figure 3.9-9b. Due to the inherent bimodal display of the pore volume distribution, the amount adsorbed shows two distinct hysteresis stages. One is attributed to the small pore range, while the other (close to the vapour pressure) is due to the larger pore range.

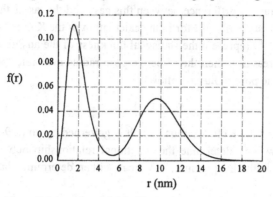

Figure 3.9-9a: Plot of the double Gamma pore volume distribution

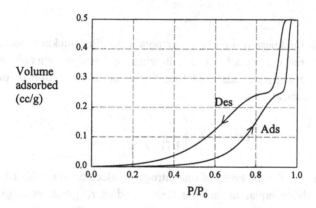

Figure 3.9-9b: Plots of the amount adsorbed versus the reduced pressure

3.9.1.6 Solids exhibiting Adsorbed Layer prior to the Condensation

The analysis so far assumes that there is no adsorption occurring on the pore wall prior to the capillary condensation or after the evaporation. That type of analysis is applicable to systems, such as water adsorption on perfect hydrophobic surface such as that of a graphitized charcoal. In this section, we will deal with the case where we will have adsorption (multilayer adsorption is allowed for) prior to the condensation, and this so-called adsorbed layer will grow with pressure (for example following the BET mechanism). This growth in the adsorbed layer will make the effective pore radius smaller, and thus making the capillary condensation to occur sooner than the case where we have no layer adsorption dealt with in the last section. Thus the difference between this case and the last is the adsorbed layer, and this layer must be allowed for in the calculation of the Kelvin radius.

If we let "t" to represent the statistical thickness of the adsorbed layer (which is a function of pressure), then the effective pore radius available for condensation is related to the true pore radius as follows:

$$r_K = r - t \qquad (3.9\text{-}24)$$

where r_K is called the Kelvin radius, which is governed by eq.(3.9-15). What needs to be done now is to determine the functional relationship between the statistical film thickness with the pressure. We can either determine that from the BET equation

$$t = \frac{V}{V_m}\sigma$$

where σ is the thickness of one layer, or obtain the film thickness measured for a nonporous reference material. The following equation is generally used as an estimate for the statistical adsorbed film thickness as a function of pressure for nitrogen adsorption at 77 K.

$$t\,(\text{nm}) = 0.354\left[\frac{5}{\ln(P_0/P)}\right]^{1/3} \qquad (3.9\text{-}25)$$

where 0.354 nm is the thickness of one nitrogen molecule. With this film thickness given in the above equation and the Kelvin radius r_K given in eq.(3.9-15), the threshold radius corresponding to a gas phase of pressure P is:

$$r(P) = \begin{cases} t + \dfrac{2v_M \sigma \cos\theta}{R_g T \ln(P_0/P)} & \text{for desorption} \\ t + \dfrac{v_M \sigma}{R_g T \ln(P_0/P)} & \text{for adsorption} \end{cases} \qquad (3.9\text{-}26)$$

If the pore volume distribution is f(r), then for a given pressure P, pores having radii less than the threshold radius (eq. 3.9-26) will be filled with adsorbate both in adsorbed form as well as condensed form, while pores having radii greater than that threshold radius will have their surfaces covered with layers of adsorbate. The amount adsorbed at a given pressure P is then:

$$V(P) = \int_0^{r(P)} f(r)\,dr + t\int_{r(P)}^{r_{max}} \frac{2f(r)}{r}\,dr \qquad (3.9\text{-}27)$$

In obtaining the above equation, we have assumed that pores are cylindrical in shape, and the film thickness is independent of the pore radius. The factor $(2/r)$ is the surface area per unit void volume. The parameter r_{max} is usually taken as the upper limit of the mesopore range, that is $r_{max} = 25$ nm.

Knowing the pore volume distribution $f(r)$, the surface area is determined from:

$$S_g = \int_0^{r_{max}} \frac{2f(r)}{r}\,dr \qquad (3.9\text{-}28)$$

Example 3.9-5: *Solids having Maxwellian pore volume distribution*

We take the case where the solid is described by the Maxwellian pore volume distribution, given as in eq.(3.9-16). With this distribution, we evaluate the amount adsorbed from eq.(3.9-27) and the result is:

$$V(P) = V_s \left\{ \left[1 - \left(1 + \frac{r}{r_0}\right)\exp\left(-\frac{r}{r_0}\right)\right] + \frac{2t}{r_0}\left[\exp\left(-\frac{r}{r_0}\right) - \exp\left(-\frac{r_{max}}{r_0}\right)\right] \right\} \qquad (3.9\text{-}29)$$

where t is given in eq.(3.9-25) and the threshold radius is given in eq.(3.9-26). For $r_0 = 5$ nm and $r_{max} = 25$ nm, Figure 3.9-10 shows the amounts adsorbed for the adsorption and desorption branches. Also plotted on the same figures are the desorption and adsorption curves for the situation where there is no adsorbed layer.

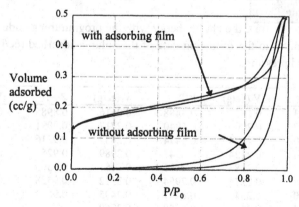

Figure 3.9-10: Plots of the amount adsorbed versus the reduced pressure

3.9.1.6.1 The Inverse Problem: Determination of the Pore Volume Distribution

The analysis presented so far can be used to solve the inverse problem, that is if we know the amount adsorbed versus pressure, the equation (3.9-27) can be used to determine the constants for the pore volume distribution provided that we know the shape of the distribution a-priori. We shall handle this inverse problem by assuming that a mesopore volume distribution can be described by the double Gamma distribution as given in eq.(3.9-22). With this form of distribution, the amount adsorbed can be calculated from eq.(3.9-27), and the result is:

$$V(P) = V_{s,1} \frac{\Gamma(p_1+1,\alpha_1 r)}{\Gamma(p_1+1)} + V_{s,2} \frac{\Gamma(p_2+1,\alpha_2 r)}{\Gamma(p_2+1)}$$

$$+ 2V_{s,1}(\alpha_1 t)\left[\frac{\Gamma(p_1,\alpha_1 r_{max})}{\Gamma(p_1+1)} - \frac{\Gamma(p_1,\alpha_1 r)}{\Gamma(p_1+1)}\right] + 2V_{s,2}(\alpha_2 t)\left[\frac{\Gamma(p_2,\alpha_2 r_{max})}{\Gamma(p_2+1)} - \frac{\Gamma(p_2,\alpha_2 r)}{\Gamma(p_2+1)}\right]$$

(3.9-30)

where the statistical thickness is calculated from eq.(3.9-25).

For a given set of experimental data of V versus P, the above equation can be used in the optimization to determine the pore volume parameters, namely

$$V_{s,1} \quad V_{s,2} \quad p_1 \quad p_2 \quad \alpha_1 \quad \alpha_2 \qquad (3.9\text{-}31)$$

There are six parameters to be determined from the optimization. Using the knowledge that the maximum volume adsorbed is the sum of $V_{s,1}$ and $V_{s,2}$, we then are left with five parameters in the optimization. A programming code MesoPSD1 is provided with the book for the optimization of these five parameters.

Example 3.9-6: To illustrate the above theory with the programming code MesoPSD1, we apply it to the following data of the volume adsorbed (cc/g) versus the reduced pressure P/P_0.

P/P_0	V (cc/g)	P/P_0	V (cc/g)	P/P_0	V (cc/g)
0.1423	0.1843	0.5488	0.2616	0.8985	0.3972
0.1977	0.2000	0.5940	0.2717	0.9091	0.4268
0.2470	0.2074	0.6474	0.2837	0.9216	0.4492
0.2983	0.2124	0.6947	0.2989	0.9259	0.4686
0.3516	0.2183	0.7481	0.3151	0.9363	0.4788
0.3988	0.2274	0.7954	0.3312	0.9528	0.4950
0.4460	0.2343	0.8407	0.3495	0.9652	0.4991
0.4954	0.2485	0.8758	0.3769		

Using the code MesoPSD1 where eq. (3.9-30) is fitted to the above experimental data, we obtain the following optimal parameters for the pore volume distribution having the form of double Gamma distribution:

$V_{s1} = 0.254$ cc/g, $V_{s2} = 0.2435$ cc/g, $p_1 = 8.119$, $p_2 = 11.88$
$\alpha_1 = 4.788$ nm^{-1}, $\alpha_2 = 2.053$ nm^{-1}

The fit between the double Gamma pore volume distribution using the above parameters with the data is shown in Figure 3.9-11a, where we see that the fit is very good. Figure 3.9-11b shows the double Gamma pore volume distribution as a function of pore radius. We note the two peaks in the distribution, suggesting that there are two distinct groups of mesopore for this solid. The mean radii for these two groups are:

$$\bar{r}_1 = (p_1 + 1)/\alpha_1 = 1.9 \text{ nm}, \quad \bar{r}_2 = (p_2 + 1)/\alpha_2 = 6.27 \text{ nm}$$

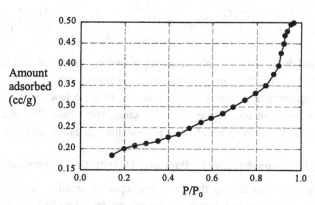

Figure 3.9-11a: The amount adsorbed versus P/P_0 (Symbol: exp. data; line: fitted curve)

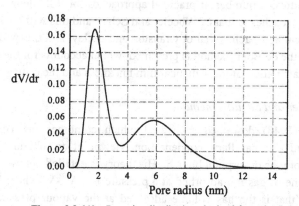

Figure 3.9-11b: Pore size distribution obtained from the fitting

The above example illustrates a simple means to determine mesopore size distribution by using the double Gamma pore volume distribution. Such a distribution is able to describe a pore volume distribution having two major peaks. Although solids having multiple peaks in pore volume distribution are rare, we will show below that the same methodology can be applied to handle this case. This time we use a combination of N Gamma distributions. The pore volume distribution can take the form:

$$f(r) = \sum_{j=1}^{N} V_{s,j} \frac{\alpha_j^{p_j+1}}{\Gamma(p_j+1)} r^{p_j} e^{-\alpha_j r} \qquad (3.9\text{-}32)$$

With this pore volume distribution, the amount adsorbed as a function of the reduced pressure is then simply the extension of eq. (3.9-30), which is:

$$V(P) = \sum_{j=1}^{N} \left\{ V_{s,j} \frac{\Gamma(p_j+1, \alpha_j r)}{\Gamma(p_j+1)} + 2V_{s,j} (\alpha_j t) \left[\frac{\Gamma(p_j, \alpha_j r_{max})}{\Gamma(p_j+1)} - \frac{\Gamma(p_j, \alpha_j r)}{\Gamma(p_j+1)} \right] \right\} \qquad (3.9\text{-}33)$$

Thus for solids having multiple peaks the above equation can be used to fit the data to optimally obtain the pore volume structural parameters p_j, α_j (j = 1, 2, 3, ..., N). From the optimization point of view, it is best to use a low value of N ≈ 2 or 3 first in the optimization and if the fit is not satisfactory then the parameter N is increased until the fit between the above equation and the data becomes acceptable.

3.10 Practical Approaches for the Pore Size Distribution Determination

The last section has shown the basic concepts of capillary condensation and how they can be utilized in the determination of pore size distribution (PSD). In this section, we address a number of practical approaches for PSD determination. One of the early approaches is that of Wheeler and Schull and this will be presented first. A more practical approach is that of Cranston and Inkley, and this will be discussed next. Finally, the de Boer method is presented, which accounts for the effect of pore shape on the calculation of the statistical film thickness and the critical pore radius.

3.10.1 Wheeler and Schull's method

Wheeler (1948) described a method which combines the BET multilayer adsorption and the capillary condensation viewpoint to obtain the pore size distribution from the desorption branch. His theory is detailed below.

The volume of gas not adsorbed at a pressure P is $V_s - V$, where V_s is the total pore volume (that is the gas volume adsorbed at the vapour pressure P_0). This volume $V_s - V$ must equal the summation of volumes of all unfilled pores of radii. Before the pore is filled its effective radius is less than the true pore radius by the thickness t of the physically sorbed multilayer. On emptying, all liquid is evaporated except the physically adsorbed layer of thickness t, that is the radius used in the Kelvin equation is (r-t) not r.

Wheeler used the length distribution $L(r)$, rather than the pore volume distribution, and proposed the following equation for the volume of gas not adsorbed at a pressure P as:

$$V_s - V = \pi \int_{r_K}^{\infty} (r-t)^2 L(r) dr \tag{3.10-1}$$

where $L(r) dr$ is the total length of pores of radii between r and $r + dr$, and r_K is the critical pore radius (corrected for physical adsorption), which means that all pores having radii less than r_K are filled and those having radii greater than r_K are empty.

The total pore volume of the solid is calculated from eq. (3.10-1) by simply setting $r_K = 0$ and $t = 0$, that is when there is no adsorbate molecule in the pore:

$$V_s = \pi \int_0^{\infty} r^2 L(r) dr \tag{3.10-2}$$

Knowing the length distribution, the specific surface area can be calculated from:

$$S = 2\pi \int_0^{\infty} rL(r) dr \tag{3.10-3}$$

The volume distribution, which has a greater physical significance than the pore length distribution, can be obtained from:

$$V(r) = \pi r^2 L(r) \tag{3.10-4}$$

3.10.1.1 The Critical Pore Radius and the Statistical Film Thickness

To evaluate the volume adsorbed from eq. (3.10-1), we need to know the critical pore radius r_K and the statistical film thickness (t). Using eq.(3.9-15) for the Kelvin radius, the critical radius is calculated from:

$$r_K = \begin{cases} t + \dfrac{2\sigma v_M \cos\theta}{R_g T \ln(P_0/P)} & \text{desorption} \\ t + \dfrac{\sigma v_M}{R_g T \ln(P_0/P)} & \text{adsorption} \end{cases} \tag{3.10-5}$$

The statistical film thickness may be calculated from eq. (3.9-25) or from the BET equation, that is:

$$t_{BET} = \frac{\theta_{BET} v_M}{a_m N_A} \tag{3.10-6}$$

where v_M is the liquid molar volume, a_m is the molecular projection area, N_A is the Avogadro number ($N_A = 6.023 \times 10^{23}$ molecules/mole), and θ_{BET} is the fractional loading relative to the monolayer coverage, calculated from the following BET equation

$$\theta_{BET} = \frac{C(P/P_0)}{(1-P/P_0)[1+(C-1)P/P_0]} \tag{3.10-7}$$

For nitrogen as the adsorbate at 77 K, we have the following values for the liquid molar volume and the molecular projection area $v_M = 34.68$ cc/mole, $a_m = 16.2$ A²/molecule. From these values, we calculate the film thickness for one adsorbed layer

$$\left(\frac{v_M}{a_m N_A}\right)_{N_2} = 0.354 \text{ nm} \tag{3.10-8}$$

Thus, the statistical film thickness for nitrogen calculated from the BET theory is:

$$t \text{ (nm)} = 0.354 \times \theta_{BET} \tag{3.10-9}$$

It is known that the BET statistical film thickness of a practical porous solid is larger than the experimental thickness for flat surfaces in the high pressure region (Schull, 1948). Figure 3.10-1 shows a plot of the statistical thickness t calculated from eq. (3.10-9) with C = 100.

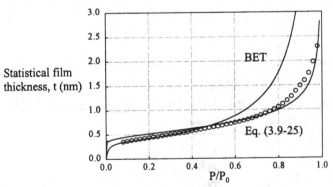

Figure 3.10-1: Plots of the statistical film thickness versus the reduced pressure

Also plotted in the same figure is the statistical thickness calculated from equation (3.9-25). The two curves deviate significantly in the high pressure region. The circle symbols on this figure are experimental data obtained on many nonporous solids (Cranston and Inkley, 1957). We see that eq. (3.9-25) agrees very well with the experimental data and it is then a better choice of equation for the calculation of the statistical film thickness.

Eq. (3.10-1) is the equation allowing us to calculate the volume adsorbed V as a function of the reduced pressure. For a given reduced pressure P/P_0, the statistical film thickness is calculated from eq. (3.9-25) and the critical radius r_K is calculated from eq. (3.10-5), and hence the volume adsorbed can be calculated by integrating the integral in eq. (3.10-1). We illustrate this with a number of examples below.

3.10.1.2 Numerical illustrations

We shall present below a number of examples to illustrate the method of Wheeler and Schull. We first use a simple Maxwellian distribution to describe the length distribution, and then consider the Gaussian distribution. Finally in the last example, we consider the double Gaussian distribution.

Example 3.10-1: *Maxwellian distribution*

Take the following Maxwellian distribution to describe the length distribution

$$L(r) = A\, r \exp\left(-\frac{r}{r_0}\right) \tag{3.10-10}$$

where A and r_0 are structural parameters. Knowing the length distribution, the total pore volume is calculated from eq. (3.10-2), that is:

$$V_s = 6\pi A r_0^4 \tag{3.10-11}$$

This equation relates V_s to the structural parameters A and r_0.

Substitution of the length distribution (eq. 3.10-10) into eq. (3.10-1) gives the following expression for the volume occupied by the adsorbate:

$$\frac{V}{V_s} = 1 - \frac{M(r_K, r_0)}{6\pi} \tag{3.10-12a}$$

where the function M is defined as follows:

$$M(r_K, r_0) = \frac{\pi}{r_0^3}\exp\left(-\frac{r_K}{r_0}\right)\left[r_K(r_K - t)^2 + 6r_0^3 + 2r_0^2(3r_K - 2t) + r_0(3r_K - t)(r_K - t)\right]$$

$$\tag{3.10-12b}$$

with r_K and t being function of pressure, given in eqs. (3.10-5) and (3.9-25). Figure 3.10-2 shows plots of the volume adsorbed versus the reduced pressure for nitrogen as adsorbate at 77 K. The total pore volume used in the plots is $V_s = 0.5$ cc/g and the characteristic pore radii r_0 are 2, 2.5 and 3 nm. In the figure, we note that the smaller is the characteristic pore radius the sooner does the capillary condensation occur.

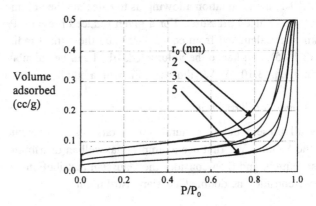

Figure 3.10-2: Volume adsorbed (cc/g) versus P/P_0 for the Maxwellian length distribution

Example 3.10-2: *Gaussian distribution*

The Maxwellian distribution is very simple and is not flexible enough to describe the length distribution as it contains only one parameter r_0 which is insufficient to characterise the shape of a practical distribution. A better distribution for this is the Gaussian distribution given by:

$$L(r) = A \exp\left[-\beta^2\left(\frac{r-r_0}{r_0}\right)^2\right] \qquad (3.10\text{-}13)$$

where A, r_0, and β are pore structural parameters. Here we see that there are two parameters r_0 and β that characterise the shape of the distribution.

The total pore volume is calculated from eq. (3.10-2) and for this Gaussian distribution, it has the following form:

$$V_s = \frac{\pi A r_0^3}{2\beta}\left[\frac{e^{-\beta^2}}{\beta} + \sqrt{\pi}\left(1 + \operatorname{erf}\beta\right) + \frac{1}{2\beta^2}\right] \qquad (3.10\text{-}14)$$

The volume adsorbed at a given pressure P is obtained by substituting eq. (3.10-13) into eq. (3.10-1) and we obtain:

$$\frac{V}{V_s} = 1 - \frac{4}{\pi} \frac{G_\beta(r_K, r_0)}{\left[\frac{e^{-\beta^2}}{\beta} + \sqrt{\pi}(1 + \text{erf } \beta) + \frac{1}{2\beta^2}\right]} \quad (3.10\text{-}15a)$$

where

$$G_\beta(r_K, r_0) = \frac{\pi}{4r_0^2}\left[\frac{r_0}{\beta}\exp(-\rho^2)\cdot(r_K - 2t + r_0) + \sqrt{\pi}\cdot\text{erfc}(\rho)\cdot(r_0 - t)^2 + \frac{1}{2}\left(\frac{r_0}{\beta}\right)^2\right]$$

$$\rho = \frac{\beta}{r_0}(r_K - r_0) \quad (3.10\text{-}15)$$

In eq. (3.10-15), the critical pore radius is calculated from eq. (3.10-5) and the statistical film thickness is calculated from eq. (3.9-25). For nitrogen as the adsorbate at 77 K, Figure 3.10-3a shows plots of the volume adsorbed versus the reduced pressure for the case where β = 5. The three mean radii, r_0, used in those plots are 2, 2.5 and 3 nm. Similar to the last example, the smaller is the pore range, the sooner does the capillary condensation occur. Figure 3.10-3b shows the effect of the variance by changing the value of β. The larger is this parameter, the sharper is the Gaussian distribution. This figure shows that the hysteresis loop is becoming vertical when the distribution is very sharp as one would expect for solids having fairly uniform pores.

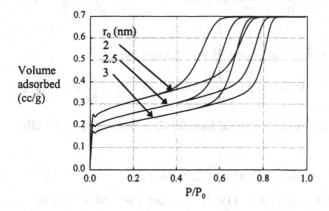

Figure 3.10-3a: Plots of the volume adsorbed versus P/P₀ for b = 5

136 Equilibria

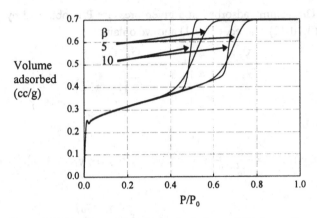

Figure 3.10-3b: Plots of the volume adsorbed versus P/P_0 for $r_0 = 2$ nm

Example 3-10-3 *Bimodal Gaussian distribution*

For solids having bimodal pore size distribution, we can use a combined Gaussian distribution as shown below.

$$L(r) = A_1 \exp\left[-\beta_1^2\left(\frac{r-r_1}{r_1}\right)^2\right] + A_2 \exp\left[-\beta_2^2\left(\frac{r-r_2}{r_2}\right)^2\right] \quad (3.10\text{-}16a)$$

which is the summation of the two Gaussian distributions.

The total pore volume is calculated from eq. (3.10-2), that is:

$$V_s = \frac{\pi A_1 r_1^3}{2\beta_1}\left[\frac{e^{-\beta_1^2}}{\beta_1} + \sqrt{\pi}\left(1+\mathrm{erf}\,\beta_1\right) + \frac{1}{2\beta_1^2}\right] + \frac{\pi A_2 r_2^3}{2\beta_2}\left[\frac{e^{-\beta_2^2}}{\beta_2} + \sqrt{\pi}\left(1+\mathrm{erf}\,\beta_2\right) + \frac{1}{2\beta_2^2}\right]$$

(3.10-16b)

The amount adsorbed V(P) is given by:

$$V_s - V = 2A_1\left(\frac{r_1^3}{\beta_1}\right)\cdot G_\beta(r_K, r_1) + 2A_2\left(\frac{r_2^3}{\beta_2}\right)\cdot G_\beta(r_K, r_2) \quad (3.10\text{-}17)$$

where the functional form for G_β is given in eq. (3.10-15b).

3.10.2 *Cranston and Inkley's (CI) method*

Cranston and Inkley (1957) presented a refined method over that of Barrett et al. (1951). Instead of using the pore length distribution and giving an equation relating the cumulative amount adsorbed up to a pressure P, Cranston and Inkley provided an equation relating the incremental adsorbed amount when the pressure is changed incrementally. Their method can provide an estimate for surface area, and can be applied to both adsorption and desorption branches of the isotherm to determine the pore size distribution.

3.10.2.1 The Statistical Film Thickness

This method, like the other methods, requires the knowledge of the statistical film thickness of the adsorbed layer on a flat surface. The experimentally determined thickness of the adsorbed layer for nitrogen on a <u>flat</u> surface was obtained by Cranston and Inkley as a function of reduced pressure as shown in Figure 3.10-1 as symbols and are tabulated in Table 3.10-1.

The data in Figure (3.10-1) was the average of isotherms of 15 nonporous materials (such as zinc oxide, tungsten powder, glass sphere, precipitated silver), obtained by dividing the volume of nitrogen adsorbed by the BET surface area. This average thickness of the adsorbed layer is a function of P/P_0:

$$t_{FLAT} = f_t(P/P^0) \qquad (3.10\text{-}18)$$

The functional form in the RHS is independent of the nature of the solid, and this is called the universal t-plot.

Table 3.10-1: Tabulation of the statistical layer thickness t(nm) versus the reduced pressure for nitrogen

P/P_0	t (nm)	P/P_0	t (nm)	P/P_0	t (nm)	P/P_0	t (nm)	P/P_0	t (nm)
0.08	0.351	0.28	0.488	0.48	0.634	0.68	0.826	0.88	1.382
0.10	0.368	0.30	0.501	0.50	0.650	0.70	0.857	0.90	1.494
0.12	0.383	0.32	0.514	0.52	0.666	0.72	0.891	0.92	1.600
0.14	0.397	0.34	0.527	0.54	0.682	0.74	0.927	0.94	1.750
0.16	0.410	0.36	0.541	0.56	0.699	0.76	0.965	0.96	1.980
0.18	0.423	0.38	0.556	0.58	0.717	0.78	1.007	0.98	2.29
0.20	0.436	0.40	0.571	0.60	0.736	0.80	1.057		
0.22	0.449	0.42	0.586	0.62	0.756	0.82	1.117		
0.24	0.456	0.44	0.602	0.64	0.777	0.84	1.189		
0.26	0.475	0.46	0.618	0.66	0.802	0.86	1.275		

3.10.2.2 The CI theory

Now back to the development of a working equation to determine the pore size distribution. Let f(r) be the pore volume distribution such that f(r)dr is the volume of pores per unit mass having radii between r and r + dr. Also let P(r) be the pressure at which pores of radii less than r will be filled with liquid, and P(r+dr) be the pressure at which pores of radii less than r+dr will be filled with liquid. The thickness of the adsorbed layer at the pressure P(r) is t(r) and that at P(r+dr) is

$$t(r + dr) = t(r) + dt \tag{3.10-19}$$

The incremental volume of nitrogen adsorbed when the pressure is increased from P(r) to P(r+dr) is given by the following equation:

$$dv = \frac{(r-t)^2}{r^2} f(r) dr + dt \int_{r+dr}^{\infty} \frac{(r-t)}{r} \frac{2f(r)}{r} dr \tag{3.10-20}$$

where r and t are functions of pressure as discussed earlier. The radius r is related to the pressure according to the Kelvin equation, while the thickness t is related to the pressure according to the universal plot shown in Figure 3.10-1 or eq. (3.9-25).

3.10.2.3 The Working Equation

Eq. (3.10-20) is the working equation, but in practice it is impossible to achieve infinitesimal small increments in pressure. Rather, only a finite increment in pressure can be made and measured, that is we have to deal with the integral form of eq.(3.10-20) rather than its current differential form. Thus, if the system pressure is increased from P_1 to P_2, the amount of nitrogen adsorbed due to this change in pressure is simply the integration of eq. (3.10-20) from r_1 to r_2, where r_1 and r_2 are the two critical radii corresponding to the pressures P_1 and P_2, respectively. This amount of nitrogen adsorbed is:

$$v_{12} = \frac{V_{12}}{r_2 - r_1} \int_{r_1}^{r_2} \frac{(r-t_1)^2}{r^2} dr + (t_2 - t_1) \int_{r_2}^{\infty} \frac{(2r - t_1 - t_2)}{r^2} f(r) dr \tag{3.10-21}$$

where V_{12} is the volume of pore having radii between r_1 and r_2. Solving for V_{12} from eq.(3.10-21), we get:

$$V_{12} = R_{12}\left[v_{12} - k_{12} \int_{r_2}^{\infty} \frac{(r - t_{12})}{2r^2} V(r) dr\right] \tag{3.10-22}$$

where

$$R_{12} = \frac{r_2 - r_1}{\int_{r_1}^{r_2}\left[(r-t_1)^2 / r^2\right] dr} \tag{3.10-23}$$

$$k_{12} = 4(t_2 - t_1), \quad t_{12} = (t_1 + t_2)/2 \tag{3.10-24}$$

Eq. (3.10-22) is the working equation for the determination of the pore size. For computational purposes, the integral in the RHS is replaced by the summation, that is:

$$V_{12} = R_{12}\left[v_{12} - k_{12}\sum_{r_2+\Delta r/2}^{r_{max}}\frac{(r-t_{12})}{2r^2}f(r)\Delta r\right] \quad (3.10\text{-}25)$$

The value of r_{max} is taken as 15 nm as it is known that surface area of pores larger than 15 nm is negligible. The further necessary equations in using the above working equation for nitrogen at 77 K are:

$$r_K = \frac{0.9399}{\ln(P_0/P)} \quad (3.10\text{-}26)$$

and

$$r = r_K(P/P_0) + t_{FLAT}(P/P_0) \quad (3.10\text{-}27)$$

Cranston and Inkley (1957) provided tables for R_{12}, k_{12} and the critical relative pressure as a function of pore size. They also tabulated $(r - t_{12})/r^2$ for the second term in the RHS of eq. (3.10-25).

3.10.2.4 *The Surface Area and the Mean Pore Radius*

The average pore radius can be calculated by assuming that the pore is of cylindrical shape:

$$\bar{r} = \frac{2V_p}{S} \quad (3.10\text{-}28)$$

Typical surface area, pore volume and mean pore size for a number of solids are given in Table 3.10-2 (Smith, 1956).

Table 3.10-2: Typical surface area, pore volume and mean pore radius of some solids

Solids	Surface area (m²/g)	Pore volume (cc/g)	Mean pore radius (A)
Activated carbon	500-1500	0.6-0.8	10-20
Silica gel	200-600	0.4	15-100
Silica-alumina	200-500	0.2-0.7	33-150
Activated clays	150-225	0.4-0.52	100
Activated alumina	175	0.39	45
Celite (kieselguhr)	4.2	1.1	11,000

3.10.3 De Boer Method

De Boer studied extensively the pore size distribution, and refined methods to determine it. Basically, he accounted for the pore shape in the calculation of the statistical film thickness as well as the critical pore radius. What we present below is the brief account of his series of papers published from early 60 to early 70.

3.10.3.1 Slit Shape Pore

For slit shape pore, the desorption branch should be used for the determination of PSD and we will discuss this further in Section 3.11. The relevant equation is then the Kelvin equation for evaporation, rewritten here in Kelvin radius versus P:

$$r_K = \frac{2\sigma v_M \cos\theta}{R_g T}\left[\ln\left(\frac{P_0}{P}\right)\right]^{-1} \tag{3.10-29}$$

where r_K is the Kelvin radius, v_M is the liquid molar volume, and σ is the surface tension. For liquid nitrogen at 77 K, the values of these parameters are:

$$\sigma = 8.72 \times 10^{-3}\, Nm^{-1}; \quad v_M = 34.68\, cm^3/mole; \quad \theta = 0 \tag{3.10-30}$$

Thus, the Kelvin equation for nitrogen adsorbate is:

$$r_K(nm) = \frac{0.405}{\log_{10}(P_0/P)} \tag{3.10-31}$$

This Kelvin equation is not accurate in its use in the calculation of pore size because the adsorption force modifies the meniscus shape. The following equation is suggested to replace the simple Kelvin equation (Anderson and Pratt, 1985; Broekhoff and de Boer, 1967):

$$r_K = (d-2t) = \begin{cases} \dfrac{0.405 + 0.2798[(1/t)-(2/d)] - 0.068(d/2-t)}{\log_{10}(P_0/P)} & \text{for } t < 1.00\,nm \\[2ex] \dfrac{0.405 + 0.3222[(1/t)-(2/d)] + 0.2966[\exp(-0.569d)-\exp(-1.137t)]}{\log_{10}(P_0/P)} & \text{for } t > 0.55\,nm \end{cases}$$

$$\tag{3.10-32}$$

where d is the width of the slit, and t is the adsorbed layer thickness. Both have units of nm. The film thickness t of the adsorbed layer is related to P/P_0 as:

$$\log_{10}(P_0/P) = \frac{0.1399}{t^2} - 0.034 \quad \text{for } t < 1\,nm$$
$$\log_{10}(P_0/P) = \frac{0.1611}{t^2} - 0.1682\exp(-1.137t) \quad \text{for } t > 0.55\,nm \tag{3.10-33}$$

Thus, for a given reduced pressure, eqs.(3.10-33) can be solved to obtain the adsorbed layer thickness t, and knowing this thickness t, eqs.(3.10-32) can then be solved to obtain "d" and thence the Kelvin radius.

The procedure of getting the pore size distribution is then a simple and straightforward manner. The volume of nitrogen in the capillary condensation range is V, which is a function of the reduced pressure (P/P_0), that is:

$$V = f(P/P_0) \quad (3.10\text{-}34)$$

But from eqs.(3.10-32) and (3.10-33), we can obtain a relationship between P/P_0 and the width d, that is:

$$P/P_0 = g(d) \quad (3.10\text{-}35)$$

Combining eqs.(3.10-34) and (3.10-35), we get the volume of nitrogen versus the width of the pore:

$$V = f[g(d)] \quad (3.10\text{-}36)$$

and then the pore size distribution is simply the derivative of the above equation with respect to d, that is:

$$\frac{dV}{d(d)} = \frac{df}{dg} \frac{dg}{d(d)} \quad (3.10\text{-}37)$$

3.10.3.2 Cylindrical Shape Pore

In the case of cylindrical pore, the layer thickness is not just a function of the reduced pressure but also on the pore diameter d. The necessary equations relating t and the reduced pressure and d are (Broekhoff and de Boer, 1967):

$$\log_{10}(P_0/P) = \frac{0.1399}{t^2} - 0.034 + \frac{0.2025}{(d/2 - t)} \quad \text{for} \quad t < 1 \text{ nm}$$
$$\log_{10}(P_0/P) = \frac{0.1611}{t^2} - 0.1682\exp(-1.137t) + \frac{0.2025}{(d/2 - t)} \quad \text{for} \quad t > 0.55 \text{ nm} \quad (3.10\text{-}38)$$

The difference between this equation and eq.(3.10-33) is the last term in the RHS of eq.(3.10-38), accounting for the curvature.

The modified Kelvin equation for this case of cylindrical pore is:

$$r_K = (d-2t) = \begin{cases} \dfrac{0.810}{\log_{10}(P_0/P)} + \dfrac{0.5596[d/2t-1-\ln(d/2t)]-0.068(d/2-t)^2}{(d/2-t)\log_{10}(P_0/P)} & \text{for } t < 1.00\text{n}\text{i} \\[2ex] \dfrac{0.810}{\log_{10}(P_0/P)} + \dfrac{0.6444[d/2t-1-\ln(d/2t)]}{(d/2-t)\log_{10}(P_0/P)} \\[1ex] \quad - \dfrac{0.5932[d/2-t-0.8795]\exp(-1.137t)}{(d/2-t)\log_{10}(P_0/P)} \\[1ex] \quad - \dfrac{0.512\exp(-0.569d)}{(d/2-t)\log_{10}(P_0/P)} & \text{for } t > 0.55\text{n}\text{i} \end{cases}$$

(3.10-39)

Solving these two equations (eqs. 3.10-38 and 3.10-39) will then yield the pore diameter, the thickness t and the Kelvin radius r_K in terms of the reduced pressure.

3.11 Assessment of Pore Shape

There are two ways of assessing the shape of a pore (Anderson and Pratt, 1985). One way is to investigate the shape of the hysteresis loop, and the other way is the shape of the amount adsorbed plotted against t.

3.11.1 Hysteresis Loop

If the hysteresis loop is vertical and the adsorption and desorption branches are parallel to each other, the pores are tubular in shape and open at both ends (type A). If the hysteresis is very flat and parallel, the pores have slit shape with parallel walls (type B). The third type is the type whereby the adsorption branch is vertical and the desorption branch is inclined (type C). This type is for systems where there is heterogeneous distribution of pores of type 1. The fourth type is that whereby the adsorption branch is flat and the desorption branch is inclined (type D). This type is exhibited by tapered slit pores. The fifth type has inclined adsorption branch and a vertical desorption branch (Type E). Schematic diagram of these hysteresis loops is shown in Figure 3.11-1.

The analysis of the hysteresis loop using the adsorption branch or the desorption branch depend on the shape of the pore. For the ink-bottle shape, once the pore is full the desorption will occur from the narrow neck and this is replenished from the larger parts of the pore; thus analysis of the desorption branch does not give information about the main part of the pore. Therefore, the analysis of the adsorption branch is necessary.

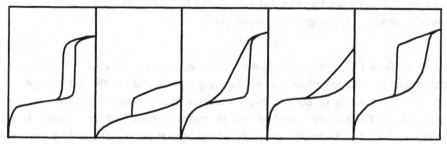

Figure 3.11-1: Various types of hysteresis loop

For slit shape pores, the pore becomes filled with adsorbate when the multilayers of the opposite walls meet; therefore, the desorption branch must be analysed.

For cylindrical pores with uniform cross section, either the adsorption branch or desorption branch could be used. For cylindrical pores of significantly varied cross section, the adsorption branch should be used as this type of pore behaves like ink-bottle pore.

3.11.2 t-Method

If the adsorption occurs in a free surface (no restriction on the number of layers which can be built up on top of the surface), the statistical thickness of the adsorbed layer is a function of the reduced pressure as seen in Figure 3.10-1 or Table 3.10-1, that is

$$t_{FLAT} = f_{FLAT}(P/P_0) \tag{3.11-1}$$

This functional form is called the universal t-plot for a <u>flat</u> surface. The statistical average thickness can also be calculated by using eq.(3.9-25).

For pores of cylindrical or slit shape, the behaviour of the calculated statistical thickness does not follow that of a flat surface as the pore shape can influence the statistical film thickness. This is explained as follows. For cylindrical pores, the solid will take up more sorbates than a free surface, that is

$$t_{CYL} > t_{FLAT} \tag{3.11-2}$$

For a pore of slit shape, adsorption will occur on two surfaces of the pore when pressure increases. As the pressure approaches a certain pressure, the adsorbed layers of the two opposing walls meet; and this pore will be no longer available for uptake when the pressure is increased further. This means that for a porous solid having a distribution of slit-shaped pores, the calculated statistical thickness will be less than that corresponding to a flat surface due to the loss in surface area resulted from the merging of two opposing layers, that is

$$t_{SLIT} < t_{FLAT} \tag{3.11-3}$$

Thus, for a free surface, the plot of the amount adsorbed versus "t" will be a straight line because the adsorption is a layering process and the area for adsorption does not change as the number of layers increases. For porous solids, however, the magnitude of the amount adsorbed can be either increased or decreased. If it increases, then it indicates the presence of cylindrical pores, ink bottle pores, or voids between closed packed spherical particles. On the other hand, if it decreases this indicates slit-shaped pores.

These influences can be studied by collecting data of the amount adsorbed versus the reduced pressure. For each value of the reduced pressure, we obtain the hypothetical t_{FLAT} from the universal t-plot. Then we finally plot the amount adsorbed versus this hypothetical t_{FLAT}. This plot will have the initial straight line passing through the origin because all surfaces are available for adsorption initially, and afterwards the slope can either increase or decrease compared to the initial slope depending on the shape of the pore as discussed in eqs. (3.11-2) and (3.11-3). The initial slope is the result of multilayering unhindered by the pore structure, and hence it can be used to calculate the total surface area. This is done as follows. Let V_m be the monolayer amount and V is the amount at a given pressure P. If this amount is to build multilayer, then the number of layer is simply V/V_m. The number of layer is also equal to t/σ, where t is the statistical film thickness of a flat surface corresponding to the amount adsorbed V and σ is the thickness of one layer. Thus

$$\frac{V}{V_m} = \frac{t}{\sigma} \tag{3.11-4a}$$

or

$$V_m = \sigma \left(\frac{V}{t}\right) \tag{3.11-4b}$$

where V/t is simply the initial slope of the plot of V versus t. Knowing the monolayer coverage, the surface area is simply calculated from

$$A = V_m a_m N_A \tag{3.11-5}$$

where a_m is the projection area per molecule and N_A is the Avogadro number. For nitrogen as the adsorbate, the relevant parameters are:

$$\sigma = 0.354 \text{ nm}, \ a_m = 0.162 \text{ nm}^2 / \text{molecule}$$

The procedure of using the t-plot is provided below.

The procedure:
1- A reference non-porous material with similar surface characteristics is chosen to obtain the information on the statistical film thickness as a function of the reduced pressure. The t values for different reference adsorbents as a function of the reduced pressure are available in literature, normally in the table form or a best fit equation.
2- The equilibrium data of the amount adsorbed versus the reduced pressure are plotted as the amount adsorbed versus the statistical thickness by using the data of nonporous material in step 1.
3- From the plot of step 2, there are usually two distinct linear regions and two straight lines can be drawn from these two regions. The slope of the first linear line passing through the origin can be used to calculate the total surface area, while the slope of the second straight line is used to calculate the external surface area. The intercept of the latter line gives the volume of the micropores.

We now illustrate the above procedure with the following example.

Example 3.11-1 *Application of t-method*

The nitrogen adsorption data at 77 K of a sample of activated carbon are presented in the following table.

Table 3.11-1: Adsorption data of a sample of activated carbon.

P/P_0 (-)	V (cc/g STP)	t (Å)	α_s	P/P_0 (-)	V (cc/g STP)	t (Å)	α_s
0.010	234.0	2.631	0.470	0.450	321.7	5.980	1.068
0.029	259.7	3.054	0.545	0.500	322.9	6.315	1.128
0.061	281.9	3.380	0.604	0.550	324.0	6.709	1.198
0.077	288.7	3.476	0.621	0.600	325.0	7.090	1.266
0.104	297.1	3.809	0.654	0.650	326.0	7.506	1.340
0.125	301.7	3.809	0.680	0.700	327.0	7.885	1.408
0.148	305.4	3.965	0.708	0.740	327.9	8.211	1.466
0.171	308.1	4.105	0.733	0.770	328.6	8.481	1.514
0.192	310.1	4.244	0.758	0.800	329.3	8.796	1.571
0.210	311.5	4.366	0.780	0.820	329.8	8.796	1.612
0.251	314.3	4.638	0.828	0.841	330.5	9.285	1.658
0.305	316.9	5.001	0.893	0.860	331.2	9.732	1.703
0.357	319.0	5.363	0.958	0.875	331.9	9.732	1.738
0.399	320.3	5.596	0.999	0.891	332.7	10.12	1.773

In the above table, V is the amount adsorbed at the relative pressure P/P_0. For each value of the reduced pressure, the statistical film thickness t can be obtained from a reference material of the same characteristics or calculated using eq. (3.9-25). This has been included in the third column of the above table by using eq. (3.9-25).

The next step is to plot the amount adsorbed V (cc/g) versus the statistical thickness t as shown in Figure 3.11-2.

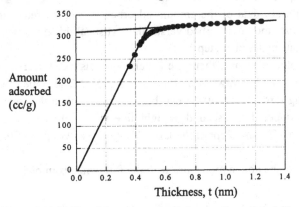

Figure 3.11-2: Plot of the amount adsorbed versus the statistical film thickness

The best fits of two straight lines representing the two linear regions of the t plot are also shown in the figure. The slope of the first straight line is 84.64 cc STP/g/Å, which is equivalent to 3.78×10^{-2} mole N_2/g/nm. Substituting this slope into eq. (3.11-5), we obtain a total surface area of 1305 m²/g. Similarly, using the slope of the second straight line of 2.65 cc STP/g/Å (or 1.183×10^{-3} mole N_2/g/nm), we obtain the surface area of the external area of 41 m²/g. Finally the intercept of the second straight line to the volume axis will provide the micropore volume. Assuming the state of adsorbate in the micropore as liquid, we can calculate the micropore volume. From Figure 3.11-2, we have an intercept of 306 cc STP/g. Knowing the liquid density of nitrogen at 77 K as 0.807 g/cc, we calculate the micropore volume as 0.47 cc/g. The following table summarises the results.

Slope 1 (cc STP/g/Å)	Slope 2 (cc STP/g/ Å)	Intercept (cc STP/g)	Total area (m²/g)	External area (m²/g)	Micropore volume (cc/g)
84.64	2.65	306	1305	41	0.47

3.11.3 The α_s Method

The t-method presented in the last section is applicable to the type IV isotherm. It requires the information of the adsorbed film thickness of a reference material as a function of the reduced pressure. If the reference material is used to compare with the test sample, there is another method, called the α_s-method, which can be used to derive information about the test sample. This α_s-method was developed by Sing and co-workers and it is done as follows. First a reference solid having the same surface characteristics as the test sample is chosen and then adsorption is carried out on that reference material. The reduced adsorbed amount ($\alpha_s = V/V_{0.4}$) is then plotted versus the reduced pressure, where $V_{0.4}$ is the volume adsorbed at a reduced pressure of 0.4. This plot is called the standard α_s-plot, and it plays a similar role to the t-plot in the t-method presented in the last section. This standard α_s-plot is then used to construct a plot of the amount adsorbed versus α_s for the test sample, that is for every value of the reduced pressure for the test sample, we obtain a value α_s from the α_s-plot. Then finally we obtain a relationship between the amount adsorbed versus α_s for the test sample, and a plot of such relationship is constructed. As the α_s-method does not assume any value for the thickness of the adsorbed layer, it can be used with any adsorbate gas. The surface area of the test sample is calculated by measuring the slope of the linear part of the α_s-plot of the test sample and that of the reference sample. By knowing the surface area of the reference, the surface area of the test sample is calculated from:

$$A_{test} = \frac{Slope_{test}}{Slope_{ref.}} \times A_{ref.} \qquad (3.11\text{-}6)$$

Similar procedure as applied in the t-method can be used to calculate the external surface area and micropore volume using the α_s values.

Example 3.11-2 *Application of α_s-method*

Using the previous example, the α_s values are shown in the fourth column of Table 3.11-1. The surface area and the slope of the reference sample is 81 m²/g and 29.42 cc STP/g. Figure 3.11-3 shows a plot of the amount adsorbed versus α_s. The slope of the first line of the test sample is 474 cc STP/g. Thus the total surface area is calculated using eq. (3.11-6) and its value is 1305 m²/g. This is exactly the same as that obtained by the t-plot. The slope of the second line is 15 cc STP/g (or 4.03 × 10²⁰ molecules/g). Knowing the projection area of nitrogen molecule as 0.162 nm²/molecule, we obtain the external surface area as 65 m²/g, compared to

41 m²/g of the t-plot. The intercept of the plot 3.11-3 is 305.4 cc STP/g. Knowing the liquid density of nitrogen at 77 K as 0.807 g/cc, we calculate the micropore volume as 0.47 cc/g which is also the same as that obtained by the t-plot.

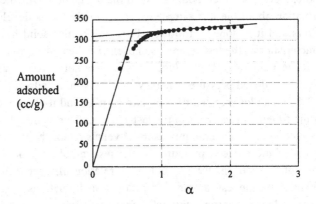

Figure 3.11-3: Plot of the amount adsorbed versus α

3.12 Conclusion

Various practical isotherm equations have been presented and they are useful in describing adsorption data of many adsorption systems. Among the many equations presented, the Toth equation is the attractive equation because of its correct behaviour at low and high loading. If the Henry behaviour is not critical then the Sips equation is also popular. For sub-critical vapours, multilayer isotherm equations are also presented in this chapter. Despite the many equations proposed in the literature, the BET equation still remains the most popular equation for the determination of surface area. When condensation occurs in the reduced pressure range of around 0.4 to 0.995, the theory of condensation put forward by Kelvin is useful in the determination of the pore size as well as the pore size distribution.

4

Pure Component Adsorption in Microporous Solids

4.1 Introduction

In Chapter 2, we discussed the fundamentals of adsorption equilibria for pure component, and in Chapter 3 we presented various empirical equations, practical for the calculation of adsorption kinetics and adsorber design, the BET theory and its varieties for the description of multilayer adsorption used as the yardstick for the surface area determination, and the capillary condensation for the pore size distribution determination. Here, we present another important adsorption mechanism applicable for microporous solids only, called micropore filling. In this class of solids, micropore walls are in proximity to each other, providing an enhanced adsorption potential within the micropores. This strong potential is due to the dispersive forces. Theories based on this force include that of Polanyi and particularly that of Dubinin, who coined the term micropore filling. This Dubinin theory forms the basis for many equations which are currently used for the description of equilibria in microporous solids.

The Dubinin equation (Dubinin, 1966, 1967, 1972, 1975) has its history in the development of theory for adsorption in activated carbon. Activated carbon has a very complex structure (see Chapter 1 for some details), with pores ranging from macropores of order of greater than 1000 Å to micropores of order of 10 Å. It is this micropore network where most of the adsorption capacity resides. Because of the pore dimension comparable to the dimension of adsorbate molecule, the adsorption mechanism in micropore is completely different from that on a surface of a large pore, where adsorption occurs by a layering process. In micropores, the mechanism is due to micropore filling because of the adsorption force field encompassing the entire volume of micropores. Such an enhancement in the adsorption potential would lead to higher heat of adsorption in micropore compared to that on a surface. For example, adsorption heat of n-hexane on activated carbon at 20 °C at a loading of 0.25 mmole/g is about 15 kcal/mole, while for the same set of conditions on nonporous carbon black, the heat of adsorption is about 10 kcal/mole (Dubinin, 1966).

4.1.1 Experimental Evidence of Volume Filling

The concept of micropore filling was demonstrated with zeolite CaA (5A) and NaX. Since zeolites have regular structure, the specific surface area of the channels (micropores) can be calculated using X-ray studies. These areas are 1640 and 1400 m² /g for CaA and NaX, respectively. The external surface area of these zeolite crystals are so low compared to the internal surface area; hence, adsorption on these external areas can be neglected. Knowing the area, the theoretical "monolayer" amount adsorbed can be calculated using the information of area occupied by one adsorbate molecule. This is shown in Table 4.1-1 (Dubinin, 1966) together with the experimental limiting amount adsorbed. Here C_μ^* is the theoretical monolayer amount, and C_μ is the experimental amount adsorbed.

Table 4.1-1: Amount adsorbed and theoretical monolayer amount for CaA and NaA zeolites

Vapor	T(°C)	CaA (1640 m²/g)			NaX (1400m²/g)		
		C_μ^* (mmole/g)	C_μ (mmole/g)	C_μ^*/C_μ	C_μ^* (mmole/g)	C_μ (mmole/g)	C_μ^*/C_μ
H_2O	20	26.5	15.50	1.71	22.6	17.95	1.26
CO	-196	16.2	8.61	1.88	14.1	9.71	1.45
N_2	-196	16.8	8.27	2.03	14.4	9.55	1.51
Ar	-196	19.7	8.58	2.30	16.8	10.27	1.64
Benzene	20	-	-	-	7.25	3.3	2.2
n-pentane	20	7.53	2.26	3.33	6.42	2.56	2.51

From Table 4.1.1, the amount calculated assuming monolayer adsorption mechanism is 2 to 3 times larger than the experimental values, indicating a micropore filling rather than surface layering mechanism. Furthermore, for a given adsorbate the ratio of the experimental amount adsorbed for the two zeolites is

$$\frac{C_\mu(CaA)}{C_\mu(NaX)} \approx 0.861$$

which is in perfect agreement with the ratio of the void volumes of the two zeolites

$$\frac{V(CaA)}{V(NaX)} = \frac{0.278 \text{ cc}/g}{0.322 \text{ cc}/g} \approx 0.863$$

rather than with the ratio of surface areas

$$\frac{S(CaA)}{S(NaX)} = \frac{1640 \text{ m}^2/g}{1400 \text{ m}^2/g} \approx 1.17$$

This shows clearly that the mechanism of adsorption is by micropore filling. This is summarised graphically in Figure 4.1-1.

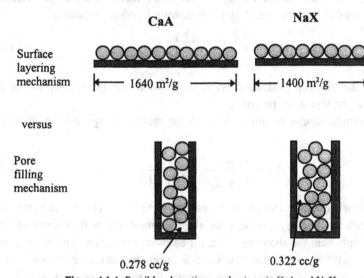

Figure 4.1-1: Possible adsorption mechanisms in CaA and NaX

4.1.2 Dispersive Forces

Forces of adsorption are electrostatic forces, the valance energy force and the cohesive energy force. The force of relevant interest here is the van der Waals force, which exists between all atoms and molecules. The van der Waals forces can be classified into three groups:

(a) dipole-dipole forces
(b) dipole-induced dipole forces: In this case one molecule having a permanent dipole will induce a dipole in a non-polar atom, such as neon
(c) dispersion forces

The dispersion force is the most important force in physical adsorption. It has an origin in the quantum mechanics. Let us take an example of a non-polar atom such as neon or helium. The time average of its dipole moment is zero. But at instant time "t" there is an asymmetry in the distribution of electrons around the nucleus and this generates a finite dipole. This so generated dipole will polarise any nearby atom (that is it distorts the electron distribution) so that the nearby atom will acquire a dipole. These two dipoles will attract, and the time average of this attractive force is finite. This is the basis of the van der Waals's dispersion force.

The dispersion interaction energy between two identical atoms or molecules having distance r apart is governed by the following London's equation:

$$\phi(r) = -\frac{3}{4}\frac{\alpha_0^2 \, I}{(4\pi\varepsilon_0)^2 \, r^6} \qquad (4.1\text{-}1)$$

where α_0 is the electron polarizability, ε_0 is the dielectric permitivity of the free space, and I is the ionisation potential.

For dissimilar atoms or molecules, the interaction energy by the dispersion force is:

$$\phi(r) = -\frac{3}{2}\frac{\alpha_{0,1}\,\alpha_{0,2}}{(4\pi\varepsilon_0)^2 \, r^6}\frac{I_1 I_2}{I_1+I_2} \qquad (4.1\text{-}2)$$

The interaction potential energy is proportional to r^{-6}. To evaluate the total interactive potential energy between an atom or molecule with a layer of solid atoms, we simply sum the above equation for pairwise interaction with respect to all atoms on the solid layer. If the distance z between an atom or molecule and the solid layer is larger than the solid atom spacing, the summation is replaced by the following integral:

$$\phi_L(z) = -\int_0^\infty \frac{3}{2}\frac{\alpha_{0,1}\,\alpha_{0,2}}{(4\pi\varepsilon_0)^2 \, r^6}\frac{I_1 I_2}{I_1+I_2} \times N \times (2\pi x)dx \qquad (4.1\text{-}3a)$$

where N is the number of surface atoms per unit area, $2\pi x\,dx$ is the differential area on the solid layer, and the inter-atom distance "r" is given by

$$r = \sqrt{z^2 + x^2} \qquad (4.1\text{-}3b)$$

Evaluating the integral (4.1-3a) gives:

$$\phi_L(z) = -\frac{3\pi N}{4}\frac{\alpha_{0,1}\,\alpha_{0,2}}{(4\pi\varepsilon_0)^2 \, z^4}\frac{I_1 I_2}{I_1+I_2} \qquad (4.1\text{-}4)$$

This is the interaction potential energy between an atom or molecule and <u>a layer</u>. It is proportional to z^{-4}. If the solid is composed of semi-infinite layers of atoms, the interaction energy is simply:

$$\phi_S(z) = \int_z^\infty \phi_L(z')dz' = -\frac{\pi N'}{4} \frac{\alpha_{0,1} \alpha_{0,2}}{(4\pi\varepsilon_0)^2 z^3} \frac{I_1 I_2}{I_1 + I_2} \qquad (4.1\text{-}5)$$

where N' is the number of atoms per unit volume. Thus the interaction potential energy between an atom or molecule with a solid of semi-infinite in extent is proportional to z^{-3}, stronger than the interaction with a single layer.

For a slit shape pore whose walls are made of semi-infinite layers of atoms, the interaction potential energy between an atom or molecule and the pore is the sum of eq.(4.1-5) for both walls, that is:

$$\phi_{PORE} = \phi_S(z) + \phi_S(H - z) \qquad (4.1\text{-}6)$$

where H is the distance between the centers of the first layers of the two walls (Figure 4.1-2). The summation in the potential results in an enhancement of the interaction energy between the atom or molecule and the pore. This is the energy source responsible for the micropore filling. We shall discuss more on this in Chapter 6.

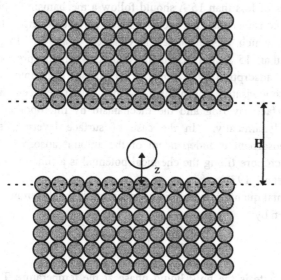

Figure 4.1-2: Configuration of a molecule residing in a slit shaped pore made up of semi-infinite layers of solid atoms

What we shall present in the next sections are the various adsorption equations developed to describe adsorption equilibria in micropores. For solids containing both micropores and mesopores, the data collected must have the contribution from the mesopores removed before the data can be used to test the micropore filling theory. For microporous solids having total surface area greater than 800 m²/g, the contribution of the mesopore toward adsorption capacity is usually negligible if its surface area is less than 10 m²/g. If the mesopore surface area is greater than about 50 m²/g, the amount adsorbed on the mesopore surface must be subtracted from the total amount, that is:

$$C_{\mu,micro} = C_{\mu,exp} - \alpha S_{meso}$$

where S_{meso} is the mesopore specific surface area and α is the value of adsorption per unit surface area, usually obtained from the adsorption of a nonporous material having similar surface characteristics as that of the microporous solid. For example, in the case of microporous activated carbon, nonporous carbon black is normally used to determine the parameter α.

4.1.3 Micropore Filling Theory

Bering et al. (1966, 1972) studied adsorption in micropores and suggested that adsorption in pores of less than 15 Å should follow a mechanism of pore filling of the adsorption space rather than the mechanism of surface coverage (formation of successive layers), which was discussed in Chapters 2 and 3. In pores having dimension larger than 15 Å and less than 1000 Å, when the pressure reaches a threshold pressure, adsorption by layering on the surface turns into volume filling by the capillary condensation mechanism. A distinction between the adsorption mechanism by surface layering and the mechanism by micropore filling can be explained thermodynamically. In the case of surface layering, the chemical potential of the adsorbent is independent of the amount adsorbed, while in the second case of micropore filling the chemical potential is a function of the amount adsorbed (Bering et al., 1972).

The fundamental quantity in the micropore filling is the differential molar work of adsorption, given by:

$$A = R_g T \ln\left(\frac{P_0}{P}\right) \tag{4.1-7}$$

The standard state is the bulk liquid phase at the temperature T, which is in equilibrium with saturated vapour. The parameter P_0 is the vapour pressure of the free liquid.

The principal feature of the micropore filling is the temperature invariance of the differential molar work of adsorption at a constant degree of filling of the adsorption space, that is:

$$\left(\frac{\partial A}{\partial T}\right)_\theta = 0 \qquad (4.1\text{-}8)$$

where θ is the fraction of the micropore volume occupied by the adsorbate. Integrating the above equation gives:

$$A = H(\theta) \qquad (4.1\text{-}9)$$

Assuming that an inverse of the above equation exists, we can write:

$$\theta = H^{-1}(A) \qquad (4.1\text{-}10)$$

This equation is the basis of the theory of micropore filling put forwards by the school of Dubinin. As the adsorption potential A has the unit of molar energy, we can scale it against a **characteristic energy**, hereafter denoted as E, and rewrite the micropore filling equation (4.1-10) as follows:

$$\theta = f(A/E, n) \qquad (4.1\text{-}11)$$

The characteristic energy is a measure of the adsorption strength between adsorbate and adsorbent. The function f is regarded as the distribution function of filling of micropores θ over the differential molar work of adsorption, and n is the parameter associated with the distribution function.

For two different adsorbates, their adsorption potentials must follow the following equation to attain the same degree of filling θ

$$\left(\frac{A}{E}\right) = \left(\frac{A}{E}\right)_0 \qquad (4.1\text{-}12)$$

if the parameter n is the same for both adsorbates. Here A and E are the adsorption potential and the characteristic energy of one adsorbate and A_0 and E_0 are those of the reference adsorbate. For activated carbon, the reference adsorbate is chosen as benzene.

Eq.(4.1-12) leads to:

$$\frac{E}{E_0} = \frac{A}{A_0} = \beta \qquad (4.1\text{-}13)$$

where the parameter β is the coefficient of similarity. This coefficient is taken as the ratio of the liquid molar volume to that of the reference vapor.

4.2 Dubinin Equations

4.2.1 Dubinin-Radushkevich (DR) Equation

The distribution f in eq. (4.1-11) is arbitrary. Dubinin and his co-workers chose the functional form of the Weibull distribution

$$f(A/E,n) = \exp\left[-(A/E)^n\right] \qquad (4.2\text{-}1)$$

The parameter n = 2 was first suggested by Dubinin and Radushkevich (1947), and hence the resulted adsorption equation is called the DR equation

$$\theta = \exp\left[-(A/E)^2\right] \qquad (4.2\text{-}2a)$$

where the adsorption potential is defined in eq. (4.1-7). Since the adsorption mechanism in the micropore is the volume filling, the degree of filling is:

$$\theta = \frac{W}{W_0} \qquad (4.2\text{-}2b)$$

where W is the volume of the adsorbate in the micropore and W_0 is the maximum volume that the adsorbate can occupy.

The temperature dependence of the isotherm equation is embedded in the vapor pressure and the term T appearing in the definition of the adsorption potential (eq. 4.1-7).

Observing the adsorption isotherm equation (4.2-2), we note that if the characteristic energy is independent of temperature, plots of the fractional loading versus the adsorption potential for different temperatures will collapse into one curve, called the characteristic curve. This nice feature of the Dubinin equation makes it convenient in the description of data of different temperatures.

The adsorption isotherm of the form (4.2-2) fits well numerous data of activated carbon. It does not perform well in solids having fine micropores such as molecular sieving carbon and zeolite. Before we address other forms of equation to deal with these solids, let us return to the adsorption potential.

4.2.1.1 The Adsorption Potential A

The value of A is equal to the difference between the chemical potentials of the adsorbate in the state of normal liquid and the adsorbed state at the same temperature. For ideal fluids, this adsorption potential is

$$A = \mu_L - \mu_a = \left[\mu^{(0)} + R_g T \ln P_0\right] - \left[\mu^{(0)} + R_g T \ln P\right] = R_g T \ln\left(\frac{P_0}{P}\right)$$

This adsorption potential A is the change of the Gibbs free energy on adsorption:

$$A = -\Delta G_{ads}$$

Thus, the Dubinin and Radushkevich equation states the distribution of the adsorption space W according to the differential molar work of adsorption. A typical plot of the adsorption potential versus the reduced pressure is shown in Figure 4.2-1 for T = {77, 273 and 473 K}. For low reduced pressure, the adsorption potential is high, while it is low for high reduced pressure. The latter means that less molar work is required for adsorption via micropore filling when the gas is approaching the vapour pressure.

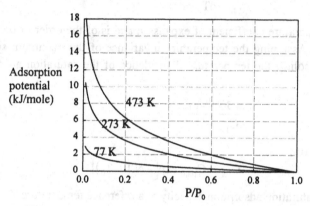

Figure 4.2-1: Plots of the adsorption potential versus the reduced pressure (T = 77, 273, 473 K)

4.2.1.2 The Molar Amount Adsorbed

Knowing the volume W taking up by the adsorbate, the number of moles adsorbed (moles/g) is simply

$$C_\mu = W / v_M \qquad (4.2\text{-}3a)$$

assuming the adsorbed phase behaves like a liquid phase. Here C_μ is the molar amount adsorbed (mole/g solid), W is the liquid volume adsorbed per unit mass of the solid, and v_M is the liquid molar volume (cc of liquid/mole).

The assumption of liquid-like adsorbed phase is made following Gurvitch (1915) who carried out experiments of various vapors on relatively large pores such as clays and earths. These solids, however, are macroporous solids; therefore, Gurvitch in fact measured the capillary condensation rather than micropore filling. It has been reported in Dubinin (1966) that the volume sorbed may exceed the limiting adsorption volume by up to a factor of 1.5, indicating that the adsorbed state in micropore is denser than the liquid state. However in the absence of the information on the state of adsorbed phase, it is reasonable to assume that it behaves like a liquid phase.

The saturation adsorption capacity corresponds to the maximum volume W_0:

$$C_{\mu s} = W_0 / v_M \qquad (4.2\text{-}3b)$$

This saturation capacity is a function of temperature as the liquid molar volume v_M is a function of temperature. This temperature dependence usually takes the form

$$\frac{1}{v_M}\frac{dv_M}{dT} = \delta$$

where δ is the temperature coefficient of expansion and is of the order of 0.003 K^{-1} for many liquids. Assuming the temperature invariance of the maximum specific volume W_0 (rigid solid), the temperature dependence of the saturation adsorption capacity is:

$$\frac{1}{C_{\mu s}}\frac{dC_{\mu s}}{dT} = -\delta \qquad (4.2\text{-}4a)$$

or in the integral form:

$$C_{\mu s} = C_{\mu s,0} \exp\left[-\delta(T - T_0)\right] \qquad (4.2\text{-}4b)$$

where $C_{\mu s,0}$ is the saturation adsorption capacity at a reference temperature T_0.

Typical values of W_0 of activated carbon are in the range of 0.25 to 0.5 cc/g. For carbon char, this value is usually lower than 0.2 cc/g.

4.2.1.3 The Characteristic Curve

Equation (4.2-2) provides what we usually call the characteristic curve, that is experimental data of various temperatures can be plotted on the same curve of lnW versus A^2.

The characteristic energy E in eq. (4.2-2) can be regarded as the average free energy of adsorption specific to a particular adsorbent-adsorbate pair. If the change in the Gibbs free energy, A, is equal to this characteristic energy, the amount adsorbed will be

$$\theta = \exp(-1) \approx 0.37$$

that is 37% of the micropore volume is occupied with adsorbate when the adsorption potential is equal to the characteristic energy.

If the adsorption potential is one third of the characteristic energy, the amount adsorbed is:

$$\theta = \exp(-1/3) \approx 0.9$$

that is 90% of the micropore volume is filled.

4.2.2 Dubinin-Astakhov Equation

The DR equation describes fairly well many carbonaceous solids with low degree of burn-off. For carbonaceous solids resulting from a high degree of burn-off during activation, the degree of heterogeneity increases because of a wider pore size distribution, and for such cases the DR equation does not describe well the equilibrium data. To this end, Dubinin and Astakhov proposed the following form to allow for the surface heterogeneity:

$$W = W_0 \exp\left[-\left(\frac{A}{E}\right)^n\right] \qquad (4.2\text{-}5a)$$

where the parameter n describes the surface heterogeneity, and the adsorption potential A is defined in eq. (4.1-7). When n = 2, the DA equation reduces to the DR equation. Rewriting eq. (4.2-5a) in terms of the characteristic energy of the reference vapor, we get

$$W = W_0 \exp\left[-\left(\frac{A}{\beta E_0}\right)^n\right] \qquad (4.2\text{-}5b)$$

The DA equation is corresponding to the choice of arbitrary value in the Weibull's distribution function (eq. 4.2-1). With this additional parameter in the adsorption isotherm equation, the DA equation provides flexibility in the description of adsorption data of many microporous solids ranging from a narrow to wide micropore size distribution. The following table shows the degree of filling when the adsorption potential is equal to some fraction of the characteristic energy.

Table 4.2-1: Degree of filling for the DA equation at some specific values of A/E

n	A/E			
	2	1	1/2	1/3
2	0.018	0.368	0.779	0.895
3	~0	0.368	0.882	0.964
4	~0	0.368	0.939	0.988
5	~0	0.368	0.969	0.996
6	~0	0.368	0.984	0.999

It is seen from the above table that the DA equation corresponding to the higher parameter n has a sharper equilibrium curve with respect to the adsorption potential. This effect of n is also shown in Figure 4.2-2 where plots of the fractional loading versus the reduced pressure for E = 10000 Joule/mole and T = 300 K are plotted.

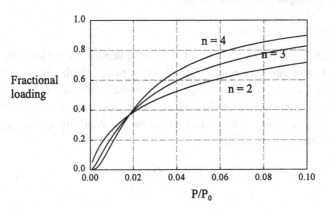

Figure 4.2-2: Plots of the DA equation versus P/P_0 for E=10kJ/mole and T = 300 K

Figure 4.2-3 shows the effect of the variation in the characteristic energy on the fractional loading versus the reduced pressure with T = 298 K and n = 2. The higher the characteristic energy, the sharper the isotherm curve, and the sharp rise occurs at lower pressure.

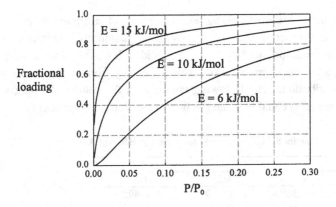

Figure 4.2-3: Plots of the DA equation versus P/P_0 for n = 2, and T = 298 K

The characteristic energy is a measure of the strength of interaction between adsorbate and adsorbent, and is different from the interaction energy in the Langmuir equation (eq. 2.2-6). The Langmuir mechanism is the monolayer type adsorption, and the interaction energy is a measure of the interaction between an adsorbate molecule and surface atoms. In the case of micropore filling, the interaction is between the adsorbent and the volume of adsorbate residing within the micropore, and this interaction is the characteristic energy.

To show the utility of the DA equation in the description of sub-critical vapors, we take the adsorption data of benzene on activated carbon at 303 K as tabulated in Table 3.2-10. The benzene vapor pressure at this temperature is 16.3 kPa. Using the program ISO_FIT1.M provided with this book, we obtain the optimized parameters as listed below:

$$C_{\mu s} = 4.96 \text{ mmole}/g, \quad E_0 = 17610 \text{ Joule}/\text{mole}, \quad n = 2.46$$

Figure 4.2-4 shows a very good fit of the DA equation with the experimental data.

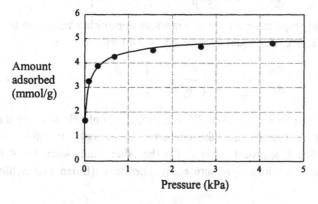

Figure 4.2-4: Fitting the benzene/ activated carbon data at 303 K with the DA equation ($P_0 = 16.3$ kPa)

4.2.2.1 *The Heterogeneity Parameter n*

Since the degree of sharpness of the adsorption isotherm versus adsorption potential or the reduced pressure increases as the parameter n increases, this parameter could be used as an empirical parameter to characterise the heterogeneity of the system. Since it is an empirical parameter, it does not point to the source of the heterogeneity. However, it can be used as a macroscopic measure of the sharpness of the micropore size distribution. For solids having narrow micropore size distribution such as the molecular sieving carbon, the DA equation with n = 3 is found to describe the data well. Therefore, if the parameter n of a given system is found to deviate from 3 (usually smaller than 3), that system is said to be heterogeneous or has a broad micropore size distribution. Typical values of n for strongly activated carbon are between 1.2 and 1.8. For zeolite having extremely narrow micropore size distribution, the parameter n is found to lie between 3 and 6.

Because n =3 is found to describe well data of solids having narrow pore size distribution, the DA equation with n = 3 is generally used as the local isotherm for the description of micropore size distribution as we shall discuss later in Section 4.4.

4.2.2.2 The incorrect Henry Law Behaviour

Thermodynamics suggests that an adsorption isotherm must exhibit the Henry law behaviour when pressure is very low. Unfortunately, the DA as well as the DR equations do not have the correct Henry law when the pressure is approaching zero. The slope of the DA adsorption isotherm equation (4.2-5) is:

$$\frac{dW}{dP} = W_0 \frac{nR_g TA^{n-1}}{E^n P} \exp\left[-(A/E)^n\right] \quad (4.2\text{-}6)$$

When n > 1, the slope is zero when the pressure approaches zero. On the other hand, when the parameter n is equal to unity, eq. (4.2-6) will become:

$$\frac{dW}{dP} = W_0 \frac{R_g T}{E} \left(\frac{P^{R_g T/E - 1}}{P_0^{R_g T/E}}\right) \quad (4.2\text{-}7)$$

Thus, for this case of n = 1, when RT > E (which is unlikely because thermal energy should not be greater than the characteristic energy), the slope of the isotherm is zero when P approaches zero. On the other hand, when RT < E, the slope will be infinity when the pressure approaches zero (Eiden and Schlunder, 1990).

4.2.2.3 Equation for Super-Critical Adsorbate

The fundamental adsorption equation presented in eq. (4.2-5) involves the vapor pressure P_0 at the adsorption temperature T. This implies that the equation is applicable only to sub-critical adsorbates. However, experiments of gas adsorption on microporous solids have shown that there is no abrupt change in the adsorption during the transition from sub-critical to super-critical conditions. This suggests that the DA isotherm equation can be empirically applied to super-critical gases as well. To do this, we need to define an effective vapor pressure P_0 for super-critical gases, sometimes called pseudo-vapour pressure, and the adsorbed phase molar volume.

The adsorbed phase molar volume for super-critical gases is estimated by the following equation (which was proposed by Ozawa et al., 1976):

$$v_M(T) = v_M(T_b) \exp[0.0025(T - T_b)] \qquad (4.2\text{-}8)$$

where $v_M(T_b)$ is the molar volume of the liquid adsorbate at the normal boiling point.

The pseudo-vapour pressure can be estimated in a number of ways. One approach is to use the following Antoine equation for the vapour pressure and extrapolate it to temperatures above the critical temperature:

$$\ln P_0 = \beta_1 - \frac{\beta_2}{T} \qquad (4.2\text{-}9)$$

Here β_1 and β_2 can be calculated from the vapour pressure information at critical condition and at the normal boiling point T_b at which $P_0 = 1$ atm (Dubinin, 1975). Another approach is to evaluate the pseudo-vapour pressure at any temperature above T_c by using the following Dubinin equation:

$$P_0 = P_c \left(\frac{T}{T_c}\right)^2 \qquad (4.2\text{-}10)$$

This approach does not render some experimental data to fall onto one characteristic curve. A generalisation of eq. (4.2-10) was proposed by Amankwah and Schwarz (1995):

$$P_0 = P_c \left(\frac{T}{T_c}\right)^k \qquad (4.2\text{-}11)$$

where k is a parameter specific to the adsorbate-adsorbent system. When this equation is used in connection with the DA equation, we have a four-parameter isotherm equation (W_0, E_0, n and k). Amankwah and Schwarz have applied this DA equation to adsorption data of methane and hydrogen on a number of carbons at temperatures well above the critical temperatures. They found that the parameter n falls in the range of 1.5 to 1.8 while the parameter k in the range of 2.1 to 4.2.

4.2.2.4 Dubinin-Astakhov for Water Adsorption

The DA equation described so far works reasonably well in describing adsorption equilibria of many vapours and gases, such as organic vapors onto microporous activated carbon and carbon molecular sieve. As we have pointed out in Section 4.2.2.2, this equation does not describe correctly the adsorption behaviour at extremely low pressure due to the zero slope at zero loading. Because of this, the DA equation has an inflexion point and its position depends on the value of the characteristic energy (Figure 4.2-3). The lower the characteristic energy, the higher the value of the reduced pressure where the inflexion point occurs. For organic vapors adsorption, the characteristic energy is high (usually greater than 15 kJoule/mole) and hence the inflexion point occurs at extremely low pressure, which is not usually manifested in the plot of the fractional loading versus the reduced pressure (see the curve corresponding to E = 15 kJoule/mole in Figure 4.2-3) unless we magnify the pressure axis.

It is interesting, however, that the zero slope at zero loading behaviour and the inflexion point exhibited by the DA equation can be utilised to describe the adsorption of water on activated carbon. Water adsorption on carbon exhibits a type V isotherm according to the BDDT classification. The curves corresponding to E = 1 and E = 2 kJoule/mole show this type of behaviour, and hence it is appropriate to use the DA equation in the description of water adsorption on carbonaceous solids. Although the DA equation with low characteristic energy E (of the order of 1 to 3 kJoule/mole) and n in the range of 2 to 8 can be used for the description of water adsorption on carbon, it is important to note that the water adsorption does not necessarily follow a micropore filling mechanism, but rather a cluster formation mechanism. This is dealt with next.

4.2.2.5 Dubinin-Serpinski Equation for Water Adsorption on Carbon

Water is known to have extremely low affinity toward the graphitic surface, and its adsorption mechanism is due to the quasi-chemisorption of water with some specific surface functional groups on the carbon surface. Once water molecules sorb onto the specific groups, they themselves act as secondary sites for further water adsorption through hydrogen bonding mechanism. Due to the finiteness of the volume space within the carbon particle, the more water adsorption occurs, the lesser is the availability of these secondary sites. Using the kinetic approach, the rate of adsorption is:

$$k_{ads}(1 - kC_\mu)(C_{\mu 0} + C_\mu)(P/P_0) \tag{4.2-12}$$

where k_{ads} is the rate constant for adsorption, C_μ is the adsorbed concentration of water, $C_{\mu 0}$ is the concentration of the primary sites, and k represents the rate of loss of the secondary sites due to the finiteness of the adsorption volume. Here $C_{\mu 0} + C_\mu$ is the total sites composed of primary and secondary sites, available for water adsorption.

The rate of desorption is proportional to the adsorbed concentration, that is

$$k_{des} C_\mu \tag{4.2-13}$$

Equating the above two rates, the following equation is obtained to describe the adsorption of water on carbonaceous materials

$$\frac{P}{P_0} = \frac{C_\mu}{c(1 - kC_\mu)(C_{\mu 0} + C_\mu)} \quad (4.2\text{-}14)$$

where c is the ratio of the rate constants. The above equation is a quadratic equation in terms of C_μ and it can be solved to obtain C_μ explicitly

$$C_\mu = \frac{-\frac{1}{k}\left(\frac{1}{cx} + kC_{\mu 0} - 1\right) + \sqrt{\frac{1}{k^2}\left(\frac{1}{cx} + kC_{\mu 0} - 1\right)^2 + \frac{4C_{\mu 0}}{k}}}{2} \quad (4.2\text{-}15)$$

where

$$x = \frac{P}{P_0}$$

This adsorption equation, known as the Dubinin-Serpinsky (D-Se) equation, exhibits a type V isotherm as shown in Figure 4.2-5 for c = 3, $C_{\mu 0}$ = 1 mmole/g and three values of k = 0.005, 0.01 and 0.02. The following ranges for $C_{\mu 0}$, c and k are typical for water adsorption in many different sources of carbon.

$$C_{\mu 0} \in [0.05 - 5 \text{ mmole} / \text{g}]$$

$$c \in [1 - 3]$$

$$k \in [0.005 - 0.05 \text{ g} / \text{mmole}]$$

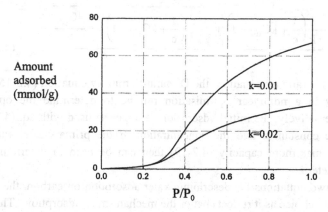

Figure 4.2-5: Plot of the Dubinin-Serpinski equation versus P/P_0 with c = 3, $C_{\mu 0}$ = 1 mmol/g

The concentration of the primary site increases with the degree of oxidation of the carbon surface. This is mainly due to the increase in the oxygen-carrying functional groups. The ratio of the two rate constants, c, is in the rather narrow range (1 - 3), and it may reflect the intrinsic association of the water molecule with the functional group to form a cluster. The parameter k represents the loss of the secondary sites with the progress of adsorption, and hence it affects strongly the maximum amount of water which can be adsorbed by the solid and it does not influence the initial adsorption behaviour as shown in Figure 4.2-5. The initial behaviour is due to the association of water molecule with the functional group. This characteristic can be described by the truncated Dubinin-Serpinsky equation:

$$C_\mu = C_{\mu 0} \frac{cx}{1-cx} \qquad (4.2\text{-}17)$$

The hyperbolic behaviour of the above equation leads to infinite adsorbed water concentration when the reduced pressure is

$$x = \frac{1}{c} \qquad (4.2\text{-}18)$$

Such behaviour is not acceptable in practical systems, and hence the parameter k is important in its provision of a constraint to limit the amount of water which can be accommodated by a solid. The smaller this parameter is, that is the slower loss of the secondary sites with adsorption, the higher is the maximum capacity for water. This physically corresponds to solids having large pore volume to accommodate large clusters. The maximum water capacity can be calculated from eq. (4.2-15) by setting x = 1, that is:

$$C_{\mu s} = \frac{-\frac{1}{k}\left(\frac{1}{c}+kC_{\mu 0}-1\right)+\sqrt{\frac{1}{k^2}\left(\frac{1}{c}+kC_{\mu 0}-1\right)^2+\frac{4C_{\mu 0}}{k}}}{2} \qquad (4.2\text{-}19)$$

Being a three parameter equation, the Dubinin-Serpinsky equation (4.2-15) can be used directly in a nonlinear optimisation routine to determine the optimal parameters. Alternatively, the initial adsorption data can be used with eq. (4.2-17) to determine the constant c and the concentration of the primary site, and the equation for the maximum capacity (4.2-19) then can be used to determine the remaining parameter k.

Among the two equations for describing water adsorption on carbon, the D-Se equation is a better choice as it reflects better the mechanism of adsorption. The DA equation, however, is the one to use if we wish to obtain the adsorption isotherm at different temperatures as the parameters E and n are almost temperature invariant (Stoeckli et al., 1994). To test these two isotherms, we use the following data of water on a sample of activated carbon.

P/P$_0$	Water adsorbed (mmole/g)	P/P$_0$	Water adsorbed (mmole/g)	P/P$_0$	Water adsorbed (mmole/g)
0.05	0.58	0.30	3.96	0.58	18.15
0.07	0.67	0.31	4.56	0.62	19.54
0.09	0.88	0.33	5.72	0.64	20.21
0.11	1.08	0.36	6.67	0.68	20.96
0.13	1.14	0.40	9.37	0.72	21.55
0.16	1.45	0.43	10.73	0.75	22.18
0.18	1.65	0.49	14.67	0.82	22.97
0.23	2.35	0.50	15.77	0.85	23.54
0.27	3.11	0.54	17.11	0.88	24.1

Figure 4.2.6 shows plots of the above data as well as fitted curves (obtained from the ISO_FIT1 program) from the Dubinin-Serpinski equation and the DA equation. Both of these equations fit the data reasonably well although the Dubinin-Serpinski equation provides a better fit in the lower range of the pressure. This could be attributed to the correct description of the water clustering in the lower pressure range.

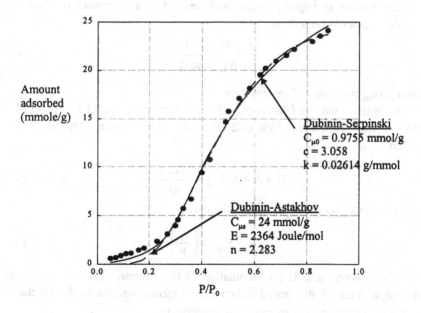

Figure 4.2-6: Fitting the Dubinin-Serpinski equation to water/activated carbon data

4.2.3 Isosteric Heat of Adsorption and Heat of Immersion

4.2.3.1 Isosteric Heat

The heat of adsorption is calculated from the following van't Hoff equation:

$$\frac{\Delta H}{R_g T^2} = -\left(\frac{\partial \ln P}{\partial T}\right)_{C_\mu} \tag{4.2-20}$$

Taking the total derivative of the DA equation (4.2-5), we have:

$$dC_\mu = \left\{\frac{dC_{\mu s}}{dT} - nC_{\mu s}\frac{A^{n-1}}{(\beta E_0)^n}\frac{\partial A}{\partial T}\right\}\exp\left[-\left(\frac{A}{\beta E_0}\right)^n\right]dT - nC_{\mu s}\frac{A^{n-1}}{(\beta E_0)^n}\frac{\partial A}{\partial P}\exp\left[-\left(\frac{A}{\beta E_0}\right)^n\right]dP \tag{4.2-21a}$$

where $\partial A/\partial T$ and $\partial A/\partial P$ are given by:

$$\frac{\partial A}{\partial T} = R_g \ln\left(\frac{P_0}{P}\right) + R_g T\frac{d\ln P_0}{dT} \quad ; \quad \frac{\partial A}{\partial P} = -\frac{R_g T}{P} \quad ; \tag{4.2-21b}$$

The temperature variation of the maximum capacity is given in eq. (4.2-4a).

The change in vapor pressure with respect to temperature is given by the Clausius-Clapeyron equation:

$$\frac{1}{P_0}\frac{dP_0}{dT} = \frac{\Delta H_{vap}}{R_g T^2} \tag{4.2-22}$$

where ΔH_{vap} is the heat of vaporization.

At constant loading (that is $dC_\mu = 0$), substitution of eqs. (4.2-21) and (4.2-22) into the van't Hoff's equation (4.2-20) yields the following expression for the isosteric heat:

$$-\Delta H = A + \Delta H_{vap} + \frac{(\beta E_0)^n \delta T}{nA^{n-1}} \tag{4.2-23}$$

where A is the adsorption potential, given by

$$A = R_g T \ln\left(\frac{P_0}{P}\right) \tag{4.2-24}$$

Thus the isosteric heat is the summation of three terms. The first is due to the adsorption potential, the second is the heat of vaporization and the third is due to the change of the maximum capacity with temperature.

To express the isosteric heat of adsorption in terms of loading, we use the definition of the adsorption equation (eq. 4.2-5) and obtain the following expression:

$$-\Delta H = \Delta H_{vap} + \beta E_0 \left(\ln\frac{1}{\theta}\right)^{1/n} + \frac{(\beta E_0)\delta T}{n}\left(\ln\frac{1}{\theta}\right)^{-(n-1)/n} \quad ; \quad \theta = \frac{C_\mu}{C_{\mu s}} \qquad (4.2\text{-}25)$$

4.2.3.2 Net Differential Heat

The net differential heat of adsorption is defined as the isosteric heat of adsorption minus the heat of vaporization:

$$q^{net} = \beta E_0\left(\ln\frac{1}{\theta}\right)^{1/n} + \frac{(\beta E_0)\delta T}{n}\left(\ln\frac{1}{\theta}\right)^{-(n-1)/n} \qquad (4.2\text{-}26)$$

The net differential heat of adsorption is infinite at zero loading. This, however, has no physical basis as the DA equation is not valid in the very small neighborhood of zero loading. Figure 4.2-7 shows a typical plot of the net differential heat of adsorption versus fractional loading for n = 2 and T = 300 K. The heat curve decreases with loading and when saturation is approached the net differential heat of adsorption increases rapidly. Such a rapid increase close to saturation is due to the change of the saturation capacity $C_{\mu s}$ with respect to temperature. If this maximum capacity is temperature-invariant, then the heat curve decreases monotonically with loading and approaches zero when saturation is reached.

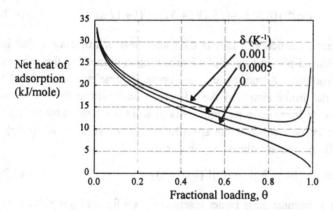

Figure 4.2-7: Net heat of adsorption versus fractional loading with E = 15 kJ/mol, n = 2, T = 300K

Figure 4.2-8 shows the effect of the heterogeneity parameter n on the heat curve. The higher the value of n, the more uniform the heat curve. This is in accordance with what was said earlier about the behaviour of the adsorption isotherm curve versus the reduced pressure when n is increased, that is the system is more homogeneous.

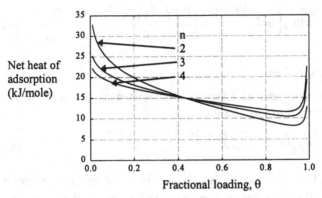

Figure 4.2-8: Net heat of adsorption versus fractional loading with E=15 kJ/mol, $\delta = 5 \times 10^{-4}$ K^{-1}, T=300K

4.2.3.3 Enthalpy of Immersion

Integration of the net heat of adsorption versus the fractional loading from 0 to 1 would give the molar enthalpy of immersion into the corresponding liquid (Stoeckli and Krahenbuhl, 1981, 1989), that is:

$$\Delta h_i = \int_0^1 q^{net}(\theta) d\theta = -(\beta E_0)(1+\delta T) \cdot \Gamma(1+1/n) \quad (4.2\text{-}27)$$

where Γ is the Gamma function. This is the molar heat of immersion, that is the amount of heat released per unit mole of adsorbate upon immersing the solid into a liquid pool of adsorbate. Assuming the micropore volume would be filled with adsorbate during the immersion and the adsorbate state within the micropore is liquid-like, the heat of immersion per unit mass of solid is obtained by multiplying the molar heat of immersion with the specific micropore volume W_0 (cc/g) and dividing it by the liquid molar volume, v_M, that is:

$$-\Delta H_i = (\beta E_0) W_0 (1+\delta T) \Gamma(1+1/n) / v_M \quad (4.2\text{-}28)$$

This specific heat of immersion is rather insensitive to n for n ranging from 2 to 6, and it is more sensitive toward the variation of the parameter E and W_0. For DR equation, that is n = 2, the specific enthalpy of immersion is:

$$\Delta H_i = -(\beta E_0) W_0 (1+\delta T) \cdot \sqrt{\pi} / (2 v_M) \quad (4.2\text{-}29)$$

The specific enthalpy of immersion is proportional to the characteristic energy, and hence it provides a convenient check on the parameter obtained by fitting the DA equation with the experimental data.

Eq. (4.2-28) or (4.2-29) are directly applicable to strictly microporous solids. For solids having high mesopore surface area, the specific enthalpy of immersion must allow for the heat associated with the open surface, that is

$$-\Delta H_i = (\beta E_0) W_0 (1 + \delta T) \Gamma(1 + 1/n) / v_M + h_i S_e \qquad (4.2\text{-}30)$$

where h_i is the specific enthalpy of immersion on open surface.

4.3 Theoretical Basis of the Potential Adsorption Isotherms

The DA equation (4.2-5) is obtained by assuming the temperature invariance of the adsorption potential at constant loading and a choice of the Weibull's distribution to describe the filling of micropore over the differential molar work of adsorption. It can be shown to be a special case of an isotherm equation derived from the statistical mechanical principles when the loading is appreciable (Chen and Yang, 1994). They derived the following isotherm

$$\ln\left(\frac{\rho^s}{\rho^g \Lambda}\right) - \frac{7}{8} \ln(1 - \eta) + \frac{2\eta}{(1-\eta)} + \frac{9}{8} \frac{\eta}{(1-\eta)^2} - \alpha(T)\eta + \frac{\Phi}{kT} = 0 \qquad (4.3\text{-}1)$$

for a given pore dimension and geometry, with a mean force field Φ. In this equation, ρ^s and ρ^g are number densities of the adsorbed and fluid phases, respectively, Λ is the de-Broglie thermal wavelength, and η is defined as:

$$\eta = \frac{\pi}{4} \sigma^2 \rho^s; \qquad \rho^s = \frac{N}{V^s} \qquad (4.3\text{-}2)$$

where N is the number of molecules exposed to the force field, V^s is the volume of the system and σ is the diameter of the hard disk. The variable η is the fraction of the surface covered.

For appreciable adsorption, they reduce the above isotherm to:

$$\ln \theta = -\left[\frac{R_g T}{K N_A \Phi} \ln\left(\frac{P}{P_0}\right)\right]^n \qquad (4.3\text{-}3)$$

where N_A is the Avogradro number and K is a constant. If we define

$$K N_A \Phi = \beta E_0 \qquad (4.3\text{-}4)$$

the above will become the DA equation. When n=2, it then reduces further to the DR equation. With the above definition of the characteristic energy (4.3-4), we see that it is proportional to the mean potential energy of the pore.

4.4 Modified Dubinin Equations for Inhomogeneous Microporous Solids

Adsorption isotherms of many microporous solids do not usually conform to the simple DR equation. Even with the adjustable parameter n in the DA equation, it also can not describe well many experimental data. This inability to fit the data is attributed to the heterogeneity of the system, that is the characteristic energy varies with the different regions in the solid.

There are a number of approaches to deal with such heterogeneous solids. One simple approach is to assume that the solid is composed of two distinct regions and there is no interaction between these two. With such an assumption, the DR or the DA equation can be applied to each region, and the overall isotherm is simply the summation of the two simple DRs or DAs equations. This is dealt with in Section 4.4.1. Another approach is to assume a continuous distribution of the characteristic energy. The distribution of the characteristic energy is completely arbitrary and its form is chosen in such a way that an analytical solution can be obtained from the averaging of a local isotherm over that distribution. Due to the arbitrary choice of the distribution function, the resulting equation is at best empirical and the parameters resulting from the averaging must be treated as empirical constants. Attempts to assign physical meaning to these constants must be treated with care and be checked with independent experimental means.

4.4.1 Ideal Inhomogeneous Microporous Solids

The simplest way of describing the adsorption isotherm of an inhomogeneous microporous solid is to use the equation given below:

$$C_\mu = \frac{W_{0,1}}{v_M} \exp\left[-\left(\frac{A}{\beta E_{0,1}}\right)^{n_1}\right] + \frac{W_{0,2}}{v_M} \exp\left[-\left(\frac{A}{\beta E_{0,2}}\right)^{n_2}\right] \qquad (4.4\text{-}1)$$

where v_M is the liquid molar volume. The micropore volume has been assumed to be divided into two independent regions and each region follows either the DR or DA equation. Here the exponent n for the two regions can be either the same or different. It is usually taken to be the same to reduce the number of parameter by one. The distinction between the two regions is then the difference in the characteristic energy. Figure 4.4-1 shows a typical plot of the amount adsorbed of eq. (4.4-1) versus the reduced pressure for E_1 = 10 kJoule/mole and E_2 = 5 kJoule/mole. Here α is the fraction of the region having E_1 = 10 kJ/mole.

This approach is one of the simplest approaches to deal with inhomogeneous solids. Practical solids, however, do not easily break down into two distinct regions having two distinct characteristic energies, but rather they possess a distribution of energy. This is dealt with in the next section.

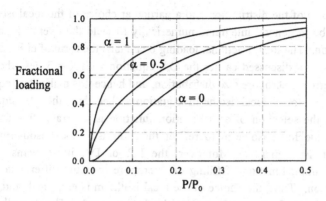

Figure 4.4-1: Plots of the dual DA equation versus P/P_0 with $E_1 =10$ kJ/mole, $E_2 =5$ kJ/mole, n=2, T=300K

4.4.2 Solids with Distribution in Characteristic Energy E_0

With the concept of distribution of the characteristic energy, the adsorption isotherm equation can be written in general as follows:

$$W = W_0 \int_{\Omega(E_0)} f(E_0)\, \theta(A/\beta E_0, n)\, dE_0 \qquad (4.4\text{-}2)$$

where W_0 is the maximum specific volume of the micropore, $f(E_0)$ is the distribution of the characteristic energy based on the reference vapour, and θ is the local fractional loading corresponding to the characteristic energy βE_0. The parameter n may vary with the variation of the characteristic energy, but for simplicity it is always taken as a constant.

It is noted in eq. (4.4-2) that the characteristic energy of the reference vapour is used as the distributed variable, and therefore the functional form of the distribution $f(E_0)$ would reflect the heterogeneity of the solid. Hence the parameters of such distribution function are the solid heterogeneity parameters.

The range of integration, $\Omega(E_0)$, strictly speaking should reflect that of a micropore system, that is the proper range should be between E_{min} and E_{max}. The lower limit E_{min} is the characteristic energy associated with the largest pore in the micropore system, and the upper limit E_{max} associates with the smallest accessible micropore. However, the range of integration is usually extended beyond its range of validity to facilitate the analytical integration of the integral (4.4-2). For example, the lower limit could be set to zero and the upper limit could be set to infinity.

With any form of the distribution and a particular choice of the local isotherm, eq. (4.4-2) can be in general integrated numerically to yield the overall adsorption isotherm equation. The local fractional loading θ can take the form of either the DR or DA equation. As discussed earlier the DA equation with n = 3 describes well solids having narrow micropore size distribution, and hence this makes this equation a better candidate for a local isotherm equation rather than the DR equation. However, since the selection of a distribution function is arbitrary, this does not strictly enforce the local isotherm to reflect the "intrinsic" local isotherm for a specific characteristic energy. Moreover, the DR or DA itself stems from a Weibull's distribution function of filling of micropore over the differential molar work of adsorption. Thus, the choice of the local isotherm is empirical, and in this sense the procedure of eq. (4.4-2) is completely empirical. The overall result, however, provides a useful means to describe the equilibrium data in microporous solids.

Because of the way that the characteristic energy appears in the DR or DA equation, instead of using the distributed variable as E_0 it will be chosen as

$$\left(\frac{1}{E_0}\right)^2 \tag{4.4-3a}$$

or

$$\left(\frac{1}{E_0}\right) \tag{4.4-3b}$$

depending on the choice of the distribution function to facilitate the analytical integration of the integral equation. Thus the overall adsorption isotherm is evaluated from the following equation:

$$W = W_0 \int_{\Omega(E_0)} f\left(\frac{1}{E_0^2}\right) \theta(A/\beta E_0, n) \, d\left(\frac{1}{E_0^2}\right) \tag{4.4-4a}$$

or

$$W = W_0 \int_{\Omega(E_0)} f\left(\frac{1}{E_0}\right) \theta(A/\beta E_0, n) \, d\left(\frac{1}{E_0}\right) \tag{4.4-4b}$$

The distribution functions of the Gaussian and Gamma form are tabulated in the following table (Table 4.4-1).

Table 4.4-1: The functional form of the distribution function

Distribution function	Functional form	Eq. No.
Gaussian	$f\left(\dfrac{1}{E_0^2}\right) = \dfrac{1}{\sigma\sqrt{2\pi}} \exp\left[-\left(\dfrac{1}{E_0^2} - \dfrac{1}{\overline{E_0^2}}\right)^2 \Big/ 2\sigma^2\right]$	(4.4-5)
Shifted Gamma	$f\left(\dfrac{1}{E_0^2}\right) = \dfrac{q^{n+1}}{\Gamma(n+1)} \left(\dfrac{1}{E_0^2} - \dfrac{1}{E_{0,max}^2}\right)^n \exp\left[-q\left(\dfrac{1}{E_0^2} - \dfrac{1}{E_{0,max}^2}\right)\right]$	(4.4-6)
Gamma	$f\left(\dfrac{1}{E_0}\right) = \dfrac{3q^n}{\Gamma(n/3)} \left(\dfrac{1}{E_0}\right)^{n-1} \exp\left[-\left(\dfrac{q}{E_0}\right)^3\right]$	(4.4-7)

The parameter E_0 in eq. (4.4-5) is the mean characteristic energy of the solid toward the reference vapor, and σ is the variance in the Gaussian distribution (mole^2/Joule^2). The parameters q and n are distribution parameters, and they affect the mean characteristic energy and the variance of the distribution. For the shifted Gamma distribution (4.4-6), the mean and the variance are:

$$\overline{\left(\dfrac{1}{E_0^2}\right)} = \dfrac{1}{\overline{E_0^2}} + \dfrac{(n+1)}{q} \tag{4.4-8a}$$

$$\sigma = \dfrac{\sqrt{n+1}}{q} \tag{4.4-8b}$$

and for the Gamma distribution (4.4-7) the corresponding mean and variance are:

$$\overline{\left(\dfrac{1}{E_0}\right)} = \dfrac{\Gamma(n/3 + 1/3)}{q\,\Gamma(n/3)} \tag{4.4-9a}$$

$$\sigma = \dfrac{\sqrt{\Gamma(n/3)\,\Gamma(n/3 + 2/3) - [\Gamma(n/3 + 1/3)]^2}}{q\,\Gamma(n/3)} \tag{4.4-9b}$$

The choice of the Gaussian and Gamma distribution in Table 4.4-1 is sufficient to illustrate the use of distribution function to describe adsorption equilibria in inhomogeneous solid. Other distributions could be used, but the lack of the physical basis of the choice of the functional form has deterred the use of other functions. Let us now address the distribution functions in Table 4.4-1 with the local isotherm DR or DA equation.

4.4.2.1 Gaussian Distribution and local DR Equation

This section will address the DR local isotherm and the Gaussian distribution of the form given in eq. (4.4-5). The volume of the micropore occupied by the adsorbate at a given adsorption potential A is:

$$W = W_0 \int_0^\infty \frac{1}{\sigma\sqrt{2\pi}} \exp\left[-\left(\frac{1}{E_0^2} - \frac{1}{\overline{E}_0^2}\right)^2 \bigg/ 2\sigma^2\right] \times \exp\left[-\left(\frac{A}{\beta E_0}\right)^2\right] d\left(\frac{1}{E_0^2}\right) \quad (4.4\text{-}10)$$

The range of the integration is between 0 and ∞. The lower limit corresponds to infinite characteristic energy, and the upper limit corresponds to zero characteristic energy. Since these two limits are not physically feasible, the variance in the Gaussian equation has to be narrow. Integration of the above equation gives the following expression for the adsorption isotherm:

$$W = \frac{W_0}{2} \exp\left[-\left(\frac{A}{\beta \overline{E}_0}\right)^2\right] \times \exp\left[\frac{\sigma^2}{2}\left(\frac{A}{\beta}\right)^4\right] \times \left\{1 - \mathrm{erf}\left[\frac{\sigma^2 (A/\beta)^2 - 1/\overline{E}_0^2}{\sigma\sqrt{2}}\right]\right\} \quad (4.4\text{-}11)$$

This form of equation was first derived by Stoeckli (1977). Figure 4.4-2 shows plots of the above equation versus the reduced pressure with the variance as the varying parameter.

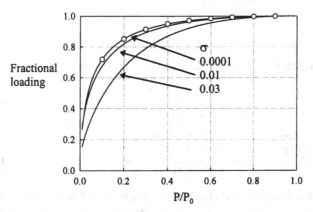

Figure 4.4-2: Plots of eq. (4.4-11) versus P/P_0 with $E = 10 \text{kJ/mol}$, $\beta = 1$, $T = 300\text{K}$
(Symbol o is from the DR equation)

Here we see that the more heterogeneous is the solid, the broader is the isotherm curve, that is it has a sharper rise in the low pressure range and a slower approach to saturation at high pressure. The sharper rise at very low pressure is due to the filling of pores having high characteristic energy and the slow approach to saturation is due to the progressive filling of lower characteristic energy sites.

Stoeckli (1977) applied the above equation to experimental data of nitrogen, xenon, SF_6, N_2O and benzene onto activated carbon, and found that all gases leading to similar value of the variance σ for a given solid, supporting the assumption that it is a structural parameter.

Eq. (4.4-11) is applicable to heterogeneous solid and the heterogeneity is reflected in the distribution function, and to be more specific, in the variance of the distribution function. In other words, the larger is the variance the more heterogeneous is the solid. We recall earlier the parameter n in the DA equation is treated as a measure for the solid heterogeneity. Thus, the parameter n in a sense is related to the variance σ. The equivalence of these parameters was found by Dubinin (1979) to take the following relationship

$$n = 2 - (1.9 \times 10^6)\sigma$$

This relation states that when the variance of the distribution is zero, the exponent n in the DA equation is 2, suggesting that DR equation describes homogeneous solid, which is the assumption used in the derivation of eq. (4.4-11). When the variance is positive, the exponent n is less than 2, and this is consistent with the experimental data of many activated carbon systems where the exponent n is found to fall between 1.2 and 1.8.

4.4.2.1.1 Micropore Size Distribution

If eq. (4.4-11) is accepted as the proper equation to describe the amount adsorbed versus the reduced pressure, it can be used to match with the experimental data to extract the mean characteristic energy and the variance. Having these parameters, the distribution describing the heterogeneity is given in eq. (4.4-5), but it does not point to the source of the heterogeneity. Dubinin and Zaverina (1949) have shown that the characteristic energy and the micropore size are related to each other. They found that the characteristic energy decreases with the activation of carbon with carbon dioxide at high temperature. This process enlarges the micropore, and by using the small angle X-ray scattering, the following approximate equation relates the characteristic energy and the micropore half width, assuming micropores are slit shape pores:

$$\frac{1}{E_0^2} = Mx^2 \qquad (4.4\text{-}12a)$$

where M is a constant and x is the half width of the pore. For E_0 taking the units Joule/mole and x taking the units nm, the constant M takes the value:

$$M = 6.944 \times 10^{-9} \; J^{-2} \, nm^{-2} \tag{4.4-12b}$$

The distribution of the micropore volume in terms of $1/E_0^2$ is:

$$\frac{dW_0}{d(1/E_0^2)} = \frac{W_0^0}{\sigma\sqrt{2\pi}} \exp\left[-\left(\frac{1}{E_0^2} - \frac{1}{\overline{E}_0^2}\right)^2 \bigg/ 2\sigma^2\right] \tag{4.4-13}$$

where W_0^v is the specific total micropore volume.

Replacing the characteristic energy with the pore half width (eq. 4.4-12a) in eq.(4.4-13), we obtain the pore volume distribution in terms of the pore half width:

$$\frac{dW_0}{dx} = \frac{2M \times W_0^0}{\sigma\sqrt{2\pi}} \exp\left[-\frac{M\left(x^2 - \overline{x}^{-2}\right)}{2\sigma^2}\right] \tag{4.4-14}$$

Figure 4.4-3 shows distribution of pore volume with the variance as the varying parameter.

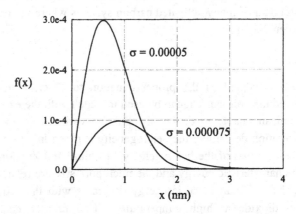

Figure 4.4-3: Micropore size distribution from eq. (4.4-14) with E_0 = 10 kJ/mole

Thus, with the use of a distribution in terms of the characteristic energy (eq. 4.4-5) and the relationship between the characteristic energy and the micropore half width (eq. 4.4-12a), the method allows an indirect means to calculate the micropore volume distribution. We must however treat this volume distribution as an

approximation to the intrinsic distribution as there are a number of assumptions embedded in the derivation of such pore volume distribution:
(a) the use of DR as the local isotherm
(b) the use of the distribution function in the form of eq. (4.4-5)
(c) the relationship between the characteristic energy and the pore half width

4.4.2.1.2 Geometrical Surface Area

Assuming the micropores having parallel sided slits with a half width of x, the geometric surface area of the micropore walls is:

$$S_g = \frac{W_0}{x} \tag{4.4-15a}$$

This equation provides a good estimation of the geometric surface area for homogeneous solids. For inhomogeneous solids, the differential increase in the geometric surface area is related to the differential increase in the micropore volume as follows:

$$dS_g = \frac{dW_0}{x} \tag{4.4-15b}$$

assuming all slits have the same length. Using eq. (4.4-14) for the distribution of pore volume with half-width, we obtain the following equation for the geometrical surface area:

$$S_g = \frac{2MW_0^0}{\sigma\sqrt{2\pi}} \int_0^\infty \exp\left[-\frac{M(x^2 - x_0^2)^2}{2\sigma^2}\right] dx \tag{4.4-16}$$

4.4.2.2 DA and Gaussian distribution

The last section shows the analysis of an inhomogeneous solid when the local isotherm taking the form of the DR equation and the distribution in terms of $1/E_0^2$ taking the Gaussian distribution (eq. 4.4-5). In this section, the local isotherm is taking the form of DA (n = 3) as discussed earlier it describes narrow pore solids better than the DR equation, and hence it should be a better choice for the local isotherm. But this DA, like the DR equation, does not truly represent the "intrinsic" local isotherm for the reason that the DA comes from the Weibull's distribution of micropore filling over the differential molar work of adsorption.

Using the DA equation and the Gaussian distribution (eq. 4.4-5), the following equation is obtained

$$W = W_0 \int_0^\infty \frac{1}{\sigma\sqrt{2\pi}} \exp\left[-\left(\frac{1}{E_0^2} - \frac{1}{\overline{E}_0^2}\right)^2 \Big/ 2\sigma^2\right] \times \exp\left[-\left(\frac{A}{\beta E_0}\right)^n\right] d\left(\frac{1}{E_0^2}\right) \quad (4.4\text{-}17)$$

or

$$W = W_0 \exp\left[\frac{\sigma^2}{2}\left(\frac{A}{\beta}\right)^{2n} - B_0\left(\frac{A}{\beta}\right)^n\right] \times \frac{\mathrm{erfc}\left[\frac{\sigma}{\sqrt{2}}\left(\frac{A}{\beta}\right)^n - \frac{B_0}{\sigma\sqrt{2}}\right]}{\mathrm{erfc}\left(-B_0/\left(\sigma\sqrt{2}\right)\right)} \quad (4.4\text{-}18)$$

where

$$B_0 = \left(\frac{1}{\overline{E}_0}\right)^n \quad (4.4\text{-}19)$$

The above solution was first obtained by Rozwadowski and Wojsz (1984). This equation is rarely used because of its complicated form.

4.4.2.3 DR Local Isotherm and Shifted Gamma Distribution

The Gaussian distribution used in the last two sections does not give a zero value when the micropore size is zero. The choice of the shifted Gamma distribution function as in eq. (4.4-6) overcomes this deficiency. In eq. (4.4-6), $E_{0,\max}$ is the maximum characteristic energy that the solid possesses. Using DR as the local isotherm, the following equation is obtained:

$$\theta = \exp\left[-B_0(A/\beta)^2\right]\left[\frac{q}{q+(A/\beta)^2}\right]^{n+1} \quad (4.4\text{-}20)$$

where

$$B_0 = \frac{1}{E_{0,\max}^2} \quad (4.4\text{-}21)$$

The above equation can take a simpler form if the maximum characteristic energy is taken to be infinity, that is:

$$\theta = \frac{W}{W_0} = \left[\frac{q}{q+(A/\beta)^2}\right]^{n+1} \quad (4.4\text{-}22)$$

This adsorption isotherm equation has three parameters, W_0^o, q and n, with the last two being the structural parameters (hence will be independent from adsorbate and temperature). By fitting this adsorption equation with the experimental data, preferably with different adsorbates and temperatures, the three parameters can be obtained.

As before, the characteristic energy E_0 is related to the half width of the micropore according to eq. (4.4-12). Therefore, the micropore size distribution, $J(x)$, is given by:

$$J(x) = \frac{2xM^{n+1}q^{n+1}}{\Gamma(n+1)}\left(x^2 - x_0^2\right)^n \exp\left[-Mq\left(x^2 - x_0^2\right)\right] \quad (4.4\text{-}23)$$

where x_0 is the micropore half width corresponding to $E_{0,max}$.

4.4.2.4 DA (n=3) Local Isotherm and Shifted Gamma Distribution

The DA equation with n = 3

$$\theta = \exp\left[-\left(\frac{A}{\beta E_0}\right)^3\right] \quad (4.4\text{-}24)$$

is a better description of adsorption in activated carbon having fairly uniform micropores. Taking this as the local isotherm for a heterogeneous solid instead of the DR equation, the overall fractional loading can be calculated according to the following equation:

$$\theta = \int_0^\infty \exp\left[-z^3\left(\frac{A}{\beta}\right)^3\right] F(z)dz; \quad z = \frac{1}{E_0} \quad (4.4\text{-}25)$$

The variable z is the inverse of the characteristic energy, and is related to the micropore half-width. There are a number of correlations available in the literature (Dubinin and Stoeckli, 1980; Stoeckli et al., 1989) and they are listed below:

$$x = 15z + 2852.5z^3 + \frac{0.014}{z} - 0.75 \quad (4.4\text{-}26a)$$

$$x = z\left(13.028 - \frac{1.53 \times 10^{-5}}{z^{3.5}}\right) \quad (4.4\text{-}26b)$$

$$x = 12z \quad (4.4\text{-}26c)$$

where x is in nm and z is in mole/kJoule. Figure 4.4-5 shows plots of the micropore half-width versus the characteristic energy obtained from the above three correlations.

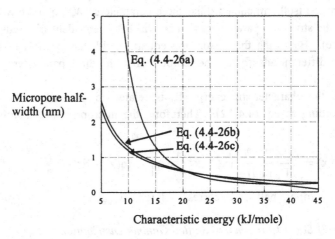

Figure 4.4-4: Plot of the micropore half-width versus the characteristic energy

Knowing the energy distribution F(z) the pore size distribution is:

$$J(x) = F(z)\frac{dz}{dx} \qquad (4.4\text{-}27)$$

The choice of the function form for the energy distribution F(z) is arbitrary, and again it is chosen to allow the integration to be performed analytically. The following Gamma distribution satisfies this requirement:

$$F(z) = \frac{3q^n}{\Gamma(n/3)} z^{n-1} \exp\left[-(qz)^3\right] \qquad (4.4\text{-}28)$$

where n > -1, and q > 0. Using this form of distribution function, eq. (4.4-25) can be analytically integrated to give:

$$\theta = \frac{1}{\left[1+(A/\beta q)^3\right]^{n/3}} \qquad (4.4\text{-}29)$$

By fitting the above adsorption isotherm equation with equilibrium experimental data, the parameters n and q can be obtained; hence the energy distribution F(z) is obtained, from which one could readily determine the micropore size distribution from eq. (4.4-27).

Knowing the parameters n and q, the differential adsorption potential distribution X(A) is obtained as:

$$X(A) = -\frac{d\theta}{dA} = n(\beta q)^{-3} A^2 \left[1+(A/\beta q)^3\right]^{(n/3)-1} \qquad (4.4\text{-}30)$$

This distribution is useful to calculate enthalpy and entropy of adsorption in the micropores (Jaroniec, 1987) and the heat of immersion (Jaroniec and Madey, 1988).

To test eq. (4.4-29), we fit it to the adsorption data of benzene on activated carbon at 30 °C (Table 3.2-10) and the result is shown in Figure 4.4-5. Knowing the parameters n and q from the isotherm fit, the micropore size distribution is calculated from eq. (4.4-27) and it is shown graphically in Figure 4.4-6.

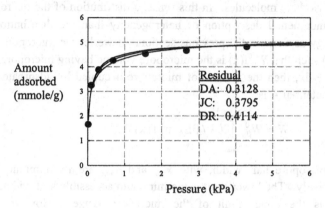

Figure 4.4-5: Fitting of the benzene/activated carbon data of 30 °C with DR, DA, and Jaroniec-Choma eqs.

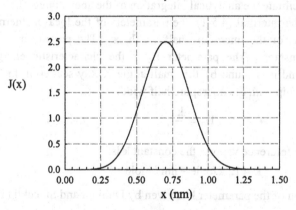

Figure 4.4-6: Pore size distribution calculated from the benzene/activated carbon fitting with JC equation

4.5 Solids with Micropore Size Distribution

The above section assumes that the solid has a distribution in terms of the structural parameter $B = 1/E_0^2$, and the micropore distribution is then calculated based on the information of this f(B) distribution by using the relationship between B and the micropore half width, x. Such a choice of distribution function in terms of E_0 is based solely on mathematical convenience rather than on any physical reasoning.

Since adsorption of many adsorbates in micropores of carbonaceous solids is due to the dispersion force, the micropore size is therefore playing a major role in the attraction of adsorbate molecules. In this sense, a distribution of the micropore size is a more fundamental description of heterogeneity than the distribution in characteristic energy as done in the previous section. If we let the micropore size distribution as f(x) such that $W_0^0 f(x)$ is the micropore volume having micropore size between x and x + dx, then the volume of micropore occupied by adsorbate at a given adsorption potential A is:

$$W = W_0^0 \int_{x_{min}}^{x_{max}} \theta(A/\beta E_0, n) f(x) \, dx \qquad (4.5\text{-}1)$$

where x is the micropore half width, with x_{min} and x_{max} are its minimum and maximum, respectively. The lower limit is the minimum accessible half width, and the upper limit is the upper limit of the micropore range. However, for mathematical convenience, those limits are usually replaced by 0 and ∞, respectively, to facilitate the analytical integration of the above integral.

To evaluate the intergral (4.5-1), the parameters of the local isotherm must be expressed in terms of the micropore half width, x. The parameter n is usually regarded as a constant. The parameter E_0 is the characteristic energy of the reference vapor, and it is found by the small angle X-Ray scattering to follow the following relationship with the micropore half width

$$E_0 = \frac{k}{x} \qquad (4.5\text{-}2)$$

For benzene as the reference vapour, the constant k is:

$$k = 12 \text{ kJoule} - \text{nm} / \text{mole} \qquad (4.5\text{-}3)$$

Another description of the parameter k is given by Dubinin and Stoeckli (1980)

$$k = 13.028 - (1.53 \times 10^{-5}) E_0^{3.5} \qquad (4.5\text{-}4)$$

Figure 4.4-4 shows plots of the characteristic energy versus the micropore half width using these two forms for k. The difference between the two is insignificant except when the micropore half width is less than 0.4 nm. Thus for solids having large micropores, such as the well developed activated carbon, the simple form for k (eq. 4.5-3) is adequate.

4.5.1 DR Local Isotherm and Gaussian Distribution

The micropore size distribution can be assumed to take the following Gaussian distribution:

$$\frac{dW_0}{dx} = \frac{W_0^0}{\delta\sqrt{2\pi}} \exp\left[-\frac{(x-x_0)^2}{2\delta^2}\right] = W_0^0 \, f(x) \qquad (4.5\text{-}5)$$

where W_0^0 is the total micropore volume, x_0 is the half-width of a slit shaped micropore which corresponds to the maximum of the distribution curve, and δ is the variance.

Using the relationship between the characteristic energy and the micropore half width (eq. 4.5-2), the adsorption equation of DR written in terms of this half width x is:

$$W = W_0 \exp\left(-mx^2 A^2\right) \qquad (4.5\text{-}6)$$

where

$$m = \frac{1}{(\beta k)^2} \qquad (4.5\text{-}7)$$

To obtain an equation for a heterogeneous solid, the adsorption equation for a micropore volume element dW is:

$$dW = dW_0 \exp\left(-mx^2 A^2\right) \qquad (4.5\text{-}8)$$

Combining eq. (4.5-5) and (4.5-6) yields the following equation for the volume occupied by the adsorbate:

$$W = \frac{W_0^0}{\delta\sqrt{2\pi}} \int_0^\infty \exp\left[-\frac{(x-x_0)^2}{2\delta^2}\right] \exp\left(-mx^2 A^2\right) dx \qquad (4.5\text{-}9)$$

Evaluation of the above integral gives:

$$W = \frac{W_0^0}{2\sqrt{1+2m\delta^2 A^2}} \exp\left(-\frac{mx_0^2 A^2}{1+2m\delta^2 A^2}\right) \left[1 + \mathrm{erf}\left(\frac{x_0}{\delta\sqrt{2}\sqrt{1+2m\delta^2 A^2}}\right)\right] \qquad (4.5\text{-}10)$$

This equation was first obtained by Dubinin and Stoeckli, and is hereafter called the D-S equation. Using this equation to fit experimental data, three parameters can be extracted from this fitting process, namely W_0^0, x_0 and δ. Knowing these parameters, the micropore size distribution in terms of volume then can be calculated from eq. (4.5-5).

Figures 4.5-1 and 4.5-2 show the plot of eq. (4.5-10) versus the reduced pressure with the variance and the mean pore half width as the parameters, respectively.

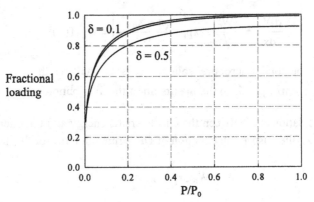

Figure 4.5-1: Plots of the DS equation versus P/P_0 with T=300 K, x_0 = 1 nm and δ = {0.1, 0.3, 0.5}

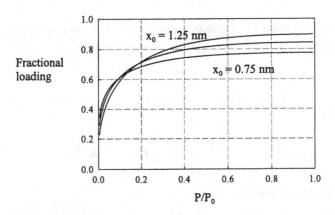

Figure 4.5-2: : Plots of the DS equation versus P/P_0 with T=300 K, δ = 1 nm and x_0 = {0.75, 1, 1.25 nm}

4.5.1.1 Geometrical Surface Area of Micropore

If the micropore volume is described by a Gaussian distribution given as eq. (4.5-5), the geometrical surface of the micropore walls under the assumption of slit shaped pore is:

$$S_g = \frac{W_0^0}{\delta\sqrt{2\pi}} \int_c^\infty \frac{1}{x} \exp\left[-\frac{(x-x_0)^2}{2\delta^2}\right] dx \qquad (4.5\text{-}11)$$

Here the parameter c is the minimum pore half width below which adsorbate molecules can not penetrate due to the steric effect, that is only pores having spacing greater than or equal to the diameter of the adsorbate molecule will allow the adsorption to proceed.

4.5.2 DA Local Isotherm and Gamma Micropore Size Distribution

Using DA as the local adsorption isotherm, and $E_0 = k/x$, Jaroniec et al. (1988) assumed the following micropore size distribution:

$$J(x) = \frac{nq^\upsilon}{\Gamma(\upsilon/n)} x^{\upsilon-1} \exp\left[-(qx)^n\right] \qquad (4.5\text{-}12)$$

to obtain the following overall isotherm equation:

$$\theta = \left[1 + (A/\beta q k)^n\right]^{-\upsilon/n} \qquad (4.5\text{-}13)$$

The distribution has the following properties:

$$\bar{x} = \frac{\Gamma\left(\frac{\upsilon+1}{n}\right)}{q\Gamma\left(\frac{\upsilon}{n}\right)}; \quad \sigma = \left[\frac{\Gamma\left(\frac{\upsilon+2}{n}\right)}{\Gamma\left(\frac{\upsilon}{n}\right)} - \frac{\Gamma^2\left(\frac{\upsilon+1}{n}\right)}{\Gamma^2\left(\frac{\upsilon}{n}\right)}\right]^{1/2} q^{-1}; \quad x_m = \left(\frac{\upsilon-1}{n}\right)^{1/n} q^{-1} \qquad (4.5\text{-}14)$$

Due to the simplicity of the overall isotherm equation (4.5-13), this equation has been used to fit many experimental data to obtain the structural parameters n, q and ν. Knowing these parameters, the micropore size distribution can be calculated from eq. (4.5-12).

Like other isotherm equations developed in this section, the micropore size is assumed to take a certain form and then the isotherm equation derived from such assumed pore size distribution is fitted with equilibrium data to obtain the relevant structural parameters. This method must be used with some precaution because if the true micropore size distribution does not conform to the assumed form the derived micropore size distribution can be erroneous.

4.6 Other Approaches

Since the development of the DR equation in 1947, many versions of Dubinin type equations have appeared in the literature, such as the DA equation which allows for the solid heterogeneity and the DS equation which allows for heterogeneity in terms of micropore size distribution. These equations, however, suffer from the disadvantage of zero Henry constant at zero pressure. This difficulty in incorrectly describing the adsorption behaviour at low pressure has been addressed by a number of workers. Here we will present a number of studies.

4.6.1 Yang's Approach

Adsorption isotherm in microporous solids such as activated carbon is described well by the now well known Dubinin equation, such as the Dubinin-Radushkevich equation or its generalised version the Dubinin-Astakhov equation. These equations are suitable to describe the isotherm of solids having micropore size ranging from 4 to about 40 Å. Although they are popular in their use to describe numerous systems with activated carbon in the literature, they suffer from one problem, that is it does not exhibit the Henry law isotherm at very low pressure, a requirement demanded often by thermodynamics (Talu and Myers, 1988). Therefore, when using the Dubinin equations to predict multicomponent adsorption equilibria using the solution thermodynamics models, large errors may result. Recognising the potential of the Dubinin equations in their description of isotherms in microporous solids, various researchers have attempted to modify it, usually in an empirical way, to enforce the Henry law behaviour at very low pressure. For example, Kapoor *et al.* (1989) used the following empirical equation:

$$\theta = \beta_1 \exp\left[-\left(\frac{R_g T \ln(P_0/P)}{\beta E_0}\right)^n\right] + \beta_2 K P \quad (4.6\text{-}1)$$

Implicit in this equation is that the two modes of adsorption mechanisms, pore filling (Dubinin) and site adsorption (Henry law) are operative simultaneously at all range of pressure. Clearly this equation has a Henry law slope as the contribution of the Dubinin at zero pressure is zero. The coefficients β_1 and β_2 are obtained empirically by Kapoor *et al.* (1989) and Kapoor and Yang (1989):

$$\beta_1 = 1 - \exp\left(-\alpha \frac{P}{P_0}\right) \quad (4.6\text{-}2a)$$

$$\beta_2 = \exp\left(-\alpha \frac{P}{P_0}\right) \quad (4.6\text{-}2b)$$

where α is a constant determined by fitting eq. (4.6-1) with experimental data. This value of α was found to fall in a very large range, ranging from 15 to about 10000, a range too large to deduce any significant meaning of that parameter.

4.6.2 Schlunder's Approach

Schlunder and co-workers proposed a theory based on a DA equation and they allowed for the end of micropore filling by introducing a limiting potential A_{Gr}. Eiden and Schlunder (1990) modified the DA equation as follows:

$$W = W_0 \exp\left[-\left(\frac{A - A_{Gr}}{\beta E_0}\right)^n\right] \quad (4.6-3)$$

The advantage of this model is that for A approaching A_{Gr} the limiting volume is reached (at pressure less than the vapor pressure), before the capillary condensation is setting in. This is because the DA equation is not meant to cover such capillary condensation.

This limiting potential, A_{Gr}, is independent of adsorbent but is a function of adsorbate. Using the capillary condensation theory, the limiting pore radius r_{Gr} corresponding to A_{Gr} is calculated from:

$$r_{Gr} = \frac{2\sigma v_m \cos\theta}{A_{Gr}} + d_{mol} \quad (4.6-4)$$

In this equation, the radius is corrected for the thickness of one preadsorbed layer.

4.6.3 Modified Antoine Equation

Hacskaylo and LeVan (1985) adapted the concept of micropore filling of Dubinin in microporous solids and with the assumption that the state of adsorbed molecule in the pores is in liquid form, they utilised the Antoine equation, which expressed the vapour pressure versus temperature, to express the isotherm as a function of pressure with the constants assumed to be dependent on loading. The form is chosen so that it has the correct limit of Henry law at low loading.

The Antoine equation for the vapor pressure above a flat liquid surface is:

$$\ln P_0 = A - \frac{B}{C + T} \quad (4.6-5)$$

where A, B and C are constants specific to the species. This equation is valid for completely filled pores. For a partially filled pore, the modified adsorption isotherm equation will be:

$$\ln P = A'(\theta) - \frac{B'(\theta)}{C'(\theta) + T} \tag{4.6-6}$$

where A', B' and C' are function of loading and they take the form such that A' = A, B' = B and C' = C when the pores are saturated. With this form of equation, the isosteric heat is evaluated as (from the Van Hoff equation):

$$\Delta H = -R_g T^2 \left.\frac{\partial \ln P}{\partial T}\right|_\theta = -R_g T^2 \frac{B'(\theta)}{[C'(\theta) + T]^2} \tag{4.6-7}$$

In the limit of low loading, the modified equation must reduce to the Henry law form $\theta = KP$, that is:

$$\ln P = \ln \theta - \ln K, \qquad K = K_\infty \exp\left(\frac{Q}{R_g T}\right) \tag{4.6-8}$$

Comparing this equation with the above modified Antoine equation, the parameter A' must contain a logarithm term, and the parameters B' and C' must be bounded. The choice of these parameters now rests with the asssumption on how the isosteric heat would vary with loading.

For the heat of adsorption to be independent of loading, B' = B and C' = C, and

$$A' = A + \ln \theta \tag{4.6-9}$$

If the heat of adsorption increases linearly with loading, we have:

$$A' = A + \ln \theta \quad ; \quad B' = B + b(1-\theta) \quad ; \quad C' = C \tag{4.6-10}$$

If the dependence on loading varies in a complex way, we can have:

$$A' = A + \ln \theta \quad ; \quad B' = B + b(1-\theta) \quad ; \quad C' = C + c(1-\theta) \tag{4.6-11}$$

The parameter A' given in eq. (4.6-9) can take another form and still satisfies the Henry law behaviour at zero loading, for example $A' = A + \ln \theta - \ln(1-\theta)$.

4.7 Concluding Remarks

This chapter has addressed a number of isotherm equations developed mainly for microporous carbon type solids. The concept of micropore filling is the basis for the derivation. The simplest equations, like the DR and DA, are the first choice for describing the isotherm data, and when the information on the micropore size distribution is required, eq. (4.5-10) is a convenient equation for the determination of the micropore size distribution. Although those equations have been developed for carbonaceous type solids, they can be used to describe other microporous solids such as zeolite. For zeolitic particle, other approaches developed from the statistical thermodynamics approach (Ruthven, 1984) are also applicable.

5
Multicomponent Adsorption Equilibria

5.1 Introduction

The last three chapters deal with the fundamental and empirical approaches of adsorption isotherm for pure components. They provide the foundation for the investigation of adsorption systems. Most, if not all, adsorption systems usually involve more than one component, and therefore adsorption equilibria involving competition between molecules of different type is needed for the understanding of the system as well as for the design purposes. In this chapter, we will discuss adsorption equilibria for multicomponent system, and we start with the simplest theory for describing multicomponent equilibria, the extended Langmuir isotherm equation. This is then followed by a very popularly used IAS theory. Since this theory is based on the solution thermodynamics, it is independent of the actual model of adsorption. Various versions of the IAS theory are presented, starting with the Myers and Prausnitz theory, followed by the LeVan and Vermeulen approach for binary systems, and then other versions, such as the Fast IAS theory which is developed to speed up the computation. Other multicomponent equilibria theories, such as the Real Adsorption Solution Theory (RAST), the Nitta et al.'s theory, the potential theory, etc. are also discussed in this chapter.

5.2 Langmuirian Multicomponent Theory

5.2.1 Kinetic approach

The treatment of the extended Langmuir isotherm of a binary system is due to Markham and Benton (1931). Here, we present the general treatment for a multicomponent gaseous system containing N species. The assumptions made by

Langmuir in the treatment of pure component systems in Chapter 2 are applicable to multicomponent systems as well, that is each site occupying only one molecule, no mobility on the surface, and constant heat of adsorption. We assume that the rate constants for adsorption and desorption for all species are invariant.

The rate of adsorption of the species i onto the solid surface is

$$R_{a,i} = k_{a,i} P_i \left(1 - \sum_{j=1}^{N} \theta_j \right) \qquad (5.2\text{-}1a)$$

where θ_j is the fractional coverage of the species j and $k_{a,i}$ is the rate constant for adsorption and from the kinetic theory of gases it is given by

$$k_{a,i} = \frac{\alpha_i}{\sqrt{2\pi M_i RT}} \qquad (5.2\text{-}1b)$$

in which we have assumed that the rate of adsorption is due to the rate of collision of molecules towards the surface. Here α_i is the sticking coefficient of the species i.

Here, the rate of adsorption of the species i is proportional to the partial pressure of that species and to the fraction of the vacant sites on the surface, namely

$$\left(1 - \sum_{j=1}^{N} \theta_j \right) \qquad (5.2\text{-}1c)$$

The rate of desorption of the species i is proportional to its fractional loading

$$R_{d,i} = k_{d,i} \theta_i \qquad (5.2\text{-}1d)$$

Here the rate of desorption of a species is unaffected by the presence of all other species. In general, one would expect that the rate constant for desorption $k_{d,i}$ is a function of fractional loadings of all other species. However, it is treated as a constant in this analysis because of the one molecule per site and the no lateral interaction assumptions in the Langmuir mechanism.

At equilibrium, the rate of adsorption of the species i (eq. 5.2-1a) is equal to the rate of desorption of that species (eq. 5.2-1d):

$$b_i P_i (1 - \theta_T) = \theta_i \qquad (5.2\text{-}2a)$$

where b_i is the ratio of the rate constant for adsorption to that for desorption, and θ_T is the sum of fractional loadings of all species:

$$b_i = \frac{k_{a,i}}{k_{d,i}} \qquad (5.2\text{-}2b)$$

$$\theta_T = \sum_{j=1}^{N} \theta_j \qquad (5.2\text{-}2c)$$

Summing eq. (5.2-2a) with respect to i over all species, we can solve for the total fractional coverage in terms of the partial pressures of all components:

$$\theta_T = \frac{\sum_{j=1}^{N} b_j P_j}{1 + \sum_{j=1}^{N} b_j P_j} \qquad (5.2\text{-}3a)$$

and hence the fraction of the vacant sites is:

$$1 - \theta_T = \frac{1}{1 + \sum_{j=1}^{N} b_j P_j} \qquad (5.2\text{-}3b)$$

Knowing this total fractional coverage (eq. 5.2-3), the fractional coverage contributed by the species i is then obtained from eq. (5.2-2a), that is

$$\theta_i = \frac{b_i P_i}{1 + \sum_{j=1}^{N} b_j P_j} \qquad (5.2\text{-}4a)$$

or written in terms of molar concentration, we get

$$\frac{C_{\mu,i}}{C_{\mu s,i}} = \frac{b_i P_i}{1 + \sum_{j=1}^{N} b_j P_j} \qquad (5.2\text{-}4b)$$

This equation is known in the literature as the extended Langmuir isotherm equation, which gives the adsorbed concentration of the species "i" in the multicomponent system. For a binary system, Figure 5.2-1 shows plots of the fractional coverage of the component 1 versus $b_1 P_1$ with the parameter $b_2 P_2$ as the varying parameter. We see that the presence of the additional component 2 causes a decrease in the surface concentration of the component 1 and vice versa due to the competition of the two species. The reduction in the adsorbed concentration of the species 1 is

$$\frac{1 + b_1 P_1}{1 + b_1 P_1 + b_2 P_2}$$

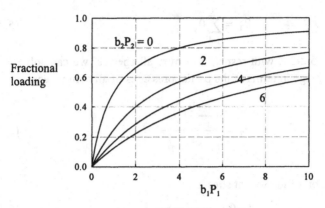

Figure 5.2-1: Plot of the fractional loading of the component 1 versus b_1P_1

The selectivity of species "i" in relation to the species "j" is defined as:

$$S_{i,j} = \frac{(C_{\mu,i}/P_i)}{(C_{\mu,j}/P_j)} = \frac{b_i}{b_j}$$

which is equal to the ratio of the affinities of the corresponding two species.

Eq. (5.2-4b) has assumed that each species maintains its own surface area (i.e. the area covered by one species is not affected by the others). The thermodynamic consistency of eq. (5.2-4b) is only possible when the monolayer capacity is the same for all species, that is

$$C_{\mu s,i} = C_{\mu s,j}$$

Discussion of this thermodynamic consistency is given in Section 5.3.6.

For binary systems where $C_{\mu s,1} \neq C_{\mu s,2}$, the following formula might be used to calculate the monolayer capacity:

$$\frac{1}{C_{\mu s}} = \frac{x_1}{C_{\mu s,1}} + \frac{x_2}{C_{\mu s,2}} \quad (5.2\text{-}5a)$$

where x_1 and x_2 are mole fractions of the adsorbed species, defined as

$$x_1 = \frac{C_{\mu,1}}{C_{\mu,1} + C_{\mu,2}} \quad (5.2\text{-}5b)$$

$$x_2 = \frac{C_{\mu,2}}{C_{\mu,1} + C_{\mu,2}} \quad (5.2\text{-}5c)$$

It is reminded that the extended Langmuir isotherm is derived assuming the surface is ideal and the adsorption is localised, and one molecule can adsorb onto one adsorption site.

5.2.2 Equilibrium Approach

Although the extended Langmuir equations can be derived from the kinetic approach presented in the last section, it can be derived from thermodynamics, which should be the right tool to study the equilibria (Helfferich, 1992). The two approaches, however, yield identical form of the equilibrium equation. What to follow is the analysis due to Helfferich.

At constant temperature and pressure, the change in the free energy must be zero at equilibrium:

$$dG = \sum_{j=1}^{N} \frac{\partial G}{\partial n_j} dn_j = 0 \quad \text{at equilibrium} \quad (5.2\text{-}6)$$

The rate of change of the free energy with respect to the number of mole of the species i is the chemical potential of that species, that is

$$\frac{\partial G}{\partial n_j} = \mu_j = \mu_j^0 + R_g T \ln C_j \quad \text{for ideal systems} \quad (5.2\text{-}7a)$$

Let us consider a reacting system

$$\sum_{j=1}^{N} v_j A_j = 0$$

with the convention that the stoichiometry of product is positive and that of reactant is negative. The above reaction would require that the change in the number of mole of the species i is proportional to the stoichiometry coefficient, that is

$$dn_i \propto v_i \quad (5.2\text{-}7b)$$

Substituting eqs. (5.2-7) for the chemical potential and the stoichiometry coefficient into eq. (5.2-6) we get:

$$\sum_{j=1}^{N} v_j \mu_j = \sum_{j=1}^{N} v_j \mu_j^0 + R_g T \sum_{j=1}^{N} v_j \ln C_j = 0 \quad (5.2\text{-}8)$$

Recognising the following property of logarithm

$$\sum_{j=1}^{N} v_j \ln C_j = \sum_{j=1}^{N} \ln C_j^{v_j} = \ln\left(\prod_{j=1}^{N} C_j^{v_j}\right) \quad (5.2\text{-}9)$$

eq. (5.2-8) can be rewritten as:

$$\prod_{j=1}^{N} C_j^{v_j} = K \equiv \text{constant} \tag{5.2-10}$$

since the standard chemical potentials μ_j° are constant. This is basically the mass action law, commonly used in chemical reaction equilibrium.

In this thermodynamic formulation, there is no mention of the nature of the chemical bonds and the status of the species involved. Therefore, this thermodynamic derivation is applicable to adsorption as well. As a matter of fact, it is applicable to any process in which something is converted into something else.

Now, we apply this thermodynamics approach to the adsorption system where there is a reversible reaction between the free adsorbate and the bound species as shown below:

$$C_j + S \Leftrightarrow [C_j - S] \tag{5.2-11}$$

where S denotes the adsorption site, C_j is the free adsorbate and $[C_j - S]$ is the bound adsorbate.

We rewrite this adsorption equilibrium equation in the standard reaction equation format as follows:

$$[C_j - S] - C_j - S = 0 \tag{5.2-12}$$

Applying the thermodynamic equation (eq. 5.2-10) to the above specific equation, we have:

$$\theta_i \, C_i^{-1} \left[1 - \sum_{j=1}^{N} \theta_j \right]^{-1} = b_i \, (\text{constant}) \tag{5.2-13}$$

where θ_i is the fractional loading of species i, and $\left[1 - \sum \theta_j\right]$ is the fraction of the free sites.

Rewrite eq. (5.2-13) as follows:

$$\theta_i = b_i C_i \left[1 - \sum_{j=1}^{N} \theta_j \right]$$

which is basically the statement of the equality between the rate of adsorption and the rate of desorption obtained in Section 5.2.1 (eq. 5.2-2a). Solving for the fractional coverage of the species i, we obtain:

$$\theta_i = \frac{b_i C_i}{1+\sum_{j=1}^{N} b_j C_j} \tag{5.2-14}$$

The concentration of the adsorbed species i will be proportional to the fractional coverage occupied by that species, that is:

$$C_{\mu,i} = C_{\mu s,j} \frac{b_i C_i}{1+\sum_{j=1}^{N} b_j C_j} \tag{5.2-15}$$

which is identical to the formula obtained by the kinetic approach of Langmuir. This lends support to the fundamental Langmuir equation, which can be derived from either the kinetic approach or the thermodynamic approach. Although it is fundamentally sound, the assumptions associated with its derivation, such as one molecule - one site interaction, no interaction among adsorbate molecules, and constant heat of adsorption with loading are too restrictive and do not readily apply to practical systems and hence very often that it does not describe well multicomponent adsorption equilibria. Since there are no particular models which are superior in the description of equilibria, empirical approaches are frequently applied especially in the dynamics studies of adsorber columns such as fixed bed or PSA column. These empirical approaches are usually based on the extended Langmuir equations by allowing some parameters to relax in order to fit multicomponent equilibrium data. One approach is to vary the affinity constant and the other is to adopt the exponent on the partial pressure. This is done in the next section.

5.2.3 Empirical Approaches Based on the Langmuir Equation

When there are lateral interactions between adsorbed species, which are generally different from the self interactions among molecules of the same type, the following empirical equation is recommended (Yang, 1987):

$$C_{\mu,i} = C_{\mu s,j} \frac{(b_i/\eta_i) P_i^{1/n_i}}{1+\sum_{j=1}^{N} (b_j/\eta_j) P_j^{1/n_j}} \tag{5.2-17}$$

The parameter η_i allows for interaction among molecules of the same type, and n_i allows for other effects of heterogeneity. Eq.(5.2-17) can be viewed as the generalization of the Langmuir-Freundlich (Sips) isotherm.

5.3 Ideal Adsorption Solution Theory (IAS)

5.3.1 The Basic Thermodynamic Theory

Recognizing the deficiency of the extended Langmuir equation, despite its sound theoretical footing on basic thermodynamics and kinetics theories, and the empiricism of the loading ratio correlation, other approaches such as the ideal adsorbed solution theory of Myers and Prausnitz, the real adsorption solution theory, the vacancy solution theory and the potential theory have been proposed. In this section we will discuss the ideal adsorbed solution theory and we first develop some useful thermodynamic equations which will be used later to derive the ideal adsorbed solution model.

Sircar and Myers (1973) have shown that the theories associated with the ideal solution are based on the same principle, that is the assumption of ideal adsorbed solution. The difference among them is simply the choice of the standard states.

The Gibbs free energy of the mixture per mole of adsorbate is:

$$g = \sum_{j=1}^{N} x_j g_j^0 + g^m \qquad (5.3\text{-}1a)$$

where g_j^0 is the molar Gibbs free energy of the pure j-th component measured at the same value of the intensive variable I, to be defined later, and g^m is the Gibbs free energy of mixing. The variable I defines the standard states of the pure adsorbate. The free energy per unit mole, g, is

$$g = \frac{G}{\sum_{j=1}^{N} n_j'} \qquad (5.3\text{-}1b)$$

with n_j' being the number of moles of the adsorbate, and

$$G = U - TS \qquad (5.3\text{-}1c)$$

The isothermal Gibbs free energy of mixing per mole, g^m, is:

$$g^m = R_g T \sum_{j=1}^{N} x_j \ln(\gamma_j x_j) \qquad (5.3\text{-}1d)$$

where γ_j is the activity coefficient (=1 for ideal solution).

The starting thermodynamic equation is (Bering et al., 1970)

$$U - TS - \phi m - \sum_{j=1}^{N} \mu_j n_j' = 0 \qquad (5.3\text{-}2)$$

The above equation mimics the following thermodynamics equation for the bulk phase:

$$U - TS + PV - \sum_{j=1}^{N} \mu_j n_j' = 0$$

Here U is the internal energy, S is the entropy, m is the mass of adsorbent, ϕ is the surface potential of the adsorbed phase per unit mass of the adsorbent, and μ_j is the chemical potential, related to the fugacity as follows:

$$\mu_j = \mu_j^* + R_g T \ln f_i \qquad (5.3\text{-}3)$$

where μ_j^* is the standard chemical potential at 1 atm, and f_i is the fugacity of the species i.

Combining eqs. (5.3-1) we get:

$$\frac{U - TS}{\sum_j n_j'} = \sum_j x_j g_j^0 + R_g T \sum_j x_j \ln(\gamma_j x_j) \qquad (5.3\text{-}4)$$

Replacing U − TS of the above equation by that of the thermodynamic equation (5.3-2) yields the following relationship between the surface potential and the chemical potential of all species

$$\frac{\phi m}{\sum_j n_j'} + \sum_j x_j \mu_j = \sum_j x_j g_j^0 + R_g T \sum_j x_j \ln(\gamma_j x_j) \qquad (5.3\text{-}5)$$

Note that g_j^0 is the molar Gibbs free energy of the pure component j.

It is convenient from the measurement point of view that the chemical potential is expressed in terms of the fugacity. Thus by combining eqs. (5.3-5) and (5.3-3) we obtain

$$\frac{\phi}{n} + R_g T \sum_j x_j \ln\left(\frac{f_j}{\gamma_j x_j}\right) + \sum_j x_j \left(\mu_j^0 - g_j^0\right) = 0 \qquad (5.3\text{-}6a)$$

where

$$n = \sum_j n_j \qquad (5.3\text{-}6b)$$

$$n_j = \frac{n'_j}{m} \tag{5.3-6c}$$

Here n is the total number of moles per unit mass of the adsorbent. Before arriving at the desired equation, we now need to express the molar Gibbs free energy in terms of the fugacity of the pure component. We have:

$$g_j^0 = \frac{G_j^0}{n_j'^0} = \frac{U^0 - TS}{n_j'^0} = \frac{\phi_j^0 m}{n_j'^0} + \mu_j^0 = \frac{\phi_j^0}{n_j^0} + \mu_j^0 \tag{5.3-7}$$

where n_j^0 is the number of mole of pure component j per unit mass of the adsorbent and ϕ_j^0 is the surface potential of the pure component j.

Combining eqs. (5.3-6) and (5.3-7) we obtain the following fundamental equation for the mixture

$$R_g T \sum_{j=1}^{N} x_j \ln\left(\frac{f_j}{f_j^0 \gamma_j e^{z_j x_j}}\right) + \phi\left[\frac{1}{n} - \sum_{j=1}^{N} \frac{x_j}{n_j^0}\right] = 0 \tag{5.3-8}$$

where f_j^0 is the fugacity of the pure component j at the reference state, n is the number of moles per unit mass of adsorbent, n_j^- is the number of moles of pure j-th component per unit mass at the same reference state, and z_j is defined as:

$$z_j = -\frac{\phi - \phi_j^0}{n_j^0 R_g T} \tag{5.3-9}$$

where ϕ_j^0 is the surface potential of the pure j-th component at the reference state.

One of the possible solutions to the fundamental equation (5.3-8) is to force each term in the LHS of that equation to zero, and we get:

$$f_j = f_j^0 \exp(z_j) x_j \tag{5.3-10a}$$

$$\frac{1}{n} = \sum_{j=1}^{N} \frac{x_j}{n_j^0} \tag{5.3-10b}$$

This special solution in fact defines an <u>ideal adsorbed solution</u>.

The fugacity in the adsorbed phase is equal to the fugacity of the gas phase and when the gas phase pressure is low the fugacity can be written in terms of the total pressure and the mole fraction as follows (this is reasonable unless the gas phase pressure is very high):

$$f_j = Py_j \tag{5.3-11}$$

Combining eqs. (5.3-10a) and (5.3-11), we get:

$$Py_j = P_j^0 x_j \exp(z_j) \tag{5.3-12}$$

Eqs. (5.3-10b) and (5.3-12) define the ideal adsorbed solution theory. The surface potential, ϕ, is calculated from the Gibbs adsorption isotherm equation (Bering et al., 1970):

$$-\frac{1}{nR_g T} d\phi = \sum_{j=1}^{N} x_j d\ln(Py_j) \tag{5.3-13}$$

For pure components, the surface potential of the pure component i is:

$$\frac{\phi_i^0}{R_g T} = -\int_0^{P_i^0} n_i^0 d\ln P_i^0 \tag{5.3-14}$$

Thus, if the pure component isotherm $n_i^0(P_i^0)$ is known, the surface potential of the pure component i can be obtained from the integral given by eq. (5.3-14).

5.3.2 Myers and Prausnitz Theory

The theory presented in the previous section involves a standard state. This standard state can be defined in a way that the surface potential of the mixture is the same as the surface potentials of all pure components, that is:

$$\frac{\phi}{R_g T} = \frac{\phi_j^0}{R_g T} = -\int_0^{P_j^0} \frac{n_j^0}{P_j^0} dP_j^0 \tag{5.3-15}$$

for all j. If this is the case, $z_j = 0$ (defined in eq. 5.3-9), and hence eq. (5.3-12) will become:

$$Py_j = P_j^0 x_j \tag{5.3-16}$$

which basically mimics the Raoult's law in vapor-liquid equilibrium. The mole fractions in the gas phase and in the adsorbed phase must satisfy:

$$\sum_{j=1}^{N} y_j = 1 \tag{5.3-17a}$$

$$\sum_{j=1}^{N} x_j = 1 \tag{5.3-17b}$$

For clarity sake, we rewrite below the basic equations of the Myers and Prausnitz theory

$$\frac{1}{n} = \sum_{j=1}^{N} \frac{x_j}{n_j^0} \qquad (5.3\text{-}18a)$$

$$\frac{\phi}{R_g T} = \frac{\phi_j^0}{R_g T} = -\int_0^{P_j^0} \frac{n_j^0(P_j^0)}{P_j^0} dP_j^0 \qquad \text{for all } j \qquad (5.3\text{-}18b)$$

$$Py_j = P_j^0 x_j \qquad \text{for all } j \qquad (5.3\text{-}18c)$$

$$\sum_{j=1}^{N} x_j = 1 \qquad (5.3\text{-}18d)$$

The number of equations of eqs. (5.3-18) are shown in Table 5.3-1.

Table 5.3-1. The number of equations in the Myers-Prausnitz theory

Eq. no.	Number of equations
(5.3-18a):	1
(5.3-18b):	N-1
(5.3-18c):	N
(5.3-18d):	1
	Total = 2N + 1

We have a total of 2N+1 equations in the ideal adsorption solution theory. Let us now apply this ideal adsorption solution theory to the usual case of adsorption equilibria, that is we specify the total pressure and the mole fractions in the gas phase and wish to determine the properties of the adsorbed phase which is in equilibrium with the gas phase. For such a case the number of unknowns that we wish to obtain is given in the following table:

Table 5.3-2. Total number of unknowns when the gas phase is specified

Variable	Number of variable
Pure component pressure, P_j^0:	N
Adsorbed phase mole fraction, x_j	N
Total adsorbed phase concentration, n	1
	Total unknown variable = 2N + 1

The total number of unknown variables is 2N+1, which is the same as the number of equations given by the ideal adsorption solution theory; thus the problem is properly posed. Once the total adsorbed amount is determined, the adsorbed phase concentration of the component i is:

$$n_i = n\, x_i \qquad (5.3\text{-}19)$$

In this theory of Ideal Adsorption Solution, adsorption isotherm equation for pure components can take any form which fits the data best (Richter et al., 1989). Two isotherms commonly used are Toth and Unilan equations (Chapter 3), although Toth equation is the preferable equation from the computational point of view because it usually gives faster convergence than the Unilan equation does. When DR equation is used to describe pure component data, the application of the IAST in this case has some special features (Richter et al., 1989) and it was used by Lavanchy et al. (1996) in the description of equilibria of chlorobenzene and tetrachloride on activated carbon.

The IAS theory is thermodynamically consistent and exact at the limit of zero pressure (Valenzuela and Myers, 1989). The success of the calculations of IAS depends on how well the single component data are fitted, especially in the low pressure region as well as at high pressure region where the pure hypothetical pressure lies. An error in these regions, in particular the low pressure region, can cause a large error in the multicomponent calculations..

5.3.3 Practical Considerations of The Myers-Prausnitz IAS Equations

The IAS theory developed in the previous section is now recast here in the form convenient for the purpose of computation. For a system containing N species, the IAS equations are given below.

The analog Raoult's law for an ideal adsorption system is:

$$P y_i = P_i = x_i P_i^0(\pi) \qquad \text{for} \quad i = 1, 2, \cdots, N \qquad (5.3\text{-}20a)$$

and

$$\sum_{j=1}^{N} x_j = 1 \qquad (5.3\text{-}20b)$$

where $P_i^0(\pi)$ is the hypothetical pressure of the pure component that gives the same spreading pressure on the surface, that is:

$$\frac{A\pi}{R_g T} = \int_0^{P_1^0} \frac{C_{\mu 1}}{P_1}\, dP_1 = \int_0^{P_2^0} \frac{C_{\mu 2}}{P_2}\, dP_2 = \cdots = \int_0^{P_N^0} \frac{C_{\mu N}}{P_N}\, dP_N \qquad (5.3\text{-}21)$$

The spreading pressure is the negative of the surface potential. For a given total pressure (P) and the mole fraction in the gas phase (y_i), eqs. (5.3-20) provides N+1 equations, and eq. (5.3-21) gives N equations, a total of 2N+1 equations. With this set of 2N+1 equations, there are 2N+1 unknowns.

1. N values of mole fractions in the adsorbed phase (x_i),
2. one value of the spreading pressure π, and
3. N values of the hypothetical pressure of the pure component, P_j^0, which gives the same spreading pressure as that of the mixture.

Thus, solving numerically eqs. (5.3-20) and (5.3-21) will give a set of solution for the adsorbed phase mole fractions and a solution for the spreading pressure.

Knowing the mole fractions in the adsorbed phase (x_j) and the hypothetical pressure of the pure component (P_j^0) that gives the same spreading pressure as that of the mixture (π), the total amount adsorbed can be calculated from the equation:

$$\frac{1}{C_{\mu T}} = \sum_{j=1}^{N} \frac{x_j}{C_{\mu j}^0} \qquad (5.3\text{-}22a)$$

where $C_{\mu j}^0$ is the adsorbed amount of pure component j at the hypothetical pressure P_j^0, that is

$$C_{\mu j}^0 = f^0\left(P_j^0\right) \qquad (5.3\text{-}22b)$$

Knowing the total amount adsorbed ($C_{\mu T}$), the amount adsorbed contributed by the component "i" is given by:

$$C_{\mu i} = x_i C_{\mu T} \qquad (5.3\text{-}23)$$

Eqs. (5.3-20) to (5.3-23) form a set of powerful equations for the IAS theory, that is if the total pressure and the mole fractions in the gas phase are given, the adsorbed phase mole fractions, the spreading pressure, the pure component pressure that gives the same spreading pressure as the mixture, the total amount adsorbed and the component amount adsorbed can be calculated. The inverse problem of this situation is that if the amounts of the adsorbed phase are given, the total pressure as well as the mole fraction in the gas phase can also be calculated.

Although the set of eqs. (5.3-20 to 5.3-23) define the fluid-solid equilibria, there is another way to determine this equilibria. If the spreading pressure is known explicitly as a function of the gaseous partial pressures, $P_1, P_2, ..., P_N$, then the multicomponent equilibria can be determined by noting the Gibbs adsorption equation for a mixture:

$$d\pi = \sum_{i=1}^{N} \frac{C_{\mu i}}{P_i} dP_i \qquad (5.3-24)$$

which means that

$$C_{\mu i} = P_i \left(\frac{\partial \pi}{\partial P_i} \right)_{P_j, j \neq i} \qquad (5.3-25)$$

This is the approach used by LeVan and Vermeulen (1981) which we will discuss in Section 5.5.

5.3.4 The Lewis Relationship

Before we proceed further with the computational illustration of the IAS theory, it is worthwhile to have another look at eq. (5.3-22a). For a binary system, we can rewrite that equation in the following form:

$$\sum_{j=1}^{2} \frac{C_{\mu j}}{C_{\mu j}^0} = 1 \qquad (5.3-26a)$$

where

$$C_{\mu j}^0 = f^0\left(P_j^0\right) \quad \text{for } j = 1 \text{ and } 2 \qquad (5.3-26b)$$

Eq. (5.3-26a) is known as the Lewis relationship. It has been tested with a number of systems involving silica gel and activated carbon. This equation relates the amounts adsorbed in multicomponent system ($C_{\mu j}$) to those for pure component systems ($C_{\mu j}^0$) evaluated at the hypothetical pressure P_j^0.

Although eq. (5.3-26) were derived from thermodynamic analysis (Section 5.3-2), the Lewis relationship can be derived by the following alternative approach, which assumes the adsorption is by a micropore filling mechanism. The derivation of this equation is as follows. The maximum volumetric capacity for micropore filling is W_0, which is assumed the same for all sorbates. Therefore, the maximum number of moles of adsorbed species is:

$$C_{\mu T} = \frac{W_0}{v_m} \qquad (5.3-27)$$

where v_m is the molar volume of mixed adsorbate. If there is no volume change during mixing in the adsorbed phase (ideal adsorbed solution), the partial molar volume is thence additive, that is

$$v_m = \sum_{j=1}^{N} x_j v_j \qquad (5.3\text{-}28)$$

Substituting eq. (5.3-28) into eq. (5.3-27), we get:

$$C_{\mu T} = \frac{W_0}{\sum_{j=1}^{N} x_j v_j} \qquad (5.3\text{-}29)$$

By noting that

$$C_{\mu j} = C_{\mu T} x_j \qquad (5.3\text{-}30a)$$

and

$$C_{\mu j}^0 = \frac{W_0}{v_j} \qquad (5.3\text{-}30b)$$

eq. (5.3-29) will become

$$\sum_{j=1}^{N} \frac{C_{\mu j}}{C_{\mu j}^0} = 1,$$

which is the Lewis equation given in eq.(5.3-22a). This equation is valid provided that adsorbates are adsorbed by the micropore filling mechanism and the adsorbed phase is ideal, that is no mixing between species. It has been found to be applicable for solids such as activated carbon, zeolites, and silica gel.

We have completed the theoretical description of the IAS theory, now we turn to the algorithm for the purpose of computation with the various forms of pure component isotherm. We start first by considering the problem where the gas phase conditions (P and y) are specified, and the adsorbed phase conditions are required. Next, we will consider the inverse problem, that is the adsorbed amount and the mole fractions x are given, and the gas phase conditions are required.

5.3.5 General IAS Algorithm: Specification of P and y

The set of equations presented in the previous section (5.3.3) in general can not be solved analytically; hence it must be solved numerically. Even if the spreading pressure equation (5.3-21) can be integrated analytically, the inverse of the hypothetical pure component pressure versus spreading pressure is not generally available in analytical form with the exception of the Langmuir, Freundlich and Sips equations (see Table 5.3-3).

For general case, the IAS equations must be solved numerically and this is quite effectively done with standard numerical tools, such as the Newton-Raphson method for the solution of algebraic equations and the quadrature method for the evaluation of integral. We shall develop below a procedure and then an algorithm for solving the equilibria problem when the gas phase conditions (P, y) are given (Myers and Valenzuela, 1986).

Eqs. (5.3-20) can be combined together to yield the following equation

$$F(z) = \sum_{j=1}^{N} x_j - 1 = \sum_{j=1}^{N} \frac{Py_j}{P_j^0(z)} - 1 = 0 \qquad (5.3\text{-}31a)$$

where z is called the reduced spreading pressure

$$z = \frac{\pi A}{R_g T} \qquad (5.3\text{-}31b)$$

Eq. (5.3-31a) is a function of only the reduced spreading pressure as the hypothetical pressure $P_j^0(z)$ is a function of the spreading pressure and it is calculated from eq. (5.3-21), from which the hypothetical pressure can be obtained either analytically or numerically depending on the isotherm equation used to describe the pure component equilibria.

To solve eq. (5.3-31a) for the reduced spreading pressure we need to resort to a numerical procedure. An effective tool to meet this goal is the Newton-Ralphson method, which is an iterative method to obtain z. The iteration formula for the reduced spreading pressure is given below:

$$z^{(k+1)} = z^{(k)} - \frac{F(z^{(k)})}{F'(z^{(k)})} \qquad (5.3\text{-}32)$$

where the upperscript denotes for the iteration number, and F and F' = dF/dz are given below:

$$F(z^{(k)}) = \sum_{j=1}^{N} \frac{Py_j}{P_j^0(z^{(k)})} - 1 \qquad (5.3\text{-}33a)$$

$$F'(z^{(k)}) = \left[-\sum_{j=1}^{N} \frac{Py_j}{\left[P_j^0(z)\right]^2} \frac{dP_j^0(z)}{dz} \right]_{z=z^{(k)}} = \left[-\sum_{j=1}^{N} \frac{Py_j}{\left[P_j^0(z)\right] C_{\mu j}^0} \right]_{z=z^{(k)}} \qquad (5.3\text{-}33b)$$

In obtaining eq.(5.3-33b), we have used the following formula

208 Equilibria

$$\frac{dP_j^0(z)}{dz} = \frac{P_j^0(z)}{C_{\mu j}^0} \qquad (5.3\text{-}34)$$

which is obtained by differentiating eq. (5.3-21).

The algorithm 5.3-1 presented below is for the evaluation of the reduced spreading pressure (eq. 5.3-31a):

Algorithm 5.3.1

Step #	Action	
1	1a.	First the input parameters are supplied: the parameters for the single component isotherm, the total gaseous pressure and the mole fractions in the gas phase.
2	2a.	Estimate the reduced spreading pressure as the molar average of the following integral $$z = \frac{\pi A}{R_g T} = \sum_{j=1}^{N} y_j \int_0^P \frac{C_{\mu j}}{P_j} dP_j \qquad (5.3\text{-}35)$$ Note that the total pressure is used as the upper limit of the integral.
	2b.	The RHS of the above equation can be evaluated because all variables (y, P, and the single component isotherm equations) are known. If the pure component isotherm can be approximated by a Langmuir equation, then the initial estimate of the spreading pressure can be taken as: $$z = C_{\mu s} \ln\left(1 + \sum_{i=1}^{N} b_i P_i\right) \qquad (5.3\text{-}36)$$ where $C_{\mu s}$ can be taken as the average of the maximum adsorbed concentration of all species.
3	3a.	Knowing the estimated reduced spreading pressure from step 2, evaluate the pure component pressure P_j^- that gives that reduced pressure (using eq. 5.3-21), and then evaluate the amount adsorbed for the single component from the single component isotherm at that hypothetical pressure $P_j^{0.}$
	3b.	Next evaluate $F(z^{(k)})$ and $F'(z^{(k)})$ from eq. (5.3-33) and thence calculate the reduced spreading pressure for the next iteration step from eq. (5.3-32).
4		Continue step 3 until the method converges

In step 3, we need to evaluate the pure component pressure from the reduced spreading pressure. This is fine if the integral for the reduced spreading pressure (eq. 5.3-21) can be obtained analytically, for example when the single component isotherm takes the form of Langmuir equation:

$$C_{\mu j} = C_{\mu s j} \frac{b_j P_j}{1 + b_j P_j} \qquad (5.3\text{-}37a)$$

The spreading pressure for the pure component isotherm of Langmuir form is:

$$z = C_{\mu s j} \ln\left(1 + b_j P_j^0\right) \qquad (5.3\text{-}37b)$$

Hence, the pure component pressure can be obtained explicitly from the above equation as:

$$P_j^0 = \frac{\exp(z/C_{\mu s j}) - 1}{b_j} \qquad (5.3\text{-}37c)$$

Such an explicit expression for the pure component pressure is only possible with Langmuir and very limited number of other isotherms (see Table 5.3-3). When the pure component isotherm takes a general form, the pure component pressure, P_j^0, as a function of z must be obtained iteratively. Myers and Valenzuela (1986) presented an algorithm to perform this task iteratively. For a given spreading pressure, eq. (5.3-21) can be put in the form of Newton-Raphson format as follows:

$$G\left(P_j^0\right) = \int_0^{P_j^0} \frac{C_{\mu j}}{P_j} dP_j - z = 0 \qquad (5.3\text{-}38)$$

The iteration formula for the solution of the hypothetical pressure is then:

$$P_j^{0(k+1)} = P_j^{0(k)} - \frac{G\left(P_j^{0(k)}\right)}{G'\left(P_j^{0(k)}\right)} \qquad (5.3\text{-}39a)$$

where the function G is given in eq. (5.3-38) and $G' = dG/dP_j^0$ is given by:

$$G'\left(P_j^0\right) = \frac{C_{\mu j}^0}{P_j^0} \qquad (5.3\text{-}39b)$$

The integral in eq. (5.3-38) must be evaluated numerically for a general pure component isotherm. The algorithm to calculate the pure component pressure for a given reduced spreading pressure is as follows:

Algorithm 5.3.2

Step #	Action
1	Input the parameters for the single component isotherm equation
2	2a. First estimate the pure component pressure $$P_j^0 = \frac{C_{\mu s j}}{H_j}\left[\exp\left(\frac{z}{C_{\mu s j}}\right) - 1\right] \quad (5.3\text{-}40)$$ where H_j is the Henry constant, $C_{\mu,j} = H_j P_j^0$. This estimate basically comes from the form of Langmuir isotherm equation, that is $z = C_{\mu s j} \ln(1 + b_j P_j^0)$
3	Next, calculate $C_{\mu j}^0(P_j^0)$, and then the next estimate for the pure component pressure from eq. (5.3-39a). Repeat this step until the method converges

Once z is obtained numerically from the algorithm 5.3-1, the hypothetical pressure P_j^0 is then calculated from eq. (5.3-21) or the algorithm 5.3-2. Once this is done the mole fractions in the adsorbed phase are calculated from eq. (5.3-20a) and the total amount adsorbed is calculated from eq. (5.3-22a); hence the individual adsorbed amount is from eq. (5.3-23).

The following table 5.3-3 shows the various formula for the spreading pressure and the pure component hypothetical pressure for various commonly used isotherms. Some isotherms such as Langmuir, Freundlich, LRC have analytical expressions for the spreading pressure as well as the pure component hypothetical pressure. Other isotherms, such as O'Brien & Myers, Ruthven, Toth and Nitta have analytical expression for the spreading pressure, but the pure component hypothetical pressure expressed in terms of the reduced pressure must be determined from a numerical method. For other general isotherms, such as Unilan, Aranovich, Dubinin-Radushkevich, Dubinin-Astakhov, Dubinin-Stoeckli, Dubinin-Jaroniec, one must resort to a numerical method to obtain the spreading pressure as well as the pure component hypothetical pressure.

Table 5.3-3a: Isotherm equations and their respective Henry constant.

Isotherm	Isotherm expression	Henry law
Langmuir	$C_\mu = C_{\mu s} \dfrac{bP}{1+bP}$	$C_\mu = C_{\mu s} bP$
Freundlich	$C_\mu = KP^{1/n}$	-
Sips	$C_\mu = C_{\mu s} \dfrac{(bP)^{1/n}}{1+(bP)^{1/n}}$	-
Dual Langmuir	$C_\mu = C_{\mu s1} \dfrac{b_1 P}{1+b_1 P} + C_{\mu s2} \dfrac{b_2 P}{1+b_2 P}$	$C_\mu = C_{\mu s1} b_1 P + C_{\mu s2} b_2 P$
O'Brien and Myers	$C_\mu = C_{\mu s}\left[\dfrac{bP}{1+bP} + \dfrac{\sigma^2 bP(1-bP)}{2(1+bP)^3}\right]$	$C_\mu = C_{\mu s} b\left(1+\dfrac{\sigma^2}{2}\right)P$
Toth	$C_\mu = C_{\mu s} \dfrac{bP}{\left[1+(bP)^t\right]^{1/t}}$	$C_\mu = C_{\mu s} bP$
Nitta	$nbP = \dfrac{\theta}{(1-\theta)^n} \;;\; \theta = \dfrac{C_\mu}{C_{\mu s}}$	$C_\mu = C_{\mu s} nbP$
Ruthven	$C_\mu = C_{\mu s} \dfrac{\sum_{j=1}^{M} A_j (bP)^j}{1+\sum_{j=1}^{M} A_j (bP)^j / j!}$	$C_\mu = C_{\mu s} bP$
Unilan	$C_\mu = C_{\mu s}\left[\dfrac{1}{2s}\ln\left(\dfrac{1+be^s P}{1+be^{-s}P}\right)\right]$	$C_\mu = \left(C_{\mu s} b \dfrac{\sinh(s)}{s}\right)P$

Note that the reduced spreading pressure has the units of the adsorbed concentration and it increases logarithmically with the pure component hypothetical pressure.

Table 5.3-3b: Formula for Spreading Pressure and Pure Component Pressure

Isotherm	Spreading pressure	Pure component pressure
Langmuir	$z = C_{\mu s} \ln(1 + bP^0)$	$P^0 = \dfrac{1}{b}\left[\exp\left(\dfrac{z}{C_{\mu s}}\right) - 1\right]$
Freundlich	$z = nK(P^0)^{1/n}$	$P^0 = \left(\dfrac{z}{nK}\right)^n$
Sips	$z = nC_{\mu s} \ln\left[1 + (bP^0)^{1/n}\right]$	$P^0 = \dfrac{1}{b}\left[\exp\left(\dfrac{z}{nC_{\mu s}}\right) - 1\right]^n$
Dual Langmuir	$z = C_{\mu s1} \ln(1 + b_1 P^0) + C_{\mu s2} \ln(1 + b_2 P^0)$	numerical
O'Brien and Myers	$z = C_{\mu s}\left[\ln(1 + bP^0) + \dfrac{\sigma^2 bP}{2(1+bP)^2}\right]$	numerical
Toth	$z = C_{\mu s}\left[-\left(\dfrac{1}{t}-1\right)\theta - \dfrac{\theta^{1/t}\ln(1-\theta)}{t} - \left(\dfrac{1}{t}-1\right)\sum_{k=1}^{\infty}\dfrac{\theta^{(1+kt)}}{(1+k)(1+kt)} \right]$ $\theta = C_\mu / C_{\mu s}$	numerical
Nitta	$z = C_{\mu s}\left[(n-1)\theta + (2-n)\ln\left(\dfrac{1}{1-\theta}\right)\right]$	numerical
Ruthven	$z = C_{\mu s} \ln\left[1 + \sum_{j=1}^{M} A_j (bP)^j / j!\right]$	numerical
Unilan	numerical	numerical

For Unilan equation, the spreading pressure must be evaluated numerically. The integration to evaluate the reduced pressure in eq. (5.3-21) converges slowly (Myers, 1984). Since the Unilan equation is the result of the integration of a local Langmuir equation over a uniform energy distribution (see Chapter 6 for more detail of energy distribution concept), one could interchange the order of the integration by taking the spreading pressure of the local Langmuir isotherm and then taking the integration of the local spreading pressure over the energy distribution as done below. Thus, instead of

$$z = \frac{\pi A}{RT} = \int_0^{P^0} \frac{C_\mu^0}{P} dP = \frac{C_{\mu s}}{2s} \int_0^{P^0} \frac{1}{P} \ln\left(\frac{1+be^s P}{1+be^{-s}P}\right) dP \qquad (5.3\text{-}42)$$

we can use the following integral

$$z = \frac{C_{\mu s}}{2s} \int_{-s}^{s} \ln(1+be^z P) dz \qquad (5.3\text{-}43)$$

The last equation converges rapidly. Myers (1984) used the Unilan equation to fit the data of Ziegler and Rogers, and the predictions of the binary data were done with the IAS theory. The prediction is reasonable, but they are not very good when the loading is more than 50% of the maximum saturation capacity (occurred when the total pressure increases). Myers postulated a number of reasons:

1. The data are incorrect
2. Imperfection in the vapor phase, since ideal gas law was used in high pressure data
3. No adsorbate-adsorbate interaction
4. Heterogeneity is ignored.

Among these reasons, the first two are dismissed. The third is not so easily dismissed. The fourth seems to be the reason for the discrepancy between the prediction and the data. In IAS theory the surface is assumed uniform in composition, although it is recognized that the surface is heterogeneous, with different selectivities on different parts of the surface. It has been shown by Myers (1983) that when surface heterogeneity is ignored some apparent deviations are observed. However, it is still difficult to point exactly to the source of discrepancies until more experimental data are carefully collected and the IAS is fully tested.

5.3.6 Thermodynamic Justification of the Extended Langmuir Equation

We have shown in the last few sections the IAS theory as well as its computation implementation to obtain multicomponent adsorption isotherm. Since this theory is based on solution thermodynamics it can be applied to prove the thermodynamic consistency of the extended Langmuir equation.

Consider a system containing N components and the pure component isotherm of each component can be described by the following Langmuir equation

$$C^0_{\mu,i} = C_{\mu s} \frac{b_i P^0_i}{1 + b_i P^0_i} \tag{5.3-44}$$

All components have the same saturation capacity $C_{\mu s}$.

At the mixture reduced spreading pressure, the hypothetical pure component pressure is given by (Table 5.3-3)

$$P^0_i = \frac{1}{b_i}\left[\exp\left(\frac{z}{C_{\mu s}}\right) - 1\right] \tag{5.3-45}$$

Substituting this hypothetical pure component pressure into eq. (5.3-31a) to solve for the reduced spreading pressure, we get

$$\exp\left(\frac{z}{C_{\mu s}}\right) - 1 = \sum_{j=1}^{N} b_j P_j \tag{5.3-46}$$

Knowing this reduced spreading pressure, eq. (5.3-45) can be solved for the hypothetical pure component pressure in terms of the gas phase partial pressures

$$P^0_i = \frac{1}{b_i}\sum_{j=1}^{N} b_j P_j \tag{5.3-47}$$

Substituting this hypothetical pure component pressure into the Raoult's law equation (5.3-20a), we get the mole fraction of the adsorbed phase

$$x_i = \frac{b_i P_i}{\sum_{j=1}^{N} b_j P_j} \tag{5.3-48}$$

The total amount adsorbed is obtained by substituting eqs. (5.3-47), (5.3-48), (5.3-44) into eq. (5.3-22a):

$$C_{\mu T} = C_{\mu s} \frac{\sum_{j=1}^{N} b_j P_j}{1 + \sum_{j=1}^{N} b_j P_j} \tag{5.3-49}$$

Then the adsorbed amount contributed by the component i is

$$C_{\mu i} = C_{\mu T} x_i = C_{\mu s} \frac{b_i P_i}{1 + \sum_{j=1}^{N} b_j P_j} \tag{5.3-50}$$

which is the extended Langmuir isotherm. Thus, the extended Langmuir isotherm is only thermodynamically correct when the saturation capacities of all species are the same.

Example 5.3-1: *Multicomponent Sips equation*

The above conclusion for the Langmuir equation does not readily apply to other isotherms. For example, if the pure component isotherm is described by the Sips equation

$$C_{\mu,i}^0 = C_{\mu s} \frac{(b_i P_i^0)^{1/n}}{\left[1 + (b_i P_i^0)^{1/n}\right]} \qquad (5.3\text{-}51)$$

the multicomponent adsorption isotherm does not necessarily follow the form

$$C_{\mu i} = C_{\mu s} \frac{(b_i P_i)^{1/n}}{1 + \left(\sum_{j=1}^{N} b_j P_j\right)^{1/n}} \qquad (5.3\text{-}52)$$

even though $C_{\mu s}$ and n are the same for all components we shall show this below.

It is not difficult to show from eq. (5.3-31a) and the corresponding equation for Sips isotherm in Table 5.3-3 that the hypothetical pure component pressure is given by

$$P_i^0 = \frac{1}{b_i} \sum_{j=1}^{N} b_j P_j \qquad (5.3\text{-}53)$$

which has the same form as that in the case of Langmuir isotherm (eq. 5.3-47).

From the Raoult'law, we get the mole fraction for the adsorbed phase:

$$x_i = \frac{b_i P_i}{\sum_{j=1}^{N} b_j P_j} \qquad (5.3\text{-}54)$$

Combining eqs. (5.3-51), (5.3-53) and (5.3-54) into eq. (5.3-22a) gives the following solution for the total adsorbed amount.

$$C_{\mu T} = C_{\mu s} \frac{\left(\sum b_k P_k\right)^{1/n}}{1+\left(\sum b_k P_k\right)^{1/n}} \tag{5.3-55}$$

Hence the adsorbed amount contributed by the component "i" is:

$$C_{\mu,i} = C_{\mu s} \frac{b_i P_i \left(\sum b_k P_k\right)^{1/n-1}}{1+\left(\sum b_k P_k\right)^{1/n}} \tag{5.3-56}$$

which is not the same form as one would conjecture in eq. (5.3-52). Further applications of the Sips equation to multicomponent systems are detailed in Rudzinski et al. (1995).

5.3.7 Inverse IAS Algorithm: Specification of $C_{\mu T}$ and x_i:

The last sections show that the IAS as given in eqs. (5.3-20) to (5.3-23) can be used to determine the adsorbed phase compositions and amounts from the information of the gas phase conditions (namely the total pressure and the gas phase mole fractions). In this section, we will show that the same set of equations can be used to solve the inverse problem, that is what are the total pressure and the gas phase mole fractions when the adsorbed phase information ($C_{\mu T}$ and x) are given.

We write eq. (5.3-22a) as follows:

$$F(z) = \left(\sum_{j=1}^{N} \frac{x_j}{C_{\mu j}^0}\right) - \frac{1}{C_{\mu T}} = 0 \tag{5.3-57}$$

The pure component adsorbed amount is a function of P_j^0, but P_j^0 is a function of the reduced spreading pressure, z, according to eq. (5.3-21). Thus, eq. (5.3-57) is a function of the reduced spreading pressure.

Applying the Newton-Raphson formula to eq. (5.3-57), we obtain the following iteration formula for the reduced spreading pressure

$$z^{(k+1)} = z^{(k)} - \left[\frac{F(z)}{F'(z)}\right]_{z=z^{(k)}} \tag{5.3-58}$$

where F(z) is given in eq. (5.3-57), and F'(z) = dF / dz is:

$$F'(z) = -\sum_{j=1}^{N} \frac{x_j P_j^0}{\left(C_{\mu j}^0\right)^3} \frac{dC_{\mu j}^0}{dP_j^0} \tag{5.3-59}$$

The algorithm for solving this inverse problem now is given as follows.

Algorithm 5.3-3

Step #	Action
1	Estimate z
2	Calculate the pure component pressure from eq. (5.3-21) and hence obtain $C_{\mu j}^0$ and $dC_{\mu j}^0 / dP_j^0$. The calculation of the pure component pressure is carried out with the algorithm 5.3-2.
3	Obtain the next estimate for z from eq. (5.3-58). If convergence is satisfied, go to Step 4; else go back to Step 2.
4	4a. Knowing the pure component pressure of Step 2, the total pressure is calculated from eq. (5.3-20) by noting that $\sum y_j = 1$, that is $$P = \sum_{j=1}^{N} P_j^0 x_j \qquad (5.3\text{-}60)$$ 4b. The mole fraction of the gas phase is then calculated from eq. (5.3-20a): $$y_j = \frac{P_j^0 x_j}{P} \qquad (5.3\text{-}61)$$

5.3.8 Numerical Example of the IAS Theory

In this section, we apply the IAS theory to calculate the multicomponent adsorption equilibria using only pure component data. The data are taken from Szepesy and Illes (1963).

Example 5.3-2: *Ethane/ ethylene /activated carbon*
Using the pure component data of ethane and ethylene at 293 K tabulated in Valenzuela and Myers (1989), we fit the data with the Toth equation and the following table lists the optimized parameters.

	Ethane	Ethylene
$C_{\mu s}$ (mmole/g)	8.327	9.823
b (kPa^{-1})	0.06942	0.04426
t (-)	0.3996	0.3716
Residual	0.1366	0.1243

The MatLab code IAS provided with this book is used for the calculation of binary data, and the following table tabulates the reduced spreading pressure, the pure component hypothetical pressure, the adsorbed phase mole fraction, and the adsorbed concentrations for a given gas phase condition (total pressure and gas mole fractions).

P_T	y_1	π/RT	P^0_1	P^0_2	x_1	$C_{\mu1}$	$C_{\mu2}$	$C_{\mu T}$
101	0.115	5.768	73.5	106.5	0.1584	0.4674	2.4835	2.951
101	0.197	5.857	75.8	109.8	0.2621	0.7807	2.1980	2.979
101	0.288	5.973	79.0	114.3	0.3692	1.1133	1.9023	3.016
101	0.388	6.087	82.1	118.8	0.4783	1.4591	1.5917	3.051
101	0.478	6.186	85.0	122.8	0.5696	1.775	1.3262	3.081
101	0.563	6.280	87.7	126.4	0.6500	2.0208	1.0880	3.109
101	0.646	6.371	90.2	130.3	0.7249	2.2723	0.8625	3.135
101	0.708	6.435	92.2	133.0	0.7777	2.4528	0.7010	3.154
101	0.766	6.494	94.0	135.6	0.8253	2.6169	0.5542	3.171
101	0.805	6.533	95.2	137.3	0.8562	2.7248	0.4576	3.182

Figure 5.3-1a shows a plot of the calculated adsorbed mole fraction versus the gas mole fraction of ethane as continuous line, and Figure 5.3-1b shows the total adsorbed concentration versus the gas mole fraction. Experimental data are also included in the figure as symbols. The agreement between the theory and the data is very good due to the similar nature of the two adsorbates used. Figure 5.3-2a presents the reduced spreading pressure vs the gas mole fraction of ethane, and Figure 5.3-2b shows the hypothetical pure component pressures vs the gas mole fraction.

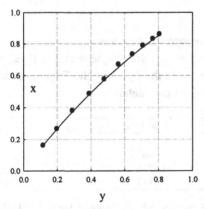

Figure 5.3-1a: Plot of the adsorbed amount versus gas mole fraction

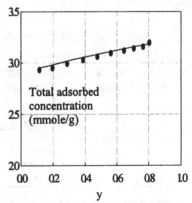

Figure 5.3-1b: Plot of the total adsorbed concentration versus the gas mole fraction

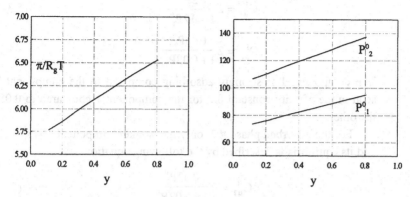

Figure 5.3-2a: Plot of the reduced spreading pressure versus the gas mole fraction

Figure 5.3-2b: Plots of the hypothetical pure component pressure vs gas mole fraction

The IAS theory is a convenient tool to calculate the multicomponent adsorption equilibria, but its predictability is limited, which is mainly due to the assumption of treating the adsorbed phase as one thermodynamic entity. It is this reason that the IAS theory can not predict the azeotropic behaviour commonly encountered in practice, especially systems involving hydrocarbons and carbon oxides in zeolitic adsorbents. One simple way of treating the azeotropic behaviour is to treat the adsorbed phase as a combination of two indendent different adsorbed phases, and the IAS is applied each adsorbed phase. We demonstrate this concept in the following example.

Example 5.3-3: *Azeotropic behaviour*

We consider an adsorption of two components on a solid, and the adsorbed phase is composed of two indendent phases. The adsorption of pure adsorbate 1 occurs only in the adsorbed phase 1, while the adsorption of pure adsorbate 2 occurs in both adsorbed phases.

In the adsorbed phase # 1, the adsorbate 1 is preferentially adsorbed to the other component. The pure component adsorption of these two adsorbates on the adsorbed phase # 1 is described by the following equations:

$$C_{\mu 1}^{I} = 2\frac{0.2P_{1}^{0}}{1+0.2P_{1}^{0}} \tag{5.3-62a}$$

$$C_{\mu 2}^{I} = 2\frac{0.08P_{2}^{0}}{1+0.08P_{2}^{0}} \tag{5.3-62b}$$

The component 1 is strongly adsorbing compared to the component 2 as seen in the affinity constant 0.2 for the component 1 compared to 0.08 for the other.

For the adsorbed phase # 2, only the weaker component 2 is adsorbing and its equilibria is described by the following equation:

$$C_{\mu 2}^{II} = 2\frac{0.02P_{2}^{0}}{1+0.02P_{2}^{0}} \tag{5.3-63}$$

Note that the affinity of this component 2 on the adsorbed phase # 2 is even weaker than its affinity towards the adsorbed phase # 1.

Now to calculate the binary adsorption equilibria for this system with the partial pressures of the two components being P_1 and P_2, we apply the IAS theory to the adsorbed phase # 1. It has been shown in the Section 5.3.6 that if the saturation capacities of the adsorbates are the same, the adsorption of these adsorbates in the binary mixtures can be described by the extended Langmuir equation, that is:

$$C_{\mu 1}^{I} = 2\frac{0.2P_{1}}{1+0.2P_{1}+0.08P_{2}}, \tag{5.3-64a}$$

$$C_{\mu 1}^{I} = 2\frac{0.08P_{2}}{1+0.2P_{1}+0.08P_{2}} \tag{5.3-64b}$$

For the adsorption phase # 2, the adsorption equilibria is just that of the component 2, that is:

$$C_{\mu 2}^{II} = 2\frac{0.02P_{2}}{1+0.02P_{2}} \tag{5.3-65}$$

Thus the total number of mole adsorbed in the adsorbed phase is:

$$C_{\mu T} = 2\frac{0.2P_{1}+0.08P_{2}}{1+0.2P_{1}+0.08P_{2}} + 2\frac{0.02P_{2}}{1+0.02P_{2}} \tag{5.3-66}$$

Therefore the mole fraction of the component 1 in the adsorbed phase is the ratio of eqs.(5.3-64) to (5.3-66), that is:

$$x = \frac{0.2y}{0.2y + 0.08(1-y) + 0.02(1-y)\left[1 + 0.2Py + 0.08P(1-y)\right]/\left[1 + 0.02P(1-y)\right]} \quad (5.3\text{-}67)$$

in which we have used $P_1 = Py$ and $P_2 = P(1-y)$, with P being the total pressure. Figure 5.3-3 shows plots of the adsorbed phase mole fraction versus the gas mole fraction with the total pressure being the varying parameter. At low pressure P=10, the solid preferentially adsorbs the component 1. This is expected because at this low pressure the adsorption mainly occurs in the adsorbed phase # 1, where the component 1 is more favourable. When the pressure increases, the adsorbed phase # 2 begins to contribute to the total adsorption capacity and since only component 2 exists in the adsorbed phase # 2 we start to see the azeotropic behaviour, that is the selectivity switches from the component 1 to the component 2. Any further increase in the pressure will result in the system more preferentially favourable towards the component 2.

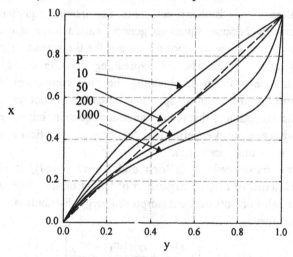

Figure 5.3-3: Plot of the adsorbed phase mole fraction versus gas mole fraction

If the adsorbed phase is treated as one entity instead of two distinct adsorbed phases, and the IAS theory is applied the x-y plot will not exhibit the azeotropic behaviour. Rather it will exhibit the component 1 preferential behaviour at low pressures and the component 2 preferential behaviour at high pressures, and no azeotropic is observed. We shall leave this calculation to the reader.

5.4 Fast IAS Theory

The IAS theory presented in Section 5.3 provides a convenient means to calculate the multicomponent equilibrium. It has a number of attractive features, that is the theory does not require any mixture data, it is independent of the actual model of physical adsorption since the IAS theory is an application of the solution thermodynamics. The form of adsorption isotherm equation for pure component data is arbitrary in the IAS theory; hence any equations that best fit the data can be used. Furthermore, different solutes can have different forms of the isotherm equation.

Despite the versatility of the IAS theory, one disadvantage is the evaluation of the reduced spreading pressure given in eq. (5.3-21), rewritten below for clarity

$$\frac{A\pi}{R_g T} = \int_0^{P_1^0} \frac{C_{\mu 1}}{P_1} dP_1 = \int_0^{P_2^0} \frac{C_{\mu 2}}{P_2} dP_2 = \cdots = \int_0^{P_N^0} \frac{C_{\mu N}}{P_N} dP_N$$

With the exception of few isotherm equations, this integral equation is generally evaluated numerically because equations generally used for best describing pure component data do not allow the above equation for the spreading pressure to be evaluated analytically. Even when the spreading pressure can be analytically evaluated, the inverse problem of obtaining the pure component "hypothetical" pressure in terms of the reduced spreading pressure is not always available analytically. The exception to this is the Langmuir equation, but unfortunately that equation rarely describes well experimental data of many practical systems because it contains only two fitting parameters.

An isotherm, introduced by O'Brien and Myers (1984), is obtained as a truncation to two terms of a series expansion of the adsorption integral equation in terms of the central moments of the adsorption energy distribution. The isotherm equation takes the form:

$$C_\mu = C_{\mu s} \left[\frac{bP}{1+bP} + \frac{\sigma^2 \, bP(1-bP)}{2(1+bP)^3} \right] \qquad (5.4\text{-}1a)$$

where σ^2 is a measure of the width of the adsorption energy distribution, made dimensionless using the thermal energy parameter $R_g T$. This isotherm equation is a three-parameter model, like the Toth and Unilan equations, and is capable of describing many experimental data fairly well. The behaviour of the O'Brien and Myers equation at zero loading is

$$C_\mu = C_{\mu s} b \left(1 + \sigma^2 / 2\right) P \qquad (5.4\text{-}1b)$$

The temperature dependence of the isotherm parameter b is:

$$b = b_\infty \exp\left(\frac{Q}{R_g T}\right) \tag{5.4-1c}$$

The advantage of this model over the Toth and Unilan from the IAS theory point of view is that the reduced spreading pressure can be obtained analytically, which is:

$$z = C_{\mu s}\left[\ln(1+bP) + \frac{\sigma^2 bP}{2(1+bP)^2}\right] \tag{5.4-2}$$

Any isotherm that has explicit expression for the spreading pressure can be used in the FastIAS formalism, which we will describe in details below. Langmuir, O'Brien and Myers, Sips, and dual Langmuir equations fall to this class (Table 5.3-3). Another isotherm, called the statistical model obtained by Ruthven (Ruthven and Goddard, 1986) also belongs to this class:

$$C_\mu = C_{\mu s} \frac{\sum_{j=1}^{M} A_j (bP)^j}{1 + \sum_{j=1}^{M} A_j (bP)^j / j!} \quad \text{with} \quad A_1 = 1 \tag{5.4-3}$$

This model was developed for zeolite, and the parameter M of the model is the number of molecules that can fit into a cavity of the zeolite. The corresponding reduced spreading pressure of the Ruthven model is:

$$z = C_{\mu s} \ln\left[1 + \sum_{j=1}^{M} A_j (bP)^j / j!\right] \tag{5.4-4}$$

5.4.1 Original Fast IAS Procedure

With the isotherm equations, such as the Langmuir, Myers-O'Brien and Ruthven equations, we shall present below the fast IAS theory first proposed by O'Brien and Myers (1985) and later refined by the same authors in 1988. Although the refined version is more efficient from the computational point of view, it is worthwhile to show here the original version of the Fast IAS theory so that readers can see the improvement.

The method essentially makes use of the analytical expression for the reduced spreading pressure in terms of the hypothetical pure component pressures. Since the reduced spreading pressure, z, is the same for all components, we let:

$$z = f_i(\eta_i) \tag{5.4-5a}$$

where

$$\eta_i = b_i P_i^0 \tag{5.4-5b}$$

The functional form f_i takes any form given in the reduced spreading pressure column of Table 5.3-3. It is interesting to note the logarithm dependence of all the isotherm equations in the table. Eq. (5.4-5a) represents (N-1) equations in terms of N unknown $b_i P_i^0$ (i = 1, 2,..., N). We rewrite eq. (5.4-5a) as follows:

$$f_i(\eta_i) = f_{i+1}(\eta_{i+1}) \tag{5.4-6}$$

for i =1, 2, 3,..., N-1.

The fast IAS procedure is to solve firstly for the variable η_i (i = 1, 2,..., N) rather than the reduced spreading pressure in the traditional IAS theory.
Written in terms of the variable η_i, the Raoult law (eq. 5.3-20a) becomes:

$$P_i = x_i \frac{\eta_i}{b_i} \tag{5.4-7}$$

in which we have made use of eq. (5.4-5b).

Since the sum of the adsorbed phase mole fraction is unity, eq.(5.4-7) can be rearranged and summed with respect to all species, leading to the following equation:

$$\sum_{j=1}^{N} \frac{b_i P_i}{\eta_i} = 1 \tag{5.4-8}$$

This equation together with the N-1 equations (eq. 5.4-6) will form a set of N equations in terms of N unknown variables η. Once η are known, the adsorbed phase mole fractions are calculated from the Raoult's law (eq. 5.4-7), the total adsorbed concentration is calculated from eq.(5.3-22a) and the component adsorbed concentration are calculated from eq. (5.3-23).

Both the Fast IAS and IAS theories require the numerical computation for the solution. So what are the difference and the advantage of the FastIAS? The difference is that the Fast IAS involves the solution of N variables of pure component pressure, $b_i P_i^0$, while the solution for the spreading pressure is sought in the IAS theory (see Section 5.3.3). Once the spreading pressure is known in the IAS theory, the hypothetical pure component pressures can be obtained as the inverse of the integral equation as given in eq. (5.3-21). Thus, solving directly for the hypothetical pressures of pure component makes the computation using the Fast IAS faster than the IAS theory.

Putting the set of equations (5.4-6) and (5.4-8) in vector form, we have:

$$\underline{g}(\underline{\eta}) = \underline{0} \qquad (5.4\text{-}9a)$$

where the underline denotes vector of N dimension. The elements of the vector g are:

$$g_i(\underline{\eta}) = f_i(\eta_i) - f_{i+1}(\eta_{i+1}) \qquad \text{for } i = 1, 2, 3, \dots, N\text{-}1 \qquad (5.4\text{-}9b)$$

$$g_N(\underline{\eta}) = \sum_{i=1}^{N} \frac{b_i P_i}{\eta_i} - 1 \qquad (5.4\text{-}9c)$$

Solution for this set of equations (5.4-9) is obtained by the Newton-Raphson method, the algorithm of which is given below.

5.4.1.1 Algorithm for the Original Fast IAS Method:

The iteration formula for the pure component pressures obtained by the application of the Newton-Raphson method to eq. (5.4-9a) is

$$\underline{\eta}^{(k+1)} = \underline{\eta}^{(k)} - \underline{\delta}^{(k)} \qquad (5.4\text{-}10a)$$

where the upperscript k denotes the k-th iteration step, and $\underline{\delta}$ is determined from the following linear equation:

$$\underline{\underline{\Phi}} \cdot \underline{\delta} = \underline{g} \qquad (5.4\text{-}10b)$$

The matrix $\underline{\underline{\Phi}}$ is the Jacobian matrix obtained by differentiating the vector $\underline{g}$ with respect to vector $\underline{\eta}$. It is given by the following equation

$$\Phi_{i,i} = f_i'(\eta_i) \qquad \text{for } i = 1, 2, \dots, N\text{-}1 \qquad (5.4\text{-}11a)$$

$$\Phi_{i,i+1} = -f_{i+1}'(\eta_{i+1}) \qquad \text{for } i = 1, 2, 3, \dots, N\text{-}1 \qquad (5.4\text{-}11b)$$

$$\Phi_{N,i} = -\frac{b_i P_i}{(\eta_i)^2} \qquad \text{for } i = 1, 2, 3, \dots, N \qquad (5.4\text{-}11c)$$

The derivatives of the function f for a number of isotherm equations are given in Table 5.4-1.

Table 5.4-1 Analytical reduced spreading pressures and their derivatives with respect to $\eta=bP$ for a number of adsorption equilibrium models

Isotherm equation	$z = f(\eta)$	$f'(\eta)$
Langmuir $$C_\mu = C_{\mu s}\frac{\eta}{1+\eta}$$	$z = C_{\mu s}\left[\ln(1+\eta)\right]$	$\dfrac{C_{\mu s}}{(1+\eta)}$
O'Brien-Myers $$C_\mu = C_{\mu s}\left[\frac{\eta}{1+\eta} + \frac{\sigma^2\eta(1-\eta)}{2(1+\eta)^3}\right]$$	$z = C_{\mu s}\left[\ln(1+\eta) + \dfrac{\sigma^2\eta}{2(1+\eta)^2}\right]$	$C_{\mu s}\left[\dfrac{1}{1+\eta} + \dfrac{\sigma^2(1-\eta)}{2(1+\eta)^3}\right]$
Sips $$C_\mu = C_{\mu s}\frac{\eta^{1/n}}{1+\eta^{1/n}}$$	$z = nC_{\mu s}\left[\ln\left(1+\eta^{1/n}\right)\right]$	$\dfrac{C_{\mu s}}{(1+\eta^{1/n})\eta^{1-1/n}}$
Dual Langmuir $$C_\mu = C_{\mu s,1}\frac{\eta}{1+\eta} + C_{\mu s,2}\frac{(b_2/b_1)\eta}{1+(b_2/b_1)\eta}$$	$z = C_{\mu s,1}\left[\ln(1+\eta)\right] + C_{\mu s,2}\left[\ln(1+b_2\eta/b_1)\right]$	$\dfrac{C_{\mu s,1}}{(1+\eta)} + \dfrac{C_{\mu s,1}(b_2/b_1)}{(1+b_2\eta/b_1)}$
Ruthven $$C_\mu = C_{\mu s}\frac{\sum_{j=1}^{M}A_j(\eta)^j}{1+\sum_{j=1}^{M}A_j(\eta)^j/j!}$$ $A_1 = 1$	$z = C_{\mu s}\ln\left[1+\sum_{j=1}^{M}A_j(\eta)^j/j!\right]$	$\dfrac{C_{\mu s}\sum_{j=1}^{M}A_j(\eta)^{j-1}/(j-1)!}{1+\sum_{j=1}^{M}A_j(\eta)^j/j!}$

The Jacobian matrix $\underline{\Phi}$ is zero everywhere, except the diagonal terms, the superdiagonal terms and the last row as shown in Figure 5.4-1. Such a matrix can be utilized numerically to speed up the computation.

$$\underline{\underline{\Phi}} = \begin{bmatrix} x & x & 0 & 0 & 0 & \cdots & 0 & 0 \\ 0 & x & x & 0 & 0 & \cdots & 0 & 0 \\ 0 & 0 & x & x & 0 & \cdots & 0 & 0 \\ 0 & 0 & 0 & x & x & \cdots & 0 & 0 \\ \vdots & \vdots & \vdots & \vdots & \ddots & \ddots & \vdots & \vdots \\ 0 & 0 & 0 & 0 & 0 & \cdots & x & 0 \\ 0 & 0 & 0 & 0 & 0 & \cdots & x & x \\ x & x & x & x & x & & x & x \end{bmatrix} \quad (5.4\text{-}12)$$

Figure 5.4-1 Schematics of the Jacobian structure of the original Fast IAS method

This is the essential idea behind the FastIAS theory of Myers and O'Brien. What we shall present next is their modified version of the FastIAS theory.

5.4.2 Modified Fast IAS Procedure

Although the matrix $\underline{\underline{\Phi}}$ shown in eq. (5.4-12) is sparse (that is many of its components are zero), its form could be made more amenable to simpler computation by choosing the proper form of the vector $\underline{g}$. Instead of choosing the vector $\underline{g}$ as defined in eq. (5.4-9), another form could be chosen as follows (O'Brien and Myers, 1988)

$$\underline{g}(\underline{\eta}) = \underline{0} \quad (5.4\text{-}13a)$$

where

$$g_i(\underline{\eta}) = f_i(\eta_i) - f_N(\eta_N) \quad \text{for } i=1, 2, 3,..., N\text{-}1 \quad (5.4\text{-}13b)$$

$$g_N(\underline{\eta}) = \sum_{i=1}^{N} \frac{b_i P_i}{\eta_i} - 1 \quad (5.4\text{-}13c)$$

The difference between this modified method and the original one is the definitions of equations (5.4-9) and (5.4-13). In the modified version, they define g_i as the difference between the function f_i and the function f_N, that is the spreading pressure of the component i is compared with that of the N-th component. While in the original version this function g_i is the difference between f_i and f_{i+1}.

Having defined the function $\underline{g}$ in terms of N pure component hypothetical pressures P_i^0 (or $\eta_i = b_i P_i^0$, eq. (5.4-13) is readily subject to the Newton-Raphson method to solve for the hypothetical pressures. This is done in an algorithm presented below.

5.4.2.1 Algorithm for the Modified Fast IAS Method:

For a set of nonlinear algebraic equations (eq. 5.4-13), the Newton-Raphson method can be applied, and the iteration formula for the pure component pressures at the (k+1)-th iteration is

$$\underline{\eta}^{(k+1)} = \underline{\eta}^{(k)} - \underline{\delta}^{(k)} \qquad (5.4\text{-}14)$$

where the vector $\underline{\delta}$ is determined from the following linear equation:

$$\underline{\underline{\Phi}} \cdot \underline{\delta} = \underline{g} \qquad (5.4\text{-}15a)$$

Here the vector $\underline{g}$ is given in eq. (5.4-13), and $\underline{\underline{\Phi}}$ is the Jacobian matrix (obtained by differentiating eq. (5.4-13) with respect to $\underline{\eta}$ and is given below:

$$\Phi_{i,i} = f_i{'}(\eta_i) \qquad \text{for } i = 1, 2, ..., N-1 \qquad (5.4\text{-}15b)$$

$$\Phi_{i,N} = -f_N{'}(\eta_N) \qquad \text{for } i = 1, 2, 3, ..., N-1 \qquad (5.4\text{-}15c)$$

$$\Phi_{N,i} = -\frac{b_i P_i}{(\eta_i)^2} \qquad \text{for } i = 1, 2, 3, ..., N \qquad (5.4\text{-}15d)$$

In this modified method, the Jacobian matrix $\underline{\underline{\Phi}}$ is zero everywhere, except the diagonal terms, the last column and the last row, as shown in Figure 5.4-2.

$$\underline{\underline{\Phi}} = \begin{bmatrix} x & 0 & 0 & 0 & 0 & \cdots & 0 & x \\ 0 & x & 0 & 0 & 0 & \cdots & 0 & x \\ 0 & 0 & x & 0 & 0 & \cdots & 0 & x \\ 0 & 0 & 0 & x & 0 & \cdots & 0 & x \\ \vdots & \vdots & \vdots & \vdots & \ddots & \ddots & \vdots & \vdots \\ 0 & 0 & 0 & 0 & 0 & \cdots & 0 & x \\ 0 & 0 & 0 & 0 & 0 & \cdots & x & x \\ x & x & x & x & x & x & x & x \end{bmatrix} \qquad (5.4\text{-}16)$$

Figure 5.4-2 Schematics of the Jacobian structure of the modified Fast IAS method

With the peculiar structure of the Jacobian matrix of the form in Figure 5.4-2, it can be reduced to a form such that all the last row of the reduced matrix contains zero elements, except the last element of that row. Such a reduction operation is standard in matrix algebra, and it is done by using the first row to make the first element of the last row to zero, and the second row to make the second element of the last row to zero, and so on. In so doing the last term of the last row is replaced by:

$$\Phi_{N,N}^{new} = \Phi_{N,N}^{old} - \sum_{j=1}^{N-1} \frac{\Phi_{N,j}}{\Phi_{j,j}} \Phi_{j,N} \tag{5.4-17}$$

After this transformation, the reduced Jacobian matrix will then take the form (Figure 5.4-3), which is more convenient for the computation of $\underline{\delta}$

$$\underline{\underline{\Phi}}^{NEW} = \begin{bmatrix} x & 0 & 0 & 0 & 0 & \cdots & 0 & x \\ 0 & x & 0 & 0 & 0 & \cdots & 0 & x \\ 0 & 0 & x & 0 & 0 & \cdots & 0 & x \\ 0 & 0 & 0 & x & 0 & \cdots & 0 & x \\ \vdots & \vdots & \vdots & \vdots & \ddots & \ddots & \vdots & \vdots \\ 0 & 0 & 0 & 0 & 0 & \cdots & 0 & x \\ 0 & 0 & 0 & 0 & 0 & \cdots & x & x \\ 0 & 0 & 0 & 0 & 0 & 0 & 0 & y \end{bmatrix} \tag{5.4-18}$$

Figure 5.4-3 Schematics of the reduced Jacobian structure of the modified Fast IAS method

With the transformation carried out for the last row of the Jacobian matrix $\underline{\underline{\Phi}}$, such transformation also changes the last element of the vector $\underline{g}$ and that element is replaced by

$$g_N^{new} = g_N^{old} - \sum_{j=1}^{N-1} \frac{\Phi_{N,j}}{\Phi_{j,j}} g_j \tag{5.4-19}$$

With the new reduced Jacobian matrix (eq. 5.4-18) and the vector $\underline{g}$, the solution for $\underline{\delta}$ is simply:

$$\delta_N = \frac{g_N^{new}}{\Phi_{N,N}^{new}} \tag{5.4-20a}$$

$$\delta_{N-1} = \frac{g_{N-1} - \Phi_{N-1,N}\delta_N}{\Phi_{N-1,N-1}} \quad (5.4\text{-}20\text{b})$$

$$\delta_{N-2} = \frac{g_{N-2} - \Phi_{N-2,N}\delta_N}{\Phi_{N-2,N-2}} \quad (5.4\text{-}20\text{c})$$

and so on, until the last one

$$\delta_1 = \frac{g_1 - \Phi_{1,N}\delta_N}{\Phi_{1,1}} \quad (5.4\text{-}20\text{d})$$

Having obtained the vector $\underline{\delta}$ the iteration formula (5.4-14) can be executed with some proper initial guess until the solution converges. We will show in the next section how this initial guess can be made.

5.4.3 *The Initial Guess for the Hypothetical Pure Component Pressure*

Like any iteration formula, the rate of convergence depends on a good choice of the initial guess. In this section, a choice of the initial guess is suggested and this is based on the behaviour of the isotherm at low pressures. We illustrate this with the O'Brien and Myers equation.

At low pressures, the O'Brien-Myers equation has the following limit

$$\lim_{P_i \to 0} C_{\mu,i} = C_{\mu s,i} b_i \left(1 + \frac{\sigma_i^2}{2}\right) P_i \quad (5.4\text{-}21)$$

If this is an estimate for the amount adsorbed of the component i, the total adsorbed amount is

$$C_{\mu T} \approx \sum_{i=1}^{N} C_{\mu s i} b_i \left(1 + \frac{\sigma_i^2}{2}\right) P_i \quad (5.4\text{-}22)$$

and the adsorbed phase mole fraction is

$$x_i = \frac{C_{\mu,i}}{C_{\mu T}} \approx \frac{C_{\mu s,i} b_i \left(1 + \frac{\sigma_i^2}{2}\right) P_i}{C_{\mu T}} \quad (5.4\text{-}23)$$

Using this mole fraction in the Raoult law (eq. 5.4-7), we get the following initial estimate for the reduced spreading pressure or η_i

$$\eta_i = \frac{b_i P_i}{x_i} \cong \frac{C_{\mu T}}{C_{\mu si}\left(1 + \dfrac{\sigma_i^{\,2}}{2}\right)} \tag{5.4-24}$$

Thus the initial guess for the reduced spreading pressure is

$$\eta_i^{(0)} = \frac{C_{\mu T}}{C_{\mu si}\left(1 + \dfrac{\sigma_i^{\,2}}{2}\right)} \tag{5.4-25}$$

5.4.4 The Amount Adsorbed

Once the hypothetical pure component pressures are obtained, the mole fractions in the adsorbed phase are obtained from the Raoult law (Eq. 5.4-7) and then the total adsorbed concentration can be calculated from

$$\frac{1}{C_{\mu T}} = \sum_{j=1}^{N} \frac{x_j}{C_{\mu j}^0} \tag{5.4-26}$$

where $C_{\mu j}^0$ is the adsorbed amount of the pure component j at the hypothetical pressure P_j^0, that is

$$C_{\mu j}^0 = f_j^0\!\left(P_j^0\right) \tag{5.4-27}$$

Knowing the total amount adsorbed ($C_{\mu T}$), the amount adsorbed contributed by the component "i" is given by:

$$C_{\mu i} = x_i C_{\mu T} \tag{5.4-28}$$

5.4.5 The FastIAS Algorithm

The algorithm to calculate the adsorbed concentrations can now be summarised below for the case of Myers-O'Brien isotherm equation.

Algorithm 5.4-1:

Step no.	Action
1	1a. Supply the adsorption equilibrium parameters of all components $C_{\mu s,i}$, b_i, and σ_i (i = 1, 2, ···, N)
	1b. Supply the partial pressures of all components in the gas phase P_i (i = 1, 2, ..., N)

2	Calculate the initial guess for the pure component hypothetical pressures $$\eta_i^{(0)} = \frac{C_{\mu T}}{C_{\mu s i}\left(1+\sigma_i^2/2\right)}; \text{ where } C_{\mu T} = \sum_{i=1}^{N} C_{\mu s i} b_i\left(1+\sigma_i^2/2\right)P_i$$
3	For $i = 1, 2, \ldots, N$, calculate f_i, f'_i $$f_i = C_{\mu s,i}\left[\ln(1+\eta_i) + \frac{\sigma_i^2 \eta_i}{2(1+\eta_i)^2}\right]; f'_i = C_{\mu s,i}\left[\frac{1}{(1+\eta_i)} + \frac{\sigma_i^2(1-\eta_i)}{2(1+\eta_i)^3}\right]$$ $$\begin{bmatrix} g_i = f_i - f_N & \text{for } i = 1, 2, \cdots, N-1 \\ g_N = \left(\sum_{j=1}^{N}\frac{b_j P_j}{\eta_j}\right) - 1 \end{bmatrix}$$
4	Calculation of the matrix $\underline{\underline{\Phi}}$ 4a. Initialization $\underline{\underline{\Phi}} = \underline{\underline{0}}$ 4b. Assign values to some elements of the matrix $\underline{\underline{\Phi}}$ as shown in eqs. (5.4-15)
5	Calculation of the modified matrix $\underline{\underline{\Phi}}^{NEW}$ as follows: 5a. Initialization $\underline{\underline{\Phi}}^{NEW} = \underline{\underline{0}}$ 5b. Assign values to some elements of this modified matrix $\Phi_{i,i}^{NEW} = \Phi_{i,i}$ $\Phi_{i,N}^{NEW} = \Phi_{i,N}$ $\Phi_{N,N}^{NEW} = \Phi_{N,N} - \sum_{j=1}^{N-1}\frac{\Phi_{N,j} b_j}{\Phi_{j,j}}$
6	Modify the last element of the vector g with eq. (5.4-19)
7	Solve for $\underline{\delta}$ according to eq.(5.4-20)
8	Obtain the next estimated for $\underline{\eta}$ according to eq. (5.4-14), and calculate the relative error. If the error is acceptable, go to step 9; else go back to step 3
9	Calculate the mole fraction of the adsorbed phase from eq.(5.4-7), the pure component adsorbed concentration from eq. (5.4-1a), and then the total adsorbed concentration from eq.(5.4-26).

A programming code FastIAS is provided with this book, and readers are encouranged to use the code to perform calculation of multicomponent equilibria. We illustrate this with the following example.

Example 5.4-1: *Ethane/Ethylene/Activated carbon*

We use the ethane/ethylene/ activated carbon system of Szepesy and Illes. The adsorption equilibrium data are tabulated in Valenzuela and Myers. Using the ISO_FIT1 program code, we fit the pure component data with the Myers-O'Brien equation, and the following table lists the optimally extracted parameters at 293 K:

	$C_{\mu s}$ (mmole/g)	b(kPa^{-1})	σ^2 (-)
ethane	5.808	0.01476	1.497
ethylene	5.985	0.00980	1.319

For partial pressures of ethane and ethylene of 11.64 and 89.61 kPa, the code FastIAS is used and the following results are obtained:

$$x_{ethane} = 0.162; \; C_{\mu T} = 3.027 \text{ mmol/g}$$

The experimental values for the mole fraction of ethane in the adsorbed phase and the total amount adsorbed are 0.162 and 2.928 mmol/g, respectively. The relative errors of the model predictions are 0.12% and 3.4% for these two quantities.

5.4.6 Other Cases

We have described in the last sections about the original IAS and the modified IAS for the case where the gas phase conditions are specified, that is the total pressure and the gaseous mole fractions are given. Other given conditions such as

1. The total pressure and the adsorbed phase mole fractions
2. The total amount adsorbed, and the adsorbed phase mole fractions

can also be dealt with by the Fast IAS method. More details in terms of initial guesses for those cases can be found in O'Brien and Myers (1988).

5.4.7 Summary

Although the FastIAS is attractive in terms of its utilization of the sparse matrix $\underline{\underline{\Phi}}$ to obtain the set of solution for the hypothetical pressures, the applicability of this theory is restricted to only a few equations which yield analytical expressions for the reduced spreading pressure. For other equations, the IAS theory has to be used instead.

5.5 LeVan and Vermeulen (1981) approach for binary systems

The approach of IAS of Myers and Prausnitz presented in Sections 5.3 and 5.4 is widely used to calculate the multicomponent adsorption isotherm for systems not deviated too far from ideality. For binary systems, the treatment of LeVan and Vermeulen presented below provides a useful solution for the adsorbed phase compositions when the pure component isotherms follow either Langmuir equation or Freundlich equation. These expressions are in the form of series, which converges rapidly. These arise as a result of the analytical expression of the spreading pressure in terms of the gaseous partial pressures and the application of the Gibbs isotherm equation.

The adsorption isotherm of pure component is assumed to take the form of Langmuir equation.

$$C_\mu^0 = C_{\mu s} \frac{bP^0}{1 + bP^0} \qquad (5.5\text{-}1)$$

At constant temperature, the Gibbs adsorption isotherm equation is

$$-A d\pi + \sum C_{\mu i} d\mu_i = 0 \qquad (5.5\text{-}2)$$

This is the Gibbs-Duhem equation for a two-dimensional system with volume and pressure being replaced by area and spreading pressure, respectively. For three dimensional systems where all components experience the same total pressure at equilibrium, all components in the two dimensional systems will experience the same spreading pressure.

At equilibrium, the chemical potential of the adsorbed phase of the species i must be equal to the chemical potential of the fluid phase of the same species:

$$\mu_i = \mu_i^0 + R_g T \ln P_i \qquad (5.5\text{-}3)$$

Substituting the above equation into the Gibbs isotherm equation (5.5-2) for pure component systems we have:

$$\frac{A}{R_g T} d\pi = \frac{C_{\mu i}}{P_i} dP_i \qquad (5.5\text{-}4)$$

When the gas phase pressure is zero, the spreading pressure is zero, and at a pressure P_i^0 the spreading pressure is π. Integration of eq. (5.5-4) subject to these two limits gives:

$$\frac{A\pi}{R_g T} = \int_0^{P_i^0} \frac{C_{\mu i}}{P_i} dP_i \tag{5.5-5}$$

This is the spreading pressure equation in terms of the pressure of the pure component, P_i^0.

For multicomponent systems obeying the ideal adsorption solution theory, the spreading pressure of the adsorbed mixture is π. The partial pressure of the species i in the gas phase is related to the hypothetical pure component pressure which gives the same spreading pressure π as that of the mixture according to the Raoult's law analogy:

$$P_i = x_i P_i^0(\pi) \tag{5.5-6}$$

Since the sum of mole fraction is unity, we then have:

$$\frac{P_1}{P_1^0} + \frac{P_2}{P_2^0} = 1 \tag{5.5-7}$$

This means that from the knowledge of the partial pressures P_1 and P_2 we can solve eq. (5.5-5) and (5.5-7) for the spreading pressure as a function of partial pressures P_1 and P_2. Knowing this, the adsorption isotherm of the species i in the binary mixture can be obtained by applying the Gibbs isotherm eq. (5.5-4), that is:

$$C_{\mu,i} = P_i \frac{\partial (A\pi / R_g T)}{\partial P_i} \tag{5.5-8}$$

This equation is useful to determine the amount adsorbed if the spreading pressure is known explicitly as a function of the partial pressures.

5.5.1 Pure Component Langmuir Isotherm

For Langmuir isotherm of the form (5.5-1), the integration of the spreading pressure equation (5.5-5) gives:

$$z = \frac{A\pi}{R_g T} = C_{\mu si} \ln(1 + b_i P_i^0) \tag{5.5-9}$$

from which we can obtain the pure component pressure P_i^0 in terms of the reduced spreading pressure z:

$$P_i^0 = \frac{1}{b_i}\left[\exp\left(\frac{z}{C_{\mu si}}\right) - 1\right] \tag{5.5-10}$$

Substitute this into eq. (5.5-7), we have:

$$\frac{P_1}{\frac{1}{b_1}\left[\exp\left(\frac{z}{C_{\mu s1}}\right)-1\right]} + \frac{P_2}{\frac{1}{b_2}\left[\exp\left(\frac{z}{C_{\mu s2}}\right)-1\right]} = 1 \qquad (5.5\text{-}11)$$

from which we can solve for the reduced spreading pressure, z, as a function of the gaseous phase pressures, P_1 and P_2. Knowing this function, the adsorption equilibrium isotherm is simply obtained by applying eq. (5.5-8).

Investigation of eq.(5.5-11), we note that if the saturation capacity of the component 1, $C_{\mu 1}$, is different from that of the component 2, $C_{\mu 2}$, an analytical solution for the reduced spreading pressure is not possible. For such a case, we need to resort to the IAS or FastIAS theory. When the two saturation capacities are equal, an analytical solution for the reduced spreading pressure is possible and the analysis for this case will be presented next. Also when the two saturation capacities are very close to each other, a perturbation method can be applied to obtain the asymptotic solution for the reduced spreading pressure and thence the adsorbed concentrations.

5.5.1.1 *Equal Saturation Capacities*

When the maximum saturation capacities of the two components are the same, we can solve for the reduced spreading pressure (from eq. 5.5-11) exactly as follows:

$$z = C_{\mu s}\ln(1+b_1 P_1 + b_2 P_2) \qquad (5.5\text{-}12)$$

This is the spreading pressure of the mixture. Knowing this analytical expression for z, we can determine the adsorption isotherm for each component in the mixture by using the Gibbs equation (5.5-8):

$$C_{\mu,i} = P_i \frac{\partial z}{\partial P_i} = C_{\mu s}\frac{b_i P_i}{1+b_1 P_1 + b_2 P_2} \qquad (5.5\text{-}13)$$

which is the usual looking extended Langmuir isotherm. Thus, the extended Langmuir isotherm is only thermodynamically consistent if the maximum saturation capacities of all components are the same.

Knowing the reduced spreading pressure given as in eq. (5.5-12), the pure component pressure that gives the same spreading pressure as the mixture is (from eq. 5.5-10):

$$P_i^0 = \frac{b_1 P_1 + b_2 P_2}{b_i} \qquad (5.5\text{-}14)$$

that is

$$P_1^0 = P_1 + \frac{b_2}{b_1} P_2 \qquad (5.5\text{-}15)$$

$$P_2^0 = P_2 + \frac{b_1}{b_2} P_1 \qquad (5.5\text{-}16)$$

Thus, one can see that the hypothetical pressure of a pure component which gives the same spreading pressure as that of the mixture is <u>always greater than its partial pressure</u> in the gas phase. If the partial pressures of the two components are of the same order of magnitude, the hypothetical pressure is larger than its partial pressure by a factor which is the ratio of the affinity of the other component to its affinity. This means that under the conditions of comparable partial pressures, the hypothetical pressure of the weaker component is much greater than its partial pressure while the hypothetical pressure of the stronger component is about the same as its partial pressure. The following analysis illustrates this point. If the component 1 is assumed to be stronger adsorbing species than the other component in the sense that

$$b_1 P_1 \gg b_2 P_2 \qquad (5.5\text{-}17)$$

eq. (5.5-15) and (5.5-16) give:

$$P_1^0 \approx P_1 \qquad (5.5\text{-}18)$$

$$P_2^0 \approx \left(\frac{b_1 P_1}{b_2 P_2}\right) P_2 \gg P_2 \qquad (5.5\text{-}19)$$

that is the hypothetical pure component pressure P_1^0 (strong component) is about the same as its partial pressure, while the hypothetical pure component pressure of the weaker component, P_2^0, is much greater than its partial pressure P_2. This means that the weaker component requires a much larger hypothetical pure component pressure than its partial pressure to achieve the same spreading pressure as that of the mixture.

5.5.1.2 *Very close Saturation Capacities: Two Term Expansion*

When the saturation capacities of the two components are very close to each other, eq. (5.5-11) can be expanded using Taylor series to obtain a series solution for the reduced spreading pressure in terms of the partial pressures; hence the component isotherm can be evaluated by substituting it into the Gibbs equation (5.5-8).

When two terms are kept in the series solution, the explicit solution for z in terms of gas phase partial pressures is:

$$z = \overline{C}_{\mu s} \ln(1 + b_1 P_1 + b_2 P_2) \qquad (5.5\text{-}20)$$

where

$$\overline{C}_{\mu s} = \frac{C_{\mu s 1} b_1 P_1 + C_{\mu s 2} b_2 P_2}{b_1 P_1 + b_2 P_2} \qquad (5.5\text{-}21)$$

The multicomponent adsorption isotherm for the component 1 is obtained by the application of the Gibbs equation:

$$C_{\mu 1} = \overline{C}_{\mu s} \frac{b_1 P_1}{1 + b_1 P_1 + b_2 P_2} + \Delta_{L2} \qquad (5.5\text{-}22a)$$

where

$$\Delta_{L2} = (C_{\mu s 1} - C_{\mu s 2}) \frac{b_1 b_2 P_1 P_2}{(b_1 P_1 + b_2 P_2)^2} \ln(1 + b_1 P_1 + b_2 P_2) \qquad (5.5\text{-}22b)$$

The multicomponent adsorption isotherm of the second component is obtained by simply interchanging the subscripts 1 and 2 in the above equation.

5.5.1.3 Very close Saturation Capacities: Three Term Expansion

If three terms are kept in the Taylor series expansion, the explicit expression for the spreading pressure is:

$$z = \overline{C}_{\mu s} \ln(1 + b_1 P_1 + b_2 P_2) \qquad (5.5\text{-}23)$$

where

$$\overline{C}_{\mu s} = \frac{b_1 P_1 C_{\mu s 1} + b_2 P_2 C_{\mu s 2}}{b_1 P_1 + b_2 P_2}$$

$$+ 2 \frac{(C_{\mu s 1} - C_{\mu s 2})^2}{C_{\mu s 1} + C_{\mu s 2}} \frac{b_1 b_2 P_1 P_2}{(b_1 P_1 + b_2 P_2)^2} \left[\left(\frac{1}{b_1 P_1 + b_2 P_2} + \frac{1}{2} \right) \ln(1 + b_1 P_1 + b_2 P_2) - 1 \right]$$

$$(5.5\text{-}24)$$

Next, substitute this expression for the spreading pressure into the Gibbs equation, we obtain:

$$C_{\mu,1} = \overline{C}_{\mu s} \frac{b_1 P_1}{1 + b_1 P_1 + b_2 P_2} + \Delta_{L2}(1 + \Delta_{L3}) \qquad (5.5\text{-}25a)$$

where

$$\Delta_{L3} = \frac{C_{\mu s1} - C_{\mu s2}}{C_{\mu s1} + C_{\mu s2}} \cdot \frac{1}{b_1 P_1 + b_2 P_2} \times \left[\frac{(b_2 P_2)^2 + 2(b_2 P_2) - 4(b_1 P_1) - (b_1 P_1)^2}{b_1 P_1 + b_2 P_2} \ln(1 + b_1 P_1 + b_2 P_2) + \right.$$

$$\left. \frac{3(b_1 P_1)^2 + 4(b_1 P_1) + b_1 b_2 P_1 P_2 - 2(b_2 P_2) - 2(b_2 P_2)^2}{1 + b_1 P_1 + b_2 P_2} \right]$$

$$(5.5\text{-}25b)$$

and Δ_{L2} is defined in eq. (5.5-22b).

5.5.2 Pure component Freundlich Isotherm

If the pure component isotherm takes the form of Freundlich equation:

$$C_{\mu,i} = K_i P^{n_i} \qquad (5.5\text{-}26)$$

the spreading pressure equation written in terms of the partial pressures is:

$$\left(\frac{K_1}{n_1}\right)^{1/n_1} P_1 \exp\left(-\frac{1}{n_1} \ln z\right) + \left(\frac{K_2}{n_2}\right)^{1/n_2} P_2 \exp\left(-\frac{1}{n_2} \ln z\right) = 1 \qquad (5.5\text{-}27)$$

5.5.2.1 Same Freundlich Exponent:

If the Freundlich exponent of the two components are the same ($n_1 = n_2 = n$), the spreading pressure can be obtained analytically (from eq. 5.5-27):

$$z = \left[\left(\frac{K_1}{n}\right)^{1/n} P_1 + \left(\frac{K_2}{n}\right)^{1/n} P_2 \right]^n \qquad (5.5\text{-}28)$$

Using the Gibbs equation, we can obtain the multicomponent isotherm as follows for the first component:

$$C_{\mu,1} = \frac{n \left(\frac{K_1}{n}\right)^{1/n} P_1}{\left[\left(\frac{K_1}{n}\right)^{1/n} P_1 + \left(\frac{K_2}{n}\right)^{1/n} P_2 \right]^{1-n}} \qquad (5.5\text{-}29)$$

For the second component, simply interchange 1 and 2 in the above equation.

5.5.2.2 Unequal Freundlich Exponent

When the two exponents are very close to each other, the spreading pressure equation can be solved for its solution by using the Taylor series expansion with the following small parameter:

$$\varepsilon = \frac{n_1 - n_2}{n_1 + n_2} \qquad (5.5-30)$$

The two term expansion obtained by LeVan and Vermeulen (1981) is:

$$C_{\mu,1} = \frac{\overline{n}\left(\frac{K_1}{n_1}\right)^{1/n_1} P_1}{\left[\left(\frac{K_1}{n_1}\right)^{1/n_1} P_1 + \left(\frac{K_2}{n_2}\right)^{1/n_2} P_2\right]^{1-n}} + \Delta_{F2} \qquad (5.5-31)$$

where

$$\Delta_{F2} = (n_1 - n_2) \frac{\left(\frac{K_1}{n_1}\right)^{1/n_1} P_1 \left(\frac{K_2}{n_2}\right)^{1/n_2} P_2}{\left[\left(\frac{K_1}{n_1}\right)^{1/n_1} P_1 + \left(\frac{K_2}{n_2}\right)^{1/n_2} P_2\right]^{2-n}} \ln\left[\left(\frac{K_1}{n_1}\right)^{1/n_1} P_1 + \left(\frac{K_2}{n_2}\right)^{1/n_2} P_2\right]$$

$$\qquad (5.5-32)$$

$$\overline{n} = \frac{n_1 \left(\frac{K_1}{n_1}\right)^{1/n_1} P_1 + n_2 \left(\frac{K_2}{n_2}\right)^{1/n_2} P_2}{\left(\frac{K_1}{n_1}\right)^{1/n_1} P_1 + \left(\frac{K_2}{n_2}\right)^{1/n_2} P_2} \qquad (5.5-33)$$

For the component 2, simply interchange the subscripts 1 and 2.

5.6 Real Adsorption Solution Theory (RAST)

The ideal adsorption solution theory presented in previous sections provides a useful means to determine the multicomponent adsorption equilibria. The procedure is simple and the method of calculation is also straight forward. The method, unfortunately, only works well when the adsorption systems do not behave too far from ideality. For example, adsorption of the same paraffin hydrocarbon gases on activated carbon can be described well by the IAS theory. However for systems such as hydrocarbon and carbon oxides adsorption onto zeolitic material, the IAS theory is inadequate to describe those systems, in particular in the prediction of azeotropic behaviour sometimes encountered in those systems. This deviation could be attributed to the nonideality of the adsorbed phase. The Raoult's law of eqs. (5.3-20a) is now replaced by

$$Py_i = x_i \gamma_i P_i^0 \tag{5.6-1}$$

where γ_i is the activity coefficient, accounting for the adsorbed phase nonideality. There are numerous equations available for the calculation of the activity coefficient. For example, for binary systems, the following table tabulates some equations for evaluating the activity coefficients.

Table 5.6-1: Equation for activity coefficient

Regular Solution	$\ln \gamma_1 = Ax_2^2$	(5.6-2)
	$\ln \gamma_2 = Ax_1^2$	
Two constant Margules	$\ln \gamma_1 = (2B - A)x_2^2 + 2(A - B)x_2^3$	(5.6-3)
(Glessner & Myers, 1969)	$\ln \gamma_2 = (2A - B)x_1^2 + 2(B - A)x_1^3$	
Wilson (1964)	$\ln \gamma_1 = 1 - \ln(x_1 + x_2 \Lambda_{12}) - \left[\dfrac{x_1}{x_1 + x_2 \Lambda_{12}} + \dfrac{x_2 \Lambda_{21}}{x_1 \Lambda_{21} + x_2} \right]$	
	$\ln \gamma_2 = 1 - \ln(x_2 + x_1 \Lambda_{21}) - \left[\dfrac{x_2}{x_2 + x_1 \Lambda_{21}} + \dfrac{x_1 \Lambda_{12}}{x_2 \Lambda_{12} + x_1} \right]$	(5.6-4)

Among these equations for activity coefficients, the Wilson equation is the widely used correlation and it involves only binary interaction parameters Λ_{12} and Λ_{21}. This equation can be readily extended to multicomponent mixture:

$$\ln \gamma_i = 1 - \ln\left(\sum_j x_j \Lambda_{ij} \right) - \sum_k \frac{x_k \Lambda_{ki}}{\sum_j x_j \Lambda_{kj}} \tag{5.6-5}$$

The activity coefficients so defined satisfy the following restrictions.

(a)   the activity coefficient of the component i must approach unity when the mole fraction of that species approaches 1, that is

$$\lim_{x_i \to 1} \gamma_i = 1 \tag{5.6-6}$$

(b) For low surface coverage, that is ideal condition, the spreading pressure approaches zero and the activity coefficient of all species must approach unity (Talu and Zwiebel, 1986)

$$\lim_{\pi \to 0} \gamma_i = 1 \tag{5.6-7}$$

(c) The thermodynamic consistency test of the Gibbs Duhem equation;

$$\sum x_i d\ln\gamma_i = \left(\frac{1}{C_{\mu T}} - \sum \frac{x_i}{C_{\mu i}^0}\right) d\left(\frac{\pi A}{R_g T}\right) \tag{5.6-8}$$

at constant temperature

Using the condition (c), the total amount adsorbed is calculated from:

$$\frac{1}{C_{\mu T}} = \sum_{i=1}^{n} \frac{x_i}{C_{\mu,i}^0} + \sum_{i=1}^{n} x_i \left(\frac{\partial \ln \gamma_i}{\partial z}\right)_x \tag{5.6-9}$$

where z is the reduced spreading pressure.

Thus the complete set of equations for the RIAST is written below for completeness

$$Py_i = x_i \gamma_i(\underline{x}) P_i^0(z) \tag{5.6-10a}$$

$$\sum_{i=1}^{n} x_i = 1 \tag{5.6-10b}$$

$$z = \frac{A\pi}{R_g T} = \int_0^{P_i^0(z)} \frac{C_{\mu i}^0}{P^0} dP^0 \tag{5.6-10c}$$

$$\frac{1}{C_{\mu T}} = \sum_{i=1}^{n} \frac{x_i}{C_{\mu,i}^0} + \sum_{i=1}^{n} x_i \left(\frac{\partial \ln \gamma_i}{\partial z}\right)_x \tag{5.6-10d}$$

$$C_{\mu,i} = C_{\mu T} x_i \tag{5.6-10e}$$

The real adsorption theory has been used by a number of workers with a good degree of success (Glessner and Myers, 1969; Costa et al., 1981; Talu and Zwiebel, 1986; Chen et al., 1990; Karavias and Myers, 1991; Dunne and Myers, 1994; Yun et al., 1996). The difference between the IAST and the RAST models is the introduction of the activity coefficients in the RAST model. These coefficients can be either calculated from known theoretical correlations or from binary equilibrium data. The activity coefficients are a function of the composition of the adsorbed phase and the spreading pressure. To calculate them, we need the binary experimental data:

(a) the total pressure of the gas phase and its compositions
(b) the compositions of the adsorbed phase
(c) the spreading pressure of the mixture

Knowing these activity coefficients, they are then used in some theoretical models, such as the Wilson equation presented in Table 5.6-1 to derive the binary interaction parameters of that model. After this step is done, eqs. (5.6-10a) to (5.6-10c) represent 2N+1 equations in terms of the following 2N+1 unknowns

N pure component hypothetical pressures
N mole fractions of the adsorbed phase
1 reduced spreading pressure.

Solving this set numerically, we will obtain the above 2N+1 variables. Then the amount adsorbed can be calculated from eqs.(5.6-10d) and (5.6-10e). Applications of the RAST are shown in Yun et al. (1996), Chen et al. (1990), Talu and Zwiebel (1986), Costa et al. (1981) and Glessner and Myers (1969).

5.7 Multi site Occupancy Model of Nitta et al.

In Section 2.4, we presented a model of Nitta et al. to allow for an adsorbing molecule to occupy more than one site on the surface. Localized adsorption is assumed in the Nitta et al.'s model. Using the statistical thermodynamics, Nitta derived the following equation for the description of pure component adsorption equilibria.

$$\ln(nbP) = \ln\theta - n\ln(1-\theta) - \frac{u}{kT}n\theta \qquad (5.7\text{-}1)$$

where n is the number of site occupied by one adsorbate molecule, b is the adsorption affinity, u is the interaction energy among adsorbate molecules and θ is the fractional coverage, written in terms of the adsorbed concentration

$$\theta = C_\mu / C_{\mu s} \qquad (5.7\text{-}2)$$

The saturation concentration $C_{\mu s}$ is related to the active site concentration C_0 as

$$C_{\mu s} = \frac{C_0}{n} \qquad (5.7\text{-}3)$$

When dealing with multicomponent systems, Nitta et al. suggested the following equation to describe the adsorption equilibria of the component "i"

$$\ln(n_i b_i P_i) = \ln\theta_i - n_i \ln\left(1 - \sum_j \theta_j\right) - n_i \sum \theta_j u_{ij} / kT \qquad (5.7\text{-}4)$$

The interaction energy of two different types of molecule can be calculated using the following arithmetic mean:

$$u_{ij} = \frac{u_i + u_j}{2} \qquad (5.7\text{-}5)$$

The Nitta et al.'s equation (5.7-4) works satisfactorily for activated carbon and molecular sieving carbon with adsorbates of similar physical and chemical nature such as lower order paraffin hydrocarbons. This is exemplified with the experimental data of Nakahara et al. (1974) using methane, ethane, propane and n-butane on molecular sieving carbon 5A. The following table shows some typical parameters obtained using the Nitta et al. equation (5.7-1) assuming the interaction energy to be zero.

Table 5.7-1: Parameters of methane, ethane and propane on MSC 5A (Nitta et al., 1984)

	n	b (kPa^{-1}) 5.4°C	20°C	51°C
CH_4	2.81	3.8×10^{-3}	1.78×10^{-3}	-
C_2H_6	3.22	0.199	6.05×10^{-2}	2.54×10^{-2}
C_3H_8	4.17	1.67	0.827	0.363

$C_0 = 10.9$ mmol/g

Using these parameters in eq. (5.7-4) for the prediction of the multicomponent data, the agreement is not excellent but is quite satisfactory.

This approach can be extended to heterogeneous surface. The reader is suggested to read Nitta et al. (1984, 1991) and Nitta and Yamaguchi (1991, 1993) for further details.

5.8 Mobile Adsorption Model of Nitta et al. (1991)

When the adsorbed molecules on a surface are mobile and the free area is governed by the scaled particle theory, Nitta and his co-workers have derived the following adsorption isotherm equations (see Section 2.5)

$$\ln(\phi bP) = \ln\left(\frac{\eta}{1-\eta}\right) + \frac{(3-2\eta)\eta}{(1-\eta)^2} - \frac{u\eta}{kT} \tag{5.8-1}$$

where b is the adsorption affinity, ϕ is the fugacity coefficient, η is the dimensionless surface density, u is the adsorbate-adsorbate interaction parameter defined as

$$u = \frac{2\alpha}{\beta} \tag{5.8-2}$$

The parameters α and β are defined in eqs. (2.5-4b) and (2.5-3), respectively. Applying the above equation to a multicomponent mixture, the following mixing rule is applied

$$\alpha = \sum_i \sum_j x_i x_j \alpha_{ij} \tag{5.8-3}$$

$$\beta = \sum_i x_i \beta_i \tag{5.8-4}$$

$$\alpha_{ij} = (\alpha_{ii}\alpha_{jj})^{1/2}(1-k_{ij}) \tag{5.8-5}$$

where k_{ij} is called the binary interaction parameter. The adsorption isotherm equation is given by

$$\ln(\phi_i b_i y_i P) = \ln \eta_i - \ln(1-\eta) + \frac{\eta}{1-\eta} + \left(\sum_j \frac{\beta_i \eta_j}{\beta_j}\right)\frac{(2-\eta)}{(1-\eta)^2} - \sum_j \frac{u_{ij}\eta_j}{kT} \tag{5.8-6}$$

where P is the total pressure, y_i is the gas phase mole fraction, η_i is the surface density of species i, η is the total surface density, and u_{ij} is calculated from

$$u_{ij} = \frac{2\alpha_{ij}}{\beta_j} \tag{5.8-7}$$

Further details as well as the application of this theory are given in Nitta et al. (1991) and Nitta and Yamaguchi (1992, 1993).

5.9 Potential Theory

The extension of the potential theory was studied by Bering et al. (1963), Doong and Yang (1988) and Mehta and Dannes (1985) to multicomponent systems. We shall present below a brief account of a potential theory put forward by Doong and Yang (1988). The approach is simple in concept, and it results in analytical solution for the multicomponent adsorption isotherm. The basic assumption of their model is that there is no lateral interaction between molecules of different types and pure component isotherm data are described by the DA equation. With this assumption, the parameters of the DA equation (W_0, E_0, n) of each species are not affected by the presence of the other species, but the volume available for each species is reduced. This means that the volume available for the species i is:

$$W_{0,i} - \sum_{\substack{j=1 \\ j \neq i}}^{N} W_j \qquad (5.9\text{-}1)$$

For a given volume "available" to the species "i" as given in the above equation, the volume taken up by that species at a pressure of P_i is simply:

$$W_i = \left(W_{0,i} - \sum_{\substack{j=1 \\ j \neq i}}^{N} W_j \right) \times \exp\left[-\left(\frac{A_i}{\beta E_{0,i}} \right)^{n_i} \right] \qquad (5.9\text{-}2)$$

where A_i is the adsorption potential of the species "i"

$$A_i = R_g T \ln\left(\frac{P_{0,i}}{P_i} \right) \qquad (5.9\text{-}3)$$

We now rewrite eq. (5.9-2) by adding $-W_i \exp\left[-\left(\frac{A_i}{\beta E_{0,i}} \right)^{n_i} \right]$ to both sides of that equation:

$$W_i \left\{ 1 - \exp\left[-\left(\frac{A_i}{\beta E_{0,i}} \right)^{n_i} \right] \right\} = \left(W_{0,i} - \sum_{j=1}^{N} W_j \right) \times \exp\left[-\left(\frac{A_i}{\beta E_{0,i}} \right)^{n_i} \right] \qquad (5.9\text{-}4)$$

Solving for the volume of the micropore occupied by the species "i", we get:

$$W_i = \cfrac{W_{0,i}\exp\left[-\left(\cfrac{A_i}{\beta E_{0,i}}\right)^{n_i}\right]}{\left\{1-\exp\left[-\left(\cfrac{A_i}{\beta E_{0,i}}\right)^{n_i}\right]\right\}} - \cfrac{\exp\left[-\left(\cfrac{A_i}{\beta E_{0,i}}\right)^{n_i}\right] \times \sum_{j=1}^{N} W_j}{\left\{1-\exp\left[-\left(\cfrac{A_i}{\beta E_{0,i}}\right)^{n_i}\right]\right\}} \qquad (5.9\text{-}5)$$

We now note that the LHS is the volume occupied by the species "i" (W_i) and the RHS contains the sum of volumes contributed by all species $\sum_{j=1}^{N} W_j$; thus this sum of volumes can be found by simply summing the above equation with respect to all species and we get:

$$\sum_{j=1}^{N} W_j = \left\{\sum_{i=1}^{N} \cfrac{W_{0,i}\exp\left[-\left(\cfrac{A_i}{\beta E_{0,i}}\right)^{n_i}\right]}{\left\{1-\exp\left[-\left(\cfrac{A_i}{\beta E_{0,i}}\right)^{n_i}\right]\right\}}\right\} \times \left\{1+\sum_{i=1}^{N} \cfrac{\exp\left[-\left(\cfrac{A_i}{\beta E_{0,i}}\right)^{n_i}\right]}{\left\{1-\exp\left[-\left(\cfrac{A_i}{\beta E_{0,i}}\right)^{n_i}\right]\right\}}\right\}^{-1} \qquad (5.9\text{-}6)$$

Knowing this sum of volumes contributed by all species, the volume occupied by the species "i" is given as in eq. (5.9-5). Thence the amount adsorbed is calculated by dividing the volume by the liquid molar volume, assuming the adsorbed state is liquid-like and there is no interaction between species. The following equation can be used to calculated the liquid molar volume if the values are known at the normal boiling point and the critical temperature

$$\begin{array}{ll} v_M = v_{M,nbp} & \text{for } T < T_{nbp} \\ v_M = v_{M,c} - \left(v_{M,c} - v_{M,nbp}\right)\left[(T_c - T)/(T_c - T_{nbp})\right] & \text{for } T_{nbp} < T < T_c \\ v_M = v_{M,c}\left(\cfrac{T}{T_c}\right)^{0.6} & \text{for } T > T_c \end{array} \qquad (5.9\text{-}7)$$

where T_{nbp} and T_c are normal boiling point and critical temperature, respectively. The liquid molar volume at the critical condition can be taken as the van der Waals constant $R_g T_c/8P_c$.

For super-critical gases, the following equation can be used to calculate the "vapor" pressure used in the DA equation

$$P_0 = P_c \times \exp\left[\frac{T_{nbp}}{T_c}\left(\frac{\ln P_c}{1 - T_{nbp}/T_c}\right)\left(1 - \frac{T_c}{T}\right)\right] \qquad (5.9\text{-}8)$$

Eq. (5.9-5) for N = 2 has been tested with Doong and Yang (1988) for a number of adsorption systems. The prediction of this theory is reasonable, and it is claimed to be comparable to that of the IAS theory.

5.10 Other Approaches

Beside the approaches of Langmuir, IAS, RAS, Nitta, potential theory, there are other approaches available in the literature. The vacancy solution method presented in Chapter 2 can be extended to deal with mixtures. Readers are referred to Cochran et al. (1985) for exposition of this method. Other methods such as the Grant-Manes method (1966) and its modification by Mehta et al. (1985), the multi-space adsorption model of Gusev et al. (1996), etc. These methods are not as popular as the IAST and need to be further tested before they would enjoy wide spread popularity as the IAS model.

5.11 Conclusions

Theories for adsorption equilibria in multicomponent systems are not as advanced as those for single component systems. This slow progress in this area has been due to a number of reasons: (i) lack of extensive experimental data for multicomponent systems, (ii) solid surface is too complex to model adequately. However, some good progress has been steadily achieved in this area.

6

Heterogeneous Adsorption Equilibria

6.1 Introduction

Adsorption in practical solids is a very complex process owing to the fact that the solid structure is generally complex and is not so well defined. The complexity of the system is usually associated with the heterogeneity between the solid and the adsorbate concerned. In other words, heterogeneity is not a solid characteristic alone but rather it is a characteristics of the specific solid and adsorbate pair.

The direct evidence of the solid heterogeneity is the decrease of the isosteric heat of adsorption versus loading. This is because sites of highest energy are usually taken up by adsorbate first (unless the highest energy sites are so restricted that the time scale for adsorption on such sites is much longer than that required on lower energy sites) and then sites of lower energy are progressively filled as the gas pressure is increased. A behaviour of constant heat of adsorption versus loading is not necessary, however, to indicate that the solid is homogeneous because this constant heat behaviour could be the result of the combination of the surface heterogeneity and the interaction between adsorbed molecules.

One practical approach in dealing with the problem of heterogeneity is to take some macroscopic thermodynamic quantity and impose on such quantity a statistical attribute (that is a distribution function). Once a local isotherm is chosen, the overall (observed) isotherm can be obtained by averaging it over the distribution of that thermodynamic quantity. Many local isotherms have been used, and among them the Langmuir equation is the most widely used. Topography of adsorption sites is usually taken as the patchwise model, whereby all sites having the same energy are grouped into one patch and there is no interaction between patches.

Other local isotherm equations such as Volmer, Fowler-Guggenheim, Hill-deBoer, DR, DA, and BET have also been used in the literature.

The parameters used as the distributed variable can be one of the following:

1. the interaction energy between the solid and the adsorbate molecule,
2. the micropore size
3. the Henry constant and
4. the free energy of adsorption.

Among these, the interaction energy between the solid and the adsorbate molecule is the commonly used as the distributed variable.

When the micropore size is used as the distributed variable, a relationship between the interaction energy and the micropore size has to be known, and this can be determined from the potential energy theory, or if the local isotherm used is the DR or DA equation, the relationship between the characteristic energy and the micropore size proposed by Dubinin and Stoeckli could be used (see Section 4.5).

The contribution of solid toward heterogeneity is the geometrical and energetical characteristics, such as the micropore size distribution and the functional group distribution (they both give rise to the overall energy distribution which characterizes the interaction between the solid and the adsorbate molecule), while the contribution of the adsorbate molecule is its size, shape and conformation. All these factors will affect the system heterogeneity, which is macroscopically observed in the adsorption isotherm and dynamics. Therefore, by measuring adsorption equilibrium, isosteric heat, and dynamics, one could deduce some information about heterogeneity, which is usually characterized by a so called apparent energy distribution. The inverse problem of determining this energy distribution would depend on the choice of the local adsorption isotherm, the shape of the energy distribution, and the topography of the surface (that is whether it is patchwise or random) as the observed adsorption isotherm is an integral of the local adsorption isotherm over the full energy distribution. Let us address these two factors one by one.

The choice of the local isotherm depends on the nature of the surface. If the fluctuation of energy on the surface is periodic (Figure 6.1-1) and the magnitude of this fluctuation (ΔE) is less than the thermal energy kT of the adsorbate molecule, then we talk of the mobile adsorption. If there is no lateral interaction among the adsorbed molecules, then the Volmer equation is the proper equation to describe the local adsorption; on the other hand, if there is a lateral interaction, the Hill-deBoer equation is the proper choice.

If the energy fluctuation of the solid surface is much larger than the molecular thermal energy kT, and the distance between the troughs (D) is larger than the diameter of the molecule, the active site is considered as localised (Figure 6.1-1) and the Langmuir equation is proper for the case of no lateral adsorbate interaction while the Fowler-Guggenheim equation is the one when there is interaction. If the distance between the troughs is smaller than the adsorbate molecule, then we might have the situation whereby the molecule might adsorb onto more than one active site. The multi-site mechanism proposed by Nitta et al. (1984) could describe this situation well (Section 2.4).

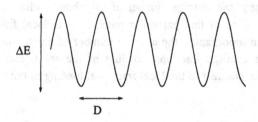

Figure 6.1-1; Fluctuation of energy on the solid surface

The surface topography (that is how surfaces of different energy are arranged between themselves) might also be an important factor in the calculation of the overall adsorption isotherm. These surfaces of different energy can distribute between the two extremes. In one extreme, the solid is composed of patches, wherein all sites of the same energy are grouped together, and there is no interaction between these patches. Here, we talk of the patchwise topography (Figure 6.1-2). The other extreme is the case where surfaces of different energy are randomly distributed. Of course, real solids would have a topography which is somewhere between these two extremes.

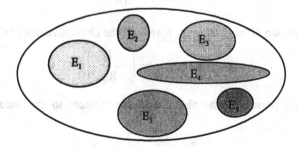

Figure 6.1-2: Schematic diagram of the surface topography composed of patches of sites, of which each patch contains sites of the same energy

If there is no interaction among the adsorbed molecules, then the surface topography is irrelevant, as the adsorption equilibria is simply the direct interaction between the adsorbate molecule and the surface atoms (vertical interaction). This means that the topography is only important when we deal with the case whereby the interaction between adsorbed molecules is important. Adsorption equations such as the Fowler-Guggenheim, the Hill-deBoer, and the Nitta et al. equations are capable of describing the adsorbate-adsorbate interaction. The energy accounting for this interaction depends on the number of neighbouring adsorbed molecules. In the patchwise topography, the average number of neighbour molecules is proportional to the fractional loading of that particular patch (that is local fractional loading), while in the random topography, the average number of neighbouring molecules is proportional to the average fractional loading of the solid, that is the observed fractional loading, in contrast to the local fractional loading in the case of patchwise topography.

6.2 Langmuir Approach

The simplest model describing the heterogeneity of solid surface is that of Langmuir (1918). He assumed that the surface contains several different regions. Each region follows the usual Langmuir assumptions of one molecule adsorbing onto one site, homogeneous surface and localized adsorption. The further assumptions are that there is no interaction between these regions, i.e. they act independently, and within each region there is no interaction between adsorbed molecules.

If there are N such regions, the adsorption equation is simply the summation of all the individual Langmuir equations for each region, that is:

$$C_\mu = \sum_{j=1}^{N} C_{\mu s,j} \frac{b_j P}{1 + b_j P} \qquad (6.2\text{-}1a)$$

where the adsorption affinity constant b_j reflects the characteristics of the patch j;

$$b_j = b_{\infty,j} \exp\left(-\frac{E_j}{R_g T}\right) \qquad (6.2\text{-}1b)$$

For low enough pressures, the above equation reduces to the usual Henry law relation, that is

$$\lim_{P \to 0} C_\mu = \sum_{j=1}^{N} C_{\mu s,j} b_j P = \sum_{j=1}^{N} H_j P = H P \qquad (6.2\text{-}2a)$$

where the overall Henry constant is the sum of all individual Henry constants

$$H = \sum_{j=1}^{N} H_j \qquad (6.2\text{-}2b)$$

If all the patches are very different in terms of energy, then the overall Henry law constant is approximately equal to that of the strongest patch, that is to say at low pressures almost all adsorbed molecules are located in the patch of highest energy.

Eq. (6.2-1) is a special case of a general case of energy distributed model which we will present later. The energy distribution for the multiple Langmuir equation (6.2-1) is:

$$F(E) = \sum_{j=1}^{N} \alpha_j \delta(E - E_j) \qquad (6.2\text{-}3a)$$

where F(E) dE is the fraction of adsorption sites having energy between E and E + dE, and the fraction of the patch j is:

$$\alpha_j = \frac{C_{\mu s,j}}{\sum_{k=1}^{N} C_{\mu s,k}} \qquad (6.2\text{-}3b)$$

When N=2 in eq. (6.2-1) we have a dual Langmuir isotherm occasionally used in the literature. This semi-empirical equation contains 4 parameters, and if the purpose is to fit experimental data the three-parameter models learnt in Chapter 3 such as the Toth equation is a better choice because it contains one less parameter and the four-parameter dual Langmuir equation very often does not provide any better fit than the three-parameter Toth equation.

6.2.1 Isosteric Heat of Adsorption

The isosteric heat for the Langmuir isotherm equation (6.2-1a) can be calculated from the van't Hoff equation:

$$\frac{\Delta H}{R_g T^2} = -\left(\frac{\partial \ln P}{\partial T}\right)_{C_\mu} \qquad (6.2\text{-}4)$$

Taking the total derivative of the Langmuir equation (eq.6.2-1), we have:

$$dC_\mu = \sum_{j=1}^{N} C_{\mu s,j} \frac{b_j dP}{(1+b_j P)^2} + P \sum_{j=1}^{N} C_{\mu s,j} \frac{b'_j dT}{(1+b_j P)^2} \qquad (6.2\text{-}5a)$$

where

$$b'_j = \frac{db_j}{dT} = -b_j \frac{E_j}{R_g T^2} \qquad (6.2\text{-}5b)$$

At constant loading ($dC_\mu = 0$), we can solve for $(\partial \ln P / \partial T)_{C_\mu}$ from eq.(6.2-5):

$$\left(\frac{\partial \ln P}{\partial T}\right)_{C_\mu} = \frac{1}{R_g T^2} \frac{\displaystyle\sum_{j=1}^{N} C_{\mu s,j} \frac{b_j E_j}{(1+b_j P)^2}}{\displaystyle\sum_{j=1}^{N} C_{\mu s,j} \frac{b_j}{(1+b_j P)^2}} \qquad (6.2\text{-}6)$$

Hence, combining eqs. (6.2-4) and (6.2-6) gives the following expression for the isoteric heat of adsorption

$$(-\Delta H) = \frac{\displaystyle\sum_{j=1}^{N} C_{\mu s,j} \frac{b_j E_j}{(1+b_j P)^2}}{\displaystyle\sum_{j=1}^{N} C_{\mu s,j} \frac{b_j}{(1+b_j P)^2}} \qquad (6.2\text{-}7)$$

The above equation shows the variation of the isosteric heat with loading as the pressure P is related to loading via the Langmuir equation (6.2-1).

The equation for the isosteric heat (6.2-7) does not show explicitly the variation with loading but it is possible to derive from such an equation the limits at low and high loadings. At low and high loadings, the isosteric heat (eq. 6.2-7) reduces to:

$$\Delta H_0 = -\frac{\displaystyle\sum_{j=1}^{N} C_{\mu s,j} b_j E_j}{\displaystyle\sum_{j=1}^{N} C_{\mu s,j} b_j} \qquad (6.2\text{-}8)$$

and

$$\Delta H_\infty = -\frac{\displaystyle\sum_{j=1}^{N} C_{\mu s,j} \frac{E_j}{b_j}}{\displaystyle\sum_{j=1}^{N} C_{\mu s,j} \frac{1}{b_j}} \qquad (6.2\text{-}9)$$

respectively. Note that b_j takes the form of eq. (6.2-1b). If patches have widely different energies, the isosteric heats given in eqs. (6.2-8) and (6.2-9) will reduce to:

$$\Delta H_0 \approx E_1 \qquad (6.2\text{-}10)$$

and

$$\Delta H_\infty \approx E_N \qquad (6.2\text{-}11)$$

where E_1 is the energy of interaction of the strongest sites, and E_N is that of the weakest sites. The above equations basically state that the adsorption process proceeds with molecules adsorbing preferentially to sites of highest energy (E_1) and then filling sites of progressingly lower energies. The pattern of isosteric heat variation with loading depends on the magnitudes of $C_{\mu s j}$, b_j and E_j. Figure 6.2-1a shows typical plot of the amount adsorbed versus $b_1 P$ (eq. 6.2-1) for a solid having two patches of energy with patch # 1 having an affinity 10 times that of the patch # 2. The relative contributions of each patch are also shown on the same figure, where it is seen that the weaker sites (patch 2) have a slower rise in adsorption density with pressure than that of the stronger sites (patch 1).

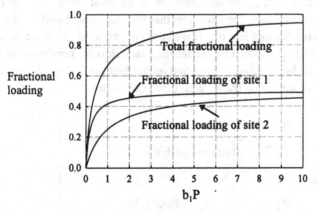

Figure 5.2-1a: Fractional loading of a dual Langmuir isotherm

The isosteric heat (6.2-7) versus loading is shown in Figure 6.2-1b, where the monotonic decrease is the indication of progressing adsorbing onto sites of lower energies. Here we use $E_1 = 40$ kJoule/mole and $E_2 = 20$ kJoule/mole.

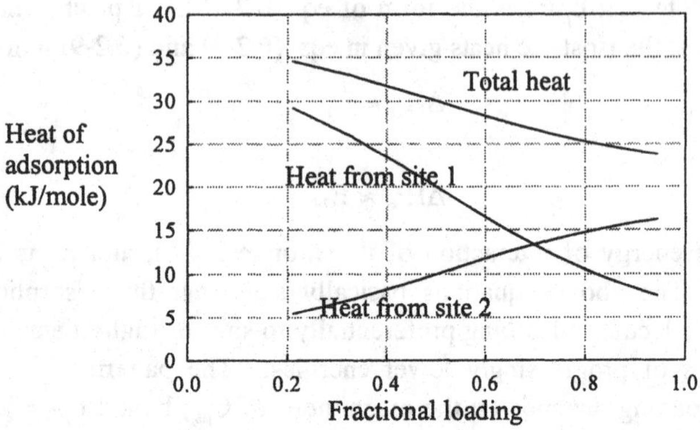

Figure 6.2-1b: Heat of adsorption versus loading (E_1 = 40 kJ/mole and E_2 = 20 kJ/mole)

Figure 6.2-2 shows the heat of adsorption versus fractional loading for a solid having three different patches of different energy. The first patch is the strongest while the third one is the weakest. The affinity of the first patch is ten times that of the second patch, which is also ten times that of the third patch. We use E_1 = 60, E_2 = 40 and E_3 = 20 kJ/mole. Here we note that the heat contributed by the first patch decreases with the loading and that contributed by the third patch increases with the loading. The heat contributed by the second patch shows a maximum as one would expect because at low loadings adsorption on the first patch dominates while at very high loadings adsorption on the weakest patch dominates.

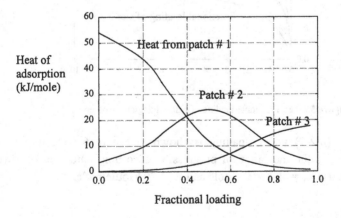

Figure 6.2-2: Heat of adsorption versus loading (E_1 = 60, E_2 = 40 and E_3 = 20 kJ/mole)

6.3 Energy Distribution Approach

Adsorption of molecule in surfaces having constant energy of interaction (one of the assumptions of the Langmuir equation) is very rare in practice as most solids are very heterogeneous. In this section, we will discuss the degree of heterogeneity by assuming that the energy of interaction between the surface and the adsorbing molecule is governed by some distribution. We first discuss the surface topography, the saturation capacity, and then the local isotherm and the form of energy distribution.

6.3.1 Random Topography

The observed adsorption equilibria, expressed as the observed fractional loading θ_{obs} (which is defined as the amount adsorbed at a given pressure and temperature divided by the maximum adsorption capacity), can be in general written in terms of the following integral, with the intergrand being the product of a local adsorption isotherm and an energy distribution:

$$\theta_{obs.} = \int_{\text{all patches}} \theta(P, T; E, u, \theta_{obs}) \, F(E) dE \tag{6.3-1}$$

where θ is the local isotherm (which is the isotherm of a homogenous patch where the interaction energy between that patch and the adsorbing molecule is E), P is the gas pressure, T is the temperature, u is the interaction energy between adsorbed molecules, and F(E) is the energy distribution with F(E)dE being the fraction of surfaces having energy between E and E+dE.

Eq. (6.3-1) is written for the case of <u>random topography</u>, where the interaction between the adsorbed molecules is characterized by the interaction energy u and the average number of neighboring adsorbed molecules, which is proportional to the overall fractional loading (note the variable θ_{obs} in the RHS of eq. 6.3-1). This is because in the case of random topography the concentration of adsorbed molecule around a particular adsorbed molecule is proportional to the overall loading rather than the local loading as is the case of patchwise topography dealt with in the next section.

6.3.2 Patchwise Topography

For patchwise topography where all sites having the same energy are grouped together in one patch and there are no interaction between patches, the extent of interaction between an adsorbed molecule and its neighbours will depend on the local fractional loading of that particular patch. In such a case, the adsorption isotherm equation is:

$$\theta_{obs.} = \int_{\text{all patches}} \theta(P,T;E,u) F(E) dE \qquad (6.3\text{-}2)$$

These equations (6.3-1 and 6.3-2) are used to serve two purposes. If the local isotherm and the energy distribution are known, eqs. (6.3-1) and (6.3-2) can be readily integrated to yield the overall adsorption isotherm. This is a direct problem, but the problem usually facing us is that very often the local isotherm is not known and neither is the energy distribution. Facing with these two unknown functions, one must carry out experiments (preferably at wider range of experimental conditions as possible), and then solve eqs. (6.3-1) and (6.3-2) as an inverse problem for the unknown integrand. Very often, we assume a form for the local isotherm and then solve the inverse problem for the energy distribution. There are a number of fundamental questions about the energy distribution obtained in this manner:

1. what physical meaning does it have?
2. does it vary with temperature?
3. does the form of the distribution vary with the adsorbate used?

Due to the associated experimental error of equilibrium data, there exists a uncertainty of the solution of this inverse problem, and hence attempts have been carried out to determine the energy distribution from a sounder theoretical footing, for example from an independent information of micropore size distribution. We shall address this point later in Section 6.10.

Misra (1970) used the Langmuir equation as the local isotherm (which is also used by many because of the simplicity of the such equation), and for some specific overall isotherms they obtained the energy distribution by using the method of Stieltjes transform to solve the inverse problem.

6.3.3 The Maximum Adsorption Capacity

Once the overall fractional loading is calculated from either eq. (6.3-1) or (6.3-2), the amount adsorbed can be obtained by multiplying it with the maximum adsorbed concentration $C_{\mu s}$. In the case of microporous solids where the adsorption mechanism is by micropore filling (that is the adsorbate inside the micropore volume is assumed to behave like a liquid), it can be obtained from the information of the solid micropore volume and the liquid state of adsorbed molecule. It means that if the micropore volume is V_μ, then the maximum saturation capacity is then given by:

$$C_{\mu s} = \frac{V_\mu}{v_M} \qquad (6.3\text{-}3)$$

where v_M is the liquid molar volume of the adsorbate. The maximum adsorption capacity in terms of moles/g of solid is not the same for all adsorbates as the liquid molar volume is different for different adsorbates. Another factor could contribute to the different maximum adsorbed amount of different species is the size exclusion, that is large molecules are excluded from entering pores having size smaller than the molecular size.

Example 6.3-1: *Langmuir local isotherm & Kronecker energy distribution*

We will first consider the simplest case of the energy distribution approach. The energy distribution takes the following form of Kronecker delta function:

$$F(E) = \sum_{j=1}^{N} \alpha_j \, \delta(E - E_j) \qquad (6.3\text{-}4a)$$

where α_j is the fraction of sites having energy E_j and

$$\sum_{j=1}^{N} \alpha_j = 1 \qquad (6.3\text{-}4b)$$

The local adsorption isotherm takes the form of the Langmuir equation:

$$\theta(E) = \frac{b(E)P}{1 + b(E)P}; \qquad b(E) = b_\infty \exp\left(\frac{E}{R_g T}\right) \qquad (6.3\text{-}5)$$

Using eqs. (6.3-4) and (6.3-5) in either eq. (6.3-1) for random topography or (6.3-2) for patchwise topography, we obtain:

$$\theta_{obs} = \sum_{j=1}^{N} \alpha_j \frac{b_j P}{1 + b_j P}; \qquad b_j = b_\infty \exp\left(\frac{E_j}{R_g T}\right) \qquad (6.3\text{-}6)$$

The reason why either eq. (6.3-1) or (6.3-2) can be used because there is no interaction between the adsorbed molecules; hence the arrangement of sites on the surface is irrelevant.

Eq. (6.3-6) is exactly the form (6.2-1) proposed by Langmuir (1918), and it is considered as the earliest isotherm equation proposed in the literature to address heterogeneous solids.

:::Example 6.3-2: *Langmuir local isotherm & uniform energy distribution*

We now consider next the case of local isotherm of the Langmuir type and a uniform energy distribution. The local fractional loading takes the form of eq. (6.3-5). The energy interaction between adsorbate and solid is assumed uniform, that is equal density of sites of all energy between E_{min} and E_{max}:

$$F(E) = \begin{cases} (E_{max} - E_{min})^{-1} & \text{for } E_{min} < E < E_{max} \\ 0 & \text{elsewhere} \end{cases} \quad (6.3\text{-}7)$$

Since there is no interaction between adsorbed molecules for this case of Langmuir local isotherm, the surface topography is irrelevant. Hence, by substituting eqs.(6.3-5) and (6.3-7) into eq.(6.3-2), we get:

$$\theta_{obs} = \int_{E_{min}}^{E_{max}} \frac{b_\infty \exp(E/R_g T) P}{1 + b_\infty \exp(E/R_g T) P} \frac{1}{(E_{max} - E_{min})} dE \quad (6.3\text{-}8)$$

Integrating this equation gives:

$$\theta_{obs} = \frac{1}{2s} \ln\left(\frac{1 + \bar{b} e^s P}{1 + \bar{b} e^{-s} P}\right) \quad (6.3\text{-}9)$$

where $\bar{E}$ is the mean energy and s is a measure of the energy variation

$$\bar{E} = \frac{E_{min} + E_{max}}{2}; \quad s = \frac{E_{max} - E_{min}}{2 R_g T} \quad (6.3\text{-}10)$$

and $\bar{b}$ is the mean adsorption affinity

$$\bar{b} = b_\infty \exp\left(\frac{\bar{E}}{R_g T}\right) \quad (6.3\text{-}11)$$

Eq. (6.3-9) has two parameters: $\bar{b}$ (mean affinity between the solid and the adsorbate) and s if it is used to correlate the experimental data at one temperature. Fitting one temperature data is not sufficient to determine the minimum and maximum energies, unless b_∞ is known a-priori. However, if the experimental data of more than one temperature are used to extract parameters we can use eq. (6.3-9) to fit those adsorption data simultaneously and if we assume that E_{max} and E_{min} are temperature independent, it is then possible to determine the parameters b_∞, $\bar{E}$, ΔE (= E_{max} - E_{min}) in addition to the maximum adsorbed concentration. Once ΔE and $\bar{E}$ are optimally extracted, the minimum and maximum energies can be readily determined using eq. (6.3-10).

Eq. (6.3-9) is known as the Unilan equation (Uniform energy distribution & Langmuir local isotherm) and is used extensively in the literature for the description of adsorption data of heterogeneous solids such as activated carbon (Valenzuela and Myers, 1989). It gives the proper Henry law constant at low loading, that is

$$\lim_{P \to 0} \theta_{obs} = H\,P \qquad (6.3\text{-}12a)$$

where H is the Henry constant given by

$$H = \bar{b}\left(\frac{\sinh s}{s}\right) \qquad (6.3\text{-}12b)$$

The Henry constant is equal to the affinity constant evaluated at the mean energy ($\bar{b}$) multiplied by a factor (sinh s/s) which characterises the energy variance (or the degree of heterogeneity). The following table shows the magnitude of this factor as a function of s

s	sinhs/s
0	1
1	1.17
2	1.81
3	3.34

It is seen that the higher is the energy variance, the larger is the heterogeneity factor and the Henry constant is contributed more by the higher energy sites. Take the case of the energy variance (s = 3), the Henry constant H is 3 times the affinity $\bar{b}$ evaluated at the mean energy. This

Henry constant is dominated by high energy sites. For comparison, the affinity of the strongest site

$$b(E_{max}) = b_\infty \exp\left(\frac{E_{max}}{R_g T}\right) = \overline{b}\, e^s$$

For this example of s = 3, the affinity of the strongest site is $\overline{b}.e^3 \cong 20\overline{b}$, about 20 times higher than the affinity evaluated at the mean energy.

It must be borne in mind that this model should be treated at best a semi-empirical model for describing the adsorption isotherm data because of the assumption of local Langmuir equation and uniform energy distribution. One would expect in general that the energy distribution is not uniform and the adsorption mechanism may not follow the assumptions of the Langmuir equation. Despite the choice of ideal local adsorption equation (Langmuir) and a simple uniform energy distribution, the resulting Unilan equation fits many data of activated carbon and zeolite reasonably well. This could be attributed to the "cancelling-out" phenomenon, that is the error associated with the choice of the local Langmuir equation cancelled out (or at least partially) the choice of the uniform energy distribution. This, however, should not be treated as a general rule.

6.3.4 Other Local Adsorption Isotherms & Energy Distribution

Recognising that the local adsorption isotherm is not Langmuir and the energy distribution is not uniform, the other forms of local isotherm as well as energy distribution can be used in eq. (6.3-1) or (6.3-2). Such a combination is infinite; however we will list below a number of commonly used local adsorption isotherm equations and energy distributions.

6.3.4.1 The Local Adsorption Isotherm

The local adsorption isotherm equations of the form Langmuir, Volmer, Fowler-Guggenheim and Hill-de Boer have been popularly used in the literature and are shown in the following Table 6.3-1. The first column shows the local adsorption equation in the case of patchwise topography, and the second column shows the corresponding equations in the case of random topography. Other form of the local isotherm can also be used, such as the Nitta equation presented in Chapter 2 allowing for the multisite adsorption.

In Table 6.3-1, the parameter w is defined as $w = zu$, where z is the coordination number and u is the interaction energy between adsorbed molecules. The choice of the local adsorption isotherm equation given in Table 6.3-1 depends on the knowledge of the adsorption mechanism of the system at hand. We list below the mechanism of adsorption behind those equations.

1. <u>Langmuir</u>: localised adsorption and no interaction between molecules
2. <u>Volmer</u>: mobile adsorption and no interaction between molecules
3. <u>Fowler-Guggenheim</u>: localised adsorption and interaction between molecules
4. <u>Hill-deBoer</u>: mobile adsorption and interaction between molecules

Table 6.3-1: Local adsorption isotherm equations for the patchwise and random topography

Local Isotherm	Patchwise topography	Random topography
L	$\theta = \left[1 + \dfrac{K_\infty}{P}\exp\left(-\dfrac{E}{R_g T}\right)\right]^{-1}$; $K_\infty = \dfrac{1}{b_\infty}$	
V	$\theta = \left[1 + \dfrac{K_\infty}{P}\exp\left(\dfrac{\theta}{1-\theta} - \dfrac{E}{R_g T}\right)\right]^{-1}$	
F-G	$\theta = \left[1 + \dfrac{K_\infty}{P}\exp\left(-\dfrac{E + w\theta}{R_g T}\right)\right]^{-1}$	$\theta = \left[1 + \dfrac{K_\infty}{P}\exp\left(-\dfrac{E + w\theta_{obs}}{R_g T}\right)\right]^{-1}$
H-dB	$\theta = \left[1 + \dfrac{K_\infty}{P}\exp\left(\dfrac{\theta}{1-\theta} - \dfrac{E + w\theta}{R_g T}\right)\right]^{-1}$	$\theta = \left[1 + \dfrac{K_\infty}{P}\exp\left(\dfrac{\theta}{1-\theta} - \dfrac{E + w\theta_{obs}}{R_g T}\right)\right]^{-1}$

For the case of no interaction among adsorbed molecules, there is no difference in form between the patchwise and random topographies as we have mentioned earlier that the surface topography is irrelevant for this case (Langmuir and Volmer); however, for the case of interaction the two topographies give rise to different local adsorption isotherm equations. The fractional loading for the interaction factor (appearing in the exponential argument in Table 6.3-1) in the case of patchwise topography is the local value of the patch under consideration (θ), while in the case of random topography, this fractional loading is the observed fractional loading (θ_{obs}). In terms of computation, for the case of patchwise topography, the local fractional loading must be obtained for every value of pressure and the energy of interaction before the integral (6.3-2) can be evaluated. On the other hand, in the case of random topography, the overall fractional loading must be assumed in the local adsorption isotherm equation and the integral (6.3-1) is then evaluated. If the integral is the same as the assumed overall fractional loading, it is the desired solution; on the other hand if it is not a new value for the overall fractional loading must be assumed and the process is repeated until the solution converges.

6.3.4.2 The Energy Distribution

The energy distribution for real solids is largely unknown apriori, and therefore the usual and logical approach is to assume a functional form for the energy distribution, such as the uniform distribution we dealt with in Example 6.3-2. Many distribution functions have been used in the literature, such as

1. Uniform distribution
2. Exponential distribution
3. Gamma distribution
4. Shifted Gamma distribution
5. Normal distribution
6. log-normal distribution
7. Rayleigh distribution

These distributions have the following form:

Distribution	Mean & variance	Eq. #
Uniform $$F(E) = \begin{cases} \dfrac{1}{E_{max} - E_{min}} & E_{min} < E < E_{max} \\ 0 & \text{elsewhere} \end{cases}$$	$\overline{E} = \dfrac{E_{max} + E_{min}}{2}$ $\sigma = \dfrac{E_{max} - E_{min}}{2\sqrt{3}}$	6.3-13
Exponential distribution $$F(E) = \dfrac{1}{E_0} \exp\left(-\dfrac{E}{E_0}\right) \quad E > 0$$	$\overline{E} = E_0$ $\sigma = E_0$	6.3-14
Gamma distribution $$F(E) = \dfrac{q^{n+1}}{\Gamma(n+1)} E^n e^{-qE} \quad E > 0$$	$\overline{E} = (n+1)/q$ $\sigma = \sqrt{n+1}/q$	6.3-15

Shifted Gamma distribution $$F(E) = \frac{q^{n+1}}{\Gamma(n+1)}(E-E_0)^n \cdot \exp[-q(E-E_0)] \quad E > E_0$$	$\overline{E} = E_0 + \dfrac{n+1}{q}$ $\sigma = \dfrac{\sqrt{n+1}}{q}$	6.3-16
Normal distribution $$F(E) = \frac{1}{s\sqrt{2\pi}} \exp\left[-\frac{(E-E_0)^2}{2s^2}\right] \quad -\infty < E < \infty$$	$\overline{E} = E_0$ $\sigma = s$	6.3-17
Log normal distribution $$F(E) = \frac{1}{m\sqrt{2\pi}E}\exp\left\{-\frac{[\ln(E/E_0)]^2}{2m^2}\right\} \quad E > 0$$	$\overline{E} = E_0 \exp\left(\dfrac{m^2}{2}\right)$ $\sigma = E_0\sqrt{e^{2m^2}-e^{m^2}}$	6.3-18
Rayleigh distribution $$F(E) = \frac{(E-E_0)}{(E_1-E_0)^2}\exp\left[-\frac{1}{2}\left(\frac{E-E_0}{E_1-E_0}\right)^2\right] \quad E_0 < E < \infty$$	$\overline{E} = E_0 + \sqrt{\dfrac{\pi}{2}}(E_1 - E_0)$ $\sigma = \sqrt{\dfrac{4-\pi}{2}}(E_1 - E_0)$	6.3-19

Any combination of the above distributions can also be used as the energy distribution in the evaluation of the overall fractional loading.

6.4 Isosteric Heat

With the observed adsorption isotherm given in eq. (6.3-1) or (6.3-2), the isosteric heat can be calculated from the van't Hoff equation:

$$\frac{\Delta H}{R_g T^2} = -\left(\frac{\partial \ln P}{\partial T}\right)_\theta \qquad (6.4\text{-}1)$$

Substitution of the adsorption isotherm equation (6.3-1 or 6.3-2) into the above van't Hoff equation yields the following expression for the isosteric heat

$$\frac{\Delta H}{R_g T^2} = \frac{\int \left(\dfrac{\partial \theta}{\partial T}\right) F(E)\, dE}{P\int \left(\dfrac{\partial \theta}{\partial P}\right) F(E)\, dE} \qquad (6.4\text{-}2)$$

which is usually evaluated numerically, except for a few special cases such as the one dealt with in the following example.

Example 8.4-1: *Isoteric heat of local Langmuir and uniform energy distribution*

We will illustrate in this example the evaluation of the isosteric heat when the local adsorption isotherm takes the form of the Langmuir equation and the energy distribution is uniform. The local Langmuir equation and the distribution are:

$$\theta = \frac{b_\infty \exp\left(\dfrac{E}{R_g T}\right) P}{1 + b_\infty \exp\left(\dfrac{E}{R_g T}\right) P} \tag{6.4-3}$$

$$F(E) = (E_{max} - E_{min})^{-1} \quad \text{for} \quad E_{min} < E < E_{max} \tag{6.4-4}$$

and $F(E)$ is zero elsewhere.

Substituting the above equations into the isosteric heat equation (6.4-2), we obtain:

$$(-\Delta H) = \frac{\int_{E_{min}}^{E_{max}} E \dfrac{bP}{(1+bP)^2} dE}{\int_{E_{min}}^{E_{max}} \dfrac{bP}{(1+bP)^2} dE} \tag{6.4-5}$$

where

$$b = b_\infty \exp\left(\frac{E}{R_g T}\right) \tag{6.4-6}$$

Evaluation of the integral of the above equation (6.4-5) gives the following expression for the isosteric heat as a function of loading

$$(-\Delta H) = \overline{E} + \frac{2(1-\theta)}{\overline{b} P}\left[\frac{\left(e^s + \overline{b}P\right)\left(e^{-s} + \overline{b}P\right)}{\left(e^s - e^{-s}\right)}\right]\frac{\Delta E}{2} - \left(\frac{2 + e^s \overline{b}P + e^{-s}\overline{b}P}{e^s - e^{-s}}\right)\frac{\Delta E}{2\overline{b}P} \tag{6.4-7}$$

where

$$\overline{b} = b_\infty \exp\left(\frac{\overline{E}}{R_g T}\right) \tag{6.4-8a}$$

$$\overline{E} = \frac{E_{max} + E_{min}}{2} \qquad (6.4\text{-}8b)$$

$$\Delta E = E_{max} - E_{min} \qquad (6.4\text{-}8c)$$

The term $\overline{b}P$ in eq. (6.4-7) is related to the overall fractional loading according to eq. (6.3-9), that is

$$\overline{b}P = \frac{e^{s\theta} - e^{-s\theta}}{e^{s(1-\theta)} - e^{-s(1-\theta)}} \qquad (6.4\text{-}9)$$

Eq. (6.4-7) for the isosteric heat is identical to eq. (3.2-27a) which was obtained by applying the van't-Hoff equation directly to the form of the Unilan equation (6.3-9).

Example 8.4-2: *Isosteric heat of multi-modal energy distribution*

Here we take an example of a solid where the local adsorption isotherm takes the form of Langmuir equation (6.4-3) and the energy distribution has the form of multi-modal distribution:

$$F(E) = \sum_{j=1}^{N} \alpha_j \delta(E - E_j)$$

Substituting the local Langmuir adsorption equation and the above distribution into isosteric heat equation (6.4-2), we obtain the following isosteric heat equation:

$$(-\Delta H) = \frac{\sum \alpha_j \frac{b_j E_j}{(1+b_j P)^2}}{\sum \alpha_j \frac{b_j}{(1+b_j P)^2}}$$

At very loading ($\theta \to 0$) the isosteric heat is equal to the heat of the strongest site while at the very high loading it is approaching the heat of the weakest site.

We have seen the example of isosteric heat (eq. 6.4-2) for the case of local Langmuir equation and an uniform distribution. For other choices of local adsorption isotherm and energy distribution, eq. (6.4-2) must be done numerically.

6.5 Brunauer, Love and Keenan Approach

We have presented a general approach of energy distribution in Section 6.3. What we shall present in this section and the next few sections are a number of approaches proposed in the literature. They are presented to show the different ways of treating the heterogeneity.

The approach dealt with in this section is that of Brunauer, Love and Keenan (1942). They assumed that the extent of heterogeneity is due to the solid itself, and is characterized by the variation of the heat of adsorption with the surface coverage. Their analysis is briefly described below.

The surface is divided into surface elements ds. For a given temperature and pressure, each of the surface elements is partly covered and the fraction coverage of each element is governed by the Langmuir equation, with the heat of adsorption varying as a linear function of s, i.e.

$$E = E_0 - \alpha s \tag{6.5-1}$$

where E_0 is the heat of adsorption at zero loading. If the local fractional loading in each surface element ds is θ, described by the Langmuir equation:

$$\theta = \frac{b_\infty \exp(E/R_g T) P}{1 + b_\infty \exp(E/R_g T) P} \tag{6.5-2}$$

then the overall adsorption isotherm is simply the summation of all the fractional loading of each element ds, that is:

$$\theta_{obs} = \int_0^1 \theta \, ds \tag{6.5-3}$$

Evaluating the above integral using eqs. (6.5-1) and (6.5-2), we obtain the following expression for the overall fractional uptake:

$$\theta_{obs} = \frac{R_g T}{\alpha} \ln\left[\frac{1 + a_0 P}{1 + a_0 \exp(-\alpha/R_g T) P}\right] \tag{6.5-4a}$$

where a_0 is the affinity at the energy E_0 (that is at zero loading) and is defined as

$$a_0 = b_\infty \exp\left(\frac{E_0}{R_g T}\right) \tag{6.5-4b}$$

This equation has a correct Henry law limit at low pressure as well as a correct saturation loading at high pressure. This is not unexpected as the overall adsorption isotherm is the average of the Langmuir local isotherm over the whole surface element (eq. 6.5-3). We shall denote eq. (6.5-4) hereafter as the BLK equation. The overall fractional loading is plotted versus a_0P for different values of $\alpha/R_gT = 1, 5, 10$, and those plots are shown in Figure 6.5-1. We see that the isotherm is more favourable when the value of α is small. This is understandable because the parameter α is a measure of the rate of decline of the adsorption energy (eq. 6.5-1). Hence when this parameter is smaller, the overall energy of interaction is higher, resulting a higher overall fractional loading.

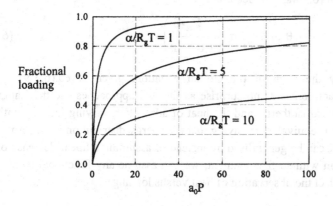

Figure 6.5-1: Plot of the BLK isotherm equation versus a_0P

Under appropriate limit (that is $a_0P \gg 1$ and $a_0P\exp(-\alpha/R_gT) \ll 1$), the above equation reduces to the following Temkin equation:

$$\theta_{obs} = \frac{R_gT}{\alpha}\ln(a_0P) \qquad (6.5-5)$$

6.5.1 BLK Equation versus the Unilan Equation

The isotherm equation of BLK presented above is in fact identical to the Unilan equation, derived in Example 6.3-2 (eq. 6.3-9) using a local Langmuir isotherm and a uniform energy distribution. To prove this, we note from the equation for the heat of adsorption (6.5-1) that the minimum and maximum heats of adsorption corresponding to maximum and minimum coverages are

$$E_{min} = E_0 - \alpha; \qquad E_{max} = E_0 \tag{6.5-5}$$

respectively. Thus, the mean heat of adsorption is:

$$\overline{E} = \frac{E_{min} + E_{max}}{2} = E_0 - \frac{\alpha}{2} \tag{6.5-6}$$

If we now define the mean affinity constant and the energy variance as

$$\overline{b} = b_\infty \exp\left(\frac{\overline{E}}{R_g T}\right); \qquad s = \frac{\alpha}{2 R_g T} = \frac{E_{max} - E_{min}}{2 R_g T} \tag{6.5-7}$$

the BLK equation (6.5-4a) will become:

$$\theta_{obs} = \frac{1}{2s} \ln\left(\frac{1 + \overline{b} \, e^s P}{1 + \overline{b} \, e^{-s} P}\right) \tag{6.5-8}$$

which is simply the form of the Unilan equation, given in eq.(6.3-9). This equivalence in fact comes as no surprise as the two approaches use the Langmuir equation as the local isotherm, and the heat of adsorption varying linearly with the surface coverage is equivalent to saying that the energy distribution is uniform. The approach of BLK can be generalised by instead of assuming a linear decrease of the heat of adsorption with surface element, we can assume any functional form other than linear to reflect the observation of heat versus loading.

6.6 Hobson Approach (1965, 1969)

Using the general approach of the Section 6.3, an overall adsorption isotherm can be calculated after the local isotherm and the distribution of energy are given. The inverse problem, which is usually the more difficult problem, is more relevant in practice as experimental adsorption isotherm is readily measured and the problem is the one to determine the energy distribution if a local adsorption isotherm is assumed. This type of approach is usually numerical by nature. Hobson (1965) proposed an approach whereby the distribution of energy function can be obtained analytically, and this section will briefly present his approach.

To obtain the energy distribution analytically, Hobson has chosen a special local adsorption isotherm, namely a step-like isotherm, to describe the local equilibrium. This step-like isotherm has its fractional loading varying linearly with respect to pressure up to a certain pressure, and then beyond which the fractional loading is equal to unity. This step-like isotherm is:

$$\theta = \begin{cases} \dfrac{p}{K} \exp\left(\dfrac{E}{R_g T}\right) & \text{for } p < P' \\ 1 & \text{for } p > P' \end{cases} \quad (6.6\text{-}1)$$

where K is the inverse of the frequency factor b_∞. Assuming a vibration time of adsorbed molecule as 10^{-12} seconds and a monolayer molecular density of 6.2×10^{14} (molecules of nitrogen/cm^2), Hobson (1965) obtained the following expression for the parameter K

$$K = 1.76 \times 10^4 \sqrt{MT} \; (\text{Torr}) \quad (6.6\text{-}2)$$

In the isotherm equtaion (6.6-1), P' is the threshold pressure above which the surface is saturated, defined as below:

$$P' = K \exp\left[-\dfrac{(E+w)}{R_g T}\right] \quad (6.6\text{-}3)$$

where w is the interaction energy between adsorbed molecules. The local isotherm equation (6.6-1) is shown graphically in Figure 6.6-1. When $w = 0$, there will be no vertical (condensation) segment in the local isotherm, that is the local isotherm is composed of a linear portion and the horizontal line, shown as the dashed line in the figure.

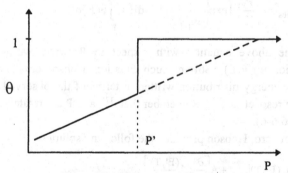

Figure 6.6-1: Hobson's step-like isotherm

The local isotherm proposed by Hobson is an approximation to the Hill-deBoer equation, where there is a sudden jump in the density of the adsorbed phase.

Having chosen the form for the local adsorption isotherm, the overall (observed) isotherm is then given by:

$$\theta_{obs} = \int_0^\infty \theta(P,T,E) \, F(E) \, dE \qquad (6.6\text{-}4)$$

Since this equation is written in terms of the interaction energy, we have to rewrite the local isotherm in terms of energy instead of pressure. From eq. (6.6-1), it is easy to convert it in terms of the energy of interaction, shown below:

$$\theta = \begin{cases} \left(\dfrac{P}{K}\right) \exp\left(\dfrac{E}{R_g T}\right) & \text{for } E < E' \\ 1 & \text{for } E > E' \end{cases} \qquad (6.6\text{-}5)$$

where E' is the threshold energy. Any sites having energy greater than this threshold energy are fully covered with adsorbate. The threshold energy is a function of pressure and is given below:

$$E' = E^0 - w; \qquad E^0 = -R_g T \ln\left(\dfrac{P}{K}\right) \qquad (6.6\text{-}6)$$

The discontinuity in the local isotherm is nicely utilized to determine the solution for the distribution function F(E). This is done as follows. Substituting the local isotherm (eq. 6.6-5) into the observed fractional loading equation (6.6-4), we get:

$$\theta_{obs} = \left(\dfrac{P}{K}\right) \int_0^{E'} \exp\left(\dfrac{E}{R_g T}\right) F(E) \, dE + \int_{E'}^\infty F(E) \, dE \qquad (6.6\text{-}7)$$

Differentiating the above equation with respect to P twice, we will obtain a differential equation for F(E). Solving such equation, Hobson obtained an explicit expression for the energy distribution written in terms of the observed loading and its derivative with respect to E'. Remember that E' and P are related to each other according to eq. (6.6-6).

When w is not zero, Hobson provided the following solution:

$$F(E) = \dfrac{A(1+A)}{R_g T} \theta_{obs}(E',T)\Big|_{E'=E} + A\left[\dfrac{\partial \theta_{obs}(E',T)}{\partial E'}\right]_{E'=E}$$
$$+ \dfrac{A^2(1+A)}{(R_g T)^2} \exp\left(\dfrac{AE}{R_g T}\right) \int_0^E \exp\left(-\dfrac{AE'}{R_g T}\right) \theta_{obs}(E',T) \, dE' - \exp\left(\dfrac{AE}{R_g T}\right) \dfrac{A(1+A)}{R_g T} [\theta_{obs}(E',T)]_{E'=0}$$

$$(6.6\text{-}8)$$

where

$$A = \frac{\exp(w/R_gT)}{1-\exp(w/R_gT)} \qquad (6.6\text{-}9)$$

When w = 0, the solution is

$$F(E') = -\frac{\partial \theta_{obs}}{\partial E'} - R_gT\frac{\partial^2 \theta_{obs}}{\partial (E')^2} \qquad (6.6\text{-}10)$$

If the experimental data are expressed in terms of energy E' by using eq. (6.6-6), the above equation can be used to determine the energy distribution. This requires numerical differentiation of the experimental curve twice, hence accurate data is absolutely critical.

6.7 DR / DA as the Local Isotherm

When the Dubinin equation is used as the local isotherm in eq. (6.3-1) or eq (6.3-2), there are a vast literature on the use of this approach. Like other local isotherm equations, Dubinin equations can be used with many forms of the energy distribution, and for each combination an isotherm equation can be obtained. In Chapter 4, we have discussed the use of Dubinin equations as the local isotherm to describe pure component adsorption in heterogeneous microporous carbon. Readers should refer to Chapter 4 for further exposition of this application.

6.8 Distribution of Henry Constant

An approach of using the Henry constant as the distributed parameter, instead of the traditional use of the interaction energy between the solid and the adsorbate, can also be used (Sircar, 1991). One notes that the Henry constant is related to the interaction energy between the solid and the adsorbate as follows:

$$b = b_\infty \exp\left(\frac{E}{R_gT}\right) \qquad (6.8\text{-}1)$$

implying that a distribution in b means a distribution in the interaction energy.

The local isotherm was assumed to follow the Langmuir equation, and by assuming a patchwise topography the overall isotherm can be readily obtained by averaging the local isotherm over the distribution of the Henry constant. One should note here that since there is no interaction between the adsorbed molecules (assumption of the Langmuir adsorption mechanism), the surface topography is irrelevant. This approach of Sircar yields an isotherm which has a finite slope at zero loading and a finite adsorption heat at zero loading. This is due to the choice of the local Langmuir equation.

The overall isotherm equation is given by the following equation:

$$\theta_{obs} = \int \theta(P,T;b) F(b) \, db \qquad (6.8\text{-}2)$$

where b is the Henry constant and F(b) is its distribution function. The local isotherm equation is:

$$\theta = \frac{bP}{1+bP} \qquad (6.8\text{-}3)$$

Thus, by assuming a form of the distribution for the Henry constant, the above equation can be integrated to give an equation for the observed fractional loading. Let us try the following uniform distribution for the Henry constant

$$F(b) = \begin{cases} \dfrac{1}{b_{max} - b_{min}} & \text{for } b_{min} < b < b_{max} \\ 0 & \text{otherwise} \end{cases} \qquad (6.8\text{-}4)$$

This distribution has the following mean and variance:

$$\bar{b} = \frac{b_{min} + b_{max}}{2}; \qquad \sigma = \frac{b_{max} - b_{min}}{2\sqrt{3}} \qquad (6.8\text{-}5)$$

A large value of σ does not necessarily mean that the system is very heterogeneous. To assess the degree of heterogeneity, we must consider the ratio of the variance to the mean as follows:

$$\psi = \frac{\sqrt{3}\sigma}{\bar{b}} = 1 - \frac{2b_{min}}{b_{max} + b_{min}} \qquad (6.8\text{-}6)$$

The higher is this value, the more heterogeneous is the solid surface. A value of zero means a homogeneous surface, while a value of unity represents the maximum degree of heterogeneity. Maximum heterogeneity is possible when b_{min} is zero or b_{max} is infinity.

By substituting the uniform distribution (6.8-4) into the overall isotherm equation (6.8-2), we obtain the following solution for the overall isotherm:

$$\theta_{obs} = \theta_H \left[1 - \frac{1-\theta_H}{\theta_H} f(z) \right] \qquad (6.8\text{-}7)$$

where the various functions are defined below:

$$\theta_H = \frac{\bar{b}P}{1+\bar{b}P} \tag{6.8-8a}$$

$$f(z) = \frac{1}{2z}\ln\left(\frac{1+z}{1-z}\right) - 1; \qquad z = \psi\theta_H \tag{6.8-8b}$$

The slope of this isotherm at zero loading (or zero pressure) is:

$$\frac{d\theta_{obs}}{dP} = \bar{b} \tag{6.8-9}$$

This states that the mean Henry constant ($\bar{b}$) is the Henry constant of the heterogeneous surface, which is expected because we use the uniform distribution of the Henry constant. This is in contrast to the case when we use the interaction energy as the distributed parameter, and when the distribution function for this energy is a uniform distribution function the Henry constant of the heterogeneous solid is equal to the Henry constant evaluated at the mean energy multiplied by the heterogeneous factor, sinh(s)/s (eq. 6.3-12b), that is the Henry constant of the heterogeneous solid is affected by the stronger energy sites.

6.8.1 The Energy Distribution

If the distribution in terms of the Henry constant is F(b), the corresponding distribution in terms of the interaction energy is evaluated from

$$G(E) = F(b)\frac{db}{dE} \tag{6.8-10}$$

where b is related to the interaction energy in eq. (6.8-1).

If the distribution of the Henry constant takes the form of uniform distribution (eq. 6.8-4), the corresponding distribution for the interaction energy is:

$$G(E) = \frac{\exp(E/R_gT)}{R_gT\left[\exp(E_{max}/R_gT) - \exp(E_{min}/R_gT)\right]} \tag{6.8-11a}$$

where

$$E_{min} = R_gT\ln\left(\frac{b_{min}}{b_\infty}\right); \qquad E_{max} = R_gT\ln\left(\frac{b_{max}}{b_\infty}\right) \tag{6.8-11b}$$

Thus, a uniform distribution in the Henry constant is corresponding to an exponential distribution in the interaction energy.

6.9 Distribution of Free Energy Approach

We have discussed the treatment of a heterogeneous solid by using an integral equation written in terms of a local adsorption isotherm and a distribution of the interaction energy between the solid and the adsorbate (eq. 6.3-1 or 6.3-2). A distribution in Henry constant is also used and was discussed in Section 6.8. In this section we will present a free energy approach developed by Aharoni and Evans (1992).

If the filling of a pore happens at a pressure determined by the free energy of the sorbate in that pore, the overall adsorption isotherm is then determined by the distribution of pore volume with respect to the free energy, rather than the interaction energy between the solid and the adsorbate. The argument for choosing the probability function for the pore volume in terms of free energy put forwarded by Aharoni and Evans stems from the fact that porous structure, when they are made such as activated carbon, is a result of an ensemble of chemical reactions, the extent to which these reactions take place is controlled by chemical potentials. Hence the appropriate distribution for the pore volume is written as follows:

$$V_P(P_a) \tag{6.9-1}$$

where P_a is the characteristic pressure, and $V(P_a)dP_a$ is the pore volume associated with a characteristic pressures between P_a and $P_a + dP_a$. The volume distribution is a function of the characteristic pressure P_a through the free energy. We write

$$V_G(G_a) \tag{6.9-2}$$

where the free energy is related to this characteristic pressure as follows:

$$G_a - G^0 = R_g T \ln\left(\frac{P_a}{P_0}\right) \tag{6.9-3}$$

Here P_0 is the saturation pressure, and G^0 is the free energy at that pressure.

Any distribution function can be used to characterize the pore volume, but we use the following Gaussian distribution as suggested by Aharoni and Evans

$$V_G(G_a) = \frac{1}{n}\exp\left[-b(G_a - G_m)^2\right] \tag{6.9-4}$$

where G_m is the free energy at which the volume distribution is maximum. The parameters of this distribution are b and n.

We shall present this approach in two separate parts. In the first part, we shall study the system of water adsorption onto activated carbon, and in the second part we shall deal with a general adsorption system. The reason for this distinction is that in the case of water adsorption on activated carbon, water does not fill the pore at pressures lower than a threshold pressure. At threshold pressure, the pore is instantaneously filled with water.

6.9.1 Water Adsorption in Activated Carbon

Let us deal with the case of water adsorption on activated carbon first. If the gas phase pressure is P, then all pores having characteristic pressure P_a less than P will be filled at the maximum density, while pores having P_a greater than P will still be empty. This is because the gas phase only have sufficient potential to fill pores with $P_a < P$, and it not strong enough to induce the instant filling of pores having $P_a > P$. Thus for a given pressure P, less than the saturation pressure, the amount of water adsorbed by the solid is:

$$M = \int_0^P D_m V_p(P_a) dP_a \qquad (6.9\text{-}5)$$

where D_m is the maximum density per unit volume. The maximum loading will occur when the gas phase pressure reaches the vapor pressure, that is:

$$M_{max} = \int_0^{P_0} D_m V_p(P_a) dP_a \qquad (6.9\text{-}6)$$

Therefore, the fractional loading at any gas phase pressure P is:

$$\theta = \frac{\int_0^P D_m V_p(P_a) dP_a}{\int_0^{P_0} D_m V_p(P_a) dP_a} \qquad (6.9\text{-}7)$$

The distribution of pore volume in terms of the characteristic pressure P_a is related to the distribution of pore volume in terms of the free energy as follows:

$$V_p(P_a) dP_a = V_G(G_a) dG_a \qquad (6.9\text{-}8)$$

where G_a is related to P_a according to eq.(6.9-3). In terms of the free energy, the fractional loading of eq. (6.9-7) will become:

$$\theta = \frac{\int_{-\infty}^{G} D_m V_G(G_a) dG_a}{\int_{-\infty}^{G^0} D_m V_G(G_a) dG_a} \qquad (6.9\text{-}9)$$

Substituting the Gaussian distribution of the volume distribution function (eq. 6.9-4) into the above equation, we get:

$$\theta = \frac{\int_{-\infty}^{\ln(P/P_m)} \exp(-\beta x^2) dx}{\int_{-\infty}^{\ln(P_0/P_m)} \exp(-\beta x^2) dx} \qquad (6.9\text{-}10)$$

where

$$x = \frac{G_a - G_m}{R_g T}; \quad \beta = b(R_g T)^2 \qquad (6.9\text{-}11)$$

Evaluating eq. (6.9-10) analytically, the fractional loading equation can be written in terms of a closed form Error function as below:

$$\theta = \frac{1 + \mathrm{erf}\left[\sqrt{b}R_g T \ln\left(\frac{P}{P_m}\right)\right]}{1 + \mathrm{erf}\left[\sqrt{b}R_g T \ln\left(\frac{P_0}{P_m}\right)\right]} \qquad (6.9\text{-}12)$$

Thus, the adsorption isotherm equation has three parameters, P_m, b and the maximum adsorption capacity. Plots of this isotherm equation for $P_0/P_m = 2$ are shown in Figure (6.9-1) for three values of $\sqrt{b}R_g T$. The fractional loading curve has an inflexion point (a characteristic of water adsorption onto activated carbon), and the slope at the inflexion point depends on the magnitude of the parameter b. The higher is the value of b, the sharper is the Gaussian distribution (eq. 6.9-4) and hence the sharper is the isotherm curve at the inflexion point. If b is infinitely large, that is the pore volume distribution is a Dirac delta function occurring at $P = P_m$, the adsorption curve will then be an ideal condensation type of which the condensation occurs at $P = P_m$, that is if the pressure is less than P_m, solid is void of any adsorbate molecule until the pressure in the gas phase reaches P_m, at which the solid is instantaneously filled.

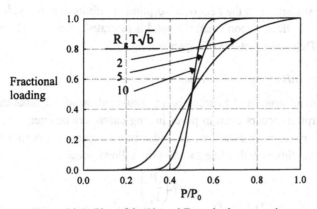

Figure 6.9-1: Plots of the Aharoni-Evans isotherm equation versus P/P_0

6.9.1.1 Slope of Isotherm

The slope of the adsorption isotherm equation (6.9-12) can be obtained by differentiating it with respect to pressure. Investigation the resulting equation we see that the slope of the adsorption equilibrium curve at zero loading is zero, commonly observed for water adsorption onto hydrophobic activated carbon. It increases with loading and finally decreases to zero when the solid is approaching saturation.

To investigate the inflexion point of the isotherm curve, we consider the second derivative and it is easy to show that the inflexion point occurs when:

$$P_{inf} = P_m \cdot \exp\left(-\frac{1}{2b(R_g T)^2}\right) \quad (6.9\text{-}13)$$

Thus the pressure at which the inflexion point occurs is less than the characteristic pressure P_m.

6.9.1.2 Isosteric Heat

To calculate the isosteric heat, we apply the van't Hoff's equation to the adsorption isotherm equation (eq. 6.9-12) and obtain the following solution:

$$(-\Delta H) = R_g T \left\{ \theta \exp\left[-(u_0)^2 + u^2\right] \ln\left(\frac{P_0}{P_m}\right) - \ln\left(\frac{P}{P_m}\right) \right\} + \theta \exp\left[-(u_0)^2 + u^2\right] \Delta H_{vap} \quad (6.9\text{-}14a)$$

where

$$u = \sqrt{\beta} \ln\left(\frac{P}{P_m}\right) = \sqrt{b} R_g T \ln\left(\frac{P}{P_m}\right); \quad u_0 = \sqrt{\beta} \ln\left(\frac{P_0}{P_m}\right) = \sqrt{b} R_g T \ln\left(\frac{P_0}{P_m}\right) \quad (6.9\text{-}14b)$$

This equation suggests that the heat of adsorption is infinite at zero loading and approaches the heat of liquefaction when the solid is saturated. Similar behaviour is observed with the Dubinin equations (Chapter 4).

6.9.2 Hydrocarbon Adsorption in Activated Carbon

Now we turn to adsorption of hydrocarbon in activated carbon. Unlike the case of water, the adsorption occurs even in pores having characteristic energy P_a lower than the gas phase pressure P; however, this adsorption occurs as a partial filling, and if we assume that this partial filling satisfies the following equation:

$$D(P) = D_m \frac{K\left(\frac{P}{P_a}\right)}{1 + (K-1)\left(\frac{P}{P_a}\right)} \qquad (6.9\text{-}15)$$

where D_m is the maximum density and K represents the degree of partial filling. If K = 0, this case will reduce to the case of water adsorption in activated carbon dealt with earlier, and if K = 1 we have the linear partial filling.

This partial density equation can be derived based on the following kinetic argument. For a given micropore having a characteristic pressure of P_a, the rate of adsorption into this pore is given by:

$$R_{ads} = k_a \left(\frac{P}{P_a}\right)(1-\theta) \qquad (6.9\text{-}16)$$

where P/P_a represents the driving force for adsorption, that is for pore of higher characteristic pressure P_a the gas phase pressure needs to be higher to fill that pore to a given extent, k_a is the rate constant for adsorption, and θ is the fractional coverage of that pore.

The rate of desorption is assumed to be proportional to the fractional coverage, and proportional to $(1-P/P_a)$, that is when the gas phase reaches the value of P_a, the gas phase attains sufficient force to stop adsorbed molecules from desorbing; thus, the rate of desorption is:

$$R_{des} = k_d \theta \left(1 - \frac{P}{P_a}\right) \qquad (6.9\text{-}17)$$

Equating these two rates of adsorption and desorption, we get the following solution for the fractional loading

$$\theta = \frac{K(P/P_a)}{1+(K-1)(P/P_a)} \qquad (6.9\text{-}18)$$

which is the same form as that postulated in eq.(6.9-15).

For pores having characteristic pressure P_a less than P, the pores are filled with adsorbate at maximum density, while pores having characteristic energy P_a greater than P will have partial filling of which the density is given in eq. (6.6-15). Thus, the adsorption isotherm equation is:

$$\theta = \frac{\int_0^P D_m V_p(P_a)dP_a + \int_P^{P_0} D(P)V_p(P_a)dP_a}{\int_0^{P_0} D_m V_p(P_a)dP_a} \qquad (6.9\text{-}19)$$

Defining x and β as in eq. (6.9-11), the fractional loading equation of eq. (6.9-19) is:

$$\theta = \frac{\frac{1}{2}\sqrt{\frac{\pi}{\beta}}\left[1+\mathrm{erf}\left(\sqrt{\beta}\ln\frac{P}{P_m}\right)\right] + K\int_{\ln(P/P_m)}^{\ln(P_0/P_m)} \frac{\exp(-\beta x^2)dx}{(K-1)+(P_m/P)\exp(x)}}{\frac{1}{2}\sqrt{\frac{\pi}{\beta}}\left[1+\mathrm{erf}\left(\sqrt{\beta}\ln\frac{P_0}{P_m}\right)\right]} \qquad (6.9\text{-}20)$$

Special case

When $K = 0$, the above equation reduces to the isotherm equation applicable for systems such as water adsorption onto activated carbon (eq. 6.9-12).

When $K = 1$ (linear partial filling), the integral in the numerator (eq. 6.9-20) can be integrated analytically, and the result is:

$$\theta = \frac{\left[1+\mathrm{erf}\left(\sqrt{\beta}\ln\frac{P}{P_m}\right)\right] + \exp\left(\frac{1}{4\beta}\right)\left(\frac{P}{P_m}\right)\left[\mathrm{erf}\left(\frac{1}{2\sqrt{\beta}}+\sqrt{\beta}\ln\frac{P_0}{P_m}\right) - \mathrm{erf}\left(\frac{1}{2\sqrt{\beta}}+\sqrt{\beta}\ln\frac{P}{P_m}\right)\right]}{\left[1+\mathrm{erf}\left(\sqrt{\beta}\ln\frac{P_0}{P_m}\right)\right]}$$

$$(6.9\text{-}21)$$

The adsorption isotherm of eq. (6.9-21) is complicated than many empirical or semi-empirical isotherm equations dealt with in Chapter 3. Because of its limited testing against experimental data, eq. (6.9-21) does not receive much applications.

6.10 Relationship between Slit Shape Micropore and Adsorption Energy

The energy distribution approach presented in the last few sections provides a useful means to describe the adsorption isotherm of heterogeneous solids. But the fundamental question still remains in that how does this energy distribution relates to the intrinsic parameters of the system (solid + adsorbate). In dealing with adsorption of some adsorbates in microporous solids where the mechanism of adsorption is resulted from the enhancement of the dispersive force, we can relate this interaction energy with the intrinsic parameter of the solid (the micropore size) and the molecular properties of the adsorbate. This approach was first introduced by Everett and Powl (1976). In this section, we will present their analysis for micropores having slit geometry. The cylindrical geometry micropores will be dealt with in Section 6.12.

6.10.1 Two Atoms or Molecules Interaction

The basic equation of calculating the potential energy of interaction between two atoms or molecules of the same type "k" is the empirical Lennard-Jones 12-6 potential equation.

$$\varphi_{kk} = 4\varepsilon_{kk}\left[\left(\frac{\sigma_{kk}}{r}\right)^{12} - \left(\frac{\sigma_{kk}}{r}\right)^{6}\right] \quad (6.10\text{-}1a)$$

where r is the distance between the nuclei of the two atoms or molecules, ε_{kk} is the depth of the potential energy minimum and σ_{kk} is the distance at which φ_{kk} is zero. The distance σ_{kk} is called the characteristic diameter or the collision diameter. The minimum of the potential occurs at $2^{1/6}\sigma_{kk}(=1.122\sigma_{kk})$.

The 12-6 potential energy equation exhibits a weak attraction at large separation (proportional to r^{-6}) and strong repulsion at small separation (proportional to r^{-12}), that is the potential energy has a very sharp decrease when the intermolecular distance r increases from 0 to $2^{1/6}\sigma_{kk}$, at which the potential energy is minimum. A further increase in r will result in an increase in the potential energy at a rate much slower than that observed when the repulsion force is operative.

Eq. (6.10-1a) is the potential energy for the two atoms or molecules of the same type. For two atoms or molecules of different type (say type 1 and type 2), the relevant 12-6 potential energy equation is:

$$\varphi_{12} = 4\varepsilon_{12}^{*}\left[\left(\frac{\sigma_{12}}{r}\right)^{12} - \left(\frac{\sigma_{12}}{r}\right)^{6}\right] \quad (6.10\text{-}1b)$$

where σ_{12} and ε_{12} are calculated using the Lorentz-Betherlot rule

$$\sigma_{12} = (\sigma_{11} + \sigma_{22})/2 \qquad (6.10\text{-}1c)$$

$$\varepsilon_{12}^* = \sqrt{\varepsilon_{11}\varepsilon_{22}} \qquad (6.10\text{-}1d)$$

Figure 6.10-1 shows a plot of the reduced potential energy $\varphi_{12}/\varepsilon_{12}^*$ versus the reduced distance r/σ_{12}. Here we see the weak attraction for large separations and the strong repulsion for short separations.

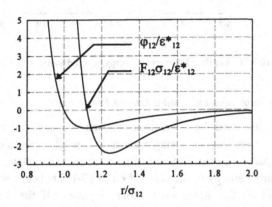

Figure 6.10-1: Plot of the 12-6 potential and force versus distance between two nuclei

The force of interaction is the change of the potential energy φ_{12} with respect to the distance. It is given by the following equation:

$$F_{12} = -\frac{d\varphi_{12}}{dr} = \frac{24\varepsilon_{12}^*}{r}\left[2\left(\frac{\sigma_{12}}{r}\right)^{12} - \left(\frac{\sigma_{12}}{r}\right)^{6}\right] \qquad (6.10\text{-}2)$$

A plot of this force versus the reduced distance (r/σ_{12}) is shown in Figure 6.10-1. This force is zero when the distance between the two atoms or molecules is equal to 1.122 times the collision diameter. It is positive when $r < 1.122\ \sigma_{12}$, implying repulsion, and it is negative when $r > 1.122\sigma_{12}$, suggesting attraction. At a distance r about three times the collision diameter, the force becomes negligible.

To show the effect of the collision diameter on the behaviour of the potential, we show on Figure 6.10-2 plots of the reduced potential energy versus r/σ_{12}°, where σ_{12}° is some reference collision diameter, and the parameter of the plot is $\sigma_{12}/\sigma_{12}^\circ = 0.95, 1, 1.1$. The smaller is the collision diameter, the closer is the equilibrium position between the two atoms or molecules.

284   Equilibria

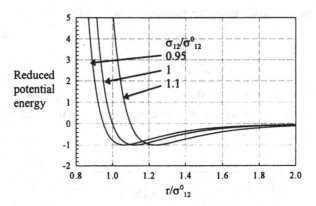

Figure 6.10-2: Plot of the 12-6 potential energy versus distance with $\sigma_{12}/\sigma^0_{12}$ as parameter

6.10.2 An Atom or Molecule and a Lattice Plane

We have discussed the interaction between two atoms or two molecules. We now turn to the problem of a single atom or a single molecule interacting with a single lattice plane. To calculate the interaction for this system, we need to sum the pairwise potential energy of the form of eq.(6.10-1b) between the molecule 1 in the gas phase and all the atoms on the lattice plane. If the distance between the molecule and the lattice plane is much larger than the distance between two adjacent lattice atoms, we can replace the summation by the integration, for which the process will yield analytical expression if the extent of the lattice plane is infinite. As a result, the potential energy will be written in terms of the shortest distance between the center of the atom or molecule and the center of the lattice plane (Figure 6.10-3).

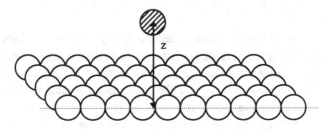

Figure 6.10-3: Interaction between a species and a single lattice plane

Let n be the number of the interacting centers per unit area of the lattice plane, the analytical expression for the potential energy between a single atom or molecule and a lattice plane is (Appendix 6.1):

$$\varphi_{1,SLP} = 4\pi n \varepsilon^*_{12} \sigma^2_{12} \left[\frac{1}{5} \left(\frac{\sigma_{12}}{z} \right)^{10} - \frac{1}{2} \left(\frac{\sigma_{12}}{z} \right)^{4} \right] \tag{6.10-3}$$

where SLP is the acronym for the single lattice plane. Note the exponents 10 and 4 in the above equation, and this is hereafter called the 10-4 potential.

The 10-4 potential energy equation has a minimum and its depth is given by:

$$\varepsilon^*_{1,SLP} = \frac{6}{5} \pi n \varepsilon^*_{12} \sigma^2_{12} \tag{6.10-4}$$

and this minimum occurs at a distance $z^* = \sigma_{12}$, which is shorter than the distance for minimum potential to occur between the two single atoms or molecules (1.122 σ_{12}). This is due to the enhancement of attraction between the atom or molecule and all the atoms on the lattice plane.

The potential energy equation, when written in terms of the minimum potential energy (eq. 6.10-4), is:

$$\varphi_{1,SLP} = \frac{10}{3} \varepsilon^*_{1,SLP} \left[\frac{1}{5} \left(\frac{\sigma_{12}}{z} \right)^{10} - \frac{1}{2} \left(\frac{\sigma_{12}}{z} \right)^{4} \right] \tag{6.10-5}$$

A plot of this 10-4 potential energy equation is shown in Figure 6.10-4a.

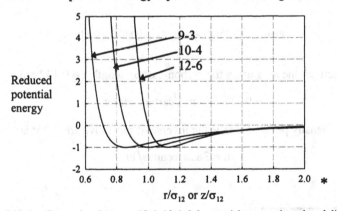

Figure 6.10-4a: Comparison between 12-6, 10-4, 9-3 potentials versus the reduced distance

It has a similar shape to that in the 12-6 potential between two single atoms or molecules (also plotted in the same figure for comparison). The difference between the case of a single atom and a lattice and the case of two single atoms is that the minimum for the former case occurs at a distance shorter than that in the case of two single atoms or molecules.

6.10.2.1 Number Density per Unit Area 'n':

Compared to the 12-6 potential, the additional parameter that the 10.4 potential has is the number density per unit area (n) of the lattice layer. In this section, we will calculate this number density per unit area for the case of graphite lattice layer. Graphite lattice layer is composed of carbon atoms arranged in the hexagonal form with the distance between two carbon centers being a = 0.142nm. The area occupied by one carbon center has a triangular shape as shown in Figure 6.10-4b, and this area is:

$$A = (a + a.\sin 30)(a \cos 30)$$

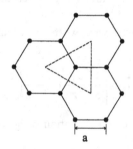

Figure 6.10-4b: Diagram of a graphitic lattice layer

Thus we calculate the area using the carbon-carbon length a = 0.142 nm

$$A = 0.026194 \text{ nm}^2/\text{carbon center}$$

The number density per unit area is the inverse of the above area, that is

$$n = 38.2 \text{ centers}/\text{nm}^2$$

Example 6.10.1: *Minimum potential energy of krypton and a graphite layers*

This example shows a calculation of the minimum potential energy of krypton and a graphite layer. We denote 1 for krypton and 2 for carbon.

The collision diameter and the energy ε for pure component are (Bird et. al., 1960):

$$\sigma_{11} = 0.3498 \text{ nm}$$

$$\sigma_{22} = 0.34 \text{ nm}$$

$$\frac{\varepsilon_{11}^*}{k} = 225 \text{ K}$$

$$\frac{\varepsilon_{22}^*}{k} = 28 \text{ K}$$

Using the Lorentz-Berthelot rule, we calculate

$$\sigma_{12} = \frac{\sigma_{11} + \sigma_{22}}{2} = \frac{0.3498 + 0.34}{2} = 0.3449 \text{ nm}$$

$$\frac{\varepsilon_{12}^*}{k} = \sqrt{\left(\frac{\varepsilon_{11}}{k}\right)\left(\frac{\varepsilon_{22}}{k}\right)} = \sqrt{(225)(28)} = 79.37 \text{ K}.$$

Having the Boltzmann constant $k = 1.38 \times 10^{-23}$ Joule/molecules/K, the characteristic energy ε_{12} is calculated as:

$$\varepsilon_{12}^* = 1.096 \times 10^{-21} \frac{\text{Joule}}{\text{molecule}}$$

For graphite layer, the number density per unit area $n = 38.2$ centers/nm^2, the minimum potential energy between a krypton atom and the graphite layer is

$$\varepsilon_{1,SLP}^* = \frac{6}{5}\pi n \varepsilon_{12}^* \sigma_{12}^2$$

$$= \frac{6}{5}\pi (38.2)(1.096 \times 10^{-21})(0.3449)^2$$

$$\varepsilon_{1,SLP}^* = 1.88 \times 10^{-20} \frac{\text{Joule}}{\text{molecule}}$$

6.10.3 An Atom or Molecule and a Slab

We have obtained the useful equation for the potential energy between the single atom or molecule and the lattice plane. Now we turn to the case whereby the solid is composed of stacked lattice layers, which we hereafter call it slab (Figure

6.10-5). The interaction between a single atom or molecule and a slab is calculated by integrating the interaction between the species and the single lattice (eq. 6.10-3) over the thickness of the slab.

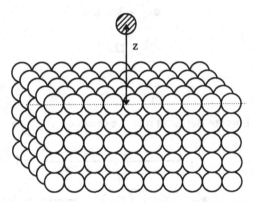

Figure 6.10-5 Distance between a species and a single slab

The result of the integration gives the following the potential equation:

$$\varphi_{1,S} = \frac{2}{3}\pi n' \varepsilon_{12}^* \sigma_{12}^3 \left[\frac{2}{15}\left(\frac{\sigma_{12}}{z}\right)^9 - \left(\frac{\sigma_{12}}{z}\right)^3 \right] \qquad (6.10\text{-}6)$$

where n' is the number of interacting centres per unit volume of solid. This potential equation is called the 9-3 potential because of the exponents 9 and 3 in the equation. The subscript 1,S denotes the interaction between the atom or molecule of species 1 and the slab object. The plot of this 9-3 potential energy equation is shown in Figure 6.10-4.

The minimum of the potential energy is obtained by taking the derivative of the above equation and setting the result to zero. This gives the minimum potential energy of:

$$\varepsilon_{1,S}^* = \frac{2\sqrt{10}}{9}\pi n' \varepsilon_{12}^* \sigma_{12}^3 \qquad (6.10\text{-}7)$$

which occurs at a distance

$$z^* = \left(\frac{2}{5}\right)^{1/6} \sigma_{12} = 0.858\,\sigma_{12} \qquad (6.10\text{-}8)$$

which is shorter than the distance where the minimum occurs for the case of interaction between the two molecules or for the case of interaction between a molecule and a single lattice plane (see Figure 6.10-4a).

Written in terms of the minimum potential energy of eq. (6.10-7), the 9-3 potential energy equation becomes

$$\varphi_{1,S} = \frac{3}{\sqrt{10}} \varepsilon^*_{1,S} \left[\frac{2}{15} \left(\frac{\sigma_{12}}{z} \right)^9 - \left(\frac{\sigma_{12}}{z} \right)^3 \right] \quad (6.10\text{-}9)$$

6.10.3.1 Number Density per Unit Volume for Graphite.

The additional parameter we have in the 9-3 potential is the number density per unit volume, n'. In this section, we calculate this value for the case of graphite. Knowing the area occupied by one carbon center in one lattice layer is 0.026194 nm^2 (Section 6.10.2.1). The volume contains one such center is simply this area multiplied by the spacing between the two adjacent graphite lattice layers. Since this spacing for graphite is 0.335nm, the volume occupied by one carbon center is

$$V = (0.026194)(0.335) = 0.008775 \text{ nm}^3/\text{center}$$

Thus, the number density of carbon center per unit volume of graphite slab is

$$n' = \frac{1}{V} = 114 \text{ centers}/\text{nm}^3$$

Example 6.10-2 *Minimum potential energy of krypton and a graphite slab*
Taking the same system as in Example 6.10-1, we calculate the minimum potential energy for a krypton atom and a graphite slab as

$$\varepsilon^*_{1,S} = \frac{2\sqrt{10}}{9} \pi n' \varepsilon^*_{12} \sigma^3_{12}$$

$$= \frac{2\sqrt{10}}{9} \pi (114)(1.096 \times 10^{-21})(0.3449)^3$$

$$\varepsilon^*_{1,S} = 1.132 \times 10^{-20} \frac{\text{Joule}}{\text{molecule}}$$

The potential energies of interaction for the three cases considered so far are summarised in Table 6.10-1 and are plotted in Figure 6.10-4. The potential energy of each case has been scaled with respect to their minimum potential energy; thus, the minimum of these plots occur at minimum reduced energy of -1. We see in the figure that due to the stronger interaction between the molecule and the slab than between two molecules, the distance where the minimum potential energy occurs will be smaller for the case of molecule and slab interaction. The distance for the case of interaction between a molecule and a lattice plane falls between the other two cases.

Table 6.10-1: Potential energy equations and their characteristics

	Potential function	Minimum Potential	Minimum Position
A	$\varphi_{1,2} = 4\varepsilon_{12}^{*}\left[\left(\dfrac{\sigma_{12}}{r}\right)^{12} - \left(\dfrac{\sigma_{12}}{z}\right)^{6}\right]$	ε_{12}^{*}	$r^{*} = 2^{1/6}\sigma_{12}$
B	$\varphi_{1,SLP} = \dfrac{10\varepsilon_{1,SLP}^{*}}{3}\left[\dfrac{1}{5}\left(\dfrac{\sigma_{12}}{z}\right)^{10} - \dfrac{1}{2}\left(\dfrac{\sigma_{12}}{z}\right)^{4}\right]$	$\varepsilon_{1,SLP}^{*} = \dfrac{6}{5}\pi n \varepsilon_{12}^{*}\sigma_{12}^{2}$	$z^{*} = \sigma_{12}$
C	$\varphi_{1,S} = \dfrac{3}{\sqrt{10}}\varepsilon_{1,S}^{*}\left[\dfrac{2}{15}\left(\dfrac{\sigma_{12}}{z}\right)^{9} - \left(\dfrac{\sigma_{12}}{z}\right)^{3}\right]$	$\varepsilon_{1,S}^{*} = \dfrac{2\sqrt{10}}{9}\pi n' \varepsilon_{12}^{*}\sigma_{12}^{3}$	$z^{*} = \left(\dfrac{2}{5}\right)^{1/6}\sigma_{12}$

<u>Case A</u>: Two single atoms or molecules
<u>Case B</u>: One single atom or molecule and a single lattice plane
<u>Case C</u>: One single atom or molecule and a slab object

6.10.4 A Species and Two Parallel Lattice Planes

We have considered the potential energy of interaction for the three cases:
1. between two molecules (12-6 potential)
2. between a molecule and a single lattice plane (10-4 potential)
3. between a molecule and a slab (9-3 potential)

We now consider the case where a molecule is located between two parallel lattice planes with a distance of 2d apart. This distance 2d is the distance between two fictitious planes passing through the centers of the surface atoms. This case represents the situation where a single nonpolar adsorbate molecule resides inside a slit-shaped micropore of infinite extent. Activated carbon is known to have slit-shaped micropore, and hence the analysis here is applicable to the adsorption of nonpolar molecule in activated carbon.

We denote z being the distance between the molecule and the central plane as shown in the Figure (6.10-6). The limit on z is $-d < z < d$. Note that the distance z defined in the last two sections is the distance between the single atom or molecule and the lattice layer or slab.

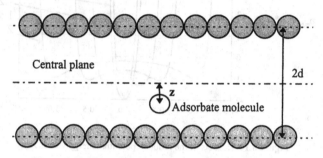

Figure 6.10-6: A molecule and two parallel lattice planes

The potential energy of interaction between a molecule and two lattice layers is simply the summation of the potential energies between the molecule and each of the two lattice layers, that is:

$$\varphi_{1,=} = \varphi_{1,SLP}(d+z) + \varphi_{1,SLP}(d-z) \quad (6.10\text{-}10a)$$

where the functional form of $\varphi_{1,SLP}$ is given in eq. (6.10-5). Rewriting the above equation explicitly, we get

$$\varphi_{1,=}(z,d) = \frac{10}{3}\varepsilon_{1,SLP}^*\left\{\frac{1}{5}\left[\left(\frac{\sigma_{12}}{d+z}\right)^{10} + \left(\frac{\sigma_{12}}{d-z}\right)^{10}\right] - \frac{1}{2}\left[\left(\frac{\sigma_{12}}{d+z}\right)^{4} + \left(\frac{\sigma_{12}}{d-z}\right)^{4}\right]\right\} \quad (6.10\text{-}10b)$$

where $\varepsilon_{1,SLP}^*$ is given in eq. (6.10-4). Figure 6.10-7 shows plots of the above potential energy versus z/σ_{12} with the reduced half-width d/σ_{12} as the varying parameter. We see that when the distance between the two lattice planes is far apart ($d/\sigma_{12} = 2$) we have two local minima and the minimum of the potential energy is essentially the same as the minimum for the case of single lattice plane (that is no enhancement in energy because of the large separation between the two lattice planes). However, when the two planes are getting closer, the two minima coalese and the minimum of the potential energy is enhanced. This enhancement is maximum when the half width of the parallel planes is equal to the collision diameter σ_{12} (that is $d/\sigma_{12} = 1$). At this maximum enhancement the minimum potential energy is twice the minimum of the potential energy for the case of a single lattice plane. We now discuss the properties of this case in more details.

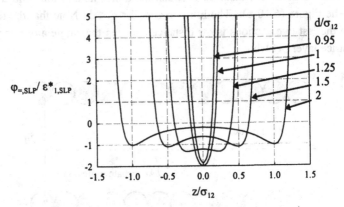

Figure 6.10-7: Plots of the reduced potential energy versus z/σ_{12}

6.10.4.1 Central Potential Energy

First we consider the potential energy at the central plane, and this is obtained by setting $z = 0$ in eq. (6.10-10b) and we get:

$$\frac{\varphi_{1,=}(0,d)}{\varepsilon^*_{1,SLP}} = -\frac{20}{3}\left(\frac{\sigma_{12}}{d}\right)^4\left[\frac{1}{2} - \frac{1}{5}\left(\frac{\sigma_{12}}{d}\right)^6\right] \tag{6.10-11}$$

The central potential energy is a function of the pore half width d. Figure 6.10-8 shows the plot of the reduced central potential energy

$$\frac{\varphi_{1,=}(0)}{\varepsilon^*_{1,SLP}}$$

versus the reduced half width d/σ_{12}. It shows a minimum of -2 and approaches zero when the spacing is large ($d/\sigma_{12} > 3$). There are four particular half-widths that define the characteristics of the central potential energy. The first two half widths are those at which the central potential energies are zero and $-\varepsilon^*_{1,SLP}$, respectively. The third is the half-width at which the central potential energy is minimum, and the fourth is the half-width at which the two local minima coalesce into one. The latter happens when the second derivative of the central potential energy is zero. We now further elaborate these four particular half-widths.

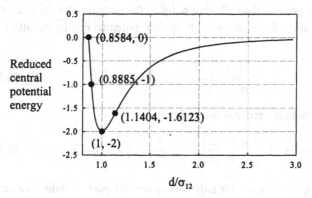

Figure 6.10-8: Plot of the reduced central potential energy for the two parallel lattice layers

1. When the central potential energy is zero, the pore half width is denoted as d_0:

$$\frac{d_0}{\sigma_{12}} = \left(\frac{2}{5}\right)^{1/6} \approx 0.8584 \qquad (6.10\text{-}12a)$$

2. When the central potential energy is equal to the minimum potential energy corresponding to a single lattice layer, $-\varepsilon^*_{1,SLP}$, the pore half width is denoted as d_1/σ_{12} and is a solution of (obtained by setting eq. 6.10-11 to -1)

$$-1 = -\frac{20}{3}\left(\frac{\sigma_{12}}{d_1}\right)^4 \left[\frac{1}{2} - \frac{1}{5}\left(\frac{\sigma_{12}}{d_1}\right)^6\right]$$

Solving this nonlinear algebraic equation, we get

$$\frac{d_1}{\sigma_{12}} = 0.8885 \qquad (6.10\text{-}12b)$$

3. The half width at which the central potential energy is minimum is denoted as d_2/σ_{12}. It is obtained by setting the first derivative of eq. (6.10-11) to zero, and we have:

$$\frac{d_2}{\sigma_{12}} = 1 \qquad (6.10\text{-}12c)$$

The central potential energy at this pore half-width is $-2\varepsilon^*_{1,SLP}$.

4. The last pore half width is the half width at which the two local minima of the potential energy profile (eq. 6.10-10b) coalesce into one at the central plane of the pore. This point is the inflexion point of the central potential energy. Thus by setting the second derivative of the central potential energy equation to zero, we get:

$$\frac{d_3}{\sigma_{12}} = \left(\frac{11}{5}\right)^{1/6} \approx 1.1404 \qquad (6.10\text{-}12d)$$

The central potential energy at this point is

$$\frac{\varphi_{1,=}}{\varepsilon^*_{1,SLP}} = \left(\frac{10}{3}\right)\left[\frac{2}{5}\left(\frac{5}{11}\right)^{5/3} - \left(\frac{5}{11}\right)^{2/3}\right] \cong -1.6123 \qquad (6.10\text{-}12e)$$

The following table lists these four half-widths and the corresponding values of the central potential energies.

d/σ_{12}	$\varphi_{1,=}(0,d)/\varepsilon^*_{1,SLP}$
0.8584	0
0.8885	-1
1.0000	-2
1.1404	-1.6123

6.10.4.2 Local Minimum of Potential Energy Profile.

What we have analysed on the central potential energy provides useful information on the subsequent determination of the local minimum of the potential energy profile (eq. 6.10-10b).

1. When the pore half width is less than the "inflexion" pore half width d_3/σ_{12} (eq. 6.10-12d), the minimum of the potential energy profile of eq. (6.10-10b) occurs at the center of the pore, and the minimum potential energy is simply the central potential energy. Thus if the interaction energy (E) between a molecule and the pore is taken as the negative of the minimum potential energy, we have

$$\frac{E}{\varepsilon^*_{1,SLP}} = \frac{20}{3}\left(\frac{\sigma_{12}}{d}\right)^4\left[\frac{1}{2} - \frac{1}{5}\left(\frac{\sigma_{12}}{d}\right)^6\right] \qquad (6.10\text{-}12f)$$

for $\left(\dfrac{d}{\sigma_{12}}\right) \leq \left(\dfrac{d_3}{\sigma_{12}}\right) = 1.1404$

2. When the pore half width is greater than the "inflexion" pore half width d_3/σ_{12}, there are two local minima and these minima must be found by solving

numerically eq. (6.10-10b). This is done with an iterative produce, such as the Newton-Ralphson method. The initial guess for the position where the minimum occurs is done as follows. Since the minima will occur approximately at a distance of σ_{12} from the wall of the pore, the initial guess for z in searching for the local minimum is

$$z^{(0)} = d - \sigma_{12}$$

With this, we will get a tabular relationship between the minimum potential energy and the pore half width. We shall discuss more about this in Section 6.10.6.

6.10.4.3 Enhancement in the Interaction Energy.

If the energy of interaction between the adsorbent (represented as two parallel planes) and the adsorbate molecule is taken as the negative of the minimum of the potential energy, this energy of interaction will have values ranging between $\varepsilon^*_{1,SLP}$ (when the distance between the two planes is far apart) and $2\varepsilon^*_{1,SLP}$ (when the distance is twice the collision diameter), that is:

$$\varepsilon^*_{1,SLP} < E < 2\varepsilon^*_{1,SLP}$$

To illustrate the enhancement in energy, we compare the affinity of a molecule towards a single lattice layer and that in the case of two parallel lattice layers having a half spacing equal to the collision diameter σ_{12}. The interaction energy between the adsorbate molecule and the single lattice plane is $\varepsilon^*_{1,SLP}$ and that for the case of two parallel lattice layers is $2\varepsilon^*_{1,SLP}$. The adsorption affinity takes the following form:

$$b_\infty \exp\left(\frac{E}{kT}\right)$$

Thus the ratio of the two affinities for these two cases is

$$\frac{\exp\left(\dfrac{2\varepsilon^*_{1,SLP}}{kT}\right)}{\exp\left(\dfrac{\varepsilon^*_{1,SLP}}{kT}\right)} = \exp\left(\frac{\varepsilon^*_{1,SLP}}{kT}\right)$$

296 Equilibria

To illustrate the enhancement in adsorption, we take the example of krypton in Example 6.10-1. We have the following values:

$$\varepsilon^*_{1,SLP} = 1.88 \times 10^{-20} \text{ Joule/molecule}$$
$$T = 300 \text{ K}$$

The ratio of the two affinities calculated as 94, an enhancement of about 100 fold in adsorption density within two parallel layers compared to that of a single lattice layer. This enhancement is even more significant at lower temperatures. If we take T = 200K, the enhancement is about 900 fold. This is only correct when the density of adsorbate within the two lattice layers is very low, that is there is no interaction between adsorbate molecules and when the interaction between the adsorbate and the two lattice planes is much stronger than that between adsorbate molecules.

6.10.5 A Species and Two Parallel Slabs

For the case of interaction between a molecule and two parallel slabs, the potential energy of interaction is calculated from the following equation, which is simply the sum of the two potential equations between a single atom or molecule and a single slab (from eq. 6.10-9).

$$\varphi_{1,SS}(z,d) = \frac{3}{\sqrt{10}} \varepsilon^*_{1,S} \left\{ \frac{2}{15}\left[\left(\frac{\sigma_{12}}{d+z}\right)^9 + \left(\frac{\sigma_{12}}{d-z}\right)^9\right] - \left[\left(\frac{\sigma_{12}}{d+z}\right)^3 + \left(\frac{\sigma_{12}}{d-z}\right)^3\right] \right\} \quad (6.10\text{-}13)$$

where z is the distance between the atom or molecule and the central plane, and d is the half spacing. Figure 6.10-9 shows plots of the reduced potential energy $\varphi_{1,SS} / \varepsilon^*_{1,S}$ versus the reduced distance z/σ_{12}. The behaviour of this potential energy is very similar to that of the case of a molecule and two parallel planes dealt with in Section 6.10-4, except that the maximum of the enhancement is achieved at a shorter distance ($d/\sigma_{12} = 0.858$) compared to $d/\sigma_{12} = 1$. We now first study the behaviours of the central potential energy, and then study the minimum of the potential energy profile (6.10-13).

6.10.5.1 Central Potential Energy

The potential energy at the central plane is obtained by setting z to zero in eq.(6.10-13), and we get:

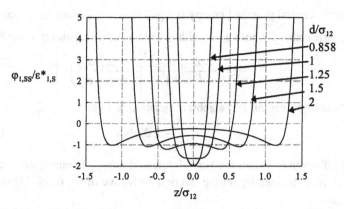

Figure 6.10-9: Plots of the reduced potential energy versus z/σ_{12} for two parallel slabs

$$\frac{\varphi_{1,SS}(0,d)}{\varepsilon_{1,S}^*} = -\frac{6}{\sqrt{10}}\left(\frac{\sigma_{12}}{d}\right)^3\left[1-\frac{2}{15}\left(\frac{\sigma_{12}}{d}\right)^6\right] \quad (6.10\text{-}14)$$

Figure 6.10-10 shows a plot of the central potential energy versus the reduced pore half-width d/σ_{12}. Four particular pore half-widths of interest are listed below

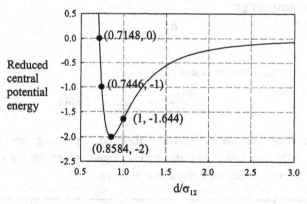

Figure 6.10-10: Plot of the reduced central potential energy for two parallel slabs

1. When the central potential energy is zero, the pore half width is

$$\frac{d_0}{\sigma_{12}} = \left(\frac{2}{15}\right)^{1/6} \approx 0.7148 \quad (6.10\text{-}15a)$$

2. When the central potential energy is equal to $-\varepsilon^*_{1,S}$ (minimum potential energy of a single slab), the pore half width is denoted as d_1/σ_{12} and is a solution of

$$-1 = -\frac{6}{\sqrt{10}}\left(\frac{\sigma_{12}}{d_1}\right)^3\left[1-\frac{2}{15}\left(\frac{\sigma_{12}}{d_1}\right)^6\right]$$

Solving this nonlinear algebraic equation, we get:

$$\frac{d_1}{\sigma_{12}} \cong 0.7446 \quad (6.10\text{-}15b)$$

3. The half width at which the central potential energy is minimum is denoted as d_2/σ_{12}. It is obtained by setting the first derivative of eq. (6.10-14) to zero, and we have:

$$\frac{d_2}{\sigma_{12}} = 0.8584 \quad (6.10\text{-}15c)$$

The central potential energy at this pore half-width is $-2\varepsilon^*_{1,S}$.

4. The last pore half-width of interest is the pore half width at which the two local minima of the potential energy profile (eq. 6.10-13) coalesce into one. This occurs at the central plane, and this pore half width is the inflexion point of the central potential energy (eq. 6.10-14). By setting the second derivative of eq. (6.10-14) to zero, we get

$$\frac{d_3}{\sigma_{12}} = 1 \quad (6.10\text{-}15d)$$

The central potential energy at this point is

$$\frac{\varphi_{1,SS}}{\varepsilon^*_{1,S}} = -\frac{26}{5\sqrt{10}} \approx -1.6444$$

We summarise these four half-widths and their corresponding central potential energies in the following table. We also include those values for the two parallel lattice layers in the same table for comparison purpose.

Two parallel lattice layers		Two parallel slabs	
d/σ_{12}	$\varphi_{1,=}(0,d)/\varepsilon^*_{1,SLP}$	d/σ_{12}	$\varphi_{1,SS}(0,d)/\varepsilon^*_{1,S}$
0.8584	0	0.7148	0
0.8885	-1	0.7446	-1
1	-2	0.8584	-2
1.1404	-1.6123	1	-1.6444

6.10.5.2 Local Minimum of Potential Energy Profile.

What we have done in the analysis of the central potential energy provides useful information for the determination of the local minimum of the potential energy profile (eq. 6.10-13).

1. When the pore half width is less than the inflexion pore half width d_3/σ_{12}, the minimum of the potential energy of eq. (6.10-13) occurs at the center of the pore, and the minimum potential energy is simply the central potential energy. Thus if the interaction energy (E) between a molecule and the pore is taken as the negative of the minimum potential energy, we have

$$\frac{E}{\varepsilon_{1,S}^*} = \frac{6}{\sqrt{10}} \left(\frac{\sigma_{12}}{d}\right)^3 \left[1 - \frac{2}{15}\left(\frac{\sigma_{12}}{d}\right)^6\right] \qquad (6.10\text{-}15e)$$

for

$$\left(\frac{d}{\sigma_{12}}\right) < \frac{d_3}{\sigma_{12}} = 1$$

2. When the pore half width is greater than the inflexion pore half width d_3/σ_{12}, there are two local minima and these must be found by solving numerically eq. (6.10-13). This is done with an interative procedure, such as the Newton-Raphson method. The initial guess for the position at which the minimum occurs is done as follows. Since the minima will occur at a distance about $0.858\ \sigma_{12}$ from the slab surface (see eq. 6.10-8), the initial guess for z in searching for the local minimum is

$$z^{(0)} = d - 0.858\sigma_{12}$$

Solving this, we will get a relationship between the minimum potential energy and the pore half width for any pore half-width greater than d_3.

What we will do in the next section is to apply what we have learnt so far in the last two sections to study the adsorption behaviour inside a slit-shape pore.

6.10.6 Adsorption Isotherm for Slit Shape Pore

We have addressed the potential energy for a number of cases in the last sections (Sections 6.10-1 to 6.10-5). The analysis of the last two cases: (i) a molecule and two parallel lattice planes and (ii) a molecule and two parallel slabs, are particularly useful for the study of adsorption of nonpolar molecules in slit-shaped micropore solids, such as activated carbon.

Using the results obtained in Sections 6.10-4 and 6.10-5, an overall adsorption isotherm can be obtained from the local adsorption isotherm and a micropore size distribution. Let us denote the micropore size distribution as f(r) such that

$$V_\mu \int_{r_{min}}^{r} f(r)dr \qquad (6.10\text{-}16)$$

is the volume of the micropores having half width from r_{min} to r, where V_μ is the micropore volume. The minimum micropore half width r_{min} is defined as the minimum micropore size accessible to the adsorbate, hence it is a function of the adsorbate.

If the local adsorption in a micropore having a half width of r is denoted as

$$\theta(E(r), P, T) \qquad (6.10\text{-}17)$$

where E is the interaction energy between the adsorbent and the adsorbate (which is a function of pore half-width), then the overall adsorption isotherm is taken in the following form:

$$C_\mu = \frac{V_\mu}{v_M} \int_{r_{min}}^{r_{max}} \theta(E(r), P, T) \, f(r) \, dr \qquad (6.10\text{-}18)$$

where r_{max} is the maximum half width of the micropore region. In writing eq.(6.10-18), we have assumed that the state of the adsorbate in the micropore is liquid-like with v_M being the liquid molar volume (m³/mole). We could write eq. (6.10-18) as follows:

$$C_\mu = C_{\mu s} \int_{r_{min}}^{r_{max}} \theta(E(r), P, T) \, f(r) \, dr \qquad (6.10\text{-}19)$$

The integral in eq.(6.10-18) or (6.10-19) can only be evaluated if the relationship between the energy of interaction E and the pore half width r is known. This relationship is possible with the information learnt in Section 6.10-4 for two parallel lattice planes and Section 6.10-5 for two parallel slabs. The depth of the potential minimum is the interaction energy between the micropore and the adsorbate.

6.10.6.1 Interaction Energy versus the Half-Width for two Parallel Lattice Planes

Interaction energy between a molecule and the pore formed by two parallel lattice layers is taken as the negative of the minimum potential energy. When the half-width is less than 1.1404 σ_{12}, the interaction energy is calculated from eq. (6.10-12f), and when it is greater than 1.1404 σ_{12} the interaction energy is obtained numerically from eq. (6.10-10b). Figure 6.10-11 shows a plot of the reduced interaction energy

$$\frac{E}{\varepsilon_{1,SLP}^*} = -\left(\frac{\varphi_{1,=}}{\varepsilon_{1,SLP}^*}\right)_{min} \qquad (6.10\text{-}20)$$

versus the reduced half width, d/σ_{12}. Results are also tabulated in Table 6.10-2.

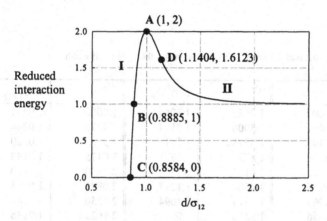

Figure 6.10-11: Plot of the reduced interaction energy versus d/σ_{12} for two parallel lattice layers

When the two parallel planes are widely separated ($d/\sigma_{12} > 2$), the energy of interaction is $\varepsilon_{1,SLP}^*$. This energy of interaction increases when the distance between the two lattice planes is getting smaller, and it reaches a maximum of 2 $\varepsilon_{1,SLP}^*$ when the half width is equal to the collision diameter σ_{12} (point A). When the distance decreases below σ_{12}, the energy of interaction decreases due to the repulsion and it has a value of $\varepsilon_{1,SLP}^*$ when the half width is 0.8885 σ_{12} (point B), and when the half width is 0.8584 σ_{12} the energy of interaction is zero (point C).

For the purpose of computing the adsorption isotherm later, it is convenient to obtain an analytical expression for the reduced interaction energy versus the reduced half width, d/σ_{12}. When the reduced half width is less than 1.14043, the reduced interaction energy is simply the negative of the central potential energy given as in eq. (6.10-12f), that is

$$\frac{E}{\varepsilon^*_{1,SLP}} = \frac{10}{3}\left[\left(\frac{\sigma_{12}}{d}\right)^4 - \frac{2}{5}\left(\frac{\sigma_{12}}{d}\right)^{10}\right] \quad (6.10\text{-}21)$$

for $d/\sigma_{12} < 1.14043$.

For $d/\sigma_{12} > 1$, the following approximate solution reasonably describes the reduced interaction energy as a function of the reduced half-width.

$$\frac{E}{\varepsilon^*_{1,SLP}} = 2 - 1.0567619\left(1 - \frac{\sigma_{12}}{d}\right)^3 \exp\left[\frac{2.7688274}{(d/\sigma_{12}) - 0.6165728}\right] \quad (6.10\text{-}22)$$

Table 6.10-2 Values of the scaled energy of interaction versus the scaled half width

d/σ_{12}	$E/\varepsilon^*_{1,SLP}$	d/σ_{12}	$E/\varepsilon^*_{1,SLP}$	d/σ_{12}	$E/\varepsilon^*_{1,SLP}$
0.8885	1.0000	1.2625	1.3068	1.9000	1.0271
0.8996	1.2496	1.3000	1.2534	1.9375	1.0244
0.9108	1.4498	1.3375	1.2112	1.9750	1.0220
0.9220	1.6086	1.3750	1.1773	2.0125	1.0199
0.9331	1.7323	1.4125	1.1500	2.0500	1.0180
0.9442	1.8268	1.4500	1.1277	2.0875	1.0164
0.9554	1.8965	1.4875	1.1094	2.1250	1.0149
0.9666	1.9456	1.5250	1.0943	2.1625	1.0136
0.9777	1.9774	1.5625	1.0817	2.2000	1.0125
0.9888	1.9947	1.6000	1.0711	2.2375	1.0114
1.0000	2.0000	1.6375	1.0622	2.2750	1.0105
1.0375	1.9542	1.6750	1.0546	2.3125	1.0097
1.0750	1.8491	1.7125	1.0482	2.3500	1.0089
1.1125	1.7170	1.7500	1.0427	2.3875	1.0082
1.1500	1.5775	1.7875	1.0379	2.4250	1.0076
1.1875	1.4627	1.8250	1.0338	2.4625	1.0070
1.2250	1.3748	1.8625	1.0302		

6.10.6.2 Interaction Energy versus the Half Width for two Parallel Slabs

When the pore half-width is less than σ_{12}, the interaction energy is calculated from eq. (6.10-15e), and when it is greater than σ_{12} the interaction energy is calculated as the negative of the minimum of the potential energy profile (eq. 6.10-13). The results are plotted graphically in Figure 6.10-12 as a plot of the reduced interaction energy

$$\frac{E}{\varepsilon_{1,S}^*} \tag{6.10-23}$$

versus the reduced half width, d/σ_{12}.

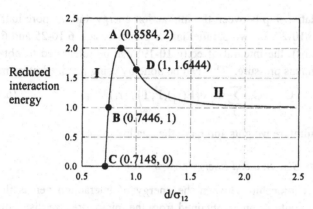

Figure 6.10-12: Plot of the reduced interaction energy versus d/σ_{12} for two parallel slabs

The behaviour of this plot is very similar to that for the case of two parallel lattice planes. The maximum enhancement in the potential energy occurs at

$$\frac{d}{\sigma_{12}} = \left(\frac{2}{5}\right)^{1/6} \cong 0.8584 \tag{6.10-24}$$

below which the enhancement starts to dissipate very rapidly due to the strong repulsion of the two slabs onto the molecule.

For the purpose of computation of adsorption isotherm, we obtain the relationship between the reduced interaction energy versus the reduced half width. When the half width is less than σ_{12}, the interaction energy is given by eq. (6.10-15e) or written again below for completeness.

$$\frac{E}{\varepsilon_{1,S}^*} = \frac{6}{\sqrt{10}}\left[\left(\frac{\sigma_{12}}{d}\right)^3 - \frac{2}{15}\left(\frac{\sigma_{12}}{d}\right)^9\right] \tag{6.10-25}$$

for $d/\sigma_{12} < 1$.

For $d/\sigma_{12} > 1$, the following approximate solution can be used to describe the relationship between the reduced interaction energy versus the reduced half width:

$$\frac{E}{\varepsilon_{1,S}^*} = 2 - 1.04417153\left[1 - \left(\frac{2}{5}\right)^{1/6}\frac{\sigma_{12}}{d}\right]^3 \exp\left[\frac{2.41519301}{(d/\sigma_{12}) - 0.4954486}\right] \quad (6.10\text{-}26)$$

6.10.6.3 Adsorption Isotherm

Having the relationship between the interaction energy and the pore half width (eqs. 6.10-21 and 6.10-22 for two parallel lattice planes or eqs. 6.10-25 and 6.10-26 for two parallel slabs), the integral of eq.(6.10-19) can be integrated to obtain the amount adsorbed versus pressure. This can be done by using the quadrature method

$$C_\mu = C_{\mu s}\sum_j w_j\, \theta\!\left(E(r_j), P, T\right) f(r_j) \quad (6.10\text{-}27)$$

where w_j is the quadrature weights and r_j are the quadrature points.

6.10.6.4 Micropore Size-Induced Energy Distribution

Knowing the relationship between the energy of interaction versus the pore size, the energy distribution can be obtained from the micropore size distribution by using the following formula:

$$F(E)dE = f(r)dr \quad (6.10\text{-}28)$$

where $F(E)dE$ is the fraction of the micropore volume having energy of interaction between E and E+dE. Thus, we have:

$$F(E) = \frac{f(r)}{dE/dr} \quad (6.10\text{-}29)$$

where dE/dr is the slope of the curve in Figure 6.10-11 or Figure 6.10-12. Since for a given value of the energy of interaction E, there are two values of r and hence two values of the slope dE/dr. One slope is positive while the other is negative. To deal with this problem, we consider the case of two parallel lattice planes (as similar procedure will apply to the case of two parallel slabs) and split the integral of eq. (6.10-19) into three integrals as follows:

$$\frac{C_\mu}{C_{\mu s}} = \int_{r_{min}}^{r_{max}}\theta(E,P,T)\cdot f(r)dr = \int_{r_{min}}^{0.8885\sigma_{12}}\theta(E,P,T)\cdot f(r)dr + \int_{0.8885\sigma_{12}}^{\sigma_{12}}\theta(E,P,T)\cdot f(r)dr + \int_{\sigma_{12}}^{r_{max}}\theta(E,P,T)\cdot f(r)dr$$

$$(6.10\text{-}30)$$

The first integral covers the range of micropore size where the interaction energy is less than that of a single surface. The contribution of this range is usually very small and we could neglect its contribution. Hence, eq (6.10-30) becomes:

$$\frac{C_\mu}{C_{\mu s}} = \int_{0.8885\sigma_{12}}^{\sigma_{12}} \theta(E,P,T) \cdot f(r)\,dr + \int_{\sigma_{12}}^{r_{max}} \theta(E,P,T) \cdot f(r)\,dr \qquad (6.10\text{-}31)$$

Within the limits of the first integral, the interaction energy increases with r while within the limits of the second integral it decreases with an increase in r. Applying the chain rule of differentiation to these two integrals, we get:

$$\frac{C_\mu}{C_{\mu s}} = \int_{0.8885\sigma_{12}}^{\sigma_{12}} \theta(E,P,T) \left(\frac{f(r)}{dE/dr}\right)_I dE + \int_{\sigma_{12}}^{r_{max}} \theta(E,P,T) \left(\frac{f(r)}{dE/dr}\right)_{II} dE \qquad (6.10\text{-}32)$$

where the subscripts I and II denote the branches I and II as shown in Figure 6.10-11, respectively. Now the integrals are written in terms of the energy of interaction, the lower and upper limits of these two integrals are replaced by values as shown in the following table.

Pore size	Corresponding energy of interaction
$0.8885\,\sigma_{12}$	$\varepsilon_{1,SLP}^{*}$
σ_{12}	$2\varepsilon_{1,SLP}^{*}$
r_{max}	E_{min}

Here E_{min} is the energy of interaction corresponding to the pore size r_{max}, which is read from the branch II of Figure 6.10-11. Thus, eq.(6.10-32) will become:

$$\frac{C_\mu}{C_{\mu s}} = \int_{\varepsilon_{1,SLP}^{*}}^{2\varepsilon_{1,SLP}^{*}} \theta(E,P,T) \left(\frac{f(r)}{dE/dr}\right)_I dE + \int_{2\varepsilon_{1,SLP}^{*}}^{E_{min}} \theta(E,P,T) \left(\frac{f(r)}{dE/dr}\right)_{II} dE \qquad (6.10\text{-}33)$$

We note that $(dE/dr)_{II}$ in the second integral is negative, and the upper limit E_{min} is smaller than $2\varepsilon_{1,SLP}^{*}$, we then can rewrite the above equation as follows:

$$\frac{C_\mu}{C_{\mu s}} = \int_{\varepsilon_{1,SLP}^{*}}^{2\varepsilon_{1,SLP}^{*}} \theta(E,P,T) \left(\frac{f(r)}{dE/dr}\right) dE + \int_{E_{min}}^{2\varepsilon_{1,SLP}^{*}} \theta(E,P,T) \left(\frac{f(r)}{|dE/dr|}\right)_{II} dE \qquad (6.10\text{-}34)$$

If the maximum micropore size is sufficiently large (that is the half width of the largest micropore size is twice larger the collision diameter such that E_{min} is approximately equal to $\varepsilon_{1,SLP}^*$, eq.(6.10-34) will be reduced to:

$$\frac{C_\mu}{C_{\mu s}} = \int_{\varepsilon_{1,SLP}^*}^{2\varepsilon_{1,SLP}^*} \theta(E,P,T)\left[\left(\frac{f(r)}{dE/dr}\right)_I + \left(\frac{f(r)}{|dE/dr|}\right)_{II}\right]dE \qquad (6.10\text{-}35)$$

This equation can be written in terms of the energy distribution as:

$$\frac{C_\mu}{C_{\mu s}} = \int_{\varepsilon_{1,SLP}^*}^{2\varepsilon_{1,SLP}^*} \theta(E,P,T)\cdot F(E)dE \qquad (6.10\text{-}36)$$

where the energy distribution takes the form

$$F(E) = \left[\left(\frac{f(r)}{dE/dr}\right)_I + \left(\frac{f(r)}{|dE/dr|}\right)_{II}\right] \qquad (6.10\text{-}37)$$

having a domain of energy between $\varepsilon_{1,SLP}^*$ and $2\varepsilon_{1,SLP}^*$. Eq. (6.10-37) is the desired energy distribution induced by the micropore size distribution. It is obtained as follows. For a given value of the interaction energy E between $\varepsilon_{1,SLP}^*$ and $2\varepsilon_{1,SLP}^*$, there will be two values of pore half-width, obtained from eqs. (6.10-21) and (6.10-22) (or from Figure 6.10-11). Knowing these two values of r, the slopes dE/dr can be calculated and two values for f(r) can be obtained from the micropore size distribution. Substituting these values into eq. (6.10-37), we then obtain the energy distribution.

A typical plot of the energy distribution versus energy is shown in the following figure. This distribution exhibits a pattern (Figure 6.10-13) such that it is infinite at $E = 2\varepsilon_{1,SLP}^*$ because at this energy the slope (dE/dr) is zero.

6.10.6.5 Evaluation of the Adsorption Isotherm

To evaluate the adsorption isotherm for a microporous solid having a known micropore size distribution, we use eq.(6.10-18) or (6.10-19). Here, for the purpose of illustration, we use the following Gamma distribution to describe the micropore size distribution

$$f(r) = \frac{\alpha^{n+1} r^n e^{-\alpha r}}{\Gamma(n+1)} \qquad (6.10\text{-}38a)$$

which has the following mean and variance as:

$$\bar{r} = \frac{(n+1)}{\alpha}; \quad \sigma = \frac{\sqrt{n+1}}{\alpha} \tag{6.10-38}$$

and we use the Langmuir isotherm as the local isotherm as given below:

$$\theta(E,P,T) = \frac{b_\infty \cdot \exp\left[E(r)/R_g T\right] P}{1 + b_\infty \cdot \exp\left[E(r)/R_g T\right] P} \tag{6.10-39}$$

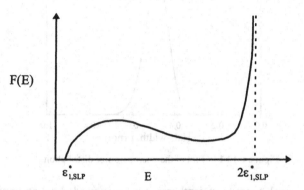

Figure 6.10-3: Typical plot of the energy distribution derived from

Eq.(6.10-19) then can be written as:

$$C_\mu = C_{\mu s} \int_{r_{min}}^{r_{max}} \left\{ \frac{b_\infty \cdot \exp\left[E(r)/R_g T\right] P}{1 + b_\infty \cdot \exp\left[E(r)/R_g T\right] P} \right\} \times \left[\frac{\alpha^{n+1} r^n \exp(-\alpha r)}{\Gamma(n+1)} \right] \cdot dr \tag{6.10-40}$$

This equation contains two parameters for the micropore size distribution, and three parameters for the adsorbate-adsorbent interaction, $C_{\mu s}$, b_∞ and $\varepsilon^*_{1,SLP}$. Thus by fitting the above equation to adsorption equilibrium data, we will obtain the optimal values for the above mentioned parameters. Using the adsorption data of propane on an activated carbon at 303 K tabulated in Table 3.2-1, we carry out the optimization and obtain the following optimal parameters:

α	20 nm^{-1}
n	89
$C_{\mu s}$	6 mmol/g
b_∞	9×10^{-7} kPa^{-1}
$\varepsilon^*_{1,SLP}$	16.7 kJoule/mole

Using these values of α and n in the above table, we plot the micropore size distribution as shown in Figure 6.10-14 where we see that the mean micropore size is 9 A, which falls in the range determined by experimental methods for many samples of activated carbon.

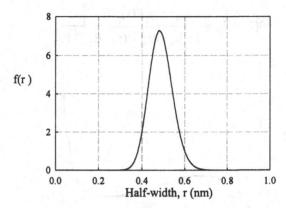

Figure 6.10-14: Plot of the micropore size distribution

Concerning the local adsorption isotherm, we have used the Langmuir equation. Other fundamental equations discussed in Chapter 2 can be used as the local isotherm. For example, Jagiello and Schwartz (1992) used Hill-de Boer as the local isotherm equation. Equation such as the Nitta equation can be used as it allows for the multiple sites adsorption and adsorbate-adsorbate interaction.

6.10.7 An Atom or Molecule and two Parallel Lattice Planes with sub-lattice layers

In Sections 6.10-2 to 6.10-5, we have dealt with cases of interaction between a species and a lattice plane, a slab, two parallel lattice planes and two parallel slabs. Here, we will extend to the case of two parallel lattice planes with sublayers underneath each lattice layer. This case represents the case of activated carbon where the walls of slit-shaped micropore are made of many lattice layers. Although real micropore configuration is more complex than this, this configuration is the closest to describe activated carbon micropore structure. Before we address molecular interacts with two lattice layers with sub-lattice layers underneath, we consider first the interaction between one atom or molecule with one lattice layer with sub-lattice layers.

6.10.7.1 An Atom or Molecule and one Lattice Plane with Sub-Lattice Layers

We denote z as the distance from the molecule to the solid surface, and Δ is the distance between the two planes passing through the nuclei of the surface atoms of the lattice layer and those of the adjoining sublayer (Figure 6.10-15).

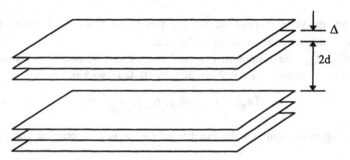

Figure 6.10-15: Schematic diagram of two lattice layers with sub-lattice layers

Starting with the fluid-fluid interaction potential as given in eq.(6.10-1), Steele (1972) obtained the fluid-solid interaction potential between a species "k" and a lattice layer with its sublayers as given below:

$$\varphi_{k,\Delta} = \varphi_w \left[\frac{1}{5}\left(\frac{\sigma_{ks}}{z}\right)^{10} - \frac{1}{2}\left(\frac{\sigma_{ks}}{z}\right)^4 - \frac{\sigma_{ks}^4}{6\Delta(z+0.61\Delta)^3} \right] \quad (6.10\text{-}44a)$$

where σ_{ks} is the collision diameter between the molecule of type k and the surface atom, and the wall potential energy parameter φ_w is given by

$$\varphi_w = \left(\frac{10}{3}\right)\left(\frac{6}{5}\pi\rho_s \sigma_{ks}^3 \varepsilon_{ks}\right) \quad (6.10\text{-}44b)$$

Here ρ_s is the number density of surface center per-unit volume and ε_{ks} is the Lennard-Jones well-depth of the molecule and the surface atom.

The potential equation of the form in eq. (6.10-44a) is called the 10-4-3 potential. This potential energy exhibits a behaviour similar to that of the 10-4 or 9-3 potential energy equations. The minimum potential energy of the 10-4-3 is obtained by taking the first derivative to zero, that is

$$\frac{d\varphi_{k,\Delta}}{dz} = \frac{\varphi_w}{\sigma_{ks}}\left[-2\left(\frac{\sigma_{ks}}{z}\right)^{11} + 2\left(\frac{\sigma_{ks}}{z}\right)^5 + \frac{\sigma_{ks}^5}{2\Delta(0.61\Delta+z)^4} \right] = 0 \quad (6.10\text{-}45)$$

from which we can solve for the distance from the lattice layer, at which the potential energy is minimum

$$\left(\frac{z_{min}}{\sigma_{ks}}\right) = f\left(\frac{\Delta}{\sigma_{ks}}\right) \qquad (6.10\text{-}46)$$

There is no analytical expression for the functional form f. It must be found numerically from the solution of eq. (6.10-45).

Knowing z_{min}, the minimum potential energy is obtained by substituting eq. (6.10-46) into the potential energy equation (6.10-44a), we get

$$\varphi_{k,\Delta}(z_{min}) = \varphi_w \, g(\Delta/\sigma_{ks}) \stackrel{def}{\equiv} \varphi_\infty \qquad (6.10\text{-}47)$$

The following table tabulates the functions f and g as a function of the parameter Δ/σ_{ks}.

Δ/σ_{ks}	$f(\Delta/\sigma_{ks})$	$g(\Delta/\sigma_{ks})$
0.2	0.9159	-0.9744
0.25	0.9336	-0.7809
0.5	0.9750	-0.4546
0.75	0.9884	-0.3727
1	0.9940	-0.3402
2	0.9991	-0.3076
3	0.9998	-0.3025
∞	1	-0.30

The special case ($\Delta \to \infty$) of the 10-4-3 potential is simply the 10-4 potential we have dealt with in Section 6.10-2.

6.10.7.2 An atom or a Molecule and two Lattice Planes with Sub Lattice Layers

Having understood the behaviour of the single lattice layer with sub-lattice layers, we now turn to the case where there are two lattice layers and sub-lattice layers underneath each of those layers. An atom or a molecule residing inside the slit pore has to interact with two surface layers including their sub-lattice layers. In this case, the potential energy of interaction is the sum of the potentials for each surface, that is

$$\varphi_{k,\Delta\Delta} = \varphi_{k,\Delta}(d+z) + \varphi_{k,\Delta}(d-z) \qquad (6.10\text{-}48a)$$

where d is the half-spacing between the two lattice layers (Figure 6.10-15) and z is the distance of the atom or the molecule from the central plane between the two surface layers. Writing eq. (6.10-48a) explicitly we get

$$\varphi_{k,\Delta\Delta}(z;d) = \varphi_w \left\{ \frac{1}{5}\left[\left(\frac{\sigma_{ks}}{d+z}\right)^{10} + \left(\frac{\sigma_{ks}}{d-z}\right)^{10}\right] - \frac{1}{2}\left[\left(\frac{\sigma_{ks}}{d+z}\right)^{4} + \left(\frac{\sigma_{ks}}{d-z}\right)^{4}\right] \right.$$
$$\left. - \left[\frac{\sigma_{ks}^4}{6\Delta(d+z+0.61\Delta)^3} + \frac{\sigma_{ks}^4}{6\Delta(d-z+0.61\Delta)^3}\right]\right\} \quad (6.10\text{-}48b)$$

Like the case of two parallel lattice layers (Section 6.10-4) and two parallel slabs (Section 6.10-5), we now investigate the behaviour of the potential energy equation (6.10-48) by first studying the central potential energy. By setting $z = 0$ into eq. (6.10-48b), we obtain the following central potential energy equation

$$\varphi_{k,\Delta\Delta}(0;d) = \varphi_w \left[\frac{2}{5}\left(\frac{\sigma_{ks}}{d}\right)^{10} - \left(\frac{\sigma_{ks}}{d}\right)^{4} - \frac{\sigma_{ks}^4}{3\Delta(0.61\Delta+d)^3}\right] \quad (6.10\text{-}49)$$

This central potential energy is a function of the pore half width d, and it has a minimum when the half width is equal to some threshold value. But first let us consider the situation when the central potential energy is zero. We denote this half width as d_0, that is

$$\frac{2}{5}\left(\frac{\sigma_{ks}}{d_0}\right)^{10} - \left(\frac{\sigma_{ks}}{d_0}\right)^{4} - \frac{\sigma_{ks}^4}{3\Delta(0.61\Delta+d_0)^3} = 0 \quad (6.10\text{-}50)$$

of which the solution is

$$\frac{d_0}{\sigma_{ks}} = f_0(\Delta/\sigma_{ks}) \quad (6.10\text{-}51)$$

When the central potential energy is equal to the minimum potential energy corresponding to the case of single lattice layer and its associated sub-layers (eq. 6.10-47), the half width is denoted as d_1 and it is a solution of

$$\varphi_{k,\Delta\Delta}(0;d_1) = \varphi_w \left[\frac{2}{5}\left(\frac{\sigma_{ks}}{d_1}\right)^{10} - \left(\frac{\sigma_{ks}}{d_1}\right)^{4} - \frac{\sigma_{ks}^4}{3\Delta(0.61\Delta+d_1)^3}\right] = \varphi_w\, g(\Delta/\sigma_{ks}) \quad (6.10\text{-}52)$$

Solving the above equation gives:

$$\frac{d_1}{\sigma_{ks}} = f_1(\Delta/\sigma_{ks}) \qquad (6.10\text{-}53)$$

The central potential energy has a minimum with respect to the half width. This is obtained by setting the first derivative of eq. (6.10-49) with respect to the half-width d to zero, that is

$$\frac{d\varphi_{k1\Delta\Delta}(0;d)}{d(d)} = \frac{2\varphi_w}{\sigma_{ks}}\left[-2\left(\frac{\sigma_{ks}}{d}\right)^{11} + 2\left(\frac{\sigma_{ks}}{d}\right)^5 + \frac{\sigma_{ks}^5}{2\Delta(0.61\Delta+d)^3}\right] = 0 \qquad (6.10\text{-}54)$$

We note that the solution for the minimum of the central potential energy (eq. 6.10-54) is identical to the solution for the minimum of the potential energy for the case of a single lattice layer (eq. 6.10-45). Thus we write

$$\left(\frac{d_2}{\sigma_{ks}}\right) = f_2\left(\frac{\Delta}{\sigma_{ks}}\right) \qquad (6.10\text{-}55)$$

where d_2 is the half width that gives the minimum central potential energy and f_2 is the same functional form as f in eq. (6.10-46). The minimum central potential energy at $d = d_2$ is

$$\varphi_{k,\Delta\Delta}(0;d_2) = 2\varphi_w\, g(\Delta/\sigma_{ks}) \qquad (6.10\text{-}56)$$

The function g has the same functional form as that defined in eq. (6.10-47). This means that the minimum of the central potential energy is <u>twice</u> the minimum potential energy corresponding to the single lattice layer.

Another half width of interest is the one at which the two local minima of the potential energy profile (eq. 6.10-48b) coalesce. This coalescence occurs at the central plane of the pore, and when this occurs the second derivative of the central potential energy with respect to the pore half-width is zero, that is

$$\frac{d^2\varphi_{k,\Delta\Delta}(0;d)}{d(d)^2} = \frac{\varphi_w}{\sigma_{ks}^2}\left[44\left(\frac{\sigma_{ks}}{d}\right)^{12} - 20\left(\frac{\sigma_{ks}}{d}\right)^6 - \frac{4\sigma_{ks}^6}{\Delta(0.61\Delta+d)^5}\right] = 0 \qquad (6.10\text{-}57)$$

Solving the above equation, we get

$$\frac{d_3}{\sigma_{ks}} = f_3(\Delta/\sigma_{ks}) \qquad (6.10\text{-}58)$$

The central potential energy at this half width is

$$\varphi_{k,\Delta\Delta}(0;d_3) = \varphi_w \left[\frac{2}{5}\left(\frac{\sigma_{ks}}{d_3}\right)^{10} - \left(\frac{\sigma_{ks}}{d_3}\right)^4 - \frac{\sigma_{ks}^4}{3\Delta(0.61\Delta + d_3)^3} \right] \stackrel{\text{def}}{=} \varphi_w \cdot g_3(\Delta/\sigma_{ks}) \quad (6.10\text{-}59)$$

The following table tabulates the functional values of f_0, f_1, f_2, f_3 and g_3 as a function of Δ/σ_{ks}.

Δ/σ_{ks}	f_0	f_1	f_2	f_3	g_3
0.3	0.8038	0.8337	0.9467	1.0871	-1.0848
0.4	0.8203	0.8505	0.9642	1.1057	-0.8615
0.5	0.8308	0.8611	0.9750	1.1169	-0.7422
0.7	0.8427	0.8730	0.9867	1.1286	-0.6236
1	0.8506	0.8810	0.9940	1.1355	-0.5520
2	0.8570	0.8872	0.9991	1.1399	-0.4970
∞	0.8584	0.8885	1	1.1404	-0.4837

Like the last two cases of parallel lattice layers and parallel slabs, what we have done so far provides useful information on the determination of the local minimum of the potential energy profiles (eq. 6.10-48b).

1. When the pore half-width is less than the inflexion pore half-width d_3, the minimum of the potential energy equation (6.10-48b) occurs at the center of the pore. Hence the minimum potential energy is simply the central potential energy. If we take the energy of interaction as the negative of the minimum potential energy, it is then given by:

$$E = \varphi_w \left[-\frac{2}{5}\left(\frac{\sigma_{ks}}{d}\right)^{10} + \left(\frac{\sigma_{ks}}{d}\right)^4 + \frac{\sigma_{ks}^4}{3\Delta(0.61\Delta + d)^3} \right] \quad (6.10\text{-}60)$$

for $d < d_3$, where d_3 is given by eq. (6.10-58).

2. When the pore half-width is greater than the inflexion pore half-width d_3, there are two local minima for the potential energy profile (eq. 6.10-48b). These minima must be found numerically by some iterative method, such as the Newton-Raphson method. The initial guess for the position where the minimum occurs is:

$$z^{(0)} = d - \sigma_{ks} f(\Delta/\sigma_{ks}) \quad (6.10\text{-}61)$$

where $\sigma_{ks} f(\Delta/\sigma_{ks})$ is the distance from the layer where the minimum occurs if there is only one lattice layer and its associated sub-layers (see eq. 6.10-46). Once the position where the minimum is found from the iteration process, the local minimum potential energy is obtained from eq. (6.10-48b) and the energy of interaction is the negative of that minimum potential energy.

Figure 6.10-16 shows schematically the energy of interaction as a function of the reduced pore half-width (d/σ_{ks}). This relationship between the interaction energy and the pore half-width can then be used to calculate the adsorption isotherm exactly the way we have done for the two parallel layers and two parallel slabs in Sections 6.10.6.4 and 6.10.6.6.

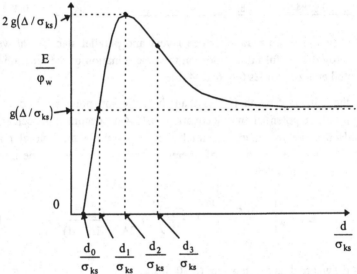

Figure 6.10-16: The energy of interaction between a molecule and two layers with sub-lattice layers (10-4-3 potential)

Activated carbon is made of many graphite-like microcrystalline units and in each unit there exists many graphite layers. The spacing between these units is usually small enough to form micropore space for adsorption. Because of the existence of these graphite layers that the 10-4-3 potential is usually used with good success to describe the adsorption of many gases or vapours in activated carbon.

6.11 Horvath and Kawazoe's Approach on the Micropore Size Distribution

In the last sections, we presented the Lennard-Jones potential method, where we can relate the interaction energy to the micropore size. Knowing this information and the local isotherm, we can either calculate the overall adsorption isotherm or with the given experimental equilibrium data we can obtain the information about the micropore size distribution.

In this section, we show another approach but very similar to what we did in Section 6.10, that of Hovarth and Kawazoe (1983), to obtain the average potential energy for a slit shape pore, from which a method is derived to determine the micropore size distribution from the information of experimental isotherm data. What to follow is the brief description of the theory due to Horvarth and Kawazoe.

6.11.1 The Basic Theory

Starting with the equation for the molar integral change of the free energy of adsorption at constant temperature

$$\Delta G = \Delta H - T\Delta S \tag{6.11-1}$$

Hovarth and Kawazoe obtained the following equation relating the gas phase pressure to the potential functions as follows:

$$R_g T \ln\left(\frac{P}{P_0}\right) = U_0 + P_a \tag{6.11-2}$$

where U_0 is the potential function describing the adsorbent-adsorbate interaction, and P_a describes the adsorbate-adsorbate-adsorbent interactions.

The potential function of a gas molecule over a graphite layer of infinite extent is:

$$\phi = 3.07\phi^*\left[\left(\frac{\sigma}{z}\right)^{10} - \left(\frac{\sigma}{z}\right)^{4}\right] \tag{6.11-3}$$

where σ is the distance between the gas molecule and the surface which gives zero interaction energy, and z is the distance from the gas molecule and the surface. Eq.(6.11-3) was derived in Section 6.10 (see eq. 6.10-5). The distance where the interaction energy is zero is obtained by setting eq.(6.10-5) to zero, and we obtain:

$$\sigma = \left(\frac{2}{5}\right)^{1/6} \sigma_{12} \tag{6.11-4}$$

Hence using this parameter σ into eq.(6.10-5), it will become:

$$\phi = \frac{10}{3} \times \frac{5^{2/3}}{2^{5/3}} \phi^* \left[\left(\frac{\sigma}{z}\right)^{10} - \left(\frac{\sigma}{z}\right)^4 \right] \qquad (6.11\text{-}5a)$$

which is the equation (6.11-3) obtained above by Hovarth and Kawazoe. Here ϕ^* is the minimum energy, and is given in terms of molecular parameters:

$$\phi^* = \frac{N_2 A_2}{3.07 (2\sigma)^4} \qquad (6.11\text{-}5b)$$

where N_2 is the number of atoms per unit area of surface, A_2 is the constant in the Kirkwood-Muller equation.

The potential function between one adsorbate molecule and the two parallel planes is given by (cf. eq. 6.10.10)

$$\phi = \frac{N_2 A_2}{(2\sigma)^4} \left[\left(\frac{\sigma}{d+z}\right)^{10} + \left(\frac{\sigma}{d-z}\right)^{10} - \left(\frac{\sigma}{d+z}\right)^4 - \left(\frac{\sigma}{d-z}\right)^4 \right] \qquad (6.11\text{-}6)$$

where σ is defined as in eq.(6.11-4), $2d$ is the distance between the two nuclei of the two parallel layers, and z is the distance of the gas molecule from the central plane.

Knowing the potential energy between the parallel planes and one molecule, the potential function between the two parallel layers filled with adsorbates is:

$$\phi = \frac{N_1 A_1 + N_2 A_2}{(2\sigma)^4} \left[\left(\frac{\sigma}{d+z}\right)^{10} + \left(\frac{\sigma}{d-z}\right)^{10} - \left(\frac{\sigma}{d+z}\right)^4 - \left(\frac{\sigma}{d-z}\right)^4 \right] \qquad (6.11\text{-}7)$$

where N_1 is the number of molecules per unit area of the adsorbate and A_1 and A_2 are defined as follows:

$$A_2 = \frac{6mc^2 \alpha_1 \alpha_2}{\frac{\alpha_1}{\chi_1} + \frac{\alpha_2}{\chi_2}} \quad ; \quad A_1 = \frac{3mc^2 \alpha_1 \chi_1}{2} \qquad (6.11\text{-}8a)$$

Here m is the electron mass, c is the speed of light, α_2 is the polarizability and χ_2 is the magnetic susceptibility of an adsorbent atom, α_1 and χ_1 are the polarizability and magnetic susceptibility of an adsorbate molecule.

The potential given as in eq. (6.11-7) varies with the distance z away from the central plane. Averaging this potential over the available distance within the slit (Figure 6.11-1), and substituting the result into eq. (6.11-2), we get:

$$R_g T \ln\left(\frac{P}{P_0}\right) = K \frac{N_1 A_1 + N_2 A_2}{(2\sigma)^4 [2d - (\sigma_1 + \sigma_2)]} \int_{-z^*}^{z^*} \left[\left(\frac{\sigma}{d+z}\right)^{10} + \left(\frac{\sigma}{d-z}\right)^{10} - \left(\frac{\sigma}{d+z}\right)^4 - \left(\frac{\sigma}{d-z}\right)^4\right] dz$$

(6.11-9a)

where K is the Avogadro's number, and z^* is given by

$$d - \frac{(\sigma_1 + \sigma_2)}{2}$$

(6.11-9b)

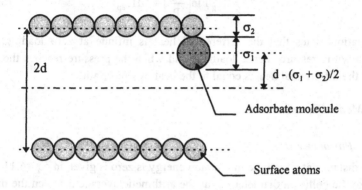

Figure 6.11-1: Schematic diagram of a slit composed of two parallel lattice planes

Integration of eq.(6.11-9a) gives:

$$R_g T \ln\left(\frac{P}{P_0}\right) = K \frac{N_1 A_1 + N_2 A_2}{\sigma^4 [2d - (\sigma_1 + \sigma_2)]} \times \left[\frac{\sigma^{10}}{9\left(\frac{\sigma_1+\sigma_2}{2}\right)^9} - \frac{\sigma^4}{3\left(\frac{\sigma_1+\sigma_2}{2}\right)^3} - \frac{\sigma^{10}}{9\left(2d - \frac{\sigma_1+\sigma_2}{2}\right)^9} + \frac{\sigma^4}{3\left(2d - \frac{\sigma_1+\sigma_2}{2}\right)^3}\right]$$

(6.11-10)

where 2d should be greater than $(\sigma_1 + \sigma_2)$.

Eq.(6.11-10) relates the gas phase pressure in terms of the width of the slit-shape pore (2d). Thus, by measuring the adsorption isotherm as a function of the reduced pressure

$$C_\mu = C_{\mu s} \times f(P/P^0)$$

(6.11-11)

we can derive the relationship between the amount adsorbed and the effective pore width by using the relationship between the relative pressure versus the effective pore width (2d-σ_2) of eq. (6.11-10).

6.11.2 Differential Heat

Assuming the state of adsorbed phase as liquid, the differential heat of adsorption is obtained from the following formula:

$$q^{diff} = R_g T \ln\left(\frac{P_0}{P}\right) + \Delta H_{vap} \qquad (6.11\text{-}12)$$

This equation states that the differential heat is infinite at zero loading, and it decreases with pressure (hence loading) and when the pressure reaches the vapour pressure the differential heat is equal to the heat of vaporisation.

6.11.3 Model Parameters

6.11.3.1 Parameter σ

The distance at which the interaction energy is zero is given in eq. (6.11-4), and if we take the collision diameter σ_{12} as the arithmetic average between the diameter of the adsorbate molecule and the adsorbent atom, we will get the following expression for the parameter σ.

$$\sigma = \left(\frac{2}{5}\right)^{1/6} \frac{\sigma_1 + \sigma_2}{2} \qquad (6.11\text{-}13)$$

For carbonaceous adsorbents, the diameter of a carbon atom is 0.34 nm, and if the adsorbate is nitrogen the diameter is 0.3 nm; thus, we have:

$$\sigma = 0.2747 \text{ nm}$$

6.11.3.2 Parameters N_1 and N_2

From Walker et al. (1966), we have:

$$N_1 = 6.7 \times 10^{14} \text{ molecules/cm}^2 \qquad (6.11\text{-}14a)$$

$$N_2 = 3.845 \times 10^{15} \text{ molecules/cm}^2 \qquad (6.11\text{-}14b)$$

Eq.(6.11-14a) is calculated from the liquid nitrogen density, which is taken as 0.808 g/cm^3 (Hovarth and Kawazoe, 1983).

6.11.3.3 Polarizability and Magnetic Susceptibility

Values for the polarizability and the magnetic susceptibility for carbon and nitrogen are tabulated in the following table (Table 6.11-1).

Table 6.11-1 The polarizability and magnetic susceptibility of carbon and nitrogen

	Polarizability (cm^3)	Magnetic susceptibility (cm^3)	Reference
Carbon	1.02×10^{24}	13.5×10^{29}	Sams et al. (1960)
Nitrogen	1.46×10^{24}	2×10^{29}	Sansonov (1968)

6.11.3.4 The working Equation for Nitrogen at 77 K

Substituting eqs.(6.11-13) and (6.11-14) into (6.11-10), we get:

$$\ln\left(\frac{P}{P_0}\right) = \frac{62.38}{2d - 0.64} \times \left[\frac{1.895 \times 10^{-3}}{(2d - 0.32)^3} - \frac{2.7087 \times 10^{-7}}{(2d - 0.32)^9} - 0.05014\right] \quad (6.11\text{-}15)$$

where d is in nm. This working equation allows us to relate the measured pressure with the effective pore width, from which the amount adsorbed can be plotted directly versus the pore width.

6.11.3.5 Tabulation of Relative Pressure versus Effective Pore Width

Eq.(6.11-15) is evaluated for a range of relative pressure, and the results are tabulated in Table 6.11-2. The effective pore width has the units of nm.

Table 6.11-2 Tabulation of the effective width versus the relative pressure

P/P_0	$2d-\sigma_2$	P/P_0	$2d-\sigma_2$	P/P_0	$2d-\sigma_2$	P/P_0	$2d-\sigma_2$
1×10^{-8}	0.3511	5×10^{-6}	0.4790	2×10^{-3}	0.7638	3×10^{-1}	2.8939
2×10^{-8}	0.3636	1×10^{-5}	0.4986	5×10^{-3}	0.8573	4×10^{-1}	3.7110
5×10^{-8}	0.3801	2×10^{-5}	0.5201	1×10^{-2}	0.9512	5×10^{-1}	4.8109
1×10^{-7}	0.3928	5×10^{-5}	0.5525	2×10^{-2}	1.0766	6×10^{-1}	6.4220
2×10^{-7}	0.4060	1×10^{-4}	0.5805	5×10^{-2}	1.3280	7×10^{-1}	9.0687
5×10^{-7}	0.4246	2×10^{-4}	0.6125	7×10^{-2}	1.4625	8×10^{-1}	14.3165
1×10^{-6}	0.4396	5×10^{-4}	0.6627	1×10^{-1}	1.6473	9×10^{-1}	29.9860
2×10^{-6}	0.4557	1×10^{-3}	0.7086	2×10^{-1}	2.2370		

Figure 6.11-2 shows the log-log plot of the effective width versus the relative pressure. Note the increase in the effective pore width with the pressure. The limit

of the pressure beyond which the Hovarth-Kawazoe method is not applicable should be taken with care because at high pressure the effective micropore half width is reaching the mesopore limit where the Horvarth-Kawazoe may not be applicable.

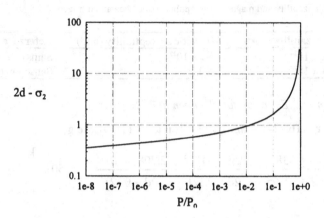

Figure 6.11-2: Plot of the effective width versus P/P_0 using the Hovarth-Kawazoe method

6.11.4 Applications

Hovarth and Kawazoe (1983) carried out adsorption and desorption of nitrogen on a carbon sample at liquid nitrogen temperature of -196 °C. The characteristics of this sample are shown in the following table.

Table 6.11-3

Properties	Units	Values
Particle density	g/cc	0.90
True density	g/cc	1.80
Micropore volume	cm^3/g	0.21
Mean micropore size	nm	0.50

The range of pressure measured is between 2.3×10^{-4} and 740 Torr, and the maximum amount adsorbed in the micropore is taken as the amount adsorbed at a relative pressure of 0.9, which is 173.2 mg nitrogen/g of carbon, and the following table tabulates the relative amount adsorbed versus the relative pressure. The effective half width is then calculated using eq.(6.11-15).

Table 6.11-4: Adsorption data of Hovarth and Kawazoe (1983) and the effective width.

$C_\mu/C_{\mu s}$	P/P_0	Effective width (nm)
0.019	3.03×10^{-7}	0.4143
0.039	5.00×10^{-7}	0.4246
0.079	1.07×10^{-6}	0.4411
0.108	1.45×10^{-6}	0.4481
0.222	2.76×10^{-6}	0.4636
0.301	4.61×10^{-6}	0.4769
0.550	1.33×10^{-5}	0.5072
0.658	5.66×10^{-4}	0.6703
0.814	2.11×10^{-3}	0.7686
0.880	1.32×10^{-2}	0.9968
0.907	6.18×10^{-2}	1.4090
0.942	1.08×10^{-1}	1.6948
0.963	2.17×10^{-1}	2.3413
0.986	5.13×10^{-1}	4.9845
0.992	7.04×10^{-1}	9.2111
0.999	8.42×10^{-1}	18.4870

The micropore size distribution is shown in Figure 6.11-3, and it shows a very sharp distribution exhibited by this sample of carbon. The mean pore width is approximately 0.5 nm.

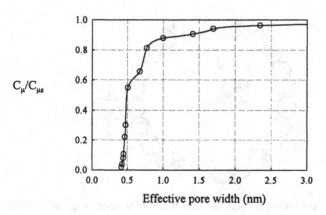

Figure 6.11-3: Micropore size distribution of a carbon sample

6.12 Cylindrical pores

In the last two sections (6.10 and 6.11), we see the analysis of the interaction energy and adsorption isotherm analysis for slit-shaped pores. We started with the potential energy between

1. two molecules or two atoms,
2. a molecule and a single lattice plane,
3. a molecule and a slab,
4. a single molecule and two parallel lattice planes,
5. a single molecule and two parallel slabs, and
6. a single molecule and one lattice layer with its associated sub-layers
7. a single molecule and two lattice layers with their associated sub-layers

The analysis of the cases 4, 5 and 7 are utilised in the study of adsorption isotherm of a nonpolar adsorbate in a microporous solid having <u>slit-shaped</u> micropores, such as activated carbon. To complete the potential theory analysis, we now deal with solids having <u>cylindrical</u> pores of molecular dimension.

6.12.1 A Molecule and a Cylindrical Surface

Here, we will consider the case of a molecule confined in a cylindrical pore having a radius R (Figure 6.12-1).

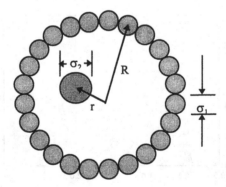

Figure 6.12-1: An adsorbate molecule in a cylindrical pore made of a single lattice layer

Let r be the distance of the center of the molecule from the center of the pore. Everett and Powl (1976) presented the following equation for the potential energy between a molecule and the surface of the cylindrical pore:

$$\varphi_{O,SL}(r) = \frac{5}{2}\pi\varepsilon^*_{1,SLP}\left[\frac{21}{32}\left(\frac{\sigma_{12}}{R}\right)^{10}\sum_{k=0}^{\infty}\alpha_k\left(\frac{r}{R}\right)^{2k} - \left(\frac{\sigma_{12}}{R}\right)^{4}\sum_{k=0}^{\infty}\beta_k\left(\frac{r}{R}\right)^{2k}\right] \quad (6.12\text{-}1a)$$

where $\varepsilon^*_{1,SLP}$ is the minimum potential of a single lattice plane, defined in eq.(6.10-4). Written in terms of the molecular parameters, it is:

$$\varepsilon^*_{1,SLP} = \frac{3}{10}\frac{N_1 A_1 + N_2 A_2}{\sigma_{12}^4} \quad (6.12\text{-}1b)$$

The parameters A_1 and A_2 are the Kirkwood-Muller dispersion constants, and are given in eqs. (6.11-8). Here σ_{12} is the collision diameter (can be taken as the arithmetic mean between the diameter of the adsorbate molecule and the surface atom), and α_k and β_k are defined as follows:

$$\alpha_k = \left[\frac{\Gamma(-4.5)}{\Gamma(-4.5-k)\cdot\Gamma(k+1)}\right]^2 \quad (6.12\text{-}2a)$$

$$\beta_k = \left[\frac{\Gamma(-1.5)}{\Gamma(-1.5-k)\cdot\Gamma(k+1)}\right]^2 \quad (6.12\text{-}2b)$$

Figure 6.12-2 shows the plot of the reduced potential energy $\varphi_{O,SL}/\varepsilon^*_{1,SLP}$ versus the reduced radius (r/σ_{12}). The parameter of the plot is (R/σ_{12}).

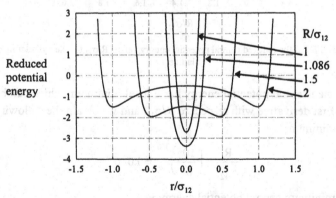

Figure 6.12-2: Plots of the reduced potential energy versus r/σ_{12} for a cylinder with a single lattice layer

We note from this figure that depending on the magnitude of (R/σ_{12}) the potential energy profile has two minima and when (R/σ_{12}) is less than $(231/64)^{1/6}$ the two minima coalesce. Before discussing any further on the behaviour of this potential energy profile we consider the potential energy at the centre of the cylindrical pore.

6.12.1.1 Central Potential Energy

From eq.(6.12-1), we obtain the potential energy at the center of the cylindrical pore by setting $r = 0$ as (see Figure 6.12-3):

$$\frac{\varphi_{O,SL}(0;R)}{\varepsilon^*_{1,SLP}} = -\frac{5\pi}{2}\left(\frac{\sigma_{12}}{R}\right)^4\left[1-\frac{21}{32}\left(\frac{\sigma_{12}}{R}\right)^6\right] \tag{6.12-3}$$

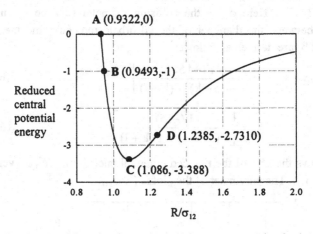

Figure 6.12-3: Plot of the reduced central potential energy versus R/σ_{12} for the cylinder made by a lattice

This reduced central potential energy has a local minimum, which can be found by taking its first derivative with respect to R/σ_{12}, and we obtain the following value for the local minimum

$$\frac{R}{\sigma_{12}} = \left(\frac{105}{64}\right)^{1/6} = 1.086 \tag{6.12-4a}$$

at which the minimum central potential energy is:

$$\left(\frac{\varphi_{O,SL}(0;R)}{\varepsilon^*_{1,SLP}}\right)_{min} = -\frac{3\pi}{2}\left(\frac{64}{105}\right)^{2/3} = -3.3877 \tag{6.12-4b}$$

When the reduced radius

$$\frac{R}{\sigma_{12}} = \left(\frac{21}{32}\right)^{1/6} \cong 0.9322$$

the central potential energy is zero, and when this reduced radius $R/\sigma_{12} = 0.94932$, the central potential energy is equal to the minimum potential of a single lattice plane, that is:

$$\left(\frac{\varphi_{O,SL}(0;R)}{\varepsilon^*_{1,SLP}}\right)_{R/\sigma_{12}=0.94932} = -1 \qquad (6.12\text{-}5)$$

This means that when R/σ_{12} increases from 0.9322 to 1.086, the reduced central potential energy decreases from 0 to - 3.388 and then increases when the radius of the reduced pore radius increases beyond 1.086.

To find the inflexion point of eq.(6.12-3), we take its second derivative with respect to R/σ_{12}, and obtain:

$$\left(\frac{R}{\sigma_{12}}\right)_{INFLEXION} = \left(\frac{231}{64}\right)^{1/6} = 1.2385 \qquad (6.12\text{-}6a)$$

which is basically the position at which the two local minima of the potential energy distribution (eq. 6.12-1) coalese into one minimum positioned at the center of the cylinder. The central potential energy at this value is:

$$\frac{\varphi_{O,SL}(0;R)}{\varepsilon^*_{1,SLP}} = -\frac{45\pi}{22}\left(\frac{64}{231}\right)^{2/3} = -2.731 \qquad (6.12\text{-}6b)$$

The following figure (Figure 6.12-4) shows the typical plot of the negative of the minimum potential energy versus R/σ_{12}.

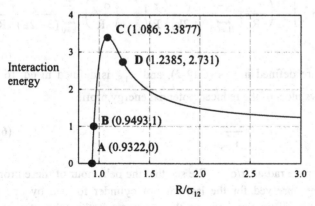

Figure 6.12-4: Plot of the interaction energy versus R/σ_{12} for a cylinder of single lattice layer

When R/σ_{12} is between 0.9322 and 1.2385, the minimum potential energy occurs at $r = 0$, and the minimum potential energy is then calculated from eq.(6.12-3). When R/σ_{12} is greater than 1.2385, the minimum potential energy must be found by minimizing eq.(6.12-1).

The interaction energy is the negative of the minimum potential energy for the case of a nonpolar molecule and a cylindrical pore. For the purpose of computation, this interaction energy is

$$\frac{E}{\varepsilon^*_{1,SLP}} = \frac{5\pi}{2}\left(\frac{\sigma_{12}}{R}\right)^4\left[1-\frac{21}{32}\left(\frac{\sigma_{12}}{R}\right)^6\right] \qquad (6.12\text{-}7a)$$

for $R/\sigma_{12} < 1.2385$.

For $R/\sigma_{12} > 1.2385$, the following approximate solution may be used to relate the interaction energy versus the pore radius

$$\frac{E}{\varepsilon^*_{1,SLP}} = 3.3877 - a\left(1-\frac{1.086}{R/\sigma_{12}}\right)^n \cdot \exp\left(\frac{b}{R/\sigma_{12}-c}\right) \qquad (6.12\text{-}7b)$$

6.12.2 A Molecule and a Cylindrical Slab

The last section considers the case when the cylinder is made of one single layer. In this section, we will consider a cylinder whose thickness is semi-infinite in extent and all atoms of the cylinder will exert interaction with an atom or a molecule confined in the pore space. For this case, the potential energy is given by the following equation (Everett and Powl, 1976).

$$\varphi_{O,S}(r) = \frac{27}{2\sqrt{10}}\pi\varepsilon^*_{1,S}\left[\frac{21}{32}\left(\frac{\sigma_{12}}{R}\right)^9\sum_{k=0}^{\infty}\frac{\alpha_k}{(9+2k)}\left(\frac{r}{R}\right)^{2k} - \left(\frac{\sigma_{12}}{R}\right)^3\sum_{k=0}^{\infty}\frac{\beta_k}{(3+2k)}\left(\frac{r}{R}\right)^{2k}\right]$$

$$(6.12\text{-}8)$$

where α_k and β_k are defined in eqs (6.12-2), and $\varepsilon^*_{1,S}$ is defined in eq. (6.10-7). Figure 6.12-5 shows plots of the reduced potential energy profile

$$\frac{\varphi_{O,S}}{\varepsilon^*_{1,S}} \qquad (6.12\text{-}9)$$

versus the reduced pore radius (r/σ_{12}). We see that the behaviour of these profiles is very similar to that observed for the last case of cylinder formed by one single lattice layer. The fine distinction between the two cases in the enhancement is the potential energy due to the interaction of the confined molecule with more atoms of the cylindrical pore.

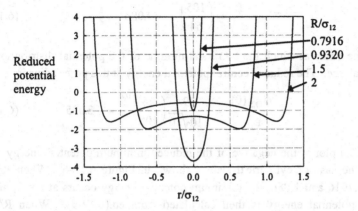

Figure 6.12-5: Plots of the reduced potential energy versus R/σ_{12} for a cylinder of semi-infinite thickness

The central potential energy is obtained by setting $r = 0$ in eq. (6.12-8):

$$\frac{\varphi_{0,S}(0)}{\varepsilon_{1,S}^*} = -\frac{9\pi}{2\sqrt{10}}\left(\frac{\sigma_{12}}{R}\right)^3\left[1 - \frac{7}{32}\left(\frac{\sigma_{12}}{R}\right)^6\right] \quad (6.12\text{-}9)$$

This central potential energy has a minimum at the position

$$\frac{R}{\sigma_{12}} = \left(\frac{21}{32}\right)^{1/6} = 0.932 \quad (6.12\text{-}10)$$

at which the reduced central potential energy is:

$$\frac{\varphi_{0,S}(0)}{\varepsilon_{1,S}^*} = -\frac{3\pi}{\sqrt{10}}\left(\frac{32}{21}\right)^{1/2} = -3.679 \quad (6.12\text{-}11)$$

At a distance $R/\sigma_{12} = 0.7916$, the central potential energy is the same as the minimum energy corresponding to the infinite slab, that is:

$$\left(\frac{\varphi_{0,S}(0)}{\varepsilon_{1,S}^*}\right)_{R/\sigma_{12}=0.7916} = -1 \quad (6.12\text{-}12)$$

Finally, by taking the second derivative of eq.(6.12-9) with respect to R/σ_{12}, and setting to zero we find

$$\frac{R}{\sigma_{12}} = \left(\frac{105}{64}\right)^{1/6} = 1.086 \qquad (6.12\text{-}13)$$

which is the value at which the two local minima of the potential distribution (eq. 6.12-8) coalesce. The reduced central potential energy at this pore radius is

$$\frac{\varphi_{0,S}(0)}{\varepsilon^*_{1,S}} = -\frac{39\pi}{10\sqrt{10}}\left(\frac{64}{105}\right)^{1/2} = -3.025 \qquad (6.12\text{-}14)$$

A typical plot of the negative of the reduced minimum potential energy versus R/σ_{12} for the case of cylindrical slab is shown in Figure 6.12-6. When R/σ_{12} is between 0.7916 and 1.086, the minimum potential energy occurs at r = 0, and the minimum potential energy is then calculated from eq.(6.12-9). When R/σ_{12} is greater than 1.086, the minimum potential energy must be found by minimizing eq.(6.12-8).

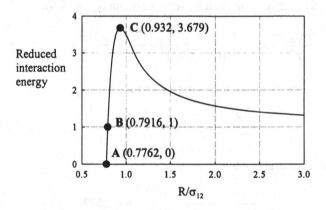

Figure 6.12-6: Plot of the reduced interaction energy versus R/σ_{12} for a cylinder of semi-infinite thickness

6.12.3 Adsorption in a Cylindrical Pore

Using the approach of Hovarth and Kawazoe (1983) for the case of adsorption in a slit shape pore, Saito and Foley (1991) solved for the case of cylindrical pore. The following assumptions are made in the analysis.

1. Pore is a uniform cylinder of infinite extent
2. The wall is made up of a single layer of atoms
3. Only dispersive force is allowed for, and the interaction is only between the adsorbent and adsorbate.

With these assumptions, the potential energy distribution is given in eq.(6.12-1). The free energy of adsorption is taken as the average of the intermolecular potential. The method of averaging depends on the way molecules move about in the cylindrical pore. If this movement is restricted to that along the pore radius, we have a line-averaged free energy, while if the movement is two dimensional, that is molecules are free to move across the pore, we have an area-averaged free energy.

The average free energies for these two cases are:

$$\langle \varphi \rangle = \frac{\int_0^{R-\sigma_{12}} \varphi(r)\,dr}{\int_0^{R-\sigma_{12}} dr} \tag{6.12-15a}$$

and

$$\langle \varphi \rangle = \frac{\int_0^{R-\sigma_{12}} 2\pi r \varphi(r)\,dr}{\int_0^{R-\sigma_{12}} 2\pi r\,dr} \tag{6.12-15b}$$

respectively. If the free energy of adsorption is assumed to be equal to the net energy of interaction, we get:

$$R_g T \ln\left(\frac{P}{P_0}\right) = N_{AV} \times \langle \varphi \rangle \tag{6.12-16}$$

where N_{AV} is the Avogadro's number.

Substituting eqs.(6.12-15) into eq.(6.12-16), we obtain the following expressions written in terms of system parameters:

$$R_g T \ln\left(\frac{P}{P_0}\right) = \frac{3\pi N_{AV}}{4} \frac{N_1 A_1 + N_2 A_2}{\sigma_{12}^4} \left[\frac{21}{32}\left(\frac{\sigma_{12}}{R}\right)^{10} \sum_{k=0}^{\infty} \frac{\alpha_k}{(2k+1)}\left(1-\frac{\sigma_{12}}{R}\right)^{2k} - \left(\frac{\sigma_{12}}{R}\right)^4 \sum_{k=0}^{\infty} \frac{\beta_k}{(2k+1)}\left(1-\frac{\sigma_{12}}{R}\right)^{2k}\right]$$

and

$$R_g T \ln\left(\frac{P}{P_0}\right) = \frac{3\pi N_{AV}}{4} \frac{N_1 A_1 + N_2 A_2}{\sigma_{12}^4} \left[\frac{21}{32}\left(\frac{\sigma_{12}}{R}\right)^{10} \sum_{k=0}^{\infty} \frac{\alpha_k}{(k+1)}\left(1-\frac{\sigma_{12}}{R}\right)^{2k} - \left(\frac{\sigma_{12}}{R}\right)^4 \sum_{k=0}^{\infty} \frac{\beta_k}{(k+1)}\left(1-\frac{\sigma_{12}}{R}\right)^{2k}\right]$$

(6.12-17)

for the line averaged and area averaged free energy, respectively.

6.12.3.1 Applications

Using the above equations, Saito and Foley (1991) applied them to the system of adsorption of argon on surface oxide ion. The relevant parameters for this system are given in the following table.

Parameter	Symbol	Units	Oxide ion	Argon
Diameter	σ	nm	0.276	0.336
Polarizability	α	cm^3	2.5×10^{-24}	1.63×10^{-24}
Magnetic susceptibility	χ	cm^3	1.3×10^{-29}	3.25×10^{-29}
Density	N	molecules/cm^2	1.31×10^{15}	8.52×10^{14}

The collision diameter σ_{12} is the mean of the diameter of the adsorbate molecule and the surface atom, that is:

$$\sigma_{12} = \frac{\sigma_1 + \sigma_2}{2} = \frac{0.276 + 0.336}{2} = 0.306 \quad (6.12\text{-}18)$$

and the relevant working equations for argon adsorbate at 87 K are:

$$\ln\left(\frac{P}{P_0}\right) = 36.48 \left[\sum_{k=0}^{\infty} \frac{1}{(2k+1)}\left(1-\frac{0.306}{R}\right)^{2k} \times \left\{\frac{21}{32}\alpha_k\left(\frac{0.306}{R}\right)^{10} - \beta_k\left(\frac{0.306}{R}\right)^4\right\}\right] \quad (6.12\text{-}19a)$$

and

$$\ln\left(\frac{P}{P_0}\right) = 36.48 \left[\sum_{k=0}^{\infty} \frac{1}{(k+1)}\left(1-\frac{0.306}{R}\right)^{2k} \times \left\{\frac{21}{32}\alpha_k\left(\frac{0.306}{R}\right)^{10} - \beta_k\left(\frac{0.306}{R}\right)^4\right\}\right] \quad (6.12\text{-}19b)$$

for the line averaged and area averaged approach. The radius R is in nm. The effective diameter of the cylindrical pore is (2R-0.276) nm. Applications of eq. (6.12-19) are detailed in Saito and Foley (1991).

6.13 Adsorption-Condensation Theory of Sircar (1985)

We have presented in previous sections the analysis of the intermolecular interaction, from which the interaction energy is derived in terms of the size of the pore. The method is particularly applicable to micropore. In the next section we shall consider a mesoporous solid where pore volume is distributed. The mechanism of adsorption is the surface adsorption and when the gas pressure is sufficiently high, capillary condensation will occur. The analysis presented below is basically the synthesis of those adsorption mechanisms, and this was done by Sircar and we will present the theory below.

6.13.1 Mesoporous Solid

6.13.1.1 Pore volume and surface area distribution

Consider a mesoporous solid whose the pore volume is described by the following Gamma distribution function in terms of pore radius r:

$$\frac{dV}{dr} = \frac{\overline{V}\,\alpha^{p+1}}{\Gamma(p+1)} r^p\, e^{-\alpha r} \qquad (6.13\text{-}1)$$

where dV is the volume of pores per unit mass of sorbent having radii between r and r + dr, p > 0 and α > 0 are parameters of the distribution, and $\overline{V}$ is the total pore volume.

Given a differential pore volume, the differential surface area within the pores having a radius between r and r + dr is

$$dS = \frac{2dV}{r} = \frac{2\overline{V}\,\alpha^{p+1}}{\Gamma(p+1)} r^{p-1}\, e^{-\alpha r} \qquad (6.13\text{-}2)$$

in which pores are assumed to be cylindrical in shape. The total surface area is then:

$$\overline{S} = \frac{2\alpha}{p}\overline{V} \qquad (6.13\text{-}3)$$

For a given pore volume distribution and a surface area distribution as given in the above equations, we calculate the cumulative specific pore volume and the cumulative specific surface area as shown below:

$$V(r) = \overline{V}\left[1 - \frac{\Gamma(p+1,\theta)}{\Gamma(p+1)}\right], \qquad (6.13\text{-}4)$$

$$S(r) = \overline{S}\left[1 - \frac{\Gamma(p,\theta)}{\Gamma(p)}\right] \qquad (6.13\text{-}5)$$

where $\Gamma(p+1,\theta)$ is the incomplete gamma function, and the parameter θ (dimensionless) is defined below:

$$\theta = \alpha r \qquad (6.13\text{-}6)$$

We have completed the presentation of the pore volume and surface area. Now we will study how these volume and surface area can be used in the analysis of adsorption. The volume is utilised in the capillary condensation while the surface area is for the surface adsorption.

6.13.1.2 Capillary Condensation and Surface Adsorption

We first consider the capillary condensation. The proposed capillary condensation theory of adsorption assumes that for a given adsorbate, pressure P and temperature T, certain pores of the adsorbent $(0 \leq r < r^*)$ are filled with the liquid adsorbate according to the Kelvin model of capillary condensation, while the remainder of the pores $(r^* < r < \infty)$ provides surfaces for physical adsorption, the extent of which depends on P and T (Figure 6.13-1).

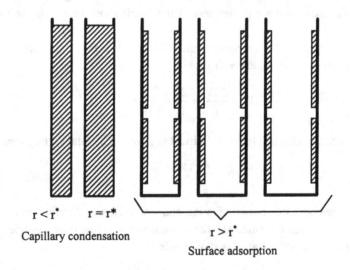

Figure 6.13-1: Capillary condensation of mesopores

According to the Kelvin's capillary condensation theory, the vapour pressure depressed by the curvature of a fluid residing in a cylindrical pore is:

$$\ln\left(\frac{P}{P_0}\right) = -\frac{2\sigma v_M \cos\theta}{R_g T \cdot r_K^*} = -\frac{C}{r_K^*} \qquad (6.13\text{-}7)$$

where P_0 is the vapour pressure of a flat liquid surface, σ is the surface tension, θ is the contact angle, v_M is the liquid molar volume and r_K^* is the effective free radius of the pore which is filled at the relative pressure (P/P_0), which is equal to the pore radius minus the thickness of the adsorbed layer:

$$r_K^* = r^* - t^* \qquad (6.13\text{-}8)$$

where t^* is the average thickness of the physically adsorbed layer on the surface of the critical pore of radius r^*.

The surface adsorption layer thickness is calculated as follows. It is assumed that the extent of physical adsorption on the pore surfaces is given by the following Langmuir-like model

$$C_s = C_{sm} \frac{bx}{1+bx} \qquad (6.13\text{-}9a)$$

where x is the reduced pressure defined as:

$$x = P/P_0 \qquad (6.13\text{-}9b)$$

Here C_s (mole/cm^2) is the amount of vapours adsorbed per unit surface area, C_{sm} (mole/cm^2) is the monolayer adsorption capacity, b represent the vapour-solid adsorption interaction parameter, defined as follows:

$$b = b_\infty \exp\left[(Q-L)/R_g T\right] \qquad (6.13\text{-}10)$$

where Q is the isosteric heat of adsorption and L is the heat of vaporization. The appearance of the heat of liquefaction in eq. (6.13-10) is due to the usage of the reduced pressure scaled against the vapour pressure.

If we assume that the physically adsorbed vapour has a liquid like density, then the average thickness t^* is given by

$$t^* = C_s v_M, \qquad (6.13\text{-}11)$$

By combining eqs. (6.13-7), (6.13-8) and (6.13-11), we obtain a relation between the critical pore radius with respect to the reduced pressure

$$r^* = -\frac{C}{\ln x} + C_{sm}v_M \frac{bx}{1+bx} \qquad (6.13\text{-}12)$$

Knowing this critical radius, we then can calculate the pore filling by capillary condensation in pores having radii less than r^*, while for pores having radii greater than r^* the surface adsorption is the sole mechanism of adsorption.

The volume of pore less than r^* is filled with adsorbed species, that is the amount adsorbed in those pores is:

$$C_{\mu 1} = \frac{V(r^*)}{v_M} \qquad (6.13\text{-}13)$$

where $V(r^*)$ is the cumulative volume of pores having radii between 0 and r^*.

The surface area of pores having radii greater than r^* is covered with physically sorbed species, that is:

$$C_{\mu 2} = [\overline{S} - S(r^*)] C_{sm} \frac{bx}{1+bx} \qquad (6.13\text{-}14)$$

Thus the total amount adsorbed is the sum of the above two concentrations:

$$C_\mu = \frac{V(r^*)}{v_M} + [\overline{S} - S(r^*)] C_{sm} \frac{bx}{1+bx} \qquad (6.13\text{-}15)$$

Written the above equation explicitly in terms of the structural parameters, we have the following needed adsorption isotherm equation:

$$\frac{C_\mu}{C_{\mu s}} = 1 + 2\left(\frac{\alpha}{p}\right)(C_{sm}v_M) \frac{\Gamma(p,\theta^*)}{\Gamma(p)} \cdot \frac{bx}{1+bx} - \frac{\Gamma(p+1,\theta^*)}{\Gamma(p+1)} \qquad (6.13\text{-}16a)$$

where

$$C_{\mu s} = \overline{V}/v_M; \qquad \theta^* = \alpha r^* \qquad (6.13\text{-}16b)$$

Here $C_{\mu s}$ is the maximum capacity when all pores are filled with liquid adsorbate.

Eq. (6.13-16) represents an adsorption isotherm for vapours based on the capillary condensation theory of adsorption. This new isotherm has the correct limits at low pressure as well as when pressure approaches the vapour pressure.

- When $x \to 0$, $C_\mu \to (C_{sm} b \overline{S}) x$ \qquad (6.13-17a)
- When $x \to 1$, $C_\mu \to C_{\mu s}$ \qquad (6.13-17b)

6.13.1.3 Model Parameters

The isotherm model equation has five parameters, and they are the monolayer surface capacity C_{sm}, the interaction affinity constant b, the capillary condensation parameter C, and the pore structural parameters α and p. If the total pore volume and the total surface area are known, the parameters α and p are related to each other according to eq. (6.13-3).

- The monolayer surface capacity C_{sm} can be calculated by assuming that the adsorbed molecules have the same packing as the molecules of the condensed phase in their plane of closest packing, that is:

$$C_{sm} = 1.085 \times 10^{-8} \, v_M^{-2/3} \, (\text{mole/cm}^2) \qquad (6.13\text{-}18)$$

where the units of the liquid molar volume are cm^3/mole.

- The parameter $b\overline{S}$ can be estimated from slope of Henry law region (eq. 6.13-17a)

- The capillary condensation parameter C can be estimated from the Kelvin equation, assuming the contact angle between the liquid and the pore wall is zero

$$C = 2\sigma v_M / R_g T \qquad (6.13\text{-}19)$$

- The total pore volume can be calculated from the maximum adsorption capacity, that is:

$$\overline{V} = C_{\mu s} v_M \qquad (6.13\text{-}20)$$

This now leaves only two adjustable parameters left in the optimisation procedure, and these two parameters are either the set of α and $\overline{S}$ or the set of p and $\overline{S}$, since α and p are related to each other, according to eq. (6.13-3).

6.13.2 Micropore-Mesoporous solids (Activated carbon type)

The last section presents the analysis of the capillary condensation and surface adsorption in mesoporous solids. This analysis can be easily extended to the class of microporous solids, where micropores exist. The pore size distribution of this class of solid contains micropores (less than 20A) and mesopores and macropores (greater than 20A). In micropore, the micropore filling is the main adsorption mechanism. The mechanisms for adsorption in meso- and macropores are those discussed in the last section. Readers are referred to Sircar (1991) for further exposition of the analysis of this microporous solid class.

6.14 Conclusion

This chapter has addressed a number of approaches to deal with heterogeneous solids. The energy distribution is general but the choice of the energy distribution is sometimes questionable in the sense whether it does reflect the true distribution for the interaction between the solid and the adsorbate concerned. Nevertheless, it does provide a convenient way of correlating data of many practical solids. For solids such as activated carbon and non-polar adsorbates, the approach using the Lennard-Jones potential theory is useful to relate the adsorption affinity to the size of the micropore, from which the overall adsorption isotherm can be derived. This approach has been used successfully on many systems such as paraffins on activated carbon. Research in the area is very fruitful and students are encouraged to pursue this area and develop their own thoughts for better ways of relating the adsorption affinity in terms of adsorbate properties and adsorbent's characteristics, such as their structural properties and surface chemistry.

7

Fundamentals of Diffusion and Adsorption in Porous Media

7.1 Introduction

In the last five chapters (Chapters 2 to 6), you have learnt about the various aspects of adsorption equilibria. This is the foremost information that you must have to gain an understanding of an adsorption system. To properly design an adsorber, however, you need to know additional information beside the adsorption equilibria. That necessary information is the adsorption kinetics. The reason for this is simply that most practical solids used in industries are porous and the overall adsorption rate is limited by the ability of adsorbate molecules to diffuse into the particle interior. Diffusion processes in porous media, such as adsorbents, are reasonably understood. This chapter and the next few chapters will address the process of diffusion and its influence on the overall adsorption rate.

Transport of gases and liquids in capillaries and porous media can be found in numerous applications, such as capillary rise, flow of gases into adsorbents, flow of gases into porous catalysts, flow of underground water, just a few examples of significant importance in chemical engineering. Before we deal with a porous medium, it is important to consider a simpler medium, a straight cylindrical capillary, whereby several types of flow can be more readily identified. These are:

- <u>Free molecular diffusion (Knudsen):</u> This type of diffusion is sometimes called molecular streaming. This flow is induced by collision of gaseous molecules with the pore wall of the capillary (that is when the mean free path is greater than the capillary diameter). Because of the collision of gaseous molecules with the wall of the capillary being the driving force for the Knudsen diffusion, transport of molecules of different type are independent of each other.

- <u>Viscous flow (streamline flow):</u> This is also called the Poiseuille flow. This flow is driven by a total pressure gradient, and as a result the fluid mixture moves through the capillary without separation because all species move at the same speed.
- <u>Continuum diffusion</u>: This flow is resulting from the collisions among molecules of different type, not of the same type because there is no net momentum change due to the collisions among molecules of the same type. This situation happens when the mean free path is much less than the diameter of the capillary.
- <u>Surface diffusion</u>: Different molecules have different mobility on the surface of the capillary due to their different extent of interaction with the surface. Hence a binary mixture can be separated using this type of flow, like the Knudsen diffusion.

The pore structure of a real solid is so complex that one has to model (idealise) the structure so that it can be represented mathematically. The model has to be simple so that the mathematics of diffusion and adsorption is tractable, but at the same time is reasonably complex enough to bring out the features of the solid and their influence on diffusion and adsorption. One of the simplest ways is to assume that the pore structure is a bundle of parallel capillaries running through the medium in the direction of the flow. Let us assume that all capillaries have the same length, L_c. Remember that this length is not the same as the length of the medium, L, because of the random orientation of the capillary. As a result the capillary length is generally longer than the medium thickness. To account for this, we introduce a tortuosity, defined as follows (Epstein, 1989):

$$\tau = \frac{L_c}{L} \geq 1 \qquad (7.1\text{-}1)$$

Thus by definition, the tortuosity, τ, is greater than unity.

7.1.1 Historical Development

Due to the significant importance of this topic, it was researched by many scientists and engineers for more than a century. The following table (Table 7.1-1) shows the historical development of major discoveries relating to mass transfer and related areas (Kaviany, 1991; Cunningham and Williams, 1980).

Table 7.1-1 Historical development of diffusion

Year	Authors	Achievement
1827	Navier	Momentum equation
1829	Graham	Mass diffusion in gas
1831	Graham	Diffusion law
1839	Hagen	Linear flow in pipe
1840	Poiseuille	Linear flow in pipe
1845	Stokes	Momentum equation
1846	Graham	Effusion law
1855	Fick	Law of mass transport
1856	Darcy	Empirical flow equation
1859	Maxwell	Distribution of velocity in gases
1860	Maxwell	Binary gas diffusivity & model in porous media
1870	Kelvin	Capillary condensation theory
1871	Stefan	Transport equations
1875	Knudt and Warburg	Slip flow between Knudsen and viscous flow
1878	Gibbs	Thermodynamics treatment of interface
1885	Boltzmann	General transport equation
1904	Buckingham	Diffusion cell
1905	Einstein, Smoluchowski	Random walk diffusion model
1909	Knudsen	Flow of rarified gases and observation of slip flow
1916	Langmuir	Theory of monolayer adsorption
1920	Dayne	Time lag method
1927	Kozeny	Permeability equation
1937	Carman	Permeability equation
1939	Thiele	Theory of diffusion and reaction in porous solids
1952	Ergun	Inertia effect to the Darcy equation
1953	Taylor	Hydrodynamics dispersion in tubes
1955	Hoogschagen	Rediscovery of Graham law of diffusion
1956	Aris	Hydrodynamics dispersion in tubes
1961	Evans, Watson, Mason	Constitutive equations for diffusion in porous media
1968	Luikov	Theories of heat and mass transfer
1977	Jackson	Stefan-Maxwell equations for reaction problems
1990	Krishna	Postulation of equation for surface diffusion

7.2 Devices used to Measure Diffusion in Porous Solids

To characterize diffusion in porous media, many experimental devices have been proposed and used successfully to determine transport diffusion coefficients. Some of these devices which are of historical significance are discussed here.

7.2.1 Graham's System

In the Graham experiment, a tube containing gas B is immersed in a water bath with the porous media mounted at the upper end of the tube (Figure 7.2-1).

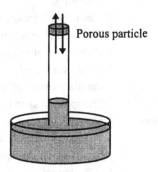

Figure 7.2-1: First Graham's system

Gas B diffuses out of the tube and the bulk gas A in the surrounding diffuses into the tube. Because the net transport of gas is not zero (that is the molar rate of A entering the tube is not the same as that of B exiting the tube), the water level inside the tube will either rise or fall. If gas B is heavier, the water level will fall, and if the gas B is lighter the level will rise. Take the latter case, because gas B is lighter, its molar transport out of the tube will be faster than the molar rate of A coming into the tube; hence the pressure in the tube will drop below the atmospheric, causing a rise in the water level. To maintain the constant pressure in the tube during the course of the experiment, Graham adjusted the tube so that the water level inside the tube is always the same as that of the water bath. He observed that the ratio of the two molar fluxes is equal to the inverse of the square root of the ratio of molecular weights of the two diffusing gases, that is

$$\frac{N_A}{N_B} = -\sqrt{\frac{M_B}{M_A}} \qquad (7.2\text{-}1)$$

The minus sign in eq. (7.2-1) is because A and B are diffusing in the opposite direction. Eq. (7.2-1) is known as the Graham's law of diffusion for open systems.

7.2.2 Hoogschagen's System

Unaware of the experiments of Graham, Hoogschagen (1953), more than a century later!, developed an experiment to rediscover the square root of molecular weight dependence. His experimental set up is shown in Figure (7.2-2).

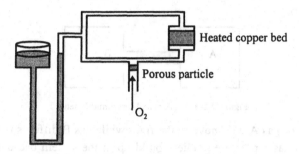

Figure 7.2-2: Hoogschagen schematic diagram

This set up was designed to perform steady state experiments. Binary diffusion of oxygen and other gases, such as nitrogen, helium and carbon dioxide were used. Oxygen is supplied at the bottom of the porous plug, and diffuses upward through the plug in exchange for the other gas inside the loop. Oxygen molecules entering the loop will be taken up by the copper bed maintained at 480 °C, at which temperature copper will react with oxygen to form solid copper oxides, removing oxygen from the atmosphere. The circulation of gas within the loop is caused by the thermal convection due to the heat generated by the copper bed. The pressure inside the loop is maintained atmospheric by raising or lowering the burette connected to the loop. The flux of the outgoing gas from the loop is calculated from the change in the liquid level of the burette, and the incoming oxygen flux is measured by weighing the copper before and after the experiment. Except for oxygen-carbon dioxide/ activated carbon system where the surface diffusion is possible, all other systems studied by Hoogschagen follow the relation given in eq. (7.2-1), confirming the observation made by Graham more than a century ago.

The porous plug used by Hoogschagen was made by compressing granules in the size range of 1 to 8 microns, which would give pores much larger than the mean free path at atmospheric pressure. Only a few experiments were carried out by Hoogschagen. Further experimental evidence was provided by Wicke and Hugo (1961), Wakao and Smith (1962) and Remick and Geankoplis (1973, 1974). Knaff and Schlunder later (1985) carried out experiments to further confirm the Graham's law up to a pore size of 2 microns.

7.2.3 Graham and Loschmidt's Systems

In this experiment of Graham and Loschmidt, the two bulbs are joined together by a tube containing either a porous medium or a capillary (Figure 7.2-3). The left bulb contains gas A, while the right bulb contains gas B, having the same pressure as that of gas A.

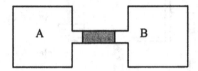

Figure 7.2-3: Graham and Loschmidt's setup 1

The flow of gas A will move to the right while gas B diffuses to the left. This type of set up has a pressure gradient build-up in the system because the counter-current flows of A and B are not equimolar, that is the molar flow of A to the right is not the same as that of B to the left. Let us take an example where A is the heavier gas, the left bulb pressure will increase while the right bulb pressure will decrease because the diffusion rate of A through the plug or capillary is slower than the diffusion rate of B. The resulting pressure gradient will then cause a viscous flow from the left to the right retarding the rate of molar flux of B to the left. This induced viscous flow will complicate the study of diffusion phenomena.

To overcome this pressure problem, they later developed another system shown in Figure (7.2-4) where the two bulbs are connected to each other through a small tube containing a drop of oil acting as a "frictionless" piston, that is it will move if there is a small pressure gradient across the two faces of the oil droplet. Take the last case where gas B is the lighter gas; hence the molar diffusion flux of A is less than the flux of B, leading the increase in pressure in the left bulb. Due to this increase in pressure in the left bulb, the oil droplet moves to the right, resulting in a balance in the pressures of the bulbs. The rate of movement of this oil piston provides the net flux of A and B through the porous plug.

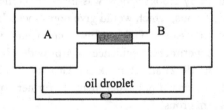

Figure 7.2-4: Graham and Loschmidt's setup 2

7.2.4 Stefan Tube

A simple device used by Stefan (Cunningham and Williams, 1980) to study the molecular diffusion is shown in Figure 7.2-5.

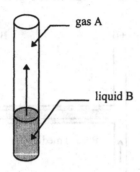

Figure 7.2-5: Stefan tube

This device is simply a tube partially filled with liquid B. This liquid B evaporates and the evaporating molecules diffuse through A up the tube. If the gas A is neither soluble in B nor reacts with B, the flux of A relative to a fixed frame of coordinate will be zero, that is the diffusive flux of A down the tube (because of its concentration gradient) is balanced by its convective flux up the tube caused by the transport of B. The flux of B is given by (Bird et al., 1960):

$$N_B = \frac{PD_{AB}}{LR_g T} \ln\left(\frac{1-p_{B2}/P}{1-p_{B1}/P}\right) \qquad (7.2-2)$$

where p_{B2} is the partial pressure of B at the top of the tube (usually zero if B is removed sufficiently fast by a moving stream of A across the tube), and p_{B1} is the partial pressure of B at the liquid gas interface, which is simply the vapor pressure of B. The variable P is the total pressure and L is the length of the gas space above the liquid surface to the top of the tube. Thus by measuring this flux by either analysing the flowing gas A for the B concentration or measuring the drop in liquid B level, eq. (7.2-2) can be readily used to calculate the binary diffusivity. This example of the Stefan tube is a semi-open system.

The semi-open system of Stefan has been used to calculate the binary diffusivity. Lee and Wilke (1954) criticised this method. However, with the exception of solvents of high volatility, the Stefan method is one of the convenient methods for calculating the binary diffusivity.

7.2.5 Diffusion Cell

In the diffusion cell configuration, introduced by Buckingham in 1904 and later exploited by Wicke (1940) and Wicke and Kallanbach (1941), gases A and B diffuse across the porous medium as shown in Figure 7.2-6.

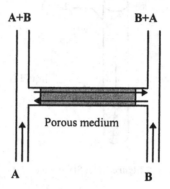

Figure 7.2-6: Wicke-Kallanbach's diffusion cell

In the left side, gas A flows in the direction as shown and picks up B due to the diffusion of B from the other side of the porous medium. Similarly, B at the other side will pick up A from its diffusion through the porous medium. The flow rates of the two sides can be carefully adjusted to give zero pressure gradient across the media (that is the total pressure is uniform throughout the porous medium). The concentrations of gases A and B are analysed by detectors, such as the thermal conductivity cell, and then the diffusive fluxes of A and B can be calculated. This is the steady state method. Recently, this method was extended to allow for transient operation such as a step change or square pulse in one chamber and the response is monitored in the other chamber. With this transient operation, the contribution from the dead end pore can be studied. This contribution is not seen by the steady state method, but its advantage is the ease of operation under isothermal operations. Detailed analysis of diffusion cell under steady state and transient conditions is provided in Chapter 13.

7.3 Modes of Transport

As we have discussed in the introduction, there are basically four modes of transport of molecules inside a porous medium. They are free molecular diffusion (Knudsen), viscous flow, continuum diffusion and surface diffusion.

- Free molecule (Knudsen): This mode of diffusion is due to Knudsen, who observed the transport of molecules as colliding and bouncing of molecules back from the wall of the porous medium (Figure 7.3-1). The driving force for this transport is the concentration gradient and the parameter characterizing this transport is the so-called Knudsen diffusivity $D_{K,i}$ for the species i or the Knudsen parameter K_0 (will be defined in Section 7.4). The subscript K denotes for Knudsen. The Knudsen flux depends on the molecular weight of the diffusing species. Molecules with smaller molecular weight travel faster than the ones with higher molecular weight under the same concentration gradient. Thus, separation of mixtures is possible with this mechanism.

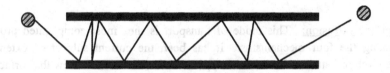

Figure 7.3-1: Knudsen diffusion mechanism

- Viscous flow: This mode of transport is due to a total pressure gradient of a continuum fluid mixture (Figure 7.3-2). Hence, there is no separation of species due to the viscous flow. The driving force is the total pressure gradient and the parameter characterizing the transport is the mixture viscosity, μ, and the viscous flow parameter, B, which is a function of solid properties only. The flow inside the pore is assumed laminar, hence the velocity profile is parabolic in shape.

P_1 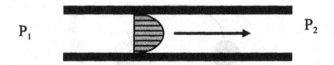 P_2

Figure 7.3-2: Viscous flow mechanism

- Continuum diffusion: The third mode of transport is the continuum diffusion, where the molecule-molecule collision is dominant over the collisions between molecules and the wall (Figure 7.3-3). In this mode of transport, different species move relative to each other. The parameter characterising this relative motion between species of different type is the binary diffusion coefficient, D_{ij}, where the subscripts i and j denote the species i and j, respectively. Because of the dependence of this parameter on the collision between molecules, the binary diffusivity is a function of the total pressure and temperature.

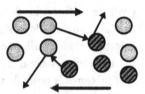

Figure 7.3-3: Continuum diffusion mechanism

- Surface diffusion: This mode of transport is the most complicated process among the four mechanisms. It has been the current subject of extensive research to better understand this type of diffusion. We can view the surface as a flat surface with specific sites, at which adsorbed molecules are located. Assuming the energy depth of these sites to be larger than the thermal energy of a molecule (kT, where k is the Boltzmann constant), molecules at each site must attain enough energy to move from one site to the next vacant site (Figure 7.3-4). This is the simple picture of the hopping mechanism, commonly used in the literature to describe the surface transport. Another mechanism called the hydrodynamics model was proposed by Gilliland et al. (1958), but the hopping model is more appropriate for many gas phase adsorption systems, where adsorption usually occurs below the capillary condensation region.

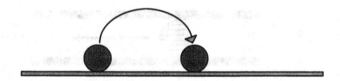

Figure 7.3-4: Surface diffusion mechanism

The three modes of transport, Knudsen, viscous, and continuum diffusion were described by Graham in 1863. A combination of viscous flow and Knudsen flow leads to a phenomenon called viscous slip. This was observed experimentally by Knudt and Warburg (1875). Another combination between viscous flow and continuum diffusion leads to diffusive slip phenomenon (Kramers and Kistemaker, 1943).

When we deal with diffusion through a porous medium, we need to describe the medium. How do we describe the medium mathematically? The simplest picture of accounting for the solid structure is absorbing all structural properties into transport coefficients or into constants of proportionality, such as the tortuosity factor. Another approach is to model fundamentally the solid structure and then incorporate it into the diffusion model. This approach is complex, and is not a subject of this book. Approaches such as the Monte-Carlo simulation (Abbasi et al., 1983), dusty gas model (Evans et al., 1961), effective medium theory (Burganos and Sotirchos, 1987; Sotirchos and Burganos, 1988; Kapoor and Yang, 1990), periodic capillary model (Gavalas and Kim, 1981), multiscale approach (Chang, 1983), percolation model (Moharty et al., 1982; Reyes and Jensen, 1985, 1986), averaging theorem (Ochoa-Tapia et al., 1993), random structure model (Pismen, 1974), random capillary structure (Tomakadis and Sotirchos, 1993; Deepak and Bhatia, 1994) and stochastic model (Bhatia, 1986, 1988) have been proposed in the literature.

In the simplest approach of lumping structure characteristics into transport coefficients, the three parameters characterising the medium for the three transport mechanisms (bulk, Knudsen, viscous) are:

- The Knudsen flow parameter, K_0
- The viscous flow parameter, B_0
- The porosity and tortuosity factor, ε/q, for continuum diffusion

For a long circular capillary of radius r, these three structural parameters take the following form:

$$K_0 = \frac{r}{2} \qquad (7.3\text{-}1a)$$

$$B_0 = \frac{r^2}{8} \qquad (7.3\text{-}1b)$$

$$\frac{\varepsilon}{q} = 1 \qquad (7.3\text{-}1c)$$

We now discuss these transport mechanisms one by one, and then discuss the transport due to the combinations of these mechanisms. First let us address the Knudsen diffusion.

7.4 Knudsen Diffusion

This mode of transport is due to Knudsen in the early 1900 (Knudsen, 1950). In this mode the mean free path is much larger than the diameter of the channel in which the diffusing molecules reside. This normally occurs at very low pressure and channels of small size, usually of order of 10 nm to 100 nm. When this happens, the molecules bounce from wall to wall, rather than colliding with themselves. The parameter which defines the conditions for this mode of transport is the Knudsen number (Levenspiel, 1984). This number is defined as:

$$Kn = \frac{\text{mean free path}}{\text{diameter of channel}} = \frac{\lambda}{d} \tag{7.4-1}$$

Depending on the magnitude of this parameter, we have three different regimes:

1. When this number is much less than unity, we have the usual viscous flow, where the Poiseuille flow is applicable. In this viscous flow, the Newton law of viscosity holds (that is the shear stress is proportional to the velocity gradient, and the proportionality constant is the viscosity), and the velocity at the wall is zero.
2. When it is of the order of unity, the velocity at the wall is not zero, that is it is not negligible compared to the overall velocity. The reason for the non-zero velocity at the wall is due to the fact of molecules bouncing to and from the wall.
3. When this number is much larger than unity, we have the molecular flow induced by the bouncing of the molecules to and from the wall (Knudsen flow). Since the collision among molecules is negligible, the concept of viscosity is no longer applicable in molecular flow.

The mean free path is a function of pressure. It is obtained from the following equation (Mulder, 1991):

$$\lambda = \frac{kT}{\pi d^2 P \sqrt{2}} \tag{7.4-2}$$

where d is the diameter of the molecule, and P is the total pressure. We note from the above equation that as the pressure decreases, the mean free path increases. The following table shows the magnitude of the mean free path for a few values of total pressure. At 760 Torr, the mean free path is 68 nm, thus pores need to be less than 50 nm for the Knudsen mechanism to operate. Many commercial porous solids have pores of dimension in the order of 50 nm; therefore, the Knudsen diffusion mechanism is usually the dominant mechanism.

Table 7.4-1 Mean free path magnitude versus the total pressure

Pressure, P (Torr)	Mean free path, λ (m)
760	6.8×10^{-8} (0.068 micron)
1	5.2×10^{-5} (52 micron)
0.1	5.2×10^{-4} (520 micron)

The first studies of Knudsen flow is restricted to small holes on a very thin plate as shown in Figure 7.4-1, with some gas density on the left hand side of the plate and vacuum on the other side.

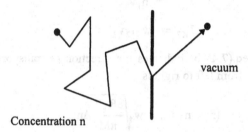

Figure 7.4-1: Knudsen diffusion through a hole

The choice of small holes in a thin plate assures that molecules do not collide with each other during their transport through the hole. Thus, the Knudsen transport is due to the collision of molecules with the pore wall (periphery of the hole), and as a result the movement of different molecules are independent of each other. The flux of species from one side having the molecular density n and vacuum at the other side is given by the following equation:

$$J_K = w\, n\, v_T \quad \left[\frac{\text{molecules}}{\text{cm}^2 - \text{sec}}\right] \qquad (7.4\text{-}3)$$

where w is the probability of a molecule that enters the hole and gets all the way to the other side, n is the molecule density (molecule/m³) and v_T is the mean thermal molecular speed (m/sec), which is given by the following equation:

$$v_T = \sqrt{\frac{8 k_B T}{\pi m}} \qquad (7.4\text{-}4)$$

Here k_B is the Boltzmann constant (1.38×10^{-23} Joule/K/molecule), T is in K, and m is the molecular mass. The thermal molecular speed increases with temperature and decreases with molecular mass. To have a feel of how fast a molecule moves, we take the case of helium at ambient temperature (T = 298 K), the mean thermal molecular speed is 1250 m/s, showing the very high random thermal velocity of molecules.

The probability factor w in eq. (7.4-3) depends on the geometry of the system. This will be discussed later.

Eq. (7.4-3) is the flux equation if the molecular concentration on one side is n and the other side is zero (vacuum). If the molecule concentration on one side is n_1 and the other side is n_2, the flux equation is simply the sum of the following two flux equations:

$$J_{K1} = w\, n_1 v_T \tag{7.4-5a}$$

$$J_{K2} = -w\, n_2 v_T \tag{7.4-5b}$$

The negative sign of eq.(7.4-5b) indicates the direction of transport from right to left. Thus, the net flux from left to right is:

$$J_K = w(n_1 - n_2)v_T = w\sqrt{\frac{8R_g T}{\pi M}}\Delta n \tag{7.4-6}$$

Eq. (7.4-6) defines the flux, written in terms of measurable quantities and the probability w. We now consider two cases: the thin orifice and the capillary.

7.4.1 Thin Orifice

If the system is a very thin orifice plate, the probability has been found as 1/4 (Graham, 1826; Knudsen, 1909), that is a quarter of molecules striking the orifice per unit area of the orifice will get through to the other side of the plate. The flux of molecules through the orifice is:

$$J_K = \frac{1}{4}\left(\sqrt{\frac{8R_g T}{\pi M}}\right) n \tag{7.4-7}$$

We note that the flux equation is independent of the size as well as the thickness of the orifice! The only requirement is that the size of the orifice is smaller than the mean free path of the diffusing gas. Let us show the applicability of the above flux equation in the following two examples.

Example 7.4-1: *Mass transfer rate through an orifice*

The mass transfer is the product of the flux times the orifice cross-sectional area. Thus it is proportional to the square of the orifice radius and to the difference in concentration. Let us take the case of helium diffusing through an orifice of diameter 25 micron and the pressure on one side is 1 Torr and that on the other side is at vacuum condition. The mass transfer through the orifice at ambient temperature is:

$$M = \frac{\pi d^2}{4}\left(\frac{1}{4}\sqrt{\frac{8R_g T}{\pi M}}\right)\Delta C$$

$$M = 2.93 \times 10^{-9}\ \frac{\text{mole}}{\text{sec}} = 1.77 \times 10^{15}\ \frac{\text{molecules}}{\text{sec}}$$

Example 7.4-2: *Thermal diffusion across an orifice*

Let us consider two vessels connected together by an orifice and the molecular densities in those vessels to be n_1 and n_2, respectively. The corresponding temperatures are T_1 and T_2. The condition of this closed system is that the net flow of molecule between these two vessels must be zero, that is

$$J_{K_{1\to2}} = \frac{1}{4}\sqrt{\frac{8R_g T_1}{\pi M}}\,n_1 = J_{K_{2\to1}} = \frac{1}{4}\sqrt{\frac{8R_g T_2}{\pi M}}\,n_2$$

Simplifying this equation, we get

$$n_1\sqrt{T_1} = n_2\sqrt{T_2}$$

But

$$n_1 = \frac{P_1}{R_g T}\ \text{and}\ n_2 = \frac{P_2}{R_g T}$$

we have:

$$\frac{P_1}{\sqrt{T_1}} = \frac{P_2}{\sqrt{T_2}}$$

Even when the pressures of the two vessels are the same, there is a net flow from the cold vessel to the hot vessel. This phenomenon is called the thermal diffusion.

The geometry of the plate does not bear much relevance to real porous solid. Thus, the above equation is not of much use in diffusion of adsorption systems. To this end, we will consider the Knudsen diffusion through a long cylindrical capillary in the next section.

7.4.2 Cylindrical Capillary

For a capillary tube, the probability that a molecule can find its way to the other end of the capillary is (Kiesling et al., 1978; Clausing, 1932):

$$w = \frac{1}{4} \frac{1}{(1 + 3L/8r)} \tag{7.4-8}$$

where r is the capillary radius and L is the length. Here, it is noted that the capillary size as well as the length of the capillary affect the probability of entry.

For a very short capillary (i.e. orifice, $L \ll 20r$), the probability becomes 1/4 as we have mentioned in Section 7.4.1.

For a long straight circular capillary (that is $r \ll L$), the above equation for the probability factor w is reduced to (Knudsen, 1909):

$$w = \frac{2}{3} \frac{r}{L} \tag{7.4-9a}$$

Hence, the flux equation for a very long capillary ($L \gg r$) is:

$$J_K = \frac{2r}{3} \sqrt{\frac{8 R_g T}{\pi M}} \cdot \frac{\Delta n}{L} \tag{7.4-9b}$$

The form of the above flux equation can be derived from the simple momentum balance of molecules in a capillary. We show this via the following example.

Example 7.4-3: *Derivation of the Knudsen equation from momentum balance*

Let n be the molecular density in a capillary. The rate of incidence of molecules per unit inner surface area is

$$\frac{1}{4} n v_T$$

where v_T is the thermal velocity. If the capillary has a perimeter of P and a length of L, the number of molecule collision at a surface area of PL is

$$\frac{1}{4}nv_T.(PL)$$

or a mass rate arriving at the capillary surface is

$$\frac{1}{4}nv_T(PL)m$$

where m is the molecular mass.

If v is the velocity of the species in the axial direction of the capillary, the rate of axial momentum by the collision is:

$$\frac{1}{4}nv_T(PL)m.v$$

If the surface is assumed to be a diffusive reflector (molecules leaving the wall uniformly in all directions regardless of the directional distribution with which they hit the surface) that is the mean axial momentum of the reflected molecules is zero, then the net change of the rate of axial momentum is

$$\left(\text{net change of rate of momentum}\right) = \frac{1}{4}nv_T(PL)mv.$$

This net change in the rate of momentum must be equal to the net force acting on the control volume. This net force is

$$\frac{1}{4}nv_T(PL)mv = -AL\frac{dp}{dz}$$

Simplifying the above two equations we get the molecular flux

$$j = nv = -\left(\frac{A\pi}{2P}\right)\sqrt{\frac{8k_BT}{\pi m}}\cdot\frac{dn}{dz}$$

For a cylindrical capillary, the proportionality constant of the above equation is:

$$\frac{A\pi}{2P} = \frac{(\pi r^2)\pi}{2(2\pi r)} = \frac{\pi r}{4}$$

Hence the molecular flux is finally

$$j = -\left(\frac{\pi}{4}\right)r\sqrt{\frac{8k_BT}{\pi m}}\frac{dn}{dz}$$

Comparing the above equation with the Knudsen flux equation given in eq. (7.4-9b), we see that they are identical in form except the numerical values: ($\pi/4 \approx 0.78$) in the above equation versus ($2/3 \approx 0.67$) in eq. (7.4-9b). Thus, the Knudsen flux in a capillary can be derived from a simple argument of momentum balance.

If we define the Knudsen diffusivity as:

$$D_K = \frac{2}{3}rv_T = \frac{2r}{3}\sqrt{\frac{8R_gT}{\pi M}} \qquad (7.4\text{-}10)$$

then the Knudsen equation (7.4-9b) can be written in terms of this Knudsen diffusivity and the concentration gradient (instead of the integral concentration difference):

$$J_K = -D_K \frac{dC}{dz} = -D_K \frac{1}{R_gT}\frac{dp}{dz} \qquad (7.4\text{-}11)$$

We note that the Knudsen diffusivity is proportional to the pore radius (capillary property), to the square root of temperature and to the inverse of the square root of the molecular weight (diffusing molecule property). It is independent of the total pressure and independent of other diffusing species.

The Knudsen diffusivity can be written in terms of the operating conditions and the capillary parameter, K_0, as follows:

$$D_K = \frac{4}{3}K_0\sqrt{\frac{8R_gT}{\pi M}} \qquad (7.4\text{-}12)$$

where K_0 is called the Knudsen flow parameter. The Knudsen flow through a complex porous medium is characterised by this parameter. It depends only on the geometry of the hole and the gas-surface scattering law (Mason and Malinauskas, 1983). Thus, it is a function of a given porous solid medium. For a long capillary of radius r, this Knudsen flow parameter is r/2 (comparing eq. 7.4-10 and 7.4-12). For a complex solid, this parameter K_0 is treated as the fitting parameter.

The following table (Table 7.4-2) presents two working formulas used to calculate the Knudsen diffusivity (Smith, 1970).

Table 7.4-2: Equations for the calculation of Knudsen diffusivity

Equation	D (units)	r (units)	T (units)	M (units)
$D_K = 9700\,r\sqrt{T/M}$	cm²/sec	cm	K	g/mole
$D_K = 3.068\,r\sqrt{T/M}$	m²/sec	m	K	kg/mole

The Knudsen flux equations, written in terms of concentration or pressure or mole fraction, are shown in the following table (Table 7.4-3). There is no interaction between species diffusing according to the Knudsen mechanism, owing to the fact that the mean free path is larger than the capillary dimension.

Table 7.4-3: Knudsen flux equation for capillary

Mole fraction gradient	Partial pressure gradient	Concentration gradient
$N_{K,j} = -D_{K,j} \dfrac{P}{R_g T} \dfrac{dx_j}{dz}$	$N_{K,j} = -D_{K,j} \dfrac{1}{R_g T} \dfrac{dp_j}{dz}$	$N_{K,j} = -D_{K,j} \dfrac{dC_j}{dz}$

At steady state, the Knudsen flux through a capillary is constant. Integrating eq. (7.4-11) with constant conditions at two ends of the capillary, we get:

$$J_K = \frac{D_K}{R_g T} \frac{\Delta P}{L} = \frac{8}{3} \frac{r}{\sqrt{2\pi M R_g T}} \frac{\Delta P}{L} \qquad (7.4\text{-}13)$$

Thus for a given pressure gradient across the capillary, the steady state flux under the Knudsen diffusion mechanism is proportional to the pore radius and is inversely proportional to the molecular weight. Viscosity does not affect the Knudsen flow as it does not have any meaning at very low pressures when continuum is no longer valid.

The proportionality constant in eq. (7.4-13) is the permeability

$$B = \frac{8}{3} \frac{r}{\sqrt{2\pi M R_g T}}$$

Experimentally we can check whether the flow is under the Knudsen flow regime by calculating $B\sqrt{MT}$ for various gases and temperatures. If it does not vary with temperature or the type of gas, the mechanism is due to the Knudsen diffusion. Adzumi (1937) studied the flow of hydrogen, acetylene and propylene through a glass capillary of radius and length of 0.0121 and 8.7 cm, respectively, and found that at low pressures (less than 0.03 torr) the permeability is independent of the mean pressure.

It is worthwhile at this point to remind the reader that the above conclusion for Knudsen diffusion is valid as long as the pressure is low or the capillary size is very small. When the capillary size is larger or the pressure is higher, the viscous flow will become important and the flow will be resulted due to the combination of the Knudsen and viscous flow mechanisms. This will be discussed in Sections 7.5 and 7.6.

Example 7.4-4: *Steady state diffusion of methane through a capillary*

To have a feel about the magnitude of the Knudsen diffusivity and the flux, we take an example of methane gas (M = 16 g/mole) diffusing through a capillary having the following properties:

$$\bar{r} = 20 \text{ Å}; \quad L = 1 \text{ cm}$$

The operating conditions are P = 0.1 atm, and T = 25 °C, and the mole fractions at two ends of the capillary are

$$x_0 = 1 \quad \text{and} \quad x_L = 0$$

The Knudsen diffusivity is calculated using the equation in Table 7.4-2:

$$D_K = 9700 \, r \sqrt{\frac{T}{M}} = 9700 \, (20 \times 10^{-8}) \sqrt{\frac{298}{16}} = 8.37 \times 10^{-3} \text{ cm}^2/\text{sec}$$

The steady state Knudsen flux then can be readily calculated using eq. (7.4-13) as

$$J_K = \frac{D_K}{L} \frac{\Delta P}{R_g T}$$

$$J_K = \frac{8.37 \times 10^{-3} (\text{cm}^2/\text{sec})}{1(\text{cm})} \frac{0.1 \text{ atm}}{(82.057 \text{ atm}-\text{cm}^3/\text{mole}/\text{K}) \times (273 + 25)\text{K}}$$

$$J_K = 3.42 \times 10^{-8} \frac{\text{moles}}{\text{cm}^2 \text{ sec}}$$

To have a better feel of this magnitude, we convert this molar flux to volume of methane gas at STP per unit hour, we get

$$\dot{V} = 3.42 \times 10^{-8} \left(\frac{\text{mole}}{\text{cm}^2 \text{ sec}}\right) \times 22{,}400 \left(\frac{\text{cm}^3 \text{STP}}{\text{mole}}\right) \times 3{,}600 \left(\frac{\text{sec}}{\text{hour}}\right) = 2.76 \frac{\text{cm}^3 \text{ STP}}{\text{cm}^2 \text{hr}}$$

Thus, methane gas diffuses through such capillary at a rate of about 3 cc of gas at standard condition per unit hour per unit square cm of the capillary.

We have addressed in the example 7.4-4 the steady state flux due to the Knudsen diffusion mechanism, but the question which is of significant interest is how long does it take for the system to response from some initial conditions to the final steady state behaviour. This is important to understand the "pure" diffusion time in a capillary. By "pure" diffusion time, we mean the diffusion time in the absence of adsorption. In the presence of adsorption, the time to approach equilibrium from some initial state is longer than the pure diffusion time due to the mass taken up by the surface. We shall discuss in more details about the adsorption time in Chapter 9.

7.4.2.1 Transient Knudsen Flow in a Capillary

Let us consider a case where a capillary is connected between two reservoirs with their respective partial pressures of p_0 and p_L. The end at $z = L$ is initially closed with a valve. At time $t=0^+$, the valve is opened and the diffusion process starts. We would like to determine the time it takes to achieve 95% of the steady state flux.

The transient mass balance equation inside the capillary is obtained by setting a shell balance in the capillary:

$$\frac{1}{R_g T} \frac{\partial p}{\partial t} = -\frac{\partial N_K}{\partial z} \qquad (7.4\text{-}14)$$

Substituting the constitutive flux equation in Table 7.4-3 into the above mass balance equation, we then have the following equation describing the distribution of partial pressure in the capillary:

$$\frac{\partial p}{\partial t} = D_K \frac{\partial^2 p}{\partial z^2} \qquad (7.4\text{-}15a)$$

subject to the following boundary conditions:

$$z = 0; \quad p = p_0 \qquad (7.4\text{-}15b)$$

$$z = L; \quad p = p_L \qquad (7.4\text{-}15c)$$

The initial condition in the capillary is assumed to be the same as the concentration at x=0, that is:

$$t = 0; \quad p = p_0 \qquad (7.4\text{-}15d)$$

The mass balance equation (7.4-15a) subject to the initial and boundary conditions (7.4-15b to d) has the following solution for the pressure as a function of time as well as position along the capillary (obtained by Laplace transform or separation of variables method)

$$\frac{p - p_0}{p_L - p_0} = \frac{z}{L} - \frac{2}{\pi} \sum_{n=1}^{\infty} \frac{-\cos(n\pi)}{n} \sin\left(\frac{n\pi z}{L}\right) \exp\left(-\frac{n^2 \pi^2 D_K t}{L^2}\right) \qquad (7.4\text{-}16)$$

Knowing the pressure distribution in the capillary, the fluxes at $z = 0$ and $z = L$ can be evaluated from:

$$N_0(t) = -D_K \frac{1}{R_g T} \frac{\partial p}{\partial z}\bigg|_0 \tag{7.4-17a}$$

$$N_L(t) = -D_K \frac{1}{R_g T} \frac{\partial p}{\partial z}\bigg|_L \tag{7.4-17b}$$

Substitute the pressure distribution (eq. 7.4-16) into the above flux equations, we obtain the fluxes at the entrance and exit:

$$N_0(t) = D_K \frac{1}{R_g T} \frac{(p_0 - p_L)}{L} \left[1 - 2\sum_{n=1}^{\infty}(-\cos(n\pi))\exp\left(-\frac{n^2\pi^2 D_K t}{L^2}\right)\right]$$

$$N_L(t) = D_K \frac{1}{R_g T} \frac{(p_0 - p_L)}{L} \left[1 - 2\sum_{n=1}^{\infty}\exp\left(-\frac{n^2\pi^2 D_K t}{L^2}\right)\right] \tag{7.4-17c}$$

The term in front of the square brackets is the steady state flux, obtained earlier in eq.(7.4-13), that is at steady state the flux entering the capillary is the same as the flux leaving the capillary, as expected.

For the flux at the exit of the capillary to achieve 95% of the steady state flux, we must have:

$$1 - 2\sum_{n=1}^{\infty}\exp\left(-\frac{n^2\pi^2 D_K t_{0.95}}{L^2}\right) = 0.95$$

Keeping only the first term in the series since all other terms in the series decay rapidly compared to the first term, we obtain the time required

$$t_{0.95} = 0.374 \frac{L^2}{D_K} = 0.352 \frac{L^2 \sqrt{M}}{r \sqrt{R_g T}} \tag{7.4-18}$$

which states that the diffusion time is proportional to the square of capillary length, the square root of the molecular weight and inversely proportional to the capillary radius and the square root of temperature.

Example 7.4-5: *Time scale of diffusion of hydrogen in a capillary*

We take the following example of hydrogen diffusing in a capillary having the pore radius of 100 Å, and the length of 10 cm. The temperature of the system is 298 K. The Knudsen diffusivity (Table 7.4-2) is calculated as 1.45 cm^2/sec. Substituting these values into eq. (7.4-18), we get the following time required by the system to reach 95% of the equilibrium:

$$t_{0.95} = 0.374\frac{(10)^2}{1.45} = 26 \text{ sec}$$

Thus, the time required for the system to attain steady state with the Knudsen mechanism is in the order of 30 seconds. It is emphasised at this point that this is the time required for the pure diffusion mechanism to reach steady state. In the presence of adsorption along the capillary, the time required will be longer because the adsorption process retards the penetration of concentration front through the capillary, that is more mass is supplied for the adsorption onto the capillary wall and hence more time is needed for the attainment of steady state. We will discuss this when we deal with diffusion and adsorption in Chapter 9.

Knowing the flux entering and leaving the capillary, the amounts per unit capillary area entering and leaving up to time t are given by:

$$Q_0(t) = \int_0^t N_0(t)dt \quad ; \quad Q_L(t) = \int_0^t N_L(t)dt \qquad (7.4\text{-}19)$$

The capillary assumed so far is cylindrical in shape and its size is uniform along the tube. If the capillary size is not uniform, but either converging or diverging, the Knudsen diffusivity and flux given in eqs. (7.4-10) and (7.4-11) are still valid and they hold for every point along the capillary, provided that the diffusion process is still dominated by Knudsen mechanism. Evaluation of the Knudsen flux for such a capillary is dealt with in the next section.

7.4.3 Converging or Diverging Capillary

Consider a diverging capillary as shown in Figure 7.4-2a. The entrance pore radius is r_1, the length is L and the angle of the pore is α. The pore radius at any position z (the origin of which is at the entrance of the pore) is:

$$r = r_1 + (\tan\alpha)\, z \qquad (7.4\text{-}20a)$$

and the pore radius at the exit is:

$$r_2 = r_1 + (\tan\alpha)\, L \qquad (7.4\text{-}20b)$$

The volume of the capillary is:

$$V = \pi \int_0^L r^2(z)\, dz = \frac{\pi}{3\tan\alpha}\left(r_2^3 - r_1^3\right) \qquad (7.4\text{-}20c)$$

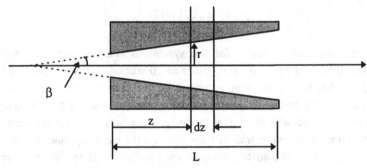

Figure 7.4-2a Diverging capillary

Having obtained the dimensions of the system, we now turn to setting up the mass balance equation describing the concentration distribution along the capillary. We consider a thin shell of thickness dz at the position z as shown in Figure 7.4-2a, and carry out the mass balance around that shell to obtain:

$$\frac{d}{dz}(\pi r^2 N) = 0 \tag{7.4-21a}$$

Note that the radius r in the above mass balance equation is a function of z as given in eq. (7.4-20a). Before solving this mass balance equation, we need to specify the flux expression and appropriate boundary conditions. We assume the flow is due to the Knudsen mechanism (that is the mean free path is longer than any radius along the diverging capillary), and hence the flux equation is given by (Table 7.4-3):

$$N = -\frac{D_{K0}}{R_g T}\left(\frac{r}{r_0}\right)\frac{dp}{dz} \tag{7.4-21b}$$

where D_{K0} is the Knudsen diffusivity at some reference pore radius r_0. The boundary conditions are assumed to be constant at two ends of the capillary, that is:

$$z = 0; \quad p = p_1 \tag{7.4-21c}$$

$$z = L; \quad p = p_2 \tag{7.4-21d}$$

Integration of eq. (7.4-21a) gives $\pi r^2 N = K$ (constant). Substituting N in eq. (7.4-21b) into that resulted equation and then integrating with respect to z, we obtain the following result for the molar rate (moles/sec) through the capillary:

$$\pi r^2 N = \frac{2 \tan\alpha}{\left[r_1^{-2} - (r_1 + L\tan\alpha)^{-2}\right]} \frac{\pi D_{K0}}{R_g T r_0}(p_1 - p_2) \tag{7.4-22}$$

7.4.3.1 An Equivalent Uniform Capillary

If we now define an equivalent capillary as a uniform capillary having a radius r_0 and the same length as that of the diverging capillary. The radius r_0 of this equivalent uniform capillary is determined by taking its volume to be the same as the volume of the diverging capillary (eq. 7.4-20c), that is:

$$\pi r_0^2 L = \frac{\pi}{3\tan\alpha}\left[(r_1 + L\tan\alpha)^3 - r_1^3\right] \qquad (7.4\text{-}23a)$$

Solving for this radius r_0 of the equivalent capillary, we have:

$$\frac{r_0}{r_1} = \sqrt{\frac{1}{3\beta}\left[(1+\beta)^3 - 1\right]} \qquad (7.4\text{-}23b)$$

where

$$\beta = \frac{L\tan\alpha}{r_1} \qquad (7.4\text{-}23c)$$

The parameter β is a measure of the degree of divergence of the capillary. If this parameter is zero, the capillary is uniform in size. The larger is this parameter, the more diverging is the capillary.

For the same constant pressures imposed at two ends of the equivalent uniform capillary, the molar rate through this capillary is:

$$\pi r_0^2 N_0 = \frac{\pi r_0^2 D_{K0}}{R_g T L}(p_1 - p_2) \qquad (7.4\text{-}23d)$$

If we now define a geometry factor I as the ratio of the molar rate of the equivalent uniform capillary (eq. 7.4-23d) to the molar rate of the diverging capillary (eq. 7.4-22), we obtain:

$$I = \frac{(r_1/r_0)^{-2} - (r_1/r_0 + \beta)^{-2}}{2\beta} \qquad (7.4\text{-}24)$$

where (r_1/r_0) is defined in eq. (7.4-23b). The utility of this geometrical factor I is as follows. If we wish to calculate the molar rate of a diverging capillary all we need to do is to calculate the molar rate of an equivalent uniform capillary (a simpler geometry) and then divide it by the geometrical factor I to account for the pore divergence of the system, that is

$$\pi r^2 N = \frac{\pi r_0^2 N_0}{I} \qquad (7.4\text{-}25)$$

This is equivalent to saying that the "apparent" Knudsen diffusivity is calculated from:

$$D_{K,app} = \frac{D_{K0}}{I} \tag{7.4-26}$$

To show the effect of the geometrical factor on the calculation of the apparent Knudsen diffusivity, we plot I of eq. (7.4-24) versus β (the degree of divergence) as shown in Figure 7.4-2b. Here we see that except for a small decrease when β is less than 1 the geometrical factor I is increased with an increase in the degree of divergence. This shows that the pore geometry has an influence on the calculation of the molar rate if an equivalent uniform capillary is used in the calculation.

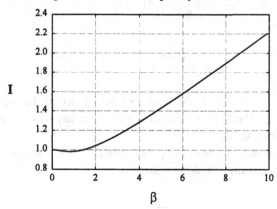

Figure 7.4-2b: Plot of I (eq. 7.4-24) versus β

7.4.4 Porous Solids:

We have dealt with Knudsen flow in a straight capillary and in a converging/diverging pore, and we see that the shape of a pore can contribute to the flux calculation. Now we turn to dealing with practical solids where pore orientation is rather random.

7.4.4.1 Parallel Capillaries Model

For an actual porous medium, the Knudsen flow parameter K_0 (eq. 7.4-12) can be used as a fitting parameter so that solid properties such as mean pore size, its variance can be embedded in such a parameter, or we can assume a model for the porous medium as a bundle of nonoverlapping capillaries of length L_c and of radius r (Figure 7.4-3). We show below the analysis of bundle of capillaries due to Epstein (1989).

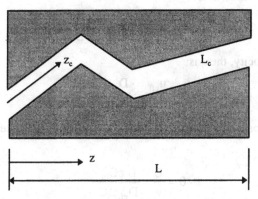

Figure 7.4-3: Capillary in a porous medium

The coordinate along the porous medium is z and that along the capillary is z_c. The steady state Knudsen flux along the coordinate of the capillary is (the upperscript c is for capillary):

$$J_K^c = D_K \frac{\Delta C}{L_c} \qquad (7.4\text{-}24)$$

Let the mean concentration in the capillary is C_m. The average velocity through the capillary is:

$$u = \frac{J_K^c}{C_m} = \frac{D_K}{C_m} \frac{\Delta C}{L_c}$$

Hence the time that a diffusing molecule spends in the capillary of length L_c is:

$$\theta = \frac{L_c}{u} = \frac{L_c^2 C_m}{D_K \Delta C} \qquad (7.4\text{-}25)$$

Now we consider an analysis of the porous medium. The steady state flux per unit total cross-sectional area of the porous medium is defined as:

$$J_K = D_{eff} \frac{\Delta C}{L}$$

where D_{eff} is the effective diffusivity based on the total cross sectional area. Given the mean concentration of C_m, the superficial velocity of the diffusing molecule through the porous medium is:

$$u_{eff} = \frac{J_K}{C_m} = \frac{D_{eff}}{C_m}\frac{\Delta C}{L}$$

The interstitial velocity, then, is:

$$v = \frac{u_{eff}}{\varepsilon} = \frac{D_{eff}}{\varepsilon C_m}\frac{\Delta C}{L}$$

where ε is the porosity. Thus the time that the diffusing molecule spent inside the porous medium is:

$$\theta = \frac{L}{v} = \frac{L^2 \varepsilon C_m}{D_{eff}\Delta C} \tag{7.4-26}$$

Since the residence times of diffusing molecules in the model capillaries and the porous medium must be the same, we derive the following expression for the effective diffusivity:

$$D_{eff} = \frac{\varepsilon D_K}{(L_c/L)^2} = \frac{\varepsilon D_K}{\tau^2} = \frac{\varepsilon D_K}{q} \tag{7.4-27a}$$

The tortuosity factor q is the square of the tortuosity. Thus the Knudsen flux equation for a porous medium obtained from the parallel capillaries model is given by:

$$J_K = -\frac{\varepsilon D_K}{q}\frac{\partial C}{\partial z} \left[\frac{\text{mole}}{\text{total area of the porous medium - time}} \right] \tag{7.4-27b}$$

where the pore diffusivity D_K is evaluated at the mean pore size of the medium. Eq. (7.4-27b) is the simplest Knudsen equation for a porous medium and it is valid when the pore size distribution is relatively narrow. For a porous medium having a pore size distribution f(r), the following equation was proposed by Wang and Smith (1983):

$$J_K = -\frac{\varepsilon}{q}\left[\int_0^\infty D_K(r)f(r)dr\right]\frac{dC}{dz} \tag{7.4-28}$$

where f(r)dr is the fraction of pore volume having pore size between r and r+dr. Here, they have assumed that the concentration gradient is the same in all pores;

hence the average Knudsen diffusion coefficient is simply the average of diffusivities of all the pores as shown in the square bracket in eq.(7.4-28).

7.4.4.2 Unconsolidated Media

The last model assumes that porous media can be idealised as parallel capillaries along the direction of flow. Porous media such as adsorbents and catalysts are usually formed by compressing small grains into pellet, and for such particles the model for unconsolidated media will be particularly useful. There are a number of equations available in the literature to describe the Knudsen flow through a unconsolidated medium. They are identical in form and differ only in the numerical proportionality coefficient.

Derjaguin (1946) used a statistical method to follow the motion of a free molecule flowing through a random pore space to obtain the following equation for the Knudsen flux based on total cross-sectional area of the medium:

$$J_K = -B \frac{\partial p}{\partial z} \tag{7.4-29}$$

where the permeability coefficient B is expressed in terms of medium porosity ε, grain diameter d_p, molecular weight of the diffusing species and temperature as follows:

$$B = \frac{1}{\sqrt{2\pi MR_g T}} \frac{\varepsilon^2 d_p}{3(1-\varepsilon)} \frac{24}{13} \tag{7.4-30}$$

Carman (1956) used the equivalent capillary model to derive the following equation for the permeability coefficient:

$$B = \frac{1}{\sqrt{2\pi MR_g T}} \frac{\varepsilon^2 d_p}{3(1-\varepsilon)} \frac{1}{3k'} \tag{7.4-31}$$

where k' is a constant.

Using a two-sided Maxwellian velocity distribution functions and Maxwell's transport equation, Asaeda et al. (1974) obtained the following equation for the permeability coefficient in a packed column containing spherical particles:

$$B = \frac{1}{\sqrt{2\pi MR_g T}} \frac{\varepsilon^2 d_p}{3(1-\varepsilon)} \frac{4}{\Phi q} \tag{7.4-32}$$

where $\Phi = 2.18$ and q is the tortuosity factor which was found experimentally as 1.41. Comparing the permeability coefficients of Derjaguin and Asaeda et al., we find

$$\frac{B_{Derjaguin}}{B_{Asaeda}} = 1.42 \tag{7.4-33}$$

Thus, with the exception of the numerical proportionality coefficient, the various models for unconsolidated media yield the same dependence on the grain diameter d_p, the porosity function:

$$\frac{\varepsilon^2}{(1-\varepsilon)} \tag{7.4-34}$$

and the property of the diffusing species (molecular weight M) and temperature

$$\frac{1}{\sqrt{2\pi MR_g T}} \tag{7.4-35}$$

In general, we then can express the Knudsen flux of a porous medium as:

$$J_K = -\frac{\alpha}{\sqrt{2\pi MR_g T}} \frac{\varepsilon^2 d_p}{3(1-\varepsilon)} \frac{\partial p}{\partial z} \tag{7.4-36}$$

where α is 24/13 or 1.3 for the Derjaguin or Asaeda et al.'s model. If we express the above equation in terms of the specific surface area S_0 (surface area per unit volume of grains), the Knudsen flux equation has the form:

$$J_K = -\frac{\alpha}{\sqrt{2\pi MR_g T}} \frac{2\varepsilon^2}{(1-\varepsilon)S_0} \frac{\partial p}{\partial z} \tag{7.4-37}$$

If the grain is spherical in shape

$$S_0 = \frac{\pi d_p^2}{\pi d_p^3 / 6} = \frac{6}{d_p} \tag{7.4-38}$$

eq. (7.4-37) reduces to eq. (7.4-36).

Example 7.4-6: *Steady state diffusion through a porous medium*

For constant pressures imposing at two ends of a porous medium containing spherical grain of diameter d_p, the steady state Knudsen flux is:

$$J_K = \frac{\alpha}{\sqrt{2\pi MR_g T}} \frac{\varepsilon^2 d_p}{3(1-\varepsilon)} \frac{\Delta p}{L} \tag{7.4-39}$$

This equation can be checked experimentally for its validity by measuring the steady state fluxes at various operating conditions T, ΔP/L on a number of porous media made by compressing the same grains with different compression pressures so that those porous media have different porosity. Then the dependence of $\dfrac{J_K \sqrt{2\pi M R_g T}}{\Delta p / L}$ on $\dfrac{\varepsilon^2 d_p}{3(1-\varepsilon)}$ can be checked.

7.4.4.3 Consolidated Media

For consolidated media such as concrete, sandstone, the Knudsen flux equation such as eq. (7.4-27b) or (7.4-28) can be used. Ash and Grove (1960) studied the flow of inert gases, hydrogen, nitrogen, oxygen, ethane, carbon dioxide and sulfur dioxide through a porous porcelian ceramic and found that for pressures less than 20 Torr the flow is described well by the Knudsen equation.

7.4.5 Graham's Law of Effusion

We have discussed so far the Knudsen diffusivity and Knudsen flux for capillary as well as for porous medium. We have said that the flow of one species by the Knudsen mechanism is independent of that of the other species. The question now is in a constant total pressure system, is there a relationship that relates fluxes of all the species when the partial pressures are constrained by the constant total pressure condition. Let us now address this issue.

Consider a system having n species and each species diffuses according to the Knudsen mechanism. The flux equation in a capillary for the component "j" is:

$$J_{K,j} = -D_{K,j} \dfrac{\partial C_j}{\partial z} \qquad (7.4\text{-}40)$$

If we divide the equation by the Knudsen diffusivity and sum the result for all components, we then have:

$$\sum_{j=1}^{N} \dfrac{J_{K,j}}{D_{K,j}} = -\sum_{j=1}^{N} \dfrac{dC_j}{dz} = -\dfrac{d\sum_{j=1}^{N} C_j}{dz} = -\dfrac{dC}{dz} = -\dfrac{1}{R_g T} \dfrac{dP}{dz} = 0 \qquad (7.4\text{-}41)$$

in which dP/dz=0 by the virtue of constant total pressure. But since the Knudsen diffusivity is inversely proportional to the square root of molecular weight, the above equation is written as follows:

$$\sum_{j=1}^{N} \sqrt{M_j}\, J_{K,j} = 0 \qquad (7.4\text{-}42)$$

This is the famous <u>Graham's law of effusion</u> for a multicomponent systems at constant pressure. For a binary system, the above equation will become:

$$\frac{J_{K,1}}{J_{K,2}} = -\sqrt{\frac{M_2}{M_1}} \qquad (7.4\text{-}43)$$

which is experimentally observed in Graham's experiments and Hoogschagen's experiments discussed in Section 7.2. The negative sign in eq. (7.4-43) means that the flux of species 1 is always in the opposite direction to the flux of species 2, and the lighter species will move faster than the heavy one. Let us take the case of helium and oxygen experiments carried out by Hoogschagen (1955). The theoretical ratio of the two fluxes is:

$$\frac{J_{K,He}}{J_{K,oxygen}} = -\sqrt{\frac{32}{4}} = -2.83$$

The experiments done by Hoogschagen have the following values 3.03, 2.66 and 2.54 in three runs carried out by him. Thus the Graham' law of effusion is experimentally confirmed. Any deviation from this law would point to an additional transport of an adsorbed surface layer. Using two commercial adsorbents with large internal surface area, this effect was detected (Table 7.4-4).

Table 7.4-4: Results of Hoogschagen's experiments

Adsorbent	$-N(CO_2)/N(O_2)$	$-N(N_2)/N(O_2)$
Activated carbon (1400m^2/g; 13.8A)	1.40	1.09
Silica gel (1020 m^2/g; 8.6A)	1.81	1.06

From the Graham' law (eq. 7.4-43), the "theoretical" Knudsen diffusion flux of carbon dioxide should be about 0.85 times that of oxygen, but yet the experiment with activated carbon shows that carbon dioxide flux is 1.4 times higher than that of oxygen. This is due to the fact that carbon dioxide is appreciably adsorbed and *additional transport* occurs in the adsorbed layer. For the pairs of nitrogen and oxygen, the theoretical ratio of nitrogen flux to oxygen flux calculated from the Graham' s law is 1.07. This is confirmed by the experiments with two adsorbents, activated carbon and silica gel. More about surface diffusion is discussed in Section 7.9.

7.5 Viscous Flow

When the flow of gas is induced by a driving force of total pressure gradient, the flow is called the viscous flow. For typical size of most capillaries, we can ignore the inertial terms in the equation of motion, and turbulence can be ignored. The flow is due to the viscosity of the fluid (laminar flow or creeping flow) and the assumption of no slip at the surface of the wall.

7.5.1 Viscous Flux in a Capillary

Starting with the Navier-Stokes equation (Bird et al., 1960):

$$\rho\left(\frac{\partial v_z}{\partial t} + v_r \frac{\partial v_z}{\partial r} + \frac{v_\theta}{r}\frac{\partial v_z}{\partial \theta} + v_z \frac{\partial v_z}{\partial z}\right) = -\frac{\partial P}{\partial z} + \mu\left[\frac{1}{r}\frac{\partial}{\partial r}\left(r\frac{\partial v_z}{\partial r}\right) + \frac{1}{r^2}\frac{\partial^2 v_z}{\partial \theta^2} + \frac{\partial^2 v_z}{\partial z^2}\right] + \rho g_z \quad (7.5\text{-}1)$$

and assuming steady flow and no variation with respect to z and angle θ, and no slip at the capillary wall we obtain the Poiseuille equation for the volumetric flow rate through a capillary tube of radius r under a pressure gradient dP/dz:

$$Q = -\frac{\pi r^4}{8\mu}\frac{dP}{dz} = -\frac{\pi d^4}{128\mu}\frac{dP}{dz} \quad \left(\frac{m^3}{\sec}\right) \quad (7.5\text{-}2)$$

Thus the molar flux of mixture is obtained by multiplying this volumetric rate with the total molar concentration, C, and then dividing the result by the area of the capillary. In so doing, we obtain the following equation for the viscous flux of mixture:

$$J_{vis} = -C\frac{r^2}{8\mu}\frac{dP}{dz} \quad \left(\frac{moles}{m^2 - \sec}\right) \quad (7.5\text{-}3a)$$

where the total molar concentration is related to the total pressure as follows

$$C = \frac{P}{R_g T} \quad (7.5\text{-}3b)$$

If we define the viscous flow parameter B_0 as $B_0 = r^2/8$ (which is the parameter characterising the capillary), the viscous flux equation then becomes:

$$J_{vis} = -C\frac{B_0}{\mu}\frac{dP}{dz} \quad \left(\frac{moles}{m^2 - \sec}\right) \quad (7.5\text{-}4a)$$

For a porous medium, the viscous flow parameter characterizes the properties of the porous medium, such as radius, its distribution, orientation, and overlapping, etc. Eq. (7.5-4a) when written in terms of pressure becomes:

$$J_{vis} = -\frac{B_0}{\mu}\left(\frac{P}{R_g T}\right)\frac{dP}{dz}\left(\frac{\text{moles}}{m^2 - \sec}\right) \tag{7.5-4b}$$

At steady state, integrating eq. (7.5-4b) with respect to z and using constant boundary conditions at two ends of the capillary, we get the following steady state equation for the viscous flux through a capillary:

$$J_{vis} = \left(\frac{B_0}{\mu}\right)\frac{(P_0^2 - P_L^2)}{(2R_g T)L} = \left(\frac{B_0}{\mu}\right)\left(\frac{\bar{P}}{R_g T}\right)\frac{(P_0 - P_L)}{L}\left(\frac{\text{moles}}{m^2 - \sec}\right) \tag{7.5-4c}$$

where $\bar{P}$ is the mean pressure in the capillary:

$$\bar{P} = \frac{(P_0 + P_L)}{2}$$

Here $\bar{P}/R_g T$ is the mean molar concentration in the capillary. A plot of permeability $B = [J_{vis}/(\Delta P/L)]$ versus the mean pressure will give a straight line with a slope of $(B_0/\mu RT)$. Thus, knowing the fluid viscosity, one could readily calculate the structural viscous flow parameter B_0, from simple steady state experiment, provided that the mechanism of flow is viscous and there is no slip on the surface. Recall the Knudsen mechanism (eq. 7.4-13) that a plot of $J_K/(\Delta P/L)$ versus pressure will give a horizontal line, indicating that there is no pressure dependence of the Knudsen permeability coefficient. Thus the dependence of the experimental permeability coefficient on the pressure can delineate which mechanism controls the flow in a capillary.

The viscous flux given in eq. (7.5-4) is the flux of the mixture, that is the mixture moves as a whole under the total pressure gradient. All species moves at the same speed (that is no separation), and the individual viscous flux of species "i" caused by the total pressure gradient is:

$$J_{vis,i} = x_i J_{vis} \tag{7.5-5}$$

where x_i is the mole fraction of species "i".

Equations obtained so far are for the condition of no molecular slip at the wall. In the case of slip, the *extra* flow can be regarded as a free molecular flow. For large pores, the extra flow due to molecular slip is negligible compared to the viscous flow. However, when the pore gets smaller the molecular slip becomes more important even though the viscous flux decreases with a decrease in capillary size. This is so because the viscous flux decreases like r^2 while the molecular slip flux decreases like r. The extra molecular slip flow will be discussed in Section 7.6.

Example 7.5-1: *Viscous flux of helium in a capillary*

Consider a capillary having a radius 0.1 μm and a length of 1 cm. Helium gas is flowing from one end of the capillary of pressure 5000 Torr to the other end of 4000 Torr. Calculate the viscous flux at 298 K. The viscosity of helium at 298K is 1.97×10^{-4} g/cm/sec.

First, we calculate the viscous flow parameter for the capillary

$$B_0 = \frac{r^2}{8} = \frac{(0.1 \times 10^{-4})^2}{8} = 1.25 \times 10^{-11} \text{ cm}^2$$

The mean concentration in the capillary is

$$\overline{C} = \frac{\overline{P}}{R_g T} = \frac{(P_0 + P_L)/2}{R_g T} = \frac{[(5000+4000)/2]/760}{82.057 \times 298} = 2.42 \times 10^{-4} \frac{\text{mole}}{\text{cm}^3}$$

The pressure gradient is:

$$\frac{P_0 - P_L}{L} = \frac{(5000-4000) \text{ Torr}}{1 \text{ cm}} \left(\frac{1 \text{ atm}}{760 \text{ Torr}}\right)\left(\frac{1.0133 \times 10^6 \text{ g cm}^{-1} \text{ sec}^{-2}}{1 \text{ atm}}\right)$$

$$\frac{P_0 - P_L}{L} = 1.333 \times 10^6 \frac{\text{g}}{\text{cm}^2 \text{s}^2}$$

Substitute these values into the viscous flux equation (7.5-4c)

$$J_{vis} = \frac{B_0}{\mu} \overline{C} \frac{(P_0 - P_L)}{L}$$

we get

$$J_{vis} = \frac{(1.25 \times 10^{-11} \text{cm}^2)}{(1.97 \times 10^{-4} \text{g cm}^{-1}\text{s}^{-1})} \left(2.42 \times 10^{-4} \frac{\text{mol}}{\text{cm}^3}\right)\left(1.333 \times 10^6 \frac{\text{g}}{\text{cm}^2 \text{s}^2}\right)$$

$$J_{vis} = 2.05 \times 10^{-5} \frac{\text{mole}}{\text{cm}^2 \text{s}}$$

For this example, we calculate the Knudsen flux using eq. (7.4-13). The Knudsen diffusivity is $D_K = 0.837$ cm²/sec and the Knudsen flux is 4.5 $\times 10^{-5}$ mole/cm²/sec. We see that the viscous flux in this case is very comparable to the Knudsen flux, and they must be accounted for in the calculation of the total flux. The reason for this significant contribution of the viscous flow is that the pressures used in this example are very high. For low pressure systems, especially those operated under sub-ambient pressure, the Knudsen mechanism is always dominating.

Example 7.5-2: *Apparent diffusivity for the viscous flow*

Under the transient conditions in a capillary with viscous flow, the following equation describes the mass balance:

$$\frac{\partial (P/R_g T)}{\partial t} = -\frac{\partial}{\partial z}(J_{vis})$$

By applying eq. (7.5-4b) for the viscous flux J_{vis} and simplifying the result we get:

$$\frac{\partial P}{\partial t} = \frac{B_0}{\mu}\frac{\partial}{\partial z}\left(P\frac{\partial P}{\partial z}\right) = \frac{B_0}{2\mu}\frac{\partial^2 (P^2)}{\partial z^2}$$

This equation describes the transient flow of an ideal gas in a porous medium. The above equation has a similar form to the standard Fickian diffusion equation

$$\frac{\partial P}{\partial t} = \frac{\partial}{\partial z}\left(D\frac{\partial P}{\partial z}\right)$$

with the "apparent" diffusion coefficient being a linear function of pressure

$$D = \frac{B_0}{\mu}P$$

that is, the higher is the pressure, the higher is the "apparent" diffusion coefficient.

7.5.2 Porous Media: Parallel Capillaries Model

We have obtained equations for the viscous flow in a capillary. For a porous medium, the viscous flow parameter B_0 can be treated as a fitting structural parameter or it can be calculated from the assumption of a model for the solid structure as it is a function of only solid properties. Like before, if we assume that the solid can be idealised as a bundle of parallel capillaries of equal diameter and length L_c running through the medium in the direction of flow, the average velocity in the capillary is (from eq. 7.5-2):

$$u_c = \frac{Q}{\pi r^2} = -\frac{r^2}{8\mu}\frac{dP}{dz_c} \tag{7.5-6}$$

where z_c is the coordinate along the capillary. The coordinate along the capillary is not the same as that along the porous medium. Now we need to convert the above equation in terms of the coordinate z and the interstitial velocity of the porous medium (Epstein, 1989).

The residence time of the fluid to travel from one end to the other end of the capillary is simply the length of the capillary divided by the average velocity in the capillary:

$$\theta = \frac{L_c}{u_c} \qquad (7.5\text{-}7)$$

If the interstitial velocity in the porous medium is denoted as v, the residence time of the fluid in the porous medium is (L/v), which must be the same as the residence time that the fluid spent in the capillary, that is:

$$\theta = \frac{L}{v} = \frac{L_c}{u_c} \qquad (7.5\text{-}8a)$$

Therefore, the interstitial velocity in the porous medium is:

$$v = \left(\frac{L}{L_c}\right) u_c \qquad (7.5\text{-}8b)$$

Eq. (7.5-8b) suggests that the interstitial velocity in the porous medium, v, is less than the velocity in the capillary as the medium length L is smaller than the capillary length.

Hence, from eqs. (7.5-6) and (7.5-8), the interstitial velocity of the porous medium is given by:

$$v = -\frac{r^2}{8\mu}\frac{dP}{dz}\left(\frac{L}{L_c}\right)^2 = -\frac{r^2}{8\mu\tau^2}\frac{dP}{dz} \qquad \left(\frac{m}{\sec}\right) \qquad (7.5\text{-}9)$$

in which we have used $dz_c/dz = L_c/L$. The parameter τ is the tortuosity of the medium (eq. 7.1-1). Hence, if the total concentration is C, the viscous molar rate per unit void area is:

$$J_{vis} = -\frac{r^2}{8\mu\tau^2} C \frac{dP}{dz} = -\frac{r^2}{8\mu\tau^2}\frac{P}{R_g T}\frac{dP}{dz} \left(\frac{moles}{m^2 - \sec}\right) \qquad (7.5\text{-}10)$$

When dealing with a porous medium, the viscous flux is usually expressed in terms of the total cross sectional area. Thus if the porosity of the medium is ε, the desired flux expression is:

$$J_{vis} = -\frac{\varepsilon r^2}{8\mu\tau^2}\frac{P}{R_g T}\frac{dP}{dz} \qquad (7.5\text{-}11)$$

If we now introduce the specific permeability coefficient B_0 (viscous flow parameter) of a porous medium as follows (Carman, 1956):

$$B_0 = \frac{\varepsilon r^2}{8\tau^2} \qquad (7.5\text{-}12)$$

that is, B_0 involves only properties of the porous medium
- radius r
- porosity ε
- tortuosity factor $q = \tau^2$

the viscous flux based on the total cross sectional area is

$$J_{vis} = -\left(\frac{B_0}{\mu}\right)\frac{P}{R_g T}\frac{dP}{dz}\left(\frac{\text{moles}}{\text{m}^2 - \text{sec}}\right) \qquad (7.5\text{-}13)$$

and the average velocity also based on the total cross sectional area is:

$$u = \frac{J_{vis}}{P/R_g T} = -\frac{B_0}{\mu}\frac{dP}{dx} \qquad (7.5\text{-}14)$$

Eq. (7.5-14) is the famous Darcy equation with B being the specific permeability coefficient. The Darcy equation implies that the flow mechanism is by viscous drag and the fluid is inert to the porous medium, that is the effects of chemical, adsorptive, electrical, electrochemical interactions between fluid molecules and the capillary are absent (Carman, 1956).

If the pressures at two ends of the porous medium are constant, the molar flux J_{VIS} is a constant. Thus, by integrating eq. (7.5-13) with the following boundary conditions

$$x = 0; \quad P = P_0$$
$$x = L; \quad P = P_L$$

we obtain the following equation for the molar viscous flux

$$J_{vis} = \frac{B_0}{\mu}\frac{\overline{P}}{R_g T}\frac{(P_0 - P_L)}{L} \qquad (7.5\text{-}15)$$

where $\overline{P}$ is the mean pressure. Thus, by rearranging eq.(7.5-15), we get:

$$\frac{J_{vis}}{(\Delta P/L)} = \left(\frac{B_0}{\mu}\right)\frac{\overline{P}}{R_g T} \tag{7.5-16}$$

If we plot the LHS of eq.(7.5-16) versus the mean pressure of the capillary, we will get a straight line with a slope being the ratio of the structural parameter to the viscosity. This is only true as long as the flow is due to the viscous mechanism. Such a plot in the regime of Knudsen flow would give a horizontal line instead (see eq. 7.4-13). This difference can be used as the criterion for the distinction between those two mechanisms. The viscous mechanism is operative at high pressure while the Knudsen mechanism dominates at lower pressure. At pressures falling between these two extremes, these two mechanisms are simultaneously operating (Section 7.6). The other distinction between these mechanisms is the viscosity (bulk property) in the viscous flow and the molecular weight (molecular property) in the Knudsen flow.

The following table (Table 7.5-1) summarizes formulas for the viscous flow.

Table 7.5-1: Equations for viscous flux

$J_{vis} = -\dfrac{B_0}{\mu}\dfrac{P}{R_g T}\dfrac{dP}{dz}$	$J_{vis} = -\dfrac{B_0}{\mu}R_g TC\dfrac{dC}{dz}$	$J_j = x_j J_{vis}$

where

$B_0 = \dfrac{r^2}{8}$ for capillary

$B_0 = \dfrac{\varepsilon r^2}{8\tau^2}$ for porous media containing non-overlapping capillaries of the same size

For porous media containing nonoverlapping capillaries of varying size, we could follow the approach of Wang and Smith (1983), that is:

$$J_{vis} = -\frac{\overline{B_0}}{\mu}\frac{P}{R_g T}\frac{dP}{dz} \tag{7.5-17}$$

where

$$\overline{B_0} = \int_0^\infty \frac{\varepsilon r^2}{8\tau^2} f(r) dr \tag{7.5-18}$$

with f(r)dr being the fraction of pore volume having size between r and r+dr.

7.5.3 Porous Media: Unconsolidated Packed Bed Model

Carman (1956) pointed out that the model of bundle of parallel and non-overlapping capillaries has a defect that the permeability in the direction perpendicular to the capillaries is zero. A more realistic model should allow for interconnection among capillaries, but this will add to the complexity of the model. What we will do instead is to assume that a porous medium is composed of random packing of spherical primary particles having size much smaller than the size of the medium.

Superficial velocity of fluid flowing in a porous medium is less than the interstitial velocity according to the Dupuit relation:

$$u = \frac{v}{\varepsilon} \tag{7.5-19}$$

where ε is the porosity of the medium. This equation is obtained assuming the pore space to be isotropically and randomly distributed. It does not apply to regular packing as for regular packing, the porosity is 0.215 at the plane passing through the centers of spherical particles and it is unity at plane passing through the contact points between particles (Figure 7.5-1). For random packing the porosity is about 0.38.

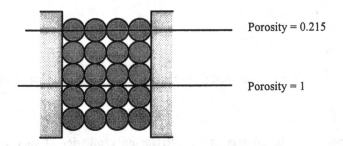

Figure 7.5-1: Diagram of a packed bed with spatial variation of porosity

According to the viscous mechanism with no slip at the wall, the molar viscous flux based on the total cross sectional area is (eq. 7.5-14):

$$J_{vis} = -\frac{\varepsilon r^2}{8\mu\tau^2} \frac{P}{R_g T} \frac{dP}{dx} = -\frac{\varepsilon d^2}{32\mu\tau^2} \frac{P}{R_g T} \frac{dP}{dx} \tag{7.5-20}$$

if the pore space is assumed as an ensemble of parallel capillaries of the <u>same diameter d</u>.

If the porous medium is a result of a random packing of spherical particles having particle size of d_P, the mean <u>pore diameter</u> is proportional to the <u>particle diameter</u> (d_P), that is:

$$d = \alpha(\varepsilon) \cdot d_P \qquad (7.5\text{-}21)$$

where the proportionality constant is a function of porosity. The variation of this proportionality constant $\alpha(\varepsilon)$ with porosity can be seen in Figure 7.5-2, where we see that the larger is the porosity the larger is the proportionality constant, meaning larger pore diameter. We note that this proportionality must become infinite when $\varepsilon \to 1$

$$\lim_{\varepsilon \to 1} \alpha(\varepsilon) = \infty$$

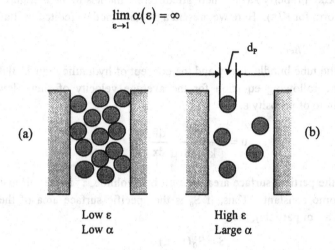

Figure 7.5-2: Packed bed of different porosity

Substituting eq. (7.5-21) into eq. (7.5-20), we obtain the molar viscous flux written in terms of the size of packing particles:

$$J_{vis} = -\frac{\varepsilon [\alpha(\varepsilon)]^2 d_P^2}{32 \mu \tau^2} \frac{P}{R_g T} \frac{dP}{dx} \qquad (7.5\text{-}22)$$

The tortuosity factor τ^2 is also a function of porosity. It is expected that a bed of low porosity will have higher tortuosity (Figure 7.5-2a) compared to a bed of high porosity (Figure 7.5-2b). Since it is difficult to distinguish the relative contribution of $\alpha(\varepsilon)$ and τ, it is customary to combine them into a so-called porosity function as:

$$J_{vis} = -\frac{d_p^2}{K(\varepsilon)\mu} \frac{P}{R_g T} \frac{dP}{dx} \tag{7.5-23a}$$

where the porosity function $K(\varepsilon)$ is:

$$K(\varepsilon) = \frac{32\,\tau^2}{\varepsilon\,\alpha(\varepsilon)} \tag{7.5-23b}$$

We have expressed the molar viscous flux in terms of separate contribution from the pressure gradient, the fluid property (viscosity μ), the solid property (d_p) and the packing property $K(\varepsilon)$. There are different models in the literature about the functional form for $K(\varepsilon)$. Here we present a popular model credited to Blake.

7.5.3.1 Blake's Theory

Using the tube bundle theory and the concept of hydraulic radius, Blake (1922) proposed the following equation for the average velocity of fluid flowing in a porous medium of porosity ε:

$$u = -\frac{\varepsilon^3}{k_0 \tau^2 S^2 \mu} \frac{dP}{dx} \tag{7.5-24}$$

where S is the particle surface area per unit bed volume, τ^2 is the tortuosity factor, and k_0 is some constant. Thus, if S_0 is the specific surface area of the packing particle (m^2/m^3 of particle), then

$$S = S_0(1-\varepsilon) \tag{7.5-25}$$

Hence the average velocity is

$$u = -\frac{\varepsilon^3}{k_0 \tau^2 \mu S_0^2 (1-\varepsilon)^2} \frac{dP}{dx} \tag{7.5-26}$$

For spherical particle, the specific surface area per unit particle volume is

$$S_0 = \frac{6}{d_p} \tag{7.5-27}$$

Thus, the equation for the viscous flux based on the total cross-sectional area of the medium is:

$$J_{vis} = -\left[\left(\frac{1}{36k}\right)\frac{\varepsilon^3}{(1-\varepsilon)^2}\right]d_p^2 \frac{1}{\mu}\frac{P}{R_g T}\frac{dP}{dx} \qquad (7.5\text{-}28a)$$

where

$$k = k_0 \tau^2 \qquad (7.5\text{-}28b)$$

Comparing eq. (7.5-28a) with eq. (7.5-14), we see that the viscous structural parameter B_0 is now written in terms of the primary particle diameter d_p and the particle porosity:

$$B_0 = \left(\frac{1}{36k}\right)\frac{\varepsilon^3}{(1-\varepsilon)^2}d_p^2 \qquad (7.5\text{-}29)$$

Also when we compare eq. (7.5-28a) with eq. (7.5-23a), we obtain the following equation for the porosity function:

$$K(\varepsilon) = \frac{36k(1-\varepsilon)^2}{\varepsilon^3}$$

For particles having shape different from sphere, d_P is the equivalent diameter of sphere having the same specific area as the particle. Empirically, the constant k was found to be about 5 to 5.6 (Carman, 1956; Dullien, 1979). Comparing the flux equation for unconsolidated media (eq. 7.5-28a) and that for the bundle of parallel capillaries model (eq. 7.5-20), we get the following relationship between the "equivalent" capillary diameter d and the primary particle diameter

$$d = \frac{1}{\sqrt{k}}\frac{\varepsilon\tau}{(1-\varepsilon)}d_P \qquad (7.5\text{-}30)$$

If the packing follows a random fashion, $\tau \approx 1/\varepsilon$, and $k = 5$ the equivalent capillary diameter is related to d_p as:

$$d \approx \frac{0.45}{(1-\varepsilon)}d_P \qquad (7.5\text{-}31)$$

For a porosity of $\varepsilon = 0.38$, the mean capillary diameter is about 0.7 d_P, which is in the same order as the dimension of the packing particle..

The following example illustrates the viscous flux calculation for a tube containing particles.

Example 7.5-3: *Viscous flux of helium in a packed column*

Calculate the viscous flux of helium through a packed column containing particle of 1 μ in size at 298K. The length of the column is 1 cm, and the pressures at two ends of the column are 500 and 10 Torr, respectively. The bed porosity is 0.34, and the viscosity of helium at 298K is 1.97×10^{-4} g/cm/sec.

Using eq. (7.5-28a) with k = 5, we calculate the viscous flow parameter.

$$B_0 = \left(\frac{1}{180}\right)\frac{\varepsilon^3}{(1-\varepsilon)^2}d_p^2 = \left(\frac{1}{180}\right)\frac{(0.34)^3}{(1-0.34)^2}(1\times 10^{-4}\,\text{cm})^2$$

$$B_0 = 5.013 \times 10^{-12}\,\text{cm}^2$$

The mean concentration in the capillary is

$$\overline{C} = \frac{\overline{P}}{R_g T} = \frac{(P_0 + P_L)/2}{R_g T} = \frac{[(500+10)/2]/760}{82.057 \times 298} = 1.37 \times 10^{-5}\,\frac{\text{mole}}{\text{cm}^3}$$

The pressure gradient is:

$$\frac{\Delta P}{L} = \frac{(500-10)\,\text{Torr}}{1\,\text{cm}}\left(\frac{1\,\text{atm}}{760\,\text{Torr}}\right)\left(\frac{1.0133 \times 10^6\,\text{g cm}^{-1}\,\text{sec}^{-2}}{1\,\text{atm}}\right)$$

$$\frac{\Delta P}{L} = 6.533 \times 10^5\,\frac{\text{g}}{\text{cm}^2\,\text{sec}^2}$$

Substituting these values into the flux equation

$$J_{vis} = \frac{B_0}{\mu}\overline{C}\cdot\frac{\Delta P}{L}$$

we get

$$J_{vis} = 2.28 \times 10^{-7}\,\frac{\text{moles}}{\text{cm}^2 - \text{sec}}$$

7.6 Transition between the Viscous Flow and Knudsen Flow

The Knudsen and viscous flows were presented separately in the last two sections. The Knudsen flow is more dominant at low pressure while the viscous flow is more dominant at high pressure. At intermediate pressures, the two mechanisms are expected to control the transport. This will be discussed in this section.

The experiments of Kundt and Warburg (1875) have shown that if the pressure of a gas flowing in a capillary is reduced to the extent that the mean free path is comparable to the capillary radius the rate of flow exceeds that predicted by the Poiseuille law, using eq (7.5-3). Figure 7.6-1 typically shows the flux versus the mean pressure in a capillary. The linear asymptote of the flux curve at high pressure when extrapolated to the flux axis yields a non-zero intercept. This extra flow at low pressure is attributed to slippage at the wall of the capillary, that is the fluid on the wall has a certain velocity instead of the zero slip assumption used in the derivation of the Poiseuille equation. If the pressure is reduced further until the mean free path is larger than the capillary diameter, viscosity will lose its meaning since molecules only collide with capillary walls, and not with each other.

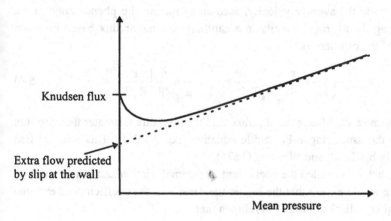

Figure 7.6-1: Plot of flux versus mean pressure

The phenomenon observed by Kundt and Warburg (1875) is called the slip flow. The slip flux is a transition flux between the viscous flux and the Knudsen flux. Hence it can be treated as the extension of either of them.

7.6.1 Extension from Viscous Flow Analysis:

If we consider the slip flow as an extension of the viscous flux, we can solve the Hagen-Poiseuille problem with the slip boundary condition, instead of the traditional zero velocity boundary condition. The slip boundary condition is (Bird et al., 1960):

$$\beta v = -\mu \frac{dv}{dr'} \quad \text{at} \quad r' = r \text{ (radius of the tube)} \qquad (7.6\text{-}1)$$

where β is called the slip friction coefficient, which determines the fluid velocity at the wall ($\beta \to \infty$ means no slip at the wall). Solving the momentum balance equation

$$\mu \left[\frac{1}{r} \frac{d}{dr} \left(r \frac{dv}{dr} \right) \right] = \frac{dP}{dz} \qquad (7.6\text{-}2)$$

with the boundary condition (7.6-1), the following average velocity can be obtained:

$$\bar{u} = -\left(\frac{r^2}{8\mu} + \frac{r}{2\beta} \right) \frac{dP}{dz} = -\left(\frac{d^2}{32\mu} + \frac{d}{4\beta} \right) \frac{dP}{dz} \qquad (7.6\text{-}3)$$

of which the first term in the RHS is the Poiseuille equation. Thus, we see that there is an extra term in the average velocity, accounting for the slip phenomenon at the wall. Knowing the average velocity in a capillary, the molar flux based on cross section area of the capillary is:

$$J_{vis} = -\left(\frac{r^2}{8\mu} + \frac{r}{2\beta} \right) \frac{P}{R_g T} \frac{dP}{dz} = -\left(\frac{d^2}{32\mu} + \frac{d}{4\beta} \right) \frac{P}{R_g T} \frac{dP}{dz} \qquad (7.6\text{-}4)$$

The second term in the RHS is the slip flux, and the total flux is greater than the flux predicted by the usual Hagen-Poiseuille equation (eq. 7.5-3a). This was verified experimentally by Kundt and Warburg (1875).

The parameter β is called the coefficient of external friction and is a function of density as one would expect that the higher the density is, the coefficient of external friction is higher. Millikan (1923) has shown that

$$\beta = \frac{1}{2} \rho v_T \frac{f_1}{2 - f_1} \qquad (7.6\text{-}5)$$

where ρ is the gas density, f_1 is the fraction of molecules which undergoes diffuse reflexion at the capillary walls (Figure 7.6-2), and v_T is the thermal velocity

$$v_T = \sqrt{\frac{8 R_g T}{\pi M}} \qquad (7.6\text{-}6)$$

But the ideal gas density is related to pressure according to:

$$\rho = \frac{PM}{R_g T} \qquad (7.6\text{-}7)$$

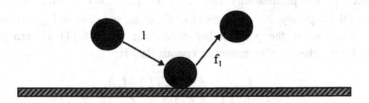

Figure 7.6-2: Deflection of molecules after hitting the surface

the coefficient of external friction will become:

$$\beta = P\sqrt{\frac{2M}{\pi R_g T}}\left(\frac{f_1}{2-f_1}\right) \quad (7.6\text{-}8)$$

Thus, the coefficient is proportional to pressure. The higher is the pressure, the larger is the value of the coefficient, implying that the slippage at the wall is not as important as the bulk flow under high pressure conditions.

Substitution of eq. (7.6-8) into eq. (7.6-4) leads to the viscous flux with slip at the wall of a capillary:

$$J_{vis} = -\frac{d^2}{32\mu}\frac{P}{R_g T}\frac{dP}{dz} - \frac{d}{4}\sqrt{\frac{\pi R_g T}{2M}}\left(\frac{2-f_1}{f_1}\right)\frac{1}{R_g T}\frac{dP}{dz}\left(\frac{\text{moles}}{\text{m}^2 - \text{sec}}\right) \quad (7.6\text{-}9)$$

The first term is the Poseuille flux, while the second term is the slippage flux.

7.6.2 Steady State Flow when Viscous and Slip Mechanisms are Operating

Having the viscous flow equation with slip at the wall (eq. 7.6-9), we integrate that equation for the case of constant boundary conditions at two ends of the capillary to obtain the following steady state flux:

$$J_{VIS} = \frac{d^2}{32\mu}\frac{\overline{P}}{R_g T}\frac{(P_0 - P_L)}{L} + \frac{d}{4}\sqrt{\frac{\pi R_g T}{2M}}\left(\frac{2-f_1}{f_1}\right)\frac{1}{R_g T}\frac{(P_0 - P_L)}{L}\left(\frac{\text{moles}}{\text{m}^2 - \text{sec}}\right) \quad (7.6\text{-}10)$$

Thus, the viscous flux per unit pressure gradient (or the permeability coefficient B) is given by:

$$B = \frac{J_{VIS}}{(\Delta P/L)} = \frac{d^2}{32\mu}\frac{\overline{P}}{R_g T} + \frac{d}{4}\sqrt{\frac{\pi R_g T}{2M}}\left(\frac{2-f_1}{f_1}\right)\frac{1}{R_g T} \quad (7.6\text{-}11)$$

A plot of the permeability coefficient of eq. (7.6-11) versus the average pressure of the capillary, $\overline{P}$, would give a straight line as shown in Figure (7.6-1).

The intercept of the permeability coefficient (eq. 7.6-11) at zero pressure predicted by this theory of extension of viscous flow is:

$$\lim_{P \to 0} \frac{J_{VIS}}{(\Delta P/L)} = \frac{d}{4}\sqrt{\frac{\pi R_g T}{2M}}\left(\frac{2-f_1}{f_1}\right)\frac{1}{R_g T} \qquad (7.6\text{-}12)$$

But the Knudsen flux, which holds at low pressure (eq. 7.4-13), is given by:

$$\frac{J_K}{(\Delta P/L)} = \frac{d}{3}\sqrt{\frac{8R_g T}{\pi M}}\frac{1}{R_g T} \qquad (7.6\text{-}13)$$

Thus, the ratio of the Knudsen flux to the slip flux predicted from the extension of the viscous flow analysis is:

$$\frac{J_K}{\lim_{P \to 0} J_{VIS}} = \frac{16}{3\pi}\frac{f_1}{2-f_1} \qquad (7.6\text{-}14)$$

The factor f_1 was measured experimentally (Carman, 1956) to fall between 0.8 and 1. Taking the average value of 0.9, the ratio of the Knudsen flux to the slip flux is about 1.4, meaning that the Knudsen flux under the molecular flow regime is higher than the flux predicted by the extra slip on the viscous flow. The reason for this is because the slip equation is an extension of the viscous flow, that is the fluid is still in its bulk state to induce viscosity. Therefore, it does not predict correctly the observed flux when $Kn \gg 1$, that is when the true molecular flow is dominant.

7.6.3 Semi-Empirical Relation by Knudsen:

To account for the full transition between the viscous and Knudsen regime, Knudsen (1909) proposed the following semi-empirical formula (later put on sounder theoretical basis by Weber, 1954) (Cunningham and Williams, 1980)

$$J = -\left(a^K P + b^K \frac{1+c_1^K P}{1+c_2^K P}\right)\frac{1}{R_g T}\frac{dP}{dz} \qquad (7.6\text{-}15)$$

This equation has been used to explain the data collected by Knudsen (Figure 7.6-1).

According to the above semi-empirical equation, the asymptote at high pressure would predict a slope of a^K and an intercept of $b^K c_1^K / c_2^K$. If we compare this with the extension of the Hagen-Poiseuille equation (7.6-4), we see that:

$$a^K = \frac{r^2}{8\mu} \tag{7.6-16}$$

$$b^K \frac{c_1^K}{c_2^K} = \frac{r\overline{P}}{2\beta} \tag{7.6-16}$$

suggesting that the slip friction coefficient β must be a function of pressure as was the case in eq. (7.6-8).

At low pressure, the above equation reduced to the Knudsen flux; thus, the value b^K is the **Knudsen diffusivity**, that is

$$b^K = D_K = \frac{2r}{3}\sqrt{\frac{8R_gT}{\pi M}} \tag{7.6-17}$$

Knudsen predicted theoretically that (Cunningham and Williams, 1980)

$$c_1^K = 2.00\sqrt{\frac{8}{\pi}} \frac{r}{\mu \overline{v}} \tag{7.6-18}$$

$$c_2^K = 2.47\sqrt{\frac{8}{\pi}} \frac{r}{\mu \overline{v}} \tag{7.6-18}$$

Hence

$$\frac{c_1^K}{c_2^K} = 0.81 \tag{7.6-18}$$

Adzumi (1937, 1939) calculated this ratio experimentally for several gases and capillaries of different materials (Table 7.6-1), and he found that this ratio is very close to 0.8, predicted theoretically by Knudsen.

Table 7.6-1: Experimental data of Adzumi

Gas	Capillary material	c_1/c_2
Hydrogen	Silver	.89
Hydrogen	Aluminum	.81
Hydrogen	Copper	.79
Hydrogen	Iron	.73
Hydrogen	Glass	.91
Acetylene	Glass	.88
Propane	Glass	.90

7.6.4 Porous Media

To describe the transition flux when the Knudsen and viscous flow mechanisms are operating inside a unconsolidated medium, we assume that the flux is additive (Kraus et al., 1953), that is the transition flux is the sum of the viscous flux (eq. 7.5-28a) and the Knudsen flux (eq. 7.4-36):

$$J = -B\frac{dP}{dz} \tag{7.6-19a}$$

where the permeability coefficient takes the form:

$$B = \frac{24}{13}\frac{1}{\sqrt{2\pi MR_gT}}\frac{\varepsilon^2 d_p}{3(1-\varepsilon)} + \frac{\varepsilon^3 d_p^2}{180(1-\varepsilon)^2 \mu}\frac{P}{R_gT} \tag{7.6-19b}$$

Eq. (7.6-19) is useful to study the properties of the porous medium. For example, by measuring the flux versus the pressure gradient for a range of mean pressure within the medium, we can obtain the permeability as a function of the mean pressure as indicated in eq. (7.6-19b). Thus, by plotting the permeability versus the mean pressure, we can obtain the slope and the intercept:

$$\text{Intercept} = \frac{24}{13}\frac{1}{\sqrt{2\pi MR_gT}}\frac{\varepsilon^2 d_p}{3(1-\varepsilon)}; \quad \text{Slope} = \frac{\varepsilon^3 d_p^2}{180(1-\varepsilon)^2 \mu}\frac{1}{R_gT} \tag{7.6-19b}$$

from which we can obtain the porosity and the packing diameter.

Kraus et al. (1953) have used this equation to study steady state flow of helium and nitrogen in packed bed containing glass microspheres, $BaSO_4$, $PbCrO_4$, TiO_2 and CuO. The BET surface area of these microparticles ranges from 0.3 to 8 m²/g. They compared these areas with the flow areas ($S_0 = 6/d_p$) and found that the flow areas calculated from the Poiseuille term (the second term in eq. 7.6-19b) is smaller than that calculated from the Knudsen term (the first term). The flow area obtained from the Knudsen term is quite comparable to the BET surface area. It was explained that in the Poiseuille flow most of it is in the center of the pores due to the zero velocity at the particle surface; thus the surface roughness can not be felt by the Poiseuille flow. On the other hand, in the Knudsen flow due to the nature of the collision between molecules and all parts of the surface, the flow areas calculated from the Knudsen flow are comparable to the BET surface area. Naturally, with the exception of particles where there is a significant amount of dead end pores, the Knudsen flow area will be less than the BET surface area.

7.7 Continuum Diffusion

This section deals with another mode of transport: continuous diffusion, resulting from the collision between molecules of one type with molecules of another type. As a result of this collision, there is a net momenta exchange between the two species. The continuum diffusion is more complicated than the free diffusion (Knudsen). Here we deal with the diffusion flux of a species within the movement of mixture (Figure 7.7-1). Take a control volume as shown in the figure, and this control volume moves at some velocity. If we choose a moving coordinate system as that moving with the control volume, the diffusion fluxes of species 1 and 2 within that mixture (assuming constant pressure) are given by:

$$J_{D,1} = -D_{12}\frac{dC_1}{dz} \quad ; \quad J_{D,2} = -D_{21}\frac{dC_2}{dz} \tag{7.7-1}$$

where D_{12} and D_{21} are binary diffusivities, characterizing the continuum diffusion. We use the symbol J to denote the diffusive flux, that is the flux relative to the moving coordinate.

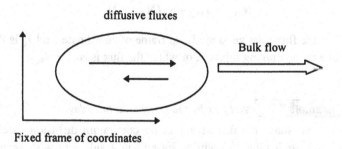

Figure 7.7-1: Frames of coordinate for continuum diffusion

Adding these two equations in eq. (7.7-1), and noting that the total fluxes of the two species within the moving coordinate is zero (as there is no net loss or gain of material from the control volume), i.e.

$$J_{1D} + J_{2D} = 0 = D_{12}\frac{dC_1}{dz} + D_{21}\frac{dC_2}{dz} \tag{7.7-2}$$

But $C_1 + C_2 = C$ (constant), the above equation will become:

$$(D_{12} - D_{21})\frac{dC_1}{dz} = 0 \tag{7.7-3}$$

Since the gradient of species 1 is generally nonzero, the necessary requirement is:

$$D_{12} = D_{21} \tag{7.7-4}$$

suggesting that the binary diffusivities are symmetric.

One of the first devices used to measure binary diffusivity is the Stefan tube (Figure 7.2-5) and the Loschmidt device (Figure 7.2-3). The latter has two bulbs connected by a capillary. Gas A is loaded into the left bulb, and gas B is in the other bulb. These two gases will diffuse in the opposite direction once the valve connecting these two bulbs is opened. Assuming gas A is lighter than gas B. The light gas travels through the capillary faster than the heavier gas does (note the Graham's law of diffusion, eg. 7.2-1), which will cause a pressure build-up in the right bulb B. This pressure build up will induce a convective flux from the right to the left to the extent that the net flux will be zero, by the virtue of the closed system, that is the sum of two fluxes relative to the fixed frame of coordinate will be zero. To allow for the total flux, the flux equations written with respect to the fixed frame of coordinate are:

$$\begin{aligned} N_{D,1} &= J_{D,1} + x_1(N_{D,1} + N_{D,2}) \\ N_{D,2} &= J_{D,2} + x_2(N_{D,1} + N_{D,2}) \end{aligned} \tag{7.7-5}$$

where N_D is the flux relative to the fixed frame of coordinate, and J_D is the diffusive flux relative to the moving mixture, of which the flux is $N_{D,1} + N_{D,2}$.

Example 7.7-1: *Net flux by the Loschmidt tube device*

To study the diffusive flux by continuum diffusion mechanism, the Loschsmidt tube is modified by adding a side arm with an oil piston as shown in Figure 7.2-4. Take the case where we have light gas in the left bulb A. The pressure in the bulb B will build up because of the faster flow of light gas. To equalize the pressure, the "frictionless" oil piston automatically is moved to the left to balance the pressure. The rate of movement of this oil piston is the net flux of the two gases, suggesting that the net flux is not zero. The net flux $N_{D,1} + N_{D,2}$ is balanced by the rate of travel of the oil piston, that is:

$$\frac{P}{R_g T} A_{side} \frac{d\ell}{dt} = (N_{D,1} + N_{D,2}) A_c \tag{7.7-6}$$

where A_{side} is the cross-sectional area of the oil side arm, and A_c is the cross sectional area of the diffusing capillary. Thus, by measuring the rate of oil travels, one could readily determine the total flux, according to eq. (7.7-6).

For the fluxes with respect to the **fixed frame of coordinates**, Graham has found experimentally in 1829 for an open system that:

$$\frac{N_{D,1}}{N_{D,2}} = -\sqrt{\frac{M_2}{M_1}} \qquad (7.7\text{-}7)$$

The generalization of this formula for a multicomponent mixture is:

$$\sum_{j=1}^{N} \sqrt{M_j}\, N_{D,j} = 0 \qquad (7.7\text{-}8)$$

This is called the Graham law of diffusion, similar in form to the Graham law of effusion eq. (7.4-42) obtained earlier for the free molecular flow (Knudsen flow).

7.7.1 Binary Diffusivity

Before we proceed further, we will discuss in this section briefly about correlation used to calculate binary diffusivity. The equation commonly used to calculate this diffusivity is derived from the Chapman-Enskog theory (Bird et al., 1960):

$$D_{12} = 0.0018583 \frac{\sqrt{T^3 \left(\frac{1}{M_1} + \frac{1}{M_2}\right)}}{P\, \sigma_{12}^2\, \Omega_{D,12}} \qquad (7.7\text{-}9)$$

where the binary diffusivity is in cm^2/sec, the total pressure P is in atm, the temperature T is in Kelvin, σ_{12} is the collision diameter in Angstrom, and $\Omega_{D,12}$ is a dimensionless function of temperature and the intermolecular potential field for one molecule of species 1 and one molecule of species 2. The potential field can be calculated for non-polar molecules by the 12-6 Lennard-Jones potential energy function

$$\phi_{12} = 4\varepsilon_{12}\left[\left(\frac{\sigma_{12}}{r}\right)^{12} - \left(\frac{\sigma_{12}}{r}\right)^{6}\right]$$

where ϕ_{12} is the potential energy, σ_{12} is the distance between molecule 1 and molecule 2 such that the potential energy of interaction is zero, and ε_{12} is the minimum of the potential energy. These Lennard-Jones parameters are determined from experimental determination of D_{12} over a range of temperature. However, for nonpolar, noninteracting molecules, the Lennard-Jones parameters of species 1 and 2 can be used:

$$\sigma_{12} = \frac{\sigma_{11}+\sigma_{22}}{2}; \qquad \varepsilon_{12} = \sqrt{\varepsilon_{11}\varepsilon_{22}} \qquad (7.7\text{-}10)$$

These parameters σ, ε can be found in Table B-1 of Bird et al. (1960). Since the dimensionless function $\Omega_{D,12}$ is a function of temperature, ranging from about 0.5 at high temperature to about 2.7 at low temperature (for rigid sphere this function is unity), the binary diffusivity will increase as T^2 at low temperature and $T^{1.65}$ at very high temperature. In average, the binary diffusivity increases as $T^{1.75}$. This dimensionless function $\Omega_{D,12}$ can be found in Table B-2 of Bird et al. (1960), and it is tabulated as a function of $k_B T/\varepsilon_{12}$. Or alternatively, it can be calculated from the following equation (Neufeld et al., 1972)

$$\Omega_{D,12} = \frac{A}{\left(T^*\right)^B} + \frac{C}{\exp(DT^*)} + \frac{E}{\exp(FT^*)} + \frac{G}{\exp(HT^*)}; \qquad T^* = \frac{k_B T}{\varepsilon_{12}} \qquad (7.7\text{-}11)$$

where the constants are obtained from Reid et al. (1988):

A = 1.06036;   B = 0.15610; C = 0.19300;   D = 0.47635;
E = 1.03587;   F = 1.52996;   G = 1.76474;   H = 3.89411

Example 7.7-2: *Binary diffusivity of propane and nitrogen*

This example shows how to use eq. (7.7-9) to calculate the binary diffusivity of propane/nitrogen at 30°C and 1 atm. To use that equation, we need to know σ, and ε. These can be found in Table B-1 of Bird et al. (1960) and are listed below:

Gas	$\sigma(\text{A})$	ε/k_B (K)
nitrogen	3.681	91.5
propane	5.061	254

We calculate

$$\sigma_{12} = \frac{3.681 + 5.061}{2} = 4.371; \qquad \frac{\varepsilon_{12}}{k_B} = \sqrt{\left(\frac{\varepsilon_1}{k_B}\right)\left(\frac{\varepsilon_2}{k_B}\right)} = \sqrt{(91.5)(254)} = 152$$

Thus, we have

$$\frac{k_B T}{\varepsilon_{12}} = \frac{303}{152} = 1.987$$

Having this value of $(k_B T)/\varepsilon_{12} = 1.987$, we look up Table B-2 of Bird et al. (1960) to obtain a value for the dimensionless function $\Omega_{D,12}$ as 1.078. Or alternatively, using eq. (7.7-11) we calculate $\Omega_{D,12}$ as 1.0778. Thus, the binary diffusivity is:

$$D_{12} = 0.0018583 \frac{\sqrt{T^3 \left(\frac{1}{M_1} + \frac{1}{M_2}\right)}}{P\sigma_{12}^2 \Omega_{D,12}} = 0.0018583 \frac{\sqrt{303^3 \left(\frac{1}{28} + \frac{1}{44}\right)}}{(1)(4.371)^2 (1.078)}$$

$$\underline{D_{12} = 0.115 \text{ cm}^2/\text{sec}}$$

The gas phase binary diffusion coefficient is usually in the order of 0.1 to 1 cm²/sec for most gases at normal temperature.

7.7.2 Constitutive Flux Equation for a Binary Mixture in a Capillary

For binary systems, the constitutive flux equation relative to a fixed frame of coordinate is obtained by combining eqs. (7.7-1) and (7.7-5):

$$\frac{1}{R_g T} \frac{dp_1}{dz} = \frac{x_2 N_{D,1} - x_1 N_{D,2}}{D_{12}} \tag{7.7-12}$$

If the system is open, the flux of the component 2 is related to that of the component 1 by the Graham's law of diffusion (eq. 7.7-7), hence the above equation can be rewritten to yield an expression for the diffusion flux of the component 1:

$$-\frac{1}{R_g T} \frac{dp_1}{dz} = \frac{N_{D,1}(1 - \sigma_{12} x_1)}{D_{12}} \tag{7.7-13}$$

where

$$\sigma_{12} = 1 - \sqrt{\frac{M_1}{M_2}} \tag{7.7-14}$$

Knowing the flux of the component 1, the flux of the component 2 is calculated from the Graham's law of diffusion.

At steady state the flux equation (7.7-13) can be integrated using constant boundary conditions at two ends of the capillary to give:

392 Kinetics

$$N_{D,1} = \frac{D_{12}}{\sigma_{12}L} \frac{P}{R_gT} \ln\left(\frac{1-\sigma_{12}x_L}{1-\sigma_{12}x_0}\right) = \frac{D_{12}^0}{\sigma_{12}L} \frac{P_0}{R_gT} \ln\left(\frac{1-\sigma_{12}x_L}{1-\sigma_{12}x_0}\right) \quad (7.7\text{-}15)$$

where D_{12}^0 is the molecular diffusivity at some reference pressure P_0. The variables x_0 and x_L are the mole fractions of the component 1 at the two ends of the capillary. We note that the molar flux under the region of molecular control is independent of pressure, which is in contrast to the Knudsen regime where the flux is proportional to pressure. This is the distinctive difference between the Knudsen flow and the continuum diffusion.

We will reserve the treatment of multicomponent systems until Chapter 8, where the systematic approach of Stefan-Maxwell will be used. Now we take an example to illustrate the binary flux calculation when the bulk diffusion is operating.

Example 7.7-2: *Molecular diffusion of H_2/N_2 in an open capillary*

Take the example of hydrogen (1) and nitrogen (2) at 1 atm and 25 °C through a capillary of length 1 cm and $x_0 = 1$ and $x_L = 0$. The binary diffusivity at 1 atm and 0°C is 0.674 cm²/sec. Thus, the molecular diffusivity at 25 C is:

$$D_{12}|_{T=298K} = D_{12}|_{T=273K} \left(\frac{298}{273}\right)^{1.75} = 0.674 \left(\frac{298}{273}\right)^{1.75}$$

$$D_{12}|_{T=298K \& 1atm} = 0.786 \text{ cm}^2/\text{sec}$$

We calculate the parameter σ_{12} of eq. (7.7-14)

$$\sigma_{12} = 1 - \sqrt{\frac{M_1}{M_2}} = 1 - \sqrt{\frac{2}{28}} = 0.733$$

The hydrogen flux is calculated from eq. (7.7-15)

$$N_{D,1} = \frac{(0.786)}{(0.733)(1)} \times \frac{1}{(82.05 \times 298)} \times \ln\left(\frac{1-0.733(0)}{1-0.733(1)}\right)$$

$$N_{D,1} = 5.79 \times 10^{-5} \frac{\text{mole}}{\text{cm}^2 \text{s}}$$

The nitrogen flux is calculated from the Graham's law of diffusion (eq. 7.7-7)

$$N_{D,2} = -\sqrt{\frac{M_1}{M_2}}\, N_{D,1} = -\sqrt{\frac{2}{28}}\,(5.79 \times 10^{-5})$$

$$N_{D,2} = -1.55 \times 10^{-5}\, \frac{\text{mole}}{\text{cm}^2 - \text{sec}}$$

The negative sign indicates that nitrogen flows in the opposite direction to hydrogen. The net flux is:

$$N_{D,1} + N_{D,2} = (5.79 - 1.55) \times 10^{-5} = 4.24 \times 10^{-5}\, \frac{\text{mole}}{\text{cm}^2\, \text{sec}}$$

which is equivalent to 0.95 cc (NTP)/cm^2-sec. The net flux is positive, indicating the net flow is the same direction as the hydrogen.

7.7.3 Porous Medium

In the last section, we have considered the diffusion flux equation for a binary system in a capillary. For a porous medium, the equivalent flux equation to that for cylindrical capillary (eq. 7.7-12) is:

$$\frac{1}{R_g T} \frac{dp_1}{dz} = \frac{x_2 N_{D,1} - x_1 N_{D,2}}{D_{12,\text{eff}}} \qquad (7.7\text{-}16)$$

where the diffusion flux is based on the total cross sectional area, and the effective diffusivity $D_{12,\text{eff}}$ is related to the binary diffusivity as follows:

$$D_{12,\text{eff}} = \frac{\varepsilon}{q} D_{12} \qquad (7.7\text{-}17)$$

Here ε is the porosity of the porous medium, and q is the tortuosity factor. The inclusion of the porosity and the tortuosity factor was proved in Section 7.4 for the case of Knudsen flow.

The following correlation for the tortuosity factor is proposed by Wakao and Smith (1962) using a random pore model:

$$q = \frac{1}{\varepsilon} \qquad (7.7\text{-}18)$$

which was later refined by Abbasi et al. (1983) using the Monte Carlo simulations of gas molecule trajectories through assemblages of spheres when the length scale of molecular-molecular collision is shorter than the particle heterogeneity scale:

$$q = \frac{1}{\varepsilon} + 1.196\frac{\sigma}{\overline{d}} \qquad (7.7\text{-}19)$$

where σ is the standard deviation of pore size, and $\overline{d}$ is the mean pore size. The correlation of Abbasi et al. (1983) reduces to the Wakao and Smith's correlation when the solid has a very narrow pore size distribution.

7.8 Combined Bulk and Knudsen Diffusion

We have considered separately the necessary flux equations for the cases of Knudsen diffusion and continuum diffusion. *Knudsen diffusion usually dominates when the pore size and the pressure are small, and the continuum diffusion dominates when the pore size and pressure are large.* In the intermediate case which is usually the case for most practical systems, we would expect that both mechanisms will control the mass transport in a capillary or a porous medium. In this section, we will consider this intermediate case and present the necessary flux equations.

7.8.1 Uniform Cylindrical Capillary

We first consider the case of a straight capillary. For a binary system, the driving force to induce the flow by continuum diffusion is the partial pressure gradient (eg. 7.7-12):

$$-\frac{1}{R_g T}\left(\frac{dp_1}{dz}\right)_D = \frac{x_2 N_{D,1} - x_1 N_{D,2}}{D_{12}} \qquad (7.8\text{-}1)$$

where the subscript D denotes continuum diffusion.

The driving force to induce the flux by the Knudsen mechanism is:

$$-\frac{1}{R_g T}\left(\frac{dp_1}{dz}\right)_K = \frac{N_{K,1}}{D_{K,1}} \qquad (7.8\text{-}2)$$

The sum of these two driving forces is the total driving force inducing the flow and since the flux must be the same, induced by both mechanisms, we write:

$$N_{D,1} = N_{K,1} = N_1 \qquad (7.8\text{-}3a)$$

and

$$N_{D,2} = N_{K,2} = N_2 \qquad (7.8\text{-}3b)$$

By summing eqs. (7.8-1) and (7.8-2) we obtain the following necessary equation to describe the flux in a capillary where both mechanisms of diffusion are operative:

$$-\frac{1}{R_g T}\frac{dp_1}{dz} = \frac{N_1}{D_{K,1}} + \frac{x_2 N_1 - x_1 N_2}{D_{12}} \qquad (7.8\text{-}4)$$

This form is the form suggested by the dusty gas theory, and will be formally proved in the context of Stefan-Maxwell approach in Chapter 8.

The relationship between the two fluxes is the Graham's law of diffusion:

$$\sqrt{M_1}\,N_1 + \sqrt{M_2}\,N_2 = 0 \qquad (7.8\text{-}5)$$

which has been experimentally proved by Graham (1831), Hoogschagen (1955), Scott and Dullien (1962), Rothfield (1963) and Knaff and Schlunder (1985).

For the second component, we can write a similar equation to eq. (7.8-4) by interchanging the indices 1 and 2 and noting the symmetry of binary diffusivities $D_{21} = D_{12}$ (eq. 7.7-4):

$$-\frac{1}{R_g T}\frac{dp_2}{dz} = \frac{N_2}{D_{K,2}} + \frac{x_1 N_2 - x_2 N_1}{D_{12}} \qquad (7.8\text{-}6)$$

This equation is not independent from eq. (7.8-4). It can be proved by adding the two equations and making use of the Graham's law of diffusion. This means that eq. (7.8-4) is the only independent constitutive equation relating fluxes and concentration gradients.

Using the Graham's law of diffusion (eq. 7.8-5), we write the flux N_1 in terms of the concentration gradient as follows (from eq. 7.8-4):

$$N_1 = -\frac{1}{\left[\dfrac{(1-\sigma_{12} x_1)}{D_{12}} + \dfrac{1}{D_{K,1}(r)}\right]}\frac{\partial C_1}{\partial z} \qquad (7.8\text{-}7)$$

where σ_{12} is defined in eq. (7.7-14). This is the equation applicable for a cylindrical capillary under isobasic conditions. It can be integrated using constant boundary conditions at two ends of the capillary of length L to give the following steady state diffusion flux:

$$N_1 = \frac{D_{12} P}{L R_g T \sigma_{12}} \ln\left[\frac{1 - \sigma_{12} x_L + D_{12}/D_{K,1}(r)}{1 - \sigma_{12} x_0 + D_{12}/D_{K,1}(r)}\right] \qquad (7.8\text{-}8)$$

Knowing the flux for species 1, the flux of species 2 can be obtained from the Graham's law of diffusion (eq. 7.8-5).

7.8.2 Porous Solids

For porous solids having a pore size distribution f(r), where f(r)dr is the fraction of pore volume having pore radii between r and r+dr, and if all the pores are cylindrical in shape and oriented along the direction of flow, the steady state flux based on total cross sectional area of the component 1 can be calculated from:

$$N_1 = \frac{\varepsilon D_{12} P}{L R_g T \sigma_{12}} \int_0^{r_{max}} \ln\left[\frac{1 - \sigma_{12} x_L + D_{12}/D_{K,1}(r)}{1 - \sigma_{12} x_0 + D_{12}/D_{K,1}(r)}\right] f(r) dr \qquad (7.8\text{-}9)$$

Such an assumption of cylindrical pores oriented along the flow is so ideal because pores can be randomly oriented and pores can have different shape and size. This deviation from ideal condition can be allowed for by the introduction of the tortuosity factor. In general, this tortuosity factor is a function of pore radius, and the following equation can be used (Brown and Travis, 1983):

$$N_1 = \frac{\varepsilon D_{12} P}{L R_g T \sigma_{12}} \int_0^{r_{max}} \ln\left[\frac{1 - \sigma_{12} x_L + D_{12}/D_{K,1}(r)}{1 - \sigma_{12} x_0 + D_{12}/D_{K,1}(r)}\right] \frac{f(r)}{q(r)} dr \qquad (7.8\text{-}10)$$

When the continuum diffusion controls the mass transfer, that is in large pore solids and high pressure, the flux equation given in eq. (7.8-10) is reduced to:

$$N_1 = \frac{\varepsilon D_{12} P}{L R_g T \sigma_{12}} \ln\left[\frac{1 - \sigma_{12} x_L}{1 - \sigma_{12} x_0}\right] \int_0^{r_{max}} \frac{f(r)}{q(r)} dr \qquad (7.8\text{-}11)$$

Thus, if we define the mean tortuosity factor as

$$\bar{q} = \left[\int_0^{r_{max}} \frac{f(r)}{q(r)} dr\right]^{-1} \qquad (7.8\text{-}12)$$

we obtain the following integral flux equation for continuum diffusion regime:

$$N_1 = \frac{\varepsilon}{\bar{q}} \frac{D_{12} P}{L R_g T \sigma_{12}} \ln\left[\frac{1 - \sigma_{12} x_L}{1 - \sigma_{12} x_0}\right] \qquad (7.8\text{-}13)$$

When Knudsen diffusion mechanism controls the mass transfer, the following flux equation is obtained from eq. (7.8-10):

$$N_1 = \frac{\varepsilon D_{K,1}^0 P(x_0 - x_L)}{L R_g T}\left[\frac{1}{r_0} \int_0^{r_{max}} \frac{r\, f(r)}{q(r)} dr\right] \qquad (7.8\text{-}14)$$

where r_0 is some reference pore radius and $D_{K,1}^0$ is the Knudsen diffusivity corresponding to this pore radius.

7.8.3 Models for Tortuosity

7.8.3.1 Pellet-Grain Model

Dogu and Dogu (1991) presented a pellet-grain model for the purpose of prediction of tortuosity factor, and obtained a correlation for the tortuosity factor based on the macroporosity ε_a.

For $\varepsilon_a > 0.476$, the tortuosity factor is calculated from:

$$q_a = \frac{\varepsilon_a}{1 - \pi\left[(1-\varepsilon_a)\frac{3}{4\pi}\right]^{2/3}} \tag{7.8-15}$$

For $\varepsilon_a < 0.476$, the following approximation formula could be used:

$$\frac{\varepsilon_a}{q_a} = 1 - \pi\left(\frac{r_p}{a}\right)^2 + 2\left(\frac{r_p}{a}\right)^2 (\beta - \sin\beta); \quad \beta = 2\cos^{-1}\left(\frac{1}{2}\frac{a}{r_p}\right) \tag{7.8-16}$$

where r_p is the grain radius and a is the unit cell dimension. The parameter (r_p/a) is calculated from the following equation:

$$\varepsilon_a = 1 + \frac{\pi}{4} + \frac{8\pi}{3}\left(\frac{r_p}{a}\right)^3 - 3\pi\left(\frac{r_p}{a}\right)^2 \tag{7.8-17}$$

This approximation formula is valid when $D_{Ta}/D_i > 1000$, where D_{ta} is the combined diffusivity in the macropore and D_i is the effective micropore diffusivity. This criterion is readily satisfied in many systems as the order of magnitude of the effective macropore diffusivity is 10^{-2} while that of D_i is in the range of 10^{-5} to 10^{-7} cm^2/sec.

7.8.3.2 Monte-Carlo Simulation

Using a Monte-Carlo simulation to a system where diffusion occurs on a length scale shorter than the heterogeneity of the medium, Akanni et al. (1987) have shown that the relation obtained by Maxwell for a dilute suspension of sphere can be applicable to a wider range of porosity:

$$q = 1 + \frac{1}{2}(1-\varepsilon) \qquad (7.8\text{-}18)$$

7.8.3.3 Weissberg Model

Using the variational approach to a bed of overlapping spheres, Weissberg obtained the following relation for the tortuosity

$$q = 1 - \frac{1}{2}\ln\varepsilon \qquad (7.8\text{-}19)$$

7.8.3.4 Bruggeman Model

Another model for tortuosity is that of Bruggeman

$$q = \frac{1}{\sqrt{\varepsilon}} \qquad (7.8\text{-}20)$$

Figure 7.8-1 shows plots of ε/q versus ε for the three models of Maxwell, Weisseberg and Bruggeman. These models are close to each other and they can be used to estimate the tortuosity factor when experimental value is not available.

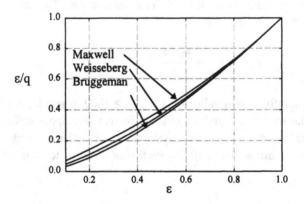

Figure 7.8-1: Plot of (ε/q) versus ε

7.9 Surface Diffusion

Surface diffusion is considered as the most important mode of transport for many sorbates as many practical sorbents have high internal surface area, such as activated carbon and silica gel. Literature data have shown that surface diffusion is important, and the mobility of adsorbed molecules varies with the loading, usually increasing sharply with loading (Carman and Raal, 1951; Carman, 1952). Aylmore and Barrer (1966) measured diffusion of several pure gases and their mixtures through a Carbolac carbon plug having surface area of 730 m^2/g. The observed extra flow of these gases when the coverage is up to 20% of the monolayer coverage was attributed to the surface diffusion.

7.9.1 Characteristics of Surface Diffusion

Surface diffusion implies a thermal motion of adsorbed molecules. It should be distinguished from "interstitial" diffusion or intracrystalline diffusion, which is more similar to solid solution than adsorption. This intracrystalline diffusion is strongly affected by the molecular size and it decreases with an increase in the molecular size. The decrease is much faster than $1/\sqrt{M}$, observed for the case of Knudsen diffusion.

In contrast to the interstitial diffusion and Knudsen diffusion, the contribution of surface diffusion increases with larger and heavier molecules because these molecules are most easily condensed and adsorbed. This is due to the higher density of adsorbed molecules.

In common with interstitial diffusion, surface diffusion is activated, possible with a smaller activation energy than in interstitial diffusion (Kirchheim, 1987).

Surface diffusion occurs even at Henry law isotherm (Barrer and Strachan, 1955). Ash, Barrer and Pope (1963) studied the surface flow of gases, such as sulphur dioxide, carbon dioxide, nitrogen, argon, and helium in microporous carbon. They observed that in many cases the surface flow dominates the transport. Even nitrogen at 190K surface flow is significantly higher than gas flow.

The microporous carbon used by Ash et al. (1963) was prepared by compressing carbon powder into a tube of 3mm internal bore. The plug has the following properties, L = 0.91cm, porosity = 0.5, and the cross sectional area of 0.07 cm^2. The electron microscopy shows the particle to have a pore size of 100 A. Each particle consists of an assembly of para-crystallites. They found that the surface diffusion in the steady state measurement increases rapidly as the monolayer coverage is approached. As the monolayer layer is exceeded the diffusion coefficient shows a minimum and then rises sharply again in the region of capillary condensation. In general, the measured surface diffusivity shows a nonlinear relationship with surface loading. It can be divided into three regions: a monolayer region, a multilayer region and a capillary condensation region (Figure 7.9-1).

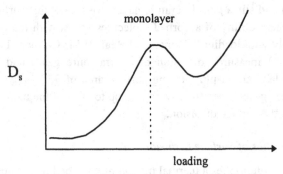

Figure 7.9-1: Surface diffusivity as a function of loading

In the monolayer region which is applicable for most adsorption systems, the effects observed are almost certainly due to surface diffusion, and the followings have been observed experimentally:

- D_S increases with loading and this is observed even at loading
- the energy of activation is one-third to one-half of the heat of adsorption
- surface diffusion can not be modelled as two dimensional gas as the activation energy of two dimensional gas is too low compared to observed values (Carman, 1956; Gilliland et al., 1974)
- surface mobility is by a hopping mechanism (random walk), that is

$$D_s = \frac{\delta^2}{4\tau} \qquad (7.9\text{-}1)$$

where δ is the distance between adjacent sites and τ is the average time that a site is occupied between jumps. The factor 4 in the above equation is for two dimensional systems and 6 for three-dimensional systems. The time τ is related to the period of vibration τ_0 at a given site by the following equation:

$$\tau = \tau_0 \exp\left(\frac{E_a}{R_g T}\right) \qquad (7.9\text{-}2)$$

where E_a is the energy of activation needed for a jump. The period of vibration τ_0 can be equated roughly to the time needed to cover a distance δ at the thermal velocity, i.e.

$$\tau_0 = \frac{\delta}{v_T} \qquad (7.9\text{-}3)$$

Thus, with this approximation, the surface diffusivity (7.9-1) will take the form:

$$D_s = \frac{1}{4} v_T \delta \exp\left(-\frac{E_s}{R_g T}\right) = D_{s0} \exp\left(-\frac{E_s}{R_g T}\right) \qquad (7.9\text{-}4)$$

Carman and Raal (1951) have shown that when less than a monolayer coverage exists in a given system, the surface diffusivity increases rapidly with loading. They attributed this rise with loading to the fact that the heat of adsorption varies over the adsorbent surface and the surface diffusion is due to the most loosely bound molecules. This means that at low coverage adsorption occurs on high energy sites where sorbed molecules exhibit a very low mobility. When the surface coverage increases, adsorbed molecules occupy more onto lower energy sites and because of the loose binding these molecules will diffuse at a rate faster than that for the strongly bound molecules in the high energy sites. This is one of the explanations for the surface diffusion increase with loading. Other explanations are also possible such as the hopping model of HIO and its modified version due to Okazaki et al. (1981) and the chemical potential argument of Darken. We will address these models later but first let us consider the definition of surface flux and the temperature-dependence of surface diffusivity.

7.9.2 Flux Equation:

Before we address further the way surface diffusivity increases with surface loading, we will discuss the surface flux equation and the conventional definition of surface diffusivity.

If the surface concentration is defined as mole per unit surface area, the surface mass transfer is defined as follows (Gilliland et al., 1974):

$$M_s = -bD_s \frac{dC_s}{dx} \quad \left[\frac{\text{moles}}{\text{sec}}\right] \qquad (7.9\text{-}5a)$$

where M_s is the mass transfer in the x direction across a line of width b under the influence of a surface concentration gradient (Figure 7.9-2). Thus, the surface flux based on unit width length is M_s/b, that is:

$$J_s = -D_s \frac{dC_s}{dx} \quad \frac{\text{mole}}{\text{m sec}} \qquad (7.9\text{-}5b)$$

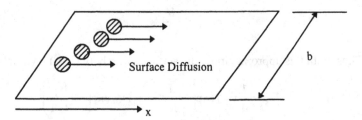

Figure 7.9-2: Schematic of surface diffusion

Adsorbed concentration is not necessarily defined in terms of mole per unit surface area. It can be defined as mole per units solid volume[1], C_μ as the surface area is not always available and moreover the adsorbed phase of practical solids is not defined as well structured surface. The adsorbed concentration is therefore conveniently defined as either mole per unit volume of the solid volume, which reflect the apparent space where adsorbed molecules reside, or mole per unit mass of the adsorbent. We shall use the definition of mole per unit solid volume as the other can be related to this by a factor of solid density.

The relationship between C_s and C_μ is obtained by equating the amount of adsorbed species in the shell of Δx as follows:

$$(\Delta x\, b\delta) C_\mu = (\Delta x\, b) C_s \tag{7.9-6}$$

where Δx is the length of the solid segment, b is the width and δ is the depth of the solid. Thus, the necessary relationship between the two concentrations is:

$$C_s = \delta C_\mu \tag{7.9-7}$$

Hence, the mass transfer equation (eq. 7.9-5a), when written in terms of the concentration based on unit solid volume, is given by:

$$M_s = -bD_s \frac{dC_s}{dx} = -b\delta D_s \frac{dC_\mu}{dx} \tag{7.9-8}$$

Thus, if we define the flux of the adsorbed species as mole transferred per unit solid cross sectional area, we will have:

$$J_s = \frac{M_s}{b\delta} = -D_s \frac{dC_\mu}{dx} \tag{7.9-9a}$$

---

[1] This volume is taken as the particle volume minus the void volume where molecules are present in free form.

This means that if the flux per unit total cross sectional area of the porous solid, instead of just the solid cross section area, is required, the necessary flux equation is:

$$J_s = -(1-\varepsilon)D_s \frac{dC_\mu}{dx} \quad \frac{\text{mole}}{\text{m}^2 \text{ sec}} \tag{7.9-9b}$$

The surface diffusion coefficient is usually a function of loading and this will be addressed with a number of models presented in Section 7.9.5.

There are many other definitions of the surface diffusion flux in the literature. For example, Smith (1970) and Schneider and Smith (1968) defined the surface diffusion flux per unit total cross sectional area as follows:

$$J_s = -D_s \rho_p \frac{dq}{dz} \tag{7.9-10}$$

where q is the surface concentration per unit mass, and ρ_p is the particle density (mass per unit total particle volume).

The surface flux equation (7.9-9b) is written in terms of the gradient of adsorbed phase concentration. It can also be written in terms of the gas phase concentration provided that there is a <u>local equilibrium</u> between the gas and adsorbed phases. By local equilibrium here, we mean that at any given point within the particle the gas and solid phases are in equilibrium with each other, despite the gradients of concentration in both phases are present. This is acceptable if the rates of adsorption and desorption at any point are much faster than the rates of diffusion in both phases. If this equilibrium is governed by the Henry law, that is

$$C_\mu = KC \tag{7.9-11}$$

where C is the gas phase concentration, the surface flux equation written in terms of the gradient of gas phase concentration is:

$$J_s = -(1-\varepsilon)D_s K \frac{dC}{dz} \tag{7.9-12a}$$

If the local equilibrium between the two phases take the general form

$$C_\mu = f(C)$$

the equivalent surface flux equation will be:

$$J_s = -(1-\varepsilon)D_s f'(C) \frac{dC}{dz} \tag{7.9-12b}$$

Various forms of the surface diffusion flux were used in the literature. The following table (Table 7.9-1) lists the popular forms used by researchers in this area of surface diffusion.

Table 7.9-1: Various forms of surface flux equations

Authors	Equation
Masamune & Smith (1964, 1965)	$J_s = -\rho_p D_{s,ps} \dfrac{dq}{dr}$; $q \equiv \dfrac{mole}{g}$, ρ_p = particle density
Rivarola & Smith (1964)	$J_s = -\phi_s \dfrac{dC_s}{dr}$; $C_s = \dfrac{mole}{m^2}$ $\phi_s = 2D_s\left[\dfrac{\varepsilon_a^2}{\bar{r}_a} + \dfrac{\varepsilon_i^2}{\bar{r}_i} + \dfrac{4\varepsilon_a(1-\varepsilon_a)}{\bar{r}_a}\right]$; $\bar{r}_a$ = macropore radius
Satterfield & Ino (1968)	$J_s = -\dfrac{D_s}{\tau_s}\rho_s(1-\varepsilon)S_g \dfrac{dC_s}{dr}$; $C_s = \dfrac{mole}{m^2}$; $\rho_s(1-\varepsilon) = \rho_p$
Schneider & Smith (1968)	$J_s = -D_{s,p}\dfrac{dC_{s,p}}{dr} = -D_{s,ps}\dfrac{dC_{s,ps}}{dr} = -D_{s,s}\dfrac{dC_{s,s}}{dr}$ $C_{s,p} = \dfrac{mole}{pore\ volume}$; $C_{s,ps} = \dfrac{mole}{particle\ volume}$; $C_{s,s} = \dfrac{mole}{solid\ volume}$ $D_{s,p} = \dfrac{\varepsilon}{\tau_s}D_s$; $D_{s,ps} = \dfrac{1}{\tau_s}D_s$; $D_{s,s} = \dfrac{(1-\varepsilon)}{\tau_s}D_s$
Reed & Butt (1971) Patel & Butt (1972)	$J_s = -D_{s,p}\dfrac{dC_{s,p}}{dr}$; $D_{s,p} = D_{s,p}^0 \exp\left[-\dfrac{E_s(\theta)}{R_g T}\right]$

7.9.3 Temperature Dependence of Surface Diffusivity:

As we have indicated before, surface diffusion is an activated process, that is the surface diffusivity follows the Arrhenius equation:

$$D_s = D_{s\infty} \exp(-E_s / R_g T) \qquad (7.9\text{-}13)$$

Let us take the case where the partition between the fluid and adsorbed phases is linear, and the Henry constant has the following temperature-dependent form

$$\dfrac{d \ln K}{dT} = \dfrac{\Delta H}{R_g T^2} \qquad (7.9\text{-}14)$$

where ΔH is the enthalpy of adsorption. In integral form, it is given by:

$$K = K_\infty \exp(-\Delta H / R_g T) \qquad (7.9\text{-}15)$$

Adsorption occurs with a decrease in surface free energy, ΔG, as well as a decrease in entropy, ΔS, because of the confining of an adsorbed molecule to a thin surface layer (certain degrees of freedom are lost). From the equation $\Delta G = \Delta H - T \cdot \Delta S$, it is clear that the enthalpy of adsorption, ΔH, is negative, that is adsorption process is endothermic.

Substituting eqs. (7.9-13) and (7.9-15) into eq. (7.9-12a), we obtain the surface flux as a function of temperature and gas phase concentration gradient:

$$J_s = -(1-\varepsilon) D_{s\infty} K_\infty \exp\left(-\frac{E_s + \Delta H}{R_g T}\right) \frac{dC}{dz} \qquad (7.9\text{-}16)$$

Heat of adsorption is usually greater the activation energy for surface diffusion, i.e.

$$-E_s - \Delta H > 0 \qquad (7.9.17)$$

Hence, for a given pressure gradient, the surface flux decreases rapidly with an increase in temperature. The pore volume diffusion flux

$$J_G = -D_{eff} \frac{dC}{dz} = -\frac{D_{eff}}{R_g T} \frac{dP}{dz}$$

on the other hand can either increase or decrease with temperature for a given pressure gradient. When it decreases with temperature, its rate is not as fast as the rate of decline of surface diffusion. Thus, the influence of the surface diffusion becomes less important as the temperature increases. One would expect this physically as when temperature is increased the amount adsorbed on the surface decreases at a rate much faster than the increase in the surface diffusivity, and hence the contribution of the surface flux decreases. This is true for linear adsorption isotherm. For highly nonlinear isotherm, however, the opposite is true because when the surface is nearly covered an increase in temperature will result in an increase in the surface diffusivity and very little reduction in the adsorbed concentration, hence an increase in the surface flux is resulted.

7.9.4 Surface Diffusion Variation with Pore Size:

As pores are getting smaller down to molecular size, the pore space ceases to have any meaning. That is, the transition from adsorption to absorption is not sharp. In an experiment of Rayleigh (1936), he pressed two optically flat glass surfaces tightly together, and measured the flow of helium from one side at 1 atm to the other at vacuum. From the flow rate measured and by assuming the flow is by Knudsen mechanism, he calculated the width between the two plates is 1 A, which is an impossible result. The actual width must be greater than 1A for helium to diffuse. Hence, the resistance must be greater than that expected for Knudsen diffusion. When air is used, and for the same pressure difference, the helium flow is 7 times higher than air, while the ratio of two fluxes predicted by the Knudsen flow is:

$$\frac{J_K(\text{helium})}{J_K(\text{air})} = \sqrt{\frac{M_{air}}{M_{helium}}} = 2.7 \qquad (7.9\text{-}18)$$

Thus, the transition from surface adsorbed molecules to interstitial sorbed molecules is marked by a profound change in the mechanism of diffusion. The important point that should be noted here is the strong dependence on the molecular size.

7.9.5 Surface Diffusivity Models

As we have pointed out before, the surface diffusivity is a strong function of surface loading. This section will present a number of models in describing this functional dependence. Various surface diffusion models have been proposed in the literature. We will start with the commonly quoted theory by Higashi et al.

7.9.5.1 Higashi et al.'s (HIO) model

Higashi et al. (1963) proposed a hopping model, in which they assumed that when a molecule adsorbed in a site jumps to one of the neighboring sites whether they are vacant or not, and if the site is occupied the molecule will not be bound to this site but rather collides with the former occupied molecule and scatters isotropically. Then the molecule continues to jump until it finds a vacant site. As θ increases the number of hopping necessary for finding a vacant site increases. The expectation value $n(\theta)$ is the summation of

(a) the probability of capture after 1st hopping = $(1-\theta)$; that is the probability is proportional to the fraction of the bare surface
(b) the probability of capture after 2nd hopping = $\theta(1-\theta)$; that is after the first hop, the chance of hitting an already occupied site is θ, and as a result the molecule hops the second time and the probability of this second hop is $(1-\theta)$. Hence, the probability of capture after the second hopping is $\theta(1-\theta)$
(c) the probability of capture after k-th hopping = $\theta^{k-1}(1-\theta)$

Therefore, the expectation number is

$$n(\theta) = \sum_{k=1}^{\infty} k\theta^{k-1}(1-\theta) = \frac{1}{1-\theta} \qquad (7.9\text{-}19)$$

Higashi et al. (1963) then finally obtained the following famous equation for the surface diffusivity:

$$D_s = \frac{D_{s\infty}}{1-\theta} \exp\left(-\frac{E_s}{R_g T}\right) \qquad (7.9\text{-}20)$$

in which they have assumed that the jumping time is negligible compared to the holding time τ, resulting in the molecule can carry out $n(\theta)$ steps of random walk.

Recognizing the surface diffusivity predicted by eq.(7.9-20) becomes infinite at monolayer coverage, the theory of Higashi et al. was later modified by Yang et al. (1973) to allow for the second layer adsorption to rid of this deficiency. Yang et al. obtained the following equation for the surface diffusivity:

$$\frac{D_s}{D_s(0)} = \frac{1}{(1-\theta) + \theta \dfrac{\upsilon_1}{\upsilon_2} \exp\left[-(\Delta E_1 - \Delta E_2)/R_g T\right]} \qquad (7.9\text{-}21)$$

where υ_1 and υ_2 are the vibration frequencies of the first and second layers, respectively, and ΔE is the effective energy of the bond. Yang et al. approximated $\Delta E_1 - \Delta E_2$ by the difference in heats of adsorption for the first and second layers. The ratio υ_1 / υ_2 is difficult to measure and they assigned a value of unity without any proof. If the second term in the denominator is very small, the Yang et al.'s theory reduces to that of Higashi et al. (1963). Figure (7.9-3) shows the variation of the surface diffusivity versus loading for the Higashi et al. theory and Yang et al. theory. Yang et al.'s theory predicts a slower rise with respect to loading, and it gives a finite limit at $\theta = 1$ while the Higashi et al. theory gives an infinite value.

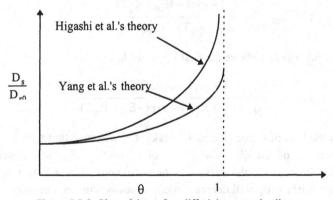

Figure 7.9-3. Plots of the surface diffusivity versus loading

Using the propane surface diffusion data on silica of Higashi et al., Yang et al. (1973) have found that $\Delta E_1 - \Delta E_2$ is 6.6 kJoule/mole.

7.9.5.2 Hopping Model of Okazaki et al. (1981)

What to follow in this section is the theory proposed by Okazaki et al., which is a refinement over the HIO model. In this model of Okazaki et al., the authors assumed molecules can hop onto already occupied sites. This will be elaborated below.

From the basis of molecular kinetic theory, it is assumed that adsorbed molecules vibrate in the perpendicular direction with a frequency τ_{s0}. For molecules in the first layer, the holding time τ_0 is:

$$\tau_0 = \tau_{s0} \int_0^{E_{a0}} f_{s0}(E)dE \bigg/ \int_{E_{s0}}^{E_{a0}} f_{s0}(E)dE \tag{7.9-22}$$

where E_{a0} is the differential heat of adsorption, E_{s0} is the potential barrier between adsorption sites, and $f_{s0}(E)$ is the distribution function of energy given by:

$$f_{s0}(E) = \frac{1}{R_g T} \exp\left(-\frac{E}{R_g T}\right) \tag{7.9-23}$$

with $\int_0^E f_{s0}(E)dE$ is the fraction of molecules having energy between 0 and E, and $\exp(-E/R_g T)$ is the fraction of molecules having energy between E and ∞.

Combining eqs. (7.9-22) and (7.9-23), we get:

$$\tau_0 = \tau_{s0} \frac{1 - \exp(-E_{a0}/R_g T)}{\exp(-E_{s0}/R_g T) - \exp(-E_{a0}/R_g T)} \tag{7.9-24}$$

Similarly for other layers above the first layer, we have:

$$\tau_1 = \tau_{s1} \frac{1 - \exp(-E_{a1}/R_g T)}{\exp(-E_{s1}/R_g T) - \exp(-E_{a1}/R_g T)} \tag{7.9-25}$$

where E_{a1} is the heat of vaporization. Okazaki et al. assumed that $\tau_{s0} = \tau_{s1}$.

The variation of the differential heat of adsorption, E_{a0}, will depend on the nature of the surface. If the surface is homogeneous, it is a constant; while on heterogeneous surface E_{a0} will decrease monotonically with the coverage.

(a) Homogeneous surface: Okazaki et al. considered four possible hopping mechanisms of an adsorbed molecule. An adsorbed molecule in one site jumps to a neighboring vacant site, an adsorbed molecule in one site jumps onto an occupied site, a molecule sitting on a former occupied molecule jumps to a neighboring vacant site, and a molecule sitting on a former occupied molecule jumps onto an occupied site. These four models are shown graphically in Figure 7.9-4.

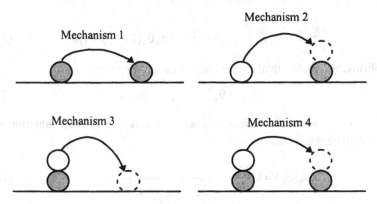

Figure 7.9-4: Four different hopping mechanisms of Okazaki et al. (1981)

In the mechanism 1, a molecule in a site hops to a neighboring vacant site. This probability is $(1-\theta_e)^2$ because the probability that the molecule is placed in vacant site and it hops to another vacant site is $(1-\theta_e)$. The holding time for this hop is

$$\tau = \tau_0 (1-\theta_e)^2 \qquad (7.9\text{-}26a)$$

where θ_e is the effective fractional coverage, which is not necessarily the same as the observed fractional loading. When multilayer adsorption occurs, the molecule under a certain molecule does not move; hence the effective fractional coverage θ_e is reduced to the coverage of the first layer. This means that if the adsorption isotherm is Langmuir, $\theta_e = \theta$, and if the isotherm follows a BET equation, $\theta_e = \theta(1-x)$, where $x = P/P_0$, with P_0 being the vapor pressure.

In the mechanism 2, a molecule in a site hops onto an occupied neighboring site. This probability is $(1-\theta_e)\theta_e$, and hence the holding time is:

$$\tau = \tau_0 \theta_e (1-\theta_e) \qquad (7.9\text{-}26b)$$

In the mechanism 3, a molecule in an occupied site hops to vacant site. The probability is $\theta_e(1-\theta_e)$, and the holding time is:

$$\tau = \tau_1\theta_e(1-\theta_e) \tag{7.9-26c}$$

Finally, the holding time for the mechanism 4 is:

$$\tau = \tau_1\theta_e^2 \tag{7.9-26d}$$

Thus, the expected holding time is the sum of these four individual holding times, that is:

$$\tau = \tau_0(1-\theta_e)^2 + \tau_0(1-\theta_e)\theta_e + \tau_1\theta_e(1-\theta_e) + \tau_1\theta_e^2 \tag{7.9-27}$$

In simplifing, we get the final expression for the holding time:

$$\tau = (1-\theta_e)\tau_0 + \theta_e\tau_1 \tag{7.9-28}$$

By assuming surface diffusion as a random walk, the Einstein equation can be used to calculate the surface diffusivity

$$D_s(\theta_e) = C\frac{\delta^2}{\tau} = \frac{C\delta^2}{\tau_0\left[1-\theta_e\left(1-\dfrac{\tau_1}{\tau_0}\right)\right]} \tag{7.9-29}$$

where C is a constant, and δ is the distance between adsorption sites. The ratio τ_1/τ_0 is obtained from eqs. (7.9-24) and (7.9-25), which is given below:

$$\frac{\tau_1}{\tau_0} = \frac{\left[\exp(-aE_{a0}/R_gT) - \exp(-E_{a0}/R_gT)\right]\left[1-\exp(-E_{a1}/R_gT)\right]}{\left[\exp(-E_{s1}/R_gT) - \exp(-E_{a1}/R_gT)\right]\left[1-\exp(-E_{a0}/R_gT)\right]} \tag{7.9-30}$$

in which we have assumed that the energy barrier between adsorption sites is proportional to the differential heat of adsorption with $a < 1$

$$E_{s0} = aE_{a0} \tag{7.9-31}$$

The surface flux is defined in terms of the effective surface concentration as:

$$J_s = -D_s(\theta_e)\frac{dC_{\mu e}}{dx} \tag{7.9-32}$$

where $C_{\mu e}$ is the effective surface concentration. But the surface flux is traditionally defined in terms of the total adsorbed concentration as:

$$J_s = -D_s(\theta)\frac{dC_\mu}{dx} \tag{7.9-33}$$

Equating these two equations, we obtain the following equation for the observed surface diffusivity:

$$D_s = D_{s0}\left(\frac{\theta_e}{\theta}\right)\frac{\exp(-aE_{a0}/R_gT) - \exp(-E_{a0}/R_gT)}{\left[1 - \exp(-E_{a0}/R_gT)\right]\left[1 - \theta_e\left(1 - \frac{\tau_1}{\tau_0}\right)\right]} \qquad (7.9.34)$$

Estimation of parameters:

The effective fractional coverage is equal to the observed fractional coverage if the adsorption isotherm is described by the Langmuir equation, and it is equal to $\theta(1-P/P_0)$ if the isotherm follows the BET equation.

The parameter E_{a1} is the heat of vaporization and can be obtained readily from the literature. The parameter E_{s1} is the activation energy of the first layer, it can be estimated from the viscosity

$$\mu_L = KT\exp(E_{s1}/R_gT) \qquad (7.9\text{-}35)$$

as the first layer would behave like a liquid. E_{s1} is of the order of 5-10 kJoule/mole.

The parameter E_{a0} is calculated from the isosteric heat, which is calculated from the Van Hoff's equation:

$$E_{st} = R_gT^2\left(\frac{\partial \ln P}{\partial T}\right)_\theta \qquad (7.9\text{-}36)$$

For Langmuir isotherm, $E_{a0} = E_{st} - R_gT$; while for BET isotherm $E_{a0} = (E_{st} - R_gT - xE_{a1})/(1-x)$, where $x = P/P_0$.

The parameter a is found in many experiments to lie between 0.4 and 0.6. DeBoer (1953) assumed adsorbed molecules lie at the center of a square of four surface atoms and showed that a is approximately 0.5.

The parameter D_{s0} is found experimentally, and is of the order of 10^{-3} to 10^{-2} cm²/sec.

We note that the model of Okazaki et al. reduces to that of Higashi et al. when the following criteria are satisfied:

$$\tau_1 \ll \tau_0$$
$$1 \gg \exp(-E_{s0}/R_gT) \gg \exp(-E_{a0}/R_gT) \qquad (7.9\text{-}37)$$
$$\theta_e = \theta$$

(b) Heterogeneous surface:

In heterogeneous surface, the differential heat of adsorption decreases with loading. This is because adsorption occurs progressively from high energy site to sites of lower energy. Okazaki et al. proposed the following equation for the surface diffusivity on heterogeneous surface:

$$\frac{D_s}{D_{s0}} = \frac{\left(\frac{\theta_e}{\theta}\right)\int_{E_{a0}^0}^{E_{a0}} \frac{\exp(-aE/R_gT) - \exp(-E/R_gT)}{[1-\exp(-E/R_gT)][1-\theta_e(1-\tau_1/\tau_0)]} g(E)dE}{\int_{E_{a0}^0}^{E_{a0}} g(E)dE} \quad (7.9\text{-}38)$$

where

$$\frac{\tau_1}{\tau_0} = \frac{\left[\exp(-aE/R_gT) - \exp(-E/R_gT)\right]\left[1-\exp(-E_{a1}/R_gT)\right]}{\left[\exp(-E_{s1}/R_gT) - \exp(-E_{a1}/R_gT)\right]\left[1-\exp(-E/R_gT)\right]} \quad (7.9\text{-}39)$$

Here g(E) is the number of molecules adsorbed which have heat of adsorption between E and E+dE. Thus, g(E) can be evaluated from the relation between E_{a0} and the amount adsorbed. The parameters E_{a0} and E_{a0}^0 are the heat of adsorption at θ_e and $\theta_e = 0$, respectively.

7.9.5.3 Darken Model

Darken (1948) in his study of diffusion of ions in metallic systems has assumed that the force acting on a particle in a potential field is the negative gradient of its potential energy. This potential energy is assumed to be the Gibbs chemical potential.

The surface diffusion flux is proportional to the product of the concentration and the gradient of chemical potential, that is:

$$J_s = -LC_\mu \frac{d\mu}{dx} \quad (7.9\text{-}40)$$

where μ is the chemical potential. Assuming this chemical potential is the same as that in the ideal gas phase (that is equilibrium between the two phases), we can write

$$\mu = \mu_0 + R_g T \ln P \quad (7.9\text{-}41)$$

where μ_0 is the reference chemical potential. Combining eqs. (7.9-40) and (7.9-41), we get:

$$J_s = -D_s^* C_\mu \frac{d \ln P}{dx} \qquad (7.9\text{-}42\text{a})$$

where D_s^* is called the corrected diffusivity

$$D_s^* = L R_g T \qquad (7.9\text{-}42\text{b})$$

But the usual definition of the surface flux is defined in terms of the gradient of adsorbed concentration, that is:

$$J_s = -D_s \frac{dC_\mu}{dx} \qquad (7.9\text{-}43)$$

Equating eqs. (7.9-42) and (7.9-43), we have:

$$D_s = D_s^* \frac{\partial \ln P}{\partial \ln C_\mu} \qquad (7.9\text{-}44)$$

This is now known as the Darken relation, and it depends on the equilibrium isotherm between the two phases. It basically states that the surface diffusivity at any loading is equal to the value at zero loading multiplied by a thermodynamic correction factor

$$\frac{\partial \ln P}{\partial \ln C_\mu}$$

The following table shows the thermodynamic correction factor for a few isotherms (Barrer, 1978).

Table 7.9.1: Thermodynamic correction factor for a number of isotherms

Model	Expression	$\partial \ln P / \partial \ln \theta$
Henry law	$bP = \theta$	1
Langmuir	$bP = \theta/(1-\theta)$	$1/(1-\theta)$
Fowler-Guggenheim	$bP = [\theta/(1-\theta)] \exp(c\theta)$	$1/(1-\theta) + c\theta$
Volmer	$bP = [\theta/(1-\theta)] \exp[\theta/(1-\theta)]$	$1/(1-\theta)^2$
Hill-deBoer	$bP = [\theta/(1-\theta)] \exp[\theta/(1-\theta) + c\theta]$	$1/(1-\theta)^2 + c\theta$

The thermodynamic correction factor of all these model isotherms reduces to $\partial \ln P / \partial \ln \theta = 1$ when $\theta \ll 1$.

We now see that except for the case of Henry isotherm, all other isotherms exhibit an increase in the thermodynamic correction factor versus loading.

7.9.5.4 Chen and Yang's Model

Chen and Yang (1991) recognized the erratic behavior of surface diffusivity versus loading and proposed a model based on "activated" adsorbed species. They obtained the following formula for the surface diffusivity

$$\frac{D_s}{D_s(0)} = \frac{1 - \theta + (\lambda/2)\theta(2-\theta) + H(1-\lambda)(1-\lambda)(\lambda/2)\theta^2}{(1 - \theta + \lambda\theta/2)^2} \quad (7.9\text{-}49)$$

where H is the Heaviside step function, defined as

$$H(x) = 1 \text{ if } x > 0; \text{ else } H(x) = 0 \quad (7.9\text{-}50)$$

and the parameter λ is the ratio of two rate constants:

$$\lambda = \frac{k_b}{k_m} \quad (7.9\text{-}51)$$

When $\lambda = 0$, this equation is reduced to the Higashi et al. model as well as the Darken model with Langmuir isotherm. Chen and Yang argued that this case is the case corresponding to surface diffusion, where no blockage is expected due to the unlimited space. The parameter λ then describes the degree of blockage by another adsorbed molecule.

It is interesting to note that the Higashi et al. model as well as Chen and Yang's model do not involve the equilibrium between the gas and solid phases. The Darken model, on the other hand, requires such an information.

7.10 Concluding Remarks

We have presented in this chapter an account on the development of diffusion theory. Various modes of flow are identified: Knudsen, viscous, continuum diffusion and surface diffusion. Constitutive flux equations are presented for all these flow mechanisms, and they can be readily used in any mass balance equations for the solution of concentration distribution and fluxes. Treatment of systems containing more than two species will be considered in a more systematic approach of Stefan-Maxwell in the next chapter.

8
Diffusion in Porous Media: Maxwell-Stefan Approach

8.1 Introduction

In Chapter 7, we have discussed the various diffusional processes for mass transfer in a capillary and a porous medium. Those discussions are sufficient for the understanding of mass transfer processes as well as the calculation of fluxes into a capillary and a porous medium for binary systems.

In this chapter, we will re-examine these processes, but from the approach developed by Maxwell and Stefan. This approach basically involves the concept of force and friction between molecules of different types. It is from this frictional concept that the diffusion coefficient naturally arises as we shall see. We first present the diffusion of a homogeneous mixture to give the reader a good grasp of the Maxwell-Stefan approach, then later account for diffusion in a porous medium where the Knudsen diffusion as well as the viscous flow play a part in the transport process. Readers should refer to Jackson (1977) and Taylor and Krishna (1994) for more exposure to this Maxwell-Stefan approach.

8.2 Diffusion in Ideal Gaseous Mixture

Consider a solution containing two components 1 and 2. If there is a difference between the concentrations of any two points within the solution there will be a net diffusion process from one point to the other. This is the macroscopic picture of diffusion, induced by a difference in concentration, or strictly speaking by a concentration gradient.

From a microscopic point of view, diffusion is the intermingling of atoms or molecules of more than one species. It is a result of the random motion of individual molecules that are distributed in space. A brief discussion of collision of two objects is presented in Appendix 8.1, where we show that the momentum transfer from the type 1 molecules to the type 2 molecules is proportional to the difference in their velocities before the collision.

8.2.1 Stefan-Maxwell Equation for Binary Systems

Accepting the concept of momentum transfer between molecules of different type, the necessary flux equation based on the frictional force is developed in this section.

Because of the collision between molecules of different type, the molecule of type 2 will interfere with the movement of the type 1 molecule. It will exert a drag on the movement of the type 1 molecule. The analogy of this is the drag exerted on the moving fluid within a pipe by the wall, that is the wall interferes with the movement of the fluid, just like the way the type 2 molecules exerts a drag on the movement of type 1 molecules.

To start with the analysis, we use the following force balance equation at steady state:

Rate of momentum in - Rate of momentum out + Sum of all forces = 0   (8.2-1)

or to put it in another form:

$$\begin{pmatrix} \text{The sum of Forces} \\ \text{acting on the system} \end{pmatrix} = \begin{pmatrix} \text{The rate of change} \\ \text{of the momentum of the system} \end{pmatrix} \quad (8.2\text{-}2)$$

Now let us apply this momentum balance on a control volume containing two types of molecule as shown in Figure 8.2-1 We will consider the force balance in one direction, say the z direction. The results for the other two directions will be identical in form, except that the z dependence will be replaced by x and y dependence, respectively.

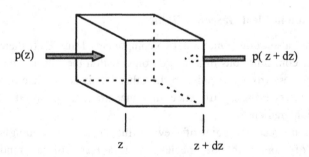

Figure 8.2-1: Control volume for the momentum balance

Momentum can enter or leave this volume due to the motion of molecules across the boundary walls. If the whole control volume moves with a velocity v, then the flow of molecules into this control volume across one boundary will be exactly balanced by the same amount of flow leaving the other side of the boundary. Thus, there is no net gain or loss of momentum due to the movement of the whole control volume.

Within the control volume, however, the type 1 molecules may lose (or gain) the momentum each time they collide with molecules of the other type. Let us consider the case where the pressure is constant inside the control volume, that is there is no pressure force acting on the control volume. The rate at which the collision occurs between the two types of molecule will depend on the molecular density of the molecule of type 1 as well as that of the molecular density of type 2 (similar to the mass action law kinetics between two reactants). Let the numbers of molecule of type 1 and type 2 per unit volume be:

$$c_1 = c_t y_1 \quad \text{and} \quad c_2 = c_t y_2 \quad (8.2\text{-}3)$$

where y_1 and y_2 are molecular fractions of the two species, and c_t is the total molecular density in the control volume:

$$c_t = \frac{P}{kT} \quad (8.2\text{-}4)$$

Here, P is the total pressure and k is the Boltzmann constant.

The number of collision between the two types of molecule will be proportional to y_1 and y_2, that is:

$$\begin{pmatrix} \text{The number of collision between species} \\ \text{1 and 2 per unit volume per unit time} \end{pmatrix} \propto y_1 y_2 \quad (8.2\text{-}5)$$

The rate of change of the momentum is equal to the average momentum transferred between one molecule of type 1 and one molecule of type 2 multiplied by the number of collision per unit volume per unit time, that is:

$$\begin{pmatrix} \text{The rate of change} \\ \text{of momentum of} \\ \text{type 1 molecule} \end{pmatrix} \propto y_1 y_2 (u_1 - u_2) \quad (8.2\text{-}6)$$

We have determined the rate of change of the momentum. Now we turn to determining the force term in the momentum balance equation (eq. 8.2-2). The force acting on the surface at z by the component 1 is

$$Ap_1\big|_z \tag{8.2-7a}$$

and the force acting on the opposite surface at $z+\Delta z$ by the same component is

$$Ap_1\big|_{z+\Delta z} \tag{8.2-7b}$$

Thus, the net force acting on the component 1 in the control volume in the z direction is:

$$A\left(p_1\big|_z - p_1\big|_{z+\Delta z}\right) \tag{8.2-8}$$

Dividing this net force by the volume $A\Delta z$ and then taking the limit when the volume is infinitesimally small, we get:

$$\begin{pmatrix} \text{Net force acting on type 1} \\ \text{molecule per unit volume} \\ \text{in the z direction} \end{pmatrix} = \lim_{\Delta z \to 0} \frac{\left(p_1\big|_z - p_1\big|_{z+\Delta z}\right)}{\Delta z} = -\frac{dp_1}{dz} \tag{8.2-9}$$

Substituting eqs. (8.2-6) and (8.2-9) into the force balance equation (eq. 8.2-2), we finally get the following momentum balance equation:

$$-\frac{dp_1}{dz} \propto y_1 y_2 (u_1 - u_2) \tag{8.2-10}$$

which simply states the net force acting on the species 1 in the control volume is balanced by the change of the momentum of the species 1 in that control volume.

Introducing the proportionality coefficient f_{12} to eq.(8.2-10), we have:

$$-\frac{dp_1}{dz} = f_{12} y_1 y_2 (u_1 - u_2) \tag{8.2-11}$$

This equation is written for one dimension. Written in three dimensional format, we have the following generalization:

$$-\nabla p = f_{12} y_1 y_2 (\underline{u}_1 - \underline{u}_2) \tag{8.2-12}$$

where we can view that ∇p_1 is the actual force exerted per unit volume of the mixture trying to move the type 1 molecule past (through) the molecules of type 2 at a relative velocity $(\underline{u}_1 - \underline{u}_2)$. The factor $y_1 y_2$ is the concentration weight factor, and the coefficient f_{12} can be viewed as the friction factor. This friction factor is expected to increase when the total molar density is high (that is when the total pressure is high). We shall take the frictional factor as a linear function of pressure:

$$f_{12} = \frac{P}{D_{12}} \tag{8.2-13}$$

The constant of proportionality is $1/D_{12}$, which we will see later that D_{12} is the binary diffusivity in a mixture containing the components 1 and 2.

Substitute eq. (8.2-13) into eq. (8.2-12) and rewrite the result into the following form:

$$d_1 = \frac{1}{P}\nabla p_1 = -\frac{y_1 y_2 (u_1 - u_2)}{D_{12}} \tag{8.2-14}$$

where d_1 can be viewed as the driving force for the diffusion of the species 1 in an ideal gas mixture at **constant pressure and temperature**. This equation is the Maxwell-Stefan equation (credited to the Scottish physicist James Clerk Maxwell and the Austrian scientist Josef Stefan), and D_{12} is called the Maxwell-Stefan binary diffusion coefficient.

8.2.1.1 Alternative Derivation

Another way of deriving the fundamental momentum balance equation (8.2-14) is as follows. The net force per unit volume exerted on the component 1 is

$$-\frac{dp_1}{dz}$$

therefore, the <u>net force per unit number of moles</u> of that species is:

$$-\frac{1}{c_1}\frac{dp_1}{dz} = -R_g T \frac{d \ln p_1}{dz} \tag{8.2-15}$$

where c_1 is the molar concentration (mole/m^3) of the component 1. In eq. (8.2-15) we have used the ideal gas law

$$c_1 = \frac{p_1}{R_g T} \tag{8.2-16}$$

This net force per unit number of moles of the species 1 is balanced by the rate of change of the momentum caused by the friction between the diffusing species 1 and 2. This friction force is proportional to the velocity difference and the mole fraction of the species 2 (Figure 8.2-2), that is:

$$f_{12} y_2 (u_1 - u_2) \tag{8.2-17}$$

where f_{12} is the friction coefficient.

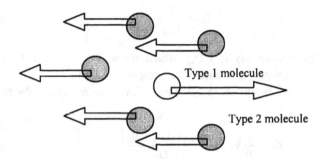

Figure 8.2-2: Schematic diagram of friction caused by molecule 1 moving through a stream of component 2

Equating eqs. (8.2-15) and (8.2-17) yields the following momentum balance equation:

$$-R_g T \frac{d \ln p_1}{dz} = R_g T y_2 \frac{(u_1 - u_2)}{D_{12}} \qquad (8.2\text{-}18)$$

where we have taken the friction coefficient to have the form:

$$f_{12} = \frac{R_g T}{D_{12}} \qquad (8.2\text{-}19)$$

Rearranging eq. (8.2-18) and noting $p_1 = y_1 P$, we finally obtain the following necessary equation:

$$d_1 = \frac{1}{P} \nabla p_1 = -\frac{y_1 y_2 (u_1 - u_2)}{D_{12}} \qquad (8.2\text{-}20)$$

which is the same equation we obtained before (eq. 8.2-14), but in this formulation we see that the net force per <u>unit moles of species 1</u> is the rate of change of the chemical potential of that component (eq. 8.2-15). Note the definition of the chemical potential of an ideal gas of species "i" is (we shall discuss the nonideal gas in Section 8.3):

$$\mu_i = \mu_{i0} + R_g T \ln p_i \qquad (8.2\text{-}21)$$

of which the gradient along the z direction is

$$\frac{d \mu_i}{dz} = R_g T \frac{d \ln p_i}{dz} \qquad (8.2\text{-}22)$$

Comparing eq. (8.2-22) with eq. (8.2-15), we note that the change in the chemical potential is simply the net force acting on one mole of the species 1. Thus, if we write the momentum balance (eq. 8.2-18) in terms of the chemical potential as the driving force, we have the more general momentum balance equation:

$$-\frac{d\mu_1}{dz} = R_g T\, y_2 \frac{(u_1 - u_2)}{D_{12}} \qquad (8.2\text{-}23)$$

Multiplying this equation by y_1/RT, we obtain the necessary flux equation written in terms of the chemical potential gradient driving force:

$$-d_1 = -\frac{y_1}{R_g T}\frac{d\mu_1}{dz} = y_1 y_2 \frac{(u_1 - u_2)}{D_{12}} \qquad (8.2\text{-}24)$$

Similarly, we can write the diffusion flux equation for the species 2 as (by interchanging the subscripts 1 and 2 in eq. 8.2-24):

$$-d_2 = -\frac{y_2}{R_g T}\frac{d\mu_2}{dz} = y_1 y_2 \frac{(u_2 - u_1)}{D_{21}} \qquad (8.2\text{-}25)$$

Summing these two constitutive flux equations (8.2-24 and 8.2-25), and noting that $p_1 + p_2 = P$ (pressure in the control volume is constant), we then have:

$$\frac{y_1 y_2 (u_1 - u_2)}{D_{12}} + \frac{y_1 y_2 (u_2 - u_1)}{D_{21}} = 0 \qquad (8.2\text{-}26a)$$

or

$$y_1 y_2 (u_1 - u_2)\left(\frac{1}{D_{12}} - \frac{1}{D_{21}}\right) = 0 \qquad (8.2\text{-}26b)$$

In general, y_1, y_2, and $(u_1 - u_2)$ are not zero, therefore eq. (8.2-26b) will result in the following important result for the binary diffusivity:

$$D_{12} = D_{21} \qquad (8.2\text{-}27)$$

Thus, the Maxwell-Stefan binary diffusion coefficients are symmetric.

8.2.2 Stefan-Maxwell Equation for Ternary Systems

Having obtained the necessary flux equations for binary systems, we can easily generalise the result to ternary systems. Similar to eq. (8.2-20) applicable for binary systems, we can write the flux equation for the species 1 in a ternary mixture as follows:

$$d_1 = \frac{1}{P}\nabla p_1 = -\frac{y_1 y_2 (u_1 - u_2)}{D_{12}} - \frac{y_1 y_3 (u_1 - u_3)}{D_{13}} \qquad (8.2\text{-}28a)$$

which simply states that the driving force for diffusion of the component 1 is balanced by the friction between molecules of type 1 and molecules of types 2 and 3, which are described by the first and second terms in the RHS of eq. (8.2-28a), respectively.

We can also write the diffusion flux equations for the species 2 and 3 as shown below:

$$d_2 = \frac{1}{P}\nabla p_2 = -\frac{y_2 y_1 (u_2 - u_1)}{D_{21}} - \frac{y_2 y_3 (u_2 - u_3)}{D_{23}} \qquad (8.2\text{-}28b)$$

$$d_3 = \frac{1}{P}\nabla p_3 = -\frac{y_3 y_2 (u_3 - u_2)}{D_{32}} - \frac{y_3 y_1 (u_3 - u_1)}{D_{31}} \qquad (8.2\text{-}28c)$$

Of the above three equations, only two are independent because of the constant pressure condition given below:

$$\nabla p_1 + \nabla p_2 + \nabla p_3 = \nabla P = 0 \qquad (8.2\text{-}29)$$

Adding eqs. (8.2-28), we get:

$$y_1 y_2 (u_1 - u_2)\left(\frac{1}{D_{21}} - \frac{1}{D_{12}}\right) + y_1 y_3 (u_1 - u_3)\left(\frac{1}{D_{31}} - \frac{1}{D_{13}}\right) + y_2 y_3 (u_3 - u_2)\left(\frac{1}{D_{32}} - \frac{1}{D_{23}}\right) = 0$$
$$(8.2\text{-}30)$$

The symmetry of binary diffusivities in the previous section (eq. 8.2-27) is applied here in eq. (8.2-30), and we see that the LHS of this equation is indeed zero. Thus, we can write in general the following equation for multicomponent systems:

$$D_{ij} = D_{ji} \qquad (8.2\text{-}31)$$

8.2.3 Stefan-Maxwell Equation for the N-Multicomponent System

Having shown the constitutive flux equation for binary and ternary systems, we can readily generalise the result to a multicomponent system containing "n" species by simply writing down the following diffusion flux equation for the component i:

$$d_i = \frac{1}{P}\nabla p_i = -\sum_{j=1}^{n}\frac{y_i y_j (u_i - u_j)}{D_{ij}} \qquad (8.2\text{-}32)$$

for i = 1, 2, ..., n, where the LHS is the net force acting on the component "i" and the RHS is the momenta exchange between the component "i" and the other components.

Note that the term corresponding to j = i in the summation of eq. (8.2-32) is zero. This equation is the general form of the diffusion equation written in terms of velocities of all species, but it is not too useful to scientists and chemical engineers as they are more comfortable in dealing with flux rather than velocity. To achieve this, we use the following definition of flux (Bird et al., 1960):

$$N_i = c_i u_i \qquad (8.2\text{-}33)$$

which is the flux relative to a fixed frame of coordinates, defined as moles transferred per unit time and per unit area perpendicular to the flow direction.

Rewrite the Stefan-Maxwell equation (eq. 8.2-32) in terms of the driving force ∇p_i and the fluxes of all components, we have:

$$d_i = \frac{1}{P}\nabla p_i = \sum_{j=1}^{n} \frac{(y_i N_j - y_j N_i)}{cD_{ij}} \qquad \text{for } i=1, 2, ..., n \qquad (8.2\text{-}34)$$

Eq. (8.2-34) is the fundamental constitutive flux equation written in terms of $\underline{N}$ relative to a fixed frame of coordinate. One could write a similar flux equation in terms of the diffusive flux, which is defined as the flux relative to the moving mixture. This diffusive flux is defined as follows:

$$J_i = c_i(u_i - u) \qquad (8.2\text{-}35a)$$

where u is the velocity of the mixture, which is defined as the average of all component velocities:

$$u = \sum_{i=1}^{n} y_i u_i \qquad (8.2\text{-}35b)$$

From this definition of the diffusion flux (eq. 8.2-35a), we see that the sum of all n diffusive fluxes is zero, that is:

$$\sum_{i=1}^{n} J_i = 0 \qquad (8.2\text{-}35c)$$

Making use of eq. (8.2-33) into eq. (8.2-35a), we obtain the following relationship between the diffusive flux $\underline{J}$ and the flux $\underline{N}$:

$$J_i = N_i - y_i N_T \qquad (8.2\text{-}36a)$$

where N_T is the total flux of all species

$$N_T = \sum_{i=1}^{n} N_i \qquad (8.2\text{-}36b)$$

Substitution of eq. (8.2-36a) into the constitutive Stefan-Maxwell equation (8.2-34), we get the following equation written in terms of the diffusive flux

$$d_i = \frac{1}{P}\nabla p_i = \sum_{j=1}^{n} \frac{(x_i J_j - x_j J_i)}{cD_{ij}} \qquad \text{for } i = 1, 2, ..., n. \qquad (8.2\text{-}37)$$

We note that the equation written in terms of the diffusive flux J (eq. 8.2-37) is identical in form to eq. (8.2-34) written in terms of the flux N relative to a fixed frame of coordinates. We further note that the n equations of these two sets are not independent as the following restriction of constant pressure:

$$\nabla p_1 + \nabla p_2 + \cdots + \nabla p_n = 0$$

that is, only (n-1) of them are independent. This is so because we are dealing with the *relative* motion of n different molecules, that is there are only (n-1) relative velocities.

8.2.3.1 The Physical Constraint

To solve for the flux $\underline{N} = [N_1 \ N_2 \ ... \ N_n]^T$, we only have n-1 equations provided by eq. (8.2-34). Thus, another equation must be found, and this is feasible by resorting to a physical constraint condition. This physical constraint is specific to the physical problem at hand, and here we shall deal with three situations, which most diffusion and adsorption problems will fall into.

8.2.3.1.1 The Open System

The first situation is the open system, where the fluxes of all species are related to each other according to what is well known as the Graham's law of diffusion (See Chapter 7).

$$\sum_{j=1}^{n} \sqrt{M_j} \times N_j = 0 \qquad (8.2\text{-}38)$$

that is the flux of the n-th component can be expressed in terms of all the other fluxes as:

$$N_n = -\sum_{j=1}^{n-1} v_j N_j \qquad (8.2\text{-}39a)$$

where

$$v_j = \sqrt{\frac{M_j}{M_n}} \qquad (8.2\text{-}39b)$$

This means that n-1 equations of eq. (8.2-34) coupling with eq. (8.2-39a) will provide the necessary n equations for the n unknown fluxes $\underline{N}$.

8.2.3.1.2 The Constant Pressure Closed System

For a constant pressure closed system, the necessary physical constraint condition is simply that the sum of all fluxes must be zero:

$$\sum_{j=1}^{n} N_j = 0 \qquad (8.2\text{-}40)$$

Thus, the flux of the n-th component in this constant pressure closed system is:

$$N_n = -\sum_{j=1}^{n-1} N_j \qquad (8.2\text{-}41)$$

8.2.3.1.3 The Stefan Tube

For a Stefan tube problem where a liquid containing n-1 components evaporating into a gas space containing the n-th component which is insoluble in liquid, the flux of the n-th component is simply zero, that is:

$$N_n = 0 \qquad (8.2\text{-}42)$$

With the three examples considered, the physical constraint can be written in general as:

$$N_n = -\sum_{j=1}^{n-1} v_j N_j \qquad (8.2\text{-}43)$$

where v_j are defined in the following table for the three cases.

Table 8.2-1: Definitions of v_j (j = 1, 2, ..., n-1) for eq. (8.2-43)

	Open system	Closed system	Stefan tube
v_j (j = 1, 2, ..., n-1)	$\sqrt{\dfrac{M_j}{M_n}}$	1	0

8.2.3.2 The Working Flux Equations

The n-1 equations given by eq. (8.2-34) and the physical constraint of the form (8.2-43) will form the necessary n equations for solving for the fluxes $\underline{N}$. What we shall show in this section that the Stefan-Maxwell equations can be <u>inverted</u> to obtain the useful flux expression written in terms of concentration gradients, instead of concentration gradient in terms of fluxes.

Taking the last term out of the series in the RHS of eq. (8.2-34), we get:

$$\nabla y_i = \sum_{j=1}^{n-1} \frac{(y_i N_j - y_j N_i)}{cD_{ij}} + \frac{y_i N_n - y_n N_i}{cD_{i,n}} \qquad (8.2\text{-}44)$$

for $i = 1, 2, ..., n-1$. It is reminded that only n-1 equations of eq. (8.2-34) are independent. We need to use the physical constraint condition (8.2-43) to eliminate the flux of the n-th species. Substitution of N_n given in eq. (8.2-43) into eq. (8.2-44), we get:

$$\nabla y_i = \sum_{j=1}^{n-1} \frac{(y_i N_j - y_j N_i)}{cD_{ij}} - \frac{y_i \sum_{j=1}^{n-1} v_j N_j}{cD_{i,n}} - \frac{y_n N_i}{cD_{i,n}} \qquad (8.2\text{-}45)$$

Simplifying the above equation by grouping N_i and N_j separately will yield the form such that the vector-matrix format can be used:

$$\nabla y_i = -\sum_{\substack{j=1 \\ j \neq i}}^{n-1} \left(\frac{y_i v_j}{cD_{i,n}} - \frac{y_i}{cD_{i,j}} \right) N_j - \left(\frac{y_i v_i}{cD_{i,n}} + \sum_{\substack{j=1 \\ j \neq i}}^{n} \frac{y_j}{cD_{i,j}} \right) N_i \qquad (8.2\text{-}46)$$

for $i = 1, 2, ..., n-1$. The above equation can be cast into a much more compact vector form as follows:

$$c\nabla \underline{y} = -\underline{\underline{B}} \cdot \underline{N} \qquad (8.2\text{-}47)$$

where $\underline{y}$ and $\underline{N}$ are (n-1) tupled vectors defined as:

$$\underline{y} = \begin{bmatrix} y_1 & y_2 & \cdots & y_{n-1} \end{bmatrix}^T \qquad (8.2\text{-}48a)$$

$$\underline{N} = \begin{bmatrix} N_1 & N_2 & \cdots & N_{n-1} \end{bmatrix}^T \qquad (8.2\text{-}48b)$$

The matrix $\underline{\underline{B}}$ (n-1, n-1) has the units of the inverse of the diffusion coefficient (sec/m^2), and is defined as below

$$\underline{\underline{B}} = \begin{cases} \dfrac{y_i v_i}{D_{i,n}} + \sum_{\substack{k=1\\k\neq i}}^{n} \dfrac{y_k}{D_{i,k}} & \text{for } i = j \\ y_i \left(\dfrac{v_j}{D_{i,n}} - \dfrac{1}{D_{i,j}} \right) & \text{for } i \neq j \end{cases} \tag{8.2-48c}$$

for $i, j = 1, 2, \ldots, n-1$.

Note that eq. (8.2-48c) has a summation having the index k ranging from 1 to n except $k = i$, and that

$$y_n = 1 - \sum_{k=1}^{n-1} y_k \tag{8.2-49}$$

Solving eq. (8.2-47) for the flux vector in terms of the concentration gradients of (n-1) species, we get:

$$\underline{N} = -c \left[\underline{\underline{B}}(\underline{y}) \right]^{-1} \nabla \underline{y} \tag{8.2-50}$$

where we see that the apparent diffusion coefficient matrix $\left[\underline{\underline{B}}(\underline{y}) \right]^{-1}$ is a function of all the concentrations, making the Stefan-Maxwell analysis different from the Fick's law, which assumes that the diffusion coefficient matrix is a constant diagonal matrix. The implication of eq.(8.2-50) is that the flux of a component "i" is affected by the concentration gradients of other species in the system.

Example 8.2-1: *Flux equation for infinite diluted conditions*

Let us now investigate the behaviour of the matrix $\underline{\underline{B}}$ (n-1, n-1) when the concentrations of n-1 components (solutes) are much lower than that of the n-th component (solvent). We take the limit of eq. (8.2-48c) when the mole fractions of n-1 solutes are very small:

$$\underline{\underline{B}}^{(0)} = \lim_{\underline{y} \to 0} \underline{\underline{B}}(\underline{y}) = \begin{bmatrix} \dfrac{1}{D_{1,n}} & 0 & \cdots & 0 \\ 0 & \dfrac{1}{D_{2,n}} & \cdots & 0 \\ \vdots & \vdots & \ddots & \vdots \\ 0 & 0 & \cdots & \dfrac{1}{D_{n-1,n}} \end{bmatrix} \tag{8.2-51}$$

which is a diagonal matrix. This means that for a dilute system where the n-th component acts as solvent, the flux equations for all n-1 solutes are independent of each other in the sense that they interact only with the solvent, and the diffusion coefficient matrix is a constant diagonal matrix:

$$\underline{N}^{(0)} = -c \left[\underline{\underline{B}}^{(0)}\right]^{-1} \nabla \underline{y} \qquad (8.2\text{-}52a)$$

where the diffusion coefficient matrix under infinite dilution conditions $\left[\underline{\underline{B}}^{(0)}\right]^{-1}$ is given by:

$$\left[\underline{\underline{B}}^{(0)}\right]^{-1} = \begin{bmatrix} D_{1,n} & 0 & \cdots & 0 \\ 0 & D_{2,n} & \cdots & 0 \\ \vdots & \vdots & \ddots & \vdots \\ 0 & 0 & \cdots & D_{n-1,n} \end{bmatrix} \qquad (8.2\text{-}52b)$$

Written in component form, eq. (8.2-52) has the following familiar looking Fick's law form:

$$N_i^{(0)} = -c\, D_{i,n} \nabla y_i \qquad (8.2\text{-}53)$$

for i = 1, 2, ..., n-1. Thus, Fick's law is only applicable to very dilute systems, in which one component is acting as a solvent and all the remaining species are acting as solutes having very low concentrations.

Example 8.2-2: *Non-dimensionalization of the diffusion matrix* $\underline{\underline{B}}$

Very often when we solve the mass balance equation under a transient condition, equations are more conveniently cast into nondimensional form for subsequent numerical analysis. What we will show in this example is the non-dimensional form of the diffusion matrix $\underline{\underline{B}}$. To do this, we need a characteristic length, and a characteristic diffusion coefficient.

The characteristic length is simply the dimension of the system under consideration, for example the radius of the particle. We let it be L. To nondimensionalize the diffusion coefficient matrix, we need a characteristic diffusivity. This characteristic diffusivity can be arbitrary, but it should be chosen to reflect the dynamic behaviour of the system at hand. Here we choose it to be the sum of all diagonal terms of the matrix $\left[\underline{\underline{B}}^{(0)}\right]^{-1}$, that is:

$$D_T = \sum_{j=1}^{n-1} D_{j,n} \qquad (8.2\text{-}54)$$

Readers can choose it to be either the smallest or largest value in the infinite dilution diffusion coefficient matrix.

With L being the characteristic length and D_T being the characteristic diffusivity, the general constitutive flux equation can be cast into the following nondimensional form:

$$\underline{N} = -\left(\frac{cD_T}{L}\right)\left[\underline{\underline{B}}^*(\underline{y})\right]^{-1}\nabla^*\underline{y} \qquad (8.2\text{-}55)$$

where (cD_T/L) is the characteristic flux (moles/m²/sec), and the nondimensional matrix $\underline{\underline{B}}^*$ and ∇^* are given by:

$$\underline{\underline{B}}^* = D_T\underline{\underline{B}} = \begin{cases} \dfrac{y_i v_i}{(D_{i,n}/D_T)} + \sum_{\substack{k=1 \\ k \neq i}}^{n} \dfrac{y_k}{(D_{i,k}/D_T)} & \text{for } i = j \\[2ex] y_i\left(\dfrac{v_j}{D_{i,n}/D_T} - \dfrac{1}{D_{i,j}/D_T}\right) & \text{for } i \neq j \end{cases} \qquad (8.2\text{-}56)$$

and

$$\nabla^* = \frac{\partial}{\partial x^*}\underline{i} + \frac{\partial}{\partial y^*}\underline{j} + \frac{\partial}{\partial z^*}\underline{k} \qquad (8.2\text{-}57)$$

Here, we use the upperscript * to denote for nondimensionality, and x*, y* and z* are nondimensional distances along the three principal coordinates of the Cartesian space. Readers will see eq. (8.2-55) again later when we deal with the transient analysis of mass balance equations.

8.2.3.3 Formulation of Mass Balance Equation

Having obtained the necessary equations for flux written in terms of the concentration gradients (eq. 8.2-50), these constitutive flux equations can be used in the mass balance equation to solve for concentration distributions. Once the concentrations are obtained, the fluxes of all species can then be calculated by using the constitutive flux equation (8.2-50). We will demonstrate this with the following example involving pure diffusion of n species into a medium having a slab geometry.

Example 8.2-3: *Mass balance equation in a slab geometry medium*

We take a pure diffusion system having a slab geometry, and the mass balance equation describing the concentration distribution of all species under transient conditions is given by:

$$\frac{\partial c_i}{\partial t} = -\frac{\partial N_i}{\partial z} \quad \text{for } i = 1, 2, \ldots, n \qquad (8.2\text{-}58)$$

Even though the mass balance equation (8.2-58) is written for all species, we need to solve for only n-1 components. The concentration of the n-th species is then given by eq.(8.2-49). Substituting the constitutive flux equation eq. (8.2-55) into the mass balance equation (8.2-58), we get:

$$c\frac{\partial \underline{y}}{\partial t} = \frac{c D_T}{L^2} \nabla^* \left\{ [\underline{\underline{B}}(\underline{y})]^{-1} \nabla^* \underline{y} \right\} \qquad (8.2\text{-}59)$$

By observing eq. (8.2-59), we see that the obvious choice for the characteristic time is simply the square of the characteristic length divided by the characteristic diffusivity

$$t_0 = \frac{L^2}{D_T} \qquad (8.2\text{-}60)$$

With this characteristic time, the nondimensional time is:

$$t^* = \frac{t}{t_0} = \frac{D_T t}{L^2} \qquad (8.2\text{-}61)$$

and the mass balance equation will become:

$$\frac{\partial \underline{y}}{\partial t^*} = \nabla^* \left\{ [\underline{\underline{B}}^*(\underline{y})]^{-1} \nabla^* \underline{y} \right\} \qquad (8.2\text{-}62)$$

With boundary and initial conditions appropriate for the given system, eq. (8.2-62) can be readily integrated numerically to yield solutions for the concentration of n-1 components ($y_1, y_2, \ldots, y_{n-1}$). Knowing these concentrations as a function of distance and time, the concentration of the n-th component is:

$$y_n = 1 - \sum_{j=1}^{n-1} y_j \qquad (8.2\text{-}63)$$

and the fluxes of the n-1 components at any time and any position then can be evaluated from eq. (8.2-55) by substituting the concentrations into such equation. Of importance to engineers are the values of fluxes at the boundary, and these are calculated from the following equation for the fluxes of the n-1 components

$$\underline{N}\big|_{\partial V} = -\left(\frac{cD_T}{L}\right)\left[\underline{\underline{B}}^*\left(\underline{y}\big|_{\partial V}\right)\right]^{-1}\nabla^*\underline{y}\big|_{\partial V} \qquad (8.2\text{-}64a)$$

where the subscript ∂V denotes for the boundary. The corresponding flux of the n-th component at the same boundary is:

$$N_n\big|_{\partial V} = -\sum_{j=1}^{n-1} v_j\, N_j\big|_{\partial V} \qquad (8.2\text{-}64b)$$

Knowing the fluxes of all components at the boundary, the total flux is then the summation of all the fluxes, that is:

$$N_T\big|_{\partial V} = \sum_{j=1}^{n} N_j\big|_{\partial V} \qquad (8.2\text{-}65)$$

If readers are interested in the diffusive fluxes at the boundary, they are calculated from:

$$\underline{J}\big|_{\partial V} = \underline{N}\big|_{\partial V} - \left(N_T\big|_{\partial V}\right)\underline{y}\big|_{\partial V} \qquad (8.2\text{-}66)$$

8.2.4 Stefan Tube with Binary System

Having presented the flux equations for a multicomponent system, we will apply the Stefan-Maxwell's approach to solve for fluxes in the Stefan tube at steady state. Consider a Stefan tube (Figure 8.2-3) containing a liquid of species 1. Its vapour above the liquid surface diffuses up the tube into an environment in which a species 2 is flowing across the top, which is assumed to be nonsoluble in liquid.

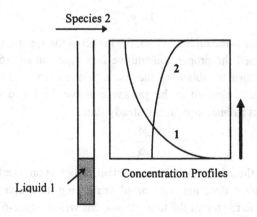

Figure 8.2-3: Concentration profile and flux directions in Stefan tube

The Stefan-Maxwell equation for the component 1 in a binary system is (obtained from eq. 8.2-34 by setting n = 2):

$$\nabla y_1 = \frac{y_1 N_2 - y_2 N_1}{cD_{12}} \qquad (8.2\text{-}67a)$$

This is the only independent constitutive flux equation. The mole fraction of the species 2 is:

$$y_2 = 1 - y_1 \qquad (8.2\text{-}67b)$$

Eq. (8.2-67a) then can be written in terms of the mole fraction of the component 1

$$N_1 = -cD_{12}\frac{dy_1}{dz} + y_1(N_1 + N_2) \qquad (8.2\text{-}68)$$

which is the familiar equation for the flux equation commonly seen in many books in transport phenomena (for example Bird et al., 1960).

Since we assume that the gas (species 2) is insoluble in the liquid (species 1), then $N_2 = 0$. Hence, eq. (8.2-68) can be rewritten in terms of flux and driving force of the component 1:

$$N_1 = -\frac{cD_{12}}{1-y_1}\frac{dy_1}{dz} \qquad (8.2\text{-}69)$$

This is the flux equation written in terms of the concentration gradient, the driving force for mass transfer with the apparent Fickian diffusivity as

$$\frac{D_{12}}{1-y_1} \qquad (8.2\text{-}70)$$

Note that the higher is the mole fraction, the higher is the apparent diffusivity.

Having obtained the proper constitutive flux equation (8.2-69) for this binary system, we now turn to obtaining the mass balance equation. Doing the mass balance across a thin element in the gas space above the liquid surface, we obtain the following mass balance equation at steady state:

$$\frac{dN_1}{dz} = 0 \qquad (8.2\text{-}71)$$

implying that the flux along the tube is constant, which is physically expected as the evaporating species 1 does not react or adsorb during its course of diffusion from the liquid surface to the top of the tube. Using this with eq. (8.2-69), we get:

$$N_1 = -\frac{cD_{12}}{1-y_1}\frac{dy_1}{dz} \equiv \text{constant} \qquad (8.2\text{-}72)$$

The boundary conditions of this problem are conditions at two ends of the tube:

$$\begin{aligned} z = 0; & \quad y_1 = y_{10} \\ z = L; & \quad y_1 = y_{1L} \end{aligned} \qquad (8.2\text{-}73)$$

The solution can be readily obtained by integrating eq. (8.2-72) subject to the conditions (8.2-73) to give the following concentration distribution of the species 1 along the tube:

$$\frac{1-y_1}{1-y_{10}} = \left(\frac{1-y_{1L}}{1-y_{10}}\right)^{z/L} \qquad (8.2\text{-}74)$$

Knowing this concentration profile, the evaporation flux can be then calculated from the flux equation (8.2-69)

$$N_1 = \frac{cD_{12}}{L}\ln\left(\frac{1-y_{1L}}{1-y_{10}}\right) \qquad (8.2\text{-}75)$$

which is indeed a constant, that is independent of distance z.

Eq. (8.2-75) is the equation yielding the evaporation rate of the species 1 per unit area of the tube. To determine the pressure and temperature dependence of this evaporation flux, we make use of the following expressions:

$$c = \frac{P}{R_g T} \qquad (8.2\text{-}76a)$$

$$D_{12} = D_{12}^0 \left(\frac{P_0}{P}\right)\left(\frac{T}{T_0}\right)^{1.75} \qquad (8.2\text{-}76b)$$

$$y_{10} = \frac{P_{vap,1}(T)}{P} \qquad (8.2\text{-}76c)$$

in which we have assumed the gas behaviour is ideal, and the molecular diffusivity D_{12} is inversely proportional to total pressure and is proportional to the temperature raised to 1.75 power. Here D_{12}^0 is the molecular diffusivity at some reference pressure P_0 and temperature T_0. The vapor pressure is assumed to follow the Antoine equation:

$$P_{vap,1} = \exp\left(A - \frac{B}{T+C}\right) \qquad (8.2\text{-}77)$$

Substitution of eqs. (8.2-76) and (8.2-77) into the evaporation flux equation (8.2-75) gives:

$$N_1 = \frac{D_{12}^0 P_0}{LR_g T_0}\left(\frac{T}{T}\right)^{0.75} \ln\left\{\frac{1-y_{1L}}{1-\exp[A-B/(T+C)]/P}\right\} \quad (8.2\text{-}78)$$

Very often the Stefan tube experiment is conducted with very high flow of gas above the tube, thus molecules of the component 1 are swept away very quickly by the gas stream, implying that the mole fraction of the component 1 can be effectively assumed as zero at the top of the tube. Thus, the evaporation written explicitly in terms of temperature and pressure is:

$$N_1 = \frac{D_{12}^0 P_0}{LR_g T_0}\left(\frac{T}{T}\right)^{0.75} \ln\left[\frac{1}{1-\exp(A-B/(T+C))/P}\right] \quad (8.2\text{-}79)$$

Example 8.2-4: *Benzene evaporation from a tube into a flowing air stream*

To have some idea of the magnitude of the evaporation rate, let us take an example of evaporation from a test tube having a diffusion length of 10 cm. Air is flowing across the tube at a rate such that the mole fraction of benzene at the top of the tube is very close to zero. The temperature and the pressure of the system are 25 °C and 760 Torr, respectively. The vapor pressure of benzene at this temperature is 100 Torr. The molecular diffusivity of benzene in air at 25 °C and 760 Torr is 0.0962 cm²/sec.

Knowing the vapor pressure at the temperature concerned, the mole fraction of benzene vapor at the gas-liquid interface is:

$$y_{10} = \frac{P_{vap,1}}{P} = \frac{100}{760} = 0.1316$$

Assuming an ideal behaviour, the total molar concentration is:

$$c = \frac{P}{R_g T} = \frac{1\,(\text{atm})}{(82.0572\text{ atm}-\text{cm}^3/\text{mole}/\text{K})\times(298\text{ K})} = 4.1\times 10^{-5}\,\frac{\text{moles}}{\text{cm}^3}$$

Substituting these values into the flux equation (8.2-75), we get the evaporation flux for benzene:

$$N_1 = \frac{cD_{12}}{L} \ln\left(\frac{1-y_{1L}}{1-y_{10}}\right) = \frac{(4.1 \times 10^{-5} \text{ moles/cm}^3)(0.0962 \text{ cm}^2/\text{sec})}{(10 \text{ cm})} \ln\left(\frac{1-0}{1-0.1316}\right)$$

$$N_1 = 2 \times 10^{-4} \frac{\text{moles}}{\text{cm}^2 - \text{hr}}$$

The flux expressed in mole/cm²/hr is hard to give us a feel about the evaporation rate. It would be useful to have it expressed in terms of liquid volume per unit time. Knowing the molecular weight of benzene is 78 g/mole, and its liquid density is 0.879 g/cc, the evaporation rate of benzene is

$$N_1 = \left(2 \times 10^{-4} \frac{\text{moles}}{\text{cm}^2 - \text{hr}}\right)\left(78 \frac{\text{g}}{\text{mole}}\right)\left(\frac{1}{0.879} \frac{\text{cm}^3 \text{ liquid}}{\text{g}}\right)$$

$$N_1 = 0.01778 \frac{\text{cm}^3 \text{ liquid}}{\text{cm}^2 - \text{hr}} \approx 0.43 \frac{\text{cm}^3 \text{ liquid}}{\text{cm}^2 - \text{day}}$$

This means that the liquid benzene level will drop by 0.43 cm per day.

Example 8.2-5: *Benzene evaporation at sub-ambient pressure*

Let us study the evaporation rate at another total pressure to investigate numerically the pressure dependence. We take a case of sub-ambient pressure with P = 300 Torr. The temperature is remained the same at 25 °C. With this new condition, the mole fraction of benzene right above the gas-liquid interface is:

$$y_{10} = \frac{P_{vap,1}}{P} = \frac{100}{300} = 0.3333$$

The total molar concentration at the new condition is:

$$c = \frac{P}{R_g T} = \frac{(300/760 \text{ atm})}{(82.0572 \text{ atm} - \text{cm}^3/\text{mole}/\text{K}) \times (298 \text{ K})} = 1.6 \times 10^{-5} \frac{\text{moles}}{\text{cm}^3}$$

At the new condition of 25 °C and 300 Torr, the binary diffusivity is calculated from (see eq. 7.7-9):

$$D_{12} @ 25°C \text{ and } 300 \text{ Torr} = \left(D_{12} @ 25°C \text{ and } 760 \text{ Torr}\right)\left(\frac{760 \text{ Torr}}{300 \text{ Torr}}\right)$$

$$= \left(0.0962 \frac{cm^2}{sec}\right)\left(\frac{760}{300}\right)$$

$$= 0.244 \frac{cm^2}{sec}$$

With these new values, the evaporation flux is readily calculated using eq. (8.2-75) as:

$$N_1 = 5.7 \times 10^{-4} \frac{moles}{cm^2 - hr}$$

Thus, the molar evaporation rate is higher when the total pressure is decreased. This is solely due to the increase in the mole fraction of benzene at the liquid interface (from 0.13 in the last example to 0.33 in this example). The increase in the molecular diffusivity as the total pressure increases is exactly compensated by the decrease in the total molar concentration.

Example 8.2-6: *Error of the Fick's calculation*

Let us now attempt to calculate the evaporation rate based on the simple Fick's law (that is, ignoring the convective contribution of the mass transfer which is the second term in the RHS of eq. 8.2-68). For this case, the evaporation flux is:

$$N_1 = cD_{12} \frac{(y_{10} - y_{1L})}{L} \quad (8.2\text{-}80)$$

Substitute the appropriate values at 25 °C and 760 Torr in Example 8.2-4 into the above Fick's law equation, we get:

$$N_1 = 1.87 \times 10^{-4} \frac{moles}{cm^2 - hr}$$

which is not much different from the value of 2×10^{-4} using the exact Maxwell-Stefan equation. This is the case because at this temperature, the convective flux of benzene vapour (which is the second term in the RHS of eq. 8.2-68) is not significant compared to the diffusion flux term.

The effect of convection will be significant if we consider a higher temperature situation. Take the case where the temperature is 60 °C and the total pressure is 1 atm. The vapor pressure of benzene at this temperature is 400 Torr. Using the Stefan-Maxwell result (eq. 8.2-75), we calculate the evaporation flux as 1.06×10^{-3} moles/cm²/hr, compared to 7.47×10^{-4} moles/cm²/hr calculated by the Fick law equation. An error of nearly 30% underpredicted by the Fick's law shows the importance of the convection term in the Stefan-Maxwell equation.

8.2.4.1 The Importance of the Convection Term

The numerical calculation of Example 8.2-6 shows the importance of the convective term in the flux equation. Let us now explore a bit further about this term. The molar flux of the component 1 is given as in eq. (8.2-75), and since the molar flux of the second component is zero, the total molar flux is simply:

$$N_T = N_1 + N_2 = \frac{cD_{12}}{L}\ln\left(\frac{1-y_{1L}}{1-y_{10}}\right) + 0 = \frac{cD_{12}}{L}\ln\left(\frac{1-y_{1L}}{1-y_{10}}\right) \quad (8.2\text{-}81)$$

The Stefan-Maxwell equation for the component 2 in a binary system is:

$$\nabla y_2 = \frac{y_2 N_1 - y_1 N_2}{cD_{12}} \quad (8.2\text{-}82)$$

that is, for one dimensional system considered here for the Stefan tube we have:

$$N_2 = -cD_{12}\frac{dy_2}{dz} + y_2(N_1 + N_2) \quad (8.2\text{-}83)$$

Since $y_2 = 1 - y_1$, we substitute eq. (8.2-74) for y_1 and eq. (8.2-81) for N_T into the above equation to get:

$$N_2 = \frac{cD_{12}}{L}\left[-(1-y_1)\ln\left(\frac{1-y_{1L}}{1-y_{10}}\right) + y_2\ln\left(\frac{1-y_{1L}}{1-y_{10}}\right)\right] \overset{!}{=} 0 \quad (8.2\text{-}84)$$

which indeed shows that the flux of the component 2 is zero. The first term in the RHS of eq. (8.2-84) is the diffusive flux of the component 2 down toward the liquid surface as a result of its concentration gradient, while the second term is the convective flux of that component up the tube carried upward by the total flux. These two contributions exactly balance each other, yielding $N_2 = 0$.

438 Kinetics

To get the magnitude of this diffusive flux, we calculate it at the liquid interface ($z = 0$) for the case of benzene evaporation in air at 30 °C and 1 atm of Example 8.2-4, we obtain the diffusive flux of air at $z = 0$ of 1.74×10^{-4} moles/cm^2/hr (compared to the benzene evaporation flux of 2×10^{-4} moles/cm^2/hr). This diffusive flux of air is balanced exactly by an amount carried by the total flow up the tube. We must note here that although the flux relative to a fixed frame of coordinate (N) is constant along the tube, the diffusive flux (J) is not, but rather a function of distance.

8.2.4.2 Determination of the Molecular Diffusivity

The flux equation (8.2-75) can be used to calculate the molecular diffusivity, that is if N_1 can be measured experimentally, the molecular diffusivity can be calculated from:

$$D_{12} = \frac{N_1 L}{c \ln\left(\frac{1-y_{1L}}{1-y_{10}}\right)} \tag{8.2-85}$$

The flux N_1 can be measured by either observing the drop of the liquid level versus time or by analysing the benzene concentration in a known constant rate of air flow. The mole fraction of benzene at the top of the tube can be controlled to zero (or close to it), and its mole fraction at the liquid interface can be calculated from the information of the vapor pressure. Thus, the molecular diffusivity is readily obtained from eq. (8.2-85).

Although such determination for the molecular diffusivity is very simple, the value obtained should be treated as the first estimate for the molecular diffusivity as in practice the flow of vapor from the liquid interface to the top is usually laminar; that is there exists a velocity distribution across the tube which the analysis dealt with here did not take into account. The readers should refer to Whitaker (1991) for detailed exposition of the influence of the velocity profile.

8.2.5 Stefan Tube for Ternary System

We have considered the Stefan tube with pure liquid in the tube. Now we consider the case whereby the liquid contains two components. These two species will evaporate and diffuse along the tube into the flow of a third component across the top of the tube. The third component is assumed to be non-soluble in the liquid. What we will consider next is the Maxwell-Stefan analysis of this ternary system, and then apply it to the experimental data of Carty and Schrodt (1975) where they used a liquid mixture of acetone and methanol. The mole fractions of acetone and methanol just above the liquid surface of the tube are 0.319 and 0.528, respectively.

The other component in the system is air, and the length of the diffusion path in the tube is 23.8 cm. The total pressure is 99.5 kPa, the temperature is 328 K and the three binary diffusivities at these conditions are:

$$D_{12} = 0.0848, \qquad D_{13} = 0.1372, \qquad D_{23} = 0.1991 \text{ cm}^2/\text{sec}$$

Here we use the numeric notation for the three diffusing species:

1 = acetone
2 = methanol
3 = air

Since air is not soluble in liquid, we write

$$N_3 = 0 \qquad (8.2\text{-}86)$$

The three Stefan-Maxwell equations for the fluxes of the three components in terms of mole fraction gradients are (obtained from eq. 8.2-34):

$$\frac{dy_1}{dz} = \frac{y_1 N_2 - y_2 N_1}{cD_{12}} + \frac{y_1 N_3 - y_3 N_1}{cD_{13}} \qquad (8.2\text{-}87a)$$

$$\frac{dy_2}{dz} = \frac{y_2 N_1 - y_1 N_2}{cD_{12}} + \frac{y_2 N_3 - y_3 N_2}{cD_{23}} \qquad (8.2\text{-}87b)$$

$$\frac{dy_3}{dz} = \frac{y_3 N_1 - y_1 N_3}{cD_{13}} + \frac{y_3 N_2 - y_2 N_3}{cD_{23}} \qquad (8.2\text{-}87c)$$

These three equations are not independent because the sum of the above equations is zero. Any two of those three equations together with eq. (8.2-86) will provide the necessary three independent equations for the three fluxes in terms of the concentration gradients.

Similar to the case of binary systems, the mass balance carried out at steady state shows that all molar fluxes are constant. We take the first two independent equations of eqs. (8.2-87), set N_3 to zero and then numerically integrate (for example, using the Runge-Kutta method) them with respect to distance z after assuming some values for N_1 and N_2. This numerical integration will yield the mole fraction profiles for the components 1 and 2. If the calculated mole fractions at the exit of the tube match the imposed boundary conditions then the assumed values for N_1 and N_2 are the correct molar fluxes required. Otherwise, another set of N_1 and N_2 values is assumed, and the integration procedure is repeated until the boundary conditions at the exit are met. This is the basis of a method called the shooting method, as suggested by Taylor and Krishna (1993). We shall present this method in the following example.

Example 8.2-7: *Shooting method for a ternary system in a Stefan tube*

Integration of eqs. (8.2-87a) and (8.2-87b) with $N_3 = 0$ and assumed values for N_1 and N_2 will yield the mole fraction of the components 1 and 2 at the top of the tube. We let these be $y_1(N_1,N_2)\big|_{z=L}$ and $y_2(N_1,N_2)\big|_{z=L}$, respectively. The trial and error procedure for the shooting method is as follows.

The trial and error method can be evaluated systematically by applying the Newton-Raphson on the following equations at the exit of the tube:

$$f_1(N_1,N_2) \stackrel{\text{def}}{=} y_1(N_1,N_2)\big|_{z=L} - y_{1L} \stackrel{?}{=} 0 \qquad (8.2\text{-}88a)$$

$$f_2(N_1,N_2) \stackrel{\text{def}}{=} y_2(N_1,N_2)\big|_{z=L} - y_{2L} \stackrel{?}{=} 0 \qquad (8.2\text{-}88b)$$

The iteration formulas for finding the solutions for N_1 and N_2 are (by application of the Newton-Raphson method):

$$\underline{N}^{(k+1)} = \underline{N}^{(k)} - \left[\underline{\underline{J}}^{(k)}(\underline{N})\right]^{-1} \underline{f}^{(k)}(\underline{N}) \qquad (8.2\text{-}89)$$

where

$$\underline{f} = [f_1 \quad f_2]^T \qquad (8.2\text{-}90a)$$

$$\underline{N} = [N_1 \quad N_2]^T \qquad (8.2\text{-}90b)$$

and $\underline{\underline{J}}$ is the Jacobian of eqs.(8.2-88) and is defined as:

$$\underline{\underline{J}} = \begin{bmatrix} \dfrac{\partial f_1}{\partial N_1} & \dfrac{\partial f_1}{\partial N_2} \\ \dfrac{\partial f_2}{\partial N_1} & \dfrac{\partial f_2}{\partial N_2} \end{bmatrix} = \begin{bmatrix} \dfrac{\partial y_1}{\partial N_1} & \dfrac{\partial y_1}{\partial N_2} \\ \dfrac{\partial y_2}{\partial N_1} & \dfrac{\partial y_2}{\partial N_2} \end{bmatrix}_{z=L} \qquad (8.2\text{-}91)$$

The elements $\dfrac{\partial y_1}{\partial N_1}$, $\dfrac{\partial y_1}{\partial N_2}$, $\dfrac{\partial y_2}{\partial N_1}$, and $\dfrac{\partial y_2}{\partial N_2}$ are obtained by appropriately differentiating eqs. (8.2-87a) and (8.2-87b). For example the first element is obtained by integrating the following equation, which is obtained by differentiating eq.(8.2-87a) with respect to N_1:

$$\frac{d\left(\frac{\partial y_1}{\partial N_1}\right)}{dz} = \frac{1}{cD_{12}}\left[N_2\frac{\partial y_1}{\partial N_1} - y_2 - N_1\frac{\partial y_2}{\partial N_1}\right] + \frac{1}{cD_{13}}\left[-N_1\frac{\partial y_3}{\partial N_1} - y_3\right] \quad (8.2\text{-}92)$$

Similarly, we can obtain the corresponding differential equations for $\frac{\partial y_1}{\partial N_2}$, $\frac{\partial y_2}{\partial N_1}$, and $\frac{\partial y_2}{\partial N_2}$ with respect to z from eqs(8.2-87a and b) by differentiating them with respect to N_1 and N_2. These four equations are then numerically integrated together with the two diffusion equations (8.2-87a and b), and then apply the Newton-Ralphson formula (eq. 8.2-89) to obtain the next estimate for the molar fluxes N_1 and N_2. This process is repeated until the convergence in the molar fluxes N_1 and N_2 is achieved to some required tolerance.

From the integration, it is found that

$$N_1 = 1.783 \times 10^{-7} \text{ moles/cm}^2\text{/sec}; \quad N_2 = 3.127 \times 10^{-7} \text{ moles/cm}^2\text{/sec}$$

The mole fraction profiles of acetone and methanol after the convergence has been achieved are shown in Figure 8.2-4. Alternatively, one could use the vector analysis presented in Appendix 8.2 to solve for the fluxes without using the shooting method. We will present the method in the next section when we deal with a Stefan tube containing "n" components, that is we have "n-1" components in the liquid and the n-th component is the non-soluble gas flowing across the tube.

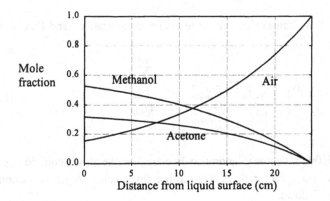

Figure 8.2-4: Methanol, acetone and air concentration profiles in Stefan tube

8.2.6 Stefan Tube with "n" Component Mixtures

We have seen the analysis of a binary system where the analytical solution for the evaporation flux is possible, and the analysis of a ternary system where we show that a numerical method called the shooting method is required to obtain the desired solutions. Here, we will analyze the same system, but this time we have (n-1) components in the liquid phase, and these (n-1) species evaporate and diffuse up the tube into the flowing stream of the n-th component, which is non-soluble in the liquid phase. In terms of solution procedure, we will present here a vector-matrix method to obtain an analytical solution. This method is elegant and is more effective in terms of computation than the shooting method shown in the last example.

For a system with n species, the relevant Stefan Maxwell equations for describing this constant pressure system are:

$$-\frac{dy_i}{dz} = \sum_{\substack{j=1 \\ j \neq i}}^{n} \frac{y_j N_i - y_i N_j}{cD_{ij}} \qquad (8.2\text{-}93)$$

for i = 1, 2, 3,..., n.

We let the n-th species be the insoluble gas flowing across the tube, then:

$$N_n = 0 \qquad (8.2\text{-}94)$$

Substituting eq.(8.2-94) into eq.(8.2-93) and making use of the relation

$$y_n = 1 - \sum_{j=1}^{n-1} y_j \qquad (8.2\text{-}95)$$

eq (8.2-93) is now written exclusively in terms of concentrations and flux of the (n-1) components:

$$-\frac{dy_i}{dz} = \sum_{\substack{j=1 \\ j \neq i}}^{n-1} \frac{y_j N_i - y_i N_j}{cD_{ij}} + \frac{1 - \sum_{j=1}^{n-1} y_j}{cD_{i,n}} \qquad (8.2\text{-}96)$$

for i = 1, 2, 3,..., n-1.

If we now define a vector $\underline{y}$ having (n-1) elements, the above equation (8.2-96) then can be cast into a vector form relating the flux vector and the concentration gradient vector as follows:

$$c\frac{d\underline{y}}{dz} = -\underline{\underline{B}}(\underline{y})\underline{N} \tag{8.2-97a}$$

where the mole fraction vector $\underline{y}$ and the flux vector $\underline{N}$ are:

$$\underline{y} = [y_1 \quad y_2 \quad \cdots \quad y_{n-1}]^T \tag{8.2-97b}$$

$$\underline{N} = [N_1 \quad N_2 \quad \cdots \quad N_{n-1}]^T \tag{8.2-97c}$$

and the matrix $\underline{\underline{B}}$ having the units of the inverse of the diffusivity is defined as follows:

$$\underline{\underline{B}} = \begin{cases} \dfrac{\left(1 - \sum_{j=1}^{n-1} y_j\right)}{D_{in}} + \sum_{\substack{j=1 \\ j \neq i}}^{n-1} \dfrac{y_j}{D_{ij}} & \text{for } i = j \\ -\dfrac{y_i}{D_{ij}} & \text{for } i \neq j \end{cases} \tag{8.2-97d}$$

The matrix $\underline{\underline{B}}$ can also be obtained directly from eq.(8.2-48c) by setting v_i ($i = 1, 2, ..., n-1$) to zero.

Alternately, we can write eq.(8.2-96) relating the concentration gradient vector and the concentration vector as follows:

$$\frac{d\underline{y}}{dz} = \underline{\underline{A}}(\underline{N})\underline{y} + \underline{\psi}(\underline{N}) \tag{8.2-98}$$

where the matrix $\underline{\underline{A}}$ and the vector $\underline{\psi}$ are defined as:

$$\underline{\underline{A}} = \begin{cases} \dfrac{N_i}{cD_{in}} + \sum_{\substack{j=1 \\ j \neq i}}^{n-1} \dfrac{N_j}{cD_{ij}} & \text{for } i = j \\ \dfrac{N_i}{cD_{in}} - \dfrac{N_i}{cD_{ij}} & \text{for } i \neq j \end{cases} \tag{8.2-99a}$$

$$\underline{\psi} = \left\{ -\dfrac{N_i}{cD_{in}} \right. \tag{8.2-99b}$$

for $i, j = 1, 2, ..., n-1$.

Solutions of eqs. (8.2-97a) and (8.2-98) are given in Appendix 8.2. For the boundary conditions of the form:

$$z = 0; \quad \underline{y} = \underline{y}_0 \tag{8.2-100a}$$

$$z = L; \quad \underline{y} = \underline{y}_L \tag{8.2-100b}$$

the solution for the fluxes is given by either

$$\underline{N} = c \left[\underline{\underline{B}}(\underline{y}_0)\right]^{-1} \underline{\underline{A}} \left[\exp(L\underline{\underline{A}}) - \underline{\underline{I}}\right]^{-1} (\underline{y}_0 - \underline{y}_L) \tag{8.2-101a}$$

or

$$\underline{N} = c \left[\underline{\underline{B}}(\underline{y}_L)\right]^{-1} \underline{\underline{A}} \exp(L\underline{\underline{A}}) \left[\exp(L\underline{\underline{A}}) - \underline{\underline{I}}\right]^{-1} (\underline{y}_0 - \underline{y}_L) \tag{8.2-101b}$$

The difference between the above two equations is the evaluation of the matrix $\underline{\underline{B}}$. In eq.(8.2-101a), it is evaluated at $z = 0$, while it is evaluated at $z = L$ in eq.(8.2-101b).

The solution for the concentration profile is (Appendix 8.2):

$$\underline{y} - \underline{y}_0 = \left[\exp(z\underline{\underline{A}}) - \underline{\underline{I}}\right] \left[\exp(L\underline{\underline{A}}) - \underline{\underline{I}}\right]^{-1} (\underline{y}_L - \underline{y}_0) \tag{8.2-102}$$

We illustrate below an example of solving the Stefan tube with "n" diffusing species by the method of Newton-Raphson.

Example 8.2-8: *Newton-Ralphson method for n-component system in a Stefan tube*

Eq.(8.2-101) is a set of (n-1) nonlinear algebraic equations in terms of the fluxes of the (n-1) diffusing components because the matrix $\underline{\underline{A}}$ is a function of fluxes $\underline{N}$ (eq. 8.2-99a). This set of equations is readily solved by the Newton-Ralphson method. This method requires an initial guess for the flux vector, and this is simply achieved by considering eq.(8.2-97a). Solving for the flux in terms of the concentration gradient, we get

$$\underline{N} = -c \left[\underline{\underline{B}}(\underline{y})\right]^{-1} \frac{d\underline{y}}{dz} \tag{8.2-103}$$

If we approximate the matrix $\underline{\underline{B}}$ by a constant matrix, which is the matrix $\underline{\underline{B}}$ evaluated at the mean concentration $\underline{y}_{ave}$, that is:

$$\underline{\underline{B}} \approx \underline{\underline{B}}\left(\underline{y}_{ave}\right) \quad (8.2\text{-}104a)$$

where

$$\underline{y}_{ave} = \frac{\underline{y}_0 + \underline{y}_L}{2} \quad (8.2\text{-}104b)$$

and approximate the concentration gradient as the linear difference as given below

$$\frac{d\underline{y}}{dz} \approx \frac{\underline{y}_L - \underline{y}_0}{L} \quad (8.2\text{-}105)$$

then the initial guess for the molar flux vector is:

$$\underline{N}^{(0)} = -c\left[\underline{\underline{B}}\left(\underline{y}_{ave}\right)\right]^{-1} \frac{\underline{y}_L - \underline{y}_0}{L} \quad (8.2\text{-}106)$$

The iteration formula for the flux vector $\underline{N}$ of dimension (n-1) is simply

$$\underline{N}^{(k+1)} = \underline{N}^{(k)} - \left[\underline{\underline{J}}(\underline{N})^{(k)}\right]^{-1} \underline{f}^{(k)} \quad (8.2\text{-}107)$$

where $\underline{\underline{J}}$ is the Jacobian of the following vector $\underline{f}$ as (from eq. 8.2-101):

$$\underline{f} = \underline{N} - c\left[\underline{\underline{B}}\left(\underline{y}_0\right)\right]^{-1} \underline{\underline{A}}\left[\exp(L\underline{\underline{A}}) - \underline{\underline{I}}\right]^{-1}\left(\underline{y}_0 - \underline{y}_L\right) \quad (8.2\text{-}108a)$$

or

$$\underline{f} = \underline{N} - c\left[\underline{\underline{B}}\left(\underline{y}_L\right)\right]^{-1} \underline{\underline{A}} \exp(L\underline{\underline{A}}) \left[\exp(L\underline{\underline{A}}) - \underline{\underline{I}}\right]^{-1}\left(\underline{y}_0 - \underline{y}_L\right) \quad (8.2\text{-}108b)$$

The Jacobian matrix is

$$\underline{\underline{J}} = \left\{\frac{\partial f_i}{\partial N_j}\right\} \quad (8.2\text{-}108c)$$

for i, j = 1, 2, ..., n-1.

The functional form of the vector function $\underline{f}$ is rather complex in terms of $\underline{N}$; therefore the Jacobian is best evaluated numerically as follows:

$$\frac{\partial f_i}{\partial N_j} \approx \frac{f_i(N_1, N_2, \cdots, N_j + \Delta N_j, \cdots, N_{n-1}) - f_i(N_1, N_2, \cdots, N_j, \cdots, N_{n-1})}{\Delta N_j} \quad (8.2\text{-}109)$$

The Newton-Raphson method usually converges within a few iterations. Once the flux vector $\underline{N}$ are obtained numerically, the concentration (mole fraction) profiles of the (n-1) diffusing species are obtained from eq.(8.2-102). The mole fraction of the n-th component at any position then can be obtained from:

$$y_n(z) = 1 - \sum_{j=1}^{n-1} y_j(z) \quad (8.2\text{-}110)$$

This is the simplest way to obtain the concentration profile of the n-th component. However, if the reader is interested just the n-th component concentration profile, we could start with the Stefan-Maxwell equation (eq. 8.2-93) written for the n-th component (note that $N_n = 0$):

$$\frac{dy_n}{dz} = y_n \left(\sum_{j=1}^{n-1} \frac{N_j}{cD_{n,j}} \right) \quad (8.2\text{-}111)$$

For constant fluxes $\underline{N}$, the above equation can be integrated from 0 to z, and we obtain:

$$y_n(z) = \left(y_n\big|_{z=0}\right) \exp\left(z \sum_{j=1}^{n-1} \frac{N_j}{cD_{n,j}} \right) \quad (8.2\text{-}112)$$

As mentioned before, the fluxes $\underline{N}$ are determined from eq.(8.2-101). Once this is known, the concentration profile of the n-th component is then given in eq.(8.2-112).

Somewhat interesting in eq.(8.2-112) is that if we evaluate it at z = L, we get:

$$y_n(L) = y_n(0) \exp\left(\frac{L}{c} \sum_{j=1}^{n-1} \frac{N_j}{D_{n,j}} \right) \quad (8.2\text{-}113)$$

This shows that the fluxes $\underline{N}$ in a special way (LHS of eq. 8.2-113) are related to the mole fractions of the n-th component at the end points of the Stefan tube.

Solution of this problem is provided with the MatLab code STEFTUBE. The user is asked by this code to supply:
1. The number of component in the system, including the nonsoluble flowing gas
2. The respective mole fractions at two ends of the tube
3. The length of the tube
4. The operating conditions, that is the temperature and pressure
5. The binary diffusivity matrix

For a typical three component system, the code takes approximately a few seconds to run on a 586/90 MHz computer to yield the fluxes as well as the concentration profiles. We take the example of Carty and Schrodt (1975) in Example 8.2-7, and execute the STEFTUBE code. The following table shows the evaporation fluxes of acetone (1) and methanol (2) as a function of the diffusion path L.

L (cm)	$N_1 \times 10^7$ (mole/cm²/s)	$N_2 \times 10^7$ (mole/cm²/s)
10	4.27	7.49
15	2.85	5.00
20	2.13	3.74
23.8*	1.79	3.14
30	1.42	2.50
35	1.22	2.14
40	1.07	1.87

* the length used by Carty and Schrodt (1975)

8.2.6.1 Drop in the Liquid Level

Since the evaporation of the species from the tube, the liquid level will drop. To calculate for this decrease in the liquid level, we need to set up a mass balance equation around the liquid body, and obtain the following mass balance equation for the species j:

$$\frac{dn_j}{dt} = -S\, N_j \qquad (8.2\text{-}114)$$

where S is the cross section area of the tube, and n_j is the number of moles in the liquid. The liquid volume contributed by the component j in the liquid is given by:

$$V_j = \frac{n_j M_j}{\rho_{L,j}} \qquad (8.2\text{-}115)$$

where $\rho_{L,j}$ is the liquid density and M_j is the molecular weight of the species j.

Combining eqs. (8.2-114) and (8.2-115), we get:

$$\frac{dV_j}{dt} = -A\frac{M_jN_j}{\rho_{L,j}} \qquad (8.2\text{-}116)$$

Summing the above equation for all evaporating species (n-1), we get:

$$\frac{dV}{dt} = -A\sum_{j=1}^{n-1}\frac{M_jN_j}{\rho_{L,j}} \qquad (8.2\text{-}117)$$

in which we have assumed that there is no change in volume upon mixing. Since $V = A(L_0 - L)$ where L_0 is the total length of the tube, eq. (8.2-117) can now be written in terms of the diffusion length L:

$$\frac{dL}{dt} = \sum_{j=1}^{n-1}\frac{M_jN_j}{\rho_{L,j}} \qquad (8.2\text{-}118)$$

Thus, knowing the fluxes in eq.(8.2-101) the diffusion length can be obtained by integrating the above equation. One must note, however, that since the evaporation rates will be different for different species, the compositions of the (n-1) components in the liquid phase will vary with time. This means that the mole fractions at z = 0 will also change with time, according to some law, for example the Raoult's law:

$$@\ z=0; \qquad Py_j = x_jP_j^0 \qquad (8.2\text{-}119)$$

where P_j^0 is the vapor pressure of the species j. However, for small change in the liquid level, the mole fractions at z=0 can be considered constant, and eq.(8.2-118) can be integrated to give:

$$\int_{L(0)}^{L}\frac{dL}{\sum_{j=1}^{n-1}\frac{M_jN_j}{\rho_{L,j}}} = t \qquad (8.2\text{-}120)$$

8.3 Transient Diffusion of Ideal Gaseous Mixtures in Loschmidt's tube:

Section 8.2.6 shows the usage of the Stefan-Maxwell equations for the steady state analysis of the Stefan tube, in which we have shown the elegance of the vector-matrix presentation in dealing with steady state diffusion problem. Now we will show the application of the constitutive Stefan-Maxwell equation to an unsteady state problem. Here we shall take a transient diffusion problem of a Loschmidt's tube, which is commonly used in the study of diffusion coefficient.

The Loschmidt tube is simply a tube with an impermeable partition separating the two sections of the tube (Figure 8.3-1). Initially, the partition is in the position that gases in the two sections do not mix with each other. Here we shall assume that the total pressure is the same in both sections of the tube, and the initial compositions are different in the two parts of the tube. At time $t = 0^+$, the partition is removed and the diffusion process is started.

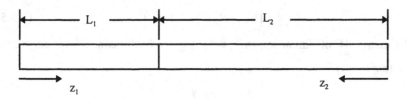

Figure 8.3-1: Schematic diagram of the Loschmidt's tube

8.3.1 The Mass Balance Equations

The constitutive Stefan-Maxwell equations for the description of fluxes of an n component system is:

$$-\frac{\partial y_i}{\partial z} = \sum_{\substack{j=1 \\ j \neq i}}^{n} \frac{(y_i N_i - y_i N_j)}{cD_{ij}} \qquad (8.3\text{-}1)$$

for i = 1, 2, ..., n. The partial derivative is used because of the time-dependent concentrations in this unsteady state problem.

Since the total pressure is constant in the tube and this will remain so throughout the course of diffusion, the sum of all fluxes must be zero, that is

$$\sum_{j=1}^{n} N_j = 0 \qquad (8.3\text{-}2)$$

from which the flux of the n-th component is related to all other fluxes as follows:

$$N_n = -\sum_{j=1}^{n-1} N_j \qquad (8.3\text{-}3)$$

Since only (n-1) equations of eq. (8.3-1) are independent, coupling these first (n-1) equations with eq. (8.3-3) will yield the following constitutive Stefan-Maxwell equations written in vector form as follows (in which all vectors have n-1 dimensions and matrices have [(n-1), (n-1)] dimensions):

$$-c\frac{\partial \underline{y}}{\partial z} = \underline{\underline{B}}(\underline{y})\underline{N} \qquad (8.3\text{-}4)$$

where

$$\underline{y} = \begin{bmatrix} y_1 & y_2 & \cdots & y_{n-1} \end{bmatrix}^T \qquad (8.3\text{-}5a)$$

$$\underline{N} = \begin{bmatrix} N_1 & N_2 & \cdots & N_{n-1} \end{bmatrix}^T \qquad (8.3\text{-}5b)$$

and the matrix $\underline{\underline{B}}$ of dimension (n-1, n-1) is given by (obtained from eq. 8.2-48c after setting $v_i = 1$):

$$\underline{\underline{B}}(\underline{y}) = \begin{cases} \dfrac{y_i}{D_{i,n}} + \sum_{\substack{k=1 \\ k \neq i}}^{n} \dfrac{y_k}{D_{i,k}} & i = j \\[6pt] y_i\left(\dfrac{1}{D_{i,n}} - \dfrac{1}{D_{i,j}}\right) & i \neq j \end{cases} \qquad (8.3\text{-}6)$$

It should be reminded here that the matrix $\underline{\underline{B}}$ is a function of mole fractions $\underline{y}$, and note the summation in eq. (8.3-6) ranges from 1 to n. Note also that this matrix $\underline{\underline{B}}$ is different from the matrix $\underline{\underline{B}}$ (eq. 8.2-97d) in the Stefan tube problem. This is because of the difference in the physical constraints of these problems. In the Stefan tube problem, the physical constraint is that the <u>flux of the n-th species is zero</u> (eq. 8.2-94), while in this problem the physical constraint is that <u>the sum of all fluxes is zero</u>.

Eq. (8.3-4) now can be inverted to obtain an expression for the flux written in terms of concentrations and concentration gradients of n-1 species:

$$\underline{N} = -c\left[\underline{\underline{B}}(\underline{y})\right]^{-1} \frac{\partial \underline{y}}{\partial z} \qquad (8.3\text{-}7)$$

Setting up the mass balance equation in either the sections of the Loschmidt's tube, we obtain the following equation for the conservation of mass

$$c\frac{\partial \underline{y}}{\partial t} = -\frac{\partial}{\partial z}(\underline{N}) \tag{8.3-8}$$

Combining eqs. (8.3-7) and (8.3-8), we get the following mass balance equation written wholly in terms of the mole fractions of the (n-1) species:

$$\frac{\partial \underline{y}}{\partial t} = \frac{\partial}{\partial z}\left\{[\underline{\underline{B}}(\underline{y})]^{-1}\frac{\partial \underline{y}}{\partial z}\right\} \tag{8.3-9}$$

This equation is the general mass balance equation. It is applicable for both sections of the tube. We now denote the upperscripts I and II for the two sections, and formally write the following mass balance equations, initial conditions and boundary conditions of the two sections as below:

Section I (LHS of the tube)	Section II (RHS of the tube)
$\frac{\partial \underline{y}^I}{\partial t} = \frac{\partial}{\partial z_1}\left\{[\underline{\underline{B}}(\underline{y}^I)]^{-1}\frac{\partial \underline{y}^I}{\partial z_1}\right\}$ (8.3-10a)	$\frac{\partial \underline{y}^{II}}{\partial t} = \frac{\partial}{\partial z_2}\left\{[\underline{\underline{B}}(\underline{y}^{II})]^{-1}\frac{\partial \underline{y}^{II}}{\partial z_2}\right\}$ (8.3-11a)
$t = 0; \quad \underline{y}^I = \underline{y}^I(0)$ (8.3-10b)	$t = 0; \quad \underline{y}^{II} = \underline{y}^{II}(0)$ (8.3-11b)
$z_1 = 0; \quad \frac{\partial \underline{y}^I}{\partial z_1} = \underline{0}$ (8.3-10c)	$z_2 = 0; \quad \frac{\partial \underline{y}^{II}}{\partial z_2} = \underline{0}$ (8.3-11c)
$z_1 = L_1; \quad \underline{y}^I\big|_{z_1=L_1} = \underline{y}^{II}\big|_{z_2=L_2}$ (8.3-10d)	$z_2 = L_2; \quad \underline{y}^{II}\big|_{z_2=L_2} = \underline{y}^I\big|_{z_1=L_1}$ (8.3-11d)
$\underline{N}^I\big|_{z_1=L_1} = -\underline{N}^{II}\big|_{z_2=L_2}$ (8.3-10e)	$\underline{N}^{II}\big|_{z_2=L_2} = -\underline{N}^I\big|_{z_1=L_1}$ (8.3-11e)

Eqs. (8.3-10d), (8.3-10e), (8.3-11d) and (8.3-11e) are simply the statement of continuity of concentrations (mole fractions) and fluxes. The minus sign in eqs. (8.3-10e) and (8.3-11e) is due to the choice of coordinates z_1 and z_2 in the opposite direction, as shown in Figure 8.3-1. There we choose the origins of the two sections at the impermeable ends of their respective sections. Note that the two sets of mass balance equation for the two sections are identical in form. They represent the mass balance at any points in their respective domains.

8.3.2 The Overall Mass Balance

The overall mass balance equation of the whole tube can be obtained by integrating eq. (8.3-10a) with respect to z_1 from 0 to L_1 and eq. (8.3-11a) with respect to z_2 from 0 to L_2, and summing the results we obtain the following equation:

$$\frac{\partial}{\partial t}\left[\int_0^{L_1} \underline{y}^I(z_1,t)\,dz_1 + \int_0^{L_2} \underline{y}^{II}(z_2,t)\,dz_2\right] = \underline{0} \qquad (8.3\text{-}12)$$

in which we have used the continuity of flux equation (eq. 8.3-10e or 8.3-11e).

Next, we integrate eq. (8.3-12) with respect to time and make use of the initial conditions (eqs. 8.3-10b and 8.3-11b) to finally obtain:

$$\int_0^{L_1} \underline{y}^I(z_1,t)\,dz_1 + \int_0^{L_2} \underline{y}^{II}(z_2,t)\,dz_2 = \underline{y}^I(0)L_1 + \underline{y}^{II}(0)L_2 \qquad (8.3\text{-}13)$$

Multiplying the above equation by Ac, where A is the cross-sectional area of the Loschmidt tube and c (the total molar concentration), we obtain what is known as the overall mass balance equation:

$$c\int_0^{V_1} \underline{y}^I(z_1,t)\,dV_1 + c\int_0^{V_2} \underline{y}^{II}(z_2,t)\,dV_2 = c\left[\underline{y}^I(0)V_1 + \underline{y}^{II}(0)V_2\right] \qquad (8.3\text{-}14)$$

where V_1 and V_2 are volumes of Sections 1 and 2, respectively. The LHS is the total number of moles of any component in the two sections at any time t, while the RHS is the total moles in the system of that component before the partition between the two sections is removed.

From eq. (8.3-14), we can easily derive the final steady state mole fraction, which must be the same in both sections when $t \to \infty$, that is

$$\underline{y}^I_\infty = \underline{y}^{II}_\infty = \frac{V_1\,\underline{y}^I(0) + V_2\,\underline{y}^{II}(0)}{V_1 + V_2} \qquad (8.3\text{-}15)$$

8.3.3 Numerical Analysis

The transient analysis of eqs. (8.3-10) and (8.3-11), which are in the form of coupled partial differential equations, must be carried out numerically. The method used by the author is the orthogonal collocation method in order to convert this set of PDEs to a larger but simpler set of ODEs with respect to time, which in turn can be numerically integrated by any standard integration routine, such as the Runge-Kutta method. Readers may choose other methods to obtain numerical solutions, with methods such as finite difference to convert PDEs into ODEs. To facilitate with the orthogonal collocation analysis, it is convenient to cast the model equations in nondimensional form as suggested in Section 8.2.3.3 (eq. 8.2-62). The necessary model equations in nondimensional form as:

Section I	Section II
$\dfrac{\partial \underline{y}^I}{\partial t^*} = \dfrac{\partial}{\partial \eta_1}\left\{\left[\underline{\underline{B}}^*(\underline{y}^I)\right]^{-1}\dfrac{\partial \underline{y}^I}{\partial \eta_1}\right\}$ (8.3-16a)	$\dfrac{\partial \underline{y}^{II}}{\partial t^*} = \left(\dfrac{L_1}{L_2}\right)^2 \dfrac{\partial}{\partial \eta_2}\left\{\left[\underline{\underline{B}}^*(\underline{y}^{II})\right]^{-1}\dfrac{\partial \underline{y}^{II}}{\partial \eta_2}\right\}$ (8.3-17a)
$t^* = 0;\quad \underline{y}^I = \underline{y}^I(0)$ (8.3-16b)	$t^* = 0;\quad \underline{y}^{II} = \underline{y}^{II}(0)$ (8.3-17b)
$\eta_1 = 0;\quad \dfrac{\partial \underline{y}^I}{\partial \eta_1} = \underline{0}$ (8.3-16c)	$\eta_2 = 0;\quad \dfrac{\partial \underline{y}^{II}}{\partial \eta_2} = \underline{0}$ (8.3-17c)
$\eta_1 = 1\ \&\ \eta_2 = 1;\quad \underline{y}^I\big|_{\eta_1=1} = \underline{y}^{II}\big|_{\eta_2=1}$	(8.3-16d)
$\eta_1 = 1\ \&\ \eta_2 = 1;\quad \left[\underline{\underline{B}}^*(\underline{y}^I)\right]^{-1}\dfrac{\partial \underline{y}^I}{\partial \eta_1}\bigg|_{\eta_1=1} = -\left(\dfrac{L_1}{L_2}\right)\cdot\left[\underline{\underline{B}}^*(\underline{y}^{II})\right]^{-1}\dfrac{\partial \underline{y}^{II}}{\partial \eta_2}\bigg|_{\eta_2=1}$	(8.3-16e)

where the nondimensional variables and parameters are defined below

$$\left[\underline{\underline{B}}^*\right]^{-1} = D_T\left[\underline{\underline{B}}\right]^{-1} \qquad (8.3\text{-}18a)$$

$$\eta_1 = \dfrac{z_1}{L_1};\qquad \eta_2 = \dfrac{z_2}{L_2};\qquad t^* = \dfrac{D_T t}{L_1^2} \qquad (8.3\text{-}18b)$$

The characteristic diffusion coefficient D_T is arbitrary. We can define that as the sum of binary diffusion coefficients of (n-1) species with respect to the n-th species, that is

$$D_T = \sum_{j=1}^{n-1} D_{j,n} \qquad (8.3\text{-}18d)$$

Readers can choose other characteristic diffusion coefficient if they so wish. For example, you can choose it as the minimum binary diffusivity or the maximum binary diffusivity.

The nondimensional matrix $\underline{\underline{B}}^*$, defined as in eq.(8.3-18a), is written explicitly below:

$$\underline{\underline{B}}^* = \begin{cases} \dfrac{y_i}{(D_{i,n}/D_T)} + \sum_{\substack{k=1 \\ k \neq i}}^{n} \dfrac{y_k}{(D_{i,k}/D_T)} & \text{for } i = j \\ y_i \left(\dfrac{1}{D_{i,n}/D_T} - \dfrac{1}{D_{i,j}/D_T} \right) & \text{for } i \neq j \end{cases} \quad (8.3\text{-}19)$$

The details of the collocation analysis are given in Appendix 8.3.

Example 8.3-1: *Methane/Argon/Hydrogen diffusion in the Loschmidt tube*

To illustrate the Stefan-Maxwell approach in the analysis of the Loschmidt diffusion tube, we apply it to the experimental results obtained by Arnold and Toor (1967). The system is a ternary mixture containing methane, argon and hydrogen. The tube length is 0.40885 m and the two sections of the tube are equal in length. The operating conditions are:

$$T = 34\ °C = 307\ K$$
$$P = 101.3\ kPa$$

We denote

1 = methane
2 = argon
3 = hydrogen

The binary diffusivities of these three species at 307K and 101.3 kPa are:

$$D_{12} = 2.157 \times 10^{-5}\ m^2/sec$$
$$D_{13} = 7.716 \times 10^{-5}\ m^2/sec$$
$$D_{23} = 8.335 \times 10^{-5}\ m^2/sec$$

Before the partition separating the two sections of the tube is removed, the mole fractions of the three components are:

Section 1	Section 2
$y_1(0) = 0$	$y_1(0) = 0.515$
$y_2(0) = 0.509$	$y_2(0) = 0.485$
$y_3(0) = 0.491$	$y_3(0) = 0$

The above parameters as well as initial conditions are supplied into a MatLab code named LOSCHMI.M. The typical number of interior collocation points used in each sections of the tube is 3 or 5. Typical execution time is about 2 to 10 minutes on a 586/90 MHz personal computer. Figures 8.3-2 and 8.3-3 show the trajectories of the average mole fractions of methane and argon in the two sections. Experimental data of Arnold and Toor are shown in Figure 8.3-3 as symbols.

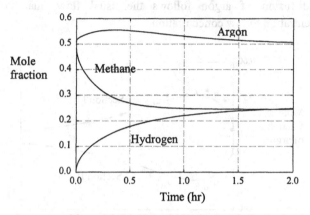

Figure 8.3-2a: Plots of mole fractions in Section I

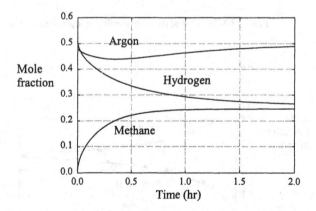

Figure 8.3-2b: Plots of mole fractions in Section II

One interesting observation is that of the diffusion of argon. Initially the argon concentration in the Section 2 of the tube is lower than that in the Section 1. When the diffusion is started we observe that argon diffuses from the Section 2 to the Section 1, <u>an uphill diffusion</u>. This uphill diffusion is sustained during the initial period of 20 minutes, after which the downhill diffusion is observed. The reason for this uphill diffusion of argon is due to the drag effect of the methane diffusion from Section 2 to the Section 1. Methane's downhill diffusion rate is in a magnitude such that it drags argon along with its direction of flow, resulting in the uphill diffusion of argon. Once the rate of diffusion of methane has decreased, the diffusion of argon follows the usual flow, that is from high concentration to low concentration.

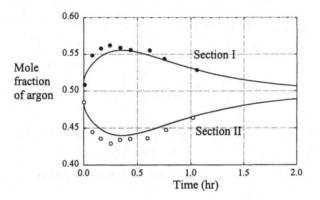

Figure 8.3-3a: Plot of the mole fraction of argon versus time

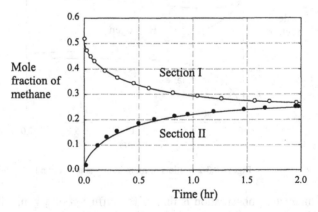

Figure 8.3-3b: Plot of the mole fraction of methane versus time

The Maxwell-Stefan analysis explains well the experimental data of Arnold and Toor (1967), justifying the importance of the Maxwell-Stefan formulation for the diffusion of multicomponent system.

8.4 Transient Diffusion of Ideal Gaseous Mixtures in Two Bulb Method

So far we have analysed the two systems using the Maxwell-Stefan approach: the steady state analysis of the Stefan tube and the transient analysis of the Loschmidt's tube, and they are conveniently used to study the diffusion characteristic of the system. Here, we consider another example which is also useful in the determination of diffusion characteristics. This system is the two bulb method, in which a small capillary tube or a bundle of capillaries is bounded by two well-mixed reservoirs as shown in Figure 7.2-3.

The constitutive Maxwell-Stefan flux equations are the same as those presented in the last section (Section 8.3) because in this case to maintain the constant pressure of the closed system the sum of all fluxes must be zero, the same requirement as that in the Loschmidt tube. The flux equations are given by eqs. (8.3-7) with the matrix $\underline{\underline{B}}$ given by eq. (8.3-6).

Setting the mass balance equation around a very thin element in the capillary between the two reservoir and making use of the constitutive flux relation (8.3-7), we have:

$$\frac{\partial \underline{y}}{\partial t} = \frac{\partial}{\partial z}\left\{\left[\underline{\underline{B}}(\underline{y})\right]^{-1} \frac{\partial \underline{y}}{\partial z}\right\} \tag{8.4-1}$$

The fluxes at two ends of the capillary tube are needed for the mass balances of the two reservoirs, and they are obtained from eq.(8.3-7) by evaluating that equation at $z = 0$ and $z = L$ (length of the capillary), respectively, that is:

$$\underline{N}\big|_{z=0} = -c\left\{\left[\underline{\underline{B}}(\underline{y})\right]^{-1} \frac{\partial \underline{y}}{\partial z}\right\}_{z=0} \quad ; \quad \underline{N}\big|_{z=L} = -c\left\{\left[\underline{\underline{B}}(\underline{y})\right]^{-1} \frac{\partial \underline{y}}{\partial z}\right\}_{z=L} \tag{8.4-2}$$

The boundary conditions for the mass balance (eq. 8.4-1) in the capillary are:

$$z = 0; \quad \underline{y} = \underline{y}_0 \tag{8.4-3a}$$

$$z = L; \quad \underline{y} = \underline{y}_L \tag{8.4-3b}$$

where $\underline{y}_0$ and $\underline{y}_L$ are the mole fractions in the left and right bulbs, respectively. These mole fractions change with time due to the mass transfer in and out of the two reservoirs. Carrying out the mass balance around the two bulbs, we get:

$$V_0 c \frac{d\underline{y}_0}{dt} = -A \, \underline{N}\big|_{z=0} \tag{8.4-4a}$$

$$V_L c \frac{d\underline{y}_L}{dt} = +A \, \underline{N}\big|_{z=L} \tag{8.4-4b}$$

where A is the cross-sectional area of the capillary, and V_0 and V_L are volumes of LHS bulb and RHS bulb, respectively.

Substituting eqs. (8.4-2) for the fluxes at two ends of the capillary into the mass balance equations for the two bulbs (eqs. 8.4-4), we get:

$$\left(\frac{V_0}{A}\right)\frac{d\underline{y}_0}{dt} = +\left[\underline{\underline{B}}(\underline{y})\right]^{-1}\frac{\partial \underline{y}}{\partial z}\bigg|_{z=0} \tag{8.4-5a}$$

$$\left(\frac{V_L}{A}\right)\frac{d\underline{y}_L}{dt} = -\left[\underline{\underline{B}}(\underline{y})\right]^{-1}\frac{\partial \underline{y}}{\partial z}\bigg|_{z=L} \tag{8.4-5b}$$

Eqs. (8.4-1), (8.4-3) and (8.4-5) completely define the behaviour of the two-bulb system after the initial state for the system is chosen. We shall assume that the partition is on the left of the capillary, that is at $z = 0$. Thus, the initial conditions of this system are:

$$t = 0; \quad \underline{y}_0 = \underline{y}_0(0); \quad \underline{y}_L = \underline{y}_L(0); \quad \underline{y} = \underline{y}_L(0) \tag{8.4-6}$$

8.4.1 The Overall Mass Balance Equation:

The overall mass balance equation can be found by integrating the mass balance for the capillary (eq. 8.4-1) with respect to z from 0 to L and combining the result with the mass balance equations for the two bulbs (eqs. 8.4-5), we get:

$$\frac{d}{dt}\left[A\int_0^L \underline{y}\,dz + V_0 \underline{y}_0 + V_L \underline{y}_L\right] = 0 \tag{8.4-7}$$

The square bracket term in the above equation is simply the total number of moles in the system, and it is invariant with respect to time as expected for a closed system.

Integrating eq.(8.4-7) with respect to time from 0 to t, and making use of the initial conditions (eq. 8.4-6) we obtain the following integral mass balance equation:

$$A \int_0^L \underline{y}\,dz + V_0 \underline{y}_0 + V_L \underline{y}_L = AL\, \underline{y}_L(0) + V_0\, \underline{y}_0(0) + V_L\, \underline{y}_L(0) \qquad (8.4\text{-}8)$$

At steady state, the mole fractions in the two bulbs and in the capillary of any component will be the same, that is:

$$\underline{y} = \underline{y}_0 = \underline{y}_L = \underline{y}(\infty) \qquad (8.4\text{-}9)$$

Substitution of eq.(8.4-9) into eq.(8.4-8) yields the following solution for the steady state mole fraction vector

$$\underline{y}(\infty) = \frac{AL\, \underline{y}_L(0) + V_0\, \underline{y}_0(0) + V_L\, \underline{y}_L(0)}{AL + V_0 + V_L} \qquad (8.4\text{-}10)$$

Usually, the volume of the capillary is much smaller than the volumes of the two bulbs. Thus, we write:

$$\underline{y}(\infty) \approx \frac{V_0\, \underline{y}_0(0) + V_L\, \underline{y}_L(0)}{V_0 + V_L} \qquad (8.4\text{-}11)$$

8.4.2 Non-Dimensionalization of the Mass Balance Equations

The system equations (eqs. 8.4-1 and 8.4-5) are coupled nonlinear partial differential equations and must be solved numerically. To facilitate with the numerical analysis, we define the following nondimensional variables

$$\eta = \frac{z}{L};\quad t^* = \frac{D_T t}{L^2} \qquad (8.4\text{-}12)$$

where the characteristic diffusion coefficient D_T is chosen as the sum of binary diffusion coefficients of (n-1) species with respect to the n-th species as given in eq. (8.3-18d). As mentioned earlier, readers can choose other characteristic diffusivity as they wish, for example the minimum binary diffusivity or the maximum binary diffusivity.

With these definitions, the nondimensional mass balance equations are:

$$\frac{\partial \underline{y}}{\partial t^*} = \frac{\partial}{\partial \eta}\left\{\left[\underline{\underline{B}}^*(\underline{y})\right]^{-1} \frac{\partial \underline{y}}{\partial \eta}\right\} \qquad (8.4\text{-}13a)$$

$$\frac{d\underline{y}_0}{dt^*} = \alpha_0 \left[\underline{\underline{B}}^*(\underline{y})\right]^{-1} \frac{\partial \underline{y}}{\partial \eta}\bigg|_{\eta=0} \quad (8.4\text{-}13b)$$

$$\frac{d\underline{y}_L}{dt^*} = -\alpha_L \left[\underline{\underline{B}}^*(\underline{y})\right]^{-1} \frac{\partial \underline{y}}{\partial \eta}\bigg|_{\eta=L} \quad (8.4\text{-}13c)$$

where

$$\alpha_0 = \frac{AL}{V_0}; \quad \alpha_L = \frac{AL}{V_L} \quad (8.4\text{-}14)$$

The boundary conditions of eq.(8.4-13a) are:

$$\eta = 0; \quad \underline{y} = \underline{y}_0 \quad (8.4\text{-}15a)$$

$$\eta = 1; \quad \underline{y} = \underline{y}_L \quad (8.4\text{-}15b)$$

The collocation analysis of this problem is given in Appendix 8.4, and the MatLab code TWOBULB.M is available for the solution of this problem.

Example 8.4-1: *Transient ternary diffusion of hydrogen, nitrogen and CO_2*

To illustrate the solution procedure of this Stefan-Maxwell approach, we apply it to the experimental results of Duncan and Toor (1962). The two well-mixed bulbs used have volumes of 7.8×10^{-5} and 7.86×10^{-5} m³, respectively, and the capillary tube has a length and a diameter of 0.086 m and 0.00208 m, respectively. The operating conditions are 35 °C and 101.3 kPa. Duncan and Toor (1962) used a ternary system of hydrogen, nitrogen and carbon dioxide in their experiment. We use the following numerical values to denote the three components used in their experiment: 1-hydrogen; 2-nitrogen and 3-carbon dioxide.

The binary diffusivities at the above operating conditions of 35 °C and 101.3 kPa are given below.

$$D_{12} = 8.33 \times 10^{-5}, D_{13} = 6.80 \times 10^{-5} \text{ m}^2/\text{sec}, D_{23} = 1.68 \times 10^{-5} \text{ m}^2/\text{sec}$$

The initial conditions for the two bulbs are given in the following table:

LHS bulb	RHS bulb
$y_1(0) = 0$	$y_1(0) = 0.50121$
$y_2(0) = 0.50086$	$y_2(0) = 0.49879$
$y_3(0) = 0.49914$	$y_3(0) = 0$

The MatLab code TWOBULB is executed and the results are shown in Figures 8.4-1 and 8.4-2 for the mole fractions of hydrogen and nitrogen.

Experimental data of Duncan and Toor (1962) are shown as symbols. The mole fraction of nitrogen in the LHS bulb is initially slightly higher than that in the RHS bulb. One would then expect to see the diffusion of nitrogen from LHS to RHS (downhill diffusion). But the experimental results as well as the computational results show the opposite, that is an uphill diffusion of nitrogen. This is due to the flow of hydrogen from the RHS to LHS bulb and its rate is such that it drags nitrogen molecules along with it, resulting in the uphill diffusion of nitrogen. Later on, the mole fraction of nitrogen in the LHS is high enough to overcome the reverse diffusion effect and hence the mole fraction of nitrogen will decrease and finally reach the steady state mole fraction.

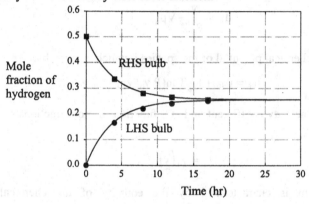

Figure 8.4-1: Mole fractions of hydrogen for LHS and RHS bulbs

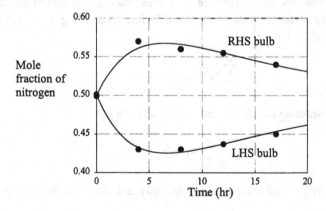

Figure 8.4-2: Mole fractions of nitrogen for LHS and RHS bulbs

8.5 Diffusion in Nonideal Fluids

8.5.1 The Driving Force for Diffusion

Nonideal fluids, such as high pressure gaseous mixture or liquid mixture, have higher density than the ideal gaseous mixtures dealt with in the previous section (Section 8.2). Thus, when dealing with such fluids, it is not possible to assume that collision is taken place by the two molecules, but three or more molecules may occur simultaneously. In this case, the Maxwell-Stefan approach still applies, but the driving force d_i is replaced by the following equation written in terms of the chemical potential:

$$d_i = \frac{y_i}{R_g T} \nabla \mu_i \qquad (8.5\text{-}1)$$

Here μ_i is the chemical potential of the species i, which is given by:

$$\mu_i = \mu_{i0} + R_g T \ln(P \gamma_i y_i) \qquad (8.5\text{-}2a)$$

where γ is the activity coefficient and is a function of mole fractions of all species, that is

$$\gamma_i = \gamma_i(\underline{y}) \qquad (8.5\text{-}2b)$$

Equilibrium is characterized by the equality of the chemical potential. Nonequilibrium is therefore induced by the gradient of the chemical potential. In the thermodynamics of irreversible processes, chemical potential gradient is the fundamentally correct driving force for diffusion. According to the Gibbs-Duhem equation (2.3-5), we have:

$$\sum_{i=1}^{n} y_i \nabla \mu_i = 0 \qquad (8.5\text{-}3)$$

therefore substituting eq. (8.5-1) into eq. (8.5-3), we get

$$\sum_{i=1}^{n} d_i = 0 \qquad (8.5\text{-}4)$$

Because of this Gibbs-Duhem equation, only n-1 Maxwell-Stefan equations are independent.

8.5.2 The Maxwell-Stefan Equation for Nonideal Fluids

The generalised Maxwell-Stefan equation for the case of nonideal fluids containing n species is given below for one dimensional problems:

$$\frac{y_i}{R_g T}\frac{d\mu_i}{dz} = \sum_{\substack{j=1\\j\neq i}}^{n} \frac{y_i N_j - y_j N_i}{cD_{ij}} \qquad (8.5\text{-}5)$$

for i = 1, 2, ..., n-1. Only (n-1) above equations are independent because of the Gibbs-Duhem restriction on the chemical potential (eq. 8.5-3). Eq.(8.5-5) is the generalized Maxwell-Stefan constitutive relation. However, such form is not useful to engineers for analysis purposes. To achieve this, we need to express the chemical potential in terms of mole fractions. This is done by using eq. (8.5-2) into the constitutive flux equation (8.5-5).

Expressing the chemical potential gradient in terms of the mole fraction gradient, we have:

$$d_i = \frac{y_i}{R_g T}\nabla\mu_i = \frac{y_i}{R_g T}\left[R_g T \cdot \nabla(\ln P + \ln\gamma_i + \ln y_i)\right] \qquad (8.5\text{-}6a)$$

For constant pressure systems, the above equation becomes:

$$d_i = \nabla y_i + y_i \nabla\left[\ln\gamma_i(\underline{y})\right] \qquad (8.5\text{-}6b)$$

Taking the total differentiation of the activity coefficient with respect to all the mole fractions, we then get:

$$d_i = \sum_{j=1}^{n} \Gamma_{ij} \nabla y_j \qquad (8.5\text{-}7)$$

where

$$\Gamma_{ij} = \delta_{ij} + y_i \frac{\partial \ln\gamma_i}{\partial y_j} \qquad (8.5\text{-}8)$$

Here δ_{ij} is the Kronecker delta function. It is equal to 1 when i = j; otherwise it takes a value of zero.

The Maxwell-Stefan constitutive equation, in terms of the mole fraction gradients, will become (by combining eqs. 8.2-7 and 8.2-5):

$$\sum_{j=1}^{n} \Gamma_{ij} \nabla y_j = \sum_{\substack{j=1 \\ j \neq i}}^{n} \frac{y_i N_j - y_j N_i}{c D_{ij}} \qquad (8.5\text{-}9)$$

for $i = 1, 2, \ldots, n\text{-}1$. We now illustrate below in Example 8.5-1 the constitutive flux equation for a non-ideal binary system.

Example 8.5-1: *Constitutive flux equation for a non-ideal binary system*

For example, for a binary system, we have the following diffusion equations for the components 1 and 2, obtained from eq. (8.5-9) for $i = 1$ and 2:

$$\left[1 + y_1 \frac{\partial \ln \gamma_1(y_1, y_2)}{\partial y_1}\right] \nabla y_1 + y_1 \frac{\partial \ln \gamma_1(y_1, y_2)}{\partial y_2} \nabla y_2 = \frac{y_1 N_2 - y_2 N_1}{c D_{12}} \qquad (8.5\text{-}10a)$$

$$y_2 \frac{\partial \ln \gamma_2(y_1, y_2)}{\partial y_1} \nabla y_1 + \left[1 + y_2 \frac{\partial \ln \gamma_2(y_1, y_2)}{\partial y_2}\right] \nabla y_2 = \frac{y_2 N_1 - y_1 N_2}{c D_{12}} \qquad (8.5\text{-}10b)$$

But for binary systems only one of the above two equations is independent, as by summing these two equations we get zero in the RHS and also zero in the LHS. The LHS equalling to zero is due to the Gibbs-Duhem relation. To illustrate this, we take the following equations for the activity coefficients (regular solution model)

$$\ln \gamma_1 = A y_2^2 \qquad (8.5\text{-}11a)$$

$$\ln \gamma_2 = A y_1^2 \qquad (8.5\text{-}11b)$$

Using these activity coefficients into the Maxwell-Stefan equations (8.5-10), we obtain:

$$\nabla y_1 + 2A y_1 y_2 \nabla y_2 = \frac{y_1 N_2 - y_2 N_1}{c D_{12}} \qquad (8.5\text{-}12a)$$

$$2A y_1 y_2 \nabla y_1 + \nabla y_2 = \frac{y_2 N_1 - y_1 N_2}{c D_{12}} \qquad (8.5\text{-}12b)$$

Indeed, we see that by adding the above two equations, we find that both sides equal to zero, indicating that the two equations are not independent.

Keeping the first equation and using the relation $y_2 = 1 - y_1$, we have the following Maxwell-Stefan equation for the binary system:

$$[1 - 2Ay_1(1-y_1)]\nabla y_1 = \frac{y_1 N_2 - y_2 N_1}{cD_{12}} \qquad (8.5\text{-}13)$$

When $A = 0$ (that is ideal system), the above equation reduces to the Maxwell-Stefan equation for ideal binary systems.

8.5.3 Special Case: Ideal Fluids

The ideal fluid is a special case of the nonideal fluid, that is when the activity coefficient is unity. For such cases, the chemical potential is:

$$\mu_i = \mu_{io} + R_g T \ln(Py_i) \qquad (8.5\text{-}14)$$

For constant pressure, we have

$$\Gamma_{ij} = \begin{cases} 1 & \text{for } i = j \\ 0 & \text{for } i \neq j \end{cases}$$

hence:

$$\frac{y_i}{R_g T} \nabla \mu_i = \nabla y_i \qquad (8.5\text{-}15)$$

we then obtain the same Stefan-Maxwell equation for ideal fluids (eq. 8.2-34), where the driving force is the gradient of the mole fraction.

8.5.4 Table of Formula of Constitutive Relations

The following table summarises the formulas for the case of bulk diffusion in nonideal systems.

Table 8.5-1 Constitutive equations for the bulk diffusion

$$\sum_{j=1}^{n} \left(\delta_{ij} + y_i \frac{\partial \ln \gamma_i}{\partial y_j} \right) \frac{dy_j}{dz} = \sum_{\substack{j=1 \\ j \neq i}}^{n} \frac{y_i N_j - y_j N_i}{cD_{ij}} \qquad \text{for } i = 1, 2, \cdots, n-1 \qquad (8.5\text{-}16a)$$

$$-\frac{P}{R_g T} \sum_{j=1}^{n} \left(\delta_{ij} + y_i \frac{\partial \ln \gamma_i}{\partial y_j} \right) \frac{dy_j}{dz} = \sum_{\substack{j=1 \\ j \neq i}}^{n} \frac{y_j N_i - y_i N_j}{D_{ij}} \qquad \text{for } i = 1, 2, \cdots, n-1 \qquad (8.5\text{-}16b)$$

These equations were obtained when the pressure is a constant. When the system's pressure and temperature are not constant, the proper equations to use are:

$$-\frac{1}{R_g T}\sum_{j=1}^{n}\left(\delta_{ij}+y_i\frac{\partial\ln\gamma_i}{\partial y_j}\right)\frac{dP_j}{dz} = \sum_{\substack{j=1\\j\neq i}}^{n}\frac{y_j N_i - y_i N_j}{D_{ij}} \quad \text{for} \quad i=1,2,\cdots,n-1 \quad (8.5\text{-}17)$$

$$-\frac{1}{R_g T}\frac{dP_i}{dz} = \sum_{\substack{j=1\\j\neq i}}^{n}\frac{y_j N_i - y_i N_j}{D_{ij}} \quad \text{for} \quad i=1,2,\cdots,n-1 \quad (8.5\text{-}18)$$

for nonideal and ideal systems, respectively. Haynes (1986) used the momentum transfer approach to derive the above two formulas. In this approach the appropriate driving force for diffusion is the gradient of the partial pressure as we have argued in Section 8.2. The driving force of concentration gradient is only applicable to isothermal systems, while the driving force of mole fraction is applicable to systems having constant pressure and temperature. When one incorrectly uses the mole fraction gradient as the driving force for a nonisobaric system, the sum of mole fractions are not equal to unity. We shall discuss in more detail about non-isobaric systems in Section 8.8.

The reason why only n-1 equations in the Stefan-Maxwell equation are independent is not surprising since these equations describe only momentum exchange between pairs of species, and they lack the necessary boundary conditions defining the rate of momentum transfer to the capillary walls (Burghdardt, 1986).

Since there are only n-1 flux equations, the other condition needed to solve the problem is tied to the specific problem, such as the Graham's law of diffusion in an open system or the flux of the n-th species is zero for the case of Stefan tube. The latter is the case that we shall deal with in the next example.

Example 8.5-2: *Nonideal binary systems in a Stefan tube*

Let us now consider the importance of the nonideality in the diffusion of a binary system in a Stefan tube. The relevant Maxwell-Stefan equation is given in eq. (8.5-10), and if the activity coefficient is given by eq. (8.5-11) the resulting constitutive flux equation is given by eq. (8.5-13).

If the flowing gas (component 2) across the tube is not soluble in liquid 1, we have $N_2 = 0$ and eq. (8.5-13) then can be rewritten:

$$N_1 = -c \frac{D_{12}[1-2Ay_1(1-y_1)]}{(1-y_1)} \nabla y_1 \tag{8.5-19}$$

from which the apparent diffusivity takes the following concentration dependence form as:

$$D_{app} = D_{12}\frac{[1-2Ay_1(1-y_1)]}{(1-y_1)} \tag{8.5-20}$$

For $A < 2$, the apparent diffusivity is positive for all range of concentration. For $A = 2$, the diffusivity is zero at the mole fraction of 0.5. What this means is that the slope of concentration gradient is infinite at this point. Figure 8.5-1 shows the plot of $y_1(1-y_1)$ and $\frac{1}{2A}$ versus y_1.

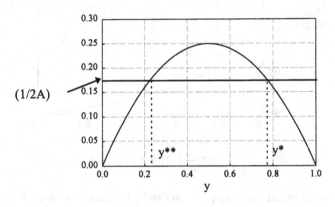

Figure 8.5-1: Plot of $y(1-y)$ versus y

When $A > 2$, we see that the line $\frac{1}{2A}$ intersects the curve $y_1(1-y_1)$ at two critical mole fractions y^* and y^{**}, given by:

$$y^* = \frac{1}{2} + \left(\frac{1}{4} - \frac{1}{2A}\right)^{\frac{1}{2}} \tag{8.5-21a}$$

$$y^{**} = \frac{1}{2} - \left(\frac{1}{4} - \frac{1}{2A}\right)^{\frac{1}{2}} \tag{8.5-21b}$$

At these two concentrations, the apparent diffusivity is zero, and for any concentrations between y^{**} and y^*, the apparent diffusivity is negative

suggesting the uphill diffusion! For concentrations either greater than y*
or less than y**, the apparent diffusivity is positive.

Let us take the case whereby the mole fraction just above the liquid surface is y_{10} greater than y*. The mole fraction will decrease from y_{10} to y* at the position z*, at which the slope is infinite due to the zero apparent diffusivity at y*. At the position $z^{*(+)}$ the mole fraction jumps to a lower value y** at which the concentration gradient is also infinite due to the zero apparent diffusivity also at that point. Figure 8.5-2 shows the typical concentration profile.

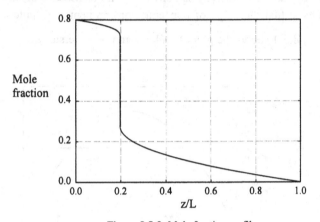

Figure 8.5-2: Mole fraction profile

For constant flux, eq. (8.5-19) can be integrated between $z = 0$ and z^* to yield the following equation:

$$N_1 z^* = -c\, D_{12} \int_{z=0}^{z=z^*} \frac{\left[1 - 2Ay_1(1-y_1)\right]}{(1-y_1)} dy_1 = -c\, D_{12} \int_{y_{10}}^{y^*} \frac{\left[1 - 2Ay_1(1-y_1)\right]}{(1-y_1)} dy_1$$

that is

$$N_1 z^* = cD_{12} \left[\ln\left(\frac{1-y^*}{1-y_{10}}\right) - A\left(y_{10}^2 - y^{*2}\right) \right] \qquad (8.5\text{-}22)$$

Similarly, we also integrate eq. (8.5-19) from $z = z^*$ to $z = L$ and get

$$(L - z^*)N_1 = cD_{12} \left[\ln\left(\frac{1-y_{1L}}{1-y^{**}}\right) - A\left(y^{**2} - y_{1L}^2\right) \right] \qquad (8.5\text{-}23)$$

The two equations (8.5-22 and 8.5-23) contain two unknowns N_1 and z^*. The evaporation flux N_1 can be obtained by summing these two equations, and we get:

$$N_1 = \frac{cD_{12}}{L}\left[\ln\left(\frac{1-y^*}{1-y_{10}} \cdot \frac{1-y_{1L}}{1-y^{**}}\right) - A\left(y_{10}^2 - y^{*2} + y^{**2} - y_{1L}^2\right)\right] \quad (8.5\text{-}24)$$

The position z^* at which the phase transition takes place is obtained by combining eqs. (8.5-22) and (8.5-24).

$$z^* = L \cdot \frac{\left[\ln\left(\frac{1-y^*}{1-y_{10}}\right) - A\left(y_{10}^2 - y^{*2}\right)\right]}{\left[\ln\left(\frac{1-y^*}{1-y_{10}} \cdot \frac{1-y_{1L}}{1-y^{**}}\right) - A\left(y_{10}^2 - y^{*2} + y^{**2} - y_{1L}^2\right)\right]} \quad (8.5\text{-}25)$$

To obtain the concentration profile for distance between 0 and z^*, we integrate eq. (8.5-19) from 0 to z, that is:

$$N_1 z = cD_{12}\left[\ln\left(\frac{1-y_1}{1-y_{10}}\right) - A\left(y_{10}^2 - y_1\right)\right] \quad (8.5\text{-}26)$$

where N_1 is given in eq. (8.5-24). Upon substitution eq. (8.5-24) into the above equation, we have the following concentration profile:

$$\frac{\ln\left(\frac{1-y_1}{1-y_{10}}\right) - A\left(y_{10}^2 - y_1^2\right)}{\ln\left(\frac{1-y^*}{1-y_{10}} \cdot \frac{1-y_{1L}}{1-y^{**}}\right) - A\left(y_{10}^2 - y^{*2} + y^{**2} - y_{1L}^2\right)} = \frac{z}{L} \quad (8.5\text{-}27)$$

for $0 \leq z < z^*$.

Similarly, to obtain the concentration profile for distance between z^* and L, we integrate eq. (8.5-19) from L to z and obtain the following result:

$$\frac{\ln\left(\frac{1-y_1}{1-y_{10}}\right) - A\left(y_1^2 - y_{1L}^2\right)}{\ln\left(\frac{1-y^*}{1-y_{10}} \cdot \frac{1-y_{1L}}{1-y^{**}}\right) - A\left(y_{10}^2 - y^{*2} + y^{**2} - y_{1L}^2\right)} = 1 - \frac{z}{L} \quad (8.5\text{-}28)$$

for $z^* < z \leq L$.

Numerical example:
We take the case of A= 2.5, the two critical mole fractions (from eqs. 8.5-21) are

$$y^* = 0.7236 \quad (8.5\text{-}29a)$$

$$y^{**} = 0.2764 \quad (8.5\text{-}29b)$$

Consider the Stefan tube having $y_{10} = 0.8$ and $y_{1L} = 0$. We compute the flux from eq. (8.5-24) as:

$$\frac{N_1 L}{cD_{12}} = \ln\left(\frac{1-0.7236}{1-0.8} \cdot \frac{1-0}{1-0.2764}\right) - 2.5(0.8^2 - 0.7236^2 + 0.2764^2 - 0^2) \quad (8.5\text{-}30)$$

$$\underline{\frac{N_1 L}{cD_{12}} = 0.165}$$

The position z^* at which the phase transition exists is (eq. 8.5-25)

$$\frac{z^*}{L} = \frac{\ln\left(\frac{1-0.7236}{1-0.8}\right) - 2.5(0.8^2 - 0.7236^2)}{0.165} \quad (8.5\text{-}31)$$

$$\underline{\frac{z^*}{L} = 0.197}$$

The concentration profile for distance between 0 and z^* is given by eq. (8.5-27), that is for this case A = 2.5, it is

$$\ln[5(1-y)] - \frac{5}{2}(0.64 - y^2) = 0.165\left(\frac{z}{L}\right) \quad (8.5\text{-}32)$$

and the concentration profile for distance between z^* and L is given by eq. (8.5-28)

$$\ln\left(\frac{1}{1-y}\right) - \frac{5}{2}y^2 = 0.165\left(1 - \frac{z}{L}\right) \quad (8.5\text{-}33)$$

The concentration profiles are given in Figure 8.5-2, where we see the abrupt change in the mole fraction at the critical position z^*.

8.6 Maxwell-Stefan Formulation for Bulk-Knudsen Diffusion in a Capillary

The last sections (Section 8.2 to 8.5) demonstrates the Maxwell-Stefan analysis of the situation when the bulk diffusion is the only diffusion mechanism, that is when the mean free path of the molecules is much shorter than the pore dimension.

However, when the mean free path is much greater than the pore dimension which is usually the case for most practical solids, the molecules will collide with the wall more frequently than they do between themselves. This is the basis of the Knudsen diffusion mechanism. When the mean free path is comparable to the pore dimension, the overall transport of molecules is due to the reflection of molecules from the wall as well as to the collisions between the diffusing molecules. This is the case where the bulk diffusion and the Knudsen diffusion will occur together. To combine these two modes of transport in a systematic way, let us adopt a dusty gas model proposed by Mason and co-workers in the late 60. In this model, the porous solid is modelled as a collection of stationary giant molecules. If we have n diffusing species in the gas phase, the solid is treated as the (n+1)-th species, uniformly distributed in space with a zero velocity. Now we apply the Maxwell-Stefan equation for the species i by balancing the force acting on a mole of species i with other species including the giant solid object, that is

$$-\frac{d\mu_i}{dz} = R_g T y_1 \frac{(v_i - v_1)}{D_{i1}} + R_g T y_2 \frac{(v_i - v_2)}{D_{i2}} + \cdots + R_g T y_n \frac{(v_i - v_n)}{D_{in}} + R_g T y_{n+1} \frac{(v_i - v_{n+1})}{D_{i,n+1}} \quad (8.6\text{-}1)$$

But the giant molecule (that is solid) is assumed stationary ($v_{n+1} = 0$), the above equation then becomes:

$$-\frac{d\mu_i}{dz} = R_g T y_1 \frac{(v_i - v_1)}{D_{i1}} + R_g T y_2 \frac{(v_i - v_2)}{D_{i2}} + \cdots + R_g T y_N \frac{(v_i - v_n)}{D_{in}} + R_g T \frac{v_i}{D_{K,i}} \quad (8.6\text{-}2a)$$

where $D_{K,i}$ is the Knudsen diffusivity defined as follows:

$$D_{K,i} = \frac{D_{i,n+1}}{y_{n+1}} \quad (8.6\text{-}2b)$$

Since the concentration distribution of the solid molecules is assumed uniform, the above Knudsen diffusivity is a constant across the domain of interest. As defined in eq. (8.6-2b), the Knudsen diffusivity does not show any explicit dependence on the solid properties as well as the diffusing species properties. However, as learnt in Chapter 7, we have shown that the Knudsen diffusivity takes the following form for a cylindrical pore of diameter d:

$$D_{K,i} = \frac{d}{3}\sqrt{\frac{8R_g T}{\pi M_i}} = \frac{2r}{3}\sqrt{\frac{8R_g T}{\pi M_i}} \quad (8.6\text{-}3a)$$

while for a porous solid with arbitrary pore cross section, it has the form

$$D_{K,i} = \frac{4K_0}{3}\sqrt{\frac{8R_gT}{\pi M_i}} \tag{8.6-3b}$$

where K_0 is the structural parameter for Knudsen diffusion.

The constitutive flux equation written as in eq. (8.6-2) is not useful for mass transfer calculation. Written in terms of the fluxes ($N_i = c_iv_i$), the desired constitutive flux equations are:

$$-\frac{y_i}{R_gT}\frac{d\mu_i}{dz} = \sum_{\substack{j=1 \\ j\neq i}}^{n} \frac{y_jN_i - y_iN_j}{cD_{ij}} + \frac{N_i}{cD_{i,K}} \tag{8.6-4}$$

for $i = 1, 2, ..., n$. For ideal gas system, where the chemical potential is related to the gas phase partial pressure

$$\mu_i = \mu_{0,i} + R_gT \ln p_i \tag{8.6-5}$$

the above equation can be written in terms of the partial pressure gradients as follows:

$$-\frac{1}{R_gT}\frac{dp_i}{dz} = \sum_{\substack{j=1 \\ j\neq i}}^{n} \frac{y_jN_i - y_iN_j}{D_{ij}} + \frac{N_i}{D_{K,i}} \tag{8.6-6}$$

for $i = 1, 2, ..., n$.

It is important to note that n of these equations are linearly independent and can be applied to determine the flux $\underline{N}$ uniquely, although in the case of bulk diffusion only (n-1) equations are linearly independent. This is due to the fact that the above equation contain the term N/D_K which characterises momentum transfer to the wall as a boundary condition (Burghardt, 1986).

8.6.1 Non-Ideal Systems:

For nonideal gas systems, the chemical potential of the component "i" is given by:

$$\mu_i = \mu_{i0} + R_gT \ln(\gamma_i p_i) \tag{8.6-7}$$

where the activity coefficient is a function of mole fractions of all species as:

$$\gamma_i = \gamma_i(\underline{y}) \tag{8.6-8}$$

Substitution of eq.(8.6-7) into eq.(8.6-4), we get:

$$-y_i \frac{d\ln(\gamma_i p_i)}{dz} = \sum_{\substack{j=1 \\ j \neq i}}^{n} \frac{y_j N_i - y_i N_j}{c D_{ij}} + \frac{N_i}{c D_{i,K}} \qquad (8.6\text{-}9)$$

or written explicitly in terms of the mole fraction gradients

$$-\frac{dy_i}{dz} - y_i \sum_{j=1}^{n} \frac{d\ln\gamma_i}{dy_j} \frac{dy_j}{dz} = \sum_{\substack{j=1 \\ j \neq i}}^{n} \frac{y_j N_i - y_i N_j}{c D_{ij}} + \frac{N_i}{c D_{i,K}} \quad \text{for } i = 1, 2, ..., n \quad (8.6\text{-}10)$$

If we define the following thermodynamic correction matrix $\underline{\underline{\Gamma}}$ as follows:

$$\Gamma_{ij} = \delta_{ij} + y_i \frac{\partial \ln\gamma_i}{\partial y_j} \qquad (8.6\text{-}11)$$

for i, j = 1, 2, ..., n, where δ_{ij} is the Kronecker delta function, eq.(8.6-10) will become:

$$-\sum_{j=1}^{n} \Gamma_{ij}(\underline{y}) \frac{dy_j}{dz} = \sum_{\substack{j=1 \\ j \neq i}}^{n} \frac{y_j N_i - y_i N_j}{c D_{ij}} + \frac{N_i}{c D_{i,K}} \qquad (8.6\text{-}12)$$

for i = 1, 2, ..., n. Eq.(8.6-12) is the general Stefan-Maxwell constitutive equation for a <u>constant pressure system</u> where the bulk diffusion and the Knudsen diffusion are simultaneously operating. The nonisobaric case will be discussed in Section 8.8.

Example 8.6-1: *Non-ideal binary systems*

Now let us apply the Stefan-Maxwell equation (8.6-12) to a binary system with the following expressions to describe the activity coefficients

$$\ln\gamma_1 = Ay_2^2, \qquad \ln\gamma_2 = Ay_1^2 \qquad (8.6\text{-}13)$$

With these forms, the Stefan-Maxwell equations of eq.(8.6-12) will become:

$$-\frac{dy_1}{dz} - 2Ay_1y_2 \frac{dy_2}{dz} = \frac{y_2 N_1 - y_1 N_2}{cD_{12}} + \frac{N_1}{cD_{K,1}} \qquad (8.6\text{-}14)$$

$$-\frac{dy_2}{dz} - 2Ay_1y_2 \frac{dy_1}{dz} = \frac{y_1 N_2 - y_2 N_1}{cD_{12}} + \frac{N_2}{cD_{K,2}} \qquad (8.6\text{-}15)$$

474 Kinetics

Summing these two equations gives (note that $y_1 + y_2 = 1$):

$$N_2 = -\sqrt{\frac{M_1}{M_2}} \, N_1 \qquad (8.6\text{-}16)$$

which is the usual looking Graham law of diffusion for a binary system.

Substituting the Graham law of diffusion into eq.(8.6-14), we get the final form for the constitutive flux equation for the component 1:

$$N_1 = -\frac{c\left[1 - 2Ay_1(1-y_1)\right]}{\left[\dfrac{1}{D_{K,1}} + \dfrac{1-\sigma_{12}y_1}{D_{12}}\right]} \frac{dy_1}{dz} \qquad (8.6\text{-}17a)$$

where

$$\sigma_{12} = 1 - \sqrt{\frac{M_1}{M_2}} \qquad (8.6\text{-}17b)$$

Knowing the flux for the component 1, the flux of the component 2 is calculated from the Graham law of diffusion equation (8.6-16).

8.6.2 Formulas for Bulk and Knudsen Diffusion case

The following table summarises the formulas for the combined bulk-Knudsen diffusion.

Table 8.6-1 Equation formulas for the case of bulk-Knudsen diffusion

$-\sum\limits_{j=1}^{n}\left(\delta_{ij} + y_i \dfrac{\partial \ln \gamma_i}{\partial y_j}\right)\dfrac{dy_j}{dz} = \sum\limits_{\substack{j=1\\ j\neq i}}^{n} \dfrac{y_j N_i - y_i N_j}{cD_{ij}} + \dfrac{N_i}{cD_{i,K}}$ for $i = 1,2,\cdots,n$	(8.6-18a)
IDEAL SYSTEMS: $-\dfrac{dy_i}{dz} = \sum\limits_{\substack{j=1\\ j\neq i}}^{n} \dfrac{y_j N_i - y_i N_j}{cD_{ij}} + \dfrac{N_i}{cD_{i,K}}$ for $i = 1,2,\cdots,n$	(8.6-18b)
$-\dfrac{P}{R_g T}\sum\limits_{j=1}^{n}\left(\delta_{ij} + y_i \dfrac{\partial \ln \gamma_i}{\partial y_j}\right)\dfrac{dy_j}{dz} = \sum\limits_{\substack{j=1\\ j\neq i}}^{n} \dfrac{y_j N_i - y_i N_j}{D_{ij}} + \dfrac{N_i}{D_{i,K}}$ for $i = 1,2,\cdots,n$	(8.6-18c)
IDEAL SYSTEMS: $-\dfrac{P}{R_g T}\dfrac{dy_i}{dz} = \sum\limits_{\substack{j=1\\ j\neq i}}^{n} \dfrac{y_j N_i - y_i N_j}{D_{ij}} + \dfrac{N_i}{D_{i,K}}$ for $i = 1,2,\cdots,n$	(8.6-18d)

As discussed in Section 8.5, the correct driving force for mass transfer in nonisothermal, non-isobaric conditions is the partial pressure gradient. Written in terms of these gradients (Appendix 8.5) the constitutive Maxwell-Stefan flux equations are:

$$-\frac{1}{R_g T}\sum_{j=1}^{n}\left(\delta_{ij}+y_i\frac{\partial \ln\gamma_i}{\partial y_j}\right)\frac{dP_j}{dz}=\sum_{\substack{j=1\\j\neq i}}^{n}\frac{y_j N_i - y_i N_j}{D_{ij}}+\frac{N_i}{D_{i,K}} \quad \text{for} \quad i=1,2,\cdots,n \quad (8.6\text{-}19a)$$

and

$$-\frac{1}{R_g T}\frac{dP_i}{dz}=\sum_{\substack{j=1\\j\neq i}}^{n}\frac{y_j N_i - y_i N_j}{D_{ij}}+\frac{N_i}{D_{i,K}} \quad \text{for} \quad i=1,2,\cdots,n \quad (8.6\text{-}19b)$$

for nonideal and ideal systems, respectively. For constant pressure conditions, the above equations reduce to eqs. (8.6-18) tabulated in Table 8.6-1.

Example 8.6-2: *Binary systems in a constant pressure capillary*

We have presented the necessary equation to relate flux and mole fraction gradient for a multicomponent system (eqs. 8.6-18) when both molecular diffusion and Knudsen diffusion are operating. Let us now treat a special case of binary systems. For such a case, the Stefan-Maxwell equations are:

$$-\frac{dc_1}{dz}=\frac{y_2 N_1 - y_1 N_2}{D_{12}}+\frac{N_1}{D_{K,1}} \quad (8.6\text{-}20a)$$

$$-\frac{dc_2}{dz}=\frac{y_1 N_2 - y_2 N_1}{D_{12}}+\frac{N_2}{D_{K,2}} \quad (8.6\text{-}20b)$$

Adding the above equations, we get

$$-\frac{dc_1}{dz}-\frac{dc_2}{dz}=\frac{N_1}{D_{K,1}}+\frac{N_2}{D_{K,2}} \quad (8.6\text{-}21)$$

For constant pressure, we have

$$\frac{dc_1}{dz}+\frac{dc_2}{dz}=0$$

and since the Knudsen diffusivity is inversely proportional to the square root of molecular weight (eq. 8.6-3), eq. (8.6-21) is reduced to:

$$N_1\sqrt{M_1} + N_2\sqrt{M_2} = 0 \qquad (8.6\text{-}22)$$

which is the usual Graham law of diffusion for a binary system.

By carrying out the mass balance for the component 1 over a thin shell in the capillary we have:

$$\frac{dN_1}{dz} = 0 \qquad (8.6\text{-}23)$$

By using the Grahams law equation (8.6-22) into the Stefan-Maxwell equation (8.6-20a), we obtain the following equation expressing the flux in terms of concentration gradient for the component 1:

$$N_1 = -\frac{c}{\left(\dfrac{1-\sigma_{12}y_1}{D_{12}} + \dfrac{1}{D_{K,1}}\right)} \frac{dy_1}{dz} \qquad (8.6\text{-}24)$$

where σ_{12} is defined as follows:

$$\sigma_{12} = 1 - \sqrt{\frac{M_1}{M_2}} \qquad (8.6\text{-}25)$$

Substitution of this equation into the mass balance equation yields $N_1 = A$ (i.e. constant flux). Integrating this equation one more time using the following two boundary conditions at two ends of the capillary:

$$\begin{aligned} z &= 0; & y_1 &= y_{10} \\ z &= L; & y_1 &= y_{1L} \end{aligned} \qquad (8.6\text{-}26)$$

we obtain the following solution for the flux of the component 1

$$N_1 = \frac{cD_{12}}{L\sigma_{12}} \ln\left(\frac{1-\sigma_{12}y_{1L} + D_{12}/D_{K,1}}{1-\sigma_{12}y_{10} + D_{12}/D_{K,1}}\right) \qquad (8.6\text{-}27a)$$

This is the steady flux passing through a capillary tube at constant pressure. This flux depends on pressure as well as the pore size. We note that D_{12} is inversely proportional to pressure while $D_{K,1}$ is proportional to pore radius. With this, we can write eq.(8.6-27a) as follows:

$$N_1 = \frac{P_0 D_{12}^0}{R_g T L \sigma_{12}} \ln\left[\frac{1-\sigma_{12}y_{1L} + \left(D_{12}^0/D_{K,1}^0\right)(P_0/P)/(r/r_0)}{1-\sigma_{12}y_{10} + \left(D_{12}^0/D_{K,1}^0\right)(P_0/P)/(r/r_0)}\right] \qquad (8.6\text{-}27b)$$

where D_{12}^0 is the binary diffusivity at some reference pressure P_0, and $D_{K,1}^0$ is the Knudsen diffusivity at some reference pore radius r_0. The following figure (8.6-1) shows the variation of the flux versus the pressure or pore radius for three values of $D_{12}^0 / D_{K,1}^0$.

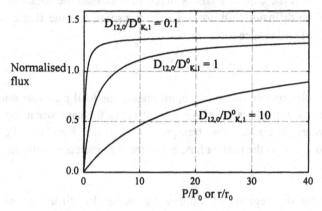

Figure 8.6-1: Plot of the normalised flux versus P/P_0 or r/r_0

Here we note that when the pressure is low or the pore radius is small, the flux increases with an increase in either pressure or radius. This is the Knudsen diffusion control regime, in which the higher is the pressure, the more collision by the molecules with the wall occurs and hence the higher flux. However, when the pressure is high or the pore radius is large, the normalised flux is independent of the pressure or the pore radius. This is so because we are in the bulk diffusion regime, and in this regime the pore radius does not have any influence on the transport of molecules. Also in this regime, an increase in pressure results in an increase in the total molar density and a decrease in the diffusivity. These two effects exactly balance each other, resulting in a constant flux with pressure.

If the capillary diameter is fairly large, that is the mean free path is smaller than the capillary size, we would expect the molecular diffusion will dominate the transport, that is $D_{12}^0 / D_{K,1}^0$ approaches zero and the flux of the component 1 in eq.(8.6-27a) will become:

$$N_1 = \frac{cD_{12}}{L\sigma_{12}} \ln\left(\frac{1-\sigma_{12}y_{1L}}{1-\sigma_{12}y_{10}}\right) \qquad (8.6\text{-}28)$$

478 Kinetics

The total concentration is proportional to the total pressure, while the molecular diffusivity is inversely proportional to the pressure; therefore the flux will be **independent of pressure** in the case of molecular diffusion as long as the mole fractions at two ends points of the capillary remain constant (Scott and Dullien, 1962; Rothfield, 1963).

When the capillary size is much smaller than the mean free path, the Knudsen diffusion will dominate the transport, and the flux equation (eq. 8.6-27a) will reduce to:

$$N_1 = \frac{cD_{K,1}}{L}(y_{10} - y_{1L}) \qquad (8.6\text{-}29)$$

The molar concentration is proportional to the total pressure; thus, the flux increases with pressure which is contrast to the case when the molecular diffusion dominates the transport. Also the Knudsen diffusivity is proportional to the pore radius, hence the flux increases with the size of the pore.

To delineate the regions of validity for molecular diffusion and Knudsen diffusion, we need to compare the mean free path (λ) to the pore size (r). The following criteria delineate those regions:

$$\begin{aligned}\lambda/2r &> 10 &&: \text{Knudsen mechanism} \\ 0.01 < \lambda/2r &< 10 &&: \text{Transition} \\ \lambda/2r &< 0.01 &&: \text{Continuum diffusion}\end{aligned} \qquad (8.6\text{-}30)$$

Example 8.6-3: *Steady state diffusion of helium-nitrogen in a capillary*

We take the following example of a binary system of helium and nitrogen diffusing in a capillary of 1 cm in length. Pure helium is at one end of the capillary while pure nitrogen is at the other end. The mean pore size (radius) of the capillary is 1 micron.

We let 1 to denote helium and 2 to denote nitrogen. The binary diffusivity of helium-nitrogen system at 25 °C and 1 atm is:

$$D_{12} = 0.769 \text{cm}^2/\text{sec} \quad @ \quad 25°C \ \& \ 1 \text{ atm}$$

From the given pore size of the capillary (1 micron), the Knudsen diffusivity is calculated from the formula (Table 7.4-2):

$$D_{K,1} = 9700 \times (1 \times 10^{-4} \text{cm})\sqrt{\frac{298K}{4\text{g/mole}}} = 8.372 \text{cm}^2/\text{sec} \qquad (8.6\text{-}31)$$

Note that the Knudsen diffusivity is about one order of magnitude higher than that of the molecular diffusivity at 25 °C and 1 atm. Therefore, one would expect that molecular diffusion will control the diffusion flux, and the proper equation to calculate the flux is eq.(8.6-28).

We have the parameter σ_{12} in eq.(8.6-25) as:

$$\sigma_{12} = 1 - \sqrt{\frac{4}{28}} = 0.622 \tag{8.6-32}$$

Let us first study the situation where the total pressure is 1 atm.

1 atm condition:

For a pressure of 1 atm and the temperature of 25°C, the total molar concentration is:

$$c = \frac{P}{R_g T} = \frac{1 \text{ atm}}{(82.057 \text{atm} - \text{cc/mole} - \text{K})(298\text{K})} = 4.0895 \times 10^{-5} \text{cm}^3/\text{sec}$$

We calculate the flux of helium from eq.(8.6-27a):

$$N_1 = 4.264 \times 10^{-5} \frac{\text{mole}}{\text{cm}^2 - \text{sec}} \tag{8.6-33}$$

This flux is equivalent to 1.043 cc of gas at 25 °C and 1 atm per cm^2 of capillary per second.

Knowing the flux of helium, the flux of nitrogen is calculated from the Graham law of diffusion (eq. 8.6-22):

$$N_2 = -N_1 \sqrt{\frac{M_1}{M_2}} = -4.264 \times 10^{-5} \sqrt{\frac{4}{28}} = -1.612 \times 10^{-5} \frac{\text{moles}}{\text{cm}^2 - \text{sec}} \tag{8.6-34}$$

A negative sign implies that nitrogen is flowing in the opposite direction to that of helium. Therefore, the total flux will be:

$$N_T = N_1 + N_2 = (4.264 - 1.612) \times 10^{-5} = 2.652 \times 10^{-5} \frac{\text{moles}}{\text{cm}^2 - \text{sec}} \tag{8.6-35}$$

which is positive, indicating that the total flow is in the same direction as that of helium.

0.1 atm condition:

Let us now consider the situation where the total pressure is reduced by a factor of 10 from the previous example, that is a new total pressure of 0.1 atm. At this pressure, the molecular diffusivity is ten times the binary diffusivity in the previous case because the binary diffusivity is inversely proportional to the total pressure, that is:

$$D_{12}(\text{@0.1 atm}) = D_{12}(\text{@1 atm})\left(\frac{1\text{ atm}}{0.1\text{ atm}}\right) = 0.769 \times 10 = 7.69 \text{cm}^2/\text{sec}$$

This binary diffusivity is now comparable to the Knudsen diffusivity (which is 8.372 cm²/sec in eq. 8.6-31). Therefore, we expect both of these diffusion mechanisms are contributing to the overall diffusion flux. For a pressure of 0.1 atm, the total molar concentration is:

$$c = \frac{P}{R_g T} = \frac{0.1 \text{atm}}{(82.057 \text{atm}-\text{cc}/\text{mole}-\text{K})(298\text{K})} = 4.0895 \times 10^{-6} \text{cm}^2/\text{sec}$$

We therefore can calculate the flux of helium as:

$$N_1 = \frac{cD_{12}}{L\sigma_{12}} \ln\left(\frac{1-\sigma_{12}y_{1L}+\frac{D_{12}}{D_{K,1}}}{1-\sigma_{12}y_{10}+\frac{D_{12}}{D_{K,1}}}\right) = N_1 = 1.98 \times 10^{-5} \frac{\text{mole}}{\text{cm}^2-\text{sec}}$$

This flux is equivalent to 4.8 cc of gas at 25 °C and 0.1 atm per cm² per second.

Knowing the flux of helium, the flux of nitrogen is calculated from the Graham law of diffusion

$$N_2 = -N_1\sqrt{\frac{M_1}{M_2}} = -1.98 \times 10^{-5}\sqrt{\frac{4}{28}} = -7.48 \times 10^{-6} \frac{\text{moles}}{\text{cm}^2-\text{sec}}$$

A negative sign implies that nitrogen is flowing in the opposite direction to that of helium. Therefore, the total flux will be

$$N_T = N_1 + N_2 = (1.98 - 0.748) \times 10^{-5} = 1.23 \times 10^{-5} \frac{\text{moles}}{\text{cm}^2-\text{sec}}$$

which is equivalent to 0.3 cc/cm²-sec at 25 °C and 1 atm, or 3 cc/cm²-sec at 25 °C and 0.1 atm.

Example 8.6-4: *Steady state ternary system in an open capillary*

Ternary systems were first dealt with by Cunningham and Geankoplis (1968) and by Remick and Geankoplis (1970). Cunningham and Geankoplis (1968) studied open systems as well as closed systems. In the closed systems where the pressure is constant, equimolar diffusion occurs. In the open systems, however, there are sources and sinks of infinite extent, which is generally the case for diffusion studies in porous solids, equimolar diffusion does not occur, but rather the fluxes will follow the Graham law.

For open systems with three diffusing species, the two independent flux equations are:

$$-\frac{dc_1}{dz} = \frac{y_2 N_1 - y_1 N_2}{D_{12}} + \frac{y_3 N_1 - y_1 N_3}{D_{13}} + \frac{N_1}{D_{K,1}}$$

$$-\frac{dc_2}{dz} = \frac{y_1 N_2 - y_2 N_1}{D_{12}} + \frac{y_3 N_2 - y_2 N_3}{D_{23}} + \frac{N_2}{D_{K,2}}$$

(8.6-36)

and the Graham law of diffusion for a ternary system is:

$$N_1 \sqrt{M_1} + N_2 \sqrt{M_2} + N_3 \sqrt{M_3} = 0 \qquad (8.6\text{-}37)$$

Note that $y_1 + y_2 + y_3 = 1$. At steady state, these three fluxes are constant, and Cunningham and Geankoplis (1968) solved this set using the boundary conditions:

$$z = 0; \quad y_j = y_{j0} \qquad \text{and} \qquad z = L; \quad y_j = y_{jL} \qquad (8.6\text{-}38)$$

They obtained solutions, but to find the flux they had to use the trial and error method, because the concentration profiles are written in terms of the flux which is not known a-priori. Differentiate the concentration profiles and evaluate the result at either end of the domain, they obtained algebraic equations written in terms of N_1 and N_2, from which they can be solved numerically. In the next section where we present a multicomponent system containing n species in a capillary, we will obtain the solution by using the vector-matrix method, which is more compact and elegant than the trial-and-error method of Cunningham and Geankoplis.

Remick and Geankoplis (1970, 1974) tested their solutions with experimental data of helium, neon and argon diffusion in an open system. These components were chosen such that they differ in molecular weights and there are no surface flow. The capillary radius is chosen such that the Knudsen diffusivity is the same order as the molecular diffusivity at 25 °C and 1 atm. Osmotic diffusion (i.e. diffusion occuring even if the concentration gradient is zero) was observed for argon in the transition and molecular regions, and osmotic diffusion does not occur in the Knudsen region. More details of their experimental set up are given in their papers.

8.6.3 Steady State Multicomponent System at Constant Pressure Conditions

We have shown the Stefan-Maxwell approach in solving for combined bulk and Knudsen diffusion in binary and ternary systems. Now we would like to present the analysis of a multicomponent system, and will show that the analysis can be elegantly presented in the form of vector-matrix format.

The neccessary constitutive equation relating the fluxes and mole fraction gradients is the following so-called modified Stefan-Maxwell equation (from eq. 8.6-6):

$$-\frac{dy_i}{dz} = \sum_{\substack{j=1 \\ j \neq i}}^{n} \frac{y_j N_i - y_i N_j}{cD_{ij}} + \frac{N_i}{cD_{K,i}} \tag{8.6-39}$$

where c is the total molar concentration. Only n-1 equations are independent because of the constant pressure condition, that is

$$\sum_{j=1}^{n} \frac{dy_j}{dz} = 0 \tag{8.6-40}$$

Eq.(8.6-39) can be summed with respect to all species to give:

$$\sum_{j=1}^{n} \frac{N_j}{D_{K,j}} = 0 \tag{8.6-41}$$

But the Knudsen diffusivity is inversely proportional to the square root of the molecular weight, hence we get

$$\sum_{j=1}^{n} N_j \sqrt{M_j} = 0 \tag{8.6-42}$$

which is the statement of the Graham's law of diffusion in open system. From this, we can solve for the flux of the n-th species in terms of the other fluxes, that is:

$$N_n = -\sum_{j=1}^{n-1} v_j N_j = 0 \tag{8.6-43a}$$

where v_j relates the molecular weight of the species j to that of the species n.

$$v_j = \sqrt{M_j / M_n} \tag{8.6-43b}$$

Taking the last term out of the second term in the RHS of eq.(8.6-39), we get

$$-\frac{dy_i}{dz} = N_i \sum_{\substack{j=1\\j\neq i}}^{n} \frac{y_j}{cD_{ij}} - y_i \sum_{\substack{j=1\\j\neq i}}^{n-1} \frac{N_j}{cD_{ij}} - y_i \frac{N_n}{cD_{i,n}} + \frac{N_i}{cD_{K,i}} \qquad (8.6\text{-}44)$$

for $i = 1, 2, ..., n-1$.

Substituting eq.(8.6-43a) for the flux of the n-th component into eq.(8.6-44), we obtain:

$$-\frac{dy_i}{dz} = N_i \sum_{\substack{j=1\\j\neq i}}^{n} \frac{y_j}{cD_{ij}} - y_i \sum_{\substack{j=1\\j\neq i}}^{n-1} \frac{N_j}{cD_{ij}} + y_i \sum_{j=1}^{n-1} \frac{v_j N_j}{cD_{i,n}} + \frac{N_i}{cD_{K,i}} \qquad (8.6\text{-}45)$$

for $i = 1, 2, ..., n-1$, or we can write it in such a way to group the terms having N_i and N_j as factor as follows:

$$-\frac{dy_i}{dz} = N_i \left(\sum_{\substack{j=1\\j\neq i}}^{n} \frac{y_j}{cD_{ij}} + \frac{y_i v_i}{cD_{i,n}} + \frac{1}{cD_{K,i}} \right) - y_i \sum_{\substack{j=1\\j\neq i}}^{n-1} \frac{N_j}{cD_{ij}} + y_i \sum_{\substack{j=1\\j\neq i}}^{n-1} \frac{v_j N_j}{cD_{i,n}} \qquad (8.6\text{-}46)$$

for $i = 1, 2, ..., n-1$.

With the form of eq.(8.6-46) where terms associated with N_i and N_j are separated, we can now rewrite it into a compact vector-matrix form as follows:

$$-c\frac{d\underline{y}}{dz} = \underline{\underline{B}}(\underline{y})\underline{N} \qquad (8.6\text{-}47)$$

where $\underline{y}$ and $\underline{N}$ are (n-1) tupled vectors defined as:

$$\underline{y} = \begin{bmatrix} y_1 & y_2 & \cdots & y_{n-1} \end{bmatrix}^T \qquad (8.6\text{-}48a)$$

$$\underline{N} = \begin{bmatrix} N_1 & N_2 & \cdots & N_{n-1} \end{bmatrix}^T \qquad (8.6\text{-}48b)$$

and the matrix $\underline{\underline{B}}$ having a dimension of (n-1, n-1) is defined as:

$$\underline{\underline{B}} = \begin{cases} \dfrac{1}{D_{K,i}} + \dfrac{v_i y_i}{D_{i,n}} + \sum_{\substack{j=1\\j\neq i}}^{n} \dfrac{y_j}{D_{i,j}} & \text{for } i = j \\[1em] \left(\dfrac{v_j}{D_{i,n}} - \dfrac{1}{D_{i,j}} \right) y_i & \text{for } i \neq j \end{cases} \qquad (8.6\text{-}49)$$

Eq.(8.6-47) expresses the relationship between the concentration gradient vector and the flux vector. We can also write eq.(8.6-39) as a relationship between the concentration and its gradient. This is done as follows.

Rewrite eq.(8.6-39) as follows:

$$-\frac{dy_i}{dz} = N_i \sum_{\substack{j=1 \\ j \neq i}}^{n-1} \frac{y_j}{cD_{ij}} + N_i \frac{y_n}{cD_{i,n}} - y_i \sum_{\substack{j=1 \\ j \neq i}}^{n} \frac{N_j}{cD_{i,j}} + \frac{N_i}{cD_{K,i}} \tag{8.6-50}$$

But

$$y_n = 1 - \sum_{j=1}^{n-1} y_j \tag{8.6-51}$$

Eq.(8.6-50) will become:

$$-\frac{dy_i}{dz} = N_i \sum_{\substack{j=1 \\ j \neq i}}^{n-1} \frac{y_j}{cD_{ij}} + \frac{N_i}{cD_{i,n}} - N_i \sum_{j=1}^{n-1} \frac{y_j}{cD_{i,n}} - y_i \sum_{\substack{j=1 \\ j \neq i}}^{n} \frac{N_j}{cD_{i,j}} + \frac{N_i}{cD_{K,i}} \tag{8.6-52a}$$

We now group the terms associated with y_i together and those associated with y_j together, and finally obtain:

$$-\frac{dy_i}{dz} = N_i \sum_{\substack{j=1 \\ j \neq i}}^{n-1} \frac{y_j}{cD_{ij}} - N_i \sum_{\substack{j=1 \\ j \neq i}}^{n-1} \frac{y_j}{cD_{i,n}} - \left(\frac{N_i}{cD_{i,n}} + \sum_{\substack{j=1 \\ j \neq i}}^{n} \frac{N_j}{cD_{ij}} \right) y_i + \frac{N_i}{cD_{i,n}} + \frac{N_i}{cD_{K,i}} \tag{8.6-52b}$$

for $i = 1, 2, ..., n-1$. Eq.(8.6-52b) can be put in a vector form as:

$$\frac{d\underline{y}}{dz} = \underline{\underline{A}}(\underline{N}) \, \underline{y} + \underline{\psi}(\underline{N}) \tag{8.6-53}$$

where $\underline{y}$, $\underline{N}$ are defined in eqs.(8.6-48), and $\underline{\underline{A}}(n-1, n-1)$ and $\underline{\psi}(n-1)$ are defined as follows:

$$\underline{\underline{A}} = \begin{cases} \dfrac{N_i}{cD_{i,n}} + \sum_{\substack{j=1 \\ j \neq i}}^{n} \dfrac{N_j}{cD_{i,j}} & \text{for } i = j \\[2ex] \left(\dfrac{1}{cD_{i,n}} - \dfrac{1}{cD_{i,j}} \right) N_i & \text{for } i \neq j \end{cases} \tag{8.6-54}$$

and

$$\underline{\psi} = \left\{ -\left(\frac{1}{cD_{i,n}} + \frac{1}{cD_{K,i}} \right) N_i \right\} \quad (8.6\text{-}55)$$

For a capillary with constant boundary conditions:

$$z = 0; \quad \underline{y} = \underline{y}_0 \quad (8.6\text{-}56a)$$

$$z = L; \quad \underline{y} = \underline{y}_L \quad (8.6\text{-}56b)$$

eqs. (8.6-47) and (8.6-53) can be integrated with constant boundary conditions to give the following solution for the fluxes (Appendix 8.2):

$$\underline{N} = c \left[\underline{\underline{B}}(\underline{y}_0) \right]^{-1} \underline{\underline{A}} \left[\exp(L\underline{\underline{A}}) - \underline{\underline{I}} \right]^{-1} \left(\underline{y}_0 - \underline{y}_L \right) \quad (8.6\text{-}57a)$$

or

$$\underline{N} = c \left[\underline{\underline{B}}(\underline{y}_L) \right]^{-1} \underline{\underline{A}} \cdot \exp(L\underline{\underline{A}}) \left[\exp(L\underline{\underline{A}}) - \underline{\underline{I}} \right]^{-1} \left(\underline{y}_0 - \underline{y}_L \right) \quad (8.6\text{-}57b)$$

The difference between eqs.(8.6-57a) and (8.6-57b) is the evaluation of the matrix $\underline{\underline{B}}$. In eq.(8.6-57a) it is evaluated at $\underline{y}_0$ while in eq.(8.6-57b) it is evaluated at $\underline{y}_L$.

Note that eq.(8.6-57a) or (8.6-57b) is a set of nonlinear algebraic equations in terms of (n-1) fluxes because the matrix $\underline{\underline{A}}$ is a function of $\underline{N}$. Thus it must be solved by a numerical method, such as the Newton-Raphson method. For using such method, the initial guess for the fluxes is

$$\underline{N}^{(0)} = c \left[\underline{\underline{B}}(\underline{y}_{av}) \right]^{-1} \frac{\left(\underline{y}_0 - \underline{y}_L \right)}{L} \quad (8.6\text{-}58a)$$

where

$$\underline{y}_{av} = \frac{1}{2} \left(\underline{y}_0 + \underline{y}_L \right) \quad (8.6\text{-}58b)$$

The code CAPILL2 written in MATLAB is available for the numerical solution of eq. (8.6-57). Knowing the fluxes of the (n-1) components, the n-th flux can be obtained from eq.(8.6-43a).

Also, knowing the fluxes, the concentration profiles are given by (Appendix 8.2)

$$\underline{y} - \underline{y}_0 = \left[\exp(z\underline{\underline{A}}) - \underline{\underline{I}} \right] \left[\exp(L\underline{\underline{A}}) - \underline{\underline{I}} \right]^{-1} \left(\underline{y}_L - \underline{y}_0 \right) \quad (8.6\text{-}59)$$

The following example illustrates the numerical evaluation of the flux solution (eq. 8.6-57) by using the code CAPILL2.

Example 8.6-5: *Steady state ternary diffusion of hydrogen/nitrogen/CO₂ through a capillary*

We take an example of diffusion of hydrogen (1), nitrogen (2) and carbon dioxide (3) through a capillary having a length of 0.086 m and a pore radius of 0.2 micron. The conditions of this system are 35 °C and 1 atm.

Knowing the temperature of the system and the pore radius, we calculate the Knudsen diffusivities of all three species (Table 7.4-2):

$$D_{K,1} = 9700 \left(2 \times 10^{-5}\right) \sqrt{\frac{35+273}{M_1}} = 2.41 \text{ cm}^2/\text{sec}$$

$$D_{K,2} = 9700 \left(2 \times 10^{-5}\right) \sqrt{\frac{35+273}{M_2}} = 0.643 \text{ cm}^2/\text{sec}$$

$$D_{K,3} = 9700 \left(2 \times 10^{-5}\right) \sqrt{\frac{35+273}{M_3}} = 0.513 \text{ cm}^2/\text{sec}$$

The binary diffusivities at the conditions of 35 °C and 1 atm are:

$$D_{12} = 0.833 \text{ cm}^2/\text{sec}, \ D_{13} = 0.680 \text{ cm}^2/\text{sec}, \ D_{23} = 0.168 \text{ cm}^2/\text{sec}$$

The mole fractions of these three components at two ends of the capillary are given in the following table

@ z = 0	@ z = L
$y_1 = 0$	$y_1 = 0.50121$
$y_2 = 0.50086$	$y_2 = 0.49879$
$y_3 = 4.9914$	$y_3 = 0$

Using the code CAPILL2, we calculate the fluxes of the three components are:

$$N_1 = -1.6 \times 10^{-6}, \ N_2 = +1.134 \times 10^{-8}, \ N_3 = +3.318 \times 10^{-7} \text{ mole/cm}^2/\text{sec}$$

The sum of these fluxes is -1.256×10^{-6} mole/cm²/sec, indicating that the bulk flow is in the direction from z = L to z = 0. We also note that the above fluxes values satisfies the Graham's law of diffusion (eq. 8.6-42).

8.7 Stefan-Maxwell Approach for Bulk-Knudsen Diffusion in Complex System of Capillaries

We have seen in Section 8.6 that the analysis using the Stefan-Maxwell approach is readily carried out for the case of a simple capillary, namely a uniformly sized capillary. In this section we will extend the analysis to more complex pore networks and will consider the three cases:

(i) bundle of parallel capillaries
(ii) two capillaries in series
(iii) network of three capillaries

to illustrate the application of the Stefan-Maxwell approach to complex configurations of capillaries.

8.7.1 Bundle of Parallel Capillaries

We first consider the case of two parallel capillaries as shown in Figure 8.7-1.

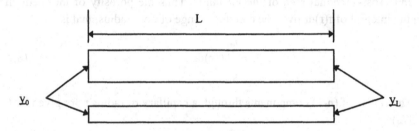

Figure 8.7-1: Bundle of parallel capillaries

In this case, the results obtained in Section 8.6 are directly applicable here as the two capillaries are acting independently, that is the fluxes passing through the two capillaries for the n-1 components are (eq. 8.6-57a):

$$\underline{N}^{(1)} = c\left[\underline{\underline{B}}^{(1)}(\underline{y}_0)\right]^{-1}\underline{\underline{A}}^{(1)}\left[\exp(L\underline{\underline{A}}^{(1)}) - \underline{\underline{I}}\right]^{-1}\left(\underline{y}_0 - \underline{y}_L\right) \quad (8.7\text{-}1a)$$

and

$$\underline{N}^{(2)} = c\left[\underline{\underline{B}}^{(2)}(\underline{y}_0)\right]^{-1}\underline{\underline{A}}^{(2)}\left[\exp(L\underline{\underline{A}}^{(2)}) - \underline{\underline{I}}\right]^{-1}\left(\underline{y}_0 - \underline{y}_L\right) \quad (8.7\text{-}1b)$$

where $\underline{N}^{(1)}$ and $\underline{N}^{(2)}$ are diffusive fluxes for the two capillaries, and $\underline{\underline{A}}^{(1)}$ and $\underline{\underline{A}}^{(2)}$ are matrices, which are function of $\underline{N}$

$$\underline{\underline{A}}^{(1)} = \underline{\underline{A}}^{(1)}(\underline{N}^{(1)}) \quad (8.7\text{-}2a)$$

$$\underline{\underline{A}}^{(2)} = \underline{\underline{A}}^{(2)}(\underline{N}^{(2)}) \tag{8.7-2b}$$

Here we use the upperscript to denote the capillary. The matrix $\underline{\underline{B}}$ is defined in eq.(8.6-49) and the matrix $\underline{\underline{A}}$ is defined in eq.(8.6-54). Eq.(8.7-1a) is a set of n-1 nonlinear algebraic equations, from which we can solve for fluxes for the n-1 components through the capillary # 1. Similarly, solving eq.(8.7-1b) will yield fluxes through the capillary # 2.

The average fluxes across the bundle of the two capillaries is then simply:

$$\underline{N} = \frac{S^{(1)}\underline{N}^{(1)} + S^{(2)}\underline{N}^{(2)}}{S^{(1)} + S^{(2)}} \tag{8.7-3}$$

where $S^{(1)}$ and $S^{(2)}$ are the cross-section areas of the two capillaries, respectively.

The same analysis can be carried out when we have a bundle of parallel capillaries with a distribution in capillary pore radius. The distribution function is f(r) with f(r)dr being the void area of capillaries having radii between r and r + dr per unit cross-sectional area of the medium. Thus the porosity of the medium is then the integral of f(r)dr over the complete range of pore radius, that is

$$\varepsilon = \int_{r_{min}}^{r_{max}} f(r)dr \tag{8.7-4}$$

The fluxes of (n-1) components through a capillary of radius r are given by (eq. 8.6-57a)

$$\underline{N} = c\left[\underline{\underline{B}}(\underline{y}_0)\right]^{-1}\underline{\underline{A}}\left[\exp(L\underline{\underline{A}}) - \underline{\underline{I}}\right]^{-1}\left(\underline{y}_0 - \underline{y}_L\right) \tag{8.7-5}$$

Note that $\underline{\underline{B}}$ is a function of the capillary radius because of its dependence through the Knudsen diffusivity (see eqs. 8.6-49 and 8.6-3). It is reminded that $\underline{\underline{A}}$ is a function of $\underline{N}$ as given in eq. (8.6-54).

The mass transfer per unit area through the capillaries having radii between r and r+dr is

$$d\underline{M} = \underline{N} \cdot f(r)dr \tag{8.7-6}$$

Thus, the total mass transfer per unit area of the medium passing through all capillaries is the integration of the above equation over the complete range of pore radius:

$$\underline{M} = \int_{r_{min}}^{r_{max}} f(r) \underline{N} \, dr \tag{8.7-7}$$

The average flux is defined as the mass transfer divided by the area available for diffusion, that is:

$$\underline{N}_{av} = \frac{\int_{r_{min}}^{r_{max}} f(r) \underline{N} \, dr}{\varepsilon} \tag{8.7-8}$$

The mass transfer rate in eq. (8.7-7) can be approximated by the following quadrature formula:

$$\underline{M} = \sum_{j=1}^{Q} w_j \underline{N}_j \, f(r_j) \tag{8.7-9}$$

where w_j is the weighting factor, and r_j is the radius at the j-th quadrature point. Here $\underline{N}_j$ is the flux vector at the radius r_j and it is solved from the algebraic equation (8.7-5).

An alternative to the above approach is to define a mean pore radius at which the matrix $\underline{\underline{B}}$ can be calculated. The mean pore radius can be calculated from (Wang and Smith, 1983), assuming a uniform cylindrical pore

$$\bar{r} = \frac{2V}{S_g} \tag{8.7-10a}$$

where V is the void volume and S_g is the internal surface area, and they can be calculated from the knowledge of the pore size distribution:

$$\bar{r} = \frac{\int_{r_{min}}^{r_{max}} f(r) \, dr}{\int_{r_{min}}^{r_{max}} \frac{f(r)}{r} \, dr} \tag{8.7-10b}$$

The approximate mass transfer per unit area of the medium is then given by

$$\underline{M}_{app} = \varepsilon \cdot c \left[\underline{\underline{B}}(\underline{y}_0, \bar{r}) \right]^{-1} \cdot \underline{\underline{A}} \cdot \left[\exp(\underline{\underline{L}}\underline{\underline{A}}) - \underline{\underline{I}} \right]^{-1} (\underline{y}_0 - \underline{y}_L) \tag{8.7-11}$$

8.7.2 Capillaries in Series

Having considered the bundle of capillaries in parallel in the previous section, we now consider two capillaries in series as shown in Figure 8.7-2. The length and radius of the section 1 are L_1 and r_1, respectively. Similarly, those for the section 2 are L_2 and r_2.

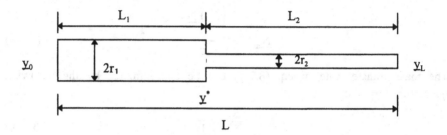

Figure 8.7-2: Two capillaries in series

To tackle this problem, we shall consider the two sections separately and then enforce the continuity of flux at the junction between the two sections to determine the desired solution. Let the mole fractions at the junction of the two sections are $\underline{y}^*$, which are unknown at this stage. The fluxes leaving the section 1 (eq. 8.6-57a) are

$$\underline{\underline{N}}^{(1)} = c\left[\underline{\underline{B}}^{(1)}(\underline{y}_0)\right]^{-1}\underline{\underline{A}}^{(1)}\left[\exp(L_1\underline{\underline{A}}^{(1)}) - \underline{\underline{I}}\right]^{-1}\left(\underline{y}_0 - \underline{y}^*\right) \qquad (8.7\text{-}12)$$

Here we use the matrix $\underline{\underline{B}}$ evaluated at $\underline{y}_0$ which is known, and

$$\underline{\underline{A}}^{(1)} = \underline{\underline{A}}^{(1)}(\underline{N}^{(1)}) \qquad (8.7\text{-}13)$$

The upperscript is used to denote the section.

The fluxes leaving the section 2 (eq. 8.6-57b) are

$$\underline{\underline{N}}^{(2)} = c\left[\underline{\underline{B}}^{(2)}(\underline{y}_L)\right]^{-1}\underline{\underline{A}}^{(2)}\cdot\exp(L_2\underline{\underline{A}}^{(2)})\left[\exp(L_2\underline{\underline{A}}^{(2)}) - \underline{\underline{I}}\right]^{-1}\left(\underline{y}^* - \underline{y}_L\right) \qquad (8.7\text{-}14)$$

The matrix $\underline{\underline{B}}^{(2)}$ is evaluated at $\underline{y}_L$ (which is also a known mole fraction) and the matrix $\underline{\underline{A}}^{(2)}$ is a function of $\underline{N}^{(2)}$, that is

$$\underline{\underline{A}}^{(2)} = \underline{\underline{A}}^{(2)}(\underline{N}^{(2)}) \qquad (8.7\text{-}15)$$

The continuity of mass at the junction between the two capillaries requires that

$$\underline{M} = S^{(1)} \underline{N}^{(1)} = S^{(2)} \underline{N}^{(2)} \tag{8.7-16}$$

which relates the flux vector of the capillary 1 to that of the capillary 2. Eq. (8.7-12), (8.7-14) and (8.7-16) are three equations in terms of $\underline{N}^{(1)}, \underline{N}^{(2)}$ and $\underline{y}*$. Thus, by solving these three equations we will obtain the solutions for the fluxes as well as the mole fractions at the junction between the two sections. What we shall do in an example below is the application of the elimination method to solve for the fluxes.

Example 8.7-1: *Two capillaries in series*

Substitution of eqs. (8.7-12) and (8.7-14) into the equation of continuity (eq. 8.7-16) gives:

$$\underline{M} = S^{(1)} c \left[\underline{\underline{B}}^{(1)}(\underline{y}_0)\right]^{-1} \underline{\underline{A}}^{(1)} \left[\exp(L_1 \underline{\underline{A}}^{(1)}) - \underline{\underline{I}}\right]^{-1} (\underline{y}_0 - \underline{y}*)$$
$$= S^{(2)} c \left[\underline{\underline{B}}^{(2)}(\underline{y}_L)\right]^{-1} \underline{\underline{A}}^{(2)} \exp(L_2 \underline{\underline{A}}^{(2)}) \left[\exp(L_2 \underline{\underline{A}}^{(2)}) - \underline{\underline{I}}\right]^{-1} (\underline{y}* - \underline{y}_L) \tag{8.7-17}$$

To simplify the notation, we let

$$\underline{\underline{C}}^{(1)} = S^{(1)} \left[\underline{\underline{B}}^{(1)}(\underline{y}_0)\right]^{-1} \underline{\underline{A}}^{(1)} \left[\exp(L_1 \underline{\underline{A}}^{(1)}) - \underline{\underline{I}}\right]^{-1} \tag{8.7-18a}$$

$$\underline{\underline{C}}^{(2)} = S^{(2)} \left[\underline{\underline{B}}^{(2)}(\underline{y}_L)\right]^{-1} \underline{\underline{A}}^{(2)} \cdot \exp(L_2 \underline{\underline{A}}^{(2)}) \cdot \left[\exp(L_2 \underline{\underline{A}}^{(2)}) - \underline{\underline{I}}\right]^{-1} \tag{8.7-18b}$$

Eq.(8.7-17) then becomes:

$$\underline{M} = c\,\underline{\underline{C}}^{(1)} (\underline{y}_0 - \underline{y}*) = c\,\underline{\underline{C}}^{(2)} (\underline{y}* - \underline{y}_L) \tag{8.7-19}$$

This equation is the required equation to solve for the mole fraction at the junction, $\underline{y}*$. We make use of the following equation:

$$\underline{y}* - \underline{y}_L = (\underline{y}_0 - \underline{y}_L) - (\underline{y}_0 - \underline{y}*) \tag{8.7-20}$$

and substitute this into eq.(8.7-19) to get

$$\underline{\underline{C}}^{(1)} (\underline{y}_0 - \underline{y}*) = \underline{\underline{C}}^{(2)} \left[(\underline{y}_0 - \underline{y}_L) - (\underline{y}_0 - \underline{y}*)\right] \tag{8.7-21}$$

Solving for $(\underline{y}_0 - \underline{y}*)$, we have:

$$(\underline{y}_0 - \underline{y}^*) = \left[\underline{\underline{C}}^{(1)} + \underline{\underline{C}}^{(2)}\right]^{-1} \underline{\underline{C}}^{(2)}(\underline{y}_0 - \underline{y}_L) \qquad (8.7\text{-}22)$$

Substitute eq.(8.7-22) into eq.(8.7-19), we obtain the following desired solution for the mass transfer rate:

$$\underline{M} = S^{(1)}\underline{N}^{(1)} = c \cdot \underline{\underline{C}}^{(1)} \cdot \left[\underline{\underline{C}}^{(1)} + \underline{\underline{C}}^{(2)}\right]^{-1} \underline{\underline{C}}^{(2)}(\underline{y}_0 - \underline{y}_L) \qquad (8.7\text{-}23)$$

written in terms of the known overall driving force. Note that $\underline{\underline{C}}^{(1)}$ is a function of $\underline{N}^{(1)}$ and $\underline{\underline{C}}^{(2)}$ is a function of $\underline{N}^{(2)}$. But $\underline{N}^{(2)}$ is related to $\underline{N}^{(1)}$ according to eq.(8.7-16). Thus, the above equation is a set of (n-1) nonlinear algebraic equations in terms of $\underline{N}^{(1)}$, which can be solved numerically.

Instead of using eq.(8.7-20), we could use:

$$\underline{y}_0 - \underline{y}^* = (\underline{y}_0 - \underline{y}_L) - (\underline{y}^* - \underline{y}_L) \qquad (8.7\text{-}24)$$

Substitution of this equation into eq.(8.7-19), we have:

$$\underline{\underline{C}}^{(1)}\left[(\underline{y}_0 - \underline{y}_L) - (\underline{y}^* - \underline{y}_L)\right] = \underline{\underline{C}}^{(2)}(\underline{y}^* - \underline{y}_L) \qquad (8.7\text{-}25)$$

from which we can solve for $\underline{y}^* - \underline{y}_L$

$$\underline{y}^* - \underline{y}_L = \left[\underline{\underline{C}}^{(1)} + \underline{\underline{C}}^{(2)}\right]^{-1} \underline{\underline{C}}^{(1)}(\underline{y}_0 - \underline{y}_L) \qquad (8.7\text{-}26)$$

Thus, substituting eq.(8.7-26) into eq.(8.7-19), we obtain another expression for the mass flux

$$\underline{M} = S^{(2)}\underline{N}^{(2)} = c \cdot \underline{\underline{C}}^{(2)}\left[\underline{\underline{C}}^{(1)} + \underline{\underline{C}}^{(2)}\right]^{-1} \underline{\underline{C}}^{(1)}(\underline{y}_0 - \underline{y}_L) \qquad (8.7\text{-}27)$$

The mass transfer rate $\underline{M}$ calculated from eqs. (8.7-27) and (8.7-23) must be the same. Thus

$$\underline{\underline{C}}^{(2)}\left[\underline{\underline{C}}^{(1)} + \underline{\underline{C}}^{(2)}\right]^{-1} \underline{\underline{C}}^{(1)} = \underline{\underline{C}}^{(1)}\left[\underline{\underline{C}}^{(1)} + \underline{\underline{C}}^{(2)}\right]^{-1} \underline{\underline{C}}^{(2)} \qquad (8.7\text{-}28)$$

which is mathematically the case (Appendix 8.6).

8.7.3 A Simple Pore Network:

Having solved the steady state mass transfer problem for two simple configurations of pore, we now turn to a more complicated pore network as shown in Figure 8.7-3. Like the previous example, we denote the mole fraction at the junction of three capillaries be $\underline{y}^*$, and then solve for the fluxes of the three capillaries separately. Once this is done, we then enforce the continuity of fluxes at the junction of these capillaries to obtain the desired solutions.

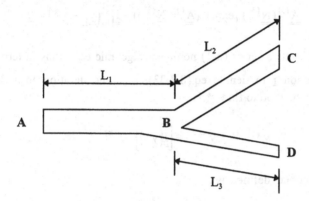

Figure 8.7-3: A simple pore network

The fluxes from A to B are given by (from eq. 8.6-57a):

$$\underline{N}^{(1)} = c\left[\underline{\underline{B}}^{(1)}(\underline{y}_0)\right]^{-1}\underline{\underline{A}}^{(1)}(\underline{N}^{(1)}).\left[\exp(L_1\underline{\underline{A}}^{(1)}(\underline{N}^{(1)}))-\underline{\underline{I}}\right]^{-1}\left(\underline{y}_0 - \underline{y}^*\right) \quad (8.7\text{-}29)$$

Similarly, the fluxes from C to B and those form D to B are (using eq. 8.6-57b rather than 8.6-57a because $\underline{y}^*$ is still unknown):

$$\underline{N}^{(2)} = c\left[\underline{\underline{B}}^{(2)}(\underline{y}_L)\right]^{-1}\underline{\underline{A}}^{(2)}(\underline{N}^{(2)}).\left[\exp(L_2\underline{\underline{A}}^{(2)}(\underline{N}^{(2)}))-\underline{\underline{I}}\right]^{-1}\left(\underline{y}_L - \underline{y}^*\right) \quad (8.7\text{-}30)$$

and

$$\underline{N}^{(3)} = c\left[\underline{\underline{B}}^{(3)}(\underline{y}_L)\right]^{-1}\underline{\underline{A}}^{(3)}(\underline{N}^{(3)}).\left[\exp(L_3\underline{\underline{A}}^{(3)}(\underline{N}^{(3)}))-\underline{\underline{I}}\right]^{-1}\left(\underline{y}_L - \underline{y}^*\right) \quad (8.7\text{-}31)$$

respectively.

Continuity of fluxes at the junction B requires that

$$\underline{\underline{S}}^{(1)}\underline{N}^{(1)} + \underline{\underline{S}}^{(2)}\underline{N}^{(2)} + \underline{\underline{S}}^{(3)}\underline{N}^{(3)} = \underline{0} \qquad (8.7\text{-}32)$$

that is:

$$\underline{\underline{S}}^{(1)}\left[\underline{\underline{B}}^{(1)}(\underline{y}_0)\right]^{-1}\underline{\underline{A}}^{(1)}(\underline{N}^{(1)}).\left[\exp(L_1\underline{\underline{A}}^{(1)}(\underline{N}^{(1)})) - \underline{\underline{I}}\right]^{-1}(\underline{y}_0 - \underline{y}^*) +$$

$$\underline{\underline{S}}^{(2)}\left[\underline{\underline{B}}^{(2)}(\underline{y}_L)\right]^{-1}\underline{\underline{A}}^{(2)}(\underline{N}^{(2)}).\left[\exp(L_2\underline{\underline{A}}^{(2)}(\underline{N}^{(2)})) - \underline{\underline{I}}\right]^{-1}(\underline{y}_L - \underline{y}^*) +$$

$$\underline{\underline{S}}^{(3)}\left[\underline{\underline{B}}^{(3)}(\underline{y}_L)\right]^{-1}\underline{\underline{A}}^{(3)}(\underline{N}^{(3)}).\left[\exp(L_3\underline{\underline{A}}^{(3)}(\underline{N}^{(3)})) - \underline{\underline{I}}\right]^{-1}(\underline{y}_L - \underline{y}^*) = \underline{0} \qquad (8.7\text{-}33)$$

Eq.(8.7-33) represents a set of (n-1) nonlinear algebraic equations in terms of $\underline{y}^*$. We let the function $\underline{f}$ to denote eq.(8.7-33), then the iteration formula for the Newton-Raphson method to find $\underline{y}^*$ is

$$\underline{y}^{*(k+1)} = \underline{y}^{*(k)} - \left[\underline{\underline{J}}(\underline{y}^*)\right]^{-1}\underline{f}^{(k)} \qquad (8.7\text{-}34)$$

where $\underline{\underline{J}}$ is the Jacobian defined as:

$$\underline{\underline{J}} = \left\{\frac{\partial \underline{f}}{\partial \underline{y}^*}\right\} \qquad (8.7\text{-}35)$$

The iteration procedure is done as follows:

(i) Choose an initial guess for $\underline{y}^{*(0)}$
(ii) For each value of $\underline{y}^{*(k)}$, $\underline{N}^{(1)}$, $\underline{N}^{(2)}$ and $\underline{N}^{(3)}$ are evaluated from eqs.(8.7-29), (8.7-30) and (8.7-31), respectively. Then evaluate $\underline{f}^{(k)}$ from eq.(8.7-33), and $\underline{\underline{J}}^{(k)}$ is evaluated by a numerical procedure.
(iii) Obtain the next iterated solution from eq.(8.7-34) and check for convergence. If convergence is not met, go back to step (ii).

8.8 Stefan-Maxwell for Bulk-Knudsen-Viscous Flow

In the last sections, you have learnt about the basic analysis of bulk flow, bulk flow and Knudsen flow using the Stefan-Maxwell approach. Very often when we deal with diffusion and adsorption system, the total pressure changes with time as well as with distance within a particle due to either the nonequimolar diffusion or loss of mass from the gas phase as a result of adsorption onto the surface of the particle. When such situations happen, there will be an additional mechanism for mass transfer: the viscous flow. This section will deal with the general case where bulk diffusion, Knudsen diffusion and viscous flow occur simultaneously within a porous medium (Jackson, 1977).

When the system is subject to a pressure variation instead of the assumption of isobaric made in the previous sections, the flux equation will contain an additional term to account for the viscous flow. Using the argument of momentum transfer, the total flux will be the sum of a flux due to the <u>diffusion mechanism</u> and a flux due to the <u>viscous mechanism</u>. The flux due to the diffusion mechanism is governed by the following modified Stefan-Maxwell equation for the component "i" (eq. 8.6-6):

$$-\frac{1}{R_gT}\frac{dp_i}{dz} = \sum_{\substack{j=1 \\ j \neq i}}^{n} \frac{y_j N_i^D - y_i N_j^D}{D_{ij}} + \frac{N_i^D}{D_{K,i}} \qquad (8.8\text{-}1)$$

The viscous flux for the component "i" is governed by the Hagen-Poiseuille type equation (see Chapter 7 for more detailed exposition of this mechanism):

$$N_i^V = -\frac{y_i}{R_gT}\frac{B_0 P}{\mu}\frac{dP}{dz} \qquad (8.8\text{-}2a)$$

where P is the total pressure, B_0 is the structural parameter of the solid characterising the viscous flow ($= r^2/8$ for cylindrical pore) and μ is the viscosity of the mixture. Here we use the uppersicpts "D" and "V" for diffusive and viscous flows, respectively. The viscosity is evaluated at the mixture condition, and it can be calculated from the semi-empirical formula of Wilke (Bird et al., 1960):

$$\mu = \sum_{i=1}^{n} \frac{y_i \mu_i}{\sum_{j=1}^{n} y_j \Phi_{ij}} \qquad (8.8\text{-}2b)$$

where

$$\Phi_{ij} = \frac{1}{\sqrt{8}}\left(1+\frac{M_i}{M_j}\right)^{-1/2}\left[1+\left(\frac{\mu_i}{\mu_j}\right)^{1/2}\left(\frac{M_j}{M_i}\right)^{1/4}\right]^2 \qquad (8.8\text{-}2c)$$

The flux of the component "i" is then the summation of the flux due to the diffusion mechanism and the flux due to the viscous mechanism, that is:

$$N_i = N_i^D + N_i^V \qquad (8.8\text{-}3)$$

Eqs.(8.8-1) to (8.8-3) are applicable for all components, that is for $i = 1, 2, 3,..., n$.

8.8.1 The Basic Equation written in terms of Fluxes N

We have presented above the three basic equations (8.8-1 to 8.8-3) for the case where bulk diffusion-Knudsen diffusion-viscous flows are simultaneously operating. What we will do in this section is to combine them to obtain a form which is useful for analysis and subsequent computation as we shall show in Chapters 9 and 10.

8.8.1.1 In Component Form

If we substitute the viscous flux equation (eq. 8.8-2a) into eq. (8.8-3) and then solve for the flux N_i^D contributed by the diffusion mechanism to get:

$$N_i^D = N_i + \frac{y_i}{R_g T} \frac{B_0 P}{\mu T} \frac{dP}{dz} \qquad (8.8\text{-}4)$$

Next, we substitute the above equation into the modified Stefan-Maxwell equation (8.8-1), we obtain the following constitutive equation to describe the flux in terms of the <u>partial pressure gradient</u> and the <u>total pressure gradient</u>:

$$\sum_{\substack{j=1 \\ j \neq i}}^{n} \frac{y_j N_i - y_i N_j}{D_{ij}} + \frac{N_i}{D_{K,i}} = -\frac{1}{R_g T}\frac{dp_i}{dz} - \frac{y_i}{R_g T}\frac{B_0 P}{\mu D_{K,i}}\frac{dP}{dz} \qquad (8.8\text{-}5)$$

for $i = 1, 2, 3, ..., n$. The first term in the RHS of eq.(8.8-5) is the driving force due to the <u>partial pressure gradient</u>, while the second term is the driving force due to the <u>total pressure gradient</u>.

If we express the partial pressure gradient in the first term in the RHS of eq.(8.8-5) in terms of the total pressure and the mole fraction

$$p_i = y_i P$$

the constitutive equation (8.8-5) can be written in terms of the mole fraction gradient and the total pressure gradient:

$$\sum_{\substack{j=1 \\ j \neq i}}^{n} \frac{y_j N_i - y_i N_j}{D_{ij}} + \frac{N_i}{D_{K,i}} = -\frac{P}{R_g T} \frac{dy_i}{dz} - \frac{y_i}{R_g T} \left(1 + \frac{B_0 P}{\mu D_{K,i}}\right) \frac{dP}{dz} \quad (8.8\text{-}6)$$

Eq. (8.8-5) or (8.8-6) is the basic equation for the case of diffusion under nonisobaric conditions. Alternative derivation of these two equations is given in Appendix 8.7. Knowing the individual fluxes obtained from eq.(8.8-6), the total flux N_T and the diffusive fluxes $\underline{J}$ are given by:

$$N_T = \sum_{j=1}^{n} N_j \quad (8.8\text{-}7a)$$

$$J_i = N_i - y_i N_T \quad (8.8\text{-}7b)$$

for $i = 1, 2, ..., n$.

Summing eq.(8.8-6) with respect to all components, we get the pressure drop equation written in terms of all fluxes:

$$\sum_{i=1}^{n} \frac{N_i}{D_{K,i}} = -\frac{dP}{dz} \frac{1}{R_g T} \left(1 + \frac{B_0 P}{\mu} \sum_{i=1}^{n} \frac{1}{D_{K,i}}\right) \quad (8.8\text{-}8)$$

This equation expresses how the total pressure changes in terms of the fluxes. Under the isobaric conditions (that is $dP/dz = 0$), the above equation reduces to:

$$\sum_{i=1}^{n} \frac{N_i}{D_{K,i}} = 0$$

which is basically the Graham law of diffusion as the Knudsen diffusivity is inversely proportional to the square root of the molecular weight. This proves that the Graham law of diffusion is only valid for constant pressure systems.

8.8.1.2 In Vector Form

With n components in the mixture, the constitutive equation (8.8-6) can be conveniently cast into more compact form by using the vector-matrix format. We define the following n-tupled vectors

$$\underline{N} = [N_1 \ N_2 \ \cdots \ N_n]; \quad \underline{y} = [y_1 \ y_2 \ \cdots \ y_n] \quad (8.8\text{-}9)$$

Eq.(8.8-6) then becomes:

$$\underline{\underline{B}}(\underline{y})\,\underline{N} = -\frac{P}{R_g T}\frac{d\underline{y}}{dz} - \frac{1}{R_g T}\frac{dP}{dz}\left(\underline{\underline{I}} + \frac{B_0 P}{\mu}\underline{\underline{\Lambda}}\right)\underline{y} \qquad (8.8\text{-}10)$$

where the matrix $\underline{\underline{B}}$ having a dimension of (n, n) is defined as below

$$\underline{\underline{B}} = \begin{cases} \dfrac{1}{D_{K,i}} + \sum_{\substack{j=1 \\ j\neq i}}^{n} \dfrac{y_j}{D_{ij}} & \text{for } i = j \\[2mm] -\dfrac{y_i}{D_{ij}} & \text{for } i \neq j \end{cases} \qquad (8.8\text{-}11a)$$

and the diagonal matrix $\underline{\underline{\Lambda}}$ is given by:

$$\underline{\underline{\Lambda}} = \left\{ \Lambda_{ii} = \frac{1}{D_{K,i}} \right. \qquad (8.8\text{-}11b)$$

and $\underline{\underline{I}}$ is an identity matrix. All the elements of $\underline{\underline{B}}$ and $\underline{\underline{\Lambda}}$ have units of inverse of the diffusion coefficients (sec/m²).

The constitutive flux equation in vector form (8.8-10) can now be inverted to yield the flux vector in terms of mole fraction gradients and the total pressure gradient. We get:

$$\underline{N} = -\frac{P}{R_g T}\left[\underline{\underline{B}}(\underline{y})\right]^{-1}\frac{d\underline{y}}{dz} - \frac{1}{R_g T}\frac{dP}{dz}\left[\underline{\underline{B}}(\underline{y})\right]^{-1}\left(\underline{\underline{I}} + \frac{B_0 P}{\mu}\underline{\underline{\Lambda}}\right)\underline{y} \qquad (8.8\text{-}12)$$

In the case of isobaric conditions, the second term in the RHS of eq.(8.8-12) becomes zero, and hence it reduces to the equation

$$\underline{N} = -c\left[\underline{\underline{B}}(\underline{y})\right]^{-1}\frac{d\underline{y}}{dz}$$

obtained earlier for isobaric conditions.

The constitutive flux equation can be written in terms of partial pressure gradients and total pressure gradient as follows:

$$\underline{N} = -\frac{1}{R_g T}\left[\underline{\underline{B}}(\underline{y})\right]^{-1}\frac{d\underline{p}}{dz} - \frac{B_0}{\mu R_g T}\frac{dP}{dz}\left[\underline{\underline{B}}(\underline{y})\right]^{-1}\underline{\underline{\Lambda}}\,\underline{p} \qquad (8.8\text{-}13a)$$

where

$$\underline{p} = \begin{bmatrix} p_1 & p_2 & \cdots & p_n \end{bmatrix}^T \tag{8.8-13b}$$

Eqs. (8.8-12) or (8.8-13) are the basic constitutive flux equations for the case where molecular diffusion, Knudsen diffusion and viscous flow are all operating. They are in the form suitable to be used in the mass balance equation, which we shall show later in this chapter as well as in Chapters 9 and 10.

Knowing the fluxes of all components, the total flux (scalar) can be obtained as by summing all elements of the flux vector as shown below:

$$N_T = \sum_{i=1}^{n} N_i \tag{8.8-14}$$

and the diffusive flux vector (relative to the moving mixture) is obtained from:

$$\underline{J} = \underline{N} - N_T \underline{y} \tag{8.8-15}$$

where

$$\underline{J} = \begin{bmatrix} J_1 & J_2 & \cdots & J_n \end{bmatrix} \tag{8.8-16}$$

It is reminded here that these n diffusive fluxes are not linearly independent as

$$\sum_{i=1}^{n} J_i = 0 \tag{8.8-17}$$

which can be easily proved from eq. (8.8-15) by summing all the elements of eq. (8.8-15) and noting that

$$\sum_{j=1}^{n} y_j = 1$$

8.8.2 The Basic Equations Written in terms of Diffusive Fluxes J

We have presented in Section 8.8.1 the basic equations (8.8-12) or (8.8-13) written in terms of the fluxes $\underline{N}$, relative to a fixed frame of coordinate. Very often that we need to solve for the diffusive fluxes $\underline{J}$ (for example in a moving mixture), and in this section we will present the basic equation written in terms of those fluxes.

8.8.2.1 In component form

Knowing the fluxes N_i (i = 1, 2, 3,..., n), the diffusive fluxes can be obtained from:

$$J_i = N_i - y_i N_T \tag{8.8-18}$$

where N_T is the total flux given in eq. (8.8-14).

Substituting eqs.(8.8-18) and (8.8-14) into eq.(8.8-6), we obtain the following equation for the diffusive flux after some algebraic manipulation:

$$\sum_{\substack{j=1 \\ j \neq i}}^{n} \frac{y_j J_i - y_i J_j}{\Delta_{ij}} = -\frac{P}{R_g T} \frac{dy_i}{dz} - \frac{y_i}{RT}\left(1 - \frac{1}{D_{K,i} \sum_{k=1}^{n} y_k / D_{K,k}}\right) \frac{dP}{dz} \qquad (8.8\text{-}19a)$$

for $i = 1, 2, 3, \ldots, (n-1)$. Detailed derivation of this equation is given in Appendix 8.8.

The parameter Δ_{ij} in the above equation is the modified binary diffusivity and is given by:

$$\frac{1}{\Delta_{ij}} = \frac{1}{D_{ij}} + \frac{1}{D_{K,i} D_{K,j} \left(\sum_{k=1}^{n} y_k / D_{K,k}\right)} \qquad (8.8\text{-}19b)$$

Note that this modified binary diffusivity is symmetric, that is

$$\Delta_{ij} = \Delta_{ji} \qquad (8.8\text{-}20)$$

Unlike the starting equation (8.8-6) which is linearly independent for n species, eq.(8.8-19) is only linearly independent for (n-1) components because summing eq.(8.8-19a) with respect to i from 1 to n will yield zero on both sides of the resulting equation.

The (n-1) equations of eq.(8.8-19a) together with the following equation

$$\sum_{i=1}^{n} J_i = 0 \qquad (8.8\text{-}21)$$

will form a complete set of n linearly independent equations for the diffusive fluxes. Knowing the diffusive fluxes, the total flux N_T and the individual fluxes are calculated from:

$$N_T = -\sum_{i=1}^{n} \frac{J_i}{\gamma D_{K,i}} - \frac{1}{R_g T}\left(\frac{B_0 P}{\mu} + \frac{1}{\gamma}\right)\frac{dP}{dz} \qquad (8.8\text{-}22a)$$

$$N_i = J_i + y_i N_T \qquad (8.8\text{-}22b)$$

where the function γ is given by

$$\gamma = \sum_{j=1}^{n}(y_j/D_{K,j}) \tag{8.8-22c}$$

8.8.2.2 In Vector Form

Like the previous Section 8.8.1.2, we now recast the diffusive flux equation (8.8-19a) in the more elegant vector form as shown below:

$$\underline{J} = -\frac{P}{R_g T}\left[\underline{\underline{B}}^{(J)}(\underline{y})\right]^{-1}\frac{d\underline{y}}{dz} - \frac{1}{R_g T}\frac{dP}{dz}\left[\underline{\underline{B}}^{(J)}(\underline{y})\right]^{-1}\underline{\underline{A}}^{(J)}(\underline{y})\underline{y} \tag{8.8-23}$$

where $\underline{J}$ and $\underline{y}$ is the (n-1)-tupled vectors defined as:

$$\underline{J} = \begin{bmatrix} J_1 \\ J_2 \\ \vdots \\ J_{n-1} \end{bmatrix} \quad \underline{y} = \begin{bmatrix} y_1 \\ y_2 \\ \vdots \\ y_{n-1} \end{bmatrix} \tag{8.8-24}$$

The matrices $\underline{\underline{B}}^{(J)}$ and $\underline{\underline{A}}^{(J)}$ are (n-1, n-1) matrices defined as follows:

$$\underline{\underline{B}}^{(J)} = \begin{cases} \displaystyle\sum_{\substack{j=1 \\ j \neq i}}^{n} \frac{y_j}{\Delta_{ij}} & \text{for } i = j \\ -\dfrac{y_i}{\Delta_{ij}} & \text{for } i \neq j \end{cases} \tag{8.8-25a}$$

$$\underline{\underline{A}}^{(J)}(\underline{y}) = \begin{cases} \Lambda_{ii}^{(J)} = 1 - \dfrac{1}{D_{K,i}\left(\displaystyle\sum_{k=1}^{n} y_k/D_{K,k}\right)} \end{cases} \tag{8.8-25b}$$

Knowing the diffusive flux of the (n-1) components, the diffusive flux of the n-th component (J_n) is given in eq.(8.8-21) and the total flux is by eq.(8.8-22b). Finally the fluxes relative to the fixed frame of coordinates are calculated from:

$$\underline{N} = \underline{J} + N_T \underline{y} \tag{8.8-26}$$

8.8.3 Another Form of Basic Equations in terms of N

We have presented in Sections 8.8.1 and 8.8.2 the two sets of basic equations written in terms of the flux $\underline{N}$ and the diffusive $\underline{J}$, respectively. Here, we present another formulation which is also useful.

Substituting eq.(8.8-18) into eq.(8.8-19a) we get

$$\sum_{\substack{j=1 \\ j \neq i}}^{n} \frac{y_j N_i - y_i N_j}{\Delta_{ij}} = -\frac{P}{R_g T} \frac{dy_i}{dz} - \frac{y_i}{R_g T} \left(1 - \frac{1}{D_{K,i} \sum_{k=1}^{n} y_k / D_{K,k}}\right) \frac{dP}{dz} \qquad (8.8\text{-}27a)$$

for $i = 1, 2, 3, \ldots, (n-1)$.

Also, substituting eq. (8.8-22b) into eq. (8.8-22a), we obtain an additional equation:

$$\sum_{i=1}^{n} \frac{N_i}{D_{K,i}} = -\frac{1}{R_g T} \left(1 + \frac{B_0 P}{\mu} \sum_{i=1}^{n} \frac{y_i}{D_{K,i}}\right) \frac{dP}{dz} \qquad (8.8\text{-}27b)$$

Eqs.(8.8-27) provide n equations for the n unknown fluxes N_i ($i = 1, 2, \ldots, n$). The total flux and the diffusive fluxes are then given in eqs. (8.8-7).

8.8.4 Limiting Cases

We have presented the three different formulations for the bulk diffusion-Knudsen diffusion-viscous flow in Sections 8.8.1 to 8.8.3. Now we will show how these formulations can be used to derive useful equations for limiting cases which are often encountered in adsorption systems.

8.8.4.1 Dilute systems

Let us consider the simplest limiting case, where all species have very low concentration except the n-th species (the solvent), that is:

$$\begin{aligned} y_n &\approx 1 \\ y_i &\ll 1 \end{aligned} \qquad (8.8\text{-}28a)$$

In this case, the modified Stefan-Maxwell equation (eq. 8.8-1) for component $i = 1, 2, \ldots, (n-1)$ is reduced to:

$$-\frac{1}{R_g T}\frac{dp_i}{dz} \cong \frac{N_i^D}{D_{i,n}} + \frac{N_i^D}{D_{K,i}} \tag{8.8-28b}$$

in which we have made use of eq.(8.8-28a). Solving for this flux contributed by diffusion mechanism, we get:

$$N_i^D \cong -\frac{D_i^0}{R_g T}\frac{dp_i}{dz} \tag{8.8-29a}$$

where

$$\frac{1}{D_i^0} = \frac{1}{D_{K,i}} + \frac{1}{D_{i,n}} \tag{8.8-29b}$$

This relation is known as the Bosanquet relation. The upperscript (0) is denoted for dilute condition.

The viscous flux is given by eq.(8.8-2) with the viscosity now being replaced by the viscosity of the n-th species as it is the dominant species in the mixture. Thus, the flux of the species i ($i = 1, 2, 3,..., n-1$) is the sum of the diffusive flux in eq. (8.8-29a) and the viscous flux:

$$N_i \cong -D_i^0 \frac{P}{R_g T}\frac{dy_i}{dz} - \frac{y_i}{R_g T}\left(D_i^0 + \frac{B_0 P}{\mu_n}\right)\frac{dP}{dz} \tag{8.8-30a}$$

where μ_n is the viscosity of the n-th component. The above equation is further approximated by

$$N_i \cong -D_i^0 \frac{P}{R_g T}\frac{dy_i}{dz} \tag{8.8-30b}$$

because the mole fraction y_i is much smaller than unity.

8.8.4.2 Knudsen Diffusion Control

When the pore size is very small or/ and the pressure is very low, the mean free path is larger than the size of the pore. In such cases the Knudsen diffusion controls the transport as the diffusing molecules are more likely to collide with the pore wall than with themselves. In this case, the Knudsen diffusivity is much smaller than the binary diffusivities D_{ij} as the Knudsen diffusivity is proportional to the pore size while the binary diffusivity is inversely proportional to the total pressure.

In this case of Knudsen diffusion control, the flux contributed by diffusion is:

$$N_i^D \cong -\frac{D_{K,i}}{R_g T}\frac{dp_i}{dz} \quad (8.8\text{-}31)$$

that is each species is diffusing under its own partial pressure gradient, independent of the presence of the others. Combining this flux with the viscous flux, we get the following constitutive relation under the limiting case of Knudsen control.

$$N_i \cong -D_{K,i}\frac{P}{R_g T}\frac{dy_i}{dz} - \frac{y_i}{R_g T}\left(D_{K,i} + \frac{B_0 P}{\mu}\right)\frac{dP}{dz} \quad (8.8\text{-}32)$$

Eq.(8.8-32) is identical in form to eq. (8.8-30) obtained under the case of dilute conditions, with a number of subtle differences shown in the following table.

Limiting case	Diffusivity	Viscosity	Validity
Dilute conditions	D^0	that of n-th species	Valid for (n-1) species
Knudsen control	D_K	that of mixture	Valid for n species

8.8.4.3 Bulk Diffusion Control

We now consider the case whereby the pore is reasonably large and the pressure is high. In this case the mean free path is much shorter than the pore size, one would expect that the molecular diffusion will control the transport as the binary diffusivity D_{ij} is much smaller than the Knudsen diffusivity. If one accepted this argument and removed the contributions of Knudsen diffusion in eq.(8.8-6), the resulting constitutive relation for the flux would be:

$$\sum_{\substack{j=1\\j\neq i}}^{n}\frac{y_j N_i - y_i N_j}{D_{ij}} \approx -\frac{P}{R_g T}\frac{dy_i}{dz} - \frac{y_i}{R_g T}\frac{dP}{dz} \quad (8.8\text{-}33)$$

that is all terms having the Knudsen diffusivity have been removed. In so doing, we would run into an inconsistency because by summing the above equation with respect to all species, we get

$$0 \approx -\frac{1}{R_g T}\frac{dP}{dz} \quad (8.8\text{-}34)$$

Under the nonisobaric conditions, the above equation is obviously wrong. Therefore, we can not ignore the contribution of the Knudsen terms as they are the terms which relate the momentum transfer from the bulk to the boundary, which is necessary for the variation in the total pressure.

So mathematically, what went wrong when we neglect the Knudsen contributions? The fact is that the small values of binary diffusivities do not mean that the molecular diffusion term will entirely dominate as the numerator in the first term of eq.(8.8-6) appeared in the form of a difference can also be a small number.

The proper starting equation to derive the reduced equation for this limiting case is eq. (8.8-19a). The dependence of the Knudsen diffusivity, the binary diffusion coefficient and the viscous flow parameter B_0 on the pore size and pressure is:

$$D_{K,i} = D_{K,i}^0 \left(\frac{r}{r_0}\right) \tag{8.8-35a}$$

$$D_{ij} = D_{ij}^0 \left(\frac{P_0}{P}\right) \tag{8.8-35b}$$

$$B_0 = B_0^0 \left(\frac{r}{r_0}\right)^2 \tag{8.8-35c}$$

where the upperscript 0 is used to denote values at pore size r_0 or the pressure P_0.

To compare the magnitude of the two terms in the RHS of eq. (8.8-19b), we investigate their ratio

$$\frac{D_{K,i}D_{K,j}\left(\sum_{k=1}^{n} y_k / D_{K,k}\right)}{D_{ij}} = \left(\frac{r}{r_0}\right)\left(\frac{P}{P_0}\right)\frac{D_{K,i}^0 D_{K,j}^0 \left(\sum_{k=1}^{n} y_k / D_{K,k}^0\right)}{D_{ij}^0} \tag{8.8-36}$$

For large pore size or large pressure, the above ratio is large, and hence the modified diffusivity Δ_{ij} is reduced to D_{ij}. Thus, the proper equation for the case of bulk diffusion control is (from eq. 8.8-19a):

$$\sum_{\substack{j=1 \\ j \neq i}}^{n} \frac{y_j J_i - y_i J_j}{D_{ij}} = -\frac{P}{R_g T}\frac{dy_i}{dz} - \frac{y_i}{R_g T}\left(1 - \frac{1}{D_{K,i}\sum_{k=1}^{n} y_k / D_{K,k}}\right)\frac{dP}{dz} \tag{8.8-37}$$

for i = 1, 2,..., (n-1). The diffusive flux of the n-th component is given by eq.(8.8-21).

The total flux is calculated from eq.(8.8-22a). However, in studying the two terms in the bracket of the second term of that equation, we consider the ratio:

$$\frac{(B_0 P/\mu)}{(1/\gamma)} \qquad (8.8\text{-}38a)$$

where γ is defined in eq.(8.5-22c). We learnt from Chapter 7 that

1. B_0 is proportional to the square of pore size
2. D_K is proportional to pore size

Thus, the ratio of eq. (8.8-38a) will become:

$$\frac{(B_0 P/\mu)}{(1/\gamma)} = \left(\frac{r}{r_0}\right)\left(\frac{P}{P_0}\right)\left[\frac{B_0^0 P_0}{\mu}\right]\sum_{k=1}^{n}\frac{y_k}{D_{K,k}^0} \qquad (8.8\text{-}38b)$$

where P_0 and r_0 are the reference pressure and pore size, respectively. We see from the above equation that under the conditions of large pore size and large pressure that ratio is a large number, suggesting that the second term in the bracket of the RHS of eq. (8.8-22a) is negligible compared to the first term. Thus, under the conditions of bulk diffusion control, the total flux equation is:

$$N_T = -\sum_{i=1}^{n}\frac{J_i}{\gamma D_{K,i}} - \frac{1}{R_g T}\frac{B_0 P}{\mu}\frac{dP}{dz} \qquad (8.8\text{-}39)$$

Knowing the diffusive fluxes and the total flux, the fluxes relative to the fixed frame of coordinates are given by eq.(8.8-22b).

The necessary equations in this limiting case are eqs.(8.8-37), (8.8-39) and (8.8-22b), and we see that even though the bulk diffusion controls the diffusion transport, the Knudsen diffusivity <u>seems</u> to play a part in the determination of fluxes. This at first looks interesting, but observing the relevant equations in this case (eqs. 8.8-37 and 8.8-39) we see that the Knudsen diffusivities appear only in ratio form, and since the Knudsen diffusivity is inversely proportional to the molecular weight

$$D_{K,i} = \frac{2r}{3}\sqrt{\frac{8R_g T}{\pi M_i}} \qquad (8.8\text{-}40)$$

eqs. (8.8-37) and (8.8-39) can now be rewritten in terms of molecular weights rather than Knudsen diffusivities as follows:

$$\sum_{\substack{j=1 \\ j \neq i}}^{n} \frac{y_j J_i - y_i J_j}{D_{ij}} = -\frac{P}{R_g T} \frac{dy_i}{dz} - \frac{y_i}{R_g T} \left(1 - \frac{\sqrt{M_i}}{\sum_{k=1}^{n} y_k \sqrt{M_k}}\right) \frac{dP}{dz} \quad (8.8\text{-}41a)$$

for i=1,2,3,...,(n-1), and

$$N_T = -\frac{\sum_{j=1}^{n} \sqrt{M_j} \cdot J_j}{\sum_{j=1}^{n} \sqrt{M_j} \cdot y_j} - \frac{1}{R_g T} \frac{B_0 P}{\mu} \frac{dP}{dz} \quad (8.8\text{-}41b)$$

where we see that there is no dependence of the above constitutive equations on the pore radius, which should be expected for the case of bulk diffusion control.

8.8.4.4 Pure Gas

For pure gas, that is n = 1, y = 1, the relevant constitutive flux equation is:

$$N_1 = -\frac{1}{R_g T}\left(D_{K,1} + \frac{B_0 P}{\mu_1}\right)\frac{dP}{dz} \quad (8.8\text{-}42)$$

This simple equation for pure gas gives us a useful tool to study the structure of a porous solid. Making use of information such as the Knudsen diffusivity is proportional to square root of temperature and inversely proportional to the molecular weight, we can carry out experiments with different gases having different molecular weights and at different temperatures to determine the value for the tortuosity in the Knudsen relation and the viscous parameter B_0.

We note from the two terms appearing in the bracket in the RHS of eq.(8.8-42) that the viscous mechanism is more important than the Knudsen mechanism at high pressures.

If the Knudsen diffusivity takes the form

$$D_{K,1} = \frac{4K_0}{3}\sqrt{\frac{8R_g T}{\pi M_1}} \quad (8.8\text{-}43)$$

where K_0 is the structural parameter for Knudsen flow, eq.(8.8-42) will become:

$$N_1 = -\frac{1}{R_g T}\left(\frac{4K_0}{3}\sqrt{\frac{8R_g T}{\pi M_1}} + \frac{B_0 P}{\mu}\right)\frac{dP}{dz} \quad (8.8\text{-}44)$$

For a uniform cylindrical pore of radius r, the structural parameters K_0 and B_0 take the following form

$$K_0 = \frac{r}{2}$$

and

$$B_0 = \frac{r^2}{8}$$

Eq.(8.8-44) shows the explicit dependence of the flux in terms of pressure, temperature, the properties of the diffusing gas (that is, molecular weight and viscosity) and the structural parameters of the solid K_0 and B_0.

Example 8.8-1: *Steady state flow of pure gas through a capillary*

For a capillary where the pressures at two ends are constant, the flux equation (8.8-44) can be integrated to give the following integral equation for the flux written in terms of the pressures at two ends:

$$N_1 = \frac{1}{R_g T}\left\{\frac{4K_0}{3}\sqrt{\frac{8R_g T}{\pi M_1}}\cdot\frac{(P_0 - P_L)}{L} + \frac{B_0}{2\mu}\cdot\frac{(P_0^2 - P_L^2)}{L}\right\} \quad (8.8\text{-}45)$$

in which we have assumed the viscosity is independent of pressure. For gases at moderate pressures, such an assumption is reasonable.

Eq.(8.8-45) can be rearranged as follows:

$$\frac{N_1 R_g T}{(P_0 - P_L)/L} = \frac{4K_0}{3}\sqrt{\frac{8R_g T}{\pi M_1}} + \frac{B_0}{\mu}\cdot\frac{(P_0 + P_L)}{2} \quad (8.8\text{-}46)$$

A plot of $N_1 R_g T/[(P_0 - P_L)/L]$ versus the mean pressure $(P_0 + P_L)/2$ would yield a straight line with a slope of

$$\frac{B_0}{\mu} \quad (8.8\text{-}47a)$$

and an intercept of

$$\frac{4K_0}{3}\sqrt{\frac{8R_g T}{M_1}} \quad (8.8\text{-}47b)$$

Knowing the viscosity of the diffusing gas and its molecular weight, the structural parameters K_0 and B_0 can be determined from such plot.

Example 8.8-2: *Effect of temperature on the steady state flow of pure gas*

To see the effect of temperature on the Knudsen flow and the viscous flow, we need only to consider the ratio of the two terms in the RHS of eq.(8.8-46):

$$\frac{\text{Knudsen}}{\text{Viscous}} = \frac{\dfrac{4K_0}{3}\sqrt{\dfrac{8R_gT}{\pi M_1}}}{\dfrac{B_0(P_0+P_L)}{\mu}\cdot\dfrac{1}{2}} = \frac{4}{3}\frac{K_0\mu}{B_0}\cdot\frac{\sqrt{\dfrac{8R_gT}{\pi M_1}}}{(P_0+P_L)/2} \tag{8.8-48}$$

For gases, the viscosity increases with temperature; thus the above ratio increases with temperature, meaning that the Knudsen flow is more important than the viscous flow at high temperatures.

Example 8.8-3: *Pure gas flow through two capillaries in series*

When two capillaries of different radii are connected in series and the pressures at two ends are kept constant, the fluxes through these two capillaries are:

$$N^{(1)} = \frac{1}{R_gT}\left\{\frac{4K_0^{(1)}}{3}\sqrt{\frac{8R_gT}{\pi M}}\frac{(P_0-P^*)}{L^{(1)}} + \frac{B_0^{(1)}}{2\mu}\cdot\frac{\left(P_0^2-P^{*2}\right)}{L^{(1)}}\right\} \tag{8.8-49a}$$

$$N^{(2)} = \frac{1}{R_gT}\left\{\frac{4K_0^{(2)}}{3}\sqrt{\frac{8R_gT}{\pi M}}\frac{(P^*-P_L)}{L^{(2)}} + \frac{B_0^{(2)}}{2\mu}\cdot\frac{\left(P^{*2}-P_L^2\right)}{L^{(2)}}\right\} \tag{8.8-49b}$$

where P* is the pressure at the junction of the two capillaries. The upperscripts (1) and (2) denotes the first and second capillaries, respectively.

Continuity of mass at the junction of the two capillaries requires that:

$$A^{(1)}N^{(1)} = A^{(2)}N^{(2)} \tag{8.8-50}$$

For a given P_0 and P_L, eqs.(8.8-49) and (8.8-50) represent three algebraic equations in terms of three unknowns $N^{(1)}$, $N^{(2)}$ and P*.

8.9 Transient Analysis of Bulk-Knudsen - Viscous Flow in a Capillary

We have presented the various constitutive flux equations for the general case of combined bulk-Knudsen and viscous flow in the last section. Now let us illustrate its application to the simple case of transient flow of a diffusing mixture in a capillary exposed to an infinite environment.

The mass balance equation describing the partial pressure variation of the component "i" in the capillary is:

$$\frac{\partial}{\partial t}\left(\frac{p_i}{R_g T}\right) = -\frac{\partial N_i}{\partial z} \tag{8.9-1}$$

for $i = 1, 2, ..., n$. Or written in terms of vector form, the above equation becomes:

$$\frac{\partial}{\partial t}\left(\frac{\underline{p}}{R_g T}\right) = -\frac{\partial \underline{N}}{\partial z} \tag{8.9-2}$$

where $\underline{p} = [p_1 \; p_2 \; \cdots \; p_n]^T$ and $\underline{N} = [N_1 \; N_2 \; \cdots \; N_n]^T$.

The molar flux N_i in eq. (8.9-2) is related to the total pressure gradient and the partial pressure gradients of all species according to eq.(8.8-13a). Substitution of this equation into eq. (8.9-2) will give the following form for the mass balance equation:

$$\frac{\partial}{\partial t}\left(\frac{\underline{p}}{R_g T}\right) = \frac{\partial}{\partial z}\left\{\frac{1}{R_g T}\left[\underline{\underline{B}}(\underline{y},P,T;r)\right]^{-1}\frac{\partial \underline{p}}{\partial z} + \frac{B_0}{\mu_m R_g T}\frac{\partial P}{\partial z}\left[\underline{\underline{B}}(\underline{y},P,T;r)\right]^{-1}\underline{\underline{\Lambda}}\,\underline{p}\right\} \tag{8.9-3}$$

where $\underline{\underline{B}}(\underline{y})$ and $\underline{\underline{\Lambda}}$ are function of mole fraction, total pressure and temperature as given in eq.(8.8-11), and the mole fractions are given by:

$$\underline{y} = \frac{\underline{p}}{P} \tag{8.9-4}$$

The parameter μ_m is the viscosity of the mixture, and for gases at moderate pressures (less than 10 atm) it is adequately calculated from the semi-empirical formula of Wilke et al. (Bird et al., 1960)

$$\mu_m = \sum_{i=1}^{n} \frac{y_i \mu_i}{\sum_{j=1}^{n} y_j \Phi_{ij}}, \tag{8.9-5}$$

where $\underline{y}$ is the mole fraction vector, $\underline{\mu}$ is the pure component viscosity vector and Φ_{ij} in the denominator is given by

$$\Phi_{ij} = \frac{1}{\sqrt{8}}\left(1+\frac{M_i}{M_j}\right)^{-1/2}\left[1+\left(\frac{\mu_i}{\mu_j}\right)^{1/2}\left(\frac{M_j}{M_i}\right)^{1/4}\right]^2 \tag{8.9-6}$$

Eq.(8.9-3) represent n partial differential equations for the n unknowns $p_1, p_2, ..., p_n$. The boundary conditions for this problem are:

$$z = 0; \quad \frac{\partial \underline{p}}{\partial z} = \underline{0} \tag{8.9-7a}$$

$$z = L; \quad \underline{p} = \underline{p}_b \tag{8.9-7b}$$

where $\underline{p}_b$ are constant partial pressures in the bulk surrounding the capillary.

The initial conditions are:

$$t = 0; \quad \underline{p} = \underline{p}_i \tag{8.9-8}$$

8.9.1 Nondimensional Equations:

The mass balance equations presented above are non-linear and must be solved numerically. We have maintained our practice that before the equations are solved numerically, they must be cast into a non-dimensional form, at least the spatial variable must be normalised as it is convenient to discretise a spatial domain having a range of (0, 1). We define the following nondimensional variables

$$\eta = \frac{z}{L}; \quad \tau = \frac{D_T t}{L^2} \tag{8.9-9}$$

where D_T is some reference diffusion coefficient. In terms of these nondimensional variables, the governing mass balance equation will become

$$\frac{\partial \underline{p}}{\partial \tau} = \frac{\partial}{\partial \eta}\left\{\left[D_T \underline{\underline{B}}(\underline{y}, P, T; r)\right]^{-1}\frac{\partial \underline{p}}{\partial \eta} + \Phi\left(\frac{\mu_0}{\mu_m}\right)\left(\frac{1}{P_0}\frac{\partial P}{\partial \eta}\right)\left[D_T \underline{\underline{B}}(\underline{y}, P, T; r)\right]^{-1}\left(D_T \underline{\underline{A}}\right)\underline{p}\right\} \tag{8.9-10}$$

where P_0 is some reference pressure, μ_0 is some reference viscosity, and the parameter Φ is a dimensionless group that characterise the strength of the viscous flow relative to the diffusive flow and is defined as follows:

$$\Phi = \frac{B_0 P_0}{\mu_0 D_T} \tag{8.9-11}$$

The role of the viscous flow is to dissipate any variation in the total pressure, that is the larger is the parameter Φ the faster is the system towards isobaric conditions. To illustrate the magnitude of this parameter, we take the following parameters typical for diffusion of most gases at ambient temperature in a capillary having a radius of 2 micron:

$\mu_0 = 1 \times 10^{-4}$ g cm^{-1} sec^{-1} (typical for most gases @ 20°C)

$P_0 = 1 \times 10^6$ g cm^{-1} sec$^{-2} \approx$ (1 atm)

$r = 2 \times 10^{-4}$ cm $\quad \Rightarrow \quad B_0 = \dfrac{r^2}{8} = 5 \times 10^{-9}$ cm^2

$D_T = 0.5$ cm^2/sec

We calculate the non-dimensional parameter Φ:

$$\Phi = \frac{(5 \times 10^{-9})(1 \times 10^6)}{(1 \times 10^{-4})(0.5)} = 100$$

This order of magnitude of 100 suggests that the viscous flow is so significant in this capillary of 1 micron, and hence any disturbance in the total pressure will be dissipated very quickly by the action of the viscous flow before any diffusive processes occur. For smaller pores, say $r = 0.2$ micron, the viscous parameter Φ is equal to 1, suggesting that the viscous flow is very comparable to the diffusion flow and it can not be neglected in the description of flow into the particle. What this means is that the total pressure variation will persist in the system while the mass transfers of all diffusing species are occurring.

The boundary and initial conditions of eqs.(8.9-10) are:

$$\eta = 0; \quad \frac{\partial \underline{p}}{\partial \eta} = \underline{0} \tag{8.9-12a}$$

$$\eta = 1; \quad \underline{p} = \underline{p}_b \tag{8.9-12b}$$

$$t^* = 0; \quad \underline{p} = \underline{p}_i \tag{8.9-13}$$

The set of non-dimensional model equations (eqs. 8.9-10 to 8.9-13) is solved numerically by a combination of the orthogonal collocation method and the Runge-Kutta method. Details of the collocation analysis are given in Appendix 8.9. A programming code ADSORB5A is provided with this book to give the reader a means to investigate the adsorption kinetics into a capillary where bulk diffusion, Knudsen diffusion and viscous flow are all operating. We illustrate this exercise with the following examples.

Example 8.9-1: *Diffusion of methane, argon and hydrogen in a capillary*

We take a system studied by Arnold and Toor (1967) where the three diffusing species are methane (1), argon (2) and hydrogen (3). The capillary is having a length of 20 cm and a pore radius of 8×10^{-5} cm. The operating conditions are 307 K and 1 atm. At these conditions, the binary diffusivities are:

$D_{12} = 0.2157$ cm^2/sec, $D_{13} = 0.7716$ cm^2/sec, $D_{23} = 0.8335$ cm^2/sec

Other properties of the three diffusing species are listed below.

	MW	μ (g/cm/sec)
Methane (1)	16	1.1×10^{-4}
Argon (2)	40	2.3×10^{-4}
Hydrogen (3)	2	0.91×10^{-4}

The Knudsen diffusivities are calculated using eq. (7.4-10):

$D_{K,1} = 3.4$ cm^2/sec, $D_{K,2} = 2.15$ cm^2/sec, $D_{K,3} = 9.61$ cm^2/sec

To simulate the problem using the code ADSORB5A, we need to specify the initial pressures and the pressures of the bulk surrounding the capillary. We use the following initial and boundary conditions:

Initial partial pressure (atm)	Bulk partial pressure(atm)
p(1) = 0.515	p(1) = 0
p(2) = 0.485	p(2) = 0.509
p(3) = 0	p(3) = 0.491

Using the code, we obtain the partial pressure profiles of all species as a function of time. Figure 8.9-1 shows the mean total pressure in the capillary versus time. Here we see that the mean total pressure in the capillary increases with time during the early stage of the diffusion. This is due to the faster diffusion of hydrogen into the capillary, more than to compensate for the slower outflux of methane. After this initial stage, the total pressure finally decays to the equilibrium total pressure of 1 atm.

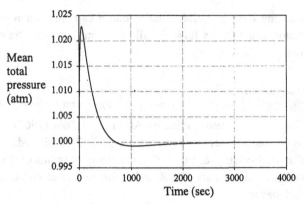

Figure 8.9-1: Plot of the mean total pressure versus time

Figure 8.9-2 shows plots of the mean partial pressures of methane, argon and hydrogen as a function of time. The behaviour of methane and hydrogen as expected, that is downhill diffusion. The behaviour of argon is interesting. Instead of going up from the initial pressure of 0.485 to a final bulk pressure of 0.509 atm, the argon mean partial pressure decreases with time during the early stage of diffusion. This behaviour is due to the outflux of methane and it drags argon to the bulk, resulting in the initial drop of argon pressure. Once the rate of outflux of methane is decreased, the argon would diffuse in a normal fashion, that is higher pressure of argon in the bulk will cause a diffusion into the capillary, resulting in an increase in the argon pressure in the capillary (see Do and Do (1998) for more simulations).

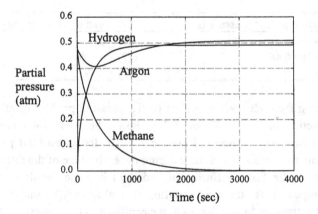

Figure 8.9-2: Plots of the partial pressures versus time

8.10 Maxwell-Stefan for Surface Diffusion

We have considered the Maxwell-Stefan approach in dealing with the bulk diffusion, Knudsen diffusion and viscous flow. Now we apply the concept of friction to surface diffusion where adsorbed molecules exhibit mobility on a surface. Here we view the surface diffusion as the hopping by the molecule from one site to the next. The hopping mechanism was considered by de Boer (1953, 1968) and Gilliland et al. (1974), and it offers a reasonable description of surface transport in systems having adsorption energies exceeding the thermal energy of the adsorbate molecule. Because of the heterogeneity of the surface the surface diffusivity increases with an increase in the surface loading, owing to the fact of progressive filling of sites of decreasing strength.

To deal with surface diffusion, it is usually assumed that (Cunningham and Williams, 1980)

1. The adsorbed molecules are located on a plane that is parallel to the solid surface and located at a distance corresponding to the minimum of the total potential energy.
2. The adsorbed molecules can move randomly over this surface.
3. The adsorbed molecules form a two dimensional rarefied gas in which only binary collisions are important.
4. The solid is at thermodynamic equilibrium
5. The phenomena of adsorption and desorption have no effect on transport properties of the adsorbed molecules.

Let $C_{s,j}$ be the surface concentration of the species j measured as mole of j per unit area. The total surface concentration is the sum of all these concentrations, i.e.

$$C_{s,T} = \sum_{j=1}^{n} C_{s,j} \tag{8.10-1}$$

The fractional of sorption sites taken up by the species j is:

$$\theta_j = \frac{C_{s,j}}{C_{sm}} \tag{8.10-2}$$

where C_{sm} is the saturation concentration. The fraction of the vacant site is:

$$\theta_V = 1 - \theta_t = 1 - \sum_{j=1}^{n} \theta_j \tag{8.10-3}$$

To formulate the Stefan-Maxwell approach for surface diffusion, we will treat the adsorption site as the pseudo-species in the mixture, a concept put forwards by Krishna (1993). If we have n species in the system, the pseudo species is denoted as the (n+1)-th species, just like the way we dealt with Knudsen diffusion where the solid object is regarded as an assembly of giant molecules stationary in space. We balance the force of the species i by the friction between that species i with all other species to obtain:

$$-\frac{d\mu_i}{dz} = R_g T \sum_{j=1}^{n} \theta_j \frac{(v_i - v_j)}{D_{ij}^s} + R_g T \theta_V \frac{(v_i - v_V)}{D_{iV}^s} \qquad (8.10\text{-}4)$$

for i = 1, 2,..., n. The coefficient $D_{i,v}$ is the single component surface diffusivity, and it describes the exchange between the sorbed species and the vacant sites. The coefficient D_{ij}^s can be regarded as the counter exchange coefficient between the species i and the species j at the same site. This counter exchange coefficient is related to the single component surface diffusivities, $D_{i,v}$ and $D_{j,v}$, and is given by (Krishna, 1990)

$$D_{i,j}^s = \left(D_{i,v}\right)^{\theta_i/(\theta_i+\theta_j)} \left(D_{j,v}\right)^{\theta_j/(\theta_i+\theta_j)} \qquad (8.10\text{-}5)$$

which is originally suggested by Vignes (1966).

The surface chemical potential is given by the equilibrium relationship with the gas phase:

$$\mu_i = \mu_i^0 + R_g T \ln f_i \qquad (8.10\text{-}6)$$

where μ_i^0 is the chemical potential at the standard state, and f_i is the fugacity of the species i in the bulk fluid mixture.

Written in terms of velocity is not useful for direct flux calculation, we now write eq. (8.10-4) in terms of the surface flux which is defined in the following equation:

$$N_i^s = C_{sm} \theta_i v_i \qquad (8.10\text{-}7)$$

It is noted that the total surface concentration is the same for all species. Using this definition, the Stefan-Maxwell equation to describe surface diffusion is:

$$-\frac{\theta_i}{R_g T} \frac{d\mu_i}{dz} = \sum_{j=1}^{n} \frac{\theta_j N_i^s - \theta_i N_j^s}{C_{sm} D_{ij}^s} + \frac{\theta_V N_i^s - \theta_i N_V^s}{C_{sm} D_{iV}^s} \qquad (8.10\text{-}8)$$

The surface chemical potential gradient can be expressed in terms of the surface concentration as follows:

$$\frac{\theta_i}{R_g T}\frac{d\mu_i}{dz} = \sum_{j=1}^{n} \Gamma_{ij}\frac{d\theta_j}{dz} \qquad (8.10\text{-}9a)$$

where

$$\Gamma_{ij} = \theta_i \frac{\partial \ln f_i}{\partial \theta_j} \qquad (8.10\text{-}9b)$$

For the Langmuir isotherm, where the fractional coverage is expressed in terms of the fugacities of all species, the coefficient Γ_{ij} is:

$$\Gamma_{ij} = \delta_{ij} + \frac{\theta_i}{\theta_V} \qquad (8.6\text{-}10)$$

Krishna (1990) used the argument that the vacancy flux must balance the sum of fluxes of all species, i.e.

$$N_V^s = -\sum_{j=1}^{n} N_j^s \qquad (8.10\text{-}11)$$

This relates the vacancy flux in terms of the other fluxes.

8.10.1 Surface Diffusivity of Single Species

The surface diffusion is related to the jump distance as well as the jump frequency as follows:

$$D_{j,V} = \lambda^2 v_j(\theta_j) \qquad (8.10\text{-}12)$$

where the jump frequency would depend on the fractional coverage. Readers interested in this approach to surface diffusion should refer to Krishna (1990) for further exposition of this topic.

8.11 Conclusion

We have presented in this chapter a systematic approach of Maxwell-Stefan approach in dealing with flow in homogeneous media as well as inside a capillary or porous media. The approach using the concept of friction is elegant, and it puts the various flow mechanisms under the same framework. For large pore space, the Maxwell-Stefan approach is comprehensive, and it is able to describe experimental results, such as the uphill diffusion, which the traditional Fick's law approach can not explain. Section 8.8 is the most general, and it provides constitutive flux equations that are applicable under general non-isobaric and non-isothermal conditions. You will find the application of this Maxwell-Stefan approach to adsorption systems in Chapters 9 and 10. The theory of surface diffusion using the Maxwell-Stefan approach is not as mature as the Maxwell-Stefan approach to the flow mechanisms in the large void space of the particle. The inherent difficulty of this approach to surface diffusion is that the surface structure is very complex, and its microstructure has a significant influence on the surface diffusion rate. Our present failure of fully understanding this microstructure has prevented us from achieving a good theory for surface diffusion. However, research in this area has achieved great progress in the last few years, and it is hoped that we will see some significant progress in the near future.

9

Analysis of Adsorption Kinetics in a Single Homogeneous Particle

9.1 Introduction

We have discussed in the last two chapters about the various transport mechanisms (diffusive and viscous flows) within a porous particle (Chapter 7) and the systematic approach of Stefan-Maxwell in solving multicomponent problems (Chapter 8). The role of diffusion in adsorption processes is important in the sense that in almost every adsorption process diffusion is the rate limiting step owing to the fact that the intrinsic adsorption rate is usually much faster than the diffusion rate. This rate controlling step has been recognized by McBain almost a century ago (McBain, 1919). This has prompted much research in adsorption to study the diffusion process and how this diffusional resistance can be minimized as the smaller is the time scale of adsorption the better is the performance of a process.

Modelling of diffusion in adsorption process is usually started with the assumption of treating solid as a unstructured homogeneous medium, that is the solid characteristics is uniform throughout the solid volume. The diffusion process is usually characterized by a Fickian like diffusion law with a parameter called the effective diffusivity, which is a function of mechanisms of flow within the particle (Figure 9.1-1). Detailed study of diffusion in Chapters 7 and 8 gives information on how this effective diffusivity can be calculated. For example, if the diffusion process is controlled by a combination of the Knudsen diffusion and the continuum diffusion, the effective diffusivity can be calculated by the formula given in eq. (8.8-29b). The advantage of this simple Fickian formulation is that analytical solutions are usually possible, allowing us to have physical insight about an adsorption process. If further details are required, more complex models must be formulated and computed usually numerically, and even in this case simple analytical solutions are valuable to confirm the numerical computation of the complex models when some limits are met. Complex models usually incorporate most or all possible phenomena occurring in the particle, and hence solution of a complex model always gives information about adsorption kinetics as a function of system parameters in greater details but at the expense of extra effort in the analysis.

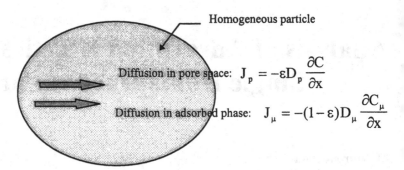

Figure 9.1-1: Schematic diagram of a homogeneous particle

In this chapter, we will present the analysis of a number of adsorption models commonly used in the literature. We will start first with a simple model. This is important to show you how adsorption kinetics would vary with parameters such as particle size, bulk concentration, pressure, temperature, pore size, and adsorption affinity. This will be studied for single component system with special adsorption isotherm such as linear isotherm and irreversible isotherm (rectangular isotherm). When dealing with nonlinear isotherm, numerical methods have to be employed to solve coupled governing equations. After understanding how single component systems behave, we will discuss multicomponent systems and apply the Stefan-Maxwell approach learnt in Chapter 8 to formulate model equations. For such cases numerical methods are used to solve the governing equations. Orthogonal collocation method is useful for this task and is chosen simply due to the author's personal taste. This book contains a number of programming codes written in MatLab language to solve some practical multicomponent systems. Students can use these interactively to understand better how a multicomponent system behaves.

After understanding the various features exhibited by a homogeneous system, the next chapter (Chapter 10) will deal with zeolitic particles and Chapter 11 will address heterogeneous systems as it is now increasingly recognized that most practical solids are not homogeneous, but rather heterogeneous in the sense that the pore structure is not uniform but rather distributed, that is the interaction energy between solid and adsorbate molecule is not constant but rather distributed. The heterogeneity in pore structure (that is the coexistence of small and large pores) is actually quite beneficial from the practical view point because large pores are good for transport, while small pores are good for capacity. Thus, the extra complexity in modelling is well worth an effort to determine how we could further utilise this heterogeneity to our advantage.

9.2 Adsorption Models for Isothermal Single Component Systems

We first discuss the various simple models, and start with linear models, favoured for the possibility of analytical solution which allows us to study the system behaviour in a more explicit way. Next we will discuss nonlinear models, and under special conditions such as the case of rectangular isotherm with pore diffusion analytical solution is also possible. Nonisothermal conditions are also dealt with by simply adding an energy balance equation to mass balance equations. We then discuss adsorption behaviour of multicomponent systems.

9.2.1 Linear Isotherms

Damkohler (1935) is perhaps one of the pioneers to develop diffusion models to understand adsorption process within a particle. In his development he assumed a parallel transport of molecules occurring in both the void space as well as the adsorbed phase (Figure 9.2-1).

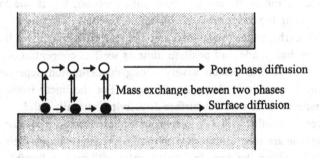

Figure 9.2-1: Schematic of parallel pore and surface diffusion

The constitutive flux equation accounting for this dual flow of molecule is:

$$J = -\varepsilon D_p \frac{\partial C}{\partial r} - (1-\varepsilon)D_s \frac{\partial C_\mu}{\partial r} \qquad (9.2\text{-}1a)$$

This combined transport is known to exist in many systems. The partition between the fluid and adsorbed phases is assumed linear as one would understand that a nonlinear partition would require numerical computation, a luxury not readily available in the early 30's. To set the consistency of the terminology used in this chapter as well as in subsequent chapters, we will define carefully all the variables in a way that there is no ambiguity and the complete list of variables as well as parameters used are listed in the nomenclature section. The mass balance equation can be obtained by carrying out mass balance around a thin shell element in the particle, and in so doing we obtain the following equation:

$$\varepsilon \frac{\partial C}{\partial t} + (1-\varepsilon)\frac{\partial C_\mu}{\partial t} = \varepsilon D_p \frac{1}{r^s}\frac{\partial}{\partial r}\left(r^s \frac{\partial C}{\partial r}\right) + (1-\varepsilon)D_s \frac{1}{r^s}\frac{\partial}{\partial r}\left(r^s \frac{\partial C_\mu}{\partial r}\right) \quad (9.2\text{-}1b)$$

where ε is the voidage of the particle (cc void volume/cc of total particle including void and solid volume), C is the fluid concentration (mole/cc of fluid), C_μ is the concentration in the adsorbed phase (mole/cc of adsorbed phase), D_p is the pore diffusivity (based on the empty cross section of the particle), D_s is the surface diffusivity (based on the solid cross section), and s is the particle shape factor (s = 0, 1 and 2 for slab, cylinder and sphere, respectively). The definition of the voidage requires some further explanation. The voidage ε is the transport void fraction, that is only the void space available for the transport is accounted for in this porosity; since only macropore and mesopore are available for transport by free molecules, this porosity can be known as the macropore porosity. The volume of the adsorbed phase in the definition of the adsorbed concentration, C_μ, is the volume of the particle excluding the transport void volume.

In eq. (9.2-1b), we have assumed that the pore diffusivity as well as the surface diffusivity are independent of position, time as well as concentration. The surface diffusivity is known to exhibit a very strong concentration dependence over the range where adsorption isotherm is nonlinear. For the linear isotherm dealt with here, the assumption of constant surface diffusivity is usually valid.

The free molecules in the pore space and the adsorbed molecules at any point within a particle are in equilibrium with each other even though their concentration gradients exist within the particle. This local equilibrium is feasible only when at any point within the particle the local adsorption kinetics is much faster than the diffusion process into the particle. This is usually the case in most practical solids. In this section, we will assume a linear partition between the two phases; thus the relationship is known as the local linear adsorption isotherm. The term local is because that particular condition is only applicable to a given position; as time approaches infinity this local adsorption isotherm will become the global adsorption isotherm (true equilibrium) as there is no gradient in concentration either in the pore space or on the surface phase at $t = \infty$. The local linear isotherm takes the form:

$$C_\mu = KC \qquad (9.2\text{-}2a)$$

where K is called the Henry constant, and is defined as

$$K = K_\infty \exp\left(\frac{Q}{R_g T}\right) \qquad (9.2\text{-}2b)$$

with K_∞ being the Henry constant at infinite temperature, and Q being the heat of adsorption. The Henry constant has the units of

$$K = \frac{(\text{moles}/\text{cc solid})}{(\text{moles}/\text{cc gas})} = \frac{(\text{cc gas})}{(\text{cc solid})} \equiv \text{dimensionless} \qquad (9.2\text{-}2c)$$

Substituting eq. (9.2-2a) into eq. (9.2-1b), we obtain the familiar looking Fickian diffusion equation written in terms of only the gas phase concentration

$$\frac{\partial C}{\partial t} = D_{app} \nabla^2 C \qquad (9.2\text{-}3a)$$

which is only applicable when the assumptions of local equilibrium and linear isotherm hold.

The parameter D_{app} is called the <u>apparent diffusivity</u> because it embeds both the two diffusion coefficients and the slope of the isotherm, K. It is defined as follows:

$$D_{app} = \frac{\varepsilon D_p + (1-\varepsilon) K D_s}{\varepsilon + (1-\varepsilon) K} \qquad (9.2\text{-}3b)$$

The apparent diffusivity describes the approach of a system response to equilibrium. The larger is this parameter, the sooner the system approaches equilibrium. A system having high mobility in both free and adsorbed phases does not necessarily mean that it would approach equilibrium quickly. It does also depend on the quantity which can be accommodated by the solid at equilibrium. The speed to approach equilibrium depends on the two factors - mobility and capacity - in the form of ratio as reflected in the definition of the apparent diffusivity. Let us demonstrate this with the following two systems.

System # 1	System # 2
$\varepsilon = 0.33$	$\varepsilon = 0.33$
$D_p = 0.01 \text{ cm}^2 \text{ s}^{-1}$	$D_p = 0.001 \text{ cm}^2 \text{ s}^{-1}$
$K = 1000$	$K = 10$
$D_s = 1 \times 10^{-5} \text{ cm}^2 \text{ s}^{-1}$	$D_s = 0 \text{ cm}^2 \text{ s}^{-1}$

The system #1 has higher mobility as reflected in the value of $\varepsilon D_p + (1-\varepsilon) K D_s$ of 0.01 cm² s⁻¹ compared to 3.3×10^{-4} for the system # 2. Despite its much higher mobility, the system # 1 takes longer time to approach equilibrium as its apparent diffusivity of 1.5×10^{-5} cm^2 s^{-1} is lower than that of the system # 2 of 4.7×10^{-5}.

Before solving the adsorption equation (9.2-3a), it is important to study the behaviour of various diffusivities with respect to temperature.

9.2.1.1 Temperature Dependence of the Apparent Diffusivity D_{app}

The apparent diffusivity of eq.(9.2-3b) involves the contribution of the pore diffusion and the surface diffusion. The relative importance of the pore and surface diffusions must be studied to have an insight on how this would vary with temperature.

9.2.1.1.1 The Relative Importance of Pore and Surface Diffusions

To investigate the relative contribution of these two processes, let us study the following ratio:

$$\delta = \frac{(1-\varepsilon)KD_s}{\varepsilon D_p} \tag{9.2-4}$$

The temperature dependence of the relevant parameters in eq.(9.2-4) is given below:

$$K = K_\infty \exp\left(\frac{Q}{R_g T}\right) \tag{9.2-5a}$$

$$D_p = D_{p0}\left(\frac{T}{T_0}\right)^\alpha \tag{9.2-5b}$$

$$D_s = D_{s\infty} \exp\left(-\frac{E_s}{R_g T}\right) \tag{9.2-5c}$$

where the exponent $\alpha = 0.5$ if Knudsen mechanism controls the pore diffusion and it is equal to about 1.75 if molecular-molecular collision mechanism controls (see Chapter 7). In eqs.(9.2-5), $D_{s\infty}$ is the surface diffusivity at infinite temperature and D_{p0} is the pore diffusivity at some reference temperature T_0.

Substitution of eqs.(9.2-5) into eq.(9.2-4) gives:

$$\delta = \frac{(1-\varepsilon)K_\infty D_{s\infty} \exp[(Q-E_s)/R_g T]}{\varepsilon D_{p0}(T/T_0)^\alpha} \tag{9.2-6}$$

Since the heat of adsorption is usually larger than the activation energy for surface diffusion ($Q > E_s$), the ratio δ will decrease as temperature increases, indicating that the surface diffusion is less important at high temperatures compared to the pore diffusion. Schneider and Smith (1968) have utilized this fact to study surface diffusion in the butane/silica gel adsorption system. Their procedure is briefly described here. Firstly experiments are conducted at high temperatures such that pore diffusion is the only transport mechanism. Analysing of these experiments allows us to understand the characteristics of pore diffusion. Once this is done, experimental results of lower temperatures are analysed to derive information about the surface diffusion rate and thence the surface diffusivity.

9.2.1.1.2 Temperature Dependence of the Combined Diffusivity εD_p + (1- ε)KD$_s$

The combined diffusivity $\varepsilon D_p + (1-\varepsilon)KD_s$ is a measure of the steady state flux as solving eq. (9.2-3a) at steady state subject to constant boundary conditions would give:

$$J_{SS} = \left[\varepsilon D_p + (1-\varepsilon)KD_s\right]\frac{\Delta C}{L} = \frac{\left[\varepsilon D_p + (1-\varepsilon)KD_s\right]}{R_g T}\frac{\Delta P}{L}$$

Thus an understanding of this combined diffusivity with temperature will give a direct information on how the steady state flux would vary with temperature for a given concentration gradient ($\Delta C/L$). For a given pressure gradient $\Delta P/L$, the temperature-dependent coefficient of the flux equation is

$$\left[\varepsilon D_p + (1-\varepsilon)KD_s\right]/R_g T \qquad (9.2\text{-}7)$$

This coefficient is known as the permeability coefficient, and its units are mole m^2 s^{-1} Joule^{-1}.

Substitution of K, D_p and D_s in eqs.(9.2-5) as function of temperature into the permeability coefficient (eq. 9.2-7) yields

$$\frac{\varepsilon D_p + (1-\varepsilon)KD_s}{R_g T} = \frac{\varepsilon D_{po}(1+\theta)^\alpha + (1-\varepsilon)K_0 D_{s0}\exp\left(-\gamma\dfrac{\theta}{1+\theta}\right)}{R_g T_0(1+\theta)} \qquad (9.2\text{-}8)$$

where K_0 and D_{s0} are the Henry constant and surface diffusivity at some reference temperature T_0, respectively, and the parameter group γ and the non-dimensional temperature θ are defined as follows:

$$\gamma = \frac{Q - E_s}{R_g T_0} \qquad (9.2\text{-}9a)$$

$$\theta = \frac{T - T_0}{T_0} \qquad (9.2\text{-}9b)$$

We define the following dimensionless permeability coefficient by scaling it against the pore diffusivity at the reference temperature T_0:

$$F(\theta) = \frac{\varepsilon D_p + (1-\varepsilon)KD_s}{\varepsilon D_{p0}} \frac{T_0}{T} = \frac{(1+\theta)^\alpha + \delta_0 \exp\left(-\gamma \frac{\theta}{1+\theta}\right)}{1+\theta} \qquad (9.2\text{-}10)$$

where δ_0 is the value of δ defined in (eq. 9.2-4) at the temperature T_0.

The range of some parameters

Before we investigate the behaviour of the combined diffusivity with respect to temperature, it is important to note the practical range of the relevant parameters. The parameter α characterises the variation of pore diffusivity with temperature. It has the following range

$$0.5 < \alpha < 1.75$$

The value of 0.5 for α corresponds to the Knudsen diffusion mechanism, i.e. molecule-wall collision dominates and $\alpha = 1.75$ corresponds to the molecule-molecule collision.

The heat of adsorption usually ranges from 10 to 60 kJoule/mole and the activation energy for surface diffusion ranges from about one third of the heat of adsorption to the heat of adsorption. Thus the parameter γ has the following practical range

$$2 < \gamma < 12$$

The parameter δ_0 is the measure of the surface diffusion flux to the pore diffusion flux. It ranges from 0 (no surface diffusion) to a very large value corresponding to a system of high mobility of the adsorbed phase. Typically

$$0 < \delta < 2$$

Now back to our dimensionless permeability coefficient (eq. 9.2-10). To investigate how $F(\theta)$ changes with θ, we investigate its first derivative with respect to θ:

$$F'(\theta) = \frac{(\alpha-1)(1+\theta)^{\alpha} - \delta_0 \left[\dfrac{\gamma}{(1+\theta)} + 1\right] \exp\left(-\gamma \dfrac{\theta}{1+\theta}\right)}{(1+\theta)^2} \qquad (9.2\text{-}11)$$

When $\alpha < 1$, the above first derivative is always negative; therefore the permeability coefficient of eq. (9.2-7) always decreases with temperature. However, when $\alpha > 1$, a stationary point is possible when $F'(\theta^*) = 0$, where θ^* is the solution of the following equation:

$$\frac{(\alpha-1)(1+\theta^*)^{\alpha+1}}{\delta_0[\gamma+(1+\theta)]} = \exp\left(-\gamma \frac{\theta^*}{1+\theta^*}\right) \qquad (9.2\text{-}12)$$

An investigation of eq.(9.2-12) reveals that there is always a solution θ^* within the physical range of temperature (that is $\theta > -1$ or $T > 0$).

To see whether the stationary point is a minimum or a maximum, we investigate the sign of the second derivative at the stationary point:

$$F''(\theta) = \alpha(\alpha-1)(1+\theta)^{\alpha-4}\left[(1+\theta^*) + \frac{\gamma}{\alpha}\frac{\gamma+2(1+\theta^*)}{\gamma+(1+\theta^*)}\right] > 0 \qquad (9.2\text{-}13)$$

Thus, the stationary point is a minimum for $\alpha > 1$. Figure 9.2-2 presents a plot of $F(\theta)$ versus θ for $\alpha = 1.75$, $\delta_0 = 1$ and $\gamma = 5$, which shows a minimum in the dimensionless coefficient $F(\theta)$. The existence of this minimum with respect to temperature suggests that for a given pressure gradient and $\alpha > 1$ the steady state flux of gases or vapours through a pellet decreases initially with temperature and when the temperature is greater than the threshold temperature θ^* the steady state flux then increases with temperature (Figure 9.2-2). This behaviour with respect to temperature can be explained as follows. The initial sharp decrease is due to rapid decrease in the amount adsorbed on the surface and hence the contribution of the surface diffusion becomes insignificant towards the total flux. As temperature increases further, the surface diffusion no longer contributes to the overall flux and the increase in the overall flux is due to the increase in the pore diffusivity with temperature, which increases slowly with T according to eq. (9.2-5b).

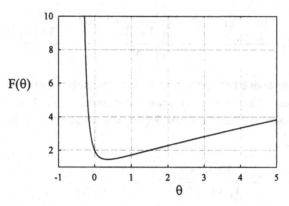

Figure 9.2-2: Plot of the non-dimensional permeability versus non-dimensional temperature

Example 9.2-1: *Temperature dependence of the combined diffusivity*

Here we illustrate the behaviour of the steady state flux with temperature using the following parameters.

Parameter	Symbol	Value	Units
Porosity	ε	0.33	-
Reference temperature T_0	T_0	298	K
Pore diffusivity at T_0	D_{p0}	0.1	cm²/sec
Pore diffusivity exponent	α	1.75	-
Surface diffusivity at T_0	D_{s0}	1×10^{-5}	cm²/sec
Henry constant	K	10000	-
Heat of adsorption	Q	40000	Joule/mole
Activation energy for surface diffusion	E_s	20000	Joule/mole

With the values of various parameters in the above table, we calculate

$$\gamma = \frac{Q - E_s}{R_g T_0} = \frac{(40000 - 20000)}{(8.314)(298)} = 8.07$$

$$\delta_0 = \frac{(1-\varepsilon)KD_{s0}}{\varepsilon D_{p0}} = \frac{(1 - 0.33)(10000)(1 \times 10^{-5})}{(0.33)(0.1)} = 2.03$$

Solving the nonlinear algebraic equation (9.2-12) for the threshold temperature θ^*, we get

$$\theta^* = 0.423$$

Thus, the temperature at which the combined diffusivity (hence the steady state flux) is minimum is:

$$T^* = (1 + \theta^*) T_0 = 424 \text{ K} = 151 \text{ }^\circ\text{C}$$

The steady state flux is calculated from the following equation:

$$J_{SS} = \frac{\left[\varepsilon D_p + (1-\varepsilon) K D_s\right]}{R_g T} \frac{\Delta P}{L}$$

Expressing this equation explicitly in terms of temperature, we get:

$$\frac{J_{SS}}{(\Delta P / L)} = \frac{\varepsilon D_{po}(1+\theta)^\alpha + (1-\varepsilon) K_0 D_{so} \exp\left(-\gamma \frac{\theta}{1+\theta}\right)}{R_g T_0 (1+\theta)}$$

Figure 9.2-3 shows the plot of the above equation versus temperature.

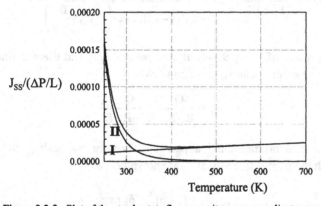

Figure 9.2-3: Plot of the steady state flux per unit pressure gradient versus temperature

Also plotted on the same figure are the contributions of the pore diffusion (curve I) and the surface diffusion (curve II) (the first and second terms of the above equation, respectively). Here we see that the surface diffusion decreases rapidly with temperature, which is due to the rapid decrease of the amount adsorbed with respect to temperature. The increase of the overall flux in the latter part of temperature range is due to the increase in the pore diffusivity with temperature.

9.2.1.1.3 The Temperature Dependence of the Apparent Diffusivity D_{app}

We have seen the behaviour of the combined diffusivity with respect to temperature, which is applicable in steady state situations. When dealing with a transient problem, the appropriate parameter which characterises the approach to equilibrium is the apparent diffusivity (eq. 9.2-3b). This apparent diffusivity is effectively the measure of the ability of the adsorbate to diffuse into the particle relative to the ability of the solid to accommodate adsorbate molecules, that is:

$$D_{app} = \frac{\varepsilon D_p + (1-\varepsilon)KD_s}{\varepsilon + (1-\varepsilon)K} \equiv \frac{\text{Ability to diffuse into the particle}}{\text{Capacity to accommodate adsorbate molecules in both phases}}$$

We consider the temperature dependence of the apparent diffusivity by substituting eq.(9.2-5) into the above equation to get

$$D_{app} = \frac{\varepsilon D_p + (1-\varepsilon)KD_s}{\varepsilon + (1-\varepsilon)K} = \frac{\varepsilon D_{p0}\left[(1+\theta)^\alpha + \delta_0 \exp\left(-\gamma\frac{\theta}{1+\theta}\right)\right]}{\varepsilon + (1-\varepsilon)K_0 \exp\left(-\gamma_H \frac{\theta}{1+\theta}\right)} \quad (9.2\text{-}15)$$

where δ_0 is the value of δ (eq. 9.2-4) at temperature T_0, γ and θ are defined in eqs. (9.2-9) and the new parameter γ_H is defined below:

$$\gamma_H = \frac{Q}{R_g T_0} > \gamma = \frac{Q-E_s}{R_g T_0} \quad (9.2\text{-}16)$$

We define a nondimensional apparent diffusivity by scaling the apparent diffusivity with the pore diffusivity at T_0 as shown below:

$$G(\theta) = \frac{D_{app}}{D_{p0}} = \frac{(1+\theta)^\alpha + \delta_0 \exp\left(-\gamma\frac{\theta}{1+\theta}\right)}{1 + \sigma \exp\left(-\gamma_H \frac{\theta}{1+\theta}\right)} \quad (9.2\text{-}17)$$

where the parameter

$$\sigma = \frac{(1-\varepsilon)K_0}{\varepsilon} \quad (9.2\text{-}18)$$

is a measure of the capacity in the adsorbed phase relative to that in the gas phase, which is usually a very large number.

We plot $G(\theta)$ versus θ as shown in Figure 9.2-4 for $\alpha = 1$, $\delta_0 = 1$, $\gamma = 5$, $\gamma_H = 10$ and $\sigma = 100$. There we see that the apparent diffusivity montonically increases with temperature, suggesting that the approach to equilibrium is faster at higher temperature. This is attributed mostly to the amount that can be accommodated by the particle being smaller at higher temperatures.

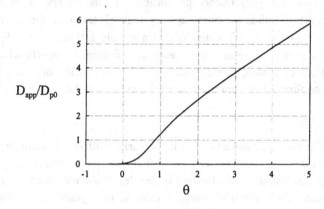

Figure 9.2-4: Plot of the reduced apparent diffusivity D_{app} versus non-dimensional temperature

9.2.1.2 Pore Diffusion Model

When the solid has low surface area or the adsorbed species is rather immobile (that is strong affinity to the surface), the contribution of surface diffusion can be neglected compared to that of pore diffusion, that is $(1-\varepsilon)D_s \ll \varepsilon D_p$, the apparent diffusivity (9.2-3b) will become:

$$D_{app} = \frac{\varepsilon D_p}{\varepsilon + (1-\varepsilon)K} \qquad (9.2\text{-}19)$$

Thus, for solids which exhibit negligible surface diffusion compared to pore diffusion, we see that when the adsorption affinity is high (that is large value of the Henry constant) the apparent diffusivity will be small; hence the system will take longer to reach equilibrium. This is the case because when the solid has high capacity the penetration of the adsorbate into the particle is retarded due to the large uptake of molecule to the surface. To further show this, we take the case of a spherical particle, and the time for the system to reach half of its equilibrium capacity, called the half time, is given by (we defer the derivation until Section 9.2.1.4):

$$t_{0.5} = 0.03055 \frac{[\varepsilon + (1-\varepsilon)K]R^2}{\varepsilon D_p} \quad (9.2\text{-}20)$$

where R is the radius of the solid. What can be learnt from this equation is that the half time is proportional to the square of the particle radius, a characteristic of a diffusion process, and is approximately proportional to the Henry constant K (because in most practical solids the pore capacity is much less than the adsorbed capacity, that is $\varepsilon \ll (1\text{-}\varepsilon)K$). This means that a double increase in the Henry constant will take the system twice as long to reach equilibrium, a property which is valid only for systems exhibiting linear isotherm. This conclusion does not hold when we deal with nonlinear isotherm as we shall see in Section 9.2.2.

9.2.1.3 Surface Diffusion Model

We now consider the other extreme, that is the surface diffusion dominates the transport. In this case adsorbate molecules in the bulk phase surrounding the particle will locally equilibrate with adsorbed molecules at the mouth of the pore, and adsorbed molecules diffuse into the particle under its own gradient. For such a case, $(1\text{-}\varepsilon)D_s \gg \varepsilon D_p$, and the apparent diffusivity (9.2-3b) becomes:

$$D_{app} = \frac{(1-\varepsilon)KD_s}{\varepsilon + (1-\varepsilon)K} \approx D_s \quad (9.2\text{-}21)$$

For most practical solids (at least in gas phase application), the Henry constant K is usually of the order of 10 to 1000, that is the adsorbed phase is approximately 10 to 1000 times more dense than the gas phase; hence the apparent diffusivity is approximately equal to the surface diffusivity. This extreme is called the solid or surface diffusion model, commonly used to describe diffusion in solids such as ion-exchange resins and zeolite crystals. The half time for the case of surface diffusion model in a spherical particle is (Section 9.2.1.4):

$$t_{0.5} = 0.03055 \frac{R^2}{D_s} \quad (9.2\text{-}22)$$

The difference between the pore diffusion model and the surface diffusion model is that in the pore diffusion model it takes longer to equilibrate the solid with higher adsorption affinity (higher K) while in the surface diffusion model the diffusion time is independent of the Henry constant. This is so because in the case of surface diffusion adsorption occurs at the pore mouth at which point the equilibrium with the bulk phase is reached and the diffusion process is simply due to the gradient of the adsorbed concentration rather than its quantity.

9.2.1.4 The Combined Diffusion Model

The linear equation (9.2-3a) was solved analytically by various means such as separation of variables, or Laplace transforms (Edeskuty and Amundson, 1952a, b). The particle is initially uniformly distributed with adsorbate molecules of concentration C_i:

$$t = 0; \qquad C = C_i \qquad (9.2\text{-}23)$$

and the boundary conditions allow for the resistance in the stagnant fluid film surrounding the particle:

$$r = 0; \qquad \frac{\partial C}{\partial r} = 0 \qquad (9.2\text{-}24a)$$

$$r = R; \qquad -\left[\varepsilon D_p + (1-\varepsilon) K D_s\right] \frac{\partial C}{\partial r}\bigg|_{r=R} = k_m \left(C|_R - C_b\right) \qquad (9.2\text{-}24b)$$

For slab object R is the half length, while for cylindrical and spherical objects, R is their respective radius. Here C_b is the concentration of the adsorbate in the bulk surrounding the particle. The first boundary condition is the usual symmetry condition at the center of the particle, and the second condition simply states that the flux into the particle at the surface is equal to that passing through the thin film surrounding the particle. The parameter characterizing the resistance to mass transfer in the film is the mass transfer coefficient, k_m. Example 9.2.2 shows how this mass transfer coefficient can be calculated.

The solution for the concentration distribution within the particle is given below in the form of an infinite series (Alternative solution to eq. 9.2-3a is obtained numerically as shown in Appendix 9.1):

$$\frac{C(x,\tau) - C_i}{C_b - C_i} = 1 - \sum_{n=1}^{\infty} a_n K_n(x) \cdot \exp\left(-\xi_n^2 \tau\right) \qquad (9.2\text{-}25)$$

where ξ_n is the eigenvalue, $K_n(x)$ is the eigenfunction corresponding to the eigenvalue ξ_n, and the non-dimensional distance and time are defined as:

$$x = \frac{r}{R}; \qquad \tau = \frac{D_{app} t}{R^2} \qquad (9.2\text{-}26)$$

Expressions for a_n, $K_n(x)$ and ξ_n for three different particle geometries are given in Table 9.2-1. Appendix 9.2 tabulates the first ten eigenvalues for slab, cylinder and spherical particles.

Table 9.2-1: Concentration distribution solutions for three particle geometries

Shape	a_n	$K_n(x)$	ξ_n	Eq. No.
Slab	$\dfrac{2\sin\xi_n}{\xi_n\left(1+\dfrac{\sin^2\xi_n}{Bi}\right)}$	$\cos(\xi_n x)$	$\xi_n \sin\xi_n = Bi\cdot\cos\xi_n$	(9.2-27)
Cylinder	$\dfrac{2}{\xi_n J_1(\xi_n)\left[1+(\xi_n/Bi)^2\right]}$	$J_0(\xi_n x)$	$\xi_n J_1(\xi_n) = Bi\cdot J_0(\xi_n)$	(9.2-27)
Sphere	$\dfrac{2(\sin\xi_n - \xi_n\cos\xi_n)}{\xi_n^2\left(1+\dfrac{\cos^2\xi_n}{Bi-1}\right)}$	$\dfrac{\sin(\xi_n x)}{x}$	$\xi_n \cos\xi_n = (1-Bi)\cdot\sin\xi_n$	(9.2-27)

Here J is the Bessel function and Bi is the non-dimensional Biot number, defined as:

$$Bi = \frac{k_m R}{\left[\varepsilon D_p + (1-\varepsilon)KD_s\right]} \qquad (9.2\text{-}27d)$$

The Biot number measures the relative resistance contributed by the stagnant film surrounding the particle to the internal diffusional resistance. For most gas phase applications, this number is of the order of 10 to 100, indicating that the internal diffusional resistance is more important than the film resistance.

Example 9.2.2: *Mass transfer coefficient and the Biot number*

The mass transfer coefficient for a stagnant film surrounding a solid adsorbent packed inside a fixed bed adsorber can be calculated from the following correlation (Wakao and Kaguei, 1982):

$$Sh = 2 + 1.1\, Re^{0.6}\, Sc^{1/3}$$

where

$$Sh = \frac{k_m(2R)}{D_{ij}}; \quad Re = \frac{u(2R)\rho}{\mu}; \quad Sc = \frac{\mu}{\rho D_{ij}}$$

Here, k_m is the mass transfer coefficient, R is the particle radius, ρ and μ are fluid density and viscosity, u is the velocity around the particle and D_{ij} is the fluid binary diffusivity.

We take an example of ethane (20%) and nitrogen (80%) at 1 atm and 303 K. The particle radius is 0.1 cm. The relevant properties of ethane and nitrogen are listed in the following table.

Species	MW (g/mol)	μ (g/cm/s)	D_{12} (cm²/s)
mitrogen	28	0.00018	0.1483
ethane	30	0.0000925	

The density of the gaseous mixture ρ is calculated from

$$\rho = \rho_1 + \rho_2$$

where

$$\rho_1 = MW_1 \frac{p_1}{RT} = 28 \times \frac{0.8}{82.057 \times 303} = 9 \times 10^{-4} \; \frac{g}{cc}$$

$$\rho_2 = MW_2 \frac{p_2}{RT} = 30 \times \frac{0.2}{82.057 \times 303} = 2.4 \times 10^{-4} \; \frac{g}{cc}$$

Thus

$$\rho = \rho_1 + \rho_2 = 11.4 \times 10^{-4} \; g/cc$$

The viscosity of the mixture is calculated from the semi-empirical formula of Wilke et al. (Bird et al., 1960)

$$\mu = \sum_{i=1}^{2} \frac{y_i \mu_i}{\sum_{j=1}^{2} y_j \Phi_{ij}}$$

$$\Phi_{ij} = \frac{1}{\sqrt{8}} \left(1 + \frac{M_i}{M_j}\right)^{-1/2} \left[1 + \left(\frac{\mu_i}{\mu_j}\right)^{1/2} \left(\frac{M_j}{M_i}\right)^{1/4}\right]^2$$

Knowing the pure component viscosity and the mole fractions, we calculate the mixture viscosity using the above equation to get:

$$\mu = 1.55 \times 10^{-4} \; g/cm/s$$

We calculate the Schmidt number

$$Sc = \frac{\mu}{\rho D_{12}} = \frac{1.55 \times 10^{-4}}{(11.4 \times 10^{-4})(0.1483)} = 0.917$$

For a range of velocity, we calculate the mass transfer coefficient using the correlation $Sh = 2 + 1.1 \, Re^{0.6} \, Sc^{1/3}$, and the results are tabulated in the following table:

u (cm/s)	Re (-)	k_m (cm/s)	Bi (-)
0	0	1.48	7.4
2	2.94	3.00	15
5	7.36	4.11	21
10	14.71	5.46	27
15	22.06	6.55	33
20	29.42	7.51	38
100	147.10	17.31	87
1000	1471.0	64.51	322

The Biot number is calculated from eq. (9.2-27d), and if the apparent diffusivity of the porous solid is 0.02 cm²/s the Biot number can be calculated and they are listed in the above table. We will show later that the mass transfer resistance in the stagnant film surrounding the particle can be neglected compared to the internal diffusion when the Biot number is greater than about 50. For this specific system, to neglect the external mass transfer resistance, a minimum velocity of 100 cm/s must be maintained.

Example 9.2-3: *More about Biot number*

Although in most gas phase applications, the Biot number is usually greater than 100, indicating that the gas film resistance is not as important as the internal diffusional resistance. In systems with fast surface diffusion, this may not be true as we shall see in this example.

Particle radius, R	0.1 cm
Particle porosity, ε	0.33
Pore diffusivity, D_p	0.32 cm²/s
Surface diffusivity, D_s	1×10^{-5} cm²/s
Henry constant, K	10,000
Mass transfer coefficient, k_m	5.5 cm/s

With the above values typical for adsorption of low hydrocarbons in activated carbon, we calculate the combined diffusivity

$$\varepsilon D_p + (1-\varepsilon)KD_s = 0.776 \quad cm^2/s$$

Thus the Biot number is calculated as Bi = 0.71. This low value of the Biot number suggests that the resistance contributed by the gas film is very much comparable to the diffusional resistance within the particle.

Now let us consider the same set of values but this time we assume that there is no surface diffusion. The Biot number in this case is Bi = 52. Thus we see that in the absence of surface diffusion the gas film resistance becomes negligible compared to the internal diffusional resistance.

9.2.1.4.1 Fractional uptake

Once the solution for the concentration distribution is known (for example, eq. 9.2-25 for the linear isotherm case), the amount uptake per unit volume of the particle can be calculated from:

$$\frac{M(t)}{V_p} = \varepsilon \overline{C}(t) + (1-\varepsilon)\overline{C}_\mu(t) \tag{9.2-28}$$

where V_p is the volume of the particle, and $\overline{C}(t)$ and $\overline{C}_\mu(t)$ are the volumetric mean concentrations in the gas and adsorbed phases, defined as follows:

$$\overline{C}(t) = \frac{1}{V}\int_V C(t,r)\,dV; \quad \overline{C}_\mu(t) = \frac{1}{V}\int_V C_\mu(t,r)\,dV \tag{9.2-29}$$

When time approaches infinity, the amount uptake by the solid per unit volume is simply:

$$\frac{M(\infty)}{V_p} = \varepsilon \overline{C}(\infty) + (1-\varepsilon)\overline{C}_\mu(\infty) \tag{9.2-30}$$

where $\overline{C}_\mu(\infty)$ is the adsorbed concentration which is in equilibrium with $\overline{C}(\infty)$, which is equal to the bulk phase concentration C_b. The fractional uptake commonly used in adsorption studies is defined as the ratio of the amount adsorbed from t = 0 to time t to the amount taken by the solid from t = 0 to time infinity, that is:

$$F = \frac{M(t) - M(0)}{M(\infty) - M(0)} = \frac{\left[\varepsilon \overline{C}(t) + (1-\varepsilon)\overline{C}_\mu(t)\right] - \left[\varepsilon \overline{C}(0) + (1-\varepsilon)\overline{C}_\mu(0)\right]}{\left[\varepsilon \overline{C}(\infty) + (1-\varepsilon)\overline{C}_\mu(\infty)\right] - \left[\varepsilon \overline{C}(0) + (1-\varepsilon)\overline{C}_\mu(0)\right]} \quad (9.2\text{-}31)$$

Eq. (9.2-31) is valid for any isotherm. For linear isotherm between the two phases (eq. 9.2-2), the above equation for the fractional uptake can be reduced to:

$$F = \frac{\left[\varepsilon + (1-\varepsilon)K\right]\overline{C}(t) - \left[\varepsilon + (1-\varepsilon)K\right]C_i}{\left[\varepsilon + (1-\varepsilon)K\right]\overline{C}(\infty) - \left[\varepsilon + (1-\varepsilon)K\right]C_i} = \frac{\overline{C}(t) - C_i}{C_b - C_i} \quad (9.2\text{-}32)$$

The solutions for the concentration distribution in the case of linear isotherm (eq. 9.2-25) are substituted into eq. (9.2-32) and we obtain the following expression for the fractional uptake:

$$F = 1 - \sum_{n=1}^{\infty} b_n \cdot \exp\left(-\xi_n^2 \tau\right) \quad (9.2\text{-}33)$$

where τ is the non-dimensional time defined in eq. (9.2-26), the coefficient b_n are tabulated in Table 9.2-2 for three different shapes of particle, and the eigenvalues ξ_n are given as solution of transcendental equations given in Table 9.2-1.

Table 9.2-2: The coefficient b_n for the fractional uptake equation (9.2-33)

Shape	b_n	Eq. No.
Slab	$\dfrac{2\sin^2 \xi_n}{\xi_n^2\left(1 + \sin^2 \xi_n / Bi\right)}$	(9.2-34a)
Cylinder	$\dfrac{4}{\xi_n^2\left[1 + (\xi_n / Bi)^2\right]}$	(9.2-34b)
Sphere	$\dfrac{6(\sin\xi_n - \xi_n \cos\xi_n)^2}{\xi_n^4\left(1 + \dfrac{\cos^2 \xi_n}{Bi - 1}\right)}$	(9.2-34c)

When the film resistance becomes negligible compared to the internal diffusional resistance, that is when Bi is very large, the fractional uptake given in eq.(9.2-33) will reduce to simpler form, tabulated in Table 9.2-3 for the three shapes of particle.

Table 9.2-3: Fractional uptake for three particle geometries when Bi → ∞

Shape	Fractional uptake	Eq. No.
Slab	$F = 1 - \dfrac{2}{\pi^2} \sum_{n=1}^{\infty} \dfrac{1}{\left(n-\frac{1}{2}\right)^2} \exp\left[-\left(n-\frac{1}{2}\right)^2 \pi^2 \tau\right]$	(9.2-35a)
Cylinder	$F = 1 - 4 \sum_{n=1}^{\infty} \dfrac{1}{\xi_n^2} \exp\left(-\xi_n^2 \tau\right); \quad J_0(\xi_n) = 0$	(9.2-35b)
Sphere	$F = 1 - \dfrac{6}{\pi^2} \sum_{n=1}^{\infty} \dfrac{1}{n^2} \exp\left(-n^2 \pi^2 \tau\right)$	(9.2-35c)

These solutions for the fractional uptake listed in Table 9.2-3 are used often in the literature, when linear isotherm is valid. The fractional uptake versus non-dimensional time τ is shown in Figure 9.2-5a for three different shapes of the particle for the case of infinite stirring in the surrounding (Bi → ∞). The computer code UPTAKEP.M written in MatLab is provided with this book to help the reader to obtain the fractional uptake versus time. As seen in Figure 9.2-5a, for the given R and D_{app} the spherical particle has the fastest dynamics as it has the highest exterior surface area per unit volume.

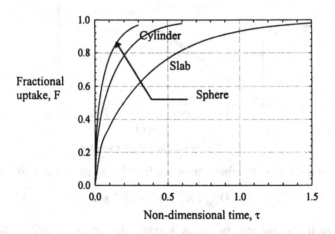

Figure 9.2-5a: Plots of the fractional uptake versus τ for slab, cylinder and sphere for Bi = ∞

Example 9.2-4: Determination of diffusivity from fractional uptake data

Figure 9.2-5a can be used to determine the diffusivity. We take the following adsorption kinetic data of fractional uptake versus time for an adsorption system with linear isotherm (the first two columns of the following table). The particle is spherical.

t (sec)	Fractional uptake	τ (from Figure 9.2-5)
60	0.2	0.005
90	0.3	0.011
160	0.4	0.020
275	0.5	0.033
425	0.6	0.051
660	0.7	0.079
1000	0.8	0.120
1600	0.9	0.190
2200	0.95	0.260

Using Figure 9.2-5a, we obtain the non-dimensional time τ for each value of the fractional uptake. This is tabulated in the third column of the above table. According to the definition of the dimensionless time (9.2-26), a plot of τ versus t (Figure 9.2-5b) gives a straight line with a slope of D_{app}/R^2.

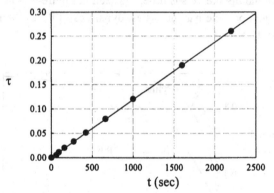

Figure 9.2-5b: Plot of τ versus t

Using the data in the above table, we find this slope as 1.2×10^{-4} sec^{-1}, i.e.

$$D_{app}/R^2 = 1.2 \times 10^{-4} \text{ sec}^{-1}$$

Thus if the particle radius is known, the apparent diffusivity can be calculated.

The effect of Biot number (film resistance relative to internal diffusion) is shown in Figure 9.2-6, where it is seen that when Bi approaches infinity the uptake curve reaches an asymptote, indicating that the system is controlled by the internal diffusion. The criterion that the gas film resistance can be neglected is Bi > 50.

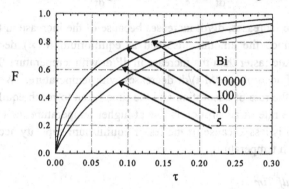

Figure 9.2-6: Plots of the fractional uptake versus τ for a spherical particle

An increase in particle radius will increase the time taken by the system to reach equilibrium (Figure 9.2-7). In fact, this time increases with the square of the particle length scale by the virtue of the definition of the non-dimensional time τ and the solution for the fractional uptake is only a function of τ.

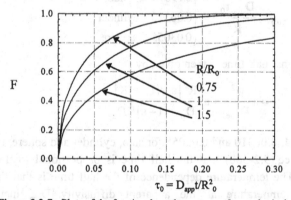

Figure 9.2-7: Plots of the fractional uptake versus τ_0 for a spherical particle

An increase in temperature also increases the speed of adsorption, that is the time to approach equilibrium, because of the two factors: the increase in diffusivity and the reduction in adsorption capacity. This has been demonstrated in Figure 9.2-4 where the apparent diffusivity increases monotonically with temperature. Although the increase in temperature makes the system to approach equilibrium faster, the rate of adsorption actually decreases with temperature. Let us write the rate of adsorption from eq. (9.2-31) as

$$\frac{dM(t)}{dt} = [M(\infty) - M(0)]\frac{dF}{dt}$$

When temperature increases, dF/dt increases because of the increase in the apparent diffusivity; but since the amount adsorbed at equilibrium $M(\infty)$ decreases with temperature at a rate faster than the increase in dF/dt with temperature the net result is that the rate of adsorption dM(t)/dt decreases with temperature. We should not confuse between the rate of adsorption and the time taken to reach equilibrium, that is even though the rate of adsorption is low at higher temperature the time taken to reach equilibrium is less because of the lesser equilibrium capacity accommodated by the solid at high temperature.

9.2.1.4.2 The Half-Time

The half time of adsorption, that is the time it takes for the solid to attain half of the equilibrium amount, can be calculated from eqs. (9.2-35) by simply setting the fractional uptake F to one half for the case of no external mass transfer resistance. When this is done, we obtain the following half times for three shapes of particle:

$$\tau_{0.5} = \frac{D_{app} t_{0.5}}{R^2} = \begin{cases} 0.19674 & \text{slab} \\ 0.06310 & \text{cylinder} \\ 0.03055 & \text{sphere} \end{cases} \quad (9.2\text{-}36a)$$

The dimensional half time is then given by

$$t_{0.5} = a\frac{R^2}{D_{app}} = a\frac{R^2[\varepsilon + (1-\varepsilon)K]}{\varepsilon D_p + (1-\varepsilon)KD_s} \quad (9.2\text{-}36b)$$

where a = 0.19674, 0.06310 and 0.03055 for slab, cylinder and sphere, respectively. It is clear from eq.(9.2-36b) that the half time is proportional to the square of particle radius. The temperature dependence of the half time is that the half time decreases with temperature as the apparent diffusivity D_{app} increases with temperature (see Figure 9.2-4).

As seen in eq. (9.2-36) that the half time for sphere is smallest and that of slab is highest. This is because the sphere has the highest external surface area per unit volume; hence it takes shorter time to reach equilibrium. If we write the above equation for half time in terms of the volume to external surface area ratio, defined as

$$R = \begin{cases} (V/S_{ext}) & \text{slab} \\ 2(V/S_{ext}) & \text{cylinder} \\ 3(V/S_{ext}) & \text{sphere} \end{cases}$$

then the half time of adsorption of eq. (9.2-36) is given by:

$$\frac{D_{app} t_{0.5}}{(V/S_{ext})^2} = \begin{cases} 0.197 & \text{slab} \\ 0.252 & \text{cylinder} \\ 0.275 & \text{sphere} \end{cases} \quad (9.2\text{-}37)$$

This means that if the half time is calculated based on the unit of volume to external surface area, the non-dimensional times defined are closer for all three shapes of particle. The small difference is attributed to the curvature effect during the course of adsorption, which can not be accounted for by the simple argument of volume to surface area. Eq. (9.2-37) is useful to calculate the half time of solid having an arbitrary shape, that is by simply measuring volume and external surface area of the solid object the half time can be estimated from eq. (9.2-37).

9.2.1.4.3 Short Time Solution

Very often in the literature short time solutions are needed to investigate the behaviour of adsorption during the initial stage of adsorption. For linear problems, this can be achieved analytically by taking Laplace transform of model equations and then considering the behaviour of the solution when the Laplace variable s approaches infinity. Applying this to the case of linear isotherm (eq. 9.2-3), we obtain the following solution for the fractional uptake at short times:

$$F \approx \frac{2(s+1)}{\sqrt{\pi}} \sqrt{\tau} \quad (9.2\text{-}38)$$

where $s = 0, 1$, and 2 for slab, cylinder and sphere, respectively, and τ is defined in eq. (9.2-26).

Written in terms of the dimensional quantities, eq.(9.2-38) becomes:

$$\frac{M(t) - M(0)}{M(\infty) - M(0)} \approx \frac{2(s+1)}{\sqrt{\pi}} \left\{ \frac{\varepsilon D_p + (1-\varepsilon)KD_s}{[\varepsilon + (1-\varepsilon)K]R^2} \right\}^{1/2} \sqrt{t} \quad (9.2\text{-}39)$$

where

$$M(\infty) = V[\varepsilon C_b + (1-\varepsilon)KC_b] \quad (9.2\text{-}40a)$$

$$M(0) = V\left[\varepsilon C_i + (1-\varepsilon)KC_i\right] \qquad (9.2\text{-}40b)$$

Thus the amount adsorbed per unit volume of the particle from $t = 0$ to t is:

$$\frac{\Delta M(t)}{V} \approx \frac{2(s+1)(C_b - C_i)}{R\sqrt{\pi}}\sqrt{\varepsilon + (1-\varepsilon)K} \cdot \sqrt{\varepsilon D_p + (1-\varepsilon)KD_s} \cdot \sqrt{t} \qquad (9.2\text{-}41)$$

This means that if one plots the change of the amount adsorbed per unit volume of the particle versus the square root of time using only initial data, a straight line should be resulted from such plot with the slope given by:

$$\text{slope} = \frac{2(s+1)(C_b - C_i)}{R\sqrt{\pi}}\sqrt{\varepsilon + (1-\varepsilon)K} \cdot \sqrt{\varepsilon D_p + (1-\varepsilon)KD_s} \qquad (9.2\text{-}42a)$$

or in terms of pressure difference

$$\text{slope} = \frac{2(s+1)(P_b - P_i)}{R\sqrt{\pi}\,R_g} \frac{\sqrt{\varepsilon + (1-\varepsilon)K} \cdot \sqrt{\varepsilon D_p + (1-\varepsilon)KD_s}}{T} \qquad (9.2\text{-}42b)$$

The slope is often utilised to determine the combined diffusivity. The nice feature about this type of plot is that we only need the initial data to quickly obtain the diffusivity.

To find the temperature dependence of this slope, we substitute the temperature dependence of the relevant parameters (eqs. 9.2-5) and obtain the following temperature dependence factor of the above slope:

$$\frac{\sqrt{1 + \sigma\exp\left(-\gamma_H \frac{\theta}{1+\theta}\right)} \cdot \sqrt{(1+\theta)^\alpha + \delta_0 \exp\left(-\gamma \frac{\theta}{1+\theta}\right)}}{(1+\theta)} \qquad (9.2\text{-}43)$$

where γ is given in eq.(9.2-9a), δ_0 in eq.(9.2-4), γ_H in eq.(9.2-16) and σ in eq.(9.2-18).

The following figure shows the dependence of the reduced slope versus non-dimensional temperature θ (Figure 9.2-8) for $\sigma = 100$, $\alpha = 1.75$, $\gamma = 5$, $\gamma_H = 10$, and $\delta_0 = 1$. We see that the slope decreases with temperature for a given pressure difference, and this decrease is due to the decrease in the amount adsorbed at high temperature (the first term in the numerator of eq. 9.2-43).

We have completed the analysis of system with linear isotherm, and local equilibrium between the two phases. The local equilibrium assumption is generally valid in many systems; the linearity between the two phases is, however, restricted to systems with very low concentration of adsorbate, usually not possible with many practical systems even with purification systems. We will discuss the analysis of nonlinear systems in the next section, and point out the behaviours which are not manifested in linear systems.

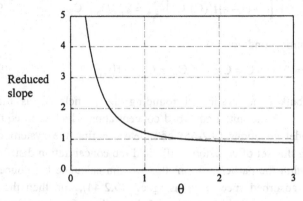

Figure 9.2-8: Plot of the reduced slope of the initial uptake versus θ

9.2.2 Nonlinear Models:

In nonlinear adsorption systems where parallel diffusion mechanisms hold, the mass balance equation given in eq. (9.2-1b) is still valid. The difference is in the functional relationship between the concentrations of the two phases, that is the local adsorption isotherm. In general, this relationship can take any form that can describe well equilibrium data. Adsorption isotherm such as Langmuir, Unilan, Toth, Sips can be used. In this section we present the mathematical model for a general isotherm and then perform simulations with a Langmuir isotherm as it is adequate to show the effect of isotherm nonlinearity on the dynamics behaviour. The adsorption isotherm takes the following functional form:

$$C_\mu = f(C) \qquad (9.2\text{-}44)$$

The mass balance equation is obtained for the case of nonlinear isotherm by substituting the isotherm equation (eq. 9.2-44) into the mass balance equation (9.2-1b)

$$\left[\varepsilon + (1-\varepsilon)f'(C)\right]\frac{\partial C}{\partial t} = \frac{1}{r^s}\frac{\partial}{\partial r}\left\{r^s\left[\varepsilon D_p + (1-\varepsilon)\,f'(C)\,D_s\right]\frac{\partial C}{\partial r}\right\} \qquad (9.2\text{-}45a)$$

subject to the following boundary conditions:

$$r = 0; \quad \frac{\partial C}{\partial r} = 0 \qquad (9.2\text{-}45b)$$

$$r = R; \quad \left[\varepsilon + (1-\varepsilon)f'(C)\,D_s\right]\frac{\partial C}{\partial r} = k_m(C_b - C) \qquad (9.2\text{-}45c)$$

and the following initial condition

$$t = 0; \quad C = C_i; \quad C_\mu = C_{\mu i} = f(C_i) \qquad (9.2\text{-}45d)$$

where C_b is the bulk concentration surrounding the particle, C_i is the initial concentration, and $C_{\mu i}$ is the initial adsorbed concentration which is in equilibrium with C_i. The model equations (9.2-45) completely define the system dynamic behaviour. Solving this set of equations will yield the concentration distribution of the free species within the particle. From this we can calculate the concentration distribution of the adsorbed species by using eq. (9.2-44), and then the amount adsorbed per unit particle volume from eq. (9.2-28) and the fractional uptake from eq. (9.2-31).

It is important to note that the mass balance equation (9.2-45a) contains the concentration-independent surface diffusivity. In general this surface diffusivity is a function of concentration when the isotherm is nonlinear. We shall postpone the treatment of concentration-dependent D_s, and assume its constancy for the study of the effects of isotherm nonlinearity on the system dynamic behaviour.

Example 9.2-5: *Non-dimensionalisation of the mass balance equations*

The model equations are eqs. (9.2-45). Because of the nonlinearity of the equations, they must be solved numerically and before this is done it is convenient to cast them into a nondimensional form. Let C_0 be some reference concentration. We define the following non-dimensional variables and parameters:

$$y = \frac{C}{C_0}; \quad \eta = \frac{r}{R}; \quad \tau = \frac{\varepsilon D_p t}{R^2} \qquad (9.2\text{-}46a)$$

$$\delta_P = \frac{(1-\varepsilon)D_s C_{\mu 0}}{\varepsilon D_p C_0}; \quad C_{\mu 0} = f(C_0) \qquad (9.2\text{-}46b)$$

$$Bi = \frac{k_m R}{\varepsilon D_p}; \quad y_b = \frac{C_b}{C_0}; \quad y_i = \frac{C_i}{C_0} \qquad (92\text{-}46c)$$

The resulting non-dimensional equations are:

$$G(y)\frac{\partial y}{\partial \tau} = \frac{1}{\eta^s}\frac{\partial}{\partial \eta}\left\{\eta^s H(y)\frac{\partial y}{\partial \eta}\right\} \qquad (9.2\text{-}47a)$$

where

$$G(y) = \varepsilon + (1-\varepsilon)f'(C_0 y) \qquad (9.2\text{-}47b)$$

$$H(y) = 1 + \delta_P \frac{C_0\, f'(C_0 y)}{C_{\mu 0}} \qquad (9.2\text{-}47c)$$

The function $G(y)$ reflects the system capacity, while the function $H(y)$ reflects the mobility. Eqs. (9.2-47) are subject to the following boundary conditions:

$$\eta = 0; \qquad \frac{\partial y}{\partial \eta} = 0 \qquad (9.2\text{-}47d)$$

$$\eta = 1; \qquad H(y)\frac{\partial y}{\partial \eta} = \text{Bi}(y_b - y) \qquad (9.2\text{-}47e)$$

and the following initial condition

$$\tau = 0; \qquad y = y_i \qquad (9.2\text{-}47f)$$

The parameter δ_P in eq. (9.2-46b) measures the strength of mobility of the adsorbed species to that of the free species. If $\delta_P = 0$, surface diffusion is absent. If it is of order of unity, both diffusion fluxes are comparable in magnitude, and if it is much greater than unity surface diffusion is more important than pore diffusion.

Solving eqs.(9.2-47) we will obtain the non-dimensional concentration y as a function of η and τ, from which we can calculate the dimensional concentrations of the free and adsorbed species:

$$C = C_0 y \qquad (9.2\text{-}48a)$$

$$C_\mu = f(C) = f(C_0 y) \qquad (9.2\text{-}48b)$$

Then the fractional uptake can be calculated from:

$$F = \frac{\left[\varepsilon\langle C\rangle(t) + (1-\varepsilon)\langle C_\mu\rangle(t)\right] - \left[\varepsilon C_i + (1-\varepsilon)C_{\mu i}\right]}{\left[\varepsilon C_b + (1-\varepsilon)C_{\mu b}\right] - \left[\varepsilon C_i + (1-\varepsilon)C_{\mu i}\right]} \qquad (9.2\text{-}48c)$$

where $C_{\mu b}$ is the adsorbed concentration in equilibrium with C_b, and $\langle \ \rangle$ denotes the volumetric average defined as:

$$\langle C \rangle = \frac{(1+s)}{R^{s+1}} \int_0^R r^s \, C \, dr \qquad (9.2\text{-}48d)$$

Eqs.(9.2-47) are in the form of nonlinear partial differential equation and are solved by a combination of the orthogonal collocation method and the Runge-Kutta method. Appendix 9.2 shows the details of the collocation analysis, and a computer code ADSORB1A.M is provided to solve the cases of Langmuir and Toth isotherms.

For Toth isotherm, the relevant functions for this model are:

$$C_\mu = f(C) = C_{\mu s} \frac{bC}{\left[1+(bC)^t\right]^{1/t}}$$

$$f'(C) = \frac{bC_{\mu s}}{\left[1+(bC)^t\right]^{(1/t)+1}}$$

$$G(y) = \varepsilon + (1-\varepsilon) \frac{bC_{\mu s}}{\left[1+(bC_0 y)^t\right]^{(1/t)+1}}$$

$$H(y) = 1 + \delta_P \frac{\left[1+(bC_0)^t\right]^{1/t}}{\left[1+(bC_0 y)^t\right]^{(1/t)+1}}$$

For Langmuir isotherm, simply replace t in the above equations by unity.

9.2.2.1 General Behaviour

Nonlinear isotherm model equations (9.2-47) are solved numerically, and a number of features from such numerical study are:

1. The fractional uptake is proportional to the square of the particle radius
2. The fractional uptake is faster with an increase in the bulk concentration
3. The fractional uptake is faster with an increase in temperature.

The points 1 and 3 are also observed for the case of linear isotherm. The second point, however, was not observed in the case of linear isotherm where the fractional uptake curve is independent of bulk concentration. This is because when the bulk concentration is double, the capacity in the solid is also doubled by the virtue of linear isotherm; thus, the times taken to reach equilibrium are the same. On the other hand, in the case of nonlinear isotherm, assuming the isotherm is convex (Langmuir isotherm is convex), a double in the bulk concentration is associated with a less than double in the adsorption capacity and as a result the time taken to reach equilibrium is shorter for the system having higher bulk concentration.

Example 9.2-6: *Dual diffusion in a particle with a Langmuir isotherm*

We illustrate this bulk concentration dependence with this example. The particle is spherical and the adsorption isotherm follows a Langmuir equation. The relevant parameters for this system are tabulated in the following table.

Particle radius, R	= 0.1 cm
Particle porosity, ε	= 0.33
Bulk concentration, C_b	= 1×10^{-6} mole/cc
Affinity constant, b	= 1×10^6 cc/mole
Saturation adsorption capacity, $C_{\mu s}$	= 5×10^{-3} mole/cc
Pore diffusivity, D_p	= 0.02 cm²/sec
Surface diffusivity, D_s	= 1×10^{-6} cm²/sec

With this bulk concentration, the equilibrium adsorbed concentration is:

$$C_{\mu b} = C_{\mu s} \frac{bC_b}{1+bC_b} = 2.5 \times 10^{-3} \frac{mole}{cc}$$

Using the program code ADSORB1A, we generate the concentration distribution as a function of time, and Figures 9.2-9a and b shows plots of the fractional uptake as well as the amount adsorbed per unit particle volume versus time (sec). Taking another case having a bulk concentration of 5×10^{-6} mole/cc, which is 5 times larger than that used in the last example. The equilibrium adsorbed concentration is:

$$C_{\mu b} = 4.17 \times 10^{-3} \frac{mole}{cc}$$

which is only 1.6 times higher than the equilibrium adsorbed concentration in the last example. Thus we see that a five fold increase in the bulk concentration results in only a 1.6-fold increase in the adsorbed concentration, suggesting that the system will approach equilibrium faster in the case of higher bulk concentration (Figure 9.2-9a). We also show on the same figure the fractional uptake versus time for another bulk concentration of 1×10^{-5} mole/cc. This illustrates the point that the fractional uptake is faster with an increase in bulk concentration.

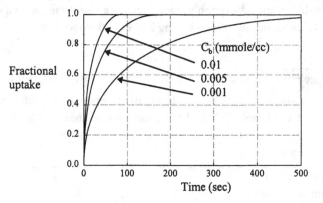

Figure 9.2-9a: Plots of the fractional uptake versus time (sec)

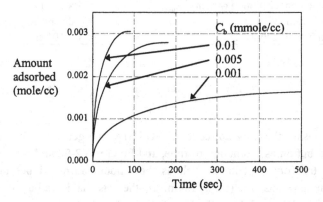

Figure 9.2-9b: Plots of the amount adsorbed per unit particle volume versus time

Another useful exercise of the computer simulation is that the internal concentration profile versus time can be studied. Figure 9.2-9c shows the time evolution of intraparticle concentrations at five specific points within the particle. The sigmoidal shape of the curve is simply due to the fact that it takes time for adsorbate molecules to reach the interior points.

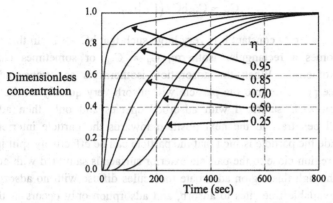

Figure 9.2-9c: Plots of the concentration at various points within the particle versus time

The effect of particle size can also be investigated numerically. Figure 9.2-9d shows plots of the amount adsorbed per unit particle volume versus time for three different particle sizes. It takes longer to reach equilibrium for larger particle size, but the equilibrium capacity is unaffected by the particle size. Note the dependence of the time scale of adsorption on the square of radius.

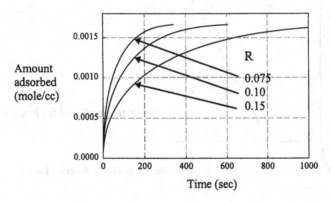

Figure 9.2-9d: Plots of the amount adsorbed per unit particle volume versus time for different particle size (R = 0.075, 0.1, 0.15 cm)

9.2.2.2 Irreversible Isotherm and Pore Diffusion

The effect of bulk concentration can be further illustrated when we consider the extreme of the Langmuir isotherm

$$C_\mu = C_{\mu s} bC / (1 + bC)$$

that is when the affinity constant b is very large such that $bC \gg 1$. In this case the isotherm becomes a rectangular isotherm, $C_\mu \approx C_{\mu s}$, or sometimes called an irreversible isotherm. When the adsorption isotherm is so strong, adsorbate molecules once penetrate into the particle will adsorb very quickly on the surface until that position is saturated with adsorbed species and only then adsorbate molecules will penetrate to the next position towards the particle interior. The behaviour inside the particle is such that the particle can be effectively split into two regions. The region close to the particle exterior surface is saturated with adsorbed species, and through this region adsorbate molecules diffuse with no adsorption as there are no available free sites to adsorb, and adsorption only occurs in the very small neighborhood around the point separating the two regions. The inner core region is, therefore, free of any adsorbate whether it is in the free form or in the adsorbed form. Figure 9.2-10 shows schematically the two regions as well as the concentration profiles of free and adsorbed species.

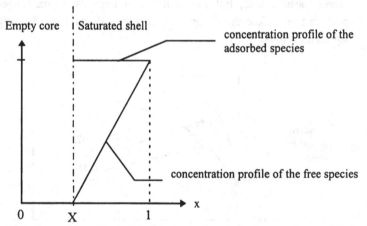

Figure 9.2-10: Concentration profiles of free and adsorbed species in the case of irreversible isotherm

By splitting the particle domain into two, for which the domain close to the particle exterior surface is saturated with the adsorbed species while the inner domain is free from any sorbates, the problem can be solved to yield analytical solution (Suzuki and Kawazoe, 1974). Using a rigorous perturbation method, Do (1986) has proved that the penetration of the adsorption front in the case of rectangular isotherm can be derived from the full pore diffusion and adsorption equations under the conditions of strong adsorption kinetics relative to diffusion, high sorption affinity and low capacity in the fluid phase compared to that in the adsorbed phase. Under these conditions, the mass balance equations are:

$$\frac{1}{r^s}\frac{\partial}{\partial r}\left(r^s \frac{\partial C}{\partial r}\right) = 0 \qquad \text{for} \quad R_f < r < R \qquad (9.2\text{-}49a)$$

$$(1-\varepsilon)C_{\mu s}\frac{dR_f}{dt} = -\varepsilon D_p \frac{\partial C}{\partial r}\bigg|_{r=R_f} \qquad (9.2\text{-}49b)$$

where $C_{\mu s}$ is the saturation capacity. The first equation is simply the pure diffusion equation describing transport in the saturated outer layer, while the second equation states that the rate of adsorption is balanced by the molar rate of supply at the front separating the two regions. Here R_f is the time-dependent position of the front demarcating the two regions.

The fractional uptake for this problem is simply the amount in the saturated region to the saturation amount of the whole particle. Written in terms of the demarcation position R_f, it is:

$$F = \frac{\left(4\pi R^3/3 - 4\pi R_f^3/3\right)(1-\varepsilon)C_{\mu s}}{\left(4\pi R^3/3\right)(1-\varepsilon)C_{\mu s}} \qquad (9.2\text{-}50a)$$

that is

$$F = 1 - \left(\frac{R_f}{R}\right)^{s+1} \qquad (9.2\text{-}50b)$$

The initial condition is that the particle is initially free of any adsorbate, that is:

$$t = 0; \quad C = 0 \qquad (9.2\text{-}51)$$

The boundary conditions are given as in eqs. (9.2-24). Solving this set of mass balance equations, we obtain the solutions for the fractional uptake for three particle shapes, tabulated in the second column of Table 9.2-4.

Table 9.2-4: Fractional uptake for the case of irreversible isotherm

Fractional uptake	Half time	Eq. No.
Slab $\left(1+\dfrac{1}{Bi}\right)F - \dfrac{1}{2}\left[1-(1-F)^2\right] = \dfrac{\varepsilon D_p C_0 t}{R^2(1-\varepsilon)C_{\mu s}}$	$t_{0.5} = \dfrac{R^2(1-\varepsilon)C_{\mu s}}{\varepsilon D_p C_0}\left(\dfrac{1}{8} + \dfrac{1}{2Bi}\right)$	(9.2-52a)
Cylinder $\left(\dfrac{1}{4}+\dfrac{1}{2Bi}\right)F + \dfrac{1}{4}(1-F)\ln(1-F) = \dfrac{\varepsilon D_p C_0 t}{R^2(1-\varepsilon)C_{\mu s}}$	$t_{0.5} = \dfrac{R^2(1-\varepsilon)C_{\mu s}}{\varepsilon D_p C_0}\left(\dfrac{1-\ln 2}{8} + \dfrac{1}{4Bi}\right)$	(9.2-52b)
Sphere $\dfrac{1}{2}\left[1-(1-F)^{2/3}\right] - \dfrac{1}{3}\left(1-\dfrac{1}{Bi}\right)F - = \dfrac{\varepsilon D_p C_0 t}{R^2(1-\varepsilon)C_{\mu s}}$	$t_{0.5} = \dfrac{R^2(1-\varepsilon)C_{\mu s}}{\varepsilon D_p C_0}\left(\dfrac{1}{3} - \dfrac{1}{2^{5/3}} + \dfrac{1}{6Bi}\right)$	(9.2-52c)

$Bi = k_m R/\varepsilon D_p$.

From the solutions for the fractional uptake in Table 9.2-4, the half time can be obtained by setting the fractional uptake F to one half, and they are tabulated in the third column of Table 9.2-4. For this case of rectangular isotherm, the half time is proportional to the square of particle radius, the maximum adsorptive capacity, and inversely proportional to the bulk concentration. The time it takes to reach equilibrium is half when the bulk concentration is doubled. This is because when the bulk concentration is doubled the driving force for mass transfer is doubled while the adsorptive capacity is remained constant (that is saturation concentration); hence the time to reach saturation will be half. Recall that when the isotherm is linear, the time scale for adsorption is independent of bulk concentration. Hence, for moderately nonlinear isotherm, the time scale would take the following form:

$$t_{0.5} = a(C_0)^\alpha \qquad (9.2-53)$$

where $\alpha = 0$ for linear isotherm, and $\alpha = -1$ for irreversible isotherm and α is between -1 and 0 for moderately nonlinear isotherm. We will show later in eq. (9.2-63) an approximate but explicit half time for the case of Langmuir isotherm.

Another feature of the irreversible isotherm case is the time taken for the particle to be completely saturated with adsorbate. This is obtained by setting F to unity in eqs. (9.2-52) and the results are tabulated in Table 9.2-5. This finite saturation time is only possible with the case of irreversible isotherm. For the cases of linear isotherm and nonlinear isotherm, the time it takes to equilibrate the particle is infinity. In such cases to obtain the magnitude of the adsorption time scale, we need to determine the time taken for the particle to reach, say 99% of the equilibrium capacity. For example, take the case of a spherical particle with linear isotherm, the fractional uptake is given in eq. (9.2-35c). By setting F = 0.99 into that equation, we obtain the following time

$$\tau_{0.99} = \frac{1}{\pi^2}\ln\left(\frac{600}{\pi^2}\right) \approx 0.42$$

which could serve as a measure of equilibration time for linear isotherm.

Table 9.2-5: Time to reach saturation

Shape	Time to reach saturation	Eq. No.
Slab	$t_1 = \dfrac{R^2(1-\varepsilon)C_{\mu s}}{\varepsilon D_p C_0}\left(\dfrac{1}{2}+\dfrac{1}{Bi}\right)$	(9.2-54a)
Cylinder	$t_1 = \dfrac{R^2(1-\varepsilon)C_{\mu s}}{\varepsilon D_p C_0}\left(\dfrac{1}{4}+\dfrac{1}{2Bi}\right)$	(9.2-54b)
Sphere	$t_1 = \dfrac{R^2(1-\varepsilon)C_{\mu s}}{\varepsilon D_p C_0}\left(\dfrac{1}{6}+\dfrac{1}{3Bi}\right)$	(9.2-54c)

Observing eqs. (9.2-54) we see that the time to reach 100% uptake in the case of cylindrical adsorbent is half of that for the case of slab adsorbent, and that in the case of spherical particle as one third of that for slab. This interesting result is explained from the ratio of particle volume to external surface area

$$\frac{V}{S_{ex}} = \begin{cases} R/3 & \text{sphere} \\ R/2 & \text{cylinder} \\ R & \text{slab} \end{cases} \quad (9.2\text{-}54d)$$

For the case of no mass transfer resistance in the film surrounding the particle, we simply set Bi to infinity in eqs. (9.2-52). For such a case, the solutions are simpler and explicit for slab and sphere and they are given in Table 9.2-6. What we note from the solution for slab is that the fractional uptake is proportional to the square root of time throughout the course of adsorption. Recall the short time solution for the case of linear isotherm (eq. 9.2-38) that the fractional uptake is also proportional to the square root of time but that is only valid for short times.

Table 9.2-6: Fractional uptake for the case of irreversible isotherm with no film resistance

Shape	Fractional uptake Expression	Eq. No.
Slab	$F = \sqrt{\dfrac{t}{t^*}}; \quad t^* = \dfrac{R^2(1-\varepsilon)C_{\mu s}}{2\varepsilon D_p C_0}$	(9.2-55a)
Cylinder	$F + (1-F)\ln(1-F) = \dfrac{t}{t^*}; \quad t^* = \dfrac{R^2(1-\varepsilon)C_{\mu s}}{4\varepsilon D_p C_0}$	(9.2-55b)
Sphere	$F = 1 - \left\{ \dfrac{1}{2} + \cos\left[\dfrac{\pi}{3} + \dfrac{1}{3}\cdot\cos^{-1}\left(1 - \dfrac{2t}{t^*}\right)\right] \right\}^3; \quad t^* = \dfrac{R^2(1-\varepsilon)C_{\mu s}}{6\varepsilon D_p C_0}$	(9.2-55c)

What has been presented so far is the analysis of the irreversible case with pore diffusion mechanism only. The concept of travelling concentration front is valid as long as there is no surface diffusion because if surface diffusion is present, which is unlikely for very strong adsorption, it will distort the sharp concentration wave front and hence the concept of two distinct regions will no longer be applicable.

9.2.2.3 Approximate Solution for Dual Diffusion Model and Langmuir Isotherm

We have discussed briefly in Chapter 7 about the complexity of the surface diffusion. It has been observed in many systems, particularly systems with large internal surface area such as activated carbon. What makes surface diffusion more difficult is that this process is highly dependent on the solid structure. It has been experimentally observed that the surface diffusivity is a strong function of loading. However, what we will do in this section is to assume that surface diffusion is described by a Fickian equation with a constant diffusion coefficient, as we did in eq. (9.2-1b), and then obtain an approximate solution to see the influence of various parameters including surface diffusion coefficient on the dynamic behaviour of a nonlinear adsorption system. Many work have assumed constant surface diffusivity, such as those of Damkohler (1935), Testin and Stuart (1966), Brecher et al. (1967a,b), Dedrick and Breckmann (1967), Nemeth and Stuart (1970), Furusawa and Smith (1973), Komiyama and Smith (1974), Neretnieks (1976), Costa et al. (1985) and Costa and Rodrigues (1985).

The surface diffusivity is usually obtained by first isolating the pore diffusion rate from the total observed rate, and then from the definition of surface flux the diffusivity can be obtained. If the surface flux is based on the concentration gradient, the surface diffusivity is called the transport surface diffusivity. The isolation of the pore diffusion can be achieved by increasing temperature to the extent that the contribution of surface diffusion is negligible (see Section 9.2.1.1). The reason why the surface diffusion becomes negligible at high temperature is that the surface diffusion flux is a product between the surface diffusivity and the surface loading. As temperature increases, the surface diffusivity increases but not as fast as the decrease of the surface loading; therefore, the surface flux decreases (Schneider and Smith, 1968; Mayfield and Do, 1991). This methodology of removing the contribution of surface diffusion at high temperature has two disadvantages. One is that the temperature beyond which the surface diffusion becomes negligible may be too high to achieve practically, and the second reason is that the surface diffusion rate is negligible compared to the pore diffusion rate as the temperature increases only when the adsorption isotherm is linear. Another possible way to overcome this problem is to use a half-time method first proposed by Do (1990) and Do and Rice (1991) to extract the surface diffusivity from a simple analysis. This is done as follows.

We demonstrate this with a spherical particle with no film resistance (that is $Bi \to \infty$), and then present the results for slab and cylindrical particles. If the pore diffusion is the only diffusion mechanism and the adsorption isotherm is linear, the half time of adsorption is (from eq. 9.2-36b)

$$\frac{\varepsilon D_p t_{0.5}}{R^2 [\varepsilon + (1-\varepsilon)K]} = 0.03055 \tag{9.2-57}$$

When the adsorption isotherm is irreversible and the pore diffusion mechanism is the controlling factor, the half time of adsorption is (from eq. 9.2-52c):

$$\frac{\varepsilon D_p C_0 t_{0.5}}{R^2 (1-\varepsilon) C_{\mu s}} = 0.01835 \tag{9.2-58}$$

Solving eq. (9.2-45) numerically for the case of Langmuir isotherm using the programming code ADSORB1A, we obtain the following approximate correlation for the half time in terms of the parameter $\lambda = bC_0$, which is a measure of the degree of nonlinearity of the isotherm

$$\frac{\varepsilon D_p t_{0.5}}{R^2 [\varepsilon + (1-\varepsilon)(C_{\mu 0}/C_0)]} = 0.03055 \left[1 - \frac{0.3\lambda}{1 + 3\lambda/4}\right] \tag{9.2-59}$$

where $C_{\mu 0}$ is the adsorbed amount in equilibrium with the bulk concentration C_0:

$$C_{\mu 0} = C_{\mu s} \frac{bC_0}{1 + bC_0} = C_{\mu s} \frac{\lambda}{1 + \lambda} \tag{9.2-60}$$

When $\lambda = 0$, eq. (9.2-59) reduces to the half time for the linear isotherm (eq. 9.2-57) since $K = C_{\mu s} b$, and when $\lambda \to \infty$ we recover the half time for the irreversible isotherm (eq. 9.2-58).

Now allowing for the pore and surface diffusion in parallel and the isotherm is linear, the half time of adsorption is then (eq. 9.2-36):

$$\frac{[\varepsilon D_p + (1-\varepsilon)KD_s]t_{0.5}}{R^2[\varepsilon + (1-\varepsilon)K]} = 0.03055 \tag{9.2-61}$$

Finally, when the surface diffusion is the controlling mechanism, the half time will be:

$$\frac{D_s t_{0.5}}{R^2} = 0.03055 \tag{9.2-62}$$

no matter what the isotherm nonlinearity is.

Thus, by combining the behaviour of the half time at various limits (eqs. 9.2-57 to 9.2-62), the following general equation for the half time for parallel pore and surface diffusion and any nonlinearity of the isotherm (Do, 1990) is obtained

$$t_{0.5} = \frac{\alpha R^2 \left[\varepsilon + (1-\varepsilon)\dfrac{C_{\mu 0}}{C_0}\right]\left(1 - \dfrac{\beta b C_0}{1+\gamma b C_0}\right)}{\varepsilon D_p + (1-\varepsilon)D_s \left(1 - \dfrac{\beta b C_0}{1+\gamma b C_0}\right)\left(\dfrac{C_{\mu 0}}{C_0}\right)} \tag{9.2-63a}$$

where

$$\frac{C_{\mu 0}}{C_0} = \frac{bC_{\mu s}}{1+bC_0} = \frac{K}{1+bC_0} \tag{9.2-63b}$$

Here C_0 is the external bulk concentration and $C_{\mu 0}$ is the adsorbed concentration which is in equilibrium with C_0 and α, β, γ are given in the following table (Table 9.2-7). Eq. (9.2-63a) is valid for three shapes of the particle. The only difference between the three particle shapes is the values of the parameters α, β, and γ.

Table 9.2-7: Parameter values for eq. (9.2-63a)

Shape of particle	α	β	γ
Slab	0.19674	0.25	0.686
Cylinder	0.06310	0.26	0.663
Sphere	0.03055	0.30	0.750

The explicit form of eq. (9.2-63) is very useful for the determination of surface diffusivity. To do so we must rely on experiments conducted <u>over the nonlinear region</u> of the isotherm. Otherwise, the contribution of the surface diffusion can not be distinguished over the linear region. Half time is measured for each bulk concentration used. By rearranging the analytical half time equation (9.2-63a), we have

$$\frac{\alpha R^2 \left[\varepsilon + (1-\varepsilon)\dfrac{C_{\mu 0}}{C_0}\right]\left(1 - \dfrac{\beta\, bC_0}{1+\gamma\, bC_0}\right)}{t_{0.5}} = \varepsilon D_p + (1-\varepsilon)D_s\left(1 - \dfrac{\beta\, bC_0}{1+\gamma\, bC_0}\right)\left(\dfrac{C_{\mu 0}}{C_0}\right) \quad (9.2\text{-}64)$$

This equation suggests that if one plots the LHS of the above equation versus

$$X = \left(1 - \frac{\beta\, bC_0}{1+\gamma\, bC_0}\right)\left(\frac{C_{\mu 0}}{C_0}\right)$$

we would obtain a straight line with the slope $(1-\varepsilon)D_s$ and the intercept εD_p. If experiments were to be carried out over the Henry's law range (i.e. $bC_0 \ll 1$ and $C_{\mu 0}/C_0 = K$, which is the slope of the isotherm), eq. (9.2-64) would become:

$$\frac{\alpha R^2[\varepsilon + (1-\varepsilon)K]}{t_{0.5}} = \varepsilon D_p + (1-\varepsilon)D_s K \quad (9.2\text{-}65)$$

Inspecting this equation, we see that the separate contribution of pore and surface diffusions can not be delineated. Instead one can only obtain the apparent diffusivity from the linear region experiments

$$D_{app} = \frac{\varepsilon D_p + (1-\varepsilon)K D_s}{\varepsilon + (1-\varepsilon)K} \quad (9.2\text{-}66)$$

To illustrate the half time dependence on bulk concentration (eq. 9.2-63), we take the following example of a spherical adsorbent particle

ε	0.31
R	0.2, 0.3 cm
b	8.4×10^5 cc/mole
$C_{\mu s}$	5×10^{-3} mole/cc
D_p	0.02 cm^2/sec
D_s	1.5×10^{-5} cm^2/sec

Figure 9.2-11 shows plots of the half time (using eq. 9.2-63) versus the bulk concentration for two particle sizes (R = 0.2 and 0.3 cm). We note that the half-time

decreases with the bulk concentration, that is the time to approach equilibrium is shorter in systems having higher bulk concentration.

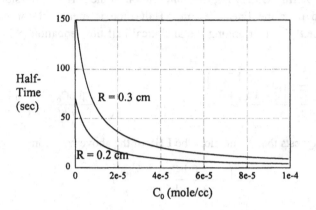

Figure 9.2-11: Plot of the half-time versus the bulk concentration C_0 (mole/cc)

Example 9.2-5: *Determination of pore and surface diffusivities*

We now illustrate the utility of eq.(9.2-64) in the determination of diffusivities in a system of n-butane adsorption on activated carbon. The adsorption isotherm can be described by a Langmuir equation:

$$C_\mu = C_{\mu s} \frac{bC}{1+bC}$$

with

$$C_{\mu s} = 4.9 \times 10^{-3} \text{ mole / cc}$$

$$b = 8.36 \times 10^5 \text{ cc / mole}$$

The particle has a slab geometry, and its half thickness and porosity are:

$$R = 0.1 \text{ cm},$$

$$\varepsilon = 0.31$$

The following table lists the experimental half-times measured at three bulk concentrations.

C_b (mole/cc)	$t_{0.5}$ (sec)
5.188×10^{-6}	60
8.285×10^{-6}	44.8
1.272×10^{-5}	36.2

Figure 9.2-12 shows a plot of the LHS of eq. (9.2-64) versus

$$X = \left(1 - \frac{\beta\, bC_0}{1+\gamma\, bC_0}\right)\left(\frac{C_{\mu 0}}{C_0}\right)$$

for this example of n-butane/activated carbon. A straight line can be plotted through the data points, and we get:

$$\text{slope} = (1-\varepsilon)D_s = 1.116 \times 10^{-5} \text{ cm}^2/\text{sec}$$

$$\text{Intercept} = \varepsilon D_p = 6.55 \times 10^{-3} \text{ cm}^2/\text{sec}$$

from which we can calculate the pore and surface diffusivities:

$$D_p = \frac{6.55 \times 10^{-3}}{0.31} = 0.021 \text{ cm}^2/\text{sec}$$

$$D_s = \frac{1.116 \times 10^{-5}}{1-0.31} = 1.62 \times 10^{-5} \text{ cm}^2/\text{sec}$$

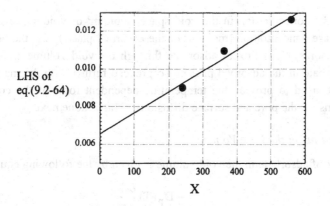

Figure 9.2-12: Plot of the LHS of eq. (9.2-64) versus X

9.3 Adsorption Models for Nonisothermal Single Component Systems

In the last section we have illustrated the essential behaviors of isothermal single component systems. In this section we will address the effect of heat release during the adsorption step or heat absorption during the desorption step on the adsorption kinetics. What we need is simply an extra equation to account for the heat balance around the particle. To make the analysis more general, we shall take adsorption isotherm being arbitrary.

Adsorption process is known to release heat upon adsorption or absorb energy upon desorption. Typical heat of adsorption of many hydrocarbons on activated carbon, alumina or silica gels ranges from 10 to 60 kJoule/mole. Such a magnitude can cause a significant change in the particle temperature if the dissipation of energy to the surrounding is not fast enough. Thus, the particle temperature increase (in adsorption) or decrease (in desorption) depends on the interplay between the rate of heat release due to adsorption and the dissipation rate of energy to the surrounding. Since the solid thermal conductivity is much larger than that of the surrounding gas, we shall assume that there is no temperature gradient within the particle. All the heat transfer resistance is in the gas film surrounding the particle.

9.3.1 Problem Formulation

The mass balance equation over a thin shell of a particle in which the dual diffusion mechanisms are operative is:

$$\varepsilon \frac{\partial C}{\partial t} + (1-\varepsilon)\frac{\partial C_\mu}{\partial t} = -\frac{1}{r^s}\frac{\partial}{\partial r}\left\{r^s\left[\varepsilon N_p + (1-\varepsilon)N_\mu\right]\right\} \qquad (9.3\text{-}1)$$

where C is the concentration in the void space (mole/m³ of void space), C_μ is the adsorbed phase concentration (mole/m³ of the adsorbed phase), ε is the porosity of the through-pores, N_p is flux of adsorbate through the void volume and N_μ is the flux of adsorbate in the adsorbed phase. To proceed further with the mass balance equation, we need to provide the temperature-dependent form of the constitutive flux equations for the pore and surface diffusions. This is done next.

9.3.1.1 The Flux in the Pore Volume

The flux of adsorbate in the void volume is given by the following equation:

$$N_p = -D_p(T)\frac{\partial C}{\partial r} \qquad (9.3\text{-}2)$$

where the pore diffusivity has the following temperature dependent form:

$$D_p(T) = D_{p0}\left(\frac{T}{T_0}\right)^\alpha \qquad (9.3\text{-}3a)$$

Here, D_{p0} is the pore diffusivity at some reference temperature T_0. The exponent α is 0.5 if Knudsen diffusion is the controlling mechanism within the void space. However, if the molecular diffusion is the controlling mechanism, this exponent takes a value of about 1.75. In the intermediate region where the Knudsen and molecular diffusions are both controlling, the pore diffusivity can be calculated from the Bosanquet relation (eq. 8.8-29b). Written explicitly in terms of temperature, the pore diffusivity is:

$$\frac{1}{D_p(T)} = \frac{1}{D_{Ko}(T/T_0)^{0.5}} + \frac{1}{D_{mo}(T/T_0)^{1.75}} \qquad (9.3\text{-}3b)$$

9.3.1.2 The Flux in the Adsorbed Phase

The fundamental driving force for the movement of adsorbed molecules is the gradient of chemical potential. The flux of adsorbate in the adsorbed phase is given by:

$$N_\mu = -L\, C_\mu \frac{\partial \mu}{\partial r} \qquad (9.3\text{-}4)$$

where L is the mobility constant, which is assumed to be independent of concentration and is dependent on temperature only, and μ is the chemical potential of the adsorbed phase. We assume that at any point within the particle local equilibrium between the gas and adsorbed phases exists, that is the chemical potential of the adsorbed phase is the same as that of the gas phase:

$$\mu = \mu_G = \mu^0 + R_g T \ln C \qquad (9.3\text{-}5)$$

With this equality in chemical potential, the diffusion flux in the adsorbed phase (eq. 9.3-4) now can be written as:

$$N_\mu = -D_\mu^0 \frac{C_\mu}{C} \frac{\partial C}{\partial r} \qquad (9.3\text{-}6)$$

in which we have assumed that there is no variation of temperature with respect to distance. In the above equation, the diffusivity D_μ^0 is called the corrected diffusivity and is given by:

$$D_\mu^0 = LR_g T \tag{9.3-7}$$

We see the flux in the adsorbed phase is governed by the gradient of the concentration in the gas phase. To write its form in terms of the gradient of the concentration of the adsorbed phase, we resort to the equilibrium relationship between the two phases:

$$C_\mu = f(C,T) \tag{9.3-8}$$

Applying the chain rule of differentiation in eq. (9.3-6), we get:

$$N_\mu = -D_\mu^0 \frac{C_\mu}{C} \frac{\partial C}{\partial r} = -D_\mu^0 \frac{d \ln C}{d \ln C_\mu} \frac{\partial C_\mu}{\partial r} \tag{9.3-9}$$

This is the flux expression for surface diffusion written in terms of the gradient of adsorbed concentration. Eq. (9.3-6) gives the same diffusion flux of the adsorbed species, but written in terms of the gradient of the fluid species concentration. Although these two equations are mathematically equivalent, eq. (9.3-6) is more efficient from the computational point of view.

The diffusion of the adsorbed species is usually activated, and hence the corrected diffusivity takes the following Arrhenius form:

$$D_\mu^0(T) = D_{\mu\infty}^0 \exp\left(-\frac{E_\mu}{R_g T}\right) \tag{9.3-10a}$$

or

$$D_\mu^0(T) = D_{\mu 0}^0 \exp\left[\frac{E_\mu}{R_g T_0}\left(1 - \frac{T_0}{T}\right)\right] \tag{9.3-10b}$$

where $D_{\mu\infty}^0$ is the corrected diffusivity at infinite temperature, and $D_{\mu 0}^0$ is that at some reference temperature T_0, and E_μ is the activation energy for surface diffusion. This activation energy is found experimentally to fall in the range of

$$\frac{Q}{3} < E_\mu < Q$$

where Q is the heat of adsorption.

9.3.1.3 The Mass and Heat Balance Equations:

Having had the necessary flux equations for the free species as well as the adsorbed species, we substitute them (eqs. 9.3-2 and 9.3-6) into the mass balance equation (9.3-1) to get:

$$\varepsilon\frac{\partial C}{\partial t}+(1-\varepsilon)\frac{\partial C_\mu}{\partial t}=\frac{1}{r^s}\frac{\partial}{\partial r}\left\{r^s\left[\varepsilon D_p(T)+(1-\varepsilon)D_\mu^0(T)\frac{C_\mu}{C}\right]\frac{\partial C}{\partial r}\right\} \quad (9.3\text{-}11a)$$

Since the two phases are in local equilibrium with each other, the adsorption isotherm equation (9.3-8) can be used to eliminate C_μ from the above equation to finally obtain the equation in terms of only concentration of the free species:

$$\left[\varepsilon+(1-\varepsilon)\frac{\partial f(C,T)}{\partial C}\right]\frac{\partial C}{\partial t}+(1-\varepsilon)\frac{\partial f(C,T)}{\partial T}\frac{\partial T}{\partial t}=\frac{1}{r^s}\frac{\partial}{\partial r}\left\{r^s\left[\varepsilon D_p(T)+(1-\varepsilon)D_\mu^0(T)\frac{f(C,T)}{C}\right]\frac{\partial C}{\partial r}\right\}$$
$$(9.3\text{-}11b)$$

The second term in the LHS accounts for the variation of the adsorbed concentration with respect to temperature. The above mass balance can, in principle, be solved for the concentration distribution if we know the temperature variation as a function of time. However, this temperature variation is governed by the interplay between the rate of mass transfer and the rate of energy dissipation. This means that mass and heat balances are coupled and their equations must be solved simultaneously.

Carrying out the heat balance around the particle, we obtain the following equation describing the variation of particle temperature with time:

$$\langle\rho C_p\rangle\frac{dT}{dt}=(1-\varepsilon)Q\frac{d\langle C_\mu\rangle}{dt}-a_H h_f(T-T_b) \quad (9.3\text{-}12a)$$

In obtaining the heat balance equation, we have assumed that the heat resistance by conduction inside the particle is insignificant compared to the heat resistance of the fluid film surrounding the particle, which is described by the second term in the RHS of eq.(9.3-12a). Because of such assumption, the temperature inside the particle is uniform. The first term in the RHS of eq.(9.3-12a) is the amount of heat released per unit volume per unit time as a result of the adsorption rate per unit volume $d<C_\mu>/dt$. The parameter h_f is the heat transfer coefficient per unit surface area, a_H is the heat transfer surface area per unit volume, Q is the molar heat of adsorption, T_b is the surrounding temperature, $<\rho C_p>$ is the mean heat capacity per unit particle volume and $<C_\mu>$ is the volumetric average concentration of the adsorbed species, defined as

$$\langle C_\mu\rangle=\frac{(1+s)}{R^{s+1}}\int_0^R r^s C_\mu\, dr \quad (9.3\text{-}12b)$$

The heat transfer surface area per unit volume a_H is written in terms of the particle size as follows:

$$a_H = \frac{1+s}{R} \qquad (9.3\text{-}12c)$$

where s is the shape factor of the particle, and it takes a value of 1, 2 or 3 for slab, cylinder and sphere, respectively.

The heat balance equation (9.3-12a) in its form is not convenient for numerical computation because of the time derivative of the average adsorbed concentration. To alleviate this problem, we proceed as follows. Taking the volume average of the mass balance equation (9.3-11a), we obtain:

$$\varepsilon \frac{\partial \langle C \rangle}{\partial t} + (1-\varepsilon) \frac{\partial \langle C_\mu \rangle}{\partial t} = \frac{1+s}{R}\left[\varepsilon D_p(T) + (1-\varepsilon)D_\mu^0(T)\frac{C_\mu}{C}\right]\frac{\partial C}{\partial r}\bigg|_{r=R} \qquad (9.3\text{-}13a)$$

which states that the rate of mass hold up in the particle is equal to the diffusion rate into the particle. Since the mass hold-up in the void volume is usually much smaller than that in the adsorbed phase

$$\varepsilon \frac{\partial \langle C \rangle}{\partial t} \ll (1-\varepsilon)\frac{\partial \langle C_\mu \rangle}{\partial t}$$

eq.(9.3-13a) is then approximated by

$$(1-\varepsilon)\frac{\partial \langle C_\mu \rangle}{\partial t} \cong \frac{1+s}{R}\left[\varepsilon D_p(T) + (1-\varepsilon)D_\mu^0(T)\frac{C_\mu}{C}\right]\frac{\partial C}{\partial r}\bigg|_{r=R} \qquad (9.3\text{-}13b)$$

Substitution of the above equation into the heat balance equation (9.3-12a), we obtain an alternative form for the heat balance equation around the particle

$$\langle \rho C_p \rangle \frac{dT}{dt} = \frac{(1+s)Q}{R}\left[\varepsilon D_p(T) + (1-\varepsilon)D_\mu^0(T)\frac{C_\mu}{C}\right]\frac{\partial C}{\partial r}\bigg|_{r=R} - a_H h_f(T-T_b) \qquad (9.3\text{-}14)$$

Eqs. (9.3-11b) and (9.3-14) are the required mass and heat balance equations. All we need to proceed with the solution of these equations is to provide the pertinent boundary and initial conditions.

9.3.1.4 The Boundary Conditions

The boundary conditions of the mass balance equation (9.3-11) are:

$$r = 0; \qquad \frac{\partial C}{\partial r} = 0 \qquad (9.3\text{-}15a)$$

$$r = R; \quad -\left[\varepsilon D_p(T) + (1-\varepsilon)D_\mu^0(T)\frac{C_\mu}{C}\right]\frac{\partial C}{\partial r}\bigg|_{r=R} = k_m\left(C|_R - C_b\right) \quad (9.3\text{-}15b)$$

The first boundary condition is the symmetry at the center of the particle, while the second boundary condition simply states that the flux at the exterior surface of the particle is equal to the flux through the stagnant film surrounding the particle. Here C_b is the bulk concentration.

9.3.1.5 The Initial Condition

The particle is assumed to be initially equilibrated with a concentration of C_i and a temperature of T_i, that is:

$$t = 0; \quad C = C_i; \quad C_{\mu,i} = f(P_i, T_i); \quad T = T_i \quad (9.3\text{-}16)$$

The initial temperature T_i may not be the same as the surrounding temperature T_b.

9.3.1.6 Non-Dimensional Equations

The set of governing equations (9.3-11b and 9.3-14) subject to the boundary and initial conditions (9.3-15) and (9.3-16) is nonlinear and must be solved numerically. We solve it by using a combination of the orthogonal collocation method and the Runge-Kutta method. Before carrying out the numerical analysis, it is convenient to cast the governing equations into non-dimensional form. The basic thing we need to achieve in this non-dimensionalisation process is to normalize the spatial variable as the orthogonal collocation method is set up for a variable having a domain of (0, 1). We need not to non-dimensionalize other variables, and this is what we shall do when we deal with complex multicomponent systems later, but for this simple single component system it is convenient to nondimensionalize all variables as we will show next.

Let C_0 be some reference concentration for the free species, and T_0 be some reference temperature. The reference concentration for the adsorbed species is taken as the adsorbed concentration which is equilibrium with the concentration C_0 and temperature T_0, that is:

$$C_{\mu 0} = f(C_0, T_0) \quad (9.3\text{-}17)$$

where f is the functional form of the adsorption isotherm.

We define the following non-dimensional variables and parameters

Non-dimensional time and distance	$\tau = \dfrac{\varepsilon D_{p0} t}{R^2}$; $\eta = \dfrac{r}{R}$	(9.3-18a)
Non-dimensional concentrations and temperature	$y = \dfrac{C}{C_0}$; $x = \dfrac{C_\mu}{C_{\mu 0}}$; $\theta = \dfrac{T - T_0}{T_0}$	(9.3-18b)
Non-dimensional bulk concentration and initial concentration	$y_b = \dfrac{C_b}{C_0}$; $y_i = \dfrac{C_i}{C_0}$	(9.3-18c)
Non-dimensional bulk temperature and initial temperature	$\theta_b = \dfrac{T_b - T_0}{T_0}$; $\theta_i = \dfrac{T_i - T_0}{T_0}$	(9.3-18d)
Surface to pore diffusion number and Biot number	$\delta_0 = \dfrac{(1-\varepsilon) D^0_{\mu 0} C_{\mu 0}}{\varepsilon D_{p0} C_0}$; $Bi = \dfrac{k_m R}{\varepsilon D_{p0}}$	(9.3-18e)
Activation energy for surface diffusion number, heat capacity group	$\gamma_\mu = \dfrac{E_\mu}{R_g T}$; $\beta = \dfrac{Q C_{\mu 0}}{\langle \rho C_p \rangle T_0}$	(9.3-18f)
Heat transfer number	$LeBi = \dfrac{a_H h_f R^2}{\varepsilon D_{p0} \langle \rho C_p \rangle}$	(9.3-18g)

With the above definitions of non-dimensional variables and parameters, the resulting non-dimensional governing equations take the following form:

$$G_1(y,\theta)\frac{\partial y}{\partial \tau} + G_2(y,\theta)\frac{\partial \theta}{\partial \tau} = \frac{1}{\eta^s}\frac{\partial}{\partial \eta}\left[\eta^s \cdot H(y,\theta)\frac{\partial y}{\partial \eta}\right] \quad (9.3\text{-}19a)$$

$$\frac{d\theta}{d\tau} = \beta\left(\frac{C_0}{C_{\mu 0}}\right)(1+s) H(y,\theta)\frac{\partial y}{\partial \eta}\bigg|_{\eta=1} - LeBi(\theta - \theta_b) \quad (9.3\text{-}19b)$$

The relevant functions G_1, G_2 and H are defined below:

$$G_1(y,\theta) = \left[\varepsilon + (1-\varepsilon)\frac{\partial f(C_0 y, T_0(1+\theta))}{\partial C}\right] \quad (9.3\text{-}20a)$$

$$G_2(y,\theta) = (1-\varepsilon)\frac{T_0}{C_0}\frac{\partial f(C_0 y, T_0(1+\theta))}{\partial T} \quad (9.3\text{-}20b)$$

$$H(y,\theta) = \left[(1+\theta)^\alpha + \delta_0 \varphi(\theta)\frac{f(C_0 y, T_0(1+\theta))}{C_{\mu 0} y}\right] \quad (9.3\text{-}20c)$$

where

$$\varphi(\theta) = \exp\left[\gamma_\mu \left(\frac{\theta}{1+\theta}\right)\right] \quad (9.3\text{-}21)$$

The boundary and initial conditions in non-dimensional form are:

$$-H(y,\theta)\frac{\partial y}{\partial \eta}\bigg|_{\eta=1} = Bi\left(y\big|_{\eta=1} - y_b\right) \quad (9.3\text{-}22)$$

$$\tau = 0; \quad y = y_i; \quad \theta = \theta_i \quad (9.3\text{-}23)$$

Solving numerically eqs.(9.3-19) subject to the boundary condition (9.3-22) and the initial condition (9.3-23), we obtain the non-dimensional concentration y as a function of distance η and time τ, and the non-dimensional temperature θ as a function of τ, that is:

$$y(\eta, \tau)$$
$$\theta(\tau)$$

From this, we can calculate the dimensional concentrations and temperature

$$C(\eta, \tau) = C_0 \, y(\eta, \tau) \quad (9.3\text{-}24a)$$

$$T(\tau) = T_0[1 + \theta(\tau)] \quad (9.3\text{-}24b)$$

$$C_\mu(\eta, \tau) = f[C(\eta, \tau), T(\tau)] \quad (9.3\text{-}24c)$$

and then obtain the amount adsorbed per unit volume of particle

$$M(\tau) = \varepsilon\langle C\rangle + (1-\varepsilon)\langle C_\mu\rangle \quad (9.3\text{-}25)$$

The fractional uptake defined as the amount adsorbed from t = 0 up to time t divided by the amount taken from t = 0 to infinite time, that is:

$$F = \frac{\left[\varepsilon\langle C\rangle + (1-\varepsilon)\langle C_\mu\rangle\right] - \left[\varepsilon C_i + (1-\varepsilon)C_{\mu i}\right]}{\left[\varepsilon C_b + (1-\varepsilon)C_{\mu b}\right] - \left[\varepsilon C_i + (1-\varepsilon)C_{\mu i}\right]} \quad (9.3\text{-}26)$$

where < > denotes the volumetric average defined as follows:

$$\langle C_\mu\rangle = (1+s)\int_0^1 \eta^s \, C_\mu \, d\eta \quad (9.3\text{-}27a)$$

and

$$C_{\mu b} = f(C_b, T_b); \quad C_{\mu i} = f(C_i, T_i) \qquad (9.3\text{-}27b)$$

9.3.1.7 The Heat Transfer Number LeBi

One of the important parameters that describe the influence of energy on the mass transfer is the heat transfer number, LeBi. This parameter is a measure of the heat transfer to the surrounding. Its physical meaning can be recognized by rearranging eq. (9.3-18g) as follows:

$$\text{LeBi} = \frac{\dfrac{R^2}{\varepsilon D_{p0}}}{\dfrac{\langle \rho C_p \rangle}{a_H h_f}} \qquad (9.3\text{-}28)$$

The numerator is the diffusion time, which is the time taken for the mass transfer to approach equilibrium. The denominator is the time which the sensible heat of the particle can be dissipated to the surrounding. Thus we expect that the larger is this parameter the closer is the system behavior towards isothermal conditions. The particle radius and the rate of stirring all affect the magnitude of this LeBi number. The dependence of the heat transfer area on the particle size is given in eq.(9.3-12c). The dependence of the heat transfer coefficient on the particle size can be shown by taking the following correlation (Wakao and Kaguei, 1982):

$$\text{Nu} = \frac{h_f(2R)}{k_f} = 2 + 1.1 \left(\frac{2 R u \rho_f}{\mu_f} \right)^{0.6} \left(\frac{C_p \mu_f}{k_f} \right)^{1/3} \qquad (9.3\text{-}29a)$$

from which the heat transfer coefficient is

$$h_f = \frac{k_f}{2R} \left[2 + 1.1 \left(\frac{2 R u \rho_f}{\mu_f} \right)^{0.6} \left(\frac{C_p \mu_f}{k_f} \right)^{1/3} \right] \qquad (9.3\text{-}29b)$$

where k_f is the thermal conductivity of the fluid surrounding the particle, u is the fluid velocity, C_p is the fluid specific heat capacity, and ρ_f and μ_f are density and viscosity of the fluid, respectively.

If the surrounding is stagnant, we have:

$$h_f \approx \frac{k_f}{R} \qquad (9.3\text{-}30)$$

that is the heat transfer coefficient is inversely proportional to the particle radius R. However, if the surrounding is highly stirred, the heat transfer coefficient will become:

$$h_f \approx 1.1 \left(\frac{k_f}{2R}\right)\left(\frac{2Ru\rho_f}{\mu_f}\right)^{0.6}\left(\frac{C_p\mu_f}{k_f}\right)^{1/3} \quad (9.3\text{-}31)$$

that is, the heat transfer coefficient is inversely proportional to $R^{0.4}$. Thus, in general

$$h_f \propto R^{-m}$$

where $0.4 < m < 1$.

Substituting the dependence of the heat transfer area (a_H) (eq. 9.3-12c) and the heat transfer coefficient on particle size into the LeBi number, we find

$$\text{LeBi} \propto R^n \quad (9.3\text{-}32)$$

where $0 < n < 0.6$, with $n = 0$ for stagnant fluid surrounding the particle and $n = 0.6$ for turbulent conditions around the particle.

Example 9.3-1: *Magnitude of the heat transfer coefficient*

The heat transfer coefficient can be calculated from a correlation given in eq. (9.3-29). In this example, we calculate the heat transfer coefficient for a spherical activated carbon particle having a diameter of 0.2cm. The gas mixture surrounding the particle contains nitrogen and ethane. The mole fractions of nitrogen and ethane are 0.8 and 0.2, respectively. The temperature of the system is 303K and the total pressure is 1 atm. We denote 1 for nitrogen and 2 for ethane.

The relevant properties of nitrogen and ethane are listed in the following table:

	MW	μ (g/cm/s)	k (W/m/K)	C_p (J/g/K)
nitrogen	28	1.8×10^{-4}	0.0265	1.05
ethane	30	0.925×10^{-4}	0.0219	1.80

where k is the fluid thermal conductivity and C_p is the specific hear capacity. The specific heat of carbon is 0.803 J/g/k, and the particle density is 0.733 g/cc. Thus the volumetric heat capacity is:

$$<\rho C_p>_P \approx 0.733 \times 0.703 = 0.515 \text{ J/cc/K}$$

in which we have ignored the heat capacity contribution of the gas within the void of the particle, which is usually about 1000 times lower than the contribution of the solid carbon.

Density of gaseous mixture:
The mixture density is the sum of component densities

$$\rho = \rho_1 + \rho_2$$

where the component densities are:

$$\rho_1 = MW_1 \cdot \frac{P_1}{R_g T} = 28 \times \frac{0.8}{(82.057)(303)} = 9 \times 10^{-4} \frac{g}{cc}$$

$$\rho_2 = MW_2 \cdot \frac{P_2}{R_g T} = 30 \times \frac{0.2}{(82.057)(303)} = 2.4 \times 10^{-4} \frac{g}{cc}$$

Thus the total density is

$$\rho = 11.4 \times 10^{-4} \ g/cc$$

Viscosity of mixture:
The viscosity of mixture can be calculated from (Bird et al., 1960)

$$\mu = \sum_{i=1}^{2} \frac{y_i \mu_i}{\sum_{j=1}^{2} y_j \Phi_{ij}}$$

where

$$\Phi_{ij} = \frac{1}{\sqrt{8}} \left(1 + \frac{M_i}{M_j}\right)^{-\frac{1}{2}} \left[1 + \left(\frac{\mu_i}{\mu_j}\right)^{\frac{1}{2}} \left(\frac{M_j}{M_i}\right)^{\frac{1}{4}}\right]^2$$

We calculate from the above equation for the viscosity of the mixture:

$$\mu = 1.55 \times 10^{-4} \ g/cm/s$$

Thermal conductivity of mixture
The thermal conductivity of the mixture is calculated from (Bird et al., 1960):

$$k = \sum_{i=1}^{n} \frac{y_i k_i}{\sum_{j=1}^{n} y_j \Phi_{ij}}$$

We calculate from the above equation for the thermal conductivity of mixture:

$$k = 0.0250 \text{ W/m-K}$$

Specific heat capacity of mixture

We calculate mass fraction:

$$w_1 = \frac{\rho_1}{\rho_1 + \rho_2} = \frac{9 \times 10^{-4}}{(9 + 2.4) \times 10^{-4}} = 0.79$$

$$w_2 = \frac{\rho_2}{\rho_1 + \rho_2} = 0.21$$

The specific heat capacity of the gaseous mixture is

$$C_p = w_1 (C_p)_1 + w_2 (C_p)_2$$

$$C_p = 1.2075 \text{ J/g-K}.$$

Heat transfer coefficient under stagnant condition

If the gas is stagnant, the heat transfer coefficient is calculated from eq. (9.3-30).

$$h_f = \frac{k_f}{R} = \frac{0.0250}{0.001} = 25 \frac{W}{m^2-K}$$

The heat transfer area per unit volume is

$$a_H = \frac{4\pi R^2}{\frac{4}{3}\pi R^3} = \frac{3}{R} = \frac{3}{0.001} = 3000 \frac{m^2}{m^3}$$

If the pore diffusivity is $\varepsilon D_{po} = 0.1 \text{ cm}^2/\text{sec}$, the LeBi number is:

$$\text{Le Bi} = \frac{R^2 a_H h_f}{<\rho C_p> \varepsilon D_{po}}$$

$$\text{Le Bi} = \frac{(0.001)^2 (3000)(25)}{(515000)(0.1\times 10^{-4})} \cong 0.015$$

Under the stagnant condition this value of Le Bi is independent of the particle size as

$$a_H \propto \frac{1}{R}$$

and

$$h_f \propto \frac{1}{R}$$

Heat transfer coefficient under flowing condition

For flowing gas, we calculate the heat transfer coefficient from eq. (9.3-29b). We calculate

$$\left(\frac{C_p \mu_f}{k_f}\right)^{1/3} = 0.908$$

and hence:

$$h_f = \frac{k_f}{2R}\left[2 + (1.1)(0.908)\left(\frac{2R u \rho}{\mu}\right)^{0.6}\right]$$

The following table shows the calculation of the heat transfer coefficient and the non-dimensional group Le Bi. The velocity of u = 10 m/sec corresponding to the case of very high velocity around the particle.

u (m/sec)	h (W/m²/K)	Le Bi
0	25	0.0146
0.02	49	0.0285
0.05	66	0.038
0.1	88	0.05
0.15	105	0.06
0.20	120	0.07
1	274	0.16
10	1018	0.59

9.3.1.8 Range of Relevant Parameters

The parameter γ_μ (eq. 9.3-18f) is the activation energy number, and for many adsorption systems it has the following range

$$2 < \gamma_\mu < 16 \tag{9.3-33}$$

The heat parameter β (eq. 9.3-18f) is a measure of the amount of heat released by adsorption process relative to the heat capacity of the solid. This parameter is called the heat capacity number. Typically it has the following range

$$0.01 < \beta < 2 \qquad (9.3\text{-}34)$$

9.3.1.9 Numerical Examples

The set of model equations (9.3-19 to 9.3-23) is solved by applying the orthogonal collocation method. The resulting equations after such application are coupled ODEs with respect to time, which can be readily integrated by any standard integration routines, such as the Runge-Kutta method. Appendix 9.4 details the orthogonal collocation analysis. A computer code ADSORB1B.M is provided to solve this set of equations.

We will illustrate this programming code with the following Langmuir adsorption isotherm:

$$C_\mu = f(C,T) = C_{\mu s} \frac{b(T)C}{1+b(T)C} \qquad (9.3\text{-}35)$$

where the adsorption affinity takes the following temperature dependent form:

$$b = b_\infty \exp\left(\frac{Q}{R_g T}\right) \qquad (9.3\text{-}36)$$

with Q being the heat of adsorption. This is the molar adsorption heat used in the heat balance equation (9.3-12a).

With the Langmuir isotherm taking the form of eq.(9.3-35), the functions $G_1(y,\theta)$, $G_2y,\theta)$ and $H(y,\theta)$ defined in eq. (9.3-20) now take the following explicit form:

$$G_1(y,\theta) = \left[\varepsilon + (1-\varepsilon)\frac{C_{\mu s} b_0 \phi(\theta)}{[1+b_0 C_0 \phi(\theta)y]^2}\right] \qquad (9.3\text{-}37a)$$

$$G_2(y,\theta) = -(1-\varepsilon)\left(\frac{Q}{R_g T_0}\right)\frac{C_{\mu s} b_0 \phi(\theta)y}{[1+b_0 C_0 \phi(\theta)y]^2 (1+\theta)^2} \qquad (9.3\text{-}37b)$$

$$H(y,\theta) = (1+\theta)^\alpha + \delta_0\, \phi(\theta)\, \phi(\theta)\frac{(1+b_0 C_0)}{[1+b_0 C_0 \phi(\theta)y]^2} \qquad (9.3\text{-}37c)$$

where b_0 is the adsorption affinity at the reference temperature T_0, and $\phi(\theta)$ is defined as:

$$\phi(\theta) = \exp\left[-\gamma_H \frac{\theta}{(1+\theta)}\right] \qquad (9.3\text{-}38a)$$

with

$$\gamma_H = \frac{Q}{R_g T_0} \qquad (9.3\text{-}38b)$$

To illustrate the heat effects on the adsorption kinetics, we take the following example typifying adsorption of light hydrocarbons onto activated carbon.

Table 9.3-1: Parameters of the base case used in the simulations

Particle radius	R	= 0.1 cm
Particle porosity	ε	= 0.33
Bulk concentration	C_b	= 1 × 10⁻⁶ mole/cc
Initial concentration	C_i	= 0
Bulk temperature	T_b	= 300 K
Initial temperature	T_i	= 300 K
Reference temperature	T_0	= 300 K
Adsorption affinity at T_0	b_0	= 1 × 10⁶ cc/mole
Saturation capacity	$C_{\mu s}$	= 5 × 10⁻³ mole/cc
Pore diffusivity at T_0	D_{p0}	= 0.02 cm²/sec
Corrected diffusivity at T_0	$D_{\mu 0}^0$	= 1 × 10⁻⁶ cm²/sec
Heat of adsorption	Q	= 30,000 Joule/mole
Activation energy for surface diffusion	E_μ	= 15,000 Joule/mole
Volumetric heat capacity	ρC_p	= 1 Joule/cc/K
Biot number for mass transfer	Bi	= ∞
Heat transfer number	LeBi	= 0.05

9.3.1.9.1 Effect of LeBi number

With the parameters listed in the above table, we study the effect of heat transfer on the mass transfer behavior as well as the particle temperature rise as a function of time. Figure 9.3-1a shows the amount adsorbed per unit particle volume (mole/cc) versus time with the LeBi as the varying parameter (LeBi = 0.01, 0.05 and 0.5). As we would expect the larger is the value of LeBi number, the faster is the adsorption kinetics due to the faster dissipation of heat from the particle. The mass transfer curve corresponding to low LeBi number of 0.01 exhibits a two-stage uptake. The first stage is due to the faster diffusion kinetics as a result of the temperature rise, and the second stage is due to the mass uptake as a result of cooling of the particle. This is substantiated by plots of the particle temperature versus time shown in Figure 9.3-1b. We see that the kink in the mass transfer curve is corresponding to the time when the particle temperature is maximum.

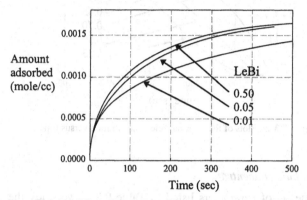

Figure 9.3-1a: Effect of the LeBi number on the amount adsorbed versus time

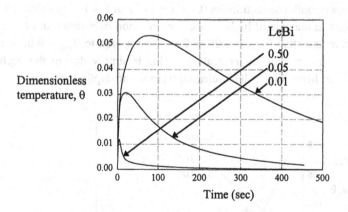

Figure 9.3-1b: Effect of LeBi number on the particle temperature evolution

The evolution of the intra-particle concentration versus time is shown in Figure 9.3-2 in which concentrations at five specific points within the particle are plotted versus time. The concentrations exhibit a sigmoidal behavior, suggesting the penetration behavior of mass into the particle.

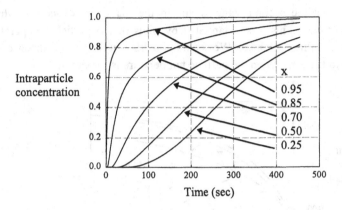

Figure 9.3-2: Plots of the intra-particle concentration versus time

9.3.1.9.2 *Effect of bulk concentration*

Using the same set of parameters listed in Table 9.3-1, we study the effect of bulk concentration on the system behavior under the condition of heat transfer of LeBi = 0.05. Figure 9.3-3 shows plots of the particle temperature as a function of time for two values of bulk concentrations ($C_b = 1 \times 10^{-6}$ and 5×10^{-6} mole/cc). The particle temperature in the case of higher bulk concentration rises faster and reaches higher maximum temperature ($\theta_{max} = 0.08$ or 24 °C) compared to $\theta_{max} = 0.03$ (or 9 °C) for the case of lower bulk concentration. This is simply due to the higher amount adsorbed and hence higher heat released from the adsorption process.

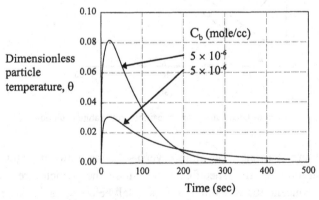

Figure 9.3-3: Plots of the particle temperature versus time

After the particle temperature has reached its maximum, the curve corresponding to higher bulk concentration decays faster, hence reaching the equilibrium faster. This is due to the effect which we have explained earlier in the isothermal analysis, i.e. due to the nonlinear convexity of the isotherm the rate of increase in the amount adsorbed due to an increase in the bulk concentration is not as fast as that in the bulk concentration, resulting in faster kinetics for the case of higher bulk concentration.

9.3.1.9.3 Effect of Particle Size

The effect of particle size on the amount adsorbed and the particle temperature is shown in Figures 9.3-4a and 9.3-4b, respectively. Three particle sizes are used in the simulations (R = 0.075, 0.1, and 0.15 cm). The value of the LeBi varies with particle size as we have discussed in Section 9.3.1.7. Assuming a turbulent environment surrounding the particle, the LeBi number is proportional to $R^{0.6}$. Taking a LeBi of 0.05 for the particle size of 0.1 cm, we calculate LeBi as 0.0421 and 0.0638 for the particle sizes of 0.075 and 0.15 cm, respectively. From Figure 9.3-4a, we see that the larger is the particle size, the slower is the mass transfer as we would expect. About the particle temperature, the rate of particle cooling is slower in the case of larger particle size because the heat transfer is limited by the smaller heat transfer area per unit volume.

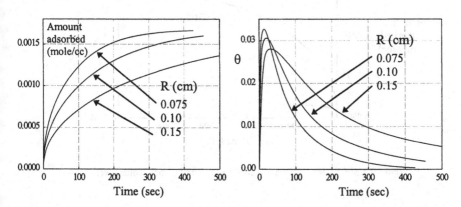

Figure 9.3-4a: Effect of particle size on the amount adsorbed versus time

Figure 9.3-4b: Effect of particle size on the particle temperature evolution

9.3.1.9.4 Effect of the Operating Temperature

We finally study how the operating temperature affects the temperature rise. We take two cases of operating temperature (T_b = 300 and 400 K). Figure 9.3-5a shows the temperature rise versus time, and we see that the rise in temperature is higher for the case of lower operating temperature. This is due to the higher amount adsorbed for the case of lower adsorption temperature T_b as shown in Figure 9.3-5b.

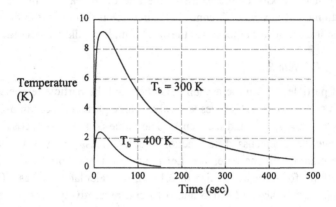

Figure 9.3-5a: Plots of the temperature change versus time

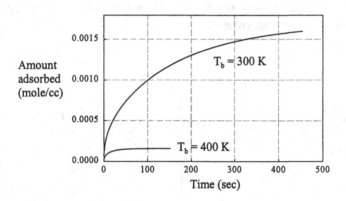

Figure 9.3-5b: Plots of the amount adsorbed per unit particle volume versus time

9.4 Finite Kinetic Adsorption Model for Single Component Systems

We have considered the analysis of single component systems under isothermal conditions as well as non-isothermal conditions. In these analyses, the local equilibrium between the fluid and adsorbed phases (eqs. 9.2-44 and 9.3-8) was invoked. This is valid when the rates of adsorption and desorption of adsorbate molecules at sorption sites are much faster than the rates of diffusion in the fluid and adsorbed phases. For well developed activated carbon where the micropore mouths are not restricted, the rate of adsorption from the fluid phase to the micropore mouth and the rate of desorption from it are expected to be faster than the diffusion rates, and hence the local equilibrium is acceptable for this situation. For solids such as carbon molecular sieve, the pore mouth is very restricted (Figure 9.4-1), and therefore the rates of adsorption and desorption at the micropore mouth are expected to be comparable to the diffusion rates, and under severe constriction conditions the rates of adsorption and desorption at the micropore mouth are in fact the rate-limiting step.

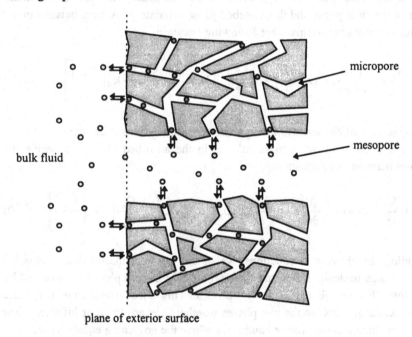

Figure 9.4-1: Finite mass exchange kinetics at the micropore mouth

To account for the finite mass exchange kinetics between the fluid and adsorbed phases, we let the number of micropore mouth (acting as sorption site) per unit area be n_μ and assume that the net rate of adsorption per pore mouth is governed by the following Langmuir kinetic equation:

$$R_{ads} = k_a C \left(1 - \frac{C_\mu}{C_{\mu s}}\right) - k_d \left(\frac{C_\mu}{C_{\mu s}}\right) \qquad (9.4\text{-}1)$$

where k_a and k_d are rate constants for adsorption and desorption, respectively. This kinetics equation implies that the equilibrium adsorption isotherm is the Langmuir equation:

$$C_\mu = C_{\mu s}\frac{bC}{1+bC}; \quad b = \frac{k_a}{k_d} \tag{9.4-2}$$

Appendix 9.5 lists a number of kinetic rate expressions which at equilibrium will give rise to Sips and Toth equations.

With the finite mass exchange between the two phases, we now do the mass balance of the fluid phase and the adsorbed phase separately. A mass balance in the fluid phase of thickness dr gives the following equation:

$$\varepsilon\frac{\partial C}{\partial t} = \varepsilon\frac{1}{r^s}\frac{\partial}{\partial r}\left\{r^s D_p \frac{\partial C}{\partial r}\right\} - \rho_p S_g n_\mu \left[k_a C\left(1-\frac{C_\mu}{C_{\mu s}}\right) - k_d\left(\frac{C_\mu}{C_{\mu s}}\right)\right] \tag{9.4-3a}$$

where ρ_p is the particle density, and S_g is the specific interior surface area.

Similarly carrying out a mass balance in the adsorbed phase, we derive the following mass balance equation:

$$(1-\varepsilon)\frac{\partial C_\mu}{\partial t} = (1-\varepsilon)\frac{1}{r^s}\frac{\partial}{\partial r}\left\{r^s D_\mu \frac{\partial C_\mu}{\partial r}\right\} + \rho_p S_g n_\mu \left[k_a C\left(1-\frac{C_\mu}{C_{\mu s}}\right) - k_d\left(\frac{C_\mu}{C_{\mu s}}\right)\right] \tag{9.4-3b}$$

Adding the above two equations, we obtain the total mass balance equation (9.2-1b), which basically states that the total hold-up in both phases is governed by the net total flux contributed by the two phases. This is true irrespective of the rate of mass exchange between the two phases whether they are finite or infinite. Now back to our finite mass exchange conditions where the governing equations are (9.4-3a) and (9.4-3b). To solve these equations, we need to impose boundary conditions as well as define an initial state for the system. One boundary is at the center of the particle where we have the usual symmetry:

$$r = 0; \quad \frac{\partial C}{\partial r} = \frac{\partial C_\mu}{\partial r} = 0 \tag{9.4-4}$$

The other boundary is at the plane of the exterior surface of the particle (Figure 9.4-1). The mass flux through the stagnant fluid film per unit total exterior surface area is:

$$R_m = k_m(C_b - C|_R) \tag{9.4-5}$$

This flux must be the same as the total flux into the particle contributed by the fluid species and the adsorbed species, that is the pertinent boundary condition is:

$$r = R; \quad k_m(C_b - C|_R) = \varepsilon D_p \frac{\partial C}{\partial r}\bigg|_{r=R} + (1-\varepsilon)D_\mu \frac{\partial C_\mu}{\partial r}\bigg|_{r=R} \tag{9.4-6}$$

Another boundary condition at the exterior surface area is simply that the finite mass exchange between the two phases at the plane of the exterior surface is equal to the diffusion flux of the adsorbed phase, that is:

$$r = R; \quad n_\mu \left[k_a C|_R \left(1 - \frac{C_\mu}{C_{\mu s}}\bigg|_R\right) - k_d \left(\frac{C_\mu}{C_{\mu s}}\bigg|_R\right) \right] = D_\mu \frac{\partial C_\mu}{\partial r}\bigg|_R \tag{9.4-7}$$

The initial state of the system is:

$$t = 0; \quad C = C_i, \quad C_\mu = C_{\mu i} = C_{\mu s} \frac{bC_i}{1 + bC_i} \tag{9.4-8}$$

Eqs. (9.4-3), (9.4-4), (9.4-6) and (9.4-7) completely define the behavior of the system. Solving these equations will yield concentration distributions in the fluid phase and the adsorbed phase. This set of equations is a superset of local equilibrium model described in Sections 9.2 and 9.3. When the rate constants k_a and k_d are very large, eqs. (9.4-3) and (9.4-7) simply lead to:

$$C_\mu(r,t) = C_{\mu s} \frac{bC(r,t)}{1 + bC(r,T)} \quad \text{for } 0 \le r \le R \tag{9.4-9}$$

which is simply the local equilibrium condition we invoked in Sections 9.2 and 9.3. The mass balance for this condition of k_a and $k_d \to \infty$ is simply equation (9.2-1b) with the adsorbed concentration relating to the fluid phase concentration according to eq. (9.4-9).

9.5 Multicomponent Adsorption Models for Porous Solids: Isothermal

You have learnt about the behaviour of adsorption kinetics of a single component in a single particle. Pore, surface diffusions and their combined diffusion have been studied in some details for linear as well as nonlinear isotherm and under isothermal as well as nonisothermal conditions. Here we will study a situation where there are more than one adsorbate present in the system and the interaction between different species will occur during diffusion as well as adsorption. Analysis of multicomponent system will require the application of the Maxwell-Stefan approach learnt in Chapter 8. To demonstrate the methodology as well as to show the essential features of how multiple species interact during the course of diffusion as well as adsorption onto adsorption sites, we will first consider a multicomponent adsorption system under isothermal conditions and dual diffusion mechanism is operating in the particle.

Carrying out mass balance over a thin shell within the particle, we obtain the following mass balance equation:

$$\frac{\varepsilon}{R_g T}\frac{\partial \underline{p}}{\partial t} + (1-\varepsilon)\frac{\partial \underline{C}_\mu}{\partial t} = -\frac{1}{r^s}\frac{\partial}{\partial r}\left\{r^s\left[\varepsilon \underline{N}_p + (1-\varepsilon)\underline{N}_\mu\right]\right\} \tag{9.5-1}$$

where r is the radial co-ordinate of the particle, $\underline{p}$ is the partial pressure vector of n components in the void space (mole/m³ of the void space) and $\underline{C}_\mu$ is the concentration vector of the adsorbed phase (mole/m³ of the adsorbed phase). They are given by:

$$\underline{p} = \begin{bmatrix} p_1(r,t) \\ p_2(r,t) \\ \vdots \\ p_n(r,t) \end{bmatrix}; \quad \underline{C}_\mu = \begin{bmatrix} C_{\mu,1}(r,t) \\ C_{\mu,2}(r,t) \\ \vdots \\ C_{\mu,n}(r,t) \end{bmatrix} \tag{9.5-2a}$$

The vectors $\underline{N}_p$ and $\underline{N}_\mu$ are fluxes in the void space and in the adsorbed phase, respectively:

$$\underline{N}_p = \begin{bmatrix} N_{p,1}(r,t) \\ N_{p,2}(r,t) \\ \vdots \\ N_{p,n}(r,t) \end{bmatrix}; \quad \underline{N}_\mu = \begin{bmatrix} N_{\mu,1}(r,t) \\ N_{\mu,2}(r,t) \\ \vdots \\ N_{\mu,n}(r,t) \end{bmatrix} \tag{9.5-2b}$$

Before further addressing the mass balance equation, we need to consider the fluxes in the void space and in the adsorbed phase.

9.5.1 Pore Volume Flux Vector $\underline{N}_p$

The flux of adsorbate in the void space is controlled by three mechanisms:
(a) Molecular diffusion
(b) Knudsen diffusion
(c) Viscous flow

Using the Maxwell-Stefan approach, Chapter 8 has addressed systematically the simultaneous contribution of those three mechanisms on the flux. The necessary equation relating the flux to the relevant driving forces is (eq. 8.8-13a):

$$\underline{N}_p = -\frac{1}{R_g T}\left[\underline{\underline{B}}(\underline{y}, p_T)\right]^{-1}\frac{\partial \underline{p}}{\partial r} - \left(\frac{B_0}{\mu_m(\underline{y}) R_g T}\right)\frac{\partial p_T}{\partial r}\left[\underline{\underline{B}}(\underline{y}, p_T)\right]^{-1}\underline{\underline{\Lambda}}\, \underline{p} \quad (9.5\text{-}3)$$

where μ_m is the viscosity of the mixture and can be calculated using a semi-empirical formula of Wilke (eq. 8.8-2), and p_T is the total pressure given by:

$$p_T = \sum_{j=1}^{n} p_j \quad (9.5\text{-}4)$$

The parameter B_0 is the viscous flow parameter and is a function of solid structure only. Since a porous solid usually exhibits a pore size distribution, this parameter must be determined experimentally.

The matrices $\underline{\underline{B}}$ and $\underline{\underline{\Lambda}}$ of (n, n) dimension are function of binary diffusivities, Knudsen diffusivities and mole fractions as follows:

$$\underline{\underline{B}}(\underline{y}, p_T) = \begin{cases} \dfrac{1}{(D_{K,i}/\tau)} + \sum_{\substack{k=1 \\ k \neq i}}^{n} \dfrac{y_k}{(D_{i,k}(p_T)/\tau)} & \text{for } i = j \\[1em] -\dfrac{y_i}{(D_{i,j}(p_T)/\tau)} & \text{for } i \neq j \end{cases} \quad (9.5\text{-}5)$$

and

$$\underline{\underline{\Lambda}} = \left\{\Lambda_{i,i} = \frac{1}{(D_{K,i}/\tau)}\right. \quad (9.5\text{-}6)$$

where τ is the solid tortuosity factor allowing for the tortuous nature of the pore (usually $2 < \tau < 6$), D_{ij} is the binary diffusivity which is inversely proportional to the total pressure:

$$D_{ij}(p_T) = D_{ij,0}\left(\frac{p_{T0}}{p_T}\right) \qquad (9.5\text{-}7a)$$

and $D_{K,i}$ is the Knudsen diffusivity defined as:

$$D_{K,i} = \frac{4K_0}{3}\sqrt{\frac{8R_g T}{\pi M_i}} \qquad (9.5\text{-}7b)$$

Here $D_{ij,0}$ is the binary diffusivity evaluated at some reference total pressure p_{T0}. The parameter K_0 is the Knudsen flow parameter and like B_0 for viscous flow it is also a function of solid structure only.

The mole fraction $\underline{y}$ in the matrix $\underline{\underline{B}}$ is defined as:

$$\underline{y} = \frac{\underline{p}}{p_T} \qquad (9.5\text{-}8)$$

The tortuosity factor is best determined experimentally. However, in the absence of this information, Abbasi et al. (1983) and Akanni et al. (1987) used the Monte Carlo simulation to obtain analytical expression for the tortuosity factor.

9.5.2 Flux Vector in the Adsorbed Phase

The surface diffusion flux is assumed to be driven by the chemical potential gradient in the adsorbed phase. The flux of a component "i" is driven by the chemical potential gradient of that species, that is:

$$N_{\mu,i} = -L_i C_{\mu,i}\frac{\partial \mu_i}{\partial r} \qquad (9.5\text{-}9)$$

where L is the mobility coefficient which is temperature dependent and is assumed to be concentration independent, $C_{\mu i}$ is the concentration of the species i in the adsorbed phase and is defined as moles per unit volume of the adsorbed phase, and μ_i is the chemical potential in the adsorbed phase.

If we make the assumption of local equilibrium between the free and adsorbed phases, the chemical potential of the adsorbed phase is the same as that of the gas phase, that is:

$$\mu_i = \mu_{G,i} = \mu_i^0 + R_g T \ln p_i \qquad (9.5\text{-}10)$$

where p_i is the partial pressure of the component i in the gas phase.

For a given set of partial pressures, $\underline{p} = \{p_1, p_2, ..., p_n\}$, there will be a set of adsorbed concentration which is in equilibrium with the gas phase, that is through the isotherm expression:

$$C_{\mu,i} = f_i(\underline{p}) \qquad (9.5\text{-}11a)$$

for i = 1, 2, 3,..., n, or written in a more compact vector form, we have:

$$\underline{C}_\mu = \underline{f}(\underline{p}) \qquad (9.5\text{-}11b)$$

Inversely, for a given set of adsorbed phase concentrations, $\underline{C}_\mu = \{C_{\mu 1}, C_{\mu 2}, ..., C_{\mu,N}\}$, there exists a set of partial pressures such that the two phases are in equilibrium with each other, that is:

$$p_i = g_i(\underline{C}_\mu) \qquad (9.5\text{-}12a)$$

or written in a vector form:

$$\underline{p} = \underline{g}(\underline{C}_\mu) \qquad (9.5\text{-}12b)$$

Substituting eq. (9.5-10) into eq.(9.5-9), we get the following expression for the flux of the species "i" written in terms of the gradient of the partial pressure p_i:

$$N_{\mu,i} = -D_{\mu,i}^0 C_{\mu,i} \frac{\partial \ln p_i}{\partial r} = -D_{\mu,i}^0 \frac{C_{\mu,i}}{p_i} \frac{\partial p_i}{\partial r} \qquad (9.5\text{-}13a)$$

where $D_{\mu,i}^0$ is the corrected diffusivity of the species "i" at zero loading

$$D_{\mu,i}^0 = L_i R_g T \qquad (9.5\text{-}13b)$$

As seen in eq.(9.5-13a) that the surface flux is driven by the gradient of the gas phase partial pressure. This is so because the two phases are in equilibrium with each other. However, if one wishes to express the surface flux in terms of the gradients of adsorbed concentration, we could make use of eq. (9.5-12), and by applying the chain rule of differentiation we get

$$\frac{\partial \ln p_i}{\partial r} = \frac{1}{p_i}\frac{\partial p_i}{\partial r} = \frac{1}{p_i}\sum_{j=1}^{n}\frac{\partial p_i}{\partial C_{\mu,j}}\frac{\partial C_{\mu,j}}{\partial r} \qquad (9.5\text{-}14)$$

where $\partial C_{\mu,j} / \partial p_i$ is simply the slope of the equilibrium isotherm of species j with respect to the partial pressure of species i.

Substituting eq.(9.5-14) into eq.(9.5-13a), we get:

$$N_{\mu,i} = -D^0_{\mu,i}\sum_{j=1}^{n} C_{\mu i}\frac{\partial \ln p_i}{\partial C_{\mu,j}}\frac{\partial C_{\mu,j}}{\partial r} \qquad (9.5\text{-}15)$$

for i = 1, 2, 3,..., n. The above equation is the flux equation for the adsorbed phase relating the surface flux to the gradients of adsorbed concentration of all species. Note that eq.(9.5-15) is mathematically equivalent to eq.(9.5-13a).

Thus, if we define the following vectors and the diffusivity matrix

$$\underline{N}_\mu = \begin{bmatrix} N_{\mu,1} \\ N_{\mu,2} \\ \vdots \\ N_{\mu,n} \end{bmatrix}; \qquad \underline{C}_\mu = \begin{bmatrix} C_{\mu,1} \\ C_{\mu,2} \\ \vdots \\ C_{\mu,n} \end{bmatrix} \qquad (9.5\text{-}16a)$$

$$\underline{\underline{D}}_\mu = \left\{ D_{\mu,ij} = D^0_{\mu,i} C_{\mu,i}\frac{\partial \ln p_i}{\partial C_{\mu,j}} \right\} \qquad (9.5\text{-}16b)$$

then eq.(9.5-15) can be put in a compact vector format:

$$\underline{N}_\mu = -\underline{\underline{D}}_\mu\left(\underline{C}_\mu, \underline{p}\right)\frac{\partial \underline{C}_\mu}{\partial r} \qquad (9.5\text{-}17)$$

The elements of the matrix $\underline{\underline{D}}_\mu$ have units of m²/sec. This matrix is called the transport diffusivity matrix and it contains the thermodynamic correction factor (eq. 9.5-16c). Although the diffusivity matrix is seen to be a function of adsorbed concentrations as well as partial pressures, it is in fact a function of either adsorbed concentrations or gas phase pressures as they are related to each other through the local equilibrium equation (9.5-11).

Eq. (9.5-17) is identical to eq. (9.5-13a), which can be written in vector form is as follows:

$$\underline{N}_\mu = -\frac{1}{R_g T}\underline{\underline{G}}\left(\underline{C}_\mu, \underline{p}, T\right)\frac{\partial \underline{p}}{\partial r} \qquad (9.5\text{-}18)$$

where

$$\underline{\underline{G}} = \left\{ D^0_{\mu,i} \frac{R_g T C_{\mu,i}}{p_i} \right\} \tag{9.5-19}$$

Either one of these two forms (eq. 9.5-17 or 9.5-18) can be used in the analysis. Computation wise, eq.(9.5-18) is a better choice as it requires shorter computation time.

We have defined the flux equations for the free species as well as the adsorbed species. Now we substitute them into the mass balance equation to obtain the necessary mass balance equation for computational analysis. This is done next.

9.5.3 The Working Mass Balance Equation

The adsorbed concentration $C_{\mu,i}$ is assumed to be in equilibrium with the gas phase concentration, that is:

$$\underline{C}_\mu = \underline{f}(\underline{p}, T) \tag{9.5-20}$$

With this functional form relating the concentrations of the two phases, the flux of the adsorbing species (eq. 9.5-18) can be written as:

$$\underline{N}_\mu = -\frac{1}{R_g T} \underline{\underline{G}}(\underline{p}, T) \frac{\partial \underline{p}}{\partial r} \tag{9.5-21}$$

where

$$\underline{\underline{G}} = \left\{ D^0_{\mu,i}(T) \frac{R_g T \, f_i(\underline{p}, T)}{p_i} \right\} \tag{9.5-22}$$

Substitution of eqs. (9.5-3), (9.5-20) and (9.5-21) into the mass balance equation (9.5-1) gives:

$$\left[\frac{\varepsilon}{R_g T} \underline{\underline{I}} + (1-\varepsilon) \underline{\underline{J}}(\underline{p}) \right] \frac{\partial \underline{p}}{\partial t} = \varepsilon \frac{1}{z^s} \frac{\partial}{\partial z} \left\{ z^s \left[\underline{\underline{B}}(\underline{y}, P_T, T) \right]^{-1} \left[\frac{1}{R_g T} \frac{\partial \underline{p}}{\partial z} + \frac{B_0}{\mu_m(\underline{y}) R_g T} \frac{\partial P_T}{\partial z} \underline{\underline{\Lambda}}^p \underline{p} \right] \right\}$$

$$+ (1-\varepsilon) \frac{1}{z^s} \frac{\partial}{\partial z} \left[z^s \underline{\underline{G}}(\underline{p}) \frac{1}{R_g T} \frac{\partial \underline{p}}{\partial z} \right]$$

(9.5-23a)

590 Kinetics

where $\underline{\underline{I}}$ is an identity matrix, and $\underline{\underline{J}}(\underline{p})$ is the Jacobian of the equilibrium vector $\underline{f}$ in terms of the partial pressure vector $\underline{p}$:

$$\underline{\underline{J}}(\underline{p},T) = \frac{\partial \underline{f}(\underline{p},T)}{\partial \underline{p}} = \begin{bmatrix} \frac{\partial f_1(\underline{p},T)}{\partial p_1} & \frac{\partial f_1(\underline{p},T)}{\partial p_2} & \cdots & \frac{\partial f_1(\underline{p},T)}{\partial p_n} \\ \frac{\partial f_2(\underline{p},T)}{\partial p_1} & \frac{\partial f_2(\underline{p},T)}{\partial p_2} & \cdots & \frac{\partial f_2(\underline{p},T)}{\partial p_n} \\ \vdots & \vdots & \ddots & \vdots \\ \frac{\partial f_n(\underline{p},T)}{\partial p_1} & \frac{\partial f_n(\underline{p},T)}{\partial p_2} & \cdots & \frac{\partial f_n(\underline{p},T)}{\partial p_n} \end{bmatrix} \quad (9.5\text{-}23b)$$

The boundary conditions of the mass balance equation (9.5-23) are:

$$r = 0; \qquad \frac{\partial \underline{p}}{\partial r} = \underline{0} \qquad (9.5\text{-}24a)$$

and

$$r = R; \qquad \underline{p} = \underline{p}_b \qquad (9.5\text{-}24b)$$

where $\underline{p}_b$ is the constant bulk partial pressure vector and R is the radius of the particle (or half length if the particle is of slab geometry).

Initially the particle is assumed to be equilibrated with a set of partial pressures $\underline{p}_i$, that is:

$$t = 0; \qquad \underline{p} = \underline{p}_i \qquad (9.5\text{-}24c)$$

9.5.4 Nondimensionalization

The mass balance equation (eqs. 9.5-23) and its boundary and initial conditions (eqs. 9.5-24) are cast into the non-dimensional form for the subsequent collocation analysis. We define the following non-dimensional variables:

$$\eta = \frac{r}{R}, \qquad t^* = \frac{D_T t}{R^2} \qquad (9.5\text{-}25)$$

where D_T is some reference diffusivity. With these definitions of non-dimensional variables, the mass balance equation (9.5-23) becomes:

$$\frac{\partial \underline{p}}{\partial t^*} = \varepsilon \underline{\underline{H}}(\underline{p}) \frac{1}{\eta^s} \frac{\partial}{\partial \eta} \left\{ \eta^s \left[D_T \underline{\underline{B}}(\underline{y}, p_T, T) \right]^{-1} \left[\frac{\partial \underline{p}}{\partial \eta} + \Phi \frac{\mu_0}{\mu_m(\underline{y})} \frac{1}{P_0} \frac{\partial P_T}{\partial \eta} (D_T \underline{\underline{\Lambda}}) \underline{p} \right] \right\}$$

$$+ (1-\varepsilon) \underline{\underline{H}}(\underline{p}) \frac{1}{\eta^s} \frac{\partial}{\partial \eta} \left\{ \eta^s \left[D_T^{-1} \underline{\underline{G}}(\underline{p}) \right] \frac{\partial \underline{p}}{\partial \eta} \right\}$$

(9.5-26)

where μ_0 is some reference viscosity and P_0 is some reference pressure and

$$\underline{\underline{H}}(\underline{p}) = \left[\varepsilon \underline{\underline{I}} + (1-\varepsilon) R_g T \underline{\underline{J}}(\underline{p}, T) \right]^{-1} \quad (9.5\text{-}27)$$

$$\Phi = \frac{B_0 P_0}{\mu_0 D_T} \quad (9.5\text{-}28)$$

The parameter Φ describes the relative strength between the viscous flow and the diffusive flow. Its significance has been discussed in Section 8.9.1.

The pertinent boundary conditions in non-dimensional form are:

$$\eta = 0; \quad \frac{\partial \underline{p}}{\partial \eta} = \underline{0} \quad (9.5\text{-}29a)$$

and

$$\eta = 1; \quad \underline{p} = \underline{p}_b \quad (9.5\text{-}29b)$$

The initial condition is:

$$\tau = 0; \quad \underline{p} = \underline{p}_i \quad (9.5\text{-}29c)$$

The set of equations (eqs. 9.5-26 to 9.5-30) is solved numerically by the orthogonal collocation method, and the details are given in Appendix 9.6. Solution of this set will yield the partial pressures of all species as a function of time and distance within the particle. Knowing the partial pressures $\underline{p}$, the adsorbed concentrations are calculated from eq. (9.5-20). The volumetric mean partial pressures and adsorbed concentration are then calculated from the following equations:

$$\langle \underline{p} \rangle = (1+s) \int_0^1 \eta^s \, \underline{p} \, d\eta \quad (9.5\text{-}30a)$$

$$\langle \underline{C}_\mu \rangle = (1+s) \int_0^1 \eta^s \, \underline{C}_\mu \, d\eta \quad (9.5\text{-}30b)$$

9.5.4.1 Fractional Uptake

The quantity of interest for the design calculation is the fractional uptake. It is defined as the ratio of the amount taken by the sample up to time t to the final amount taken at infinite time, that is:

$$F_i = \frac{M_i}{M_{\infty,i}} = \frac{V\left\{\left[\varepsilon\frac{\langle p_i \rangle}{R_g T}+(1-\varepsilon)\langle C_{\mu,i}\rangle\right]-\left[\varepsilon\frac{\langle p_i \rangle}{R_g T}+(1-\varepsilon)\langle C_{\mu,i}\rangle\right]_{t=0}\right\}}{V\left\{\left[\varepsilon\frac{\langle p_i \rangle}{R_g T}+(1-\varepsilon)\langle C_{\mu,i}\rangle\right]_{t=\infty}-\left[\varepsilon\frac{\langle p_i \rangle}{R_g T}+(1-\varepsilon)\langle C_{\mu,i}\rangle\right]_{t=0}\right\}} \quad (9.5\text{-}31)$$

for i = 1, 2, , n.

9.5.4.2 Special Case: Langmuir Isotherm

In the case of extended Langmuir isotherm, which is the simplest case of dealing with multicomponent mixtures, the isotherm expression of the component i is:

$$C_{\mu,i} = C_{\mu s,i}\frac{b_i(T)\,p_i}{1+\sum_{k=1}^{n}b_k(T)\,p_k} = f_i\!\left(\underline{p},T\right) \quad (9.5\text{-}32)$$

With this form of the isotherm, the Jacobian matrix (eq. 9.5-23b) is then given by:

$$\underline{\underline{J}}(\underline{p}) = \left\{\frac{\partial f_i}{\partial p_j}\right\} = \begin{cases} C_{\mu s,i}\dfrac{b_i(T)\left[1+\sum\limits_{\substack{k=1\\k\neq i}}^{n}b_k(T)\,p_k\right]}{\left[1+\sum\limits_{k=1}^{n}b_k(T)\,p_k\right]^2} & \text{for } i = j \\[2em] -C_{\mu s,i}\dfrac{b_i(T)\,b_j(T)\,p_i}{\left[1+\sum\limits_{k=1}^{n}b_k(T)\,p_k\right]^2} & \text{for } i \neq j \end{cases} \quad (9.5\text{-}33)$$

A computer code ADSORB5B is provided with this book to solve this problem. Readers can use and modify the code to explore interactively how the various parameters can affect the system behaviour. We shall illustrate in an example below the application of this code.

Example 9.5-1: Adsorption of ethane/propane/nitrogen in an activated carbon

We consider a system of adsorption of ethane, propane and nitrogen onto activated carbon at 303 K and a total pressure of 1 atm. At this temperature, ethane and propane adsorb onto activated carbon, and their adsorption isotherm can be adequately described by a Langmuir equation. The isotherm parameters for nitrogen, ethane and propane are:

Nitrogen:	Ethane:	Propane:
$C_{\mu s} = 0$ mole/cc	$C_{\mu s} = 5.55 \times 10^{-3}$ mole/cc	$C_{\mu s} = 5.214 \times 10^{-3}$ mole/cc
$b = 0$ atm^{-1}	$b = 3.23$ atm^{-1}	$b = 21.9$ atm^{-1}

For the computer simulation, we denote nitrogen, ethane and propane as species 1, 2 and 3, respectively.

The viscosities at 303 K and molecular weights of these three species are given in the following table.

	μ (g/cm/sec)	MW (g/mole)
Nitrogen (1)	1.8×10^{-4}	28
Ethane (2)	0.925×10^{-4}	30
Propane (3)	0.8×10^{-4}	44

The binary diffusivities are calculated from the Chapman-Enskog equation (7.7-9) are given below:

$D_{12} = 0.1483$ cm²/sec, $D_{13} = 0.1151$ cm²/sec, $D_{23} = 0.0798$ cm²/sec

The particle is of slab geometry and its half-length is 2.2 mm. The average mesopore radius is 8×10^{-5} cm, and the tortuosity factor for this sample of activated carbon is 4.7. The surface diffusivities of ethane and propane at 303 K are 5×10^{-5} and 1×10^{-5}, respectively.

For the simulation, we take the base case that the particle is initially equilibrated with pure nitrogen, and at time t = 0+ the particle is exposed to a constant environment containing 80 % of nitrogen, 10% of ethane and 10% of propane.

The above values are inputted into the ADSORB5B programming code, and the execution takes about 2 minutes on a Pentium 166 MHz to generate the concentration distributions of all three species, the total pressure as well as the fractional uptakes.

Fractional uptake: Figure 9.5-1 shows the fractional uptake of ethane and propane, and we observe the overshoot of ethane. Ethane, being a weaker adsorbing species, will penetrate the particle faster and hence adsorb more than its share of equilibrium capacity under the condition of ethane and propane presence in the system. Propane, on the other hand, is a stronger adsorbing species than ethane, and therefore it penetrates slower into the particle. When it moves into the particle interior, it displaces some of the previously adsorbed ethane molecules from the surface, resulting an overshoot in the fractional uptake of ethane. Experimental data are also shown in the same figure, and we see that the mathematical model predicts very well the fractional uptakes of all species. The model also predicts the correct time when the maximum overshoot occurs.

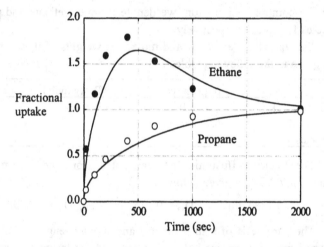

Figure 9.5-1: Fractional uptake of ethane and propane versus time

Concentration distribution: Figures 9.5-2 show the concentration profiles of ethane and propane at various times. The distribution of propane behaves in a normal fashion, and the distribution of ethane exhibits interesting behaviour which is a characteristics of a species penetrating very quickly but then being displaced by a stronger species in a later stage of the course of adsorption.

Analysis of Adsorption Kinetics in a Single Homogeneous Particle   595

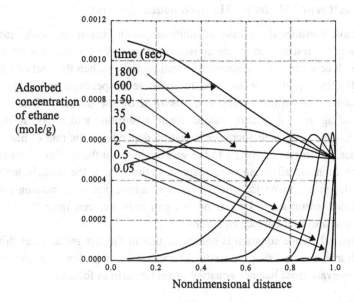

Figure 9.5-2a: Adsorbed concentration profiles of ethane at various times

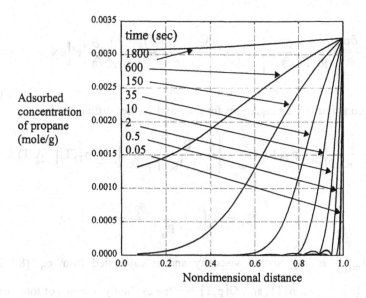

Figure 9.5-2a: Adsorbed concentration profiles of ethane at various times

9.6 Nonisothermal Model for Multicomponent Systems

We have considered the case of multicomponent adsorption under isothermal conditions in the last section. Such an isothermal condition occurs when the particle is very small or when the environment is well stirred or when the heat of adsorption is low. If these criteria are not met, the particle temperature will vary. Heat is released during adsorption while it is absorbed by the particle when desorption occurs, leading to particle temperature rise in adsorption and temperature drop in desorption. The particle temperature variation depends on the rate of heat released and the dissipation rate of energy to the surrounding. In the displacement situation, that is one or more adsorbates are displacing the others, the particle temperature variation depends also on the relative heats of adsorption of displacing adsorbates and displaced adsorbates. Details of this can only be seen from the solution of coupled mass and heat balance equations.

The mass balance equation is similar to that in the last section, but this time it must be borne in mind that the particle temperature is no longer a constant. We write the relevant mass balance equation in vector form as follows:

$$\varepsilon \frac{\partial}{\partial t}\left(\frac{p}{R_g T}\right) + (1-\varepsilon)\frac{\partial \underline{C}_\mu(r,t,\underline{p},T)}{\partial t} = -\frac{1}{r^s}\frac{\partial}{\partial r}\left\{r^s\left[\varepsilon \underline{N}_p + (1-\varepsilon)\underline{N}_\mu\right]\right\} \quad (9.6\text{-}1)$$

or

$$\varepsilon\left[\frac{1}{R_g T}\frac{\partial \underline{p}}{\partial t} - \frac{1}{R_g T^2}\frac{\partial T}{\partial t}\underline{p}\right] + (1-\varepsilon)\frac{\partial \underline{C}_\mu(r,t,\underline{p},T)}{\partial t} = -\frac{1}{r^s}\frac{\partial}{\partial r}\left\{r^s\left[\varepsilon \underline{N}_p + (1-\varepsilon)\underline{N}_\mu\right]\right\}$$

$$(9.6\text{-}2)$$

The constitutive flux equations for pore and surface diffusions are:

$$\underline{N}_p = -\frac{1}{R_g T}\left[\underline{\underline{B}}(\underline{y},p_T,T)\right]^{-1}\frac{\partial \underline{p}}{\partial r} - \left(\frac{B_0}{\mu_m(\underline{y})R_g T}\right)\frac{\partial p_T}{\partial r}\left[\underline{\underline{B}}(\underline{y},p_T,T)\right]^{-1}\underline{\underline{A}}(T)\,\underline{p} \quad (9.6\text{-}3)$$

$$\underline{N}_\mu = -\underline{\underline{G}}(\underline{p},T)\frac{1}{R_g T}\frac{\partial \underline{p}}{\partial r} \quad (9.6\text{-}4)$$

where $\mu_m(\underline{y})$ is the mixture viscosity and is calculated from eq. (8.8-2), the matrices $\underline{\underline{B}}(\underline{y},p_T,T)$, $\underline{\underline{A}}(T)$, and $\underline{\underline{G}}(\underline{p},T)$ written explicitly in terms of total pressure, temperature and mole fractions are given by:

$$\underline{B}(\underline{y}, p_T, T) = \begin{cases} \dfrac{1}{(D_{K,i}(T)/\tau)} + \sum_{\substack{k=1 \\ k \neq i}}^{n} \dfrac{y_k}{(D_{i,k}(p_T, T)/\tau)} & \text{for } i = j \\ -\dfrac{y_i}{(D_{i,j}(p_T, T)/\tau)} & \text{for } i \neq j \end{cases} \quad (9.6\text{-}5a)$$

$$\underline{\underline{\Lambda}}(T) = \left\{ \Lambda_{i,i} = \dfrac{1}{(D_{K,i}(T)/\tau)} \right\} \quad (9.6\text{-}5b)$$

$$\underline{\underline{G}}(\underline{p}, T) = \left\{ D_{\mu,i}^0(T) \dfrac{R_g T\, f_i(\underline{p}, T)}{p_i} \right\} \quad (9.6\text{-}6)$$

where $f_i(\underline{p}, T)$ is the adsorption isotherm

$$C_{\mu,i} = f_i(\underline{p}, T) \quad (9.6\text{-}7)$$

The expressions for these matrices are the same as those in the previous section when we dealt with the isothermal case, but this time all the relevant parameters are a function of temperature. The temperature dependence of the various diffusivities are given below:

$$D_{K,i} = \dfrac{4K_0}{3} \sqrt{\dfrac{8 R_g T}{\pi M_i}} = D_{K,i}(T_0) \sqrt{\dfrac{T}{T_0}} \quad (9.6\text{-}8)$$

$$D_{ij}(p_T, T) = D_{ij}(p_{T0}, T_0) \left(\dfrac{p_{T0}}{p_T} \right) \left(\dfrac{T}{T_0} \right)^{\alpha} \quad (9.6\text{-}9)$$

$$D_{\mu,i}^0(T) = D_{\mu\infty,i}^0 \exp\left(-\dfrac{E_{\mu,i}}{R_g T} \right) = D_{\mu,i}^0(T_0) \exp\left[\dfrac{E_{\mu,i}}{R_g T_0} \left(1 - \dfrac{T_0}{T} \right) \right] \quad (9.6\text{-}10)$$

where $D_{K,i}(T_0)$ is the Knudsen diffusivity of the component i at some reference temperature T_0, $D_{ij}(p_{T0}, T_0)$ is the binary diffusivity at some reference total pressure p_{T0} and temperature T_0, and $D^0{}_{\mu,i}(T_0)$ is the corrected surface diffusivity at temperature T_0.

Assuming the lumped thermal model for the particle temperature, that is uniform particle temperature and the heat transfer resistance is in the thin gas film surrounding the particle, the heat balance equation is:

$$\rho C_p \frac{dT}{dt} = (1-\varepsilon)\underline{Q} \bullet \frac{d\langle \underline{C}_\mu \rangle}{dt} - a_H h_f (T - T_b) \qquad (9.6\text{-}11)$$

where ρC_p is the volumetric heat capacity of the particle, a_H is the heat transfer area per unit volume, h_f is the heat transfer coefficient, T_b is the bulk temperature, and $\underline{Q}$ is the molar heat of adsorption vector

$$\underline{Q} = \begin{bmatrix} Q_1 \\ Q_2 \\ \vdots \\ Q_n \end{bmatrix} \qquad (9.6\text{-}12)$$

and

$$\langle \underline{C}_\mu \rangle = \frac{(1+s)}{R^{s+1}} \int_0^R r^s \, \underline{C}_\mu(r,t) \, dr \qquad (9.6\text{-}13)$$

Eq. (9.6-11) involves the time derivative of the average adsorbed concentration. Its form is not convenient for numerical computation, and what we will show is an alternative expression for the heat balance equation. We take the volumetric average of the mass balance equation (9.6-1):

$$\varepsilon \frac{\partial}{\partial t}\left\langle \frac{p}{R_g T} \right\rangle + (1-\varepsilon)\frac{\partial \langle \underline{C}_\mu \rangle}{\partial t} = -\frac{(1+s)}{R}\left[\varepsilon \underline{N}_p + (1-\varepsilon)\underline{N}_\mu \right]_{r=R} \qquad (9.6\text{-}14)$$

Since the capacity in the pore volume is much less than that in the adsorbed phase, the above equation can be approximated as follows:

$$(1-\varepsilon)\frac{d\langle \underline{C}_\mu \rangle}{dt} \cong -\frac{(1+s)}{R}\left[\varepsilon \underline{N}_p + (1-\varepsilon)\underline{N}_\mu \right]_{r=R} \qquad (9.6\text{-}15)$$

Substitution of the above equation into the heat balance equation (9.6-11) gives an alternative form for the heat balance equation:

$$\rho C_p \frac{dT}{dt} = -\frac{(1+s)}{R}\underline{Q} \bullet \left[\varepsilon \underline{N}_p + (1-\varepsilon)\underline{N}_\mu \right]_{r=R} - a_H h_f (T - T_b) \qquad (9.6\text{-}16)$$

9.6.1 The Working Mass and Heat Balance Equations

We substitute
  (i) the local isotherm equation (eq. 9.6-7)
  (ii) the pore volume flux (eq. 9.6-3)
  (iii) the surface flux (eq. 9.6-4)
into the mass balance equation (9.6-2) to finally obtain:

$$\left[\frac{\varepsilon}{R_g T}\underline{\underline{I}} + (1-\varepsilon)\underline{\underline{J}}(\underline{p},T)\right]\frac{\partial \underline{p}}{\partial t} =$$

$$\frac{\varepsilon}{R_g T^2}\frac{\partial T}{\partial t}\underline{p} - (1-\varepsilon)\frac{\partial \underline{f}(\underline{p},T)}{\partial T}\frac{\partial T}{\partial t} +$$

$$\varepsilon\frac{1}{r^s}\frac{\partial}{\partial r}\left\{r^s\left[\underline{\underline{B}}(\underline{y},P_T,T)\right]^{-1}\left[\frac{1}{R_g T}\frac{\partial \underline{p}}{\partial r} + \frac{B_0}{\mu_m(\underline{y})R_g T}\frac{\partial P_T}{\partial r}\underline{\underline{\Lambda}}(T)\,\underline{p}\right]\right\} + \quad (9.6\text{-}18)$$

$$(1-\varepsilon)\frac{1}{r^s}\frac{\partial}{\partial r}\left[r^s\underline{\underline{G}}(\underline{p})\frac{1}{R_g T}\frac{\partial \underline{p}}{\partial r}\right]$$

where $\underline{\underline{J}}$ is the Jacobian matrix $\partial \underline{f}/\partial \underline{p}$

$$\underline{\underline{J}}(\underline{p},T) = \frac{\partial \underline{f}}{\partial \underline{p}} = \left\{\frac{\partial f_i}{\partial p_j}\right\} \quad (9.6\text{-}19)$$

Similarly we substitute
(i) the pore volume flux (eq. 9.6-3)
(ii) the surface flux (eq. 9.6-4)
into the heat balance equation (9.6-16) to get

$$\rho C_p \frac{dT}{dt} = \frac{(1+s)}{R}\underline{Q}\bullet\left\{\varepsilon\left[\underline{\underline{B}}(\underline{y},P_T,T)\right]^{-1}\left[\frac{1}{R_g T}\frac{\partial \underline{p}}{\partial r} + \frac{B_0}{\mu_m(\underline{y})R_g T}\frac{\partial P_T}{\partial r}\underline{\underline{\Lambda}}(T)\,\underline{p}\right] + (1-\varepsilon)\underline{\underline{G}}(\underline{p})\frac{1}{R_g T}\frac{\partial \underline{p}}{\partial r}\right\}_{r=R}$$
$$- a_H h_f(T-T_b)$$

$$(9.6\text{-}20)$$

The pertinent boundary and initial conditions of eqs. (9.6-18) and (9.6-20) are:

$$r = 0; \quad \partial \underline{p}/\partial r = \underline{0}, \quad (9.6\text{-}21a)$$

$$r = R; \quad \underline{p} = \underline{p}_b, \quad (9.6\text{-}21b)$$

$$t = 0; \quad T = T_i\,; \quad \underline{p} = \underline{p}_i\,; \quad \underline{C}_\mu = \underline{f}(\underline{p}_i,T_i) \quad (9.6\text{-}22)$$

9.6.2 The Working Nondimensional Mass and Heat Balance Equations

We define the following non-dimensional variables and parameters

$$\eta = \frac{r}{R}, \quad \tau = \frac{D_T t}{R^2}, \quad \theta = \frac{T - T_0}{T_0} \tag{9.6-23}$$

where T_0 is some reference temperature and D_T is some reference diffusion coefficient. With these definitions, the non-dimensional mass and heat balance equations are:

$$\frac{\partial p}{\partial \tau} = \frac{d\theta}{d\tau} \underline{\underline{H}}(\underline{p}, T) \left[\frac{\varepsilon}{(1+\theta)} \underline{p} - (1-\varepsilon)(1+\theta) R_g T_0^2 \frac{\partial \underline{f}(\underline{p}, T)}{\partial T} \right]$$

$$+ \varepsilon \underline{\underline{H}}(\underline{p}, \theta) \frac{1}{\eta^s} \frac{\partial}{\partial \eta} \left\{ \eta^s \left[D_T \underline{\underline{B}}(y, P_T, T) \right]^{-1} \left[\frac{\partial \underline{p}}{\partial \eta} + \Phi \left(\frac{\mu_0}{\mu_m(y)} \right) \left(\frac{1}{P_{T0}} \frac{\partial P_T}{\partial \eta} \right) (D_T \underline{\underline{A}}(T)) \underline{p} \right] \right\}$$

$$+ (1-\varepsilon) \underline{\underline{H}}(\underline{p}, \theta) \frac{1}{\eta^s} \frac{\partial}{\partial \eta} \left[\eta^s \underline{\underline{G}}^*(\underline{p}, \theta) \frac{\partial \underline{p}}{\partial \eta} \right] \tag{9.6-24a}$$

and

$$\frac{d\theta}{d\tau} = (1+s) \frac{\beta}{(1+\theta)} \left(\frac{\underline{Q}}{Q_0} \right) \bullet$$

$$\left\{ \varepsilon \left[D_T \underline{\underline{B}}(y, P_T, T) \right]^{-1} \left[\frac{1}{P_{T0}} \frac{\partial \underline{p}}{\partial \eta} + \Phi \left(\frac{\mu_0}{\mu_m} \right) \left(\frac{1}{P_{T0}} \frac{\partial P_T}{\partial \eta} \right) (D_T \underline{\underline{A}}) \left(\frac{1}{P_{T0}} \underline{p} \right) \right] + (1-\varepsilon) \left[\underline{\underline{G}}^*(\underline{p}, \theta) \left(\frac{1}{P_{T0}} \frac{\partial \underline{p}}{\partial \eta} \right) \right] \right\}_{\eta=1}$$

$$- \text{LeBi}(\theta - \theta_b) \tag{9.6-24b}$$

where P_{T0} is some reference pressure, Q_0 is some reference heat of adsorption and

$$\underline{\underline{H}}(\underline{p}, \theta) = \left[\varepsilon \underline{\underline{I}} + (1-\varepsilon) R_g T \underline{\underline{J}}(\underline{p}, T) \right]^{-1} \tag{9.6-25a}$$

$$\underline{\underline{G}}^*(\underline{p}, \theta) = \underline{\underline{G}}(\underline{p}, T) / D_T \tag{9.6-25b}$$

$$\Phi = \frac{P_{T0} B_0}{D_T \mu_0} \tag{9.6-25c}$$

$$\beta = \frac{Q_0 (P_{T0} / R_g T_0)}{T_0 \rho C_p} \tag{9.6-25d}$$

$$\text{LeBi} = \frac{R^2 a_H h_f}{D_T \rho C_p} \qquad (9.6\text{-}25e)$$

$$\theta_b = \frac{T_b - T_0}{T_0} \qquad (9.6\text{-}25f)$$

The boundary and initial conditions are:

$$\eta = 1; \quad \underline{p} = \underline{p}_b \qquad (9.6\text{-}26a)$$

$$\tau = 0; \quad \theta = \theta_i = \frac{T_i - T_0}{T_0}, \quad \underline{p} = \underline{p}_i \qquad (9.6\text{-}26b)$$

9.6.3 Extended Langmuir Isotherm:

We shall illustrate the utility of the theory by applying to a multicomponent system in which the adsorption isotherm is described by the following extended Langmuir isotherm equation:

$$C_{\mu,i}(\underline{p},T) = f_i(\underline{p},T) = C_{\mu s,i} \frac{b_i(T) p_i}{1 + \sum_{j=1}^{n} b_j(T) p_j}, \qquad (9.6\text{-}27a)$$

where the adsorption affinity takes the following temperature dependence form:

$$b_i = b_{\infty,i} \exp\left(\frac{Q_i}{R_g T}\right) = b_{0,i} \exp\left[-\frac{Q_i}{R_g T_0}\left(1 - \frac{T_0}{T}\right)\right] \qquad (9.6\text{-}27b)$$

The Jacobian $\underline{\underline{J}}$ with respect to the partial pressures $\underline{p}$ is given by:

$$\underline{\underline{J}}(\underline{p},T) = \frac{\partial \underline{f}}{\partial \underline{p}} = \begin{cases} C_{\mu s,i} b_i \left[1 + \sum_{\substack{k=1 \\ k \neq i}}^{n} b_k p_k\right]\left[1 + \sum_{k=1}^{n} b_k p_k\right]^{-2} & \text{for } i = j \\ -C_{\mu s,i} b_i b_j p_i \left[1 + \sum_{k=1}^{n} b_k p_k\right]^{-2} & \text{for } i \neq j \end{cases} \qquad (9.6\text{-}28)$$

The change of the amount adsorbed with respect to T is:

$$\frac{\partial \underline{f}_i}{\partial T} = -\frac{C_{\mu s,i} b_i p_i}{R_g T^2} \frac{\left\{ Q_i \left[1 + \sum_{j=1}^{n} b_j p_j \right] - \sum_{j=1}^{n} Q_j b_j p_j \right\}}{\left[1 + \sum_{j=1}^{n} b_j p_j \right]^2} \qquad (9.6\text{-}29)$$

for i = 1, 2, 3, ..., n.

Substituting eqs. (9.6-28) and (9.6-29) into (the heat and mass balance equations (eqs. 9.6-24), we can solve them numerically using the combination of the orthogonal collocation method and the Runge-Kutta method. Solving these equations we obtain the partial pressures of all components at N interior collocation points and temperature as a function of time. The adsorbed concentrations at these N collocation points are then calculated from eq. (9.6-27a). Hence the volumetric average partial pressures and adsorbed concentrations are obtained from the following quadrature formula:

$$\left\langle \underline{C}_\mu \right\rangle = (1+s) \int_0^1 \eta^s \underline{C}_\mu \, d\eta = \sum_{j=1}^{N} w_j \underline{C}_{\mu j} \qquad (9.6\text{-}30a)$$

$$\left\langle \underline{p} \right\rangle = (1+s) \int_0^1 \eta^s \underline{p} \, d\eta = \sum_{j=1}^{N} w_j \underline{p}_j \qquad (9.6\text{-}30b)$$

We define the fractional uptake of the component "k" as:

$$F_k = \frac{\left[\varepsilon \frac{\langle p_k \rangle}{R_g T} + (1-\varepsilon) \langle C_{\mu,i} \rangle \right] - \left[\varepsilon \frac{\langle p_k \rangle}{R_g T_i} + (1-\varepsilon) \langle C_{\mu,i} \rangle \right]_{t=0}}{\left[\varepsilon \frac{\langle p_k \rangle}{R_g T} + (1-\varepsilon) \langle C_{\mu,i} \rangle \right]_{t=\infty} - \left[\varepsilon \frac{\langle p_k \rangle}{R_g T_i} + (1-\varepsilon) \langle C_{\mu,i} \rangle \right]_{t=0}} \qquad (9.6\text{-}31)$$

Eqs. (9.6-24) are solved numerically by the collocation method. Appendix 9.7 describes the method, and a computer code ADSORB5C is available for the simulation.

9.7 Conclusion

This chapter has presented a number of adsorption models for homogeneous particles where parallel diffusion mechanism is operating. This type of mechanism is applicable to solids such as activated carbon. We will present in the next chapter a number of models for zeolite type solids where a bimodal diffusion mechanism is operating.

10
Analysis of Adsorption Kinetics in a Zeolite Particle

10.1 Introduction

The last chapter discusses the behaviour of adsorption kinetics of a homogeneous particle, where the transport mechanism into the particle is controlled by the pore diffusion as well as the surface diffusion of adsorbed molecules along the particle length scale. Viscous flux is also allowed for in the multicomponent analysis. The pore and surface diffusion processes occur in parallel, and the assumption of local equilibrium between the two phases is generally invoked. In this chapter, we will discuss another class of solid: the zeolite-type solid. Basically the particle of this type is composed of distinct microparticles (for example zeolite crystals). Most of the adsorption capacity is accommodated in these microparticles. The void between these microparticles forms a network of large pores whose sizes are of the order of 0.1 to 1 micron. These large pores basically act as passage way for molecules to diffuse from the surrounding environment into the interior of the particle. Once they are inside the particle, these molecules adsorb at the pore mouth of the microparticles and thence these adsorbed molecules will diffuse into the interior of the microparticle. Generally, the pore size of the microparticle is of the order of molecular dimension, and therefore, molecules inside the micropores never escape the attraction potential of the pore walls. This means that molecules in the free form do not exist and only adsorbed molecules exist in the micropore.

Since there are two diffusion processes in the particle, it is possible that either of them or both of them controls the uptake, depending on the system parameters and operating conditions. We have the following three cases:

1. <u>Micropore diffusion</u>: This is the case when the diffusion into the particle interior through the large void between the microparticles is very fast. The uptake is then controlled by the diffusion of adsorbed molecules into the interior of the microparticle. This is expected for small particles or molecules having molecular dimensions close to the size of the micropore.
2. <u>Macropore diffusion</u>: This is the case when the diffusion into the microparticle is fast, and hence the uptake is controlled by the ability of the molecules to get through the macropores and mesopores. This is expected for large particles and molecules having size much smaller than the pore size of the micropore.
3. <u>Macropore-Micropore diffusion</u>: This is the case often called the bimodal diffusion model in the literature. In this case the two diffusion processes both control the uptake. This is expected when the particle size is intermediate.

The macropore diffusion case is basically dealt with in Chapter 9 where we dealt with parallel diffusion in homogeneous solids. Since surface diffusion on the exterior surface of the zeolite microparticle is almost negligible due to the very low capacity on those surfaces, we can ignore the surface diffusion contribution in the analysis of the last chapter when we apply such analysis to a zeolite-type particle.

In this chapter we will consider the micropore diffusion case and the bimodal diffusion. First, we consider the single component micropore diffusion in a single crystal under isothermal conditions, and then consider how nonisothermal conditions would affect the overall adsorption kinetics. Next, we consider the bimodal diffusion case, and study two situations. In one situation, the linear isotherm is considered, while in the other we investigate the nonlinear isotherm. Finally we consider multicomponent case where we will show how non-linear isotherm and non-isothermal conditions are incorporated into the heat and mass balance equations.

10.2 Single Component Micropore Diffusion (Isothermal)

Diffusion in micropore is assumed to be driven by the chemical potential gradient of the adsorbed species, instead of the concentration gradient. This is not a general rule, but it has been shown in many systems (Ruthven, 1984) that the chemical potential gradient is the proper description for the driving force of diffusion in zeolite, especially zeolites A, X, Y. Diffusion in other zeolites, and molecular sieve particles, there are still some discrepancies in the description of the diffusion. Solid structure and properties of the diffusing molecule may all contribute to these discrepancies.

10.2.1 The Necessary Flux Equation

The important relation used in the mass balance equation is the constitutive flux equation, which relates the flux and the concentration gradient of the adsorbed species. For the diffusion of the adsorbed species inside a micro-particle, the flux can be written in terms of the chemical potential gradient as follows:

$$J_\mu = -L\, C_\mu \frac{\partial \mu}{\partial z} \quad (10.2\text{-}1)$$

where L is the mobility coefficient which is temperature dependent, C_μ is the concentration of the species in the crystal and is defined as moles per unit volume of the crystal, and μ is the chemical potential of the adsorbed phase. Here we shall assume that the mobility coefficient is independent of concentration. It is reminded that we have used the subscript μ to denote the adsorbed phase, and this has been used throughout the book.

Let us assume that there exists a <u>hypothetical gas phase</u> such that this gas phase is in equilibrium with the adsorbed phase, that is the adsorbed phase chemical potential is the same as the chemical potential of that hypothetical gas phase. We write the following equality of chemical potentials of the two phases:

$$\mu = \mu_G = \mu^0 + R_g T \ln p \quad (10.2\text{-}2)$$

where p is the <u>hypothetical</u> partial pressure, which is in equilibrium with the adsorbed concentration C_μ, that is the following isotherm equation holds:

$$C_\mu = f(p) \quad (10.2\text{-}3)$$

where f is the functional form of the adsorption isotherm. The term hypothetical is because there is no gas phase within the crystal.

We assume here that there is one-to-one correspondence between the adsorbed concentration and the partial pressure; thus, we can write the following inverse equation relating the hypothetical pressure to the adsorbed concentration:

$$p = g(C_\mu) = f^{-1}(C_\mu) \quad (10.2\text{-}4)$$

Substituting eq. (10.2-2) into eq. (10.2-1), we get the following expression for the flux written in terms of the gradient of the hypothetical pressure p:

$$J_\mu = -(L R_g T) C_\mu \frac{\partial \ln p}{\partial z} \quad (10.2\text{-}5)$$

It is desirable, however, that we express the flux equation in terms of the adsorbed concentration instead of the gradient of the unknown partial pressure of a hypothetical gas phase. To do this, we simply apply the chain rule of differentiation to eq.(10.2-5) and obtain:

$$J_\mu = -(LR_g T) \frac{\partial \ln p}{\partial \ln C_\mu} \frac{\partial C_\mu}{\partial z} \tag{10.2-6}$$

where $\partial \ln p / \partial \ln C_\mu$ is obtained from the adsorption isotherm (10.2-3).

If we define the corrected diffusivity as (with the super-script 0 to denote the corrected diffusivity):

$$D_\mu^0 = LR_g T \tag{10.2-7}$$

which is temperature dependent, the constitutive flux equation (10.2-6) becomes:

$$J_\mu = -D_\mu \frac{\partial C_\mu}{\partial z} = -D_\mu^0 \frac{\partial \ln p}{\partial \ln C_\mu} \frac{\partial C_\mu}{\partial z} \tag{10.2-8}$$

where D_μ is called the transport diffusivity. This transport diffusivity is equal to the corrected diffusivity D_μ^0 multiplied by the term known as the thermodynamic correction factor. This factor describes the thermodynamic equilibrium between the two phases, and it is given by

$$\frac{\partial \ln p}{\partial \ln C_\mu} \tag{10.2-9}$$

When the isotherm is linear, the thermodynamic correction factor is unity, meaning that the corrected diffusivity is the transport diffusivity at zero loading conditions.

Example 10.2-1: *Thermodynamic correction factor for Langmuir isotherm case*

If the isotherm takes the form of Langmuir equation

$$C_\mu = f(P) = C_{\mu s} bP / (1 + bP)$$

the thermodynamic correction factor is:

$$\frac{\partial \ln P}{\partial \ln C_\mu} = \frac{1}{1-\theta} = \frac{1}{1 - C_\mu / C_{\mu s}}$$

Thus, for Langmuir isotherm the transport diffusivity increases as the loading inside the zeolite crystal increases. Stronger dependence with concentration can be observed if the isotherm takes the form of Volmer equation (Table 10.2-1). This stronger concentration dependence of the transport diffusivity in the case of Volmer isotherm equation is attributed to the mobility term in the Volmer equation.

Table 10.2-1: The thermodynamics correction factor for some isotherm equations

Isotherm	$\partial \ln p / \partial \ln C_\mu$
Linear	1
Langmuir	$1/(1-\theta)$
Volmer	$1/(1-\theta)^2$

Example 10.2-2 *Steady state diffusion through a zeolite membrane*

Steady state flow of molecule through a zeolite membrane is a constant. It can be obtained by integrating eq. (10.2-8) subject to constant boundary conditions at two ends of the membrane:

$$J_\mu = \frac{D_\mu^0}{L} \int_{C_{\mu,L}}^{C_{\mu,0}} \frac{\partial \ln p}{\partial \ln C_\mu} dC_\mu$$

where $C_{\mu,0}$ and $C_{\mu,L}$ are adsorbed concentrations which are in equilibrium with pressures p_0 and p_L at the two ends, respectively.

If the adsorption isotherm takes the form of Langmuir equation as given in example 10.2-1, the steady state flux equation will become:

$$J_\mu = \frac{D_\mu^0 C_{\mu s}}{L} \ln\left(\frac{1 - C_{\mu,L}/C_{\mu s}}{1 - C_{\mu,0}/C_{\mu s}}\right)$$

where

$$\frac{C_{\mu,0}}{C_{\mu s}} = \frac{bp_0}{1+bp_0}; \quad \frac{C_{\mu,L}}{C_{\mu s}} = \frac{bp_L}{1+bp_L}$$

Rewriting the above flux equation in terms of pressures, we get:

$$J_\mu = \frac{D_\mu^0 C_{\mu s}}{L} \ln\left(\frac{1+bp_L}{1+bp_0}\right)$$

10.2.1.1 Temperature Dependence of the Corrected Diffusivity D_μ^0

The diffusion process in the microparticle is assumed to be the hopping process of molecules from one low energy position to the next low energy position within the microparticle. If the energy barrier of the hopping process is E_μ, the corrected diffusivity takes the following Arrhenius relation:

$$D_\mu^0 = D_{\mu\infty}^0 \exp\left(-\frac{E_\mu}{R_g T}\right) = D_{\mu 0}^0 \exp\left[\frac{E_\mu}{R_g T_0}\left(1-\frac{T_0}{T}\right)\right] \qquad (10.2\text{-}10)$$

where $D_{\mu\infty}^0$ is the corrected diffusivity at infinite temperature, and $D_{\mu 0}^0$ is the corrected diffusivity at some reference temperature T_0. This energy barrier E_μ is usually less than the heat of adsorption. The following table lists some systems of which the heat of adsorption is greater than the activation energy E_μ.

Table 10.2-2: Adsorption heat & activation energy for micropore diffusion for some systems

Reference	Solid	Adsorbate	Heat of adsorption (kcal/mole)	Activation energy for micropore diffusion (kcal/mole)
Haq & Ruthven (1986a)	4A	CO_2	11	4.9
Haq & Ruthven (1986b)	5A	cyclo-C_3H_6	10.7	4.57
		cis-C_4H_8	11.4	6.4
Chiang et al. (1984)	5A	n-butane	10.2	4.8
Ma & Mancel (1973)	H-Mordenite	CH_4	4.4	1.8
		C_2H_6	5.8	3.8
	Na-Mordenite	C_2H_6	5.5	4.5
		C_3H_8	8.8	5.2
		n-butane	10.9	8.7

10.2.2 The Mass Balance Equation

The mass balance equation describing the concentration distribution of the adsorbed species in a microparticle of either slab, cylinder and spherical geometry takes the following form:

$$\frac{\partial C_\mu}{\partial t} = -\frac{1}{z^s}\frac{\partial}{\partial z}\left(z^s J_\mu\right) \qquad (10.2\text{-}11a)$$

Substitution of the constitutive flux equation (10.2-8) into the above equation yields:

$$\frac{\partial C_\mu}{\partial t} = \frac{1}{z^s}\frac{\partial}{\partial z}\left(z^s D_\mu^0 \frac{\partial \ln p}{\partial \ln C_\mu}\frac{\partial C_\mu}{\partial z}\right) \qquad (10.2\text{-}11b)$$

This equation is subject to the following boundary conditions:

$$z = 0; \qquad \frac{\partial C_\mu}{\partial z} = 0 \qquad (10.2\text{-}12a)$$

$$z = L; \qquad C_\mu = C_{\mu b} = f(P_b) \qquad (10.2\text{-}12b)$$

The first condition is simply the symmetry condition at the center of the micro-particle, and the second condition is the equilibrium condition at the exterior surface of the microparticle, where the gas phase pressure is maintained at P_b. Here L is the half length of the slab crystal or radius of the cylindrical and spherical crystals.

The crystal is initially assumed to be loaded with an amount which is in equilibrium with a gas phase of pressure P_i, that is:

$$t = 0; \qquad C_\mu = C_{\mu i} = f(P_i) \qquad (10.2\text{-}13)$$

Solving eq.(10.2-11) subject to the boundary conditions (10.2-12) and the initial condition (10.2-13) will give us an information on how adsorbed molecules are distributed inside the micro-particle as a function of time and the time taken for the system to relax to the new equilibrium condition. The volumetric average concentration is given by:

$$\langle C_\mu \rangle = \frac{(1+s)}{L^{s+1}}\int_0^L z^s C_\mu(z,t)\, dz \qquad (10.2\text{-}14a)$$

The quantity of interest in the analysis of adsorption kinetics is the fractional uptake, defined as the amount taken by the solid from $t = 0$ up to time t divided by the amount taken at infinite time. Mathematically, it is:

$$F = \frac{\langle C_\mu \rangle - C_{\mu i}}{C_{\mu b} - C_{\mu i}} \qquad (10.2\text{-}14b)$$

The behaviour of the fractional uptake with respect to time gives an indication of how fast a system approaches equilibrium with respect to a jump from the initial pressure P_i to some final pressure P_b.

What we will show next is to investigate the behaviour of the fractional uptake versus time to determine how the relaxation time will vary with the system

610 Kinetics

parameters and operating conditions. We first illustrate this with a linear isotherm, and then study a case of nonlinear isotherm. Langmuir isotherm equation is used as the model case as it is sufficient to study the effects of the isotherm nonlinearity on the adsorption kinetics behaviour.

10.2.2.1 *Linear Isotherm*

When the adsorption isotherm is linear (that is when the pressure is very low usually of the order of 1 Torr at ambient temperature for many low molecular weight hydrocarbons), we have the following relevant initial and boundary conditions:

$$t = 0; \quad C_\mu = C_{\mu i} = K\, P_i$$

and

$$z = L; \quad C_\mu = C_{\mu b} = K\, P_b$$

The thermodynamic correction factor for the case of linear isotherm is unity (Table 10.2-1), rendering the governing equation (10.2-11b) linear. The analytical solution for the concentration distribution is:

$$\frac{C_\mu(z,t) - C_{\mu i}}{C_{\mu b} - C_{\mu i}} = 1 - \sum_{n=1}^{\infty} a_n K_n(x) \cdot \exp\left(-\xi_n^2 \tau\right) \qquad (10.2\text{-}15a)$$

where

$$x = \frac{z}{L}; \quad \tau = \frac{D_\mu^0 \, t}{L^2} \qquad (10.2\text{-}15b)$$

The coefficient a_n, the eigenfunction $K_n(x)$ and the eigenvalue ξ_n are tabulated in the following table (Table 10.2-3) for the three micro-particle geometries. With the exception of the cylindrical geometry, the eigenvalues are in explicit form, facilitating the numerical evaluation of the concentration distribution.

Table 10.2-3: The coefficient a_n, the eigenfunction $K_n(x)$ and the eigenvalue ξ_n for the three micro-particle geometries

Shape	a_n	$K_n(x)$	ξ_n	Eq. no.
Slab	$2\sin\xi_n / \xi_n$	$\cos(\xi_n x)$	$(n-1/2)\pi$	(10.2-16a)
Cylinder	$2/[\xi_n J_1(\xi_n)]$	$J_0(\xi_n x)$	$J_0(\bullet) = 0$	(10.2-16b)
Sphere	$-2\cos\xi_n / \xi_n$	$\sin(\xi_n x)/x$	$n\pi$	(10.2-16c)

J_0 and J_1 are zero-order and first-order Bessel functions.

Knowing the concentration distribution for the case of linear isotherm as given in eq. (10.2-15a), the fractional uptake is then given in eq.(10.2-14b) and is tabulated in the following table (Table 10.2-4):

Table 10.2-4: Fractional uptake for the three microparticle geometries

Shape	Fractional uptake	Eq.no.
Slab	$F = 1 - \dfrac{2}{\pi^2} \sum_{n=1}^{\infty} \dfrac{1}{(n-1/2)^2} \exp\left[-(n-1/2)^2 \pi^2 \tau\right]$	(10.2-17a)
Cylinder	$F = 1 - 4 \sum_{n=1}^{\infty} \dfrac{1}{\xi_n^2} \exp\left(-\xi_n^2 \tau\right); \quad J_0(\xi) = 0$	(10.2-17b)
Sphere	$F = 1 - \dfrac{6}{\pi^2} \sum_{n=1}^{\infty} \dfrac{1}{n^2} \exp\left(-n^2 \pi^2 \tau\right)$	(10.2-17c)

A programming code UPTAKEP is provided with this book, and it allows the calculation of the fractional uptake versus time for three different shapes of the micro-particle. The following figure (Figure 10.2-1) shows the fractional uptake of a spherical crystal with a corrected diffusivity of 1×10^{-10} cm^2/sec for various values of microparticle sizes (1, 2 and 3 micron in radius).

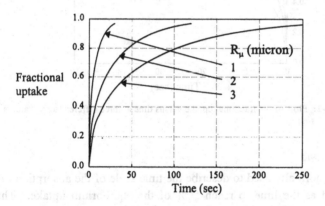

Figure 10.2-1: Plots of the fractional uptake versus time

It is seen that the larger is the micro-particle, the longer is the time required for the system to approach equilibrium. This time is in fact proportional to the square of the micro-particle radius by the virtue of the definition of the non-dimensional time

612 Kinetics

in eq. (10.2-15b). This is proved as follows. For any arbitrary fractional uptake (say F = 0.9), eq.(10.2-17) states that τ is a constant. Thus from the definition of the non-dimensional time τ in eq. (10.2-15b) the adsorption time scale is then proportional to the square of the micro-particle radius and inversely proportional to the corrected diffusivity.

Figure 10.2-2 shows plots of the fractional uptake for various temperatures for a spherical zeolite having a radius of 1 micron. The activation energy for micropore diffusion is 20 kJoule/mole, and the micropore diffusivity at 298 K is 1×10^{-10} cm^2/sec. Here we see that the higher is the temperature the faster is the uptake due to the increase in the diffusivity. What this means is that the higher temperature system will reach equilibrium faster than the lower temperature system. This, however, does not mean that the adsorption rate is higher for the higher temperature system. We shall discuss this in Section 10.2.2.1.3.

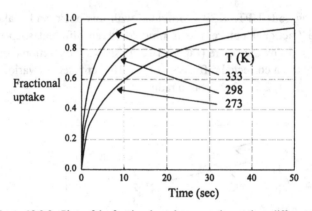

Figure 10.2-2: Plots of the fractional uptake versus time at three different temperatures

10.2.2.1.1 Half time

Half time is normally used to describe the time scale of the adsorption kinetics, and it is defined as the time to reach 50% of the equilibrium uptake. Thus, by setting F = 0.5 into eqs.(10.2-17), we get the necessary expressions for the half time written in terms of the size of the particle and the diffusivity. The following table shows the results of the half times for three shapes of the microparticle for the linear adsorption isotherm case.

Table 10.2-5: Half times of microparticles with linear isotherm

Shape	Half-time, $t_{0.5}$	Eq. no.
Slab	$0.1967 \, L^2 / D_\mu^0$	(10.2-18a)
Cylinder	$0.0631 \, L^2 / D_\mu^0$	(10.2-18b)
Sphere	$0.03055 \, L^2 / D_\mu^0$	(10.2-18c)

The half time is proportional to the square of the crystal size and inversely proportional to the diffusivity. We see that the half time is largest for the case of slab, and smallest for the spherical case. This is not surprising as the exterior surface area per unit volume is largest for the sphere and smallest for the slab; hence mass transfer per unit volume of crystal is fastest in the case of spherical crystals.

What we also see in Table 10.2-5 is that the half time is independent of $C_{\mu i}$ (the initial concentration) and $C_{\mu b}$ (the final concentration). The system is undergoing adsorption when $C_{\mu b} > C_{\mu i}$ and desorption when $C_{\mu b} < C_{\mu i}$. Thus, what this means is that the time scale for adsorption (half time) is the same for both adsorption and desorption modes. This characteristics is applicable only to linear isotherm.

10.2.2.1.2 First Statistical Moment

Half time is a convenient measure of the time scale of adsorption. Another measure of the time scale for the sorbate molecule to diffuse from the exterior surface to the interior of the micro-particle is the first statistical moment (Koricik and Zikanova, 1972), defined as the statistical mean of the adsorption rate:

$$\mu_1 = \int_0^\infty t \frac{dF}{dt} dt = \int_0^\infty (1 - F) \, dt \qquad (10.2\text{-}19)$$

Take the case of spherical microparticle as an example, we substitute the fractional uptake of eq.(10.2-17c) into the above equation and carry out the integration to obtain:

$$\mu_1 = \frac{1}{15} \frac{L^2}{D_\mu^0} \approx 0.0667 \frac{L^2}{D_\mu^0} \qquad (10.2\text{-}20a)$$

in which we have used the relation $\sum_{n=1}^{\infty}(1/n^4) = \pi^4/90$ (Jolley, 1961). This first statistical moment is compared to the half-time

$$t_{0.5} = 0.03055 \frac{L^2}{D_\mu^0} \tag{10.2-20b}$$

The fractional uptake at the first statistical moment μ_1 is

$$F = 1 - \frac{6}{\pi^2} \sum_{n=1}^{\infty} \frac{1}{n^2} \exp(-n^2\pi^2/15) \approx 0.674 \tag{10.2-21}$$

For other shapes of the particles, the first statistical moments are listed in the following table (Table 10.2-6). For comparison, the half times are also included in the table. Note that the first statistical moment is about twice as large as the half time. Readers can use either one of them as the measure for the time scale of adsorption.

Table 10.2-6: First statistical moments and half-times

Shape	First statistical moment, μ_1	Half time, $t_{0.5}$	Eq. no.
Slab	$\dfrac{1}{3}\dfrac{L^2}{D_\mu^0}$	$0.19674 \dfrac{L^2}{D_\mu^0}$	(10.2-22a)
Cylinder	$\dfrac{1}{8}\dfrac{L^2}{D_\mu^0}$	$0.0631 \dfrac{L^2}{D_\mu^0}$	(10.2-22b)
Sphere	$\dfrac{1}{15}\dfrac{L^2}{D_\mu^0}$	$0.03055 \dfrac{L^2}{D_\mu^0}$	(10.2-22c)

10.2.2.1.3 The Temperature Dependence of the Adsorption Rate

We have seen the dependence of the adsorption kinetics on the micro-particle size, that is the time scale is proportional to the square of the particle radius. Since particle size does not affect the equilibrium capacity, the adsorption rate is lower in the case of larger particle size as the time taken to reach equilibrium is longer.

To investigate the temperature dependence, we can not use the behaviour of the fractional uptake versus temperature as shown in Figure 10.2-2 because the equilibrium amount $C_{\mu b}$ is a function of temperature, and so is the diffusivity. To do this we have to investigate the amount adsorbed up to time t per unit volume of the microparticle, that is:

$$\frac{M}{V} = (C_{\mu b} - C_{\mu i}) F(\tau) \tag{10.2-22}$$

where V is the volume of the micro-particle, and M is the amount adsorbed up to time t. The temperature dependence of the various variables in the above equation is:

$$C_{\mu b} = KP_b = K_\infty \exp\left(\frac{Q}{R_g T}\right) P_b = K_0 \exp\left[-\frac{Q}{R_g T_0}\left(1 - \frac{T_0}{T}\right)\right] P_b \quad (10.2\text{-}23a)$$

$$C_{\mu i} = KP_i = K_\infty \exp\left(\frac{Q}{R_g T}\right) P_i = K_0 \exp\left[-\frac{Q}{R_g T_0}\left(1 - \frac{T_0}{T}\right)\right] P_i \quad (10.2\text{-}23b)$$

$$\tau = \frac{D_\mu^0 t}{L^2} = \frac{D_{\mu\infty}^0 \exp\left(-\frac{E_\mu}{R_g T}\right) t}{L^2} \quad (10.2\text{-}23c)$$

where Q is the heat of adsorption, K_∞ is the Henry constant at infinite temperature, and K_0 is the Henry constant at some reference temperature T_0. Expressing the adsorption rate per unit volume explicitly in terms of temperature, we have:

$$\frac{M}{V} = \left[K_0 \exp\left[-\frac{Q}{R_g T_0}\left(1 - \frac{T_0}{T}\right)\right](P_\infty - P_0)\right] \times F\left(\frac{D_\mu^0(T) t}{L^2}\right) \quad (10.2\text{-}24)$$

Eq.(10.2-24) contains two factors. The first factor (the square bracket term in the RHS) is the capacity factor and it decreases with temperature, while the second factor is the fractional uptake and it increases with temperature due to the increase in the diffusivity. The net result of these two factors is that the rate of adsorption decreases with temperature as the decrease in the capacity with respect to temperature overcompensates the increase in the fractional uptake. This is due to the higher heat of adsorption (Q) than the activation energy for diffusion (E_μ). Figure 10.2-3 shows typical plots of the adsorbed amount per unit volume as a function of time for three values of temperatures, 273, 298 and 333 K. The parameters used in generating these plots are:

L	= 1 micron
E_μ	= 20 kJoule/mole
Q	= 40 kJoule/mole
D_μ^0 @ 298 K	= 1 × 10^{-10} cm²/sec
K	= 7.5 × 10^{-4} mole/cc/kPa
P_i	= 0 kPa
P_b	= 2 kPa

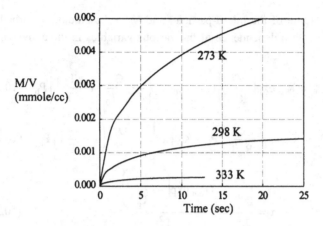

Figure 10.2-3: Plot of the amount adsorbed per unit volume versus time at T = 273, 298 and 333 K

10.2.2.1.4 Short Time Solution:

The solution for the fractional uptake given in eqs.(10.2-17) is valid for all times. However, when time is short, that is when the uptake is only a few percent of the equilibrium amount, the following solution for the fractional uptake into an initially clean microparticle can be useful

$$F \approx \frac{2(1+s)}{\sqrt{\pi}} \sqrt{\frac{D_\mu^0 \, t}{L^2}} \qquad (10.2\text{-}25)$$

The fractional uptake at short time is proportional to the square root of time, a well known characteristics of a diffusion type process. The above equation is obtained by taking the Laplace transform of the linear governing equation and finding the limit of the solution in the Laplace domain when the Laplace variable s approaches infinity.

To calculate the amount taken up by the micro-particle we multiply the fractional uptake by the amount taken up at infinite time, which is:

$$M_b = V C_{\mu b} \qquad (10.2\text{-}26)$$

where V is the volume of the micro-particle and $C_{\mu b}$ is the adsorbed concentration which is in equilibrium with P_b. Thus, the amount taken by the micro-particle at short time is given by:

$$M_t = \begin{cases} \dfrac{2}{\sqrt{\pi}}(2A)C_{\mu b}\sqrt{D_\mu^0}\cdot\sqrt{t} & \text{slab} \\ \dfrac{2}{\sqrt{\pi}}(2\pi LD)C_{\mu b}\sqrt{D_\mu^0}\cdot\sqrt{t} & \text{cylinder} \\ \dfrac{2}{\sqrt{\pi}}(4\pi L^2)C_{\mu b}\sqrt{D_\mu^0}\cdot\sqrt{t} & \text{sphere} \end{cases} \qquad (10.2\text{-}27)$$

where A is the area of one side of the slab crystal, and D is the length of the cylinder. We see that (2A) is the external area for mass transfer for slab, ($2\pi LD$) is that for cylinder and ($4\pi L^2$) is that for sphere. Thus, we could generalise the amount uptake for short time for any micro-particle shape as:

$$M_t = \frac{2}{\sqrt{\pi}}A_{ext}\,C_{\mu b}\sqrt{D_\mu^0}\cdot\sqrt{t} \qquad (10.2\text{-}28)$$

This equation is independent of the particle shape, as one would expect for short time the mass penetration is still very close to the exterior surface and the particle curvature is irrelevant. The mass transfer is only controlled by the available external surface area.

Eq.(10.2-28) is a very useful equation to measure the corrected diffusivity. One simply measures the amount uptake versus time, and then a simple plot of the initial data of M_t versus the square root of time, a straight line is resulted with a slope of

$$\text{slope} = \frac{2}{\sqrt{\pi}}A_{ext}C_{\mu b}\sqrt{D_\mu^0} = \frac{2}{\sqrt{\pi}}A_{ext}K\,P_b\sqrt{D_\mu^0} \qquad (10.2\text{-}29)$$

from which the corrected diffusivity can be obtained. The range of validity of the short time solution is that the amount adsorbed M_t is less than the equilibrium amount, that is:

$$M_t \ll M_b \qquad (10.2\text{-}30)$$

Substitution of eq.(10.2-28) and (10.2-26) into the above equation gives the following range of validity of the short time solution:

$$\sqrt{t} \ll \frac{V}{(2/\sqrt{\pi})A_{ext}\sqrt{D_\mu^0}} \qquad (10.2\text{-}31)$$

Using the temperature dependence form for the Henry constant and the corrected diffusivity

$$K = K_\infty \exp\left(\frac{Q}{R_g T}\right)$$

$$D_\mu^0 = D_{\mu\infty}^0 \exp\left(-\frac{E_\mu}{R_g T}\right)$$

the slope of the initial data of M_t versus time has the following temperature dependence form:

$$\text{slope} = \frac{2}{\sqrt{\pi}} A_{ext} K_\infty P_b \sqrt{D_{\mu\infty}^0} \times \exp\left(\frac{Q - E_\mu/2}{R_g T}\right) \quad (10.2\text{-}32)$$

which decreases as the temperature increases since the heat of adsorption (Q) is usually greater than the activation energy for diffusion (E_μ).

10.2.2.2 Langmuir Isotherm

Having understood how a linear adsorption isotherm affects the adsorption kinetics in a micro-particle, let us now turn to the case where the adsorption isotherm is nonlinear. The nonlinearity of the isotherm affects the kinetics through the thermodynamics correction factor (eq. 10.2-9). Let us start with the mass balance equation and obtain its solution by a numerical method. We shall take the case of Langmuir isotherm to illustrate the effect of the isotherm nonlinearity.

10.2.2.2.1 Mass Balance Equation

When the adsorption equilibrium between the fluid phase and the micro-particle takes the form of Langmuir equation

$$C_\mu = C_{\mu s} \frac{bP}{1 + bP} \quad (10.2\text{-}33)$$

the thermodynamic correction factor is $\left(\partial \ln P / \partial \ln C_\mu\right) = \left(1 - C_\mu / C_{\mu s}\right)^{-1}$, hence the mass balance equation (10.2-11b) will become:

$$\frac{\partial C_\mu}{\partial t} = \frac{1}{z^s} \frac{\partial}{\partial z}\left(z^s \frac{D_\mu^0}{1 - C_\mu / C_{\mu s}} \frac{\partial C_\mu}{\partial z}\right) \quad (10.2\text{-}34)$$

subject to the following boundary condition:

$$z = L; \quad C_\mu = C_{\mu b} = C_{\mu s} \frac{bP_b}{1 + bP_b} \quad (10.2\text{-}35)$$

where P_b is the pressure of the gas surrounding the micro-particle and b is the adsorption affinity.

Assuming the micro-particle is initially equilibrated with a gas phase of pressure P_i. The initial condition is:

$$t = 0; \quad C_\mu = C_{\mu i} = C_{\mu s} \frac{bP_i}{1 + bP_i} \quad (10.2\text{-}36)$$

The above equations (10.2-34 to 10.2-36) are non-linear due to the thermodynamic correction factor in the transport diffusivity term. The method we have been using in solving nonlinear partial differential equations is the orthogonal collocation method. We again apply it here, and to do so we define the following non-dimensional variables and parameters:

$$x = \frac{C_\mu - C_{\mu i}}{C_{\mu b} - C_{\mu i}}; \quad \eta = \frac{z}{L}; \quad \tau = \frac{D_\mu^0 \, t}{L^2}; \quad (10.2\text{-}37a)$$

$$\lambda_i = bP_i; \quad \lambda_b = bP_b; \quad H(x) = \left\{1 - \left[\frac{\lambda_i}{1+\lambda_i} + \left(\frac{\lambda_b}{1+\lambda_b} - \frac{\lambda_i}{1+\lambda_i}\right)x\right]\right\}^{-1} \quad (10.2\text{-}37b)$$

The above choice of the non-dimensional dependent variable x is to normalise the adsorbed concentration. Note that the function $H(x)$ is equal to $(1 + \lambda_i)$ at time $t = 0$, and will take the value of $(1 + \lambda_b)$ at $t = \infty$.

With these new variables and parameters, the non-dimensional mass balance equation is:

$$\frac{\partial x}{\partial \tau} = \frac{1}{\eta^s} \frac{\partial}{\partial \eta}\left[\eta^s \, H(x) \frac{\partial x}{\partial \eta}\right] \quad (10.2.38)$$

The boundary conditions and initial condition are:

$$\eta = 1; \quad x = 1 \quad (10.2\text{-}39)$$

$$\tau = 0; \quad x = 0 \quad (10.2\text{-}40)$$

Solving this set of equations numerically will give the non-dimensional adsorbed concentration as a function of time as well as distance. Appendix 10.1 shows the application of the collocation method to discretise eq. (10.2-38) into a set of coupled ordinary differential equations, which are then integrated with respect to time by the Runge-Kutta method.

10.2.2.2.2 Fractional Uptake

The fractional uptake is the ratio of the amount adsorbed from t = 0 to a time "t" divided by the amount adsorbed up to t = ∞. Mathematically, it is calculated from:

$$F = \frac{\langle C_\mu \rangle - C_{\mu i}}{C_{\mu b} - C_{\mu i}} \tag{10.2-41}$$

or written in terms of nondimensional variable

$$F = \langle x \rangle \tag{10.2-42}$$

where <x> is the volumetric average defined as follows:

$$\langle x \rangle = (1+s) \int_0^1 \eta^s \, x \, d\eta \tag{10.2-43}$$

The average amount uptake per unit volume of the crystal at any time t is then calculated from:

$$\langle C_\mu \rangle = C_{\mu i} + (C_{\mu b} - C_{\mu i}) \langle x \rangle \tag{10.2-44}$$

10.2.2.2.3 Size Dependence and Temperature Dependence

To investigate the dependence of the adsorption kinetics on the micro-particle size, we simply define the non-dimensional time scaled against some reference length, L_{ref}, that is:

$$\tau = \frac{D_\mu^0 \, t}{L_{ref}^2} \tag{10.2-45}$$

With this time scale the mass balance equation (10.2-34) becomes:

$$\frac{\partial x}{\partial \tau} = \left(\frac{L_{ref}}{L}\right)^2 \frac{1}{\eta^s} \frac{\partial}{\partial \eta}\left[\eta^s H(x) \frac{\partial x}{\partial \eta}\right] \tag{10.2-46}$$

subject to the boundary condition (10.2-39) and initial condition (10.2-40). This equation shows that the adsorption kinetics is proportional to the inverse of the square of the micro-particle size, similar to the linear isotherm case. This means that the isotherm nonlinearity does not change the dependence of the system behaviour on the micro-particle size.

To study the temperature dependence, we need to investigate the adsorbed amount (eq. 10.2-44) as a function of time. Without loss of generality, we can assume that the initial gas phase pressure is zero ($P_i = 0$, that is $C_{\mu i} = 0$) and hence the adsorbed amount per unit volume of the microparticle is:

$$\langle C_\mu \rangle = C_{\mu b} \langle x \rangle \tag{10.2-47}$$

where the temperature dependence of $C_{\mu b}$ is:

$$C_{\mu b} = C_{\mu s} \frac{b_\infty \exp(Q/R_g T) P_b}{1 + b_\infty \exp(Q/R_g T) P_b} \tag{10.2-48a}$$

The temperature dependence of $\langle x \rangle$ comes indirectly from the temperature dependence of λ_b and D_μ^0:

$$\lambda_b = bP_b = b_\infty \exp\left(\frac{Q}{R_g T}\right) P_b \tag{10.2-48b}$$

$$D_\mu^0 = D_{\mu\infty}^0 \exp\left(-\frac{E_\mu}{R_g T}\right) \tag{10.2-48c}$$

Numerical computation shows that as temperature increases the non-dimensional average concentration $\langle x \rangle$ increases for a given time, that is the system approaches equilibrium faster at higher temperature but the equilibrium amount $C_{\mu b}$ decreases with temperature and this decrease overcompensates the increase in $\langle x \rangle$; hence the net result is the decrease of $\langle C_\mu \rangle$ with temperature. In other words, the rate of adsorption per unit volume of micro-particle decreases with temperature.

10.2.2.2.4 Numerical Analysis

We see that in the non-dimensional form of equations, the initial and boundary conditions assume constant values (eqs. 10.2-39 and 10.2-40), irrespective of the system undergoing adsorption and desorption. Whether the system is going through adsorption or desorption depends on the relative magnitude of λ_i and λ_b, that is:

1. If $\lambda_i < \lambda_b$, the system is undergoing adsorption mode
2. If $\lambda_i > \lambda_b$, the system is undergoing desorption mode

The model equation (10.2-38) is solved numerically by a combination of the orthogonal collocation method and the Runge-Kutta method. Details of the collocation analysis is given in Appendix 10.1. The code ADSORB1C.M written in MatLab version 4.0 is provided with this book to simulate the above problem. The important parameter is the bP parameter. As this parameter is much smaller than unity, we have the case of linear isotherm; which is known to give the adsorption and desorption curves as mirror images to each other. However, when this parameter is greater than unity (non-linear range of the isotherm), we see that the time it takes to desorb molecules from the micro-particle is much longer than that taken for adsorption due to the stronger affinity of molecules towards the surface.

Figure 10.2-4 shows plots of fractional uptake versus time for adsorption of gas into an initially clean crystal of slab geometry. The gas pressures are bP = 0, 1 and 2. The curve for bP = 0 corresponds to the linear isotherm case, while the other two curves (bP = 1 and 2) correspond to the nonlinear situation. We see that as the nonlinearity of the isotherm increases, the adsorption dynamics is faster, and this is due to the higher value of the transport diffusivity under the nonlinear isotherm condition (see eq. 10.2-34).

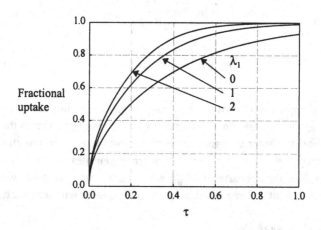

Figure 10.2-4: Fractional uptake versus τ for the case of adsorption with $\lambda_0 = 0$ and $\lambda_1 = (0,1,2)$

The same behaviour when we consider the desorption mode. First the crystal is equilibrated with a gas at various pressures (bP = 0, 1 and 2) and these loaded crystals are then desorbed into a vacuum environment. We see in Figure 10.2-5 that the nonlinearity of the adsorption isotherm also makes the desorption rate faster, but the rate is not as fast as that in the case of adsorption mode.

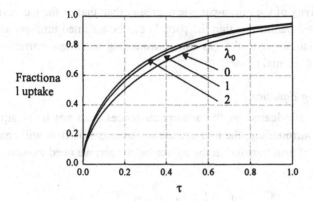

Figure 10.2-4: Fractional uptake versus τ for the case of desorption with $\lambda_1 = 0$ and $\lambda_0 = (0,1,2)$

10.3 Nonisothermal Single Component Adsorption in a Crystal

In the last section we have shown the kinetics behaviour of a microparticle when isothermal conditions prevail. A number of features have been observed for such conditions:

1. For linear isotherm, the adsorption time is the same as the desorption time; the time scale is proportional to the square of the radius of the microparticle, and the time scale is independent of the bulk concentration. With respect to temperature, the approach to equilibrium is faster when the system temperature is higher but the rate of adsorption is lower.
2. For nonlinear isotherm, the adsorption process is faster than the desorption; the time scale is still proportional to the square of microparticle radius, and the time scale is lower when the bulk concentration increases. With respect to temperature, the behaviour is the same as that for the case of linear isotherm.

Adsorption process is known to release heat upon adsorption or absorb energy upon desorption. Typical heat of adsorption of many hydrocarbons on zeolite ranges from 10 to 60 kJoule/mole, and such a magnitude can cause a significant change in the temperature of the crystal if the dissipation of energy to the surrounding is not fast enough. Thus, the crystal temperature increase (adsorption) or decrease (desorption) depends on the interplay between the rate of heat release due to adsorption or heat absorption due to desorption and the dissipation rate of energy to the surrounding. In this section we will address this interplay and investigate what way the heat transfer affects the kinetics of mass transfer. Since the thermal conductivity of the micro-particle is greater than that of the gas surrounding the particle, the temperature within the particle can be assumed uniform and all the heat transfer resistance is in the gas film surrounding the micro-particle. This is called the lumped thermal model.

10.3.1 Governing Equations

When the heat released by the adsorption process can not be dissipated fast enough into the surrounding, the temperature of the micro-particle will change. To study the effects of heat transfer on the uptake behaviour, we need to solve the mass balance equation

$$\frac{\partial C_\mu}{\partial t} = \frac{1}{z^s}\frac{\partial}{\partial z}\left(z^s D_\mu^0 \frac{\partial \ln p}{\partial \ln C_\mu}\frac{\partial C_\mu}{\partial z}\right) \tag{10.3-1a}$$

where the corrected diffusivity has the following Arrhenius temperature dependence

$$D_\mu^0 = D_{\mu\infty}^0 \exp\left(-\frac{E_\mu}{R_g T}\right) = D_{\mu 0}^0 \exp\left[\frac{E_\mu}{R_g T_0}\left(1-\frac{T_0}{T}\right)\right] \tag{10.3-1b}$$

together with the following lumped thermal model for heat balance around the micro-particle:

$$\langle \rho C_p \rangle \frac{dT}{dt} = Q \frac{d\langle C_\mu \rangle}{dt} - a_H h_f (T - T_b) \tag{10.3-2}$$

In obtaining the heat balance equation, we have assumed that the heat resistance by conduction inside the micro-particle is insignificant compared to the heat resistance outside the micro-particle, which is described by the second term in the RHS of eq.(10.3-2). Because of such assumption, the temperature inside the micro-particle is uniform, and the first term in the RHS of eq.(10.3-2) is the amount of heat released per unit volume per unit time as a result of the adsorption rate per unit volume $d\langle C_\mu \rangle/dt$. In eq.(10.3-2), h_f is the heat transfer coefficient per unit surface area, a_H is the heat transfer surface area per unit volume, Q is the molar heat of adsorption, T_b is the surrounding temperature, $\langle \rho C_p \rangle$ is the mean heat capacity per unit volume of the micro-particle, and $\langle C_\mu \rangle$ is the volumetric average concentration, defined as

$$\langle C_\mu \rangle = \frac{(1+s)}{L^{s+1}} \int_0^L z^s C_\mu dz \tag{10.3-3}$$

The heat transfer surface area per unit volume, a_H, needs some elaboration. It is not necessarily the exterior surface area of the crystal per unit volume. If the crystals agglomerates into an agglomerate and if the heat resistance within this agglomerate is negligible, then the heat transfer area per unit volume a_H is the exterior surface area of the agglomerate rather than the exterior surface area of a crystal. This means that mass transfer occurs over the crystal length scale, while the heat transfer occurs over the agglomerate length scale. If we denote the size of the agglomerate as L_a, then the exterior surface area per unit volume of this agglomerate is:

$$a_H = \frac{1+s_a}{L_a} \tag{10.3-4}$$

where s_a is the shape factor of the agglomerate, and it takes a value of 1, 2 or 3 for slab, cylinder and sphere, respectively.

The boundary conditions of eq.(10.3-1) are:

$$z = 0; \quad \frac{\partial C_\mu}{\partial z} = 0 \tag{10.3-5a}$$

$$z = L; \quad C_\mu = f(P_b, T) \tag{10.3-5b}$$

Note that at any time t the concentration at the pore mouth of the micropore is in equilibrium with the gas phase pressure P_b, and the instant temperature of the agglomerate, T. Since this temperature is time variant and is governed by the heat balance equation (10.3-2), the adsorbed concentration at the solid-fluid interface (z = L) is also time variant.

The initial conditions of the mass and heat balance equations are:

$$t = 0; \quad C_\mu = C_{\mu i} = f(P_i, T_i); \quad T = T_i \tag{10.3-6}$$

The initial temperature T_i may not be the same as the surrounding temperature T_b.

10.3.2 Non-Dimensional Equations

The set of governing equations (10.3-1 to 10.3-6) is nonlinear, and must be solved numerically. We solve it by using a combination of the orthogonal collocation method and the Runge-Kutta method. Before carrying out the numerical analysis, we cast the governing equations into non-dimensional form by defining the following non-dimensional variables

Non-dimensional time	$\tau = \dfrac{D_{\mu 0}^0 t}{L^2}$	(10.3-7a)
Non-dimensional distance	$\eta = \dfrac{z}{L}$	(10.3-7b)
Non-dimensional concentration	$x = \dfrac{C_\mu - C_{\mu i}}{C_{\mu b} - C_{\mu i}}$	(10.3-7c)
Non-dimensional temperature	$\theta = \dfrac{T - T_0}{T_0}; \ \theta_b = \dfrac{T_b - T_0}{T_0}; \ \theta_i = \dfrac{T_i - T_0}{T_0};$	(10.3-7d)

where $D_{\mu 0}^0$ is the crystal corrected diffusivity evaluated at some reference temperature T_0

$$D_{\mu 0}^0 = D_{\mu \infty}^0 \exp\left(-\frac{E_\mu}{R_g T_0}\right) \qquad (10.3\text{-}7e)$$

$C_{\mu b}$ is the adsorbed concentration in equilibrium with P_b and T_b:

$$C_{\mu b} = f(P_b, T_b) \qquad (10.3\text{-}7f)$$

and $C_{\mu i}$ is the adsorbed concentration which is in equilibrium with P_i and T_i:

$$C_{\mu i} = f(P_i, T_i) \qquad (10.3\text{-}7g)$$

The non-dimensional parameters of the system are:

$$\varphi(\theta) = \exp\left[\gamma_\mu \left(\frac{\theta}{1+\theta}\right)\right] \qquad (10.3\text{-}7h)$$

$$H(x) = \frac{\partial \ln P}{\partial \ln C_\mu} \qquad (10.3\text{-}7i)$$

$$\beta = \frac{Q(C_{\mu b} - C_{\mu i})}{\langle \rho C_p \rangle T_0} \qquad (10.3\text{-}7j)$$

$$\text{LeBi} = \frac{a_H h_f L^2}{D_{\mu 0}^0 \langle \rho C_p \rangle} \qquad (10.3\text{-}7k)$$

$$\gamma_\mu = \frac{E_\mu}{R_g T_0} \qquad (10.3\text{-}7l)$$

Some physical explanation of these parameters is necessary. The parameter γ_μ is the activation energy number, and for many zeolitic systems it has the following range

$$2 < \gamma_\mu < 20$$

The heat parameter β is a measure of the amount of heat released by adsorption process relative to the heat capacity of the solid. This parameter is called the heat capacity number. Typically it has the following range

$$0.02 < \beta < 0.2$$

The parameter LeBi is a measure of the heat transfer to the surrounding. Its physical meaning can be better seen by rearranging eq. (10.3-7k) as follows:

$$\text{LeBi} = \dfrac{\dfrac{L^2}{D_{\mu 0}^0}}{\dfrac{\langle \rho C_p \rangle}{a_H h_f}}$$

The numerator is the diffusion time, which is the time it takes for the mass transfer to approach equilibrium. The denominator is the time which the sensible heat of the micro-particle can be dissipated to the surrounding. Thus we expect that the larger is this parameter the closer is the system behaviour towards the isothermal condition.

With these definitions of non-dimensional variables and parameters, the mass and heat balance equations will become:

$$\frac{\partial x}{\partial \tau} = \frac{1}{\eta^s}\frac{\partial}{\partial \eta}\left[\eta^s\, \varphi(\theta)\, H(x)\, \frac{\partial x}{\partial \eta}\right] \qquad (10.3\text{-}8a)$$

$$\frac{d\theta}{d\tau} = \beta \frac{d\langle x \rangle}{d\tau} - \text{LeBi} \cdot (\theta - \theta_b) \qquad (10.3\text{-}8b)$$

The boundary conditions are:

$$\eta = 0; \quad \frac{\partial x}{\partial \eta} = 0 \qquad (10.3\text{-}9a)$$

$$\eta = 1; \quad x = \frac{f(P_b, T) - C_{\mu i}}{C_{\mu b} - C_{\mu i}} \qquad (10.3\text{-}9b)$$

The initial condition is:

$$\tau = 0; \quad x = 0; \quad \theta = \theta_i \qquad (10.3\text{-}10)$$

The fractional uptake defined as the additional amount adsorbed up to time t divided by the amount uptaken from t = 0 to infinite time, that is:

$$F = \frac{\langle C_\mu \rangle - C_{\mu i}}{C_{\mu b} - C_{\mu i}} = \langle x \rangle \qquad (10.3\text{-}11a)$$

where the volumetric average concentration is defined as follows:

$$\langle x \rangle = (1+s) \int_0^1 \eta^s\, x\, d\eta \qquad (10.3\text{-}11b)$$

This set of non-dimensional equations can be solved numerically to yield solutions for x and θ. The non-dimensional concentration x is a function of η and τ, and the non-dimensional temperature θ is a function of time. The fractional uptake is obtained from eq.(10.3-11), and then the amount adsorbed per unit volume of the micro-particle is:

$$\langle C_\mu \rangle = C_{\mu 0} + (C_{\mu\infty} - C_{\mu 0})\langle x \rangle$$

The micro-particle size dependence and the temperature dependence of the kinetics behaviour is not apparent in this case due to a number of factors. First the equations are coupled and nonlinear, and the dependence of some parameters such as the heat transfer coefficient h_F is subject to the conditions of the environment outside the micro-particle, and it might depend on the micro-particle size. For example, the heat transfer coefficient around the agglomerate can be described by the following correlation (Wakao and Kaguei, 1982)

$$Nu = \frac{h_f(2L_a)}{k_f} = 2 + 1.1\left(\frac{2L_a u \rho_f}{\mu_f}\right)^{0.6}\left(\frac{C_p \mu_f}{k_f}\right)^{1/3} \qquad (10.3\text{-}12a)$$

that is

$$h_f = \frac{k_f}{2L_a}\left[2 + 1.1\left(\frac{2L_a u \rho_f}{\mu_f}\right)^{0.6}\left(\frac{C_p \mu_f}{k_f}\right)^{1/3}\right] \qquad (10.3\text{-}12b)$$

If the surrounding is stagnant, we have:

$$h_f \approx \frac{k_f}{L_a}$$

that is the heat transfer coefficient is inversely proportional to the agglomerate radius L_a.

However, if the surrounding is highly stirred, the heat transfer coefficient will become:

$$h_f \approx 1.1 \left(\frac{k_f}{2L_a}\right)\left(\frac{2L_a u \rho_f}{\mu_f}\right)^{0.6}\left(\frac{C_p \mu_f}{k_f}\right)^{1/3}$$

that is, the heat transfer coefficient is inversely proportional to $L_a^{0.4}$.

Knowing the size dependence of the heat transfer coefficient and the size dependence of the exterior surface area per unit volume (eq. 10.3-4), the LeBi parameter has the following dependence on the agglomerate size (L_a) and the crystal size (L).

	h_f	a_H	LeBi
Stagnant	L_a^{-1}	L_a^{-1}	$L^2 L_a^{-2}$
Turbulent	$L_a^{-0.4}$	L_a^{-1}	$L^2 L_a^{-1.4}$

Thus we see that the detailed kinetics behaviour must be investigated numerically. However, as a first approximation, we would expect the time scale of adsorption is proportional to the square of the micro-particle radius.

10.3.3 Langmuir Isotherm

We shall study this non-isothermal case with Langmuir isotherm of the form

$$C_\mu = f(P,T) = C_{\mu s} \frac{b(T)P}{1+b(T)P} \qquad (10.3\text{-}13a)$$

where the adsorption affinity takes the following temperature dependent form:

$$b = b_\infty \exp\left(\frac{Q}{R_g T}\right) = b_0 \exp\left[-\frac{Q}{R_g T_0}\left(1-\frac{T_0}{T}\right)\right] \qquad (10.3\text{-}13b)$$

where Q is the heat of adsorption, used in the heat balance equation (10.3-2), b_∞ is the adsorption affinity at infinite temperature and b_0 is that at the reference temperature T_0.

With this form of adsorption isotherm, the thermodynamic correction factor can now take the following explicit form:

$$H(x) = \left\{1 - \left[\frac{\lambda_i}{1+\lambda_i} + \left(\frac{\lambda_b}{1+\lambda_b} - \frac{\lambda_i}{1+\lambda_i}\right)x\right]\right\}^{-1} \quad (10.3\text{-}14a)$$

where the parameters λ_i and λ_b are defined as

$$\lambda_i = bP_i; \quad \lambda_b = bP_b; \quad (10.3\text{-}14b)$$

They are function of temperature due to the temperature dependence of the affinity constant b.

The boundary condition (10.3-9b) takes the following explicit form for this case of Langmuir isotherm:

$$\eta = 1; \quad x\big|_1 = \frac{C_{\mu s} \dfrac{b_0 P_b \, \phi(\theta)}{1 + b_0 P_b \, \phi(\theta)} - C_{\mu i}}{C_{\mu b} - C_{\mu i}} \quad (10.3\text{-}15)$$

where

$$\phi(\theta) = \exp\left[-\left(\frac{Q}{R_g T_0}\right)\left(\frac{\theta}{1+\theta}\right)\right] \quad (10.3\text{-}16)$$

The model equations (10.3-8) are solved numerically by the method of orthogonal collocation (Appendix 10.2). The simulations are obtained from the code ADSORB1D.M, provided with this book to help readers with means to understand the adsorption problem better. The parameters supplied to this code are grouped as follows:

Description	Parameters
Micro-particle & agglomerate	L, L_a, a_H
Operating conditions	T_i, T_b, P_i, P_b
Adsorption characteristics	$C_{\mu s}, b_0, Q$
Diffusion characteristics	$D_{\mu 0}^0, E_\mu$
Heat parameters	$\langle \rho C_p \rangle, h_f$

The order of magnitude of some heat parameters is given in Appendix 10.3. The following parameters are used in the simulations as the base case

Micro-particle shape, s	= 2
Micro-particle size, L	= 1 micron
Initial adsorbate pressure, P_i	= 0 kPa
Bulk adsorbate pressure, P_b	= 1 kPa
Initial temperature, T_i	= 300 K
Bulk temperature, T_b	= 300 K
Reference temperature, T_0	= 300 K
Adsorption affinity at T_0, b_0	= 1 kPa^{-1}
Saturation capacity, $C_{\mu s}$	= 5 × 10^{-3} mole/cc
Corrected diffusivity at T_0, $D_{\mu 0}^0$	= 1 × 10^{-10} cm^2/sec
Activation for micropore diffusion, E_μ	= 15,000 Joule/mole
Heat of adsorption, Q	= 30,000 Joule/mole
Volumetric heat capacity, ρC_p	= 1 Joule/cc/K
Heat transfer number, LeBi	= 5

Figure 10.3-1 shows the concentration evolution of six discrete points along the micro-particle co-ordinate. We note that the non-dimensional concentration at the surface decreases from unity because of the micro-particle temperature increase, resulting in a drop in the adsorbed concentration at the surface (see eq. 10.3-15).

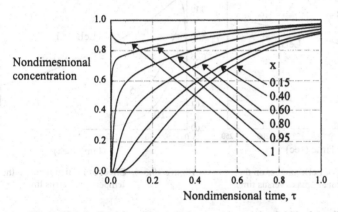

Figure 10.3-1: Concentration evolution at various points within the zeolite crystal

The temperature increase versus time is shown in Figure 10.3-2, and the time at which the temperature reaches its maximum is corresponding to the time at which the surface concentration reaches its minimum (cf. Figures 10.3-1 and 10.3-2).

632 Kinetics

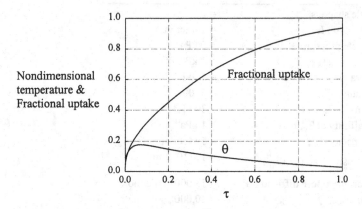

Figure 10.3-2: Plots of the non-dimensional temperature and fractional uptake versus τ

Effects of the heat transfer number, LeBi, are shown in Figure 10.3-3 and 10.3-4. The same set of parameters in the above table is used in the simulation with the following values of LeBi number {1, 5, 10}.

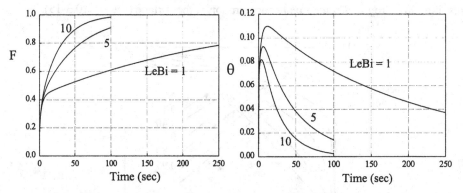

Figure 10.3-3: Effect of LeBi number on the fractional uptake versus time

Figure 10.3-4: Effect of LeBi number on the temperature versus time

As expected a reduction in the heat transfer number would cause an increase in the micro-particle temperature and as a result the fractional uptake exhibits a distinct two stage uptake when the LeBi is low. The slow second stage of the uptake is dictated by the cooling rate of the micro-particle.

Effects of the bulk pressure are shown in Figure 10.3-5. Increase in the bulk pressure causes an increase in the particle temperature because of the larger amount

Analysis of Adsorption Kinetics in a Zeolite Particle 633

adsorbed by the micro-particle. Also increase in the bulk pressure results in a faster approach to equilibrium.

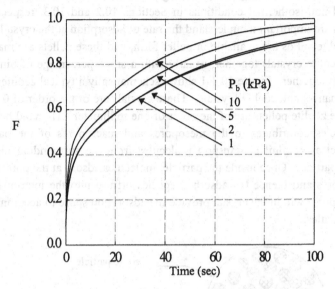

Figure 10.3-5a: Plots of the fractional uptake versus time

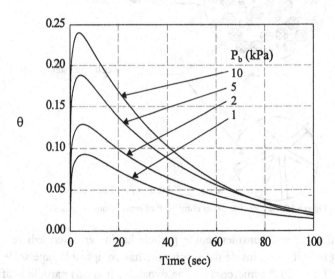

Figure 10.3-5b: Effect of the bulk pressure on the temperature response

10.4 Bimodal Diffusion Models

We have shown the analysis of a single zeolite crystal under isothermal conditions and non-isothermal conditions in Sections 10.2 and 10.3, respectively. These analyses are important to understand the rate of adsorption at the crystal level. In practice zeolite solids are available in pellet form, and these pellets are made by compressing zeolite crystals together, usually with a small percentage of binder to join the crystals together. Figure 10.4-1 shows schematically a typical zeolite pellet composed of many small zeolite crystals. These crystals are of the order of 0.1 to 1 micron, and the zeolite pellets are of the order of one millimeter. The void between the microparticles contributes to the mesopores and macropores of the particle. These pores act as conduit to transport molecules from the surrounding into the interior of the particle. Once inside the particle, molecules adsorb at the pore mouth of the micropores and thence the adsorbed species diffuse into the interior of the crystal. Micropores within the crystal provide the adsorption space to accommodate adsorbate molecules.

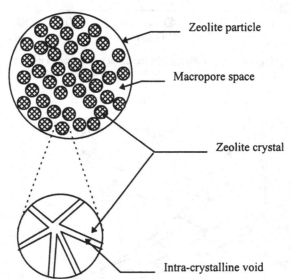

Figure 10.4-1: Zeolite pellet composed of small zeolite crystals

In this section we will consider zeolite particle having an intermediate size, in such a way that the diffusion inside the macropore has comparable time scale to the time scale of diffusion in the micropore. The dynamics into this particle is affected by the interplay between the two diffusion processes as well as the capacity accommodated by the micropores.

10.4.1 The Length Scale and The Time Scale of Diffusion

The diffusion process in the macropore and mesopore follows the combination of the molecular and Knudsen mechanisms while the diffusion process inside the zeolite crystal follows an intracrystalline diffusion mechanism, which we have discussed in Section 10.2. The length scale of diffusion in the macropore is the dimension of the particle, while the length scale of diffusion in the micropore is the dimension of the zeolite crystal; thus, although the magnitude of the intracrystalline diffusivity (in the order of 10^{-11} to 10^{-6} cm^2/sec) is very small compared to the diffusivity in the macropore the time scales of diffusion of these two pore systems could be comparable.

If the time scale of diffusion in the micropore is very short compared to that in the macropore, we will have a macropore diffusion model with the characteristic length being the particle dimension. This case is called the macropore diffusion control. The model equations of this macropore diffusion case are similar to those obtained in Chapter 9 for homogeneous-type solids. The only difference is that in the case of macropore diffusion control for zeolite particles, there is no contribution of the surface diffusion.

On the other hand, if the time scale of diffusion in the crystal is very long compared to that in the particle the system dynamics will be controlled by the intracrystalline diffusion with the characteristic length being the crystal dimension. The initial stage of the dynamics is the filling of pore space of macropore with adsorbate, which is very fast and usually of the order of one second. Furthermore, this capacity is usually very small compared to the capacity of the micropore; hence the initial stage is usually not measurable.

The two time scales of diffusion can be comparable. The relative measure between these two time scales for the case of linear isotherm is the following parameter (Do, 1983, 1990) (which will be apparent later in the formulation of mass balance equation):

$$\gamma = \frac{[\varepsilon + (1-\varepsilon)K]D_\mu R^2}{\varepsilon D_p R_\mu^2} \equiv \frac{\text{Time scale for macropore diffusion}}{\text{Time scale for micropore diffusion}} \qquad (10.4\text{-}1)$$

where ε is the particle porosity (macropore and mesopore), K is the slope of the isotherm if the isotherm is linear or the ratio of the adsorbed concentration to the bulk concentration at equilibrium for nonlinear isotherm, D_μ is the diffusivity in the micropore, D_p is the pore diffusivity, R is the particle radius and R_μ is the radius of the microparticle. This parameter is the ratio of the time scale for macropore diffusion to the time scale for micropore diffusion.

The macropore diffusion time scale	$\dfrac{R^2[\varepsilon + (1-\varepsilon)K]}{\varepsilon D_p}$	(10.4-2a)
The micropore diffusion time scale	$\dfrac{R_\mu^2}{D_\mu}$	(10.4-2b)

Thus, if γ is less than unity, meaning the time scale of diffusion in the crystal is greater than that in the macropore, we then talk about intracrystalline diffusion control. On the other hand, if γ is greater than unity, we talk about macropore diffusion control. We illustrate this criterion with the following example:

Example 10.4-1: *Controlling mechanism for a linear isotherm case*

We take the following values typical for sorption of light hydrocarbons in zeolite.

R	= 0.1 cm
R_μ	= 2 × 10^{-4} cm
D_p	= 0.01 cm²/sec
D_μ	= 1 × 10^{-9} cm²/sec
K	= 1000
ε	= 0.33

Substituting these values into eq. (10.4-1), we get

$$\gamma = \frac{2031\,\text{sec}}{40\,\text{sec}} = 50 \gg 1$$

The value in the numerator is the time scale of diffusion in macropore while that in the denominator is the time scale of diffusion in micropore. A value of 50 for γ suggests that the system overall kinetics is controlled by the macropore diffusion as the time scale it takes to diffuse along the macropore is 50 times longer than that in the micropore.

If we take the case where the intracrystalline diffusion is more restricted (10 time more restricted than the last case)

$$D_\mu = 1 \times 10^{-10} \text{ cm}^2/\text{sec}$$

compared to 1×10^{-9} cm²/sec in the last example, the value of γ is calculated as

$$\gamma = \frac{2031 \text{ sec}}{400 \text{ sec}} = 5 \equiv O(1)$$

An order of unity of this parameter suggests that the system is controlled by the diffusion in the macropore as well as the diffusion in the micropore.

10.4.2 The Mass Balance Equations

The mass balance of the bimodal particle is composed of two equations. One describes the mass balance inside the crystal, while the other describes the mass balance in the pellet, and the two are coupled through the boundary of the zeolite crystal. The coordinate framework for the zeolite pellet is shown as in Figure 10.4-2, with r being the radial distance for the pellet and r_μ being the radial distance for the crystal. We have used the notation convention that the subscript μ is for the adsorbed phase. Here the crystal is acting as the adsorbed phase.

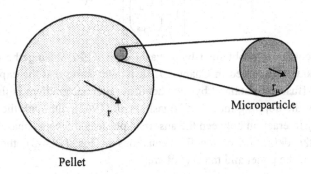

Figure 10.4-2: Frame of coordinates for zeolite pellet

Before writing down the mass balance equations, we need to define the constitutive flux equations for the macropore and the micropore. These flux equations are assumed to follow the following Fickian law equations:

$$J_p = -\varepsilon D_p \frac{\partial C}{\partial r} \qquad (10.4\text{-}3a)$$

and

$$J_\mu = -D_\mu \frac{\partial C_\mu}{\partial r_\mu} \tag{10.4-3b}$$

where C is the adsorbate concentration in the macropore and C_μ is the concentration in the microparticle. These fluxes are based on the total cross-sectional areas of their respective domains.

The intracrystalline diffusivity, D_μ, is a function of concentration in the crystal. It is a constant only when the adsorption isotherm is linear, and unlike the diffusion process in the macropore the intracrystalline diffusion is activated. Intracrystalline diffusion has been dealt with in Section 10.2, where we have shown that the transport diffusivity D_μ is related to the corrected diffusivity D_μ^0 as follows:

$$D_\mu = D_\mu^0 \left(\frac{\partial \ln P}{\partial \ln C_\mu} \right) \tag{10.4-4}$$

When the adsorption isotherm takes the form of a Langmuir equation, eq. (10.4-4) is reduced to:

$$D_\mu = \frac{D_\mu^0}{\left[1 - \left(\frac{C_\mu}{C_{\mu s}} \right) \right]} = \frac{D_\mu^0}{1-\theta} \tag{10.4-5}$$

A similar equation is obtained by Higashi et al. (1963) using the concept of the hopping model, discussed in Chapter 7. It is interesting that the expression for the transport diffusivity obtained by the chemical potential gradient is the same as that obtained by the hopping model of Higashi et al. (1963), although the latter does not involve any interaction between the adsorbed phase and the gas phase.

With the definition of the flux equations in eqs. (10.4-3), the mass balance equations for the pellet and the crystal are:

$$\varepsilon \frac{\partial C}{\partial t} + (1-\varepsilon) \frac{\partial \langle C_\mu \rangle}{\partial t} = \frac{1}{r^s} \frac{\partial}{\partial r} \left(r^s \varepsilon D_p \frac{\partial C}{\partial r} \right) \tag{10.4-6a}$$

and

$$\frac{\partial C_\mu}{\partial t} = \frac{1}{r_\mu^s} \frac{\partial}{\partial r_\mu} \left[r_\mu^s D_\mu(C_\mu) \frac{\partial C_\mu}{\partial r_\mu} \right] \tag{10.4-6b}$$

where $\langle C_\mu \rangle$ is the volumetric mean concentration, defined as follows:

$$\langle C_\mu \rangle = \frac{(1+s)}{R_\mu^{s+1}} \int_0^{R_\mu} r_\mu^s \, C_\mu \, dr_\mu \qquad (10.4\text{-}6c)$$

Note that there is no surface diffusion in the macropore as it is assumed that the adsorption on the exterior surface of the zeolite crystal is negligible compared to the adsorptive capacity within the zeolite crystal.

Having the two mass balance equations written for the two sub systems (the pellet and the crystal), we need to provide the connection (that is, boundary conditions) between these two sub-systems and the connection between them and the surrounding. The boundary condition at the surface of the microparticle is:

$$r_\mu = R_\mu \quad ; \quad C_\mu = f(C) \quad \text{(isotherm equation)} \qquad (10.4\text{-}7)$$

in which we have assumed that the local adsorption kinetics at the pore mouth of the crystal is much faster than the diffusion in the macropore as well as the diffusion in the micropore. For zeolite crystals subjected to severe hydrothermal treatment, skin barrier could be resulted and it could contribute to the mass transfer resistance at the pore mouth. However, we shall not deal with that situation here.

The connection between the pellet and the surrounding is that the mass transfer from the pellet equals to the mass transfer through the fluid film surrounding the particle, that is:

$$r = R; \quad -\varepsilon D_p \frac{\partial C}{\partial r}\bigg|_{r=R} = k_m \left(C\big|_{r=R} - C_b \right) \qquad (10.4\text{-}8)$$

where k_m is the mass transfer coefficient, and C_b is the bulk concentration.

The initial condition is:

$$t = 0; \quad C = C_i; \quad C_\mu = C_{\mu i} = f(C_i) \qquad (10.4\text{-}9)$$

in which we have assumed that the pellet is initially equilibrated with a surrounding having a concentration of C_i.

10.4.3 Linear Isotherm

To investigate the behaviour of the bimodal diffusion mechanism, we consider the linear isotherm first. In this case the mass balance equations become linear and hence they are amenable to linear analysis. With linear isotherm, the boundary condition at the zeolite pore mouth (eq. 10.4-7) becomes:

$$r_\mu = R_\mu \quad ; \quad C_\mu = KC \qquad (10.4\text{-}10)$$

The first work in the Western literature dealing with this bimodal diffusion mechanism is that of Ruckenstein et al. (1971). The model was justified as a point sink approximation of a more exact analysis (Neogi and Ruckenstein, 1980). Since then a number of papers have appeared to investigate the various aspects of adsorption using this model, for example in fixed bed chromatography (Haynes and Sharma, 1973; Kawazoe and Takeuchi, 1974; Raghavan and Ruthven, 1985), in fixed bed breakthrough curve (Rasmusen, 1982; Cen and Yang, 1986) in adsorbent particle (Dubinin et al. 1975; Lee, 1978; Ugruzov and Zolotarev, 1982; Zolotarev et al., 1982, 1984; Polte and Mersmann, 1986), and in nonisothermal adsorbent particle (Sun and Meunier, 1987).

With the linear isotherm the intracrystalline diffusivity is constant $D_\mu(C_\mu) = D_\mu^0$, and the mass balance can be solved by the method of Laplace transform or the method of generalised integral transform (Do, 1983). The solution for the fractional uptake for spherical microparticle and spherical pellet is in the form of a double series:

$$F = 1 - 36 \sum_{n=1}^{\infty} \sum_{m=1}^{\infty} a_{mn} \exp\left\{-\gamma \eta_{nm}^2 \frac{\varepsilon_M D_p t}{R^2[\varepsilon_M + (1-\varepsilon_M)K]}\right\} \qquad (10.4\text{-}11a)$$

where the coefficient a_{nm} is given by

$$a_{nm} = \frac{(\sin \eta_{nm} - \eta_{nm} \cos_{nm})^2}{\xi_n^2 \eta_{nm}^4 \left\{1 + \cos^2 \eta_{nm} / \left[(\xi_n^2/3\gamma) - 1\right]\right\}} \qquad (10.4\text{-}11b)$$

with ξ_n being the primary eigenvalue and η_{nm} being the secondary eigenvalue, determined from:

$$\xi_n = n\pi; \qquad \eta_{nm} \cot \eta_{nm} - 1 = -\frac{\xi_n^2}{3\gamma} \qquad (10.4\text{-}11c)$$

that is for a given value of n, we have one value of the eigenvalue ξ and for each value of ξ we have a spectrum of the eigenvalue η; hence a double index is associated with the eigenvalue η.

The parameter γ in eq.(10.4-11a) is the ratio of the two time scales, namely the time scale for macropore diffusion to the time scale for micropore diffusion, defined as

$$\gamma = \frac{R^2[\varepsilon + (1-\varepsilon)K]/(\varepsilon D_p)}{R_\mu^2 / D_\mu^0} \equiv \frac{\text{Time scale of macropore diffusion}}{\text{Time scale of micropore diffusion}}$$

The half time can be calculated from eq.(10.4-11a) by setting the fractional uptake to one half. Unlike the parallel pore and surface diffusion model discussed in Chapter 9 where the half time is proportional to the square of the particle radius, the half time of the bimodal diffusion model is proportional to R^α, where α is equal to 2 when macropore diffusion dominates the transport and α is equal to zero when micropore diffusion controls the uptake. An approximate expression for the half time for a bimodal diffusion model is given by Do (1990):

$$\tau_{0.5} = \frac{\varepsilon D_p t_{0.5}}{R^2[\varepsilon + (1-\varepsilon)K]} = 0.03055 \left\{ 1 + \frac{1 + 0.9\exp\left[-0.2\left(\ln\frac{\gamma}{6}\right)^2\right]}{\gamma} \right\} \quad (10.4\text{-}12)$$

The solution given in eq. (10.4-11a) reduces to simple solutions when either the macropore or micropore diffusion is the controlling mechanism, that is when γ is less than unity, that is the time scale for the diffusion in the macropore is much smaller than that in the micropore, we would then expect the micropore diffusion would control the overall adsorption kinetics. In this case, we have the following half time

$$\tau_{0.5} = \frac{\varepsilon D_p t_{0.5}}{R^2[\varepsilon + (1-\varepsilon)K]} = \frac{0.03055}{\gamma} \quad (10.4\text{-}13a)$$

or

$$t_{0.5} = 0.03055 \frac{R_\mu^2}{D_\mu^0} \quad (10.4\text{-}13b)$$

We see that the half time involves only the microparticle characteristics, which is what one would expect when micropore diffusion is the controlling mechanism.

On the other hand, when γ is greater than unity (macropore diffusion control regime), we obtain the following limit for the half time

$$\tau_{0.5} = \frac{\varepsilon D_p t_{0.5}}{R^2[\varepsilon + (1-\varepsilon)K]} = 0.03055 \quad (10.4\text{-}14)$$

It is clear that the half-time in this case involves only characteristics of the macropore.

For a bimodal solid with pellet having a cylindrical shape and the microparticle having a spherical shape, the solution was obtained by Smith (1984).

10.4.3.1 Temperature Dependence of the Parameter γ

As we have mentioned above that the parameter which demarcates the micropore diffusion and the macropore diffusion is the parameter γ. Let us now investigate its dependence on temperature to see how temperature would influence the controlling mechanism. The temperature dependence of relevant parameters in the parameter γ is given below:

$$K = K_\infty \exp\left(\frac{Q}{R_g T}\right) = K_0 \exp\left[-\frac{Q}{R_g T_0}\left(1 - \frac{T_0}{T}\right)\right] \tag{10.4-15a}$$

$$D_\mu^0 = D_{\mu\infty}^0 \exp\left(-\frac{E_\mu}{R_g T}\right) = D_{\mu 0}^0 \exp\left[\frac{E_\mu}{R_g T_0}\left(1 - \frac{T_0}{T}\right)\right] \tag{10.4-15b}$$

$$D_p = D_{p0}\left(\frac{T}{T_0}\right)^\alpha \tag{10.4-15c}$$

where D_{po} is the pore diffusivity at some reference temperature T_0, and α is between 0.5 to 1.75. It is equal to 0.5 when Knudsen diffusion controls the pore diffusion, and it is about 1.75 when molecular diffusion mechanism controls the pore diffusion. The parameter K_0 is the Henry constant at T_0, and $D_{\mu 0}^0$ is the corrected diffusivity at T_0.

Substitute these equations into eq.(10.4-1), we get:

$$\gamma = \frac{R^2\left[\varepsilon + (1-\varepsilon)K_\infty \exp\left(\frac{Q}{R_g T}\right)\right]D_{\mu\infty}^0 \exp\left(-\frac{E_\mu}{R_g T}\right)}{R_\mu^2(\varepsilon D_{p0})(T/T_0)^\alpha} \tag{10.4-16}$$

The capacity in the pore space is usually much smaller than the capacity in the adsorbed phase (that is $\varepsilon \ll (1-\varepsilon)K$); thus the above equation is simplified to:

$$\gamma = \frac{R^2(1-\varepsilon)K_\infty D_{\mu\infty}^0 \exp\left(\frac{Q-E_\mu}{R_g T}\right)}{R_\mu^2(\varepsilon D_{p0})(T/T_0)^\alpha} \tag{10.4-17a}$$

or

$$\gamma = \frac{R^2(1-\varepsilon)K_0 D_{\mu 0}^0}{R_\mu^2(\varepsilon D_{p0})} \times \frac{\exp\left[-\left(\frac{Q-E_\mu}{R_g T_0}\right)\left(1-\frac{T_0}{T}\right)\right]}{(T/T_0)^\alpha} \tag{10.4-17b}$$

Since the heat of adsorption is usually larger than the activation energy of intracrystalline diffusion, the parameter γ decreases with an increase in temperature. This means that the micropore diffusion is more important at higher temperatures and the macropore diffusion is more significant at lower temperatures. This is <u>only true for linear isotherm</u>. We will discuss this effect on nonlinear isotherm in the next section. To illustrate the temperature effect on the parameter γ for the linear isotherm case, we take the following example:

Example 10.4-2: *Effect of temperature on the controlling mechanism*

We take the following values typical for sorption of light hydrocarbons into zeolite:

R	= 0.02 cm
R_μ	= 2 × 10⁻⁴ cm
T_0	= 298 K
K_0	= 1000 @ 298 K
D_{p0}	= 0.02 cm²/sec @ 298K and 1 atm
$D_{\mu 0}^0$	= 1 × 10⁻⁹ cm²/sec @ 298K
α	= 1.5
Q	= 40000 Joule/mole
E_μ	= 20000 Joule/mole

Figure 10.4-3 shows the plot of γ versus T and we see that the demarcation temperature is about 300K. For temperature greater than 400K, micropore diffusions controls the uptake; while for temperature less than 200K, macropore diffusion controls. For temperatures between 200 and 400 K, both diffusion mechanisms control the uptake.

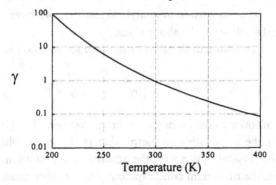

Figure 10.4-3: Plot of γ versus temperature

10.4.3.2 Pressure Dependence of the Parameter γ

The only parameter that could be affected by the total pressure is the pore diffusivity D_p. If the macropore diffusion is controlled purely by the Knudsen diffusion mechanism, the pore diffusivity is D_K and hence it is independent of total pressure, implying that the parameter γ is independent of pressure. However, if the macropore diffusion is governed by molecular-molecular collision, then the pore diffusivity is inversely proportional to the total pressure, meaning that the parameter γ increases linearly with the total pressure. This means that the system is moving toward macropore diffusion control as the total pressure increases.

10.4.4 Irreversible Isotherm

We have discussed the behaviour of the bimodal solid with a linear isotherm. Now we discuss the other extreme of the isotherm, the irreversible isotherm. What we would expect in this case is that the concentration in the macropore behaves like a wave front, that is the adsorbed concentration in the region close to the pellet exterior is very close to the maximum concentration, while the region near the core is void of adsorbate in any form, either in free or adsorbed form. The position demarcating these two regions is the adsorption (wave) front position. How this wave front penetrates into the particle depends on the rate of macropore diffusion as well as the rate of diffusion into the micropore.

What we will present below is the analysis of a zeolite pellet composed of small spherical crystals. The mass balance for the description of concentration distribution inside a zeolite crystal is:

$$\frac{\partial C_\mu}{\partial t} = D_\mu \frac{1}{r_\mu^2} \frac{\partial}{\partial r_\mu}\left(r_\mu^2 \frac{\partial C_\mu}{\partial r_\mu} \right) \quad (10.4\text{-}18)$$

To render the equation amenable to analytical analysis, we have assumed constant intra-crystalline diffusivity in the above equation.

The boundary condition of the above mass balance equation is:

$$r_\mu = R_\mu; \quad C_\mu = \begin{cases} C_{\mu s} & \text{if } C > 0 \\ 0 & \text{if } C = 0 \end{cases} \quad (10.4\text{-}19)$$

where C is the adsorbate concentration in the macropore. Eq. (10.4-19) is simply the statement of the irreversible isotherm, that is if the gas phase concentration outside the zeolite crystal is nonzero, the adsorbed concentration at the micropore mouth is equal to the maximum concentration. On the other hand, it will be zero if the gas phase concentration is zero.

Next, we obtain the mass balance in the macropore, given below

$$\varepsilon \frac{\partial C}{\partial t} + (1-\varepsilon) \frac{\partial \overline{C}_\mu}{\partial t} = \varepsilon D_p \frac{1}{r^s} \frac{\partial}{\partial r}\left(r^s \frac{\partial C}{\partial r}\right) \qquad (10.4\text{-}20a)$$

where ε is the porosity of the macropore, s is the pellet shape factor and $\overline{C}_\mu$ is the volumetric average concentration of the microparticle, defined as

$$\overline{C}_\mu = \frac{3}{R_\mu^3} \int_0^{R_\mu} r_\mu^2 \, C_\mu \, dr_\mu \qquad (10.4\text{-}20b)$$

The boundary condition of the macropore mass balance equation (10.4-20a) is:

$$r = R; \quad -\varepsilon D_p \frac{\partial C}{\partial r}\bigg|_R = k_m \left(C\big|_R - C_b\right) \qquad (10.4\text{-}21)$$

Before proceeding with the solution procedure, we define the following nondimensional variables and parameters:

$$\eta = \frac{r}{R}; \quad z = \frac{r_\mu}{R_\mu}; \quad \tau = \frac{D_\mu t}{R_\mu^2}; \quad x = \frac{C_\mu}{C_{\mu s}}; \quad y = \frac{C}{C_b} \qquad (10.4\text{-}22a)$$

$$\gamma = \frac{R^2 (1-\varepsilon)(C_{\mu s}/C_b) D_\mu}{(\varepsilon D_p) R_\mu^2}; \quad Bi = \frac{k_m R}{\varepsilon D_p} \qquad (10.4\text{-}22b)$$

and obtain the following nondimensional governing equations:

	Equation	Boundary conditions	Eq. #
Micropore	$\dfrac{\partial x}{\partial \tau} = \dfrac{1}{z^2} \dfrac{\partial}{\partial z}\left(z^2 \dfrac{\partial x}{\partial z}\right)$	$z = 1; \quad x = \begin{cases} 1 & \text{if } y > 0 \\ 0 & \text{if } y = 0 \end{cases}$	(10.4-23)
Macropore	$\gamma \dfrac{\partial \overline{x}}{\partial \tau} = \dfrac{1}{\eta^s} \dfrac{\partial}{\partial \eta}\left(\eta^s \dfrac{\partial y}{\partial \eta}\right)$	$\eta = 1; \quad -\dfrac{\partial y}{\partial \eta}\bigg|_{\eta=1} = Bi\left(y\big|_{\eta=1} - 1\right)$	(10.4-24a)

where

$$\overline{x} = 3 \int_0^1 z^2 x \, dz \qquad (10.4\text{-}24b)$$

In obtaining eq. (10.4-24a) we have ignored the hold up of the adsorbate in the macropore space compared to the hold up in the micropore. This is reasonable as the density of adsorbate in the micropore is usually about 100 to 1000 times higher than that in the macropore.

Note the definition of the parameter γ for this case of irreversible isotherm given in eq. (10.4-22b) compared to that for the case of linear isotherm in eq. (10.4-1).

$$\gamma = \begin{cases} \dfrac{R^2(1-\varepsilon)K D_\mu}{(\varepsilon D_p)R_\mu^2} & \text{for linear isotherm} \\ \dfrac{R^2(1-\varepsilon)(C_{\mu s}/C_b)D_\mu}{(\varepsilon D_p)R_\mu^2} & \text{for irreversible isotherm} \end{cases}$$

The Henry constant K in the linear isotherm case is replaced by the ratio $C_{\mu s}/C_b$ in the irreversible isotherm case. The Henry constant is a constant and therefore the parameter γ in the linear isotherm case is independent of the bulk concentration used. On the other hand, the parameter γ for the irreversible isotherm case depends on the bulk concentration in a way that an increase in the bulk concentration will result in a decrease in the parameter γ. This means that the system is moving toward the micropore diffusion control when the bulk concentration increases. This can be physically explained that as the bulk concentration increases, the mass transfer through the macropore space is faster and this renders more uniformity in the concentration distribution in the macropore; hence overall uptake is due to the other resistance, namely the micropore diffusion resistance.

10.4.4.1 Temperature Dependence of γ for the case of irreversible isotherm

The dependence of γ on temperature in the case of irreversible isotherm is opposite to what was observed for the case of linear isotherm. We have the following temperature dependence of the parameter γ:

$$\gamma = \begin{cases} \dfrac{R^2(1-\varepsilon)K_\infty D_{\mu\infty}^0 \exp\left(\dfrac{Q-E_\mu}{R_g T}\right)}{R_\mu^2(\varepsilon D_{p0})(T/T_0)^\alpha} & \text{for linear isotherm} \\ \\ \dfrac{R^2(1-\varepsilon)\left(\dfrac{C_{\mu s}}{C_0}\right)D_{\mu\infty}^0 \exp\left(-\dfrac{E_\mu}{R_g T}\right)}{R_\mu^2(\varepsilon D_{p0})(T/T_0)^\alpha} & \text{for irreversible isotherm} \end{cases}$$

In the case of irreversible isotherm, the higher is the temperature, the larger is the value of γ; while for the case of linear isotherm as discussed earlier, the higher is the temperature the smaller is the value of γ. Thus, macropore diffusion controls at high temperature for irreversible isotherm, while the micropore diffusion will control the uptake at high temperature in the case of linear isotherm.

Assuming the zeolite pellet is initially free of adsorbate, eqs. (10.4-23) and (10.4-24) have been solved by Do (1989). The solutions are summarized below for three sub-cases:

(a) Comparable macropore and micropore diffusion rates
(b) Micropore diffusion
(c) Macropore diffusion

10.4.4.2 Comparable Macropore and Micropore Diffusion Rates

For the case of comparable rates between macropore diffusion and micropore diffusion ($\gamma \equiv O(1)$), the solution for the fractional uptake is:

$$F = \begin{cases} (1+s)\int_{X(\tau)}^{1} \eta^s \bar{x}\, d\eta & \text{for } 0 < \tau < \tau_0 \quad (10.4\text{-}25a) \\ (1+s)\int_{0}^{1} \eta^s \bar{x}\, d\eta & \text{for } \tau > \tau_0 \quad (10.4\text{-}25b) \end{cases}$$

where $X(\tau)$ is the nondimensional position of the adsorption front separating the two regions: the saturated region close to the particle exterior surface and the core void of any adsorbate. It is a function of time and is obtained from the following equation:

$$\left(\frac{1}{\gamma}\right)\left(\tau + \frac{1}{15} - \frac{2}{3}\sum_{n=1}^{\infty}\frac{e^{-\zeta_n^2 \tau}}{\zeta_n^2}\right) = \begin{cases} \dfrac{X^2 - 1}{2(1-s)} - \left(\dfrac{1}{Bi} + \dfrac{1}{1-s}\right)\dfrac{(X^{s+1} - 1)}{(s+1)} & \text{for } s \neq 1 \\[2mm] \dfrac{X^2 \ln X}{2} - \left(\dfrac{1}{4} + \dfrac{1}{2Bi}\right)(X^2 - 1) & \text{for } s = 1 \end{cases} \quad (10.4\text{-}26)$$

where ζ_n are roots of the following transcendental equation

$$\zeta \cot \zeta - 1 = 0 \qquad (10.4\text{-}27)$$

The time τ_0 in eq. (10.4-25) is the time at which the adsorption front reaches the center of the particle, that is $X(\tau_0) = 0$. Setting $X = 0$ into eq. (10.4-26), we get the following implicit equation for τ_0

$$\left(\frac{1}{\gamma}\right)\left(\tau_0 + \frac{1}{15} - \frac{2}{3}\sum_{n=1}^{\infty}\frac{e^{\zeta_n^2 \tau_0}}{\zeta_n^2}\right) = \frac{1}{(s+1)}\left(\frac{1}{Bi} + \frac{1}{2}\right) \qquad (10.4\text{-}28)$$

The mean concentration in the micro-particle appearing in the integrand of eq. (10.4-25) is given by:

$$\overline{x}(\eta,\tau) = 1 - \frac{6}{\pi^2}\sum_{n=1}^{\infty}\frac{1}{n^2}\exp\left\{-n^2\pi^2\left[\tau - \tau^*(\eta)\right]\right\} \qquad (10.4\text{-}29)$$

where $\tau^*(\eta)$ is the time when the adsorption front reaches the position η and is given in eq. (10.4-26).

The half time for this case must be solved implicitly from eq. (10.4-25) after setting F to 0.5.

10.4.4.3 Micropore Diffusion Control ($\gamma << 1$)

When micropore diffusion controls the overall uptake, that is when

  (b) the pellet size is small
  (c) the temperature is low
  (d) the bulk concentration is high

we expect that the macropore is filled very quickly with adsorbate and the overall kinetics is dictated solely by the diffusion of adsorbate into the micropore. The fractional uptake is then simply

$$F = 1 - \frac{6}{\pi^2}\sum_{n=1}^{\infty}\frac{1}{n^2}\exp\left(-n^2\pi^2\tau\right) \qquad (10.4\text{-}30)$$

The half time for this micropore diffusion control is

$$\tau_{0.5} = 0.03055 \qquad (10.4\text{-}31a)$$

that is

$$t_{0.5} = 0.03055\frac{R_\mu^2}{D_\mu} \qquad (10.4\text{-}31b)$$

Thus the time scale for the case of irreversible isotherm is the same as that for the case of linear isotherm when micropore diffusion is the controlling mechanism and the intracrystalline diffusivity is a constant.

10.4.4.4 Macropore Diffusion Control ($\gamma \gg 1$)

When
(a) the pellet size is large
(b) the temperature is high
(c) the bulk concentration is low

the parameter γ will be greater than unity and hence the system kinetics is controlled by macropore diffusion. The analysis of this macropore diffusion was described in details in Section 9.2.2.2. The fractional uptake is given by

$$F = 1 - X^{s+1} \qquad (10.4\text{-}32)$$

where X is the position of the adsorption front and is determined from the following implicit equations:

$$\frac{\varepsilon D_p C_0}{R^2 (1-\varepsilon) C_{\mu s}} t = H(X) \qquad (10.4\text{-}33)$$

The functional form H depends on the shape of the zeolite pellet and it takes the following form for three shapes of the pellet

$$H(X) = \begin{cases} \left(1 + \dfrac{1}{Bi}\right)(1-X) - \dfrac{1}{2}(1-X^2) & \text{for } s = 0 \quad (10.4\text{-}34a) \\[6pt] \left(\dfrac{1}{2Bi} + \dfrac{1}{4}\right)(1-X^2) + \dfrac{1}{2} X^2 \ln X & \text{for } s = 1 \quad (10.4\text{-}34b) \\[6pt] \dfrac{1}{2}(1-X^2) + \dfrac{1}{3}\left(\dfrac{1}{Bi} - 1\right)(1-X^3) & \text{for } s = 2 \quad (10.4\text{-}34c) \end{cases}$$

The half time for this macropore diffusion control is obtained by simply setting $F = 0.5$ in eq. (10.4-32) to obtain $X_{0.5}$ and then substituting it into eq. (10.4-33), we get:

$$t_{0.5} = \frac{R^2 (1-\varepsilon) C_{\mu s}}{\varepsilon D_p C_0} \times H_{0.5} \qquad (10.4\text{-}35a)$$

where

$$H_{0.5} = \begin{cases} \dfrac{1}{8} + \dfrac{1}{2} Bi & \text{for } s = 0 \\[6pt] 0.03836 + \dfrac{1}{4} Bi & \text{for } s = 1 \\[6pt] 0.01835 + \dfrac{1}{6} Bi & \text{for } s = 2 \end{cases} \qquad (10.4\text{-}35b)$$

We see that the half-time is proportional to the square of the pellet radius for this case of macropore diffusion control while it is independent of the pellet radius in the case of micropore diffusion control (eq. 10.4-31).

10.4.5 Nonlinear Isotherm and Nonisothermal Conditions

We have dealt with the analysis of a zeolite pellet for the case of linear isotherm and the case of irreversible isotherm. These two isotherms represent the two extremes of the nonlinearity of the adsorption isotherm. In this section we will deal with the case of nonlinear isotherm and to make the formulation general we also add to it the heat balance equation to study the coupled effect of the nonlinear isotherm and the nonisothermality on the overall adsorption uptake in a zeolite pellet. Similar to the Section 10.3, we shall assume that the thermal conductivity of the zeolite pellet is high and the heat transfer resistance is due to that of the stagnant film surrounding the pellet. This means that the temperature of the pellet is uniform, and the model corresponding to this circumstance is called the lumped thermal model.

The mass balance equation in the microparticle is:

$$\frac{\partial C_\mu}{\partial t} = \frac{1}{r_\mu^{s_\mu}} \frac{\partial}{\partial r_\mu}\left(r_\mu^{s_\mu} D_\mu^0(T) \frac{\partial \ln p}{\partial \ln C_\mu} \frac{\partial C_\mu}{\partial r_\mu} \right) \qquad (10.4\text{-}36a)$$

where the corrected diffusivity takes the Arrhenius form

$$D_\mu^0 = D_{\mu\infty}^0 \exp\left(-\frac{E_\mu}{R_g T}\right) \qquad (10.4\text{-}36b)$$

The boundary condition is that at the exterior surface of the microparticle the adsorption equilibrium is established between the macropore concentration and the adsorbed concentration at the micropore mouth, that is

$$r_\mu = R_\mu; \quad C_\mu = f(C, T) \qquad (10.4\text{-}37)$$

where f is the adsorption equilibrium functional form.

The mass balance equation in the macropore is simply

$$\varepsilon \frac{\partial C}{\partial t} + (1-\varepsilon)\frac{\partial \overline{C}_\mu}{\partial t} = \varepsilon D_p(T) \frac{1}{r^s} \frac{\partial}{\partial r}\left(r^s \frac{\partial C}{\partial r} \right) \qquad (10.4\text{-}38a)$$

where the temperature dependence of the pore diffusivity is

$$D_p = D_{p0}\left(\frac{T}{T_0}\right)^\alpha \tag{10.4-38b}$$

Here T_0 is some reference temperature and D_{p0} is the pore diffusivity evaluated at that temperature. The parameter α is equal to 0.5 when Knudsen controls the pore diffusion and to 1.75 where molecular-molecular collision mechanism controls the transport. The volumetric average concentration of the adsorbed species in eq. (10.4-38a) is

$$\overline{C}_\mu = \frac{(s_\mu + 1)}{R_\mu^{s_\mu+1}} \int_0^{R_\mu} r_\mu^{s_\mu} C_\mu \, dr_\mu \tag{10.4-39}$$

The boundary condition of eq. (10.4-38a) is

$$r = R; \quad -\varepsilon D_p \frac{\partial C}{\partial r}\bigg|_R = k_m\left(C|_R - C_b\right) \tag{10.4-40}$$

The heat balance equation on the whole pellet is:

$$\langle \rho C_p \rangle \frac{dT}{dt} = Q(1-\varepsilon)\frac{d\overline{\overline{C}}_\mu}{dt} - a_H h_f(T - T_b) \tag{10.4-41}$$

where $\langle \rho C_p \rangle$ is the volumetric heat capacity of the zeolite pellet, Q is the heat of adsorption, a_H is the exterior surface area per unit volume of the pellet, h_f is the heat transfer coefficient, and T_b is the bulk temperature.

The initial condition of the problem is that the zeolite pellet is initially equilibriated with a concentration of C_i and a temperature T_i, that is

$$t = 0; \quad C = C_i; \quad C_\mu = C_{\mu i} = f(C_i, T_i); \quad T = T_i \tag{10.4-42}$$

To solve the set of governing equations (10.4-36) to (10.4-42), we transform them into a set of nondimensional governing equations. First we choose some reference concentration C_0 and temperature T_0. The reference for the concentration in the microparticle is

$$C_{\mu 0} = f(C_0, T_0) \tag{10.4-43}$$

By defining the following nondimensional variables

$$\zeta = \frac{r_\mu}{R_\mu} \; ; \; \eta = \frac{r}{R} \; ; \; \tau = \frac{R^2\left[\varepsilon C_0 + (1-\varepsilon)C_{\mu 0}\right]}{\varepsilon D_{p0} C_0} t \tag{10.4-44a}$$

$$y = \frac{C}{C_0} \ ; \ x = \frac{C_\mu}{C_{\mu 0}}, \ \theta = \frac{T - T_0}{T_0} \tag{10.4-44b}$$

$$\gamma = \frac{R^2 \left[\varepsilon C_0 + (1-\varepsilon)C_{\mu 0}\right] D_{\mu 0}^0}{(\varepsilon D_{p0}) C_0 R_\mu^2} \ ; \ H(x,\theta) = \frac{\partial \ln p}{\partial \ln C_\mu} \ ; \ F(y,\theta) = \frac{f(C,T)}{f(C_0,T_0)} \tag{10.4-44c}$$

$$\varphi(\theta) = \exp\left[\gamma_\mu \left(\frac{\theta}{1+\theta}\right)\right] \ ; \ \gamma_\mu = \frac{E_\mu}{R_g T_0} \ ; \ \beta = \frac{Q(1-\varepsilon)C_{\mu 0}}{\langle \rho C_p \rangle T_0}, \ Bi = \frac{k_m R}{\varepsilon D_{p0}} \tag{10.4-44e}$$

$$LeBi = \frac{a_H h_f R^2 \left[\varepsilon C_0 + (1-\varepsilon)C_{\mu 0}\right]}{\langle \rho C_p \rangle \varepsilon D_{p0} C_0} \tag{10.4-44g}$$

$$\sigma = \frac{\varepsilon C_0}{\varepsilon C_0 + (1-\varepsilon)C_{\mu 0}} \ ; \ \theta_b = \frac{T_b - T_0}{T_0} \ ; \ \theta_i = \frac{T_i - T_0}{T_0} \tag{10.4-44h}$$

the governing equations in nondimensional form are:

Equation	Boundary condition	Eq. #
$\dfrac{\partial x}{\partial \tau} = \gamma \dfrac{1}{\zeta^{s_\mu}} \cdot \dfrac{\partial}{\partial \zeta}\left[\zeta^{s_\mu} \varphi(\theta) H(x,\theta) \dfrac{\partial x}{\partial \zeta}\right]$	$\zeta = 1 \ ; \ x = F(y,\theta)$	(10.4-45a)
$\sigma \dfrac{\partial y}{\partial \tau} + (1-\sigma)\dfrac{\partial \overline{x}}{\partial \tau} = (1+\theta)^\alpha \dfrac{1}{\eta^s}\dfrac{\partial}{\partial \eta}\left(\eta^s \dfrac{\partial y}{\partial \eta}\right)$	$\eta = 1 \ ;$ $-(1+\theta)^\alpha \dfrac{\partial y}{\partial \eta} = Bi(y - y_b)$	(10.4-45b)
$\dfrac{d\theta}{d\tau} = \beta \dfrac{d\overline{\overline{x}}}{d\tau} - LeBi \cdot (\theta - \theta_b)$		(10.4-45c)

where

$$\overline{x}(\eta,\tau) = (s_\mu + 1)\int_0^1 \zeta^{s_\mu} x(\eta,\zeta,\tau)\, d\zeta \tag{10.4-46a}$$

$$\overline{\overline{x}}(\tau) = (s+1)\int_0^1 \eta^s \overline{x}(\eta,\tau)\, d\eta \tag{10.4-46b}$$

Analysis of eqs. (10.4-45) is carried out with the collocation method. Readers interested in this method should refer to Appendix 10.4 for further detail, and a computer code of ADSORB1E is provided with this book for the simulation purpose.

10.4.5.1 Numerical Example of Langmuir Isotherm

We shall study this non-isothermal case with Langmuir isotherm of the form

$$C_\mu = f(C,T) = C_{\mu s} \frac{b(T) C}{1 + b(T) C} \qquad (10.4\text{-}47a)$$

where the adsorption affinity takes the following temperature dependent form:

$$b = b_\infty \exp\left(\frac{Q}{R_g T}\right) = b_0 \exp\left[-\frac{Q}{R_g T_0}\left(1 - \frac{T_0}{T}\right)\right] \qquad (10.4\text{-}47b)$$

where Q is the heat of adsorption, used in the heat balance equation (10.4-41), b_∞ is the adsorption affinity at infinite temperature and b_0 is that at the reference temperature T_0.

With this form of adsorption isotherm, the thermodynamic correction factor can now take the following explicit form:

$$H(x,\theta) = \left\{1 - \left[\frac{\lambda_i}{1+\lambda_i} + \left(\frac{\lambda_b}{1+\lambda_b} - \frac{\lambda_i}{1+\lambda_i}\right)x\right]\right\}^{-1} \qquad (10.4\text{-}48a)$$

where the parameters λ_i and λ_b are defined as

$$\lambda_i = bC_i; \qquad \lambda_b = bC_b \qquad (10.4\text{-}48b)$$

They are function of temperature due to the temperature dependence of the affinity constant b.

The boundary condition (10.4-45b) takes the following explicit form for this case of Langmuir isotherm:

$$\zeta = 1; \quad x\big|_1 = \frac{C_{\mu s} \dfrac{b_0 C_0 y\, \phi(\theta)}{1 + b_0 C_0 y\, \phi(\theta)}}{C_{\mu 0}} \qquad (10.4\text{-}49)$$

where

$$\phi(\theta) = \exp\left[-\left(\frac{Q}{R_g T_0}\right)\left(\frac{\theta}{1+\theta}\right)\right] \qquad (10.4\text{-}50)$$

$$C_{\mu 0} = C_{\mu s} \frac{b_0 C_0}{1 + b_0 C_0} \qquad (10.4\text{-}51)$$

The following parameters are used as the base case in the numerical simulation.

Pellet radius, R	$= 0.1$ cm
Zeolite radius, R_μ	$= 0.0001$ cm
Porosity, ε	$= 0.33$
Bulk concentration, C_b	$= 1 \times 10^{-6}$ mole/cc
Initial concentration, C_i	$= 0$
Reference concentration, C_0	$= 1 \times 10^{-6}$ mole/cc
Bulk temperature, T_b	$= 300$ K
Initial temperature, T_i	$= 300$ K
Reference temperature, T_0	$= 300$ K
Affinity at T_0, b_0	$= 1 \times 10^6$ cc/mole
Saturation capacity, $C_{\mu s}$	$= 5 \times 10^{-3}$ mole/cc
Pore diffusivity at T_0, D_{p0}	$= 0.02$ cm²/sec
Intracrystalline diffusivity at T_0, $D_{\mu 0}^0$	$= 1 \times 10^{-10}$ cm²/sec
Heat of adsorption, Q	$= 30,000$ Joule/mole
Activation energy for intracrystalline diffusion, E_μ	$= 15,000$ Joule/mole
Biot number for mass transfer, Bi	$= \infty$
Volumetric heat capacity, $\langle \rho C_p \rangle$	$= 1$ Joule/cc/K
Exponent α for pore diffusion	$= 1$
Heat transfer number, $LeBi$	$= 5$

The effect of the heat transfer number, LeBi, is shown in Figure 10.4-4. The fractional uptake exhibits a two stage uptake and the kink in the fractional uptake occurs at the time at which the particle temperature is maximum.

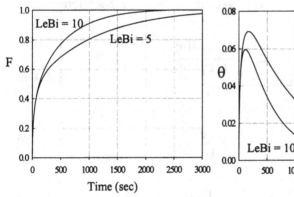
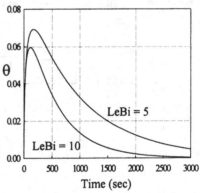

Figure 10.4-4a: Plot of the fractional uptake versus time

Figure 10.4-4b: Plot of the non-dimensional temperature versus time

The effect of crystal radius is shown in Figure 10.4-5 with the crystal radii being 1, 5 and 20 micron. The values of the parameter γ corresponding to these radii are 25, 1.02 and 0.0635. These values indicate that macropore diffusion control in zeolite pellet with crystal size of 1 micron, micropore diffusion control in zeolite pellet with crystal size of 20 micron, and both diffusions control in zeolite pellet with crystal size of 5 micron.

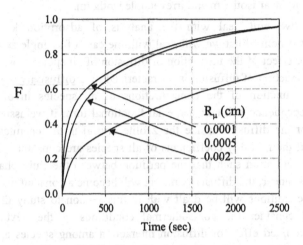

Figure 10.4-5a: Plot of the fractional uptake versus time

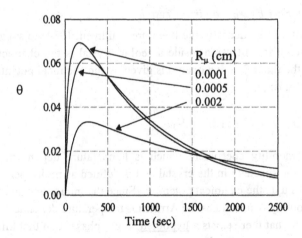

Figure 10.4-5b: Plot of the non-dimensional temperature versus time

10.5 Multicomponent Adsorption in an Isothermal Crystal

In the last sections we have addressed single component system for zeolite particle. Various aspects affecting the overall uptake have been dealt with such as

(a) isothermal and nonisothermal conditions
(b) micropore diffusion control, macropore diffusion control and a combination of them.
(c) linear, nonlinear isotherm and irreversible isotherm.

In this section we will deal with the analysis of adsorption kinetics of a multicomponent system. First we will deal with the case of a single zeolite crystal to investigate the effect of the interaction of diffusion of all species inside a zeolite crystal. This interaction of diffusion is characterized by a <u>diffusivity matrix</u>, which is in general a function of the concentrations of all species involved. This concentration dependence will take a special functional form if we assume that the driving force for the diffusion inside the zeolite crystal is the chemical potential gradient and that the mobility coefficients of all species are constant. Only in the limit of low concentration such that the partition between the fluid phase and the adsorbed phase is linear, the diffusivity matrix will become a constant matrix.

Isothermal conditions will be dealt with in this section to study the diffusion effect and will consider the nonisothermal conditions in the next section to investigate the coupled effect of diffusion interaction among species and the heat release.

10.5.1 Diffusion Flux Expression in a Crystal

There is sufficient evidence in the literature (Ruthven, 1984) to suggest that the proper driving force for diffusion inside a zeolite crystal is the chemical potential gradient. Thus, the flux of a component i is given by the chemical potential gradient of that species, that is:

$$J_{\mu,i} = -L_i C_{\mu i} \frac{\partial \mu_i}{\partial z} \tag{10.5-1}$$

where L is the mobility coefficient which is temperature dependent, $C_{\mu i}$ is the concentration of the species i in the crystal and is defined as moles per unit volume of the crystal, and μ_i is the chemical potential. Since the intracrystalline diffusion is activated, the mobility constant has the Arrhenius temperature dependent form.

Let us assume that there exists a <u>hypothetical gas phase</u> such that this gas phase is in equilibrium with the adsorbed phase within the crystal, that is the adsorbed phase chemical potential is the same as the chemical potential of that hypothetical gas phase. We have:

$$\mu_i = \mu_{G,i} = \mu_i^0 + R_g T \ln p_i \qquad (10.5\text{-}2)$$

where p_i is the hypothetical partial pressure of the species i.

We know that for a given set of partial pressures ($\underline{p} = \{p_1, p_2, ..., p_n\}$) where "n" is the number of component in the mixture, there will be a set of adsorbed concentrations which is equilibrium with the gas phase (Chapter 5), that is:

$$C_{\mu,i} = f_i(\underline{p}) \qquad (10.5\text{-}3a)$$

for i = 1, 2, 3,..., n, or written in a more compact vector form, we have:

$$\underline{C}_\mu = \underline{f}(\underline{p}) \qquad (10.5\text{-}3b)$$

Inversely, for a given set of the adsorbed phase concentrations ($\underline{C}_\mu = \{C_{\mu 1}, C_{\mu 2}, ..., C_{\mu,n}\}$), there will also exist a set of partial pressures such that the two phases are in equilibrium with each other, that is:

$$p_i = g_i(\underline{C}_\mu) \qquad (10.5\text{-}4a)$$

for i = 1, 2, 3, ..., n, or in a compact vector form, it is

$$\underline{p} = \underline{g}(\underline{C}_\mu) \qquad (10.5\text{-}4b)$$

Substituting eq. (10.5-2) into eq.(10.5-1), we get the following expression for the flux of the species "i" written in terms of the gradient of the hypothetical pressure p_i:

$$J_{\mu,i} = -(L_i R_g T) C_{\mu i} \frac{\partial \ln p_i}{\partial z} \qquad (10.5\text{-}5)$$

Written in terms of this hypothetical pressure, eq. (10.5-5) does not have direct application as we do not know the hypothetical pressure directly but rather we have to solve for them from eq. (10.5-4). It is desirable, however, that we express the flux equation in terms of the adsorbed concentration as this is known from the solution of mass balance equations in a crystal. Now that the partial pressure p_i is a function of the adsorbed concentrations of all species (eq. 10.5-4), we apply the chain rule of differentiation to get:

$$\frac{\partial \ln p_i}{\partial z} = \frac{1}{p_i}\frac{\partial p_i}{\partial z} = \frac{1}{p_i}\sum_{j=1}^{n}\frac{\partial p_i}{\partial C_{\mu,j}}\frac{\partial C_{\mu,j}}{\partial z} \qquad (10.5\text{-}6)$$

where $\partial C_{\mu,j}/\partial p_i$ is simply the slope of the equilibrium isotherm of the species j with respect to the partial pressure of the species i.

Substituting eq.(10.5-6) into the flux equation (10.5-5), we get:

$$J_{\mu,i} = -D_{\mu,i}^{0}\sum_{j=1}^{n}C_{\mu i}\frac{\partial \ln p_i}{\partial C_{\mu,j}}\frac{\partial C_{\mu,j}}{\partial z} \qquad (10.5\text{-}7a)$$

for i = 1, 2, 3, ..., n, where $D_{\mu,i}^{0}$ is the corrected diffusivity of the component i, defined as:

$$D_{\mu,i}^{0} = L_i R_g T \qquad (10.5\text{-}7b)$$

Thus, if we define the following vectors and the diffusivity matrix

$$\underline{J}_\mu = \begin{bmatrix} J_{\mu,1} \\ J_{\mu,2} \\ \vdots \\ J_{\mu,n} \end{bmatrix}; \quad \underline{C}_\mu = \begin{bmatrix} C_{\mu,1} \\ C_{\mu,2} \\ \vdots \\ C_{\mu,n} \end{bmatrix}; \quad \underline{\underline{D}}_\mu = \left\{D_{\mu,ij}(\underline{C}_\mu) = D_{\mu,i}^{0}\, H_{ij}(\underline{C}_\mu)\right\} \quad (10.5\text{-}8a)$$

where

$$H_{ij}(\underline{C}_\mu) = C_{\mu,i}\frac{\partial \ln p_i}{\partial C_{\mu,j}} \qquad (10.5\text{-}8b)$$

then eq.(10.5-7) can be put in a simple vector-matrix format as follows:

$$\underline{J}_\mu = -\underline{\underline{D}}_\mu \frac{\partial \underline{C}_\mu}{\partial z} \qquad (10.5\text{-}9)$$

The diffusivity matrix is a function of the concentrations of all species. Eq. (10.5-9) is the constitutive flux equation. The explicit functional form of the diffusivity matrix in terms of concentration depends on the choice of the adsorption isotherm. What we shall do in the next section is to show this form for the case of the extended Langmuir isotherm.

10.5.1.1 Extended Langmuir Isotherm

To see the explicit form of the diffusion matrix in terms of concentration, we take an example of the extended Langmuir equation:

$$C_{\mu,i} = C_{\mu s,i} \frac{b_i p_i}{1 + \sum_{j=1}^{n} b_j p_j} \qquad (10.5\text{-}10)$$

Eq.(10.5-10) relates the concentration of the adsorbed species "i" in terms of the set of partial pressure ($\underline{p} = \{p_1, p_2, ..., p_n\}$). Inversely, for a given set of the concentrations of the adsorbed species ($\underline{C}_\mu = \{C_{\mu 1}, C_{\mu 2}, ..., C_{\mu,n}\}$), the partial pressure of the species "i" can be written in terms of this set as:

$$p_i = \frac{1}{b_i} \frac{(C_{\mu,i}/C_{\mu s,i})}{1 - \sum_{j=1}^{n} \frac{C_{\mu,j}}{C_{\mu s,j}}} \qquad (10.5\text{-}11)$$

Knowing this relationship between the partial pressure p_i and the set of adsorbed concentrations, we can evaluate the elements of the diffusivity matrix (eq. 10.5-8) and the final result is the following equation:

$$\underline{\underline{D}}_\mu = \begin{cases} D_{\mu,i}^0 \dfrac{C_{\mu,i}}{C_{\mu s,j}\left[1 - \sum_{k=1}^{n}\left(\dfrac{C_{\mu,k}}{C_{\mu s,k}}\right)\right]} & \text{for } i \neq j \\[2em] D_{\mu,i}^0 \left[1 + \dfrac{C_{\mu,i}/C_{\mu s,i}}{1 - \sum_{k=1}^{n}\dfrac{C_{\mu,k}}{C_{\mu s,k}}}\right] & \text{for } i = j \end{cases} \qquad (10.5\text{-}12)$$

We now illustrate below an example about the relative magnitude between the elements of the above diffusion matrix.

Example 10.5-1: *Diffusivity matrix for a binary system following extended Langmuir equation*

Let us consider a binary system (n = 2) and equal saturation capacities of the two diffusing species (that is $C_{\mu s,1} = C_{\mu s,2}$), the binary diffusivity matrix of eq.(10.5-12) will become:

$$\underline{\underline{D}}_\mu = \begin{bmatrix} D^0_{\mu,1}\left[1+\dfrac{\left(C_{\mu,1}/C_{\mu s}\right)}{1-\dfrac{C_{\mu,1}}{C_{\mu s}}-\dfrac{C_{\mu,2}}{C_{\mu s}}}\right] & D^0_{\mu,1}\dfrac{\left(C_{\mu,1}/C_{\mu s}\right)}{\left(1-\dfrac{C_{\mu,1}}{C_{\mu s}}-\dfrac{C_{\mu,2}}{C_{\mu s}}\right)} \\[2em] D^0_{\mu,2}\dfrac{\left(C_{\mu,2}/C_{\mu s}\right)}{\left(1-\dfrac{C_{\mu,1}}{C_{\mu s}}-\dfrac{C_{\mu,2}}{C_{\mu s}}\right)} & D^0_{\mu,2}\left[1+\dfrac{\left(C_{\mu,2}/C_{\mu s}\right)}{1-\dfrac{C_{\mu,1}}{C_{\mu s}}-\dfrac{C_{\mu,2}}{C_{\mu s}}}\right] \end{bmatrix} \quad (10.5\text{-}13)$$

Let the species 1 be the stronger adsorbing species, which has the property that it has higher capacity but possesses lower mobility. Let us use the following values to demonstrate the magnitude of the diffusivity matrix

$$\frac{D^0_{\mu,2}}{D^0_{\mu,1}} = 10, \quad \frac{C_{\mu,1}}{C_{\mu s}} = 0.5, \quad \frac{C_{\mu,2}}{C_{\mu s}} = 0.25$$

Substitute these values into eq. (10.5-13), we get the following diffusivity matrix:

$$\underline{\underline{D}}_\mu = D^0_{\mu,1}\begin{bmatrix} 3 & 2 \\ 10 & 20 \end{bmatrix}$$

compared to the diffusivity matrix at zero loading of

$$\lim_{\underline{C}_\mu \to \underline{0}} \underline{\underline{D}}_\mu = D^0_{\mu,1}\begin{bmatrix} 1 & 0 \\ 0 & 10 \end{bmatrix}$$

We see that at finite loadings, the interaction between the diffusing species can have a substantial influence on the diffusion rate. We will see later that it is this interaction that causes some interesting behaviour in the adsorption kinetics, such as the overshoot behaviour when two species are simultaneously adsorbing into a zeolite crystal.

We have obtained the necessary flux equation in the crystal, and now turn to deriving the mass balance equation.

10.5.2 The Mass Balance Equation in a Zeolite Crystal

The mass balance equation for the species "i" inside the crystal is:

$$\frac{\partial C_{\mu,i}}{\partial t} = \frac{1}{z^s}\frac{\partial}{\partial z}\left[z^s \sum_{j=1}^{n} D_{\mu,ij}(\underline{C}_\mu)\frac{\partial C_{\mu,j}}{\partial z}\right] \qquad (10.5\text{-}14)$$

for $i = 1, 2, ..., n$. We have assumed the one dimensional diffusion inside the crystal. Here s is the shape factor with s = 0, 1, 2 for slab, cylinder and sphere, respectively. Zeolites such as mordenite, ZSM-5 and silicalite have slab-like geometry with respect to the diffusion direction, thus s = 0 is applicable in those cases, while zeolites X, Y and A will have diffusion geometry similar to sphere, and hence s = 2 is applicable.

The boundary conditions for the mass balance equation (10.5-14) are:

$$z = 0; \quad \frac{\partial C_{\mu,i}}{\partial z} = 0 \qquad (10.5\text{-}15a)$$

$$z = L; \quad C_{\mu,i} = C_{\mu b,i} = f_i(\underline{p}_b, T_b) \qquad (10.5\text{-}15b)$$

where f_i is the multicomponent isotherm of the species i, $\underline{p}_b$ is the partial pressure vector of n components in the gas surrounding the crystal, L is the characteristic length of the zeolite, and T_b is the bulk temperature.

The initial condition of this problem is:

$$t = 0; \quad C_{\mu,i} = C_{\mu i,i} = f_i(\underline{p}_i, T_i) \qquad (10.5\text{-}16)$$

for $i = 1, 2, 3, ..., n$, where the initial pressures and temperature can be different from the bulk pressures and temperature.

Eqs.(10.5-14) to (10.5-16) completely define the behaviour inside a crystal once the temperature and the partial pressures in the gas phase are given. This set of equations is nonlinear (due to the diffusivity matrix) and therefore the only tool to solve it is the numerical method. Here we use the method of orthogonal collocation, and to do so the spatial domain needs to be normalised because the orthogonal collocation method is developed for a normalised domain (0, 1). We will present in the next section the non-dimensionalisation process and the dimensionless equations.

10.5.2.1 Non-dimensionalization

The choice of the characteristic length is L, and the choice of the characteristic time is defined as the ratio of the square of a characteristic length to a characteristic diffusivity $D_{\mu T}$. This characteristic diffusivity could be chosen as the sum of all corrected diffusivities.

$$t_0 = \frac{L^2}{D_{\mu T}}; \qquad D_{\mu T} = \sum_{i=1}^{n} D^0_{\mu,i} \qquad (10.5\text{-}17)$$

Thus, by defining the following nondimensional variables and parameters

$$\eta = \frac{z}{L}; \qquad \tau = \frac{t}{t_0}; \qquad \delta_i = \frac{D^0_{\mu,i}}{D_{\mu T}}; \qquad H_{ij}(\underline{C}_\mu) = C_{\mu,i} \frac{\partial \ln p_i}{\partial C_{\mu,j}} \qquad (10.5\text{-}18)$$

we obtain the following non-dimensional equation and boundary conditions:

Equation	Boundary conditions	
$\dfrac{\partial C_{\mu,i}}{\partial \tau} = \dfrac{\delta_i}{\eta^s} \dfrac{\partial}{\partial \eta}\left[\eta^s \sum_{j=1}^{n} H_{ij}(\underline{C}_\mu) \dfrac{\partial C_{\mu,j}}{\partial \eta}\right]$ (10.5-19)	$\eta = 0;\quad \dfrac{\partial C_{\mu,i}}{\partial \eta} = 0$	(10.5-20)
	$\eta = 1;\quad C_{\mu,i} = C_{\mu b,i}$	

and the initial condition:

$$\tau = 0; \qquad C_{\mu,i} = C_{\mu i,i} \qquad (10.5\text{-}21)$$

A species undergoing adsorption or desorption depends on the adsorbed amount at time 0 and that at time infinity. This means that

1. If $C_{\mu i,i} < C_{\mu b,i}$, then the species i is undergoing a net adsorption
2. If $C_{\mu i,i} > C_{\mu b,i}$, then the species i is undergoing a net desorption

The above equation can be cast into a more elegant form of vector and matrix format, as shown below:

Equation	Boundary conditions	Initial condition	Eq. #
$\dfrac{\partial \underline{C}_\mu}{\partial \tau} = \dfrac{1}{\eta^s} \dfrac{\partial}{\partial z}\left[\eta^s \underline{\underline{\chi}}(\underline{C}_\mu) \dfrac{\partial \underline{C}_\mu}{\partial \eta}\right]$	$\eta = 0;\ \dfrac{\partial \underline{C}_\mu}{\partial \eta} = \underline{0}$ $\eta = 1;\ \underline{C}_\mu = \underline{C}_{\mu b}$	$\tau = 0;\ \underline{C}_\mu = \underline{C}_{\mu i}$	(10.5-22)

where

$$\underline{\underline{\chi}}(\underline{C}_\mu) = \{\chi_{ij} = \delta_i \, H_{ij}(\underline{C}_\mu)\} \tag{10.5-23}$$

Solving this set of equations will give the concentration $\underline{C}_\mu$ as a function of time as well as distance inside the zeolite crystal. The quantity of interest is the fractional uptake, which is defined as the amount uptaken by the zeolite crystal from time $t = 0$ up to time t divided by the maximum amount taken by the crystal, that is:

$$F_i = \frac{\langle C_{\mu,i} \rangle - C_{\mu i,i}}{C_{\mu b,i} - C_{\mu i,i}} \tag{10.5-24}$$

for $i = 1, 2, 3, \ldots, n$, where $\langle . \rangle$ is the volumetric average concentration, defined as:

$$\langle \underline{C}_\mu \rangle = (1+s) \int_0^1 \eta^s \, \underline{C}_\mu(\eta, \tau) \, d\eta \tag{10.5-25}$$

The set of equations (10.5-22) is quite readily handled by the numerical method of orthogonal collocation. Basically, the coupled partial differential equations (eq. 10.5-22) are discretized in the sense that the spatial domain η is discretized into N collocation points, and the governing equation is valid at these points. In this way, the coupled partial differential equations will become coupled ordinary differential equations in terms of concentrations at those points. These resulting coupled ODEs are function of time and are solved by any standard ODE solver. Details of the orthogonal collocation analysis are given in Appendix 10.5, and a computer code ADSORB3A is provided with this book for the readers to learn interactively and explore the simulation of this model.

10.5.2.2 Extended Langmuir Isotherm

The computation of the non-dimensional governing equations is carried out after we specify the functional form for the multicomponent isotherm. We shall do it here with the extended Langmuir isotherm (eq. 10.5-10).

For the case where the extended Langmuir isotherm can describe the multicomponent equilibria, the non-dimensional diffusivity matrix is given by:

$$\underline{\underline{\chi}}(\underline{C}_\mu) = \begin{bmatrix} \delta_1 \left[1 + \dfrac{(C_{\mu,1}/C_{\mu s})}{1 - \dfrac{C_{\mu,1}}{C_{\mu s}} - \dfrac{C_{\mu,2}}{C_{\mu s}}}\right] & \delta_1 \dfrac{C_{\mu,1}/C_{\mu s}}{\left(1 - \dfrac{C_{\mu,1}}{C_{\mu s}} - \dfrac{C_{\mu,2}}{C_{\mu s}}\right)} \\ \\ \delta_2 \dfrac{C_{\mu,2}/C_{\mu s}}{\left(1 - \dfrac{C_{\mu,1}}{C_{\mu s}} - \dfrac{C_{\mu,2}}{C_{\mu s}}\right)} & \delta_2 \left[1 + \dfrac{C_{\mu,2}/C_{\mu s}}{\left(1 - \dfrac{C_{\mu,1}}{C_{\mu s}} - \dfrac{C_{\mu,2}}{C_{\mu s}}\right)}\right] \end{bmatrix} \quad (10.5\text{-}26)$$

Example 10.5-2: *Two component diffusion in a zeolite*

To study the multicomponent effect in mass transfer, we take the following base case of two components.

Micro-particle geometry, s	= 2
Micro-particle radius, L	= 1 micron = 0.0001 cm
Corrected diffusivity, $\underline{\underline{D}}^0_\mu$	= [1×10^{-9} 1×10^{-10}] cm²/sec
Initial pressure, $\underline{p}_i$	= [0 0] kPa
Bulk pressure, $\underline{p}_b$	= [10 10] kPa
Adsorption affinity, $\underline{b}$	= [0.1 0.3] kPa⁻¹
Saturation capacity, $\underline{C}_{\mu s}$	= [5×10^{-3} 5×10^{-3}] mole/cc

With the choice of the bulk pressure and the adsorption affinity, the system is called a high-affinity system because bp >> 1. The simulation is carried out using the code ADSORB3A. Figure 10.5-1 shows the fractional uptake of the two components. We note the overshoot of the component 1 because the component 1 is a weakly adsorbing species and has a higher mobility. Its occupation of adsorption space is displaced by the slowly moving component 2, resulting in the overshoot in the fractional uptake.

By using the affinity of [0.01 0.03] instead of [0.1 0.3] kPa⁻¹, we have a low affinity system and Figure 10.5-1 also shows the fractional uptake of this low affinity system. The overshoot phenomenon is again observed but the degree of overshoot is not as significant as that in the high affinity system.

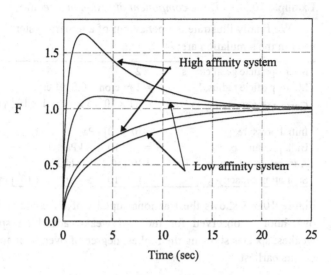

Figure 10.5-1: Plots of the fractional uptake versus time

To study the effect of the particle size, we compare the base case with a case of L = 2 micron. The results are shown in Figure 10.5-2 where we see that the larger crystal size system has a slower time scale of adsorption but the two cases have the same degree of overshoot.

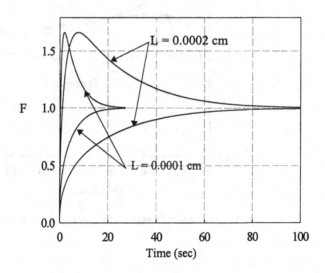

Figure 10.5-2: Effect of size on the fractional uptake versus time

Example 10.5-3: *Three component diffusion in a zeolite*

We finally illustrate the behaviour of a ternary system. The parameters used in the simulation are:

Micro-particle geometry, s	= 2
Micro-particle radius, L	= 1 micron = 0.0001 cm
Corrected diffusivity, $\underline{\underline{D}}_\mu^0$	= [1 × 10^{-9} 5 × 10^{-10} 1 × 10^{-10}] cm²/sec
Initial pressure, $\underline{p}_i$	= [0 0 0] kPa
Bulk pressure, $\underline{p}_b$	= [10 10 10] kPa
Adsorption affinity, $\underline{b}$	= [0.1 0.2 0.3] kPa^{-1}
Saturation capacity, $\underline{C}_{\mu s}$	= [5 × 10^{-3} 5 × 10^{-3} 5 × 10^{-3}] mole/cc

Figure 10.5-3 shows the fractional uptake of the three components. The overshoot is observed for the two weaker-adsorbing species with the weakest species showing the highest degree of overshoot and the overshoot occurs earliest.

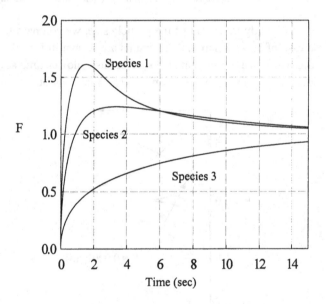

Figure 10.5-3: Plots of the fractional uptake versus time

10.6 Multicomponent adsorption in a crystal: Nonisothermal

In the last section, we dealt with the case of isothermal diffusion in a zeolite crystal. Isothermality in a zeolite crystal is usually almost assured because of the small size of the crystal (that is large surface area for heat transfer per unit volume of crystal). This isothermal condition holds as long as the crystals are separated from each other, that is there is no agglomoration of zeolite crystals. If zeolite crystals are bound in an agglomerate, the heat released by the adsorption process or taken up by the desorption process will be accumulated by the interior of the zeolite agglomorate. This agglomerate is normally of a size such that the heat transfer from the interior to the bulk is limited by the film surrounding the agglomerate due to the low ratio of the surface area for heat transfer to the agglomerate volume. This case is usually occurred in practice, and therefore to correctly describe the mass transfer kinetics, one must allow for the contribution of the heat transfer as it will affect significantly the way how the mass transfer would occur. The extent of this heat transfer contribution will depend on

1. The stirring of the surrounding environment
2. The size of the zeolite agglomerate
3. The heat of adsorption
4. The rate of diffusion into the zeolite interior which is affected by the diffusivity and the size of the crystal

10.6.1 Flux Expression in a Crystal

In this section we will derive the general expression for the constitutive flux equation written in terms of not just the concentration gradients of all species but also on the temperature gradient in the zeolite crystal. We again assume that the driving force for diffusion is the chemical potential gradient (Ruthven, 1984). Thus, the flux of a component i is given by the chemical potential gradient of that species, that is:

$$J_{\mu,i} = -L_i C_{\mu i} \frac{\partial \mu_i}{\partial z} \qquad (10.6\text{-}1)$$

Like the last sections, let us assume that there exists a hypothetical gas phase such that this gas phase is in equilibrium with the adsorbed phase at the point z, that is the adsorbed phase chemical potential is the same as the chemical potential in the gas phase. We have:

$$\mu_i = \mu_{GAS,i} = \mu_i^0 + R_g T \ln p_i \qquad (10.6\text{-}2)$$

where p_i is the hypothetical partial pressure of the species i, and T is the temperature at the point z.

For a given set of partial pressures ($\underline{p} = \{p_1, p_2, ..., p_n\}$) and temperature, there will be a set of adsorbed concentration which is equilibrium with the gas phase, that is written in a compact vector form as:

$$\underline{C}_\mu = \underline{f}(\underline{p}, T) \qquad (10.6\text{-}3)$$

Inversely, for a given set of the adsorbed phase concentration ($\underline{C}_\mu = \{C_{\mu 1}, C_{\mu 2}, ..., C_{\mu,n}\}$), there will also exist a set of partial pressure such that the two phases are in equilibrium with each other, that is:

$$\underline{p} = \underline{g}(\underline{C}_\mu, T) \qquad (10.6\text{-}4)$$

Substituting eq. (10.6-2) into eq.(10.6-1), we get the following expression for the flux of the species "i" written in terms of the gradient of the hypothetical pressure p_i:

$$J_{\mu,i} = -(L_i R_g T) C_{\mu i} \frac{\partial \ln p_i}{\partial z} \qquad (10.6\text{-}5)$$

It is desirable, however, that we express the flux equation in terms of the adsorbed concentration and temperature as p_i in the above equation is only the partial pressure of the hypothetical gas phase. Noting that the partial pressure p_i is a function of the adsorbed concentration of all species as well as temperature, we apply the chain rule of differentiation:

$$\frac{\partial \ln p_i}{\partial z} = \frac{1}{p_i} \frac{\partial p_i}{\partial z} = \frac{1}{p_i} \left(\sum_{j=1}^{n} \frac{\partial p_i}{\partial C_{\mu,j}} \frac{\partial C_{\mu,j}}{\partial z} + \frac{\partial p_i}{\partial T} \frac{\partial T}{\partial z} \right) \qquad (10.6\text{-}6)$$

Substituting eq.(10.6-6) into eq.(10.6-5), we get the following constitutive flux equation:

$$J_{\mu,i} = -D^0_{\mu,i} \left[\sum_{j=1}^{n} C_{\mu i} \frac{\partial \ln p_i}{\partial C_{\mu,j}} \frac{\partial C_{\mu,j}}{\partial z} + C_{\mu,i} \frac{\partial \ln p_i}{\partial T} \frac{\partial T}{\partial z} \right] \qquad (10.6\text{-}7)$$

for i = 1, 2, 3, ..., n, where the corrected diffusivity is:

$$D^0_{\mu,i} = L_i R_g T \qquad (10.6\text{-}8a)$$

of which its temperature dependence takes the usual Arrhenius form:

$$D_{\mu,i}^0(T) = D_{\mu\infty,i}^0 \exp\left(-\frac{E_{\mu,i}}{R_g T}\right) = D_{\mu 0,i}^0 \exp\left[\frac{E_{\mu,i}}{R_g T_0}\left(1-\frac{T_0}{T}\right)\right] \quad (10.6\text{-}8b)$$

where $D_{\mu 0,i}^0$ is the corrected diffusivity of the component i at some reference temperature T_0.

Thus, if we define the following vectors and the diffusivity matrix

$$\underline{J}_\mu = \begin{bmatrix} J_{\mu,1} \\ J_{\mu,2} \\ \vdots \\ J_{\mu,n} \end{bmatrix}; \quad \underline{C}_\mu = \begin{bmatrix} C_{\mu,1} \\ C_{\mu,2} \\ \vdots \\ C_{\mu,n} \end{bmatrix} \quad (10.6\text{-}9)$$

$$\underline{\underline{D}}_\mu = \left\{ D_{\mu,ij} = D_{\mu,i}^0(T) \times H_{ij}(\underline{C}_\mu, T) = D_{\mu,i}^0(T)\, C_{\mu,i} \frac{\partial \ln p_i}{\partial C_{\mu,j}} \right\} \quad (10.6\text{-}10a)$$

$$\underline{h} = \begin{bmatrix} D_{\mu,1}^0 C_{\mu,1} \partial \ln p_1 / \partial T \\ D_{\mu,2}^0 C_{\mu,2} \partial \ln p_2 / \partial T \\ \vdots \\ D_{\mu,n}^0 C_{\mu,n} \partial \ln p_n / \partial T \end{bmatrix} \quad (10.6\text{-}10b)$$

then eq.(10.6-7) can be put in a simple vector-matrix format:

$$\underline{J}_\mu = -\underline{\underline{D}}_\mu \frac{\partial \underline{C}_\mu}{\partial z} - \frac{\partial T}{\partial z} \underline{h} \quad (10.6\text{-}11)$$

The diffusivity matrix (eq. 10.6-10a) is a function of the adsorbed concentration as well as temperature, and the above constitutive flux is written in terms of concentration gradients as well as temperature gradient.

Example 10.6-1: *Temperature dependence form for the extended Langmuir isotherm*

For the multicomponent system satisfying the extended Langmuir isotherm (eq. 10.5-10), the affinity b_i has the following temperature dependent form

$$b_i(T) = b_{\infty,i} \exp\left(\frac{Q_i}{R_g T}\right) = b_{0,i} \exp\left[-\frac{Q_i}{R_g T_0}\left(1-\frac{T_0}{T}\right)\right] \quad (10.6\text{-}12)$$

where $b_{\infty,i}$ is the affinity of the species i at infinite temperature, $b_{0,i}$ is that at some reference temperature T_0, and Q_i is the heat of adsorption of species i.

The diffusivity matrix $\underline{\underline{D}}_\mu$ is given as in eq. (10.5-12), and the vector coefficient of the temperature gradient is

$$\underline{h} = \left\{ h_i = D^o_{\mu,i}(T) \times C_{\mu,i} \left(\frac{Q_i}{R_g T^2} \right) \right\} \tag{10.6-13}$$

10.6.2 The Coupled Mass and Heat Balance Equations

We have obtained the necessary flux equation in the crystal, and now turn to deriving the mass balance equation.

Since the zeolite crystal is now nonisothermal, we must set up the mass balance as well as heat balance equations. First, the mass balance equation for the species "i" inside the crystal is:

$$\frac{\partial C_{\mu,i}}{\partial t} = \frac{1}{z^s} \frac{\partial}{\partial z} \left\{ z^s \left[\sum_{j=1}^{n} D_{\mu,ij}(\underline{C}_\mu, T) \frac{\partial C_{\mu,j}}{\partial z} + h_i \frac{\partial T}{\partial z} \right] \right\} \tag{10.6-14a}$$

for i = 1, 2, ..., n, where n is the number of component. Usually the temperature gradient within the zeolite crystal can be neglected because the thermal conductivity of the crystal is larger than that of the gas surrounding the crystal. In such cases all the heat transfer resistances are located in the thin film surrounding the agglomerate, and the mass balance equation is simply:

$$\frac{\partial C_{\mu,i}}{\partial t} = \frac{1}{z^s} \frac{\partial}{\partial z} \left[z^s \sum_{j=1}^{n} D_{\mu,ij}(\underline{C}_\mu, T) \frac{\partial C_{\mu,j}}{\partial z} \right] \tag{10.6-14b}$$

where $\underline{\underline{D}}_\mu$ is the diffusivity matrix evaluated at the instant condition of the crystal, that is the adsorbed concentration $\underline{C}_\mu$ and the temperature T.

The boundary conditions for the mass balance equation (10.6-14) are:

$$z = 0; \quad \frac{\partial C_{\mu,i}}{\partial z} = 0 \tag{10.6-15a}$$

$$z = L; \quad C_{\mu,i} = f_i(\underline{p}_b, T) \tag{10.6-15b}$$

where f_i is the multicomponent isotherm of the species i, $\underline{p}_b$ is the partial pressure vector of the surrounding, and L is the radius of the zeolite crystal (or half length if the crystal is a slab). The surface condition (eq. 10.6-15b) is no longer a constant as is the case for the isothermal situation dealt with in Section 10.5 because in this case the temperature varies with respect to time during the course of adsorption or desorption.

The initial condition of this problem is:

$$t = 0; \quad C_{\mu,i} = C_{\mu i, i} = f_i\left(\underline{p}_i, T_i\right) \qquad (10.6\text{-}16)$$

for i = 1, 2, 3, ..., n.

The heat balance equation is obtained by carrying out the energy balance around the zeolite agglomerate, and we obtain the following equation:

$$\langle \rho C_p \rangle \frac{dT}{dt} = (1-\varepsilon)\sum_{i=1}^{n} Q_i \frac{d\langle C_{\mu,i}\rangle}{dt} - a_H h_f(T - T_b) \qquad (10.6\text{-}17)$$

where Q_i is the molar heat of adsorption of the species "i" (Joule/moles adsorbed), h_f is the heat transfer coefficient of a film surrounding the agglomerate, a_H is the heat transfer area per unit volume of the agglomerate, $\langle \rho C_p \rangle$ is the mean heat capacity per unit volume of the zeolite agglomorate and is defined as follows:

$$\langle \rho C_p \rangle = \varepsilon (\rho C_p)_G + (1-\varepsilon)(\rho C_p)_S \qquad (10.6\text{-}18)$$

with ε being the porosity of the agglomerate.

The initial condition of the heat balance is:

$$t=0; \quad T = T_i \qquad (10.6\text{-}19)$$

Eqs.(10.6-14) to (10.6-19) completely define the behaviour inside a nonisothermal zeolite agglomerate where the mass transfer is controlled by the diffusion into the crystal while the heat transfer is controlled by the film surrounding the agglomerate (Figure 10.6-1).

This set of equations is nonlinear and therefore must be solved numerically. We will present in the next section the nondimensionalization and the dimensionless equations.

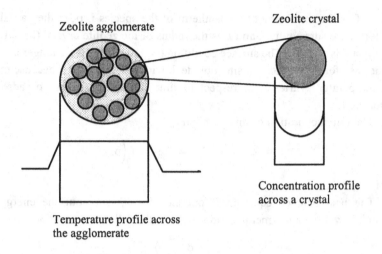

Figure 10.6-1: Temperature and concentration profiles

10.6.2.1 Nondimensionalization

The choice of the characteristic length is L, and the choice of the characteristic time is defined as the ratio of the square of the characteristic length to the characteristic diffusivity defined as follows:

$$t_0 = \frac{L^2}{D_{\mu T}}; \qquad D_{\mu T} = \sum_{i=1}^{n} D^0_{\mu 0,i} \tag{10.6-20}$$

The characteristic diffusivity is the sum of all diffusivities at zero loading evaluated at the reference temperature T_0. Thus, by defining the following nondimensional variables and parameters

$$\eta = \frac{z}{L}; \qquad \tau = \frac{t}{t_0}; \qquad \theta = \frac{T - T_0}{T_0} \tag{10.6-21a}$$

$$\delta_{0,i} = \frac{D^0_{\mu 0,i}}{D_{\mu T}} \qquad \varphi_i(\theta) = \exp\left[\gamma_{\mu,i}\left(\frac{\theta}{1+\theta}\right)\right]; \qquad \gamma_{\mu,i} = \frac{E_{\mu,i}}{R_g T_0} \tag{10.6-21b}$$

$$H_{ij}(\underline{C}_\mu, \theta) = C_{\mu,i}\frac{\partial \ln p_i}{\partial C_{\mu,j}} \qquad \theta_b = \frac{T_b - T_0}{T_0}, \qquad \theta_i = \frac{T_i - T_0}{T_0} \tag{10.6-21d}$$

$$\text{LeBi} = \frac{a_H h_f L^2}{\rho C_p D_{\mu T}}; \quad \beta_i = \frac{(1-\varepsilon)Q_i C_{\mu 0,i}}{\langle \rho C_p \rangle T_0}, \quad C_{\mu 0,i} = f_i(\underline{p}_0, T_0) \quad (10.6\text{-}21f)$$

we obtain the following non-dimensional equations:

$$\frac{\partial C_{\mu,i}}{\partial \tau} = \frac{\delta_{0,i}\, \varphi_i(\theta)}{\eta^s} \frac{\partial}{\partial \eta}\left\{\eta^s \sum_{j=1}^n H_{ij}(\underline{C}_\mu, \theta) \frac{\partial C_{\mu,j}}{\partial \eta}\right\} \quad (10.6\text{-}22a)$$

and

$$\frac{d\theta}{d\tau} = \sum_{i=1}^n \beta_i \frac{1}{C_{\mu 0,i}} \frac{d\langle C_{\mu,i}\rangle}{d\tau} - \text{LeBi}\,(\theta - \theta_b) \quad (10.6\text{-}22b)$$

subject to the following boundary conditions:

$$\eta = 0; \quad \frac{\partial C_{\mu,i}}{\partial \eta} = 0 \quad (10.6\text{-}23a)$$

$$\eta = 1; \quad C_{\mu,i} = f_i(\underline{p}_b, \theta) \quad (10.6\text{-}23b)$$

and the initial condition

$$\tau = 0; \quad C_{\mu,i} = C_{\mu i,i}; \quad \theta = \theta_i \quad (10.6\text{-}24)$$

The boundary condition at the particle exterior surface (eq. 10.6-23b) is not constant because of the variation of the temperature with respect to time.

The above equation can be cast into a more elegant form of vector and matrix format, as shown below:

$$\frac{\partial \underline{C}_\mu}{\partial \tau} = \frac{1}{\eta^s} \frac{\partial}{\partial \eta}\left\{\eta^s \underline{\underline{\chi}}(\underline{C}_\mu, \theta) \frac{\partial \underline{C}_\mu}{\partial \eta}\right\} \quad (10.6\text{-}25a)$$

$$\frac{d\theta}{d\tau} = \underline{\beta}' \bullet \frac{d\langle \underline{C}_\mu \rangle}{d\tau} - \text{LeBi}\,(\theta - \theta_b) \quad (10.6\text{-}25b)$$

$$\eta = 0; \quad \frac{\partial \underline{C}_\mu}{\partial \eta} = \underline{0} \quad (10.6\text{-}25c)$$

$$\eta = 1; \quad \underline{C}_\mu = \underline{f}(\underline{p}_b, \theta) \quad (10.6\text{-}25d)$$

$$\tau = 0; \quad \underline{C}_\mu = \underline{C}_{\mu i}; \quad \theta = \theta_i \quad (10.6\text{-}25e)$$

where • is the vector dot operation and

$$\underline{\beta}' = \left\{ \frac{\beta_i}{C_{\mu 0,i}} \right\} \tag{10.6-26}$$

and

$$\underline{\underline{\chi}} = \left\{ \delta_{0,i} \; \varphi_i(\theta) \; H_{ij} \, (\underline{C}_\mu, \theta) \right\} \tag{10.6-27}$$

Solving this set of equations (eqs. 10.6-25) will give the concentration $\underline{C}_\mu$ as a function of time as well as distance inside the zeolite crystal, and temperature as a function of time. The mean adsorbed concentration is then calculated from

$$\langle \underline{C}_\mu \rangle = (1+s) \int_0^1 \eta^s \; \underline{C}_\mu(\eta, \tau) \; d\eta \tag{10.6-28}$$

and the fractional uptake of each component is calculated from:

$$F_i = \frac{\langle C_{\mu,i} \rangle - C_{\mu i,i}}{C_{\mu b,i} - C_{\mu i,i}} \tag{10.6-29}$$

where

$$C_{\mu i,i} = f_i(\underline{p}_i, \theta_i) \tag{10.6-30a}$$

$$C_{\mu b,i} = f_i(\underline{p}_b, \theta_b) \tag{10.6-30b}$$

The set of equations (10.6-25) is quite readily handled by numerical method of orthogonal collocation. Basically, the coupled partial differential equations (eq. 10.6-25a) are discretized in the sense that the spatial domain is discretized into N collocation points, and the governing equation is valid at these points. In this way, the coupled partial differential equations will become coupled ordinary differential equations in terms of time, and together with the equation for temperature, they can be readily solved by any standard ODE solver. Details of the orthogonal collocation analysis are given in Appendix 10.6.

Example 10.6-2: Extended Langmuir isotherm

We consider the case where the extended Langmuir isotherm describes the multicomponent equilibria:

$$C_{\mu,i} = C_{\mu s,i} \frac{b_i p_i}{1 + \sum_{j=1}^{n} b_j p_j} \tag{10.6-31a}$$

with the adsorption affinity taking the form:

$$b_i = b_{\infty,i} \exp\left(\frac{Q_i}{R_g T}\right) = b_{0,i} \exp\left[-\frac{Q_i}{R_g T_0}\left(1 - \frac{T_0}{T}\right)\right] \qquad (10.6\text{-}31b)$$

where $b_{0,i}$ is the affinity of the component i evaluated at some reference temperature T_0.

The function $H_{ij}(\underline{C}_\mu, \theta)$ takes the form:

$$H_{ij}(\underline{C}_\mu, \theta) = \begin{cases} \dfrac{C_{\mu,i}}{C_{\mu s,j}\left[1 - \sum_{k=1}^{n} \dfrac{C_{\mu,k}}{C_{\mu s,k}}\right]} & \text{for } i \neq j \\[2ex] 1 + \dfrac{C_{\mu,i}/C_{\mu s,i}}{\left[1 - \sum_{k=1}^{n} \dfrac{C_{\mu,k}}{C_{\mu s,k}}\right]} & \text{for } i = j \end{cases} \qquad (10.6\text{-}32)$$

The boundary condition becomes

$$\eta = 1; \quad C_{\mu,i} = C_{\mu s,i} \frac{b_{0,i}\, \phi_i(\theta)\, p_{b,i}}{1 + \sum_{k=1}^{n} b_{0,k}\, \phi_k(\theta)\, p_{b,k}} \qquad (10.6\text{-}33)$$

where

$$\phi_k(\theta) = \exp\left[-\gamma_{Q,i}\left(\frac{\theta}{1+\theta}\right)\right] \qquad (10.6\text{-}34a)$$

$$\gamma_{Q,i} = \frac{Q_i}{R_g T_0} \qquad (10.6\text{-}34b)$$

10.7 Multicomponent Adsorption in a Zeolite Pellet. Nonisothermal

Before we close this chapter out, we would like to present a model of multicomponent adsorption in a zeolite pellet under nonisothermal conditions. Adsorption equilibrium is taken to be non-linear. The pellet is composed of many zeolite crystals, and there are a number of processes occuring when this pellet is exposed to an environment containing nc adsorbates. These processes are:

1. Diffusion in the macropore space (inter-crystalline space)
2. Diffusion in micropores (intra-crystalline space)
3. Adsorption at the pore mouth of the micropore

676 Kinetics

The macropore diffusion of nc adsorbates is described by the Maxwell-Stefan equation as learnt in Chapter 8 (Section 8.8). The micropore diffusion in crystal is activated and is described by eq. (10.6-11), and the adsorption process at the micropore mouth is assumed to be very fast compared to diffusion so that local equilibrium is established at the mouth. Adsorption and desorption of adsorbates are associated with heat release which in turn causes a rise or drop in temperature of the pellet. We shall assume that the thermal conductivity of the pellet is large such that the pellet temperature is uniform and all the heat transfer resistance is located at the thin film surrounding the pellet. How large the pellet temperature will change during the course of adsorption depends on the interplay between the rate of adsorption, the heat of adsorption and the rate of heat dissipation to the surrounding. But the rate of adsorption at any given time depends on the temperature. Thus the mass and heat balances are coupled and therefore their balance equations must be solved simultaneously for the proper description of concentration and temperature evolution.

The mass balance equation in the pellet is:

$$\varepsilon_M \frac{\partial}{\partial t}\left(\frac{p}{R_g T}\right) + (1-\varepsilon_M)\frac{\partial \langle C_\mu \rangle}{\partial t} = \frac{1}{z^s}\frac{\partial}{\partial z}\left(z^s \underline{N}_p\right) \tag{10.7-1}$$

where

$$\underline{N}_p = -\frac{1}{R_g T}\left[\underline{B}(\underline{y},p_T,T)\right]^{-1}\frac{\partial p}{\partial z} - \left(\frac{B_0}{\mu R_g T}\right)\frac{\partial p_T}{\partial z}\left[\underline{B}(\underline{y},p_T,T)\right]^{-1}\underline{A}(T)\,\underline{p} \tag{10.7-2}$$

Here μ_m is the mixture viscosity and is calculated from eq. (8.8-2). The variable $\langle \underline{C}_\mu \rangle$ is the crystal volumetric average concentration of the adsorbed species

$$\langle \underline{C}_\mu \rangle = \frac{(1+s)}{R_\mu^{s_\mu+1}}\int_0^{R_\mu} r^{s_\mu}\,\underline{C}_\mu(r,t)\,dr \tag{10.7-3}$$

The mass balance in the crystal is:

$$\frac{\partial \underline{C}_\mu}{\partial t} = \frac{1}{r^{s_\mu}}\frac{\partial}{\partial r}\left(r^{s_\mu}\underline{\underline{D}}_\mu(\underline{C}_\mu,T)\frac{\partial \underline{C}_\mu}{\partial r}\right) \tag{10.7-3}$$

where $\underline{\underline{D}}_\mu$ is given in eq. (10.6-10a).

Assuming the lumped thermal model, the heat balance equation is:

$$\langle \rho C_p \rangle \frac{dT}{dt} = (1-\varepsilon_M) \underline{Q} \bullet \frac{d\langle\langle \underline{C}_\mu \rangle\rangle}{dt} - a_H h_f (T - T_b) \qquad (10.7\text{-}5)$$

where $\langle \rho C_p \rangle$ is the pellet volumetric heat capacity, a_H is the heat transfer area per unit volume, h_f is the heat transfer coefficient, $\underline{Q}$ is the heat of adsorption and $\langle\langle \underline{C}_\mu \rangle\rangle$ is the pellet volumetric average concentration of the adsorbed species

$$\langle\langle \underline{C}_\mu \rangle\rangle = \frac{(1+s)}{R^{s+1}} \frac{(1+s_\mu)}{R_\mu^{s_\mu+1}} \int_0^R z^s \int_0^{R_\mu} r^{s_\mu} \underline{C}_\mu(r,t) \, dr \, dz \qquad (10.7\text{-}6)$$

The pertinent boundary conditions are:

$$r = 0; \qquad \frac{\partial \underline{C}_\mu}{\partial r} = \underline{0} \qquad (10.7\text{-}7a)$$

$$r = R_\mu; \qquad \underline{C}_\mu = \underline{f}(\underline{p}, T) \text{ (Adsorption isotherm)} \qquad (10.7\text{-}7b)$$

$$z = 0; \qquad \frac{\partial \underline{p}}{\partial z} = \underline{0} \qquad (10.7\text{-}8a)$$

$$z = R; \qquad \underline{p} = \underline{p}_b \qquad (10.7\text{-}8b)$$

The initial condition can take the form, in which we assume that the pellet is initially equilibrated with a set of partial pressure $\underline{p}_i$ and a temperature of T_i:

$$t = 0; \qquad \underline{p} = \underline{p}_i; \quad \underline{C}_\mu = \underline{f}(\underline{p}_i, T_i); \quad T = T_i \qquad (10.7\text{-}9)$$

The set of equations (10-7-1) to (10-7-9) can be effectively solved by a combination of the orthogonal collocation method and the Runge-Kutta method. The procedure of which has been described in Appendices 8.9, 9.6 and 9.7.

10.8 Conclusion

This chapter describes adsorption models for bidispersed solids. These models reflect a special structure of the particle, which is basically composed of many smaller grains or crystals. Diffusion through the inter-grain and intra-grain is the key in the models developed in this chapter. In the next chapter we will address general models to deal with heterogeneous solids where the heterogeneity is desribed by the distribution of interaction energy.

11

Analysis of Adsorption Kinetics in a Heterogeneous Particle

11.1 Introduction

The last two chapters have addressed the adsorption kinetics in homogeneous particle as well as zeolitic (bimodal diffusion) particle. The diffusion process is described by a Fickian type equation or a Maxwell-Stefan type equation. Analysis presented in those chapters have good utility in helping us to understand adsorption kinetics. To better understand the kinetics of a practical solid, we need to address the role of surface heterogeneity in mass transfer. The effect of heterogeneity in equilibria has been discussed in Chapter 6, and in this chapter we will briefly discuss its role in the mass transfer. More details can be found in a review by Do (1997). This is started with a development of constitutive flux equation in the presence of the distribution of energy of interaction, and then we apply it firstly to single component systems and next to multicomponent systems.

11.2 Heterogeneous Diffusion & Sorption Models:

Practical adsorbents are inherently heterogeneous, and therefore to properly account for kinetics in such adsorbents, we need to develop a mathematical model to allow for the energetic heterogeneity. First we will address single component systems to uncover the various features of the heterogeneous model.

11.2.1 Adsorption Isotherm

To deal with the heterogeneity, we assume that the surface has a patchwise topography, that is sites of the same energy of interaction with the adsorbate are grouped together in the same patch. The local isotherm on each patch is assumed to be described by the Langmuir equation. The use of such equation is valid due to the assumption of the same energy in the patch. Let the energy of interaction be E, the Langmuir equation is written as:

$$C_\mu(E) = C_{\mu s}\frac{b(E)C}{1+b(E)C} \qquad (11.2\text{-}1)$$

where $C_{\mu s}$ is the maximum adsorbed concentration, C is the fluid phase concentration and b(E) is the affinity between the sorbent and the adsorbate, which is given by:

$$b(E) = b_\infty e^{E/R_g T} \qquad (11.2\text{-}2)$$

Other forms of the local isotherm can be used to allow for the adsorbate-adsorbate interaction, such as the Fowler-Guggenheim equation:

$$C_\mu(E) = C_{\mu s}\frac{b[E;\theta(E)]\,C}{1+b[E;\theta(E)]\,C} \qquad (11.2\text{-}3a)$$

where

$$b[E;\theta(E)] = b_\infty \exp\left[\frac{E}{R_g T} + \frac{zw}{R_g T}\theta(E)\right] \qquad (11.2\text{-}3b)$$

with θ being the local fractional loading of the patch having the interaction of energy of E, w is the pairwise interaction energy and z is the number of neighboring sites.

11.2.2 Constitutive Flux Equation:

To derive the constitutive flux equation for the heterogeneous adsorbed phase, we proceed as follows. For any given point within the particle (point A in Figure 11.2-1), the chemical potential of the adsorbed phase is denoted as μ_A, which is assumed to be in equilibrium with a hypothetical gas phase having a concentration of C_A. All the patches of the adsorbed phase at that point will have the same chemical potential. At another point B, which is at a distance Δr from the point A, the chemical potential is μ_B. The gas phase concentration which would have the same chemical potential of the adsorbed phase at point B is C_B. Due to a difference in the chemical potential between the two points, the diffusion within the adsorbed phase is possible for example adsorbed molecules in patch E_1 of point A will diffuse to all the patches in point B, shown as arrows in Figure 11.2-1.

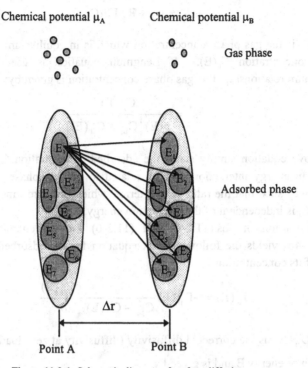

Figure 11.2-1: Schematic diagram of surface diffusion

Let the chemical potential of the point A is greater than that of the point B. The mass transfer from point A to point B along the adsorbed phase contributed by the patch having an interaction energy of E is driven by the difference in the chemical potentials:

$$J_\mu(E) = L(E) C_\mu(E) \frac{\mu_A - \mu_B}{\Delta r} \qquad (11.2\text{-}4a)$$

where $L(E)$ is the mobility constant associated with that patch. Taking the distance between the two points as small as possible, we obtain the following flux equation in differential form:

$$J_\mu(E) = -L(E) C_\mu(E) \frac{\partial \mu}{\partial r} \qquad (11.2\text{-}4b)$$

Assuming a local equilibrium between the fluid and adsorbed phases, the chemical potential of the adsorbed phase is equal to that of the fluid phase:

$$\mu_{ads} = \mu_0 + R_g T \ln C^* \tag{11.2-5}$$

where C^* is the gas phase concentration which is in equilibrium with the adsorbed phase concentration $C_\mu(E)$. If Langmuir equation is used to describe this equilibrium relationship, this gas phase concentration is given by:

$$C^* = \frac{C_\mu(E)}{b(E)\left[C_{\mu s} - C_\mu(E)\right]} \tag{11.2-6}$$

The above equation simply states the adsorbed concentration $C_\mu(E)$ of any patch having an energy interaction of E is related to the gas phase concentration in a special way such that the ratio of two terms, which both are function of interaction energy E, is independent of the interaction energy.

Substitution of eqs.(11.2-5) and (11.2-6) for the chemical potential into eq.(11.2-4b) yields the following flux equation for the adsorbed phase written in terms of its concentration

$$J_\mu(E) = -D_\mu^0(E) \frac{C_{\mu s}}{C_{\mu s} - C_\mu(E)} \frac{\partial C_\mu(E)}{\partial r} \tag{11.2-7}$$

where $D_\mu^0(E)$ is the corrected diffusivity (diffusivity at zero loading conditions) of the patch of energy E and is given by

$$D_\mu^0(E) = R_g T \times L(E) \tag{11.2-8}$$

Since the surface diffusion is activated, this corrected diffusivity has the following Arrhenius form:

$$D_\mu^0(E) = D_{\mu\infty}^0 \exp\left(-\frac{aE}{R_g T}\right) \tag{11.2-9}$$

where $D_{\mu\infty}^0$ is the diffusion coefficient at infinite temperature and zero loading condition. The coefficient "a" in the above equation means that the activation energy for adsorbed phase diffusion is "a" times the energy of interaction between the adsorbate and the adsorbent. In general, such a coefficient "a" is a function of E, but without a fundamental proof of such dependence, we shall take that as a constant. Literature on surface diffusion has shown that that constant a is between 0.33 to 1.

Knowing the flux equation contributed by the patch of energy E, the overall flux of the adsorbed phase is simply the total contribution of all patches, that is it is the integral of the local flux (eq. 11.2-7) over the complete range of the energy distribution

$$\langle J_\mu \rangle = \int_0^\infty J_\mu(E)\, F(E)\, dE = -\int_0^\infty \left[D_\mu^0(E) \frac{C_{\mu s}}{C_{\mu s} - C_\mu(E)} \frac{\partial C_\mu(E)}{\partial r} \right] F(E)\, dE \quad (11.2\text{-}10)$$

We will now address the formulation of a kinetic model for a heterogeneous particle with single component systems first to illustrate the concept of energy distribution, and then logically extend the mathematical formulation to multicomponent systems.

11.3 Formulation of the Model for Single Component Systems

The assumptions of the model for the description of kinetics in a heterogeneous adsorbent are:

- The system is isothermal
- The pore diffusivity and mass transfer coefficient are constant
- The surface diffusivity at zero loading and infinite temperature is constant

Let $C_\mu(E)$ is the adsorbed concentration in the adsorbed phase of the patch having an adsorption energy E. The energy distribution F(E) is defined such that F(E) dE is the fraction of site having energies between E and E+dE. The accumulation of the adsorbed concentration at the position r within the particle is given by:

$$\frac{\partial}{\partial t}\left[\int_0^\infty C_\mu(E)\, F(E)\, dE \right] \quad (11.3\text{-}1)$$

The flux of the adsorbed species contributed by all patches at the position r is:

$$\langle J_\mu \rangle = -\int_0^\infty \frac{D_\mu^0(E)}{\left[1 - C_\mu(E)/C_{\mu s}\right]} \frac{\partial C_\mu(E)}{\partial r} F(E)\, dE \quad (11.3\text{-}2)$$

With the above accumulation of the adsorbed species and the surface flux, the mass balance equation can be written as:

$$\varepsilon \frac{\partial C}{\partial t} + (1-\varepsilon) \frac{\partial}{\partial t} \int_0^\infty C_\mu(E) \, F(E) \, dE = \varepsilon D_p \frac{1}{r^s} \frac{\partial}{\partial r}\left(r^s \frac{\partial C}{\partial r}\right) +$$
$$(1-\varepsilon) \frac{1}{r^s} \frac{\partial}{\partial r} \left\{ r^s \int_0^\infty \frac{D_\mu^0(E)}{\left[1 - C_\mu(E)/C_{\mu s}\right]} \frac{\partial C_\mu(E)}{\partial r} F(E) \, dE \right\} \quad (11.3\text{-}3)$$

where ε is the porosity of the macropore, D_p is the pore diffusivity based on the empty cross sectional area, and s is the shape factor of the particle. The terms in the RHS of eq. (11.3-3) are the flux terms contributed by the free and adsorbed molecules. For highly immobile adsorbed species, the pore volume diffusion dominates. For most adsorption systems involving activated carbons these two terms can be comparable in magnitude because the pore diffusivity is about 100 to 1000 times larger than the surface diffusivity while the concentration of the adsorbed species is about 100 or 1000 times larger than that of the gas phase concentration.

The mass balance equation involves the gas phase and adsorbed phase concentrations. We assume equilibrium is established between the gas and surface phases; hence at any point within the particle, the adsorbed concentration at the patch of sites having an adsorption energy E is related to the gas phase concentration, C, according to eq.(11.2-6), or written in terms of the fractional loading:

$$C(r,t) = \frac{\theta(r,t;E)}{b(E)\left[1 - \theta(r,t;E)\right]}; \qquad \theta(r,t;E) = \frac{C_\mu(r,t;E)}{C_{\mu s}} \quad (11.3\text{-}4)$$

If the patches of sites having discrete adsorption energies of E_1, E_2, and so on, the adsorbed concentrations of those sites are related to each other according to:

$$C(r,t) = \frac{\theta(r,t;E_1)}{b(E_1)\left[1 - \theta(r,t;E_1)\right]} = \frac{\theta(r,t;E_2)}{b(E_2)\left[1 - \theta(r,t;E_2)\right]} = \cdots \quad (11.3\text{-}5)$$

which means that for a given gas phase concentration the sites having higher adsorption energy (that is higher affinity b) will have higher occupancy (higher fractional loading) than those sites having lower adsorption energy.

Before adsorption is proceeded the particle is equilibrated with a stream of adsorbate having a concentration of C_i, and the surface concentration is in equilibrium with this gas phase concentration, that is

$$t = 0; \quad C = C_i \quad \text{and} \quad C_\mu(E) = C_{\mu s} \frac{b(E)C_i}{1 + b(E)C_i} \quad (11.3\text{-}6)$$

The boundary condition at the center of the particle is:

$$r = 0; \quad \frac{\partial C}{\partial r} = 0 \quad \text{and} \quad \frac{\partial C_\mu(E)}{\partial r} = 0 \tag{11.3-7a}$$

The other boundary condition is that at the exterior surface of the particle, at which the total flux contributing by the pore and surface diffusions into the particle must be equal to the diffusion flux through the stagnant film surrounding the particle:

$$r = R; \quad \varepsilon D_p \frac{\partial C(R,t)}{\partial r} + (1-\varepsilon) \int_0^\infty \frac{D_\mu^0(E)}{\left[1 - \frac{C_\mu(R,t;E)}{C_{\mu s}}\right]} \frac{\partial C_\mu(R,t;E)}{\partial r} F(E) dE = k_m [C_b - C(R,t)]$$

$$\tag{11.3-7b}$$

The model is applicable for any form of the energy distribution. The energy distribution is a result of structural heterogeneity or distribution of surface defects or other surface factors (for example, surface chemistry) that cause the variation in the adsorption energy. Two extremes we could expect for the energy distribution. One is the ideal surface where the distribution is the Dirac delta function, i.e.

$$F(E) = \delta(E - \overline{E})$$

With this form, the model equations will reduce to the pore and surface diffusion model dealt with extensively in Chapter 9:

$$\varepsilon \frac{\partial C(r,t)}{\partial t} + (1-\varepsilon) \frac{\partial C_\mu(r,t;\overline{E})}{\partial t} = \varepsilon D_p \frac{1}{r^s} \left[\frac{\partial}{\partial r} \left(r^s \frac{\partial C(r,t)}{\partial r} \right) \right] +$$

$$(1-\varepsilon) \frac{1}{r^s} \frac{\partial}{\partial r} \left\{ r^s \frac{D_\mu^0(\overline{E})}{\left[1 - C_\mu(r,t;\overline{E})/C_{\mu s}\right]} \frac{\partial C_\mu(r,t;\overline{E})}{\partial r} \right\} \tag{11.3-8a}$$

$$C_\mu(r,t;\overline{E}) = C_{\mu s} \frac{b(\overline{E}) C(r,t)}{1 + b(\overline{E}) C(r,t)} \tag{11.3-8b}$$

The other extreme of energy distribution is the bimodal energy distribution, with a patch of energy E_1 occupying a fraction of α and the remainder is oocupied by the patch of energy E_2:

$$F(E) = \alpha \delta(E - E_1) + (1-\alpha) \delta(E - E_2) \tag{11.3-9}$$

The model associated with this distribution represents a dual adsorption sites model, where the diffusion into the particle is proceeded through three paths, one is the pore diffusion in the void space of the particle and the other two are diffusions in adsorbed phase of energies E_1 and E_2, respectively.

Other distributions between these two extremes often used in the literature are the uniform and Gamma distributions because of their simplicity in form. Their functional forms are:

$$F(E) = \frac{1}{E_{max} - E_{min}} \qquad (11.3\text{-}10)$$

$$F(E) = \frac{q^{n+1}}{\Gamma(n+1)}(E - E_0)^n \exp[-q(E - E_0)] \qquad (11.3\text{-}11)$$

The mean and the variance of these two distributions are:

$$\bar{E} = \frac{E_{min} + E_{max}}{2} \quad \text{and} \quad \sigma = \frac{(E_{max} - E_{min})}{2\sqrt{3}} \qquad (11.3\text{-}12)$$

$$\bar{E} = E_0 + \frac{n+1}{q} \quad \text{and} \quad \sigma = \frac{\sqrt{n+1}}{q} \qquad (11.3\text{-}13)$$

11.3.1 Simulations:

To see the effect of the energetic heterogeneity on the sorption kinetics, we will show in this section the simulations when some of the system parameters, especially the variance of the distribution, are varied.

The parameters listed in the following table are used as the base case in the simulation. These parameters are for the system of n-butane onto activated carbon at 101 kPa and 303 K. In this table, we present parameters for the two energy distributions. One is the uniform energy distribution while the other is the single site energy distribution (homogeneous solid)

Table 11.3-1: Base parameters used in the simulations

	Uniform Energy distribution	**Single Site distribution**
Particle characteristics	$\varepsilon = 0.31$; $R = 10^{-3}$ m (slab half length)	
Parameter a	a=0.5	

Energy distribution	$E_{max} = 22.7$, $E_{min} = 8.3$ kJ/mole $\overline{E} = \frac{1}{2}(E_{max} + E_{min}) = 15.5$ kJ/mole $\sigma = (2\sqrt{3})^{-1}(E_{max} - E_{min}) = 4.16$ kJ/mole	$\overline{E} = 15.5$ kJ/mole
C_0	$C_0 = 8.286 \times 10^{-3}$ kmole/m^3	
Equilibria parameters	$b_0 = 2.62$ m^3/kmole $\overline{b} = b_0 \exp(\overline{E}/R_g T) = 1231$ m^3/kmole $C_{\mu s} = 5.$ kmole/m^3	$\overline{b} = 1231$ m^3/kmole $C_{\mu s} = 5$ kmole/m^3
Dynamic parameters	$D_p = 1.24 \times 10^{-6}$ m^2/sec $D_{\mu\infty}^0 = 1.9 \times 10^{-8}$ m^2/sec $\overline{D}_\mu = D_{\mu\infty}^0 \exp\left(-a\frac{\overline{E}}{R_g T}\right) = 8.76 \times 10^{-10} \frac{m^2}{sec}$	$D_p = 1.24 \times 10^{-6}$ m^2/sec $\overline{D}_\mu = 8.76 \times 10^{-10}$ m^2/sec

The effect of the variance of the adsorption energy, σ, (a measure of the surface heterogeneity) on the adsorption and desorption dynamics is shown in Figure 11.3-1 for two values of variance σ (0 and 6.16 kJoule/mole). The adsorption kinetic curves for these variances are practically superimposed on each other. This is because the patch of strongest sites has the highest affinity, hence higher density but this patch posseses low mobility characteristics towards the adsorbate while the patch of weakest sites has the lowest density but exhibits the highest mobility. Hence, the overall kinetics of a solid having an energy distribution behaves similarly to a solid having a narrower distribution, provided that these two solids have the same mean energy and the same mean adsorption affinity. In the desorption case where the solid, initially equilibrated with a stream of adsorbate of constant concentration is desorbed into a stream of inert gas. Figure 11.3-1 shows a clear influence of the surface heterogeneity on the desorption kinetics. This is physically expected because the patch having weakest site will desorb quickly because of high mobility, while the patch having the strongest sites will desorb at a slower rate. Hence, the solids having much wider energy distribution will take longer time to desorb.

Next we study the effect of the external bulk concentration. We learn from Chapter 9 for homogeneous solids that the time scale is smaller for favourable adsorption isotherm. Here we would like to see whether such a conclusion is still valid for a heterogeneous solid. The effect of the external bulk concentration can be seen in the following two figures. Three external concentrations used in the simulation of adsorption kinetics are 10.6, 20.6 and 30.6 kPa (Figure 11.3-2). Increase in the bulk concentration results in a reduction in the adsorption time, the same conclusion obtained for homogeneous solids (Chapter 9). The reason for this decrease in the adsorption time has been given in Chapter 9. In the desorption mode, the kinetics curve is not sensitive to the adsorbate concentration with which the solid is initially equilibrated (Figure 11.3-3).

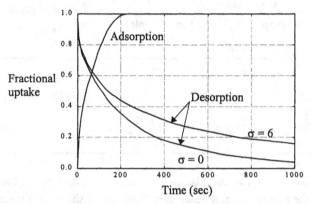

Figure 11.3-1: Effect of variance on the fractional uptake versus time

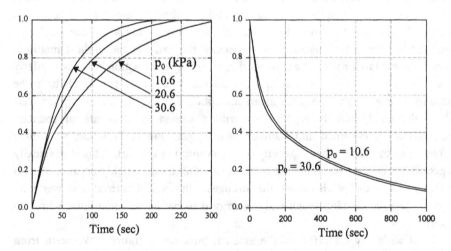

Figure 11.3-2: Effect of the bulk concentration on the adsorption kinetics

Figure 11.3-3: Effect of the bulk concentration on the desorption kinetics

We now investigate the applicability of the heterogeneous model to the experimental data of light hydrocarbons onto activated carbon (Do, 1997). First we present the experimental methods from which the adsorption equilibrium data as well as kinetics data are collected. The method for kinetic measurement is the Differential Adsorption Bed (DAB) method, and the method for equilibrium measurement is the conventional volumetric method. The DAB method has been proven to be an useful and reliable means to collect the adsorption kinetics, and is described in the next section.

11.4 Experimental Section

First we describe the adsorbent and adsorbates used to collect the kinetic data for the testing of the heterogeneous model.

11.4.1 Adsorbent and Gases

The model adsorbent we used is an activated carbon in the form of extrudate having a diameter of 1.7 mm. The properties of this activated carbon are summarised in the following table.

Table 11.4-1: Characteristics of activated carbon

Particle density	733 kg/m^3 = 0.733 g/cc
Total porosity	0.71
Macropore porosity	0.31
Micropore porosity	0.40
Average macropore diameter	8×10^{-7} m = 0.8 µm
Mesopore surface area	8.2×10^4 m^2/kg = 82 m^2/g
Micropore volume	0.44×10^{-3} m^3/kg = 0.44 cc/g
Nitrogen surface area	1.2×10^6 m^2/kg = 1200 m^2/g

The adsorbates used are the n-paraffins, ethane, propane and n-butane.

11.4.2 Differential Adsorption Bed Apparatus (DAB):

Adsorption kinetics are carried out with the DAB set up. It is designed such that a constant environment inside the adsorption cell is always maintained and the cell is kept at constant temperature. This is done by allowing a very high flow rate to pass through the bed containing a very small amount of solid. This high flow is such that the molar rate supply of the adsorbate is much greater than the adsorption rate into the particles. Also with this high flow rate heat released by the adsorption process will be dissipated very quickly.

The apparatus is consisted of four parts (Figure 11.4-1): a gas mixing system, the adsorption cell (C), the desorption bomb (B) and a gas chromatograph (or an equivalent device) for concentration analysis.

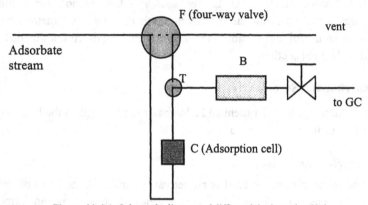

Figure 11.4-1: Schematic diagram of differential adsorption bed

The adsorption cell (C) is connected to a four-way valve (F), which is used to either isolate the cell from the flowing gas stream or allow an adsorbate stream to flow into it. The three-way valve (T) allows either the gas stream to go to the four way valve and thence to vent or the gas to flow into the reservoir, B. This reservoir is fully instrumented with a pressure gauge and a temperature sensor, from which the total number of moles in the reservoir can be calculated.

11.4.3 Differential Adsorption Bed Procedure:

In this section, we describe the operation of the DAB method. With the bed isolated by the four way valve and the three way valve set to the position such that the adsorption cell (C) is connected to the desorption bomb (B). These two are cleaned by either vacuum or heating or a combination of them.

After the bed is cleaned, the three way valve is set to the position to isolate the cell from the reservoir B, and the adsorption cell is brought to the adsorption temperature with an aid of a flowing inert gas. Once this has been done, the cell is isolated from the flowing gas by using the four way valve F.

- Step 1: With the adsorption cell isolated, mix the adsorbates to the desired concentration. While this is done, the reservoir B is evacuated and once this is done, it is isolated from the cell as well as the vacuum.

- Step 2: At time t=0, the adsorbate stream is allowed to pass through the adsorption cell by switching the four way valve. Adsorption is allowed to occur over a period of t^*, and then the adsorption cell is isolated. During the period of exposure t^*, the total amount inside the cell will be the amount adsorbed by the solid up to time t^* plus the amount in the dead volume of the cell.
- Step 3: Next, turn the three way valve T to connect the adsorption cell to the pre-evacuated reservoir B. Adsorbate molecules in the cell will desorb into the reservoir due to the total pressure driving force. To facilitate with this desorption, heat is usually applied to the adsorption cell and a small flow of pure inert gas is introduced to flush the adsorbate into the bomb. After this desorption step is completed, the three way valve is switched to isolate the reservoir. Pressure and temperature of the reservoir are then recorded; hence the total number of moles, which include that of the inert gas and that of the adsorbate, is calculated.
- Step 4: Pass the gas in the reservoir to the GC for the analysis of the adsorbate concentration, from which we can calculate the number of moles of adsorbates in the reservoir. This amount must be subtracted from the amount in the dead volume of the adsorption cell to obtain the amount adsorbed by the solid during the period of t*. The amount in the dead volume is determined experimentally by carrying the steps 2 to 4 with the adsorbents replaced by the same amount of non-porous solids.

By repeating the above steps for different exposure times, we will obtain the full uptake curve as a function of time.

We apply this DAB procedure to a sample of activated carbon whose characteristics are given in Table 11.4-1 and three n-paraffin adsorbates: ethane, propane and n-butane.

11.5 Results & Discussion:

A model system of ethane, propane and n-butane as adsorbates and an activated carbon as the adsorbent are used to validate the heterogeneous kinetic model presented in Section 11.3. The results have been discussed in Do (1997) where it was shown that the equilibria of these adsorbates are adequately described by the Unilan equation:

$$C_\mu = \frac{C_{\mu s}}{2s} \ln\left[\frac{1+\overline{b}C\,e^s}{1+\overline{b}C\,e^{-s}}\right] \qquad (11.5\text{-}1a)$$

$$s = \frac{\sqrt{3}\sigma}{R_g T} \tag{11.5-1b}$$

$$\bar{b} = b_\infty \exp\left(\frac{\bar{E}}{R_g T}\right) \tag{11.5-1b}$$

The fitting between the theory and the experimental data at three different temperatures (283, 303 and 333 K) was carried out simultaneously to extract the following parameters:
1. The saturation capacity
2. The adsorption affinity at infinite temperature
3. The minimum and maximum interaction energies
4. The mean interaction energy

The optimal parameters are tabulated in the following tables for the three adsorbates used: ethane, propane and n-butane.

Table 11.5-1: Parameters for the Unilan equation for ethane

T °C	$C_{\mu s}$ kmole/m³	b_∞ (kPa⁻¹)	E_{min} kJ/mole	E_{max} kJ/mole	$\bar{E}$ (kJ/mole)	$\bar{b}$ (kPa⁻¹)	$s = \frac{\sqrt{3}\sigma}{R_g T}$
10	13.0	8.47 × 10⁻⁷	6.26	30.75	18.5	2.20 × 10⁻³	5.20
30	13.0	8.47 × 10⁻⁷	6.26	30.75	18.5	1.31 × 10⁻³	4.86
60	13.4	8.47 × 10⁻⁷	6.26	30.75	18.5	6.77 × 10⁻⁴	4.42

Table 11.5-2: Parameters for the Unilan equation for propane

T °C	$C_{\mu s}$ kmole/m³	b_∞ (kPa⁻¹)	E_{min} kJ/mole	E_{max} kJ/mole	$\bar{E}$ (kJ/mole)	$\bar{b}$ (kPa⁻¹)	$s = \frac{\sqrt{3}\sigma}{RT}$
10	13.0	1.93 × 10⁻⁷	0	42.75	21.38	1.70 × 10³	9.08
30	13.0	1.93 × 10⁻⁷	0	42.75	21.38	9.34 × 10⁻⁴	8.48
60	13.4	1.93 × 10⁻⁷	0	42.75	21.38	4.35 × 10⁻⁴	7.72

Table 11.5-3: Parameters for the Unilan equation for n-butane

T °C	$C_{\mu s}$ (kmole/m³)	b_∞ (m³/kmole)	$\bar{E}$ (kJ/mole)	$\bar{b}$ (m³/kmole)	$s = \frac{\sqrt{3}\sigma}{RT}$
10	5.545	2.62	15.5	1780	2.89
30	5.016	2.62	15.5	1440	2.86
60	4.559	2.62	15.5	610	2.62

| 150 | 3.754 | 2.62 | 15.5 | 91.4 | 2.58 |

For ethane and propane, we show below the best fitted parameters for the Langmuir adsorption isotherm equation, which are needed for the simulations of the homogeneous model (eq. 11.3-8). This is for the comparison with the heterogeneous model. The fit of the Langmuir equation is not as good as that for the Unilan equation.

Table 11.5-4: Parameters for the Langmuir equation for ethane, propane

T °C	Ethane b (kPa^{-1})	$C_{\mu s}$ (kmole/m^3)	Propane b (kPa^{-1})	$C_{\mu s}$ (kmole/m^3)
10	0.0516	5.49	0.470	5.00
30	0.0320	5.00	0.217	4.69
60	0.0167	4.58	0.0863	4.36

In the study of the kinetics, we fit the heterogeneous model to the experimental kinetic data of ethane and propane. The only parameter required in the fitting is the surface diffusivity at infinite temperature, $D_{\mu\infty}^0$. The surface diffusivities at infinite temperature extracted from the fitting for ethane and propane are $D_{\mu\infty}^0 = 7.02 \times 10^{-7}$ and 1.34×10^{-6} m^2/sec, respectively. The following tables summarises the values of the diffusivities obtained from the fitting of the heterogeneous model as well as the homogeneous model.

Table 11.5-5: Diffusivity parameters for ethane

T °C	D_p m^2/sec	Homogeneous model D_μ (m^2/s)	Heterogeneous model $D_{\mu\infty}^0$ (m^2/s)	$\overline{D}_\mu$
10	1.51 × 10^{-6}	2.58 × 10^{-9}	702 × 10^{-9}	13.76 × 10^{-9}
30	1.68 × 10^{-6}	3.41 × 10^{-9}	702 × 10^{-9}	17.84 × 10^{-9}
60	1.96 × 10^{-6}	6.26 × 10^{-9}	702 × 10^{-9}	24.83 × 10^{-9}

Table 11.5-6: Diffusivity parameters for propane

T °C	D_p m^2/sec	Homogeneous model D_μ (m^2/s)	Heterogeneous model $D_{\mu\infty}^0$ (m^2/s)	$\overline{D}_\mu$
10	1.20 × 10^{-6}	0.789 × 10^{-9}	1340 × 10^{-9}	14.27 × 10^{-9}
30	1.30 × 10^{-6}	0.928 × 10^{-9}	1340 × 10^{-9}	19.26 × 10^{-9}
60	1.56 × 10^{-6}	1.750 × 10^{-9}	1340 × 10^{-9}	28.23 × 10^{-9}

It is generally found that the heterogeneous model fits the experimental data better than the homogeneous model. When they are used to predict the data at other conditions, the heterogeneous model predicts better than its counterpart does, implying the importance of the system heterogeneity. Although the "true" energy distribution may not be uniform as we have assumed here, its use in the model, nevertheless, points to its significance in the description of kinetic data. Furthermore, the heterogeneous model is more advantageous to the homogeneous model in the sense that the heterogeneous model requires only one diffusivity $D_{\mu\infty}^0$ for all temperature data while the homogeneous model requires one value of D_μ for each temperature data.

11.6 Formulation of Sorption Kinetics in Multicomponent Systems

We consider a solid particle exposing to an environment containing N adsorbates of constant concentrations. The surface topography is assumed to have a patchwise configuration. For the development of the model we make the following assumptions:

1. The system is isothermal
2. The particle is large enough so that the resistance to the mass transfer is due to diffusion of free and adsorbed species along the particle coordinate
3. The pore diffusivity, the film mass transfer coefficient and the surface diffusivity at infinite temperature are constant
4. Fluid phase and adsorbed phase are in local equilibrium at any point in the solid
5. Extended Langmuir isotherm is assumed to be valid at the patch level
6. The cumulative energy is the same for all species
7. The adsorbed flux is driven by the chemical potential gradient
8. At any given site, the ratio of the activation for surface diffusion and the adsorption energy is a constant
9. The adsorption energy distribution is uniform (other distributions could be used)

We first present the multicomponent equilibria and then deal with the multicomponent kinetics next.

11.6.1 Adsorption Isotherm:

The local adsorption isotherm of the species "k" takes the following extended Langmuir form:

$$C_\mu[k;E(k)] = C_{\mu s}(k) \frac{b[k;E(k)]\,C(k)}{1+\sum_{j=1}^{N} b[j;E(j)]\,C(j)} \qquad (11.6\text{-}1)$$

where $b[k;E(k)]$ is the adsorption affinity of the component k at the energy level $E(k)$. The adsorption affinity, $b[k;E(k)]$, is related to the adsorption energy and temperature according to the following equation:

$$b[k;E(k)] = b_\infty(k)\exp\left[E(k)/R_g T\right] \qquad (11.6\text{-}2)$$

where the adsorption affinity at infinite temperature, $b_\infty(k)$, is assumed to be dependent on species.

With the assumption of uniform distribution, the functional form for F(E) for the species "k" takes the form:

$$F[k;E(k)] = \begin{cases} \dfrac{1}{E_{max}(k)-E_{min}(k)} & \text{for } E_{min} < E < E_{max} \\ 0 & \text{elsewhere} \end{cases} \qquad (11.6\text{-}3)$$

with

$$\int_0^\infty F[k;E(k)]\,dE(k) = 1 \qquad (11.6\text{-}4)$$

The overall adsorption isotherm of the species "k" in a multicomponent mixture is simply the integration of the local adsorption isotherm over the complete energy distribution domain:

$$\langle C_\mu(k)\rangle = C_{\mu s}(k)\int_0^\infty \frac{b[k;E(k)]\,C(k)}{1+\sum_{j=1}^{N}b[j;E(j)]\,C(j)}\,F[k;E(k)]\,dE(k) \qquad (11.6\text{-}5)$$

The integration of the above integral can only be carried out after we relate the interaction energy of the species "j" to that of the species "k". This is achieved by assuming the matching between the cumulative energies of different species:

$$\frac{E(j)-E_{min}(j)}{E_{max}(j)-E_{min}(j)} = \frac{E(k)-E_{min}(k)}{E_{max}(k)-E_{min}(k)} \qquad (11.6\text{-}6)$$

11.6.2 Local Flux of Species k:

If the driving force for the surface flow is assumed to be the gradient of the chemical potential, we can write the local surface flux of the adsorbed species k at the energy level E(k) as:

$$J_\mu[r,t;k;E(k)] = -D_\mu^0[k;E(k)] \frac{C_\mu[r,t;k;E(k)]}{C(r,t;k)} \frac{\partial C(r,t;k)}{\partial r} \quad (11.6\text{-}7)$$

If the adsorption isotherm takes the form of the extended Langmuir equation (11.6-1), the local flux written in terms of the gradient of gas phase concentrations is:

$$J_\mu[r,t;k;E(k)] = -D_\mu^0[k;E(k)] \left\{ C_{\mu s}(k) \frac{b[k;E(k)]}{1+\sum_{j=1}^{N} b[j;E(j)] C(r,t;j)} \right\} \frac{\partial C(r,t;k)}{\partial r} \quad (11.6\text{-}8)$$

If one is interested in the expression for the flux written in terms of the gradient of the adsorbed concentrations, one can apply the total differentiation of eq. (11.6-7) to get:

$$J_\mu[r,t;k;E(k)] = -D_\mu^0[k;E(k)] \frac{C_\mu[r,t;k;E(k)]}{C(r,t)} \sum_{j=1}^{N} \frac{\partial C(r,t;k)}{\partial C_\mu[r,t;j;E(j)]} \frac{\partial C_\mu[r,t;j;E(j)]}{\partial r}$$

(11.6-9)

Using the local extended Langmuir isotherm (eq. 11.6-1), we obtain the following equation for the local flux written in terms of only the adsorbed concentrations:

$$J_\mu[r,t;k;E(k)] = -D_\mu^0[k;E(k)] \sum_{j=1}^{N} \frac{C_{\mu s}(k)}{C_{\mu s}(j)} \left\{ \delta(k,j) + \frac{\dfrac{C_\mu[r,t;k;E(k)]}{C_{\mu s}(k)}}{1-\sum_{i=1}^{N} \dfrac{C_\mu[r,t;i;E(i)]}{C_{\mu s}(i)}} \right\} \frac{\partial C_\mu[r,t;j;E(j)]}{\partial r}$$

(11.6-10)

This form is more complicated than that when written in terms of the gas phase concentration gradient (eq. 11.6-8). Computation wise, the flux equation using the gas phase concentration gradient is a better choice because it gives much more stable numerical computation.

The surface diffusion coefficient is taking the Arrhenius form:

$$D_\mu^0[k; E(k)] = D_{\mu\infty}^0(k) \exp\left[-\frac{a(k)E(k)}{R_g T}\right] \tag{11.6-11}$$

where a(k) is the ratio of the activation energy for surface diffusion to the adsorption energy. In the absence of any information regarding this parameter we will treat it as a constant for all species.

11.6.3 Mass Balance Equations:

Carrying the mass balance in the particle for the adsorbate "k", we obtain the following equation:

$$\varepsilon \frac{\partial C(r,t)}{\partial t} + (1-\varepsilon)\frac{\partial}{\partial t}\int_0^\infty C_\mu[r,t;k;E(k)]\, F[k;E(k)]\, dE(k) =$$

$$\varepsilon D_p \frac{1}{r^s}\left[\frac{\partial}{\partial r}\left(r^s \frac{\partial C(r,t)}{\partial r}\right)\right] - (1-\varepsilon)\frac{1}{r^s}\frac{\partial}{\partial r}\left\{r^s \int_0^\infty J_\mu[r,t;k;E(k)] \cdot F[k;E(k)] \cdot dE(k)\right\}$$

$$\tag{11.6-12}$$

for k = 1, 2, ..., N. Here J_μ is the surface flux determined from eq. (11.6-8) or (11.6-10). The adsorbed concentration at any point is related to the gas phase concentration at any point as given in eq. (11.6-1).

The boundary condition at the exterior surface of the particle is:

$$r = R; \quad \varepsilon D_p \frac{\partial C(R,t)}{\partial r} - (1-\varepsilon)\int_0^\infty J_\mu[R,t;k;E(k)]\, F[k;E(k)]\, dE(k) = k_m[C_b - C(R,t)]$$

$$\tag{11.6-13}$$

that is the total flux into the particle (LHS) is balanced by the flux through the stagnant film surrounding the particle (RHS).

The particle is assumed to be initially equilibrated with a stream of adsorbate at concentration of $\mathbf{C_i} = [C_i(1), C_i(2), ..., C_i(N)]$, i.e.

$$t = 0; \quad C(r,0;k) = C_i(k) \; ; \; C_\mu[r,0;k,E(k)] = C_{\mu s}(k)\frac{b[k;E(k)]\, C_i(k)}{1 + \sum_{j=1}^N b[j;E(j)]\, C_i(j)} \tag{11.6-14}$$

We have defined a set of equations which describe the sorption kinetics of a multicomponent mixture in a single particle. When N = 1, this set of equation reduces to the set for single component systems dealt with in the last section.

The model equations (11.6-2) to (11.6-4) are validated with the experimental data of binary and ternary systems of ethane, propane and n-butane onto activated carbon. All the necessary equilibrium and kinetic parameters are obtained from the single component fitting as done in the last section. In this sense the multicomponent model is the predicting tool, and it has been shown in Do (1997) that this multicomponent heterogeneous model is a good predictive model. It is capable of predicting well simultaneous adsorption, simultaneous desorption and displacement. Readers are referred to a review paper by Do (1997) for further details.

11.7 Micropore Size Distribution Induced Heterogeneity

The effect of heterogeneity was accounted for by the use of the energy distribution as shown in previous sections. For adsorption of paraffins onto activated carbon, the source of the energy distribution is assumed to be due to the micropore size distribution. This heterogeneity is called the micropore size-induced heterogeneity. The energy of interaction between the micropore and the adsorbate molecules is a strong function of the size of the adsorbate as well as the size of the micropore. The effect of micropore size distribution on adsorption equilibrium has been addressed in Chapter 6. Here we address its role in the adsorption kinetics.

The local adsorption isotherm in a pore having a width of 2r is assumed to take the form of the Langmuir equation:

$$C_\mu(E) = C_{\mu s} \frac{b(E)P}{1 + b(E)P} = C_{\mu s} \frac{b_\infty e^{E/R_g T} P}{1 + b_\infty e^{E/R_g T} P} \tag{11.7-1}$$

where the interaction energy E between the micropore and the adsorbate molecule is a function of the micropore size, that is

$$E = E(r) \tag{11.7-2}$$

This relationship can be derived from the Lennard-Jones theory as described in Chapter 6. Thus, if the micropore size distribution is known, we can write the following equation for the overall adsorbed concentration.

$$C_\mu = \int_{r_{min}}^{r_{max}} C_{\mu s} \frac{b_\infty e^{E/R_g T} P}{1 + b_\infty e^{E/R_g T} P} f(r)\, dr \tag{11.7-3}$$

where f(r) is the functional form for the micropore size distribution. This micropore size distribution is then converted into an energy distribution by the following equation:

$$f(r)\, dr = F(E)\, dE \qquad (11.7\text{-}4)$$

The micropore size-induced energy distribution is then fitted by a polynomial for the subsequent use in the dynamics calculation. The model equations for the dynamics studies are the same as those presented in Section 11.6. The only difference is that the energy distribution is deduced from the micropore size distribution, instead of the uniform energy distribution. Hu and Do (1995) have studied this and they have shown that the approach using the micropore size distribution as the source of system heterogeneity seems to provide a better description of the desorption data than the approach using the uniform energy distribution.

11.8 Conclusions

We have shown a new mathematical model, utilizing the surface energy distribution to understand the sorption dynamics into the particle. The model has shown great promise in its role as a predictive tool to understand dynamics under a wide range of conditions.

Regarding the future work, we suggest that the following aspects could be explored to further our advances in this area

- The separate role of surface energy and structural heterogeneity
- The role of pore size distribution in structural heterogeneity
- More experimental data of other binary and ternary systems
- The heat effect in the heterogeneous model
- More detailed structured models
- Effect of pore evolution on the equilibrium and dynamics parameters of the heterogeneous model

12
Time Lag in Diffusion and Adsorption in Porous Media

12.1 Introduction

Basic information of diffusion and adsorption have been discussed in some details in the last five chapters (7 to 11). In this chapter and the subsequent chapters we will present various methods devised to determine the diffusion coefficient. We start with a method of time lag, which was introduced in 1920 by Daynes. This method can be exploited for the determination of diffusion and adsorption parameters. The concept of this method is simple. Basically a porous medium is mounted between two reservoirs (Figure 12.1-1).

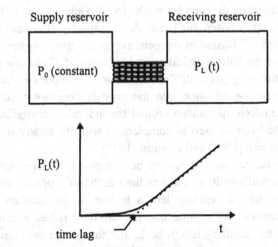

Figure 12.1-1: Time lag setup and the typical pressure response versus time

One reservoir is filled with a diffusing gas, hereafter called the supply reservoir, and the other is maintained under vacuum. The latter reservoir is called the receiving reservoir. At time t = 0, the permeation process is started and the pressure of the receiving reservoir is monitored. Since it takes time for the diffusing gas to transport through the porous media, the pressure in the receiving reservoir remains very low until the diffusing gas breaks through into it. Thus, there is a lag in the response of the pressure in the receiving reservoir; hence, the name of time lag is given. Since the time lag is a result of various kinetic processes occurring inside the porous medium, efforts have been spent to understand this time lag and to extract kinetic parameters from the time lag information.

What to follow are the basis analyses of nonadsorbing gas in Knudsen flow and in viscous flow. This is to illustrate the effect of transport mode on the time lag. Next, we will consider a special treatment due to Frisch, which is useful to obtain the time lag without getting involved in the detailed solution for the concentration distribution across the particle. Finally, we consider the time lag analysis on adsorbing gas, where the adsorption in addition to diffusion can yield many interesting behaviours owing to the fact that adsorption can occur in the dead end pores and adsorbed species could diffuse along the porous medium, processes of which do not occur for non-adsorbing gases. Other factors such as the finite nature of the receiving reservoir, the geometry of the porous medium, and the shape of the adsorption isotherm can influence the time lag behaviour.

12.2 Non-Adsorbing Gas with Knudsen flow

We will first illustrate the time lag method with a simple case of non-adsorbing gas and conditions are chosen such that the transport mechanism is due to the Knudsen mechanism. Diffusion of oxygen, nitrogen, argon, krypton, methane and ethane through inert analcite spherical crystals (Barrer, 1953) at low pressure is an example of non-adsorbing gas with Knudsen flow. Conditions of the experiments are chosen such that the diffusion into the crystals does not occur and flow is restricted to the Knudsen mechanism around the individual crystallites in the bed. The time lag method can be used to complement with the steady state method by Kozeny (1927), Carman (1948) and Adzumi (1937).

The porous solid can be assumed to be composed of many non-intersecting capillaries having equal length, longer than the length of the porous medium (Figure 12.2-1). The ratio of the capillary length to that of the medium is called the tortuosity. Furthermore, we assume that the capillary radius is uniform, and is denoted as r. Let the capillary length be L_c and the porous medium length be L. The tortuosity, which is a measure of the solid structure, is given by:

$$\tau = \frac{L_c}{L} > 1 \qquad (12.2\text{-}1)$$

which, by definition, is greater than unity.

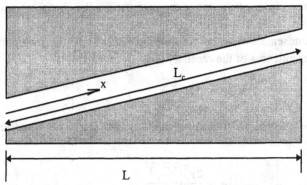

Figure 12.2-1: Schematic of an ideal capillary in the solid medium

Another model for porous solid is a model of packed bed containing very fine grains, and the inter-grain volume defines a space for diffusing molecules to pass through (Figure 12.2-2). Details of this model can be found in Chapter 7.

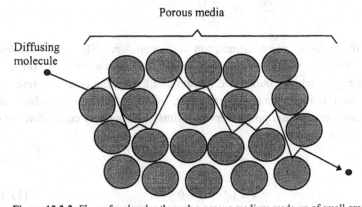

Figure 12.2-2: Flow of molecules through a porous medium made up of small grains

Since the pressures of the two ends of the medium are finite, viscous (Darcy) flow might be operating in addition to the Knudsen flow. To restrict the flow to only Knudsen diffusion, we must maintain the conditions of the experiment such that the Knudsen mechanism is dominating. This is possible when the pressure is very low or the capillary radius is very small. The condition for the validity of the Knudsen flow is P.d < 0.01 Pa-m (Levenspiel, 1984), where P is the total pressure and d is the diameter of the channel. For example, for a pore size of 1 micron, the maximum pressure for the Knudsen diffusion mechanism to occur is 10 Torr. At higher pressures, viscous flow due to the pressure gradient becomes important. This will be discussed in Section 12.4.

When Knudsen flow is the controlling mechanism, the transport flux (mole transported per unit area of the capillary cross section area) is

$$J_K = -D_K \frac{\partial C}{\partial x} \qquad (12.2\text{-}2a)$$

where x is the co-ordinate along the capillary and D_K is the Knudsen diffusivity, defined as

$$D_K = \frac{2r}{3}\sqrt{\frac{8R_g T}{\pi M}} \qquad (12.2\text{-}2b)$$

For constant capillary radius, this Knudsen diffusivity is a constant. With this constitutive flux equation, the mass balance equation describing the concentration distribution in the capillary is:

$$\frac{\partial C}{\partial t} = D_K \frac{\partial^2 C}{\partial x^2} \qquad (12.2\text{-}3)$$

where C is the diffusing gas concentration in the capillary. The porous medium is bounded by two reservoirs. The concentration in the supply reservoir is kept constant during the whole course of diffusion and that of the receiving reservoir is maintained low relative to the supply reservoir, usually of the order of a few percent of the concentration of the supply reservoir. Thus, the boundary conditions imposed at two ends of the capillary are:

$$x = 0; \quad C = C_0 \qquad (12.2\text{-}4a)$$

$$x = L_c; \quad C \approx 0 \qquad (12.2\text{-}4b)$$

Solving the mass balance equation (12.2-3) subject to the above two boundary conditions yields an analytical solution for the concentration distribution, from which system behaviours can be derived, and one such important behaviour is the time lag, which can be exploited to obtain the diffusion coefficient.

Boundary conditions of the form (12.2-4) can be readily maintained experimentally. To obtain the time evolution of the concentration distribution in the capillary, the initial state of the capillary must be defined. There are two possible initial states which are feasible experimentally. One is that the capillary is free from any adsorbate, and the other initial condition is that the concentration in the capillary is the same as the concentration in the supply reservoir. The first initial condition is possible by putting a partition between the porous medium and the supply reservoir, while to achieve the second initial condition the partition is put between the porous medium and the receiving reservoir. We now deal with these two initial conditions separately.

12.2.1 Adsorption: Medium is Initially Free from Adsorbate

When the capillary is free from any molecule of the diffusing gas, the initial condition is:

$$t = 0; \quad C = 0 \qquad (12.2\text{-}5)$$

Solving the mass balance equations (12.2-3) subject to the boundary and initial conditions (12.2-4) and (12.2-5) by methods such as the Laplace transform or the separation of variables yields the following solution for the concentration distribution along the capillary:

$$C = C_0\left(1 - \frac{x}{L_c}\right) - \frac{2C_0}{\pi}\sum_{n=1}^{\infty}\frac{1}{n}\sin\left(\frac{n\pi x}{L_c}\right)\exp\left(-\frac{D_K n^2 \pi^2 t}{L_c^2}\right) \qquad (12.2\text{-}6)$$

The first term in the RHS is simply the steady state linear concentration profile, and the second term is the transient term responsible for the evolution of the profile from the initial state to the final linear steady state profile. The linear steady state profile is due to the assumption of constant diffusion coefficient. For nonconstant diffusion coefficient as we will show later in Section 12.3, the steady state profile is no longer linear. Figure 12.2-3 shows the evolution of the concentration profile for the case of constant diffusion coefficient.

Knowing the concentration profile as given in eq. (12.2-6), the transient flux across any plane perpendicular to the capillary axial direction is calculated from the Knudsen flux equation (12.2-2), that is:

$$J(x,t) = -D_K\frac{\partial C}{\partial x} = \frac{D_K C_0}{L_c} + \frac{2D_K C_0}{L_c}\sum_{n=1}^{\infty}\cos\left(\frac{n\pi x}{L_c}\right)\exp\left(-\frac{n^2\pi^2 D_K t}{L_c^2}\right) \qquad (12.2\text{-}7)$$

of which the first term

$$J_\infty = \frac{D_K C_0}{L_c} \tag{12.2-8}$$

is the steady state flux and the second term is the transient contribution to the flux. The steady state flux is usually achieved after 2 to 3 times the time lag (Van-Amerongen, 1949), which is determined later in the next section.

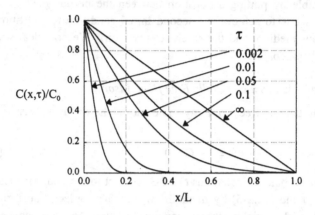

Figure 12.2-3: Evolution of the concentration profile in the porous medium

Eq. (12.2-7) is the transient flux at any point inside the capillary. Of interest to us is the flux at either end of the capillary, depending on whether we wish to measure the pressure of the receiving reservoir or that of the supply reservoir. Usually in practice the pressure of the receiving reservoir is monitored. We now investigate the behaviour of the receiving reservoir first, and then the supply reservoir next.

12.2.1.1 Properties at the Capillary Exit:

The flux at the exit of the capillary is obtained by setting $x = L_c$ into eq. (12.2-7), and we get:

$$J_L(t) = -\frac{D_K C_0}{L_c} + \frac{2 D_K C_0}{L_c} \sum_{n=1}^{\infty} \cos(n\pi) \exp\left(-\frac{n^2 \pi^2 D_K t}{L_c^2}\right) \tag{12.2-9}$$

This equation is valid for all times, but the convergence of the series is very slow for short times. For short times, a better solution can be obtained by taking Laplace transform of the governing equation and then find the asymptotic solution when $s \to$

∞, which is corresponding to $t \to 0$. The short time solution for the flux at the exit of the capillary is given (Roger et al., 1954):

$$J_L(t) = 2C_0 \sqrt{\frac{D_K}{\pi t}} \sum_{n=0}^{\infty} \exp\left[-\frac{(2n+1)^2 L_c^2}{4D_K t}\right] \qquad (12.2\text{-}10)$$

which converges faster for small values of t. Usually only the first term in the above series is sufficient for the calculation, thus the short time solution is:

$$\lim_{t \to 0} J_L(t) = 2C_0 \sqrt{\frac{D_K}{\pi t}} \exp\left(-\frac{L_c^2}{4D_K t}\right) \qquad (12.2\text{-}11)$$

Rearrange this equation, we get the following form amenable for linear plotting

$$\ln\left(J_L \sqrt{t}\right) = \ln\left(2C_0 \sqrt{\frac{D_K}{\pi}}\right) - \frac{L_c^2}{4D_K}\left(\frac{1}{t}\right) \qquad (12.2\text{-}12)$$

Thus, a plot of $\ln\left(J_L \sqrt{t}\right)$ versus $1/t$ would yield a straight line with the following slope and intercept

$$\text{Slope} = -\frac{L_c^2}{4D_K} \qquad (12.2\text{-}13a)$$

$$\text{Intercept} = \ln\left(2C_0 \sqrt{\frac{D_K}{\pi}}\right). \qquad (12.2\text{-}13b)$$

When using eq. (12.2-12) for data analysis, an accurate measurement of the flux at short time is required. This method is, therefore, recommended for very slow permeating porous media where the time lag is too long to practically measure.

Knowing the short time flux (eq. 12.2-10), the number of moles received by the receiving reservoir up to time t is:

$$Q_L(t) = A \int_0^t J_L(t)dt \qquad (12.2\text{-}14)$$

or

$$Q_L = 4AC_0 \sqrt{\frac{D_K}{\pi}} \sum_{n=0}^{\infty} \left\{\sqrt{t} \exp\left[-\frac{(2n+1)^2 L_c^2}{4D_K t}\right] - \sqrt{\frac{\pi}{4D_K}}(2n+1)L_c \,\text{erfc}\left[\frac{(2n+1)L_c}{\sqrt{4D_K t}}\right]\right\} \qquad (12.2\text{-}15)$$

where A is the capillary cross sectional area. Keeping only the first term in the above series, we get the following short time solution for the amount received by the receiving reservoir up to time t:

$$\lim_{t \to 0} Q_L = 2AC_0 \left[2\sqrt{\frac{D_K t}{\pi}} \exp\left(-\frac{L_c^2}{4D_K t}\right) - L_c \cdot \text{erfc}\left(\frac{L_c}{\sqrt{4D_K t}}\right) \right] \quad (12.2\text{-}16)$$

Knowing the amount entering the receiving reservoir, the pressure can be calculated from:

$$P_L = \frac{R_g T}{V} Q_L \quad (12.2\text{-}17)$$

Combining the above two equations we get the following measurable pressure of the receiving reservoir:

$$\lim_{t \to 0} \frac{P_L}{P_0} = 2\left(\frac{AL_c}{V}\right) \left[\sqrt{\frac{\theta}{\pi}} \exp\left(-\frac{1}{\theta}\right) - \text{erfc}\left(\frac{1}{\sqrt{\theta}}\right) \right] \quad (12.2\text{-}18)$$

where

$$\theta = \frac{4D_K t}{L_c^2} \quad (12.2\text{-}19)$$

Eq.(12.2-18) is useful to analyse data at short times. The validity of the short time solution is

$$t < t_{lag}$$

where t_{lag} is the time lag to be determined later. Eq.(12.2-18) is applicable for low permeability membranes. For highly permeable membranes, short time solution is not recommended because the range of validity for the short time is too short for accurate measurement. For this case, we use the solution for the flux valid at all time (eq. 12.2-9) and obtain the following solution for the amount collected in the receiving reservoir from t= 0 to t

$$Q_L(t) = -AD_K \int_0^t \left(\frac{\partial C}{\partial x}\right)_{x=L_c} dt = \frac{AD_K C_0}{L_c} \left[t - \frac{L_c^2}{6D_K} + \frac{2L_c^2}{\pi^2 D_K} \sum_{n=1}^{\infty} \frac{(-1)^{n+1}}{n^2} \exp\left(-\frac{D_K n^2 \pi^2 t}{L_c^2}\right) \right] (12.2\text{-}20)$$

in which the following identity has been used

$$\sum_{n=1}^{\infty} (-1)^{n+1} \frac{1}{n^2} = \frac{\pi^2}{12} \quad (12.2\text{-}21)$$

Knowing the amount diffusing into the receiving reservoir, the pressure variation with time is then obtained from eq.(12.2-17), that is:

$$\frac{P_L(t)}{P_0} = \frac{AD_K}{VL_c}\left[t - \frac{L_c^2}{6D_K} + \frac{2L_c^2}{\pi^2 D_K}\sum_{n=1}^{\infty}(-1)^{n+1}\frac{1}{n^2}\exp\left(-\frac{D_K n^2 \pi^2 t}{L_c^2}\right)\right] \quad (12.2\text{-}22)$$

This pressure increases at small times and after some time its behaviour is linear with respect to time due to the attainment of a steady state flow of molecules through the medium. The linear asymptote takes the form:

$$\lim_{t \to \infty}\frac{P_L}{P_0} = \frac{AD_K}{VL_c}\left(t - \frac{L_c^2}{6D_K}\right) \quad (12.2\text{-}23)$$

The above equation has two interesting properties. First, the slope is proportional to the Knudsen diffusivity and the linear dimension of the system. Second, the intercept of this linear asymptote with the time axis is:

$$t_{lag} = \frac{L_c^2}{6D_K} = \frac{\tau^2 L^2}{6D_K} \quad (12.2\text{-}24)$$

This is called the time lag. This time is a measure of the time scale that molecules diffuse through the medium. Since the time lag measurement is a rather straightforward task, it is a preferred method to many others to determine the Knudsen diffusivity as no further analysis is required on the data to obtain the diffusivity. Figure 12.2-4 shows a plot of $\dfrac{P_L(t)/P_0}{(AL_c/V)}$ versus $\tau = D_K t/L_c^2$. The slope of the pressure versus time at long time is obtained from eq.(12.2-23) as:

$$S_\infty = \frac{AD_K}{VL_c}P_0 \quad (12.2\text{-}25a)$$

Note that the parameter A is the cross-sectional area of the capillaries. It is related to the cross-sectional area of the pellet, A_p, according to

$$AL_c = A_p L \varepsilon.$$

Therefore, the slope S_∞ can be written as:

$$S_\infty = \frac{\varepsilon A_p L D_K}{VL_c^2}P_0 \quad (12.2\text{-}25b)$$

This equation provides the information on how quickly the pressure of the receiving reservoir rises with time at long time. This information is important for design as it is necessary to restrain the flow of molecules into the receiving reservoir such that its pressure does not exceed few percents of the inlet pressure during the course of the experiment. This constraint is necessary because of the requirement of the boundary condition (12.2-4b). To illustrate this, we take an example of benzene molecule diffusing through an inert porous medium having a capillary length of 10 cm, a diffusion area of 0.1 cm², an average pore radius of 0.5 μ, and the volume of the receiving reservoir is 1000 cm³. The operating temperature is 293 K and the inlet pressure is 1 Torr. The low pressure of the supply reservoir is necessary to ensure that the diffusion mechanism through the medium is by the Knudsen mechanism.

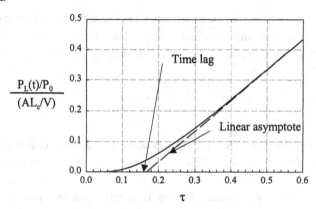

Figure 12.2-4: Plot of the reduced pressure of the receiving reservoir versus τ

The Knudsen diffusivity is calculated as (Table 7.4-2):

$$D_K = 9700\, r\, \sqrt{\frac{T}{M}} = 9700\, (5 \times 10^{-5}) \sqrt{\frac{298}{60}} = 1.08 \quad cm^2 / sec$$

The rate of pressure increase at steady state (eq. 12.2-25a) is:

$$\lim_{t \to \infty} \frac{dP_L}{dt} = S_\infty = \frac{(0.1\ cm^2)(1.08\ cm^2 / sec)(1\ Torr)}{(1000\ cm^3)(1.4)(10\ cm)} = 7.7 \times 10^{-6}\ Torr/sec = 4.63 \times 10^{-4}\ Torr/min$$

in which we have assumed a tortuosity factor of 1.4. In order for the pressure of the receiving reservoir to be maintained less than 5% of that of the supply reservoir, that is 0.05Torr, the experiment should be completed within 100 min. This stopping time must be greater than the time lag, which is calculated from eq.(12.2-24):

$$t_{lag} = \frac{(1.4 \times 10)^2}{6 \times 1.08} = 30 \text{ sec}$$

12.2.1.1.1 Parameter Determination

The parameter characterising the diffusion through the medium is the Knudsen diffusivity, which could be determined from the time lag given in eq. (12.2-24) or from the short time solution (eq. 12.2-18). The long time solution for time lag is preferrable if the experimental data exhibit a linear asymptote behaviour at long time and the constant boundary conditions (12.2-4) are maintained throughout the course of the experiment. If the medium is rather impermeable and the time lag is practically too long to measure, then the application of the short time solution is the only possible choice.

One aspect of the model is the assumption of constant Knudsen diffusivity. To validate this assumption, we can extract the diffusivity from the short time solution as it reflects the transient behaviour of the system, and also extract the diffusivity from the time lag information. The latter reflects the overall kinetic behaviour of the system. If the two extracted diffusivities are the same then the assumption of constant diffusivity holds and the system is a pure diffusion system. On the other hand, if the diffusion coefficient extracted from the short time solution is smaller than the steady state diffusivity, then there exists a sorption process occuring during the transient operation within the medium as the steady state flow is unaffected by the amount adsorbed.

What we show below is a convenient way to determine the diffusion coefficient from the short time solution without recourse to any optimization procedure. Combining the short time solution (eq. 12.2-18) and the steady state rate of pressure rise (eq. 12.2-25), we get

$$\lim_{t \to 0} \frac{P_L}{tS_\infty} \approx \frac{8}{\theta} \left[\sqrt{\frac{\theta}{\pi}} \exp\left(-\frac{1}{\theta}\right) - \text{erfc}\left(\frac{1}{\sqrt{\theta}}\right) \right] \qquad (12.2\text{-}26a)$$

This short time solution can be compared with the exact solution

$$\frac{P_L(t)}{tS_\infty} = \frac{1}{\tau}\left[\tau - \frac{1}{6} + \frac{2}{\pi^2}\sum_{n=1}^{\infty}(-1)^{n+1}\frac{1}{n^2}\exp\left(-n^2\pi^2\tau\right)\right] \qquad (12.2\text{-}26b)$$

which is obtained from eq.(12.2-22), where

$$\tau = \frac{\theta}{4} = \frac{D_K t}{L_c^2} \qquad (12.2\text{-}26c)$$

The RHS of eq. (12.2-26b) is a function of the nondimensional time τ. Plot of the RHS versus τ is shown in Figure (12.2-5).

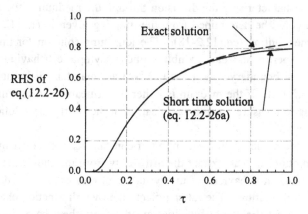

Figure 12.2-5: Plot of the RHS of eq. (12.2-26) versus τ

This plot can be used to determine the diffusion coefficient, and this is done as follows. For each experimental value $P_L(t)$, evaluate the LHS of eq. (12.2-26a), and then use the plot in Figure (12.2-5) to obtain a value of θ. After this is done for all values of $P_L(t)$, a plot of t versus θ will give a straight line according to eq. (12.2-26c). The slope of this straight line is:

$$\frac{L_c^2}{4D_K}$$

This slope has the units of time, and it is 3/2 times the time lag given in eq. (12.2-24).

We shall illustrate this with the experimental data of Barrer (1953). His system is a pure diffusion process of sulfur dioxide through a bed of analcite sphere. The bed length is 98 cm, and the pressure rise at long time as 2.234×10^{-4} cmHg/min. The following table tabulates the pressure of the receiving reservoir (P_L) versus time during the initial time period. The LHS of eq. (12.2-26a) is evaluated for every data in the table and is shown in the third column of the same table. Having these values, Figure (12.2-5) is used to obtain θ for every data and this is included in the fourth column.

Table 12.2-1: Time lag data of Barrer (1954)

time (min)	P_L (cmHg)	$P_L / S_\infty t$ (-)	θ (-)
10.3	4×10^{-5}	0.01739	0.256
15.0	1×10^{-4}	0.02985	0.294
20.0	2.5×10^{-4}	0.05593	0.355
25.3	5×10^{-4}	0.08850	0.4178
30.0	9.75×10^{-4}	0.14552	0.516

A plot of t versus θ is shown in Figure 12.2-6, and a straight line can be drawn through all the points and the slope of such line is 56 min, from which the diffusivity is calculated as:

$$D_K = \frac{(98 \text{ cm})^2}{4(56 \text{ min})(60 \text{ sec/min})} = 0.7134 \text{ cm}^2/\text{sec}$$

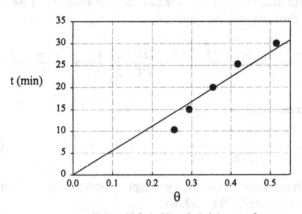

Figure 12.2-6: Plot of t (min) versus θ

Using the time lag equation (12.2-24), a value for the Knudsen diffusivity of 0.769 is obtained. The diffusion coefficient obtained from the short time analysis is 5% smaller than the time-lag diffusivity. This small difference could be attributed to the experimental error or it could be attributed to a very small adsorption of sulfur dioxide on analcite particle.

Another useful quantity for the checking of the experimental data is the zero order moment, defined as the following definite integral of the difference between the steady flux and the transient flux:

$$M = \int_0^\infty [J_\infty - J_L(t)]dt = \frac{L_c C_0}{6} \quad (12.2\text{-}27)$$

The RHS of the above zero-order moment equation is independent of the dynamic parameter and involves only known quantities: L_c and C_0. Thus the integrity of the flux data can be checked to ensure that it satisfies the zero-order moment equation before the data can be used to determine the diffusivity.

12.2.1.2 Properties at the Capillary Entrance:

We have seen that the measurements of the pressure of the receiving reservoir can be utilised to determine the diffusivity. The measurement of the pressure of the supply reservoir can also be effectively used in a similar manner to achieve the same purpose. This is done below.

Evaluating the transient flux equation (12.2-7) at the capillary entrance (that is $x = 0$), and then integrating the result with respect to time from 0 to t we obtain the following quantity diffusing into the capillary:

$$Q_0(t) = -AD_K \int_0^t \left(\frac{\partial C}{\partial x}\right)_{x=0} dt = \frac{AD_K C_0}{L_c}\left[t + \frac{L_c^2}{3D_K} - \frac{2L_c^2}{\pi^2 D_K}\sum_{n=1}^\infty \frac{1}{n^2}\exp\left(-\frac{D_K n^2 \pi^2 t}{L_c^2}\right)\right] \quad (12.2\text{-}28)$$

in which we have used the identity

$$\sum_{n=1}^\infty \frac{1}{n^2} = \frac{\pi^2}{6} \quad (12.2\text{-}29)$$

Carrying out the mass balance around the supply reservoir, the pressure of the supply reservoir is related to $Q_0(t)$ as follows:

$$P_0(t) - P_0(0) = -\frac{R_g T}{V}\int_0^t AJ_0 dt = -\frac{R_g T}{V}Q_0(t) \quad (12.2\text{-}30)$$

where V is the volume of the supply reservoir and $P_0(0)$ is its initial pressure.

A plot of $Q_0(t)$ versus τ is shown in Figure 12.2-7. There is a rapid decrease of the pressure of the supply reservoir due to the very sharp pressure gradient at the capillary entrance at very short times. At sufficiently long time, the supply reservoir pressure approaches a linear asymptote given by:

$$\lim_{t\to\infty} Q_0(t) = \frac{AD_K C_0}{L_c}\left(t + \frac{L_c^2}{3D_K}\right) \qquad (12.2\text{-}31)$$

This linear asymptote when extended to the time axis will give an intercept of

$$t_{lag} = -\frac{L_c^2}{3D_K} = -\frac{\tau^2 L^2}{3D_K} \qquad (12.2\text{-}32)$$

which is in the negative range of the time axis.

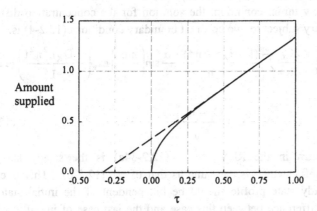

Figure 12.2-7: Plot of the amount supplied versus τ

This time lag (12.2-32) using the pressure response of the supply reservoir or the time lag (eq. 12.2-24) using the pressure response of the receiving reservoir can be used to determine the diffusivity. Thus, the time lag method provides a very convenient if not a straightforward method to determine the diffusion coefficient. The method is not restrictive to the simple Knudsen diffusion mechanism, it is also applicable to other situations, for example

(a) Knudsen and viscous flow for diffusing gas
(b) Knudsen flow for adsorbing gas following either a linear or nonlinear isotherm
(c) Knudsen and surface diffusions for adsorbing gases or vapours
(d) Knudsen flow into a finite reservoir

We shall discuss these situations in the subsequent sections, but first let us consider the pure Knudsen diffusion case whereby the capillary initially contains some diffusing molecules.

12.2.2 Medium Initially Contains Diffusing Molecules

We have considered the case whereby the capillary is initially free from any diffusing molecules. We now study the case where the capillary is filled with the diffusing and the initial concentration of the diffusing gas in the capillary is assumed to be different from that of the supply reservoir. The initial condition of eq. (12.2-3) is:

$$t = 0; \qquad C(x,0) = C_i \qquad (12.2\text{-}33)$$

where C_i is assumed uniform along the capillary.

With this new initial condition, the solution for the concentration distribution inside the capillary subject to two constant boundary conditions (12.2-4) is:

$$C = C_0\left(1 - \frac{x}{L_c}\right) - \frac{2C_0}{\pi}\sum_{n=1}^{\infty} \frac{(1+\alpha)\cos(n\pi) - \alpha}{n} \sin\left(\frac{n\pi x}{L_c}\right) \exp\left(-\frac{D_K n^2 \pi^2 t}{L_c^2}\right) \quad (12.2\text{-}34a)$$

where

$$\alpha = \frac{C_i - C_0}{C_0} \qquad (12.2\text{-}34b)$$

The first term in the RHS of eq. (12.2-34a) is the steady state linear concentration profile which is the same as that of eq. (12.2-6). This is expected because the steady state profile should be independent of the initial state of the capillary. The difference between this case and the last case of initially molecule-free capillary is the way the concentration distribution within the capillary approaches the steady state profile. Figure 12.2-8 shows this approach for the case where the capillary is free of any adsorbate and the case where the initial concentration in the capillary is the same as that of the supply reservoir.

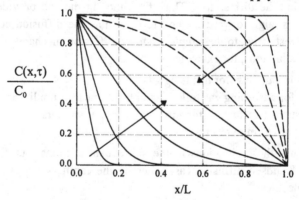

Figure 12.2-8: Evolution of the concentration profile in the porous medium

Knowing this concentration distribution as given in eq. (12.2-34), the transient fluxes at the entrance and exit of the capillary are obtained by applying the diffusion law (eq. 12.2-2), that is:

$$J_0(t) = -D_K \left.\frac{\partial C}{\partial x}\right|_0 = \frac{D_K C_0}{L_c} + \frac{2 D_K C_0}{L_c} \sum_{n=1}^{\infty} [(1+\alpha)\cos(n\pi) - \alpha] \exp\left(-\frac{n^2 \pi^2 D_K t}{L_c^2}\right) \quad (12.2\text{-}35a)$$

$$J_L(t) = -D_K \left.\frac{\partial C}{\partial x}\right|_{L_c} = \frac{D_K C_0}{L_c} + \frac{2 D_K C_0}{L_c} \sum_{n=1}^{\infty} [(1+\alpha) - \alpha \cos(n\pi)] \exp\left(-\frac{n^2 \pi^2 D_K t}{L_c^2}\right) \quad (12.2\text{-}35b)$$

Knowing the transient fluxes as given in the above equations, the amounts entering and leaving the capillary from $t = 0$ to t are:

$$Q_0 = \frac{A D_K C_0}{L_c} \left[t - \frac{\alpha L_c^2}{3 D_K} - \frac{(1+\alpha) L_c^2}{6 D_K} - \frac{2 L_c^2}{\pi^2 D_K} \sum_{n=1}^{\infty} \frac{[(1+\alpha)\cos(n\pi) - \alpha]}{n^2} \exp\left(-\frac{D_K n^2 \pi^2 t}{L_c^2}\right) \right] \quad (12.2\text{-}36a)$$

$$Q_L = \frac{A D_K C_0}{L_c} \left[t + \frac{\alpha L_c^2}{6 D_K} + \frac{(1+\alpha) L_c^2}{3 D_K} - \frac{2 L_c^2}{\pi^2 D_K} \sum_{n=1}^{\infty} \frac{[(1+\alpha) - \alpha \cos(n\pi)]}{n^2} \exp\left(-\frac{D_K n^2 \pi^2 t}{L_c^2}\right) \right] \quad (12.2\text{-}36b)$$

Plots of these quantities versus time yield a linear asymptote at long time and these linear asymptotes will cross the time axis at a time which is known as the time lag. For the pressure response of the supply reservoir, the time lag is obtained from eq. (12.2-36a) and it is

$$t_{lag} = \alpha \frac{L_c^2}{3 D_K} + (1+\alpha) \frac{L_c^2}{6 D_K} \quad (12.2\text{-}37a)$$

Similarly, the corresponding time lag for the receiving reservoir is

$$t_{lag} = -\alpha \frac{L_c^2}{6 D_K} - (1+\alpha) \frac{L_c^2}{3 D_K} \quad (12.2\text{-}37b)$$

The expressions for the time lag (eqs. 12.2-37) are applicable for whatever the initial concentration of the diffusing gas in the capillary is. For example, when the capillary is initially free of any diffusing molecules ($\alpha = -1$), the time lags for the supply and receiving reservoirs are:

$$\left(t_{lag}\right)_{\text{SUPPLY RESERVOIR}} = -\frac{L_c^2}{3 D_K} \quad (12.2\text{-}38a)$$

and

$$\left(t_{lag}\right)_{\text{RECEIVING RESERVOIR}} = \frac{L_c^2}{6D_K} \qquad (12.2\text{-}38b)$$

which are the results we have obtained earlier (eqs. 12.2-32 and 12.2-24), respectively.

When the initial concentration of the diffusing gas in the capillary is the same as that of the supply reservoir, that is $\alpha = 0$, we obtain the following time lags for the supply and receiving reservoirs:

$$\left(t_{lag}\right)_{\text{SUPPLY RESERVOIR}} = \frac{L_c^2}{6D_K} \qquad (12.2\text{-}39a)$$

and

$$\left(t_{lag}\right)_{\text{RECEIVING RESERVOIR}} = -\frac{L_c^2}{3D_K} \qquad (12.2\text{-}39b)$$

The two time lags in eqs. (12.2-38) and the two in eqs. (12.2-39) make up four time lags this method can provide. This is commonly referred to in the literature as the four time lag method, and either one of them can be used to determine the diffusivity or for the certainty of parameter determination, more than one time lag can be used.

12.3 Frisch's Analysis (1957 - 1959) on Time Lag

Section 12.2 shows the essential features about the time lag method, which is a useful tool to determine the diffusion coefficient. This is done by simply measuring the pressure of either the supply reservoir or the receiving reservoir, and then extrapolating the linear asymptotes of the long-time data to the time axis. The intercept of such extrapolation is the time lag, which is conveniently used to determine the diffusion coefficient. When the diffusion mechanism is pure Knudsen mechanism, the proper equations for the time lag are given in eqs. (12.2-38) and (12.2-39) depending the initial state of the capillary. What has been done in order to get to the expressions for the time lag is that the mass balance equation (12.2-3) must be solved subject to the constant boundary conditions (12.2-4) and some proper initial condition to first yield the solution for the concentration distribution, from which the flux can be obtained by application of the Fick's law (12.2-2a). Knowing the flux, the quantities of diffusing in and out of the capillary can be derived, and subsequently the time lags are finally obtained. Such a lengthy procedure can be avoided if our purpose is purely to obtain the expression for the time lag, rather than the evolution of the concentration distribution. Frisch (1957,

1958, 1959) was the first to provide this simple means, and hence hereafter the method is called the Frisch's method. All the method requires is the determination of the steady state concentration distribution, which is of course a much easier task than the determination of the transient concentration distribution. Furthermore, the Frisch method is very versatile as it is not applicable to the case of constant diffusion coefficient but also to many other complicated cases involving nonlinear equation as we shall show in the next few sections.

What we show in this section is the essential steps of the Frisch's method to a case of slab geometry although the method is applicable to cylindrical and spherical geometries as well. The mass balance in a slab geometry medium with a concentration dependent diffusion coefficient is:

$$\frac{\partial C}{\partial t} = \frac{\partial}{\partial x}\left[D(C)\frac{\partial C}{\partial x}\right] \qquad (12.3\text{-}1)$$

where the diffusion coefficient $D(C)$ is assumed to be a function of concentration.

12.3.1 Adsorption:

We will demonstrate the Frisch method with the case where the medium is initially free from any diffusing species (that is the initial condition is given by eq. 12.2-5). The boundary conditions are assumed constant and given in eqs. (12.2-4).

If the amount of gas (or pressure) is measured at the exit of the slab medium, there will be an initial lag in the rise of the amount versus time due to the fact that the diffusing molecules take time to transport across the medium, and then afterward the amount will increase linearly with time as a result of the steady state flow of molecules through the porous medium. Before solving for the time lag by this Frisch's method, it is necessary to consider the steady state behaviour of the system.

12.3.1.1 Steady State Consideration

At steady state, the mass balance equation (12.3-1) will become:

$$\frac{d}{dx}\left[D(C_\infty)\frac{dC_\infty}{dx}\right] = 0 \qquad (12.3\text{-}2)$$

where the subscript ∞ denotes the steady state. This steady state mass balance is solved subject to the constant boundary conditions (12.2-4), and the following solution for the steady state concentration is obtained in implicit form:

$$\int_{C_\infty(x)}^{C_0} D(u)du = \left(\frac{x}{L}\right) \int_0^{C_0} D(u)du \qquad (12.3\text{-}3)$$

from which the steady state flux can be calculated using the constitutive flux equation:

$$J_\infty = -D(C_\infty)\frac{dC_\infty}{dx} = \frac{1}{L}\int_0^{C_0} D(u)du \equiv \text{constant} \qquad (12.3\text{-}4)$$

This steady state flux is a constant as a result of the constant boundary conditions. Although it takes time for the steady state condition to be reached, if we assume this steady state flux is instantaneously attained from $t = 0$ the amount collected per unit area of the capillary from $t = 0$ to t is given by

$$Q_\infty(t) = (J_\infty)\,t = \frac{t}{L}\int_0^{C_0} D(u)du \qquad (12.3\text{-}5)$$

In fact the flux during the early stage of diffusion is lower than the steady flux due to the transient build-up of mass within the capillary; therefore the actual amount collected at time t will be less than the "theoretical" amount calculated by eq. (12.3-5).

12.3.1.2 Time Lag of the Receiving Reservoir:

Having obtained the necessary equations at steady state condition, we now turn to the derivation of the time lag for the receiving reservoir. The transient molar flux at the exit of the capillary and the amount of gas collected by the receiving reservoir per unit area of the capillary are:

$$J_L(t) = -D(C)\frac{\partial C}{\partial x}\bigg|_{x=L} \qquad (12.3\text{-}6a)$$

and

$$Q(t) = \int_0^t J_L(t)dt \qquad (12.3\text{-}6b)$$

respectively. Eq. (12.3-6b) is the quantity that we would readily obtain experimentally.

To obtain this transient amount collected by the receiving reservoir $Q_L(t)$ we need to consider the transient mass balance equation (12.3-1). First, integrating this equation with respect to x from L to x, we obtain:

$$\int_L^x \frac{\partial C}{\partial t} dx = D(C)\frac{\partial C}{\partial x} - D(C)\frac{\partial C}{\partial x}\bigg|_{x=L} = D(C)\frac{\partial C}{\partial x} + J_L(t) \qquad (12.3\text{-}7)$$

Integrating this result again with respect to x from 0 to L, and noting that the flux at the capillary exit $J_L(t)$ is not a function of x, we have:

$$\int_0^L \int_L^x \frac{\partial C(z,t)}{\partial t} dz\, dx = \int_{C_0}^0 D(u)\, du + J_L(t)\, L \qquad (12.3\text{-}8)$$

from which we can obtain the following expression for the transient flux at the capillary exit:

$$J_L(t) = \frac{1}{L}\left[\int_0^{C_0} D(u)\, du - \int_0^L \int_x^L \frac{\partial C(z,t)}{\partial t} dz\, dx\right] \qquad (12.3\text{-}9)$$

The first term in the RHS is simply the flux at steady state (eq. 12.3-4). Thus, the transient flux at any time t is less than the steady state flux due to the transient build-up of mass in the capillary (the second term in the RHS of the above equation). Only when time is sufficiently large that this transient rate of mass hold-up will be zero.

Knowing this transient flux (eq. 12.3-9), the amount collected by the receiving reservoir is calculated as in eq. (12.3-6b) and the result is:

$$Q(t) = \left[\frac{1}{L}\int_0^{C_0} D(u)du\right] t - \frac{1}{L}\int_0^L \int_x^L C(z,t)dz\,dx \qquad (12.3\text{-}10)$$

This is the actual amount collected by the receiving reservoir from t = 0 to time t. Note that the first term in the RHS of eq. (12.3-10) is the hypothetical amount collected Q_∞ if the steady state flux is instantaneously started from t = 0 (c.f eq. 12.3-5). Observing eq. (12.3-10) we note that if we know the transient concentration distribution, the integral in the RHS can be evaluated. However, this is not necessary if the time lag is what we need rather than the details of how diffusing molecules are distributed themselves within the capillary during the transient period. We take the limit of eq. (12.3-10) when time is large, and the limit is a linear asymptote. We denote this asymptote as $Q_a(t)$ and it is given by

$$\lim_{t\to\infty} Q(t) = Q_a(t) = \left[\frac{1}{L}\int_0^{C_0} D(u)du\right]\left[t - \left(t_{\text{lag}}\right)_{\text{EXIT}}\right] \qquad (12.3\text{-}11)$$

In the limit of large time, the concentration distribution $C(z,t)$ in eq. (12.3-10) is simply the steady state concentration distribution $C_\infty(z)$. Comparing eqs. (12.3-10) and (12.3-11) in the limit of large time, we get the following expression for the time lag:

$$\left(t_{lag}\right)_{EXIT} = \frac{\int_0^L \int_x^L C_\infty(z) \, dz \, dx}{C_0 \int_0^{C_0} D(u) \, du} = \frac{\int_0^L x C_\infty(x) \, dx}{C_0 \int_0^{C_0} D(u) \, du} \qquad (12.3\text{-}12)$$

This is a very useful expression for the time lag. All it needs is the concentration distribution at steady state. Thus by measuring the time lag experimentally (the LHS of eq. 12.3-12), the RHS can be evaluated to yield information about the diffusion coefficient. Let us proceed this one step further. Substituting the steady state concentration distribution given in eq. (12.3-3) into the above equation and then carrying out the integration by parts will give the following expression for the time lag written completely in terms of the diffusion coefficient function:

$$\left(t_{lag}\right)_{EXIT} = \frac{L^2 \left\{ \int_0^{C_0} w D(w) \left[\int_w^{C_0} D(u) \, du \right] dw \right\}}{\left[\int_0^{C_0} D(u) \, du \right]^3} \qquad (12.3\text{-}13)$$

Observing the above equation, we note that measuring the time lag does not necessarily provide us with the information about the functional form of the diffusion coefficient. This is usually what we would like to obtain for a given system. One could, however, choose a functional form for D containing one or more parameters, and then by measuring the time lags at various values of the concentration of the supply reservoir we can do a nonlinear optimisation to extract those parameters for the assumed functional form of the diffusion coefficient.

We note in eq. (12.3-13) that the time lag in general is a function of the inlet concentration C_0, due to the concentration dependence of the diffusion coefficient. It will become independent of C_0 only when the diffusion coefficient is a constant. Indeed, when D is a constant, the steady state concentration distribution (eq. 12.3-3) is reduced to:

$$C_\infty = C_0 \left(1 - \frac{x}{L}\right) \qquad (12.3\text{-}14)$$

and the time lag (12.3-13) will reduce to:

$$\left(t_{lag}\right)_{EXIT} = \frac{L^2}{6D} \tag{12.3-15}$$

which is the result we obtained earlier using the complete solution approach. Here we see the power of the Frisch's approach of achieving the same result without the need to obtain the transient concentration distribution.

12.3.1.3 Time Lag of the Supply Reservoir

The procedure presented in the above section can be applied to the entrance of the capillary. The only difference for this case of supply reservoir is the step in eq.(12.3-7). The limit for the integration in this case is from 0 to x, rather than L to x as in the last case for the receiving reservoir. This exercise is left to the reader. The solution for the time lag for the pressure response of the supply reservoir is:

$$\left(t_{lag}\right)_{ENTRANCE} = -\frac{\int_0^L (L-x) C_\infty(x) dx}{C_0 \int_0^{C_0} D(u) du} \tag{12.3-16}$$

When the diffusion coefficient is constant, the above equation reduces to that given in eq. (12.2-32).

12.3.2 General Boundary Conditions:

In the previous section, we dealt with the case of diffusion through a medium, which is initially free from any adsorbate. In this section, we will generalise the conditions imposed on the system. We assume that the medium is initially filled with some adsorbate, but uniform in profile, and the inlet and outlet of the medium are exposed to some constant concentrations. The method of Frisch presented in the previous section is again applicable to this general case.

The initial and boundary conditions for this general problem are:

$$t = 0 \; ; \; C = a \tag{12.3-17}$$

$$x = 0; \quad C = C_0 \tag{12.3-18a}$$

$$x = L \; ; \; C = C_1 \tag{12.3-18b}$$

Using the Frisch's method and applying the above initial and boundary conditions, the time lags obtained for the receiving and supply reservoirs are:

$$\left(t_{lag}\right)_{EXIT} = \frac{\int_0^L x\left[C_\infty(x) - a\right]dx}{\int_{C_1}^{C_0} D(u)du} \tag{12.3-19a}$$

and

$$\left(t_{lag}\right)_{ENTRANCE} = \frac{\int_0^L (L-x)\left[C_\infty(x) - a\right]dx}{\int_{C_1}^{C_0} D(u)du} \tag{12.3-19b}$$

where C_∞ is a solution of the following quadrature:

$$\int_{C_\infty(x)}^{C_0} D(u)du = \left(\frac{x}{L}\right)\int_{C_1}^{C_0} D(u)du \tag{12.3-20}$$

Substituting C_∞ of eq. (12.3-20) into eqs. (12.3-19), we obtain the following solutions for the time lag written in terms of diffusion coefficient:

$$\left(t_{lag}\right)_{EXIT} = L^2 \frac{\int_{C_1}^{C_0} D(u)(u-a)\left(\int_u^{C_0} D(w)dw\right)du}{\left[\int_{C_1}^{C_0} D(u)du\right]^3} \tag{12.3-21a}$$

$$\left(t_{lag}\right)_{ENTRANCE} = -L^2 \frac{\int_{C_1}^{C_0} D(u)(u-a)\left(\int_{C_1}^u D(w)dw\right)du}{\left[\int_{C_1}^{C_0} D(u)du\right]^3} \tag{12.3-21b}$$

Eqs. (12.3-21) provide general expressions for the time lags of the receiving and supply reservoirs when the following conditions are met:
(a) medium has slab geometry
(b) isothermal condition
(c) constant initial distribution of molecules in the capillary
(d) constant boundary conditions
(e) diffusion coefficient is a function of concentration

Example 12.3-1: *Time lag for the case of constant diffusion coefficient*

Apply the general expressions (eq. 12.3-21) for the time lag to the case of constant diffusion coefficient, we get:

$$(t_{lag})_{EXIT} = \frac{L^2}{D} \left\{ \frac{\frac{C_0}{6}(C_0 + C_1) - \frac{a}{2}(C_0 - C_1) - \frac{C_1^2}{3}}{(C_0 - C_1)^2} \right\} \quad (12.3\text{-}22a)$$

$$(t_{lag})_{ENTRANCE} = -\frac{L^2}{D} \left\{ \frac{\frac{C_0^2}{3} - \frac{C_1}{6}(C_0 + C_1) - \frac{a}{2}(C_0 - C_1)}{(C_0 - C_1)^2} \right\} \quad (12.3\text{-}22b)$$

We see that when the concentration at the exit of the medium is not zero and the initial concentration in the medium is not either zero or C_0 the time lags are dependent on the inlet concentration C_0 even though the diffusion coefficient is a constant. Thus the time lag is independent of inlet concentration only when the following conditions are satisfied:

1. diffusion coefficient is constant
2. concentration at the exit of the medium is zero
3. initial concentration is either zero or equal to the inlet conc. C_0.

Figure 12.3-1 shows plots of the reduced time lag (scaled against the time lag at a = 0 and C_1 = 0) for the case of a = 0. Here we note that the non-zero concentration at the medium exit results in an increase in the time lag, which is not unexpected because of the decrease in the overall driving force across the medium.

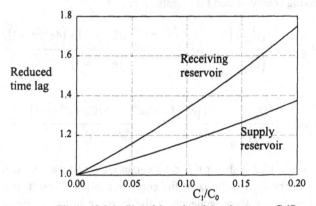

Figure 12.3-1: Plot of the reduced time lag versus C_1/C_0

Example 12.3-2: *Exponential functional form for D*

We now consider the case where the diffusion coefficient takes the following functional form:

$$D(C) = D_0 e^{\alpha C} \qquad (12.3\text{-}23)$$

where D_0 is the diffusivity at zero loading, and α is a constant. The capillary is assumed initially free of any diffusing molecules and the concentration at the capillary exit is much lower than that at the entrance, that is $C_1 \approx 0$.

Substitution of the expression for the diffusion coefficient (eq. 12.3-23) into the time lag equations (12.3-21) yields:

$$\left(t_{lag}\right)_{EXIT} = \frac{L^2}{4D_0} \frac{\left[\left(e^{\alpha C_0}\right)^2 (2\alpha C_0 - 3) + \left(4e^{\alpha C_0} - 1\right)\right]}{\left(e^{\alpha C_0} - 1\right)^3} \qquad (12.3\text{-}24a)$$

$$\left(t_{lag}\right)_{ENTRANCE} = -\frac{L^2}{4D_0} \frac{\left[e^{\alpha C_0}\left(2\alpha C_0 e^{\alpha C_0} - 4\alpha C_0 - e^{\alpha C_0} + 4\right) - 3\right]}{\left(e^{\alpha C_0} - 1\right)^3} \qquad (12.3\text{-}24b)$$

When $\alpha = 0$, the diffusion coefficient (12.3-23) becomes constant, and it is not difficult to prove using the Taylor series expansion that the time lag solutions given in eqs. (12.3-24a) and (12.3-24b) reduce to those given in eqs. (12.2-24) and (12.2-32), respectively.

To investigate the effect of concentration dependence of the diffusivity, we plot the time lag for finite α to that for $\alpha = 0$ for the receiving reservoir and the supply reservoir

$$R_{EXIT} = \left[\frac{\left(t_{lag}^{\alpha}\right)}{\left(t_{lag}^{0}\right)}\right]_{EXIT} = \frac{3}{2} \frac{\left[\left(e^{\alpha C_0}\right)^2 (2\alpha C_0 - 3) + \left(4e^{\alpha C_0} - 1\right)\right]}{\left(e^{\alpha C_0} - 1\right)^3} \qquad (12.3\text{-}25a)$$

$$R_{ENT} = \left[\frac{\left(t_{lag}^{\alpha}\right)}{\left(t_{lag}^{0}\right)}\right]_{ENT} = \frac{3}{4} \frac{\left[e^{\alpha C_0}\left(2\alpha C_0 e^{\alpha C_0} - 4\alpha C_0 - e^{\alpha C_0} + 4\right) - 3\right]}{\left(e^{\alpha C_0} - 1\right)^3} \qquad (12.3\text{-}25a)$$

Figure 12.3-2 shows this ratio versus αC_0, and we note that an increase in either the coefficient α or the concentration will result in a decrease in the time lag as a consequence of the increase in the apparent diffusivity,

$D_0 e^{\alpha C}$. Thus, if the experimental data of time lag versus concentration C_0 shows a decrease with concentration, it could point to the fact that the apparent diffusivity is a function of concentration, but it says nothing about the mechanism of transport of molecules through the medium. We will show later that the viscous flow mechanism and the adsorbing gas situations also give rise to the behaviour of a decrease of time lag for the concentration (pressure) of the supply reservoir.

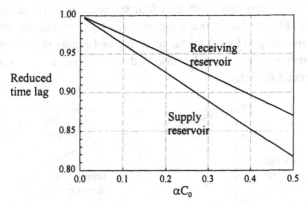

Figure 12.3-2: Plot of the reduced time lag versus αC_0

Example 12.3-3: *Darken-diffusivity*

The Darken diffusivity takes the following form:

$$D(C) = \frac{D_0}{1 - \beta C} \tag{12.3-26}$$

with $\beta C < 1$.

For the capillary initially free of any diffusing gas and the pressure at the capillary exit is maintained low during the course of diffusion, the time lags are obtained by substituting eq. (12.3-26) into eqs. (12.3-21). We get

$$(t_{lag})_{EXIT} = \frac{L^2}{2D} \frac{\left[\ln\left(\frac{1}{1-\beta C_0}\right)\right]^2 - 2\ln\left(\frac{1}{1-\beta C_0}\right) + 2\beta C_0}{\left[\ln\left(\frac{1}{1-\beta C_0}\right)\right]^3} \tag{12.3-27a}$$

$$(t_{lag})_{ENTRANCE} = \frac{L^2}{2D} \frac{\left[\ln\left(\frac{1}{1-\beta C_0}\right)\right]^2 + 2\ln\left(\frac{1}{1-\beta C_0}\right) - 2\beta C_0 \ln\left(\frac{1}{1-\beta C_0}\right) - 2\beta C_0}{\left[\ln\left(\frac{1}{1-\beta C_0}\right)\right]^3}$$

(12.3-27b)

Similarly to the example 2, the time lags for this case also decrease with an increase in either the coefficient β or the pressure of the supply reservoir. Figure 12.3-3 shows the reduced time lags (scaled against time lags corresponding to $\beta = 0$) for the receiving and supply reservoirs versus βC_0. The reduced time lags' decrease with βC_0 is due to the increase in the apparent diffusivity.

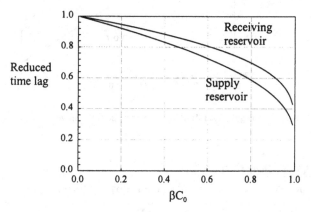

Figure 12.3-3: Plot of the reduced time lag versus βC_0

12.4 Nonadsorbing Gas with Viscous Flow

We have shown the essential features of the time lag in Section 12.2 using the simple Knudsen diffusion as an example, and a direct method of obtaining the time lag in Section 12.3. The diffusion coefficient dealt with in the Frisch's method in Section 12.3 is concentration dependent. In this section we will deal with a case where the transport through the porous medium is a combination of the Knudsen diffusion and the viscous flow mechanism. We shall see below that this case will result in an apparent diffusion coefficient which is concentration dependent, and hence it is susceptible to the Frisch's analysis as outlined in the Section 12.3. This means that the results of equations (12.3-21) are directly applicable to this case.

The viscous flow mechanism is important when the pressure of the system is reasonably high. When this is the case, the constitutive flux equation describes a combined transport of Knudsen diffusion and viscous flow as:

$$J = -\left[D_K + \left(\frac{B_0 R_g T}{\mu}\right) C\right] \frac{\partial C}{\partial x} \qquad (12.4\text{-}1)$$

where D_K is the Knudsen diffusivity, B_0 is the viscous flow parameter and μ is the viscosity. This constitutive flux equation can be rewritten as:

$$J = -D_K (1 + \alpha C) \frac{\partial C}{\partial x} \qquad (12.4\text{-}2)$$

where the parameter

$$\alpha = \frac{B_0 R_g T}{\mu D_K} \qquad (12.4\text{-}3)$$

is a measure of the relative importance of the viscous flow to the Knudsen flow.

Example 12.4-1: *Importance of the viscous flow*

To give an idea how important this parameter is, we take the following example of nitrogen at 298 K flowing through a cylindrical capillary of radius r. The parameter B_0 and the Knudsen diffusivity take the form:

$$B_0 = \frac{r^2}{8}$$

$$D_K = \frac{2r}{3}\sqrt{\frac{8 R_g T}{\pi M}}$$

and hence the parameter α is:

$$\alpha = \frac{3\pi}{16\sqrt{8}} \frac{r\sqrt{M R_g T}}{\mu}$$

Substituting the following values into the above equation for α

Viscosity, μ	$= 1.75 \times 10^{-4}$ g cm^{-1} sec^{-1}
Temperature, T	$= 298$ K
Gas constant, R	$= 8.314 \times 10^7$ g cm^2 sec^{-2} mole^{-1} K^{-1}
Molecular weight, M	$= 28$ g/mole

we get

$$\left(\frac{\alpha}{1 \text{ cm}^3 / \text{mole}}\right) = \left(5.59 \times 10^8\right)\left(\frac{r}{1 \text{ cm}}\right)$$

The following table shows the values of α for a number of capillary radius

r (micron)	α (cc/mole)
0.01	5.59×10^2
0.1	5.59×10^3
1	5.59×10^4

For a concentration of the supply reservoir of $C_0 = 1 \times 10^{-5}$ mole/cc (which is about 186 Torr), we calculate αC_0 as 0.00559, 0.0559, 0.559 for capillaries of radii 0.01, 0.1 and 1 micron, respectively. This means that viscous flux is not so important in capillaries of radii 0.01 and 0.1 micron. For larger capillary (1 micron), the viscous flux becomes more important.

The mass balance equation for describing the concentration distribution in the medium is:

$$\frac{\partial C}{\partial t} = \frac{\partial}{\partial x}\left[D_K\left(1 + \alpha C\right)\frac{\partial C}{\partial x}\right] \qquad (12.4\text{-}4)$$

This equation has exactly the form studied in Section 12.3 where the Frisch's method was illustrated. Thus, the general solutions given by eq. (12.3-21) are applicable. We shall take the case whereby the medium is initially free of any diffusing molecules and the pressure of the receiving reservoir is much less than that of the supply reservoir. The time lags for the receiving and supply reservoirs are:

$$\left(t_{lag}\right)_{EXIT} = \frac{L^2}{6D_K}\frac{\left[1 + \frac{5}{4}\alpha C_0 + \frac{2}{5}(\alpha C_0)^2\right]}{\left(1 + \alpha C_0 / 2\right)^3} \qquad (12.4\text{-}5)$$

$$\left(t_{lag}\right)_{ENTRANCE} = -\frac{L^2}{3D_K}\frac{\left[1 + \frac{9}{8}\alpha C_0 + \frac{3}{10}(\alpha C_0)^2\right]}{\left(1 + \alpha C_0 / 2\right)^3} \qquad (12.4\text{-}6)$$

A number of observations could be deduced from this analysis of the case involving viscous flow:
(a) when the viscous flow is absent, the time lag solutions are reduced to those for the case of constant Knudsen diffusivity dealt with in Section 12.2
(b) the time lag decreases when the supply pressure increases.

Figure 12.4-1 shows the effect of the viscous flow parameter αC_0 on the reduced time lag (scaled against the time lags corresponding to no viscous flow). We see that the viscous flow reduces the time lag as we would expect physically.

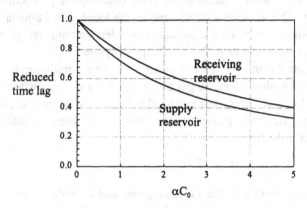

Figure 12.4-1: Plot of the reduced time lag versus αC_0

To investigate the relative contribution of the Knudsen flow and the viscous flow, let us check the temperature dependence of the parameter α. First we write the apparent diffusivity in terms of pressure instead of concentration as the measurements are done in terms of pressure:

$$D_{app} = D_K(1 + \alpha_P P) \tag{12.4-7a}$$

where

$$\alpha_P = \frac{B}{\mu D_K} \tag{12.4-7b}$$

The temperature dependence of viscosity of many gases at moderate pressures is stronger than $T^{0.5}$, while the Knudsen diffusivity is proportional to $T^{0.5}$. Thus the parameter α_P decreases with an increase in temperature, suggesting that the Knudsen mechanism is gaining its dominance at high temperature.

12.5 Time Lag in Porous Media with Adsorption

The analysis so far dealt with the time lag method for diffusing (non-adsorbing) gases. The method can be applied to adsorbing gases or vapours as well. This section and the subsequent sections will show the applicability of the time lag method to adsorption systems and how adsorption and diffusion parameters can be extracted from the analysis.

One clear distinct difference between a non-adsorbing system and an adsorbing system is the longer time lag observed in the adsorbing system. This is simply due to the accumulation of mass by the medium, hence retarding the penetration of the concentration front. The stronger is the adsorption, the longer is the time lag. Since the amount adsorbed on the medium is strongly dependent on temperature, it can be effectively used as a parameter to study the adsorption system.

We first illustrate the time lag procedure on a simple adsorption system where the partition between molecules in the gas phase and those on the surface is linear (linear isotherm). We also consider at any local point within the medium, the mass exchange between the two phases is so rapid that local equilibrium is instantaneously established, that is:

$$C_\mu = KC \qquad (12.5\text{-}1)$$

where C_μ is the concentration of the adsorbed phase and K is the Henry constant. Other factors such as nonlinear isotherm, finite mass exchange between the two phases, and non-constant diffusion coefficient will be dealt with in the subsequent sections.

12.5.1 Linear isotherm

We consider the case of linear isotherm between the gas and solid phases. The mass transport into the particle is assumed to occur by two parallel mechanisms: pore and surface diffusions. The mass balance equation describing the concentration distribution in a slab porous medium with these two parallel mechanisms is:

$$\varepsilon \frac{\partial C}{\partial t} + (1-\varepsilon)\frac{\partial C_\mu}{\partial t} = \varepsilon D_p \frac{\partial^2 C}{\partial x^2} + (1-\varepsilon)D_s \frac{\partial^2 C_\mu}{\partial x^2} \qquad (12.5\text{-}2)$$

The parameter ε is the porosity of the void space available for pore diffusion (that is mesopore and macropore voidage), C is the concentration of the free species (mole/cc of gas), and C_μ is the concentration in the adsorbed phase (mole/cc of solid phase). The diffusion coefficients for these two diffusion mechanisms are assumed constant, although the diffusion coefficient for the surface diffusion is known to have a highly concentration dependence. If we, however, restrict ourselves to low concentration (low pressure), then the adsorption isotherm will be linear and the assumption of constant surface diffusion coefficient is applicable.

The interaction between the gas phase and the adsorbed phase inside the porous medium can be very fast relative to the diffusion process or it can be comparable to the diffusion process. If the former is the case, we will have what is called the local equilibrium, that is at any time t if the gas phase concentration at a given point x is C(x,t) then the adsorbed phase will be in equilibrium with that gas phase concentration. We will first deal with the case of instantaneous adsorption and then consider the finite rate of adsorption next.

12.5.1.1 Instantaneous Adsorption

When the adsorption rate is much faster than the diffusion rate, the local adsorption equilibrium prevails:

$$C_\mu(x,t) = KC(x,t) \tag{12.5-3}$$

Substitution of this linear isotherm equation into the mass balance equation (12.5-2) yields the following equation:

$$\frac{\partial C}{\partial t} = D_{app}\frac{\partial^2 C}{\partial x^2} \tag{12.5-4}$$

where the apparent diffusivity is defined as:

$$D_{app} = \frac{\varepsilon D_p + (1-\varepsilon)D_s K}{\varepsilon + (1-\varepsilon)K} \tag{12.5-5}$$

The mass balance equation written in the form of equation (12.5-4) is identical in form to eq. (12.2-3) for the case of nonadsorbing gas operating under the Knudsen mechanism. This means that the complete analysis of Section 12.2 or the Frisch's method of Section 12.3 is applicable to this case. The time lags for the receiving and supply reservoirs when the porous medium is initially free of any molecules are:

$$\left(t_{lag}\right)_{EXIT} = \frac{L^2}{6D_{app}} \tag{12.5-6a}$$

and

$$\left(t_{lag}\right)_{ENTRANCE} = -\frac{L^2}{3D_{app}} \tag{12.5-6b}$$

respectively.

The pressure of the supply reservoir does not affect the time lags as is expected for a linear isotherm. The temperature dependence of the time lag is studied by investigating the temperature dependence of the apparent diffusivity. The dependence of relevant parameters on temperature is shown below:

$$K = K_\infty \exp\left(\frac{Q}{R_g T}\right) \tag{12.5-7a}$$

$$D_s = D_{s\infty} \exp\left(-\frac{E_s}{R_g T}\right) \tag{12.5-7b}$$

$$D_p = D_{p0} \left(\frac{T}{T_0}\right)^\alpha \tag{12.5-7c}$$

With these parameters dependence on temperature, the dependence of the apparent diffusivity has been investigated in Section 9.2.1.13 (eq. 9.2-15). This apparent diffusivity exhibits a monotonous decrease with temperature, suggesting that the time lag is smaller at higher temperature due to the combined effect of the lower amount adsorbed and the higher diffusion coefficients.

Example 12.5-1: *Magnitude of time lag for adsorbing gases*

To show an idea how long the time lag is for the case of adsorbing gas, we take an example with the following values for the relevant parameters.

Parameter	Symbol	Value
Pore diffusivity	D_p	0.5 cm²/sec
Surface diffusivity	D_s	10^{-5} cm²/sec
Henry constant	K	10000
Porosity	ε	0.33
Particle length	L	1 cm

The apparent diffusivity is calculated as:

$$D_{app} = \frac{\varepsilon D_p + (1-\varepsilon)D_s K}{\varepsilon + (1-\varepsilon)K} = \frac{(0.33)(0.5) + (1 - 0.33)(10000)(10^{-5})}{0.33 + (1 - 0.33)(10000)} = 0.000256 \frac{cm^2}{sec}$$

The time lags for the receiving and supply reservoirs are:

$$(t_{lag})_{EXIT} = \frac{L^2}{6D_{app}} = \frac{(1)^2}{6(0.000256)} = 650 \text{ sec} \approx 11 \text{ min}$$

$$(t_{lag})_{ENTRANCE} = -\frac{L^2}{3D_{app}} \approx 22 \text{ min}$$

For the same set of parameters except now that the gas is non-adsorbing, that is $K = 0$, the corresponding time lags are:

$$(t_{lag})_{EXIT} = \frac{L^2}{6D_p} = \frac{(1)^2}{6(0.5)} = 0.33 \text{ sec}$$

$$(t_{lag})_{ENTRANCE} = -\frac{L^2}{3D_p} = 0.66 \text{ sec}$$

Thus we see that the adsorbing gas increases the time lag significantly compared to the non-adsorbing gas. Conducting the time lag method to an adsorbing system exhibiting a linear isotherm is not different from that of a non-adsorbing gas except that the medium is short otherwise the time lag would be too large to practically measure

12.5.2 Finite Adsorption

One of the assumptions made in the last analysis is the local adsorption equilibrium. We would like to investigate in this section that if such assumption does not hold, that is the exchange rate between the two phases is comparable to the diffusion rates, would the time lag be affected? For this case, the mass balance equation for the concentration distribution is still governed by eq. (12.5-2). The relationship between the concentrations of the two phases is given by the following kinetic equation:

$$\frac{dC_\mu}{dt} = k_a \left(C - \frac{C_\mu}{K} \right) \qquad (12.5\text{-}8)$$

where k_a is the rate constant for adsorption. If this rate constant is very large (strictly speaking when the adsorption rate is faster than the diffusion rate), this equation is reduced to the local linear adsorption equilibrium equation (12.5-3) dealt with in the last section.

Take the case of initially free adsorbate in the medium and the constant boundary conditions (12.2-4) imposed on the system, the mass balance equations (12.5-2) and (12.5-8) can be solved by Laplace transform (Appendix 12.1) to give the following solution for the amount collected in the receiving reservoir:

$$Q_L = \frac{A(\varepsilon D_p)C_0}{L}\left\{t - \frac{L^2[\varepsilon + (1-\varepsilon)K]}{6(\varepsilon D_p)} + \sum_{n=1}^{\infty}\frac{\alpha(s_n)e^{s_n t}}{\{s \cdot \sinh[\alpha(s)L]\}'_{s_n}}\right\} \quad (12.5\text{-}9)$$

where A is the cross sectional area of the medium, and α is a function of complex variable s and is defined as

$$\alpha^2(s) = \frac{\varepsilon s + \dfrac{(1-\varepsilon)sk_a}{s + k_a/K}}{\varepsilon D_p} \quad (12.5\text{-}10)$$

and s_n are poles and are given by:

$$s_n = -u + \sqrt{u^2 - n^2\pi^2\frac{(\varepsilon D_p)k_a}{\varepsilon KL^2}} \quad \text{for } n=1,2,.... \quad (12.5\text{-}11a)$$

$$u = \frac{1}{2}\left[\frac{k_a}{K} + \frac{(1-\varepsilon)k_a}{\varepsilon} + \frac{(\varepsilon D_p)n^2\pi^2}{\varepsilon L^2}\right] \quad (12.5\text{-}11b)$$

Knowing the amount of sorbate collected in the receiving reservoir (eq. 12.5-9), the time lag can be readily calculated by observing the asymptotic behaviour of that equation (the first two terms), and we get the following expression for the time lag:

$$t_{lag} = \frac{L^2}{6}\frac{[\varepsilon + (1-\varepsilon)K]}{\varepsilon D_p} \quad (12.5\text{-}12)$$

which is <u>identical</u> to the time lag obtained for the case of local equilibrium. This means that the finite mass exchange does not affect the time lag information. In other words, the linear asymptote of the amount collected in the receiving reservoir is independent of the rate of mass interchange between the two phases, whether it is infinitely fast or finite. This rate of mass interchange *only* affects the way the transient curve approaches the linear asymptote (the third term in eq. 12.5-9). One can view the time lag as the integral of the overall diffusion process across the medium, and hence it is unaffected by the way in which mass is exchanged inside the medium.

A similar analysis was presented by Goodknight et al. (1960) and Goodknight and Fatt (1961) in the context of diffusion through a porous medium with dead end volume. The dead end pores can be viewed as the adsorption capacity sites in the context of adsorption.

12.5.2.1 Frisch's Time Lag

Instead of solving the problem by the Laplace transform, we could apply the Frisch's method to derive the time lag without the need of solving the transient concentration distribution. The Frisch's method outlined in Section 12.3 can be applied here, and the analysis presented below briefly accounts for this development.

First, we integrate the mass balance equation (12.5-2) with respect to x from x to L and then again with respect to x from 0 to L, we obtain:

$$\int_0^L \int_x^L \left[\varepsilon \frac{\partial C(x,t)}{\partial t} + (1-\varepsilon) \frac{\partial C_\mu(z,t)}{\partial t} \right] dz dx = \varepsilon D_p L \frac{\partial C}{\partial x}\bigg|_L + (1-\varepsilon) D_\mu L \frac{\partial C_\mu}{\partial x}\bigg|_L + \varepsilon D_p C_0 + (1-\varepsilon) D_\mu C_{\mu 0}$$

(12.5-13)

where $C_{\mu 0}$ is the adsorbed concentration at the entrance of the medium, which is assumed to be in equilibrium with C_0.

$$C_{\mu 0} = K C_0$$

(12.5-14)

Integrating eq. (12.5-13) again, but this time with respect to time from 0 to t, and making use of the definition of the amount collected by the receiving reservoir as:

$$Q_L(t) = A \int_0^t \left(-\varepsilon D_p \frac{\partial C}{\partial x}\bigg|_{x=L} - (1-\varepsilon) D_\mu \frac{\partial C_\mu}{\partial x}\bigg|_{x=L} \right) dt$$

(12.5-15)

we obtain the following expression for $Q_L(t)$ as a function of the concentration distribution inside the porous medium:

$$Q_L(t) = \frac{A\left[\varepsilon D_p C_0 + (1-\varepsilon)D_\mu C_{\mu 0}\right]t}{L} - \frac{A}{L}\int_0^L \int_x^L \left[\varepsilon C(z,t) + (1-\varepsilon)C_\mu(z,t)\right] dz\, dx \quad (12.5\text{-}16)$$

This is an equation for the amount collected by the receiving reservoir valid for any time, provided the concentrations of the two phases are known as function of x and t. Since the information at large time (i.e. time lag) is only needed, we take the limit of the above equation when $t \to \infty$ and note that

$$\lim_{t\to\infty} C(x,t) = C_\infty(x) = C_0\left(1 - \frac{x}{L}\right) \quad (12.5\text{-}17a)$$

$$\lim_{t\to\infty} C_\mu(x,t) = C_{\mu\infty}(x) = KC_\infty(x) \quad (12.5\text{-}17b)$$

we have

$$\lim_{t\to\infty} Q_L(t) = \frac{A\left[\varepsilon D_p C_0 + (1-\varepsilon)D_\mu C_{\mu 0}\right]t}{L} - \frac{A}{L}\int_0^L \int_x^L \left[\varepsilon C_\infty(z) + (1-\varepsilon)C_{\mu\infty}(z)\right] dz\, dx \quad (12.5\text{-}18)$$

The steady state solutions (12.5-17) were obtained by solving the mass balance equation at steady state.

Substitution of the steady state concentration distributions (eqs. 12.5-17) into the linear asymptote equation (12.5-18) yields:

$$Q_L^\infty(t) = \frac{A\left[\varepsilon D_p C_0 + (1-\varepsilon)D_\mu C_{\mu 0}\right]}{L}\left\{t - \frac{L^2\left[\varepsilon + (1-\varepsilon)K\right]}{6\left[\varepsilon D_p + (1-\varepsilon)D_\mu K\right]}\right\} \quad (12.5\text{-}19)$$

which gives the time lag

$$t_{lag} = \frac{L^2\left[\varepsilon + (1-\varepsilon)K\right]}{6\left[\varepsilon D_p + (1-\varepsilon)D_\mu K\right]} \quad (12.5\text{-}20)$$

The above time lag is independent of the rate constant for adsorption k_a. Thus, the finite mass exchange kinetics does not affect the time lag. This is true even when the finite mass exchange kinetics equation takes other form than the one shown in eq. (12.5-8) provided that the relationship between the concentrations of the two phases is linear at equilibrium.

12.5.3 Nonlinear Isotherm

The analysis in the last two sections is only applicable when the adsorption isotherm is linear, a situation where the pressure is very low. At such low pressure, the surface diffusivity is a constant, validating the use of solutions given in the last two sections. Although the linear analysis yields simple analytical solutions, its advantage disappears when we wish to learn more about the adsorption system. For example, when a time lag experiment is carried out, and if the time lag of the receiving reservoir is monitored, eq. (12.5-6a) only states that the apparent diffusivity can be calculated as:

$$D_{app} = \frac{L^2}{6 t_{lag}} \qquad (12.5\text{-}21)$$

Unless we know the Henry constant K and the pore diffusivity a-priori from some independent experiments, we have no means to calculate the surface diffusivity. This problem can be overcome if we now conduct the experiment over the nonlinear range of the isotherm. It is this isotherm non-linearity that we could delineate the separate contribution of the pore and surface diffusions.

The nonlinear isotherm is assumed to take the general form:

$$C_\mu = f(C) \qquad (12.5\text{-}22)$$

The mass balance equation describing the concentration distribution no longer takes the form of eq. (12.5-2) as over the nonlinear range of the isotherm the surface diffusivity is no longer a constant. Rather the proper mass balance equation should be:

$$\varepsilon \frac{\partial C}{\partial t} + (1-\varepsilon)\frac{\partial C_\mu}{\partial t} = \frac{\partial}{\partial x}\left[\varepsilon D_p \frac{\partial C}{\partial x} + (1-\varepsilon)D_\mu(C_\mu)\frac{\partial C_\mu}{\partial x}\right] \qquad (12.5\text{-}23)$$

Combining eqs. (12.5-22) and (12.5-23), we obtain the following mass balance equation written in terms of the fluid phase concentration:

$$G(C)\frac{\partial C}{\partial t} = \frac{\partial}{\partial x}\left[H(C)\frac{\partial C}{\partial x}\right] \qquad (12.5\text{-}24)$$

where

$$G(C) = \varepsilon + (1-\varepsilon)f'(C) \qquad (12.5\text{-}25a)$$

$$H(C) = \varepsilon D_p + (1-\varepsilon)D_\mu(f(C))f'(C) \qquad (12.5\text{-}25b)$$

The form of eq. (12.5-24) is different from the form of eq. (12.3-1) which was analysed in the section describing the Frisch's method. However, the procedure described therein is applicable here as well, and this is done as follows.

12.5.3.1 The Frisch's Method

To obtain the time lag using the Frisch's method, we must first determine the steady state concentration distribution. This is found by setting the time derivative in eq. (12.5-24) to zero, and we get:

$$\frac{d}{dx}\left[H(C_\infty)\frac{dC_\infty}{dx}\right] = 0 \qquad (12.5\text{-}26)$$

Solution of the above equation subject to the constant boundary conditions (12.2-4) is written in the following implicit form:

$$\int_{C_\infty(x)}^{C_0} H(u)\,du = \frac{x}{L}\left(\int_0^{C_0} H(u)\,du\right) \qquad (12.5\text{-}27)$$

Knowing this concentration of the free species, the concentration of the adsorbed phase at steady state is given by:

$$C_{\mu\infty} = f\bigl(C_\infty(x)\bigr) \qquad (12.5\text{-}28)$$

To obtain the time lag expression, we integrate eq. (12.5-24) with respect to x from L to x and obtain:

$$\int_L^x G(C)\frac{\partial C}{\partial t}\,dz = H(C)\frac{\partial C}{\partial x} - H(C)\frac{\partial C}{\partial x}\bigg|_L \qquad (12.5\text{-}29)$$

The second term in the RHS of the above equation is the flux entering the receiving reservoir. We denote that as $J_L(t)$. Integrating the above equation again with respect to x from 0 to L, we get:

$$\int_0^L\int_L^x G(C)\frac{\partial C}{\partial t}\,dz\,dx = \int_{C_0}^{0} H(u)\,du + \bigl[J_L(t)\bigr]L \qquad (12.5\text{-}30)$$

Finally integrating this result with respect to time from t = 0 to t, we obtain the following result for the amount collected by the receiving reservoir per unit area of the medium:

$$\frac{Q_L(t)}{A} = \int_0^t J_L(t)\, dt = \left[\frac{1}{L}\int_0^{C_0} H(u)\, du\right] t - \frac{1}{L}\int_0^L \int_L^x G(C)\, C(z,t)\, dz\, dx \qquad (12.5\text{-}31)$$

Taking the limit of the above equation when time is large, we obtain:

$$\lim_{t\to\infty}\frac{Q_L(t)}{A} = \left[\frac{1}{L}\int_0^{C_0} H(u)\, du\right]\left(t - t_{lag}\right) \qquad (12.5\text{-}32)$$

where the time lag is given by:

$$t_{lag} = \frac{\int_0^L\int_x^L G(C_\infty)\, C_\infty(z)\, dz\, dx}{\int_0^{C_0} H(u)\, du} \qquad (12.5\text{-}33a)$$

or after integration by parts to reduce the double integral to a single integral, it takes the final form:

$$t_{lag} = \frac{\int_0^L x\, G(C_\infty(x))\, C_\infty(x)\, dx}{\int_0^{C_0} H(u)\, du} \qquad (12.5\text{-}33b)$$

Changing the integration variable from x to C_∞, we finally obtain the expression for the time lag written in terms of the capacity function G and the diffusivity function H as follows:

$$t_{lag} = \frac{L^2 \int_0^{C_0} u\, G(u)\, H(u)\left(\int_u^{C_0} H(w)\, dw\right) du}{\left(\int_0^{C_0} H(u)\, du\right)^3} \qquad (12.5\text{-}34)$$

The explicit expression for this time lag depends on the choice of the adsorption isotherm f(C) and the functional form for $D_\mu(C_\mu)$.

Example 12.5-2: Langmuir isotherm and pore diffusion mechanism

Before dealing further with the time lag equation (12.5-34), let us consider the case of simple Langmuir isotherm, chosen here to illustrate the effect of the isotherm nonlinearity. The Langmuir equation is:

$$C_\mu = f(C) = C_{\mu s} \frac{bC}{1+bC} \quad (12.5\text{-}35a)$$

where $C_{\mu s}$ is the maximum saturation capacity and b is the Langmuir adsorption affinity, taking the following temperature dependence form:

$$b = b_\infty \exp\left(\frac{Q}{R_g T}\right) = b_0 \exp\left[-\frac{Q}{R_g T_0}\left(1-\frac{T_0}{T}\right)\right] \quad (12.5\text{-}35b)$$

with b_0 being the adsorption affinity at some reference temperature T_0.

To separate the contribution of the surface diffusion (which we know from Section 12.5.1 that it reduces the time lag) on the time lag, let us neglect the surface diffusion in this example and assume that pore diffusion is the only transport mechanism. For such a case, we have:

$$G(C) = \varepsilon + \frac{(1-\varepsilon)K}{(1+bC)^2} \quad (12.5\text{-}36a)$$

$$H(C) = \varepsilon D_p \quad (12.5\text{-}36b)$$

where $K = bC_{\mu s}$, which is the Henry constant at zero loading. Substitution of eqs. (12.5-36) into the general time lag equation (12.5-34), we obtain the time lag for the case of Langmuir isotherm with pore diffusion:

$$t_{lag} = \frac{L^2}{6(\varepsilon D_p)}\left[\varepsilon + (1-\varepsilon)K \cdot F(\lambda)\right] \quad (12.5\text{-}37)$$

where $F(\lambda)$ is a function of the isotherm nonlinearity (hereafter called the isotherm nonlinearity factor) and is given by:

$$F(\lambda) = \frac{6}{\lambda}\left\{\frac{1}{2} + \frac{1}{\lambda^2}[\lambda - (1+\lambda)\ln(1+\lambda)]\right\}; \quad \lambda = bC_0 \quad (12.5\text{-}38)$$

The parameter λ is a measure of the isotherm nonlinearity. The isotherm is called linear when this parameter is much less than unity, and is strongly nonlinear when it is greater than 10; otherwise it is called moderately nonlinear.

In the limit of linear isotherm, that is low concentration, the time lag given in eq. (12.5-37) is reduced to:

$$\lim_{\lambda \to 0} t_{lag} = \frac{L^2[\varepsilon + (1-\varepsilon)K]}{6(\varepsilon D_p)} \qquad (12.5\text{-}39)$$

which is the result (12.5-6a) obtained earlier in the analysis of the linear isotherm case.

To see the effect of the isotherm nonlinearity, let us investigate the reduced time lag scaled against the time lag corresponding to zero loading:

$$R = \frac{t_{lag}}{\lim_{\lambda \to 0} t_{lag}} = \frac{[\varepsilon + (1-\varepsilon)K \cdot F(\lambda)]}{\varepsilon + (1-\varepsilon)K} \approx F(\lambda) \qquad (12.5\text{-}40)$$

The approximation of the above equation is simply that the capacity of the adsorbed phase is much larger than that of the fluid phase. Figure 12.5-1 shows this ratio at three temperatures 273, 293 and 333 K plotted versus $b_0 C_0$ for $Q/R_g T_0 = 10$. Here we see that the isotherm nonlinearity reduces the time lag. This is explained as follows. A double in the pressure corresponds to less than double increase in the adsorbed phase concentration due to the convexity of the isotherm, and hence the driving force for permeating through the medium is enhanced, resulting in a lower time lag. For example, for moderate nonlinear isotherm $\lambda = 1$, we have $F(1) = 0.68$; a reduction of 32 % in the time lag.

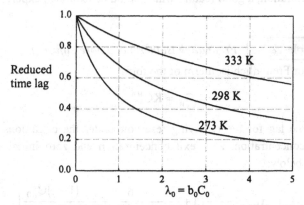

Figure 12.5-1: Plot of the reduced time lag versus $b_0 C_0$ for $Q/R_g T_0 = 10$

We now see that there are a number of factors that can cause a reduction in the time lag. These factors are:

(a) presence of the viscous flow
(b) presence of the surface diffusion
(c) nonlinearity of the isotherm

Delineation of these factors can be done with experiments carried out at different temperatures as their dependence on temperature follows different rate. The viscous flow is less important at high temperature (Section 12.4), the surface diffusivity increases quickly with temperature, and the adsorption capacity decreases with temperature, usually at a rate much faster than the rate of increase in the surface diffusivity. Thus, delineation is possible.

Let us return to the general expression (12.5-37) and investigate the situation where the adsorption isotherm is strongly nonlinear (that is $\lambda \gg 1$). For this case, we have:

$$\lim_{\lambda \to \infty} t_{lag} = \frac{L^2}{6(\varepsilon D_p)}\left[\varepsilon + (1-\varepsilon)\frac{3C_{\mu s}}{C_0}\right] \qquad (12.5\text{-}41a)$$

The second term in the RHS of the above equation is generally larger than the first term, so the time lag equation for the case of irreversible isotherm is reduced to:

$$\lim_{\lambda \to \infty} t_{lag} \approx \frac{L^2}{6(\varepsilon D_p)}(1-\varepsilon)\frac{3C_{\mu s}}{C_0} \qquad (12.5\text{-}41b)$$

Thus, we see that the time lag in this case is inversely proportional to the supply reservoir concentration, a good feature which could be exploited experimentally.

Example 12.5-3: *Freundlich isotherm and pore diffusion*

For Freundlich isotherm of the form

$$C_\mu = KC^{1/n} \qquad (12.5\text{-}42)$$

the time lag for the receiving reservoir under the conditions of constant inlet concentration, zero exit concentration and zero initial condition is given below:

$$t_{lag} = \frac{L^2}{6D_p}\left[1 + \frac{6}{(1+1/n)(2+1/n)}\frac{(1-\varepsilon)C_{\mu 0}}{\varepsilon C_0}\right] \qquad (12.5\text{-}43)$$

Example 12.5-4: Dual Langmuir isotherm and pore diffusion

For dual Langmuir isotherm of the form

$$C_\mu = C_{\mu s1} \frac{b_1 C}{1 + b_1 C} + C_{\mu s2} \frac{b_2 C}{1 + b_2 C} \tag{12.5-44}$$

the time lag for the receiving reservoir under the conditions of constant inlet concentration, zero exit concentration and zero initial condition is given below:

$$t_{lag} = \frac{L^2}{6 D_p} \left[1 + \frac{(1-\varepsilon)}{\varepsilon} K_1 F(\lambda_1) + \frac{(1-\varepsilon)}{\varepsilon} K_2 F(\lambda_2) \right] \tag{12.5-45}$$

where the functional form F is given in eq.(12.5-38) and

$$K_1 = b_1 C_{\mu s1}; \qquad K_2 = b_2 C_{\mu s2} \tag{12.5-46a}$$

Example 12.5-5: Langmuir isotherm and dual diffusion mechanism

Let us now turn to the case where both the Langmuir isotherm and surface diffusion are included. We assume that the surface diffusivity takes the following Darken relation form:

$$D_\mu = \frac{D_\mu^0}{1 - C_\mu / C_{\mu s}} \tag{12.5-47}$$

For this case, the functions G(C) and H(C) are:

$$G(C) = \varepsilon + \frac{(1-\varepsilon) K}{(1 + bC)^2}; \qquad H(C) = \varepsilon D_p + \frac{(1-\varepsilon) K D_\mu^0}{(1 + bC)} \tag{12.5-48}$$

Substitution of the above expressions for G(C) and H(C) into the expression for the time lag (eq. 12.5-34) yields the following solution:

$$t_{lag} = \frac{L^2}{D_p} \frac{\int_0^1 y \left[1 + \frac{(1-\varepsilon) K}{\varepsilon (1+\lambda y)^2} \right] \left[1 + \frac{\delta}{(1+\lambda y)} \right] \left[(1-y) + \frac{\delta}{\lambda} \ln\left(\frac{1+\lambda}{1+\lambda y}\right) \right] dy}{\left[1 + \frac{\delta}{\lambda} \ln(1+\lambda) \right]^3} \tag{12.5-49a}$$

where

$$\delta = \frac{(1-\varepsilon) K D_\mu^0}{\varepsilon D_p}; \qquad \lambda = b C_0 \tag{12.5-49b}$$

Example 12.5-6: *Toth isotherm*

We finish this section by considering another isotherm equation: the Toth equation, which is a popular equation in describing adsorption isotherm of numerous practical systems. It has the form:

$$C_\mu = f(C) = C_{\mu s} \frac{bC}{\left[1+(bC)^t\right]^{1/t}} \qquad (12.5\text{-}50)$$

For simplicity, we ignore the surface diffusion in this example. We have:

$$G(C) = \varepsilon + \frac{(1-\varepsilon)K}{\left[1+(bC)^t\right]^{1/t+1}} \; ; \qquad H(C) = \varepsilon D_p \qquad (12.5\text{-}51)$$

and the time lag then is given by eq. (12.5-34) or in implicit quadrature form below

$$t_{lag} = \frac{L^2}{\varepsilon D_p C_0^3} \int_0^{C_0} u \left[\varepsilon + \frac{(1-\varepsilon)K}{\left(1+(bu)^t\right)^{1/t+1}}\right](C_0 - u)\, du \qquad (12.5\text{-}52)$$

12.6 Further Consideration of the Time Lag Method

The time lag method has been shown to be a promising tool to study transport through a porous medium. For non-adsorbing gases, the method is effectively used to characterise the Knudsen flow as well as the viscous flow. Experimental conditions can be adjusted, for example temperature and pressure of the supply reservoir, that one of the two mechanisms dominates the overall transport. Knudsen diffusion is known to dominate the transport when the pressure is low and the temperature is high. When dealing with adsorbing gases or vapours, the nonlinearity of the isotherm results in a decrease in the time lag with an increase in the supply reservoir pressure. The surface diffusion of the adsorbed species also contributes to the decrease of the time lag.

In this section, we will address this method one step further. First we show how the steady state concentration distribution can be measured experimentally, and how the time lag can be obtained irrespective of the mechanism of transport within the medium.

12.6.1 Steady State Concentration:

Let C_T be the total concentration of the solute in the porous medium, defined as mole per unit volume of the particle, that is:

$$C_T = \varepsilon C + (1-\varepsilon)C_\mu \tag{12.6-1}$$

where C is the sorbate concentration in the void space and C_μ is the concentration in the adsorbed phase. For strong sorbates, the latter concentration is of order of 100 or 1000 times larger than the concentration in the gas phase.

The flux of the species in the medium is assumed to follow the Fick's law equation:

$$J_T = -D(C_T)\frac{\partial C_T}{\partial x} \tag{12.6-2}$$

whereby the flux equation is expressed in terms of the *total* concentration gradient, and the diffusion coefficient is a function of this total concentration.

If the mechanism of transport into the porous medium is the parallel pore and surface diffusion mechanism, the flux equation can be written in terms of the two individual concentration gradients as follows:

$$J_T = -\varepsilon D_p \frac{\partial C}{\partial x} - (1-\varepsilon)D_s \frac{\partial C_\mu}{\partial x} \tag{12.6-3}$$

To conform this flux equation (12.6-3) to the form involving the total concentration (12.6-2), we make the following transformation:

$$J_T = -\left[\varepsilon D_p + (1-\varepsilon)D_s \frac{\partial C_\mu}{\partial C}\right]\left(\frac{\partial C}{\partial C_T}\right)\frac{\partial C_T}{\partial x} = -D(C_T)\frac{\partial C_T}{\partial x} \tag{12.6-4}$$

where the total diffusion coefficient is written in terms of two individual diffusion coefficients and the isotherm behaviour:

$$D(C_T) = \left[\varepsilon D_p + (1-\varepsilon)D_s \frac{\partial C_\mu}{\partial C}\right]\left(\frac{\partial C}{\partial C_T}\right) \tag{12.6-5}$$

At steady state, the total flux across the porous medium is a constant as there is no further accumulation on the pore surface as well as the pore volume

$$J_T = -D(C_T)\frac{\partial C_T}{\partial x} \equiv \text{constant} \tag{12.6-6}$$

Integrating the flux equation with respect to x subject to the following constant boundary conditions:

$$x = 0; \quad C_T = C_{T0} \quad (12.6\text{-}7a)$$

and

$$x = L; \quad C_T \approx 0 \quad (12.6\text{-}7b)$$

we obtain the following equation for the total concentration:

$$\frac{x}{L} = \frac{\int_{C_T}^{C_{T0}} D(u)\,du}{\int_0^{C_{T0}} D(u)\,du} = \frac{\int_0^{C_{T0}} D(u)\,du - \int_0^{C_T} D(u)\,du}{\int_0^{C_{T0}} D(u)\,du} = \frac{J(C_{T0}) - J(C_T)}{J(C_{T0})} \quad (12.6\text{-}8)$$

The above equation provides a very interesting means to determine the total concentration as a function of the distance along the medium. Experimentally, the steady state flux $J(C_{T0})$ can be readily obtained for any value of C_{T0} at the inlet of the medium. In general, the following figure (12.6-1) shows a typical behaviour of the steady state flux $J(C_T)$ versus C_T. If this plot is linear, it suggests that the diffusion and adsorption properties do not change with concentration. For such cases, the steady state concentration distribution is linear. What is shown below is a procedure to determine the profile when there is a variation of diffusion and adsorption properties with concentration. Thus, the task is that what is the steady state concentration profile when the pressure of the supply reservoir is C_{T0}.

The steady state flux corresponding to C_{T0} is $J(C_{T0})$ shown as point A in the plot of Figure 12.6-1. At an arbitrary concentration C_T ($0 < C_T < C_{T0}$), we draw two lines shown as dotted lines in the figure to obtain the flux $J(C_T)$. The position at which the total concentration is C_T is calculated from eq. (12.6-8), that is:

$$x = L \left[\frac{J(C_{T0}) - J(C_T)}{J(C_{T0})} \right] \quad (12.6\text{-}9)$$

Thus, by choosing a number of concentration $C_T < C_{T0}$, the steady state concentration profile can be generated.

12.6.2 Functional Dependence of the Diffusion Coefficient

The determination of the concentration profile from the previous section also provides a useful step to determine the concentration dependence of the diffusion coefficient $D(C_T)$. Recall the flux equation as:

$$J_T = -D(C_T)\frac{\partial C_T}{\partial x} \qquad (12.6\text{-}10)$$

The steady state flux is measured experimentally (Figure 12.6-1) and the previous section shows how the concentration profile C_T versus x is obtained. Thus, the diffusion coefficient $D(C_T)$ can be calculated from eq. (12.6-10). An alternative approach to this is to use the flux curve directly (that is $J(C_T)$ versus C_T). This is done below.

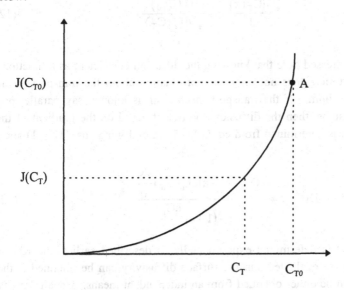

Figure 12.6-1: Plot of the flux versus the total concentration

Integrating the flux equation (12.6-10) with respect to x from x = 0 to L, we get:

$$L J_T(C_T) = \int_0^{C_T} D(u)\, du \qquad (12.6\text{-}11)$$

Differentiating this equation with respect to C_T yields the following expression for the diffusion coefficient:

$$D(C_T) = L\frac{dJ_T(C_T)}{dC_T} \qquad (12.6\text{-}12)$$

The derivative $dJ_T(C_T)/dC_T$ can be evaluated from Figure 12.6-1 and hence the diffusion coefficient is obtained from eq. (12.6-12) without a recourse to the determination of the concentration profile. For a particular experiment with the inlet total concentration of C_{T0}, we have:

$$J_T(C_{T0}) = -D(C_T(x))\frac{dC_T(x)}{dx} \qquad (12.6\text{-}13)$$

Thus, the concentration gradient at any value of C_T within the medium is obtained by combining eqs. (12.6-13) and (12.6-12):

$$\frac{dC_T(x)}{dx} = -\frac{J_T(C_{T0})}{L\dfrac{dJ_T(C_T)}{dC_T}} \qquad (12.6\text{-}14)$$

It should be stressed here that knowing the diffusion coefficient as a function of the total concentration does not infer any information about the transport mechanism within the medium. If the transport mechanism is known, say parallel pore and surface diffusions, then the diffusion coefficient based on the gradient of the total concentration is determined from eq. (12.6-5). Combining eqs. (12.6-1) and (12.6-5), we get:

$$D(C_T) = \frac{\varepsilon D_p + (1-\varepsilon)D_\mu(C_\mu)\dfrac{dC_\mu}{dC}}{\varepsilon + (1-\varepsilon)\dfrac{dC_\mu}{dC}} \qquad (12.6\text{-}15)$$

Thus if $D(C_T)$ is determined experimentally as described earlier, the RHS of eq. (12.6-15) can be evaluated and the surface diffusivity can be obtained if the pore diffusivity can be either obtained from an independent means. Usually this is done with an inert gas such as helium. Since it is non-adsorbing, its Knudsen diffusivity can be obtained from the time lag experiment as there is no surface diffusion in the case of non-adsorbing gas. Knowing the Knudsen diffusivity for helium, the Knudsen diffusivity of any adsorbate is calculated from:

$$D_p = D_{p,He}\sqrt{\frac{M_{He}}{M}} \qquad (12.6\text{-}16)$$

in which we have assumed Knudsen diffusion is the sole transport mechanism.

12.6.3 Further about Time Lag

We have discussed in some details about the time lag method and how it can be obtained directly by the method of Frisch. Before closing out this chapter, we show below some useful properties of time lag (Barrer, 1967, 1968).

Whether the flow mechanism by diffusion or viscous in nature, the conversation equation of mass requirement embodied in the following expression:

$$\frac{\partial C}{\partial t} = -\text{div}(J) \tag{12.6-17}$$

where C is the total concentration per unit volume of the medium and J is the mass transfer per unit total cross-sectional area of the medium. If the flow is uni-directional, the above mass balance equation will become:

$$\frac{\partial C}{\partial t} = -\frac{\partial J}{\partial x} \tag{12.6-18}$$

The boundary conditions and initial condition are given as in eqs. (12.2-4) and (12.2-5).

We now proceed with the Frisch method and apply it on eq. (12.6-18). Integrating the mass balance equation with respect to x from x to some plane X, where $0 < X < L$, we have:

$$\int_x^X \frac{\partial C}{\partial t} dx = -J(X,t) + J(x,t) \tag{12.6-19}$$

Integration with respect to x again over the full domain of x (that is from 0 to L) gives:

$$\int_0^L \int_x^X \frac{\partial C}{\partial t} dx = -LJ(X,t) + \int_0^L J(x,t) dx \tag{12.6-20}$$

Let $J_\infty(C_0)$ be the flux at steady state with C_0 being the concentration of the supply reservoir. Adding and subtracting the RHS of the above equation with $LJ_\infty(C_0)$, we have:

$$\int_0^L \int_x^X \frac{\partial C}{\partial t} dx = LJ_\infty(C_0) - LJ(X,t) + \int_0^L [J(x,t) - J_\infty(C_0)] dx \tag{12.6-21}$$

We define Q(X,t) as the quantity passing through the plane X from $t = 0$ up to time t per unit cross-sectional area, that is:

$$Q(X,t) = \int_0^t J(X,t) dt \tag{12.6-22}$$

Now integrating eq.(12.6-21) with respect to time from 0 to t yields:

$$\int_0^L \int_x^X C(z,t) dz dx = LtJ_\infty(C_0) - LQ(X,t) + \int_0^t \int_0^L [J(x,t) - J_\infty(C_0)] dx\, dt \tag{12.6-23}$$

Solving for Q(X,t) gives

$$Q(X,t) = tJ_\infty(C_0) - \frac{1}{L}\int_0^L\int_x^X C(z,t)dzdx - \frac{1}{L}\int_0^t\int_0^L[J_\infty(C_0) - J(x,t)]dxdt \quad (12.6\text{-}24)$$

As time is sufficiently large, the amount Q(X,t) will reach an asymptote of $Q_a(X,t)$:

$$Q_a(X,t) = tJ_\infty(C_0) - \frac{1}{L}\int_0^L\int_x^X C_\infty(z)dzdx - \frac{1}{L}\int_0^t\int_0^L[J_\infty(C_0) - J(x,t)]dxdt \quad (12.6\text{-}25)$$

where C_∞ is the steady state concentration. This linear asymptote $Q_a(X,t)$ cuts the time axis at a value, called the time lag. It is given below:

$$T_X = \frac{\int_0^L x\, C_\infty(x)\, dx - L\int_X^L C_\infty(x)\, dx + \int_0^L [Q_\infty - Q(x)]\, dx}{J_\infty(C_0)\, L} \quad (12.6\text{-}26)$$

where

$$Q_\infty - Q(x) = \int_0^\infty [J_\infty(C_0) - J(x,t)]\, dt \quad (12.6\text{-}27)$$

Eq. (12.6-26) is the time lag for any position X in the medium. To obtain the time lag at the inlet of the medium, we set X = 0 and that at the exit, replace X by L. Thus the two time lags measured experimentally are:

$$T_0 = \frac{\int_0^L (x-L)C_\infty(x)dx + \int_0^\infty\int_0^L[J_\infty(C_0) - J(x,t)]dxdt}{J_\infty(C_0)L} \quad (12.6\text{-}28a)$$

$$T_L = \frac{\int_0^L xC_\infty(x)dx + \int_0^\infty\int_0^L[J_\infty(C_0) - J(x,t)]dxdt}{J_\infty(C_0)L} \quad (12.6\text{-}28b)$$

The difference between these two time lags is:

$$\Delta T = T_L - T_0 = \int_0^L \frac{C_\infty(x)dx}{J_\infty(C_0)} \quad (12.6\text{-}29)$$

We see that this time lag difference involves the steady state concentration profile only. If we rewrite this difference as follows:

$$\Delta T = \frac{A\int_0^L C_\infty(x)\, dx}{A\, J_\infty(C_0)} \quad (12.6\text{-}30)$$

where A is the cross-sectional area of the medium. The numerator is the total amount of the adsorbate in the medium at steady state, and the denominator is the mass transfer rate at steady state. Thus the difference in time lags is the ratio of the capacity of the medium divided by the mass transfer rate, and it has a physical meaning of a hold-up time in the medium. This difference can be used to check the integrity of the time lag data.

Special case: Fick's law
If the flux $J(x,t)$ is governed by the Fick's law and the diffusion coefficient is a function of concentration only (not on distance or time)

$$J(x,t) = -D(C)\frac{\partial C}{\partial x} \qquad (12.6\text{-}31)$$

Integrating this equation with respect to x from 0 to L gives:

$$\int_0^L J(x,t)\,dx = \int_0^{C_0} D(u)\,du = L\,J_\infty(C_0) \qquad (12.6\text{-}32)$$

With the above equation, the time lags given in eqs. (12.6-28) are reduced to:

$$T_0 = \frac{\int_0^L (x-L)C_\infty(x)\,dx}{J_\infty(C_0)L} \qquad (12.6\text{-}33)$$

$$T_L = \frac{\int_0^L x C_\infty(x)\,dx}{J_\infty(C_0)L} \qquad (12.6\text{-}34)$$

Thus in this special case of diffusion coefficient being a function of only concentration, the time lags are function of only the steady state behaviour as shown in the above equations.

12.7 Other Considerations

The time lag method is shown to be a useful tool for the characterisation of a porous medium. Conditions are usually chosen in such a way that the constant boundary conditions are satisfied (12.2-4). This is usually possible but there are situations where the receiving reservoir is small and its pressure can not be maintained to satisfy the zero boundary condition (12.2-4b). In such cases, the pressure will rise and the boundary condition at the exit of the medium is replaced by:

$$x = L; \quad C = C_b(t) \tag{12.7-1}$$

where $C_b(t)$ is the time dependent concentration of the receiving reservoir, and it satisfies the mass balance around the receiving reservoir:

$$V \frac{dC_b(t)}{dt} = -A\, J(x,t)\big|_{x=L} \tag{12.7-2}$$

Here V is the volume of the reservoir and A is the cross section area of the medium. Barrie et al. (1975) have solved this problem with constant diffusion coefficient. Nguyen et al. (1992) addressed the same problem and proposed the following form for the pressure of the receiving reservoir:

$$C_b(t) = C_0 \left[1 - \sum_j a_j e^{-\beta_j t} \right] \tag{12.7-3}$$

Although mathematical solutions are always possible for time varying boundary conditions, experimental preparation should be exercised such that the constant boundary conditions hold during the course of experiment. This would then simplify the analysis and hence the ease of obtaining the diffusion coefficient, which is after all the main purpose of the time lag method.

Because of its versatility, the time lag method has been applied to many other cases, such as:
(a) Permeation with first order reaction (Ludolph et al., 1979; Leypoldt and Gough, 1980)
(b) Permeation with serial and/or parallel paths (Chen and Rosengerg, 1991; Ash et al., 1963, 1965; Jaeger, 1950)
(c) Permeation in cylindrical and spherical media (Barrer, 1941, 1944; Jaeger, 1946; Crank, 1975).

Review of some of these aspects has been given in Rutherford and Do (1997).

12.8 Conclusion

This chapter has addressed the method of time lag, and we have shown its application to a large number of diffusion and adsorption problems to show its utility in the determination of the diffusion coefficient as well as adsorption parameters. The central tool in the time lag analysis is the Frisch's method, and such a method has allowed us to obtain the expression of the time lag without any recourse to the solution of the concentration distribution within the medium. We shall present in the next few chapters other methods and they all complement each other in the determination of parameter.

13
Analysis of Steady State and Transient Diffusion Cells

13.1 Introduction

The last chapter shows the utility of the time lag method and its applications in the characterization of diffusion and adsorption of pure component systems. In this chapter we address the diffusion cell method, which is used mainly with systems containing two solutes. The process involves the counter-diffusion of these two solutes through a porous medium from one chamber to the other. Usually both of these chambers are open, but there are applications where one of the chambers is a closed chamber. There are two modes of operation of the diffusion cell method. One is the steady state diffusion cell, and the other is the transient diffusion cell.

The steady state diffusion cell was first developed by Wicke and Kallanbach in 1941. Hereafter we shall refer this method as the WK method. In their method, a pellet or many pellets are mounted in parallel between two open chambers (Figure 13.1-1). In one chamber, one component (labelled as A) is flowing into and out of the chamber by convection, and in the other chamber another component (B) is also flowing into and out of that chamber. The residence time in these two chambers are usually much smaller than the diffusion time through the pellet. The species A diffuses through the pellet in the opposite direction to the diffusion path of the species B. The pressures of the two chambers are maintained the same, and hence the counter-flows of the two species are by the mechanism of combined molecular diffusion and Knudsen diffusion. The fluxes of these solutes are calculated by simply measuring the concentrations of A and B in the exit streams of the two chambers. This steady state method of Wicke-Kallanbach is very simple to carry out and it provides a simple means to calculate steady state fluxes through the pellet. However, it does have a number of problems:

(a) dead end pores are not characterized by the measurement of the steady state flux as they do not contribute to the through-flux
(b) incorrect determination of the diffusivity if the pellet is anisotropic

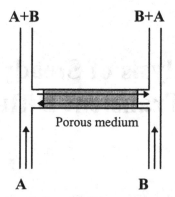

Figure 13.1-1: Wicke-Kallenbach diffusion cell

To overcome these problems, the operation of the diffusion cell in the transient mode will reveal the contribution of the dead end pores. It is important to obtain the information of these dead end pores as they are usually the pores providing most of the adsorption capacity in the pellet. The principles of the steady state and transient operations of the diffusion cell are very similar to the principles of the time lag presented in the last chapter. The dead end pores are not reflected in the time lag information. Their information must be obtained from the analysis of the transient curve describing the approach of the receiving reservoir's pressure towards steady state. Thus, in order to understand the diffusion characteristic of a pellet, both the steady state and transient operations should be carried out.

To operate the diffusion cell under transient condition, the concentration of one of the solutes is perturbed in one chamber and its concentration in the other chamber is monitored. The time-variation of that concentration will depend on the interplay of various processes occurring inside the pellet. Those processes responsible to the flow through the pellet are reflected in the response, while those processes occurring inside the pellet but not directly contributing to the through flux will be reflected in the response as a secondary level. This will be clear later when we deal with the analysis of the transient diffusion cell.

Despite the fact that the transient operation can provide additional information about the system, for example the dead end pore, it does suffer a number of disadvantages:

(a) nonisothermal behavior may be important
(b) mathematical analysis is more tedious than the steady state analysis

There are many ways that we can invoke a transient operation. The common ways used widely are as follows. A component (usually an inert gas but this is not necessary) is allowed to flow in both chambers. Once this is achieved and the pressures of both chambers are equalized, the other solute is injected into one chamber either as a pulse or a step input. In either mode of injection, the concentration of this solute is monitored at the other chamber (Figure 13.1-2).

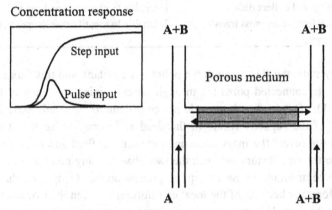

Figure 13.1-2: Transient operation of the diffusion cell

In the pulse injection method, the response curve contains information about the various processes occurring inside the pellet. The moment method is applied to analyze the response curve. The zero-order moment gives the amount injected into the system. The first order moment contains information about the processes responsible directly to the through flux, while the second-order moment contains information about the secondary processes occurring in the pellet. Similar to the pulse injection method, the response curve of the step injection method also contains information about all processes occurring in the pellet. This curve is usually analyzed by matching the time domain solution to the experimental data. The steady state of the step-injection response contains only information about the through processes while the transient part of the curve contains information of all processes.

13.2 Wicke-Kallenbach Diffusion Cell

The steady state diffusion cell is composed of a porous medium bound by two chambers. The porous medium can be either a particle or a collection of particles mounted in parallel as shown in Figure 13.1-1. The advantages and disadvantages of the steady state diffusion cell are listed in the following table.

Table 13.2-1: Advantages and disadvantages of the steady state diffusion cell

Advantages	Disadvantages
1. Easy to set up and collect data	1. Unable to characterize dead end pores
2. No heat interference on mass transfer	2. Unable to identify anisotropic pellets
3. Simple analysis	

At steady state, the flux through the pellet is a constant, and this flux is the flux through the interconnected pores (or through pores) joining two ends of the pellet (Figure 13.2-1). The dead end pores do not contribute to the steady state flux even though most of the capacity reside in the dead end pores. Also the steady state method does not reveal the mass exchange between the fluid and adsorbed phases. This is due to the equilibrium between the two phases at any point within the pellet and hence no contribution of the adsorption process on the steady state flux through the pellet. Moreover because of the local equilibrium between the two phases at any points within the particle, there is no heat release and hence isothermal condition is always ensured in the steady state operation.

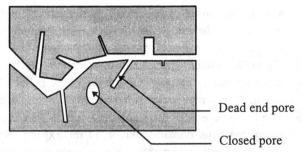

Figure 13.2-1: Through pores, dead end pores and closed pores in a particle

The steady state Wicke-Kallabach method is usually conducted with a binary system with an aim of determining the binary diffusivity and Knudsen diffusivity in the porous medium. In this binary system, one gas (A) is flowing into and out of one chamber, and the other gas (B) is flowing into and out of the other chamber.

Due to the concentration gradient imposed across the pellet, a counter-current mass transfer will occur. Let N_A be the steady state flux of the component A and N_B be that of the component B. If the total pressure is maintained constant throughout the system, the mechanism for mass transfer through the particle is due to the combined molecular and Knudsen diffusions. The steady state flux of the component A for the case of a cylindrical capillary is (Chapters 7 and 8):

$$N_A = \frac{cD_{AB}}{\sigma_{AB} L} \ln\left(\frac{1 - \sigma_{AB} y_2 + D_{AB}/D_{K,A}}{1 - \sigma_{AB} y_1 + D_{AB}/D_{K,A}}\right) \qquad (13.2\text{-}1)$$

where c is the total molar concentration ($= P/R_g T$ for ideal gas), D_{AB} is the molecular binary diffusivity, $D_{K,A}$ is the Knudsen diffusivity of the component A, L is the length of the capillary, y_1 and y_2 are mole fractions of A at two ends of the capillary, and σ_{AB} is defined as:

$$\sigma_{AB} = 1 - \sqrt{\frac{M_A}{M_B}} \qquad (13.2\text{-}2)$$

The Knudsen diffusivity for a cylindrical capillary is given by:

$$D_{K,A} = \frac{2r}{3}\sqrt{\frac{8R_g T}{\pi M_A}} \qquad (13.2\text{-}3)$$

For the case of pellet, the presence of uneven pore size as well as the interconnection of pores and tortuous path of diffusion, the flux expression for the pellet case has to be derived from that for a cylindrical capillary (eq. 13.2-1) through some model about the structure of the pellet. One such model is the parallel path pore model and in this model the "effective" flux is calculated by summing the combined Knudsen and molecular binary diffusions over each increment of pore volume. The expression for the steady state effective flux is:

$$N_{A,\text{eff}} = \frac{cD_{AB}}{q\,\sigma_{AB} L} \int_0^\infty \ln\left(\frac{1 - \sigma_{AB} y_2 + D_{AB}/D_{K,A}(r)}{1 - \sigma_{AB} y_1 + D_{AB}/D_{K,A}(r)}\right) f(r)\, dr \qquad (13.2\text{-}4)$$

where f(r) dr is the void volume having pore radii between r and r + dr per unit volume of the pellet, and it has the following property

$$\varepsilon = \int_0^\infty f(r)\, dr \qquad (13.2\text{-}5)$$

with ε being the particle porosity. The effective steady state flux (13.2-4) has units of moles transported per unit <u>total</u> cross sectional area of the pellet per unit time. In that equation, the parameter q is the tortuosity factor accounting for the tortuous path of diffusion.

The effective steady state flux is usually defined in terms of a parameter called the effective diffusivity as follows:

$$N_{A,eff} = -D_e \frac{dC}{dx} \qquad (13.2\text{-}6)$$

Integrating this equation subject to constant boundary conditions at two ends of the pellet, we get (assuming constant D_e):

$$N_{A,eff} = D_e c (y_1 - y_2) \qquad (13.2\text{-}7)$$

Comparing eqs. (13.2-7) and (13.2-4) yields the following expression for the effective diffusivity:

$$D_e = \frac{D_{AB}}{q\,\sigma_{AB}(y_1 - y_2)} \int_0^\infty \ln\left(\frac{1 - \sigma_{AB} y_2 + D_{AB}/D_{K,A}(r)}{1 - \sigma_{AB} y_1 + D_{AB}/D_{K,A}(r)}\right) f(r)\,dr \qquad (13.2\text{-}8)$$

We see that in the transition regime (that is when bulk and molecular diffusions are both operating) the effective diffusivity is not just a function of the system parameters but also on the operating conditions as well.

When pore size is very small or pressure is very low, the Knudsen diffusion will dominate the transport, eq. (13.2-8) is reduced to:

$$D_e = \frac{1}{q} \int_0^\infty D_{K,A}(r)\,f(r)\,dr \qquad (13.2\text{-}9)$$

The effective diffusivity in the regime of Knudsen diffusion is simply the average of the Knudsen diffusivity over the pore size distribution. The steady state flux in this case takes a simple form:

$$N_{A,eff} = \frac{c(y_1 - y_2)}{q\,L} \int_0^\infty D_{K,A}(r)\,f(r)\,dr \qquad (13.2\text{-}10)$$

On the other hand, when the pore size is large or the pressure is high, the molecular diffusion will control the overall flux through the pellet. For this case, eq. (13.2-8) is reduced to:

$$D_e = \left(\frac{\varepsilon}{q}\right) \frac{D_{AB}}{\sigma_{AB}(y_1 - y_2)} \ln\left(\frac{1-\sigma_{AB}y_2}{1-\sigma_{AB}y_1}\right) \qquad (13.2\text{-}11)$$

and the steady state flux is:

$$N_{A,\text{eff}} = \left(\frac{\varepsilon}{q}\right) \frac{c\, D_{AB}}{\sigma_{AB} L} \ln\left(\frac{1-\sigma_{AB}y_2}{1-\sigma_{AB}y_1}\right) \qquad (13.2\text{-}12)$$

Having the general expression for the steady state flux (eq. 13.2-8) and the two limits when either Knudsen diffusion or molecular diffusion dominates the transport (eqs. 13.2-10 and 13.2-12), experiments can be conducted to extract the necessary parameters.

For example, the experimental conditions can be adjusted such that the molecular diffusion is the dominating mechanism (high pressure). This can be experimentally confirmed by the validation of the independence of the steady state flux with respect to the total pressure because the total molar concentration c is proportional to the total pressure while the molecular diffusivity is inversely proportional to pressure (eq. 13.2-12). In this regime, eq. (13.2-12) can be used to determine the tortuosity factor if the binary diffusion coefficient D_{AB} is known.

We can also adjust the conditions, for example low pressures, such that the Knudsen diffusion is the controlling mechanism. In this case, eq. (13.2-10) can be used to describe the steady state flux and to extract the tortuosity factor if the pore volume distribution is known from other independent experiments, such as the capillary condensation experiment as learnt in Chapter 3.

The porosity ε used in the above analysis is that of the through pores, which are pores connected between the two ends of the pellet, and the tortuosity factor is for those pores. As mentioned earlier, the steady state method does not reveal any features about the dead end pores as well as the adsorptive characteristics if there are any. Usually pellets are made by compressing small grains or crystals (usually of the order of 1 micron) at very high pressure.

(a) If those grains remain discrete after the compression and they themselves have pores usually much smaller than pores formed by the inter-particle voids, these pores are called the dead end pores. Using the steady state method, the measured flux does not reflect any contribution from those dead end pores. Only pores between the grains are characterized. In this case the porosity in eq. (13.2-8) is the porosity of the through-pores, and the pore volume distribution f(r) only reflects those pores.

(b) If the grains form a continuum joining two ends of the pellet, then the diffusion through the grain will also contribute to the measured steady state flux. The diffusion through the grain is very complex. If the pores of the grain are of molecular dimension then they can only accommodate molecules in the adsorbed state and the diffusion of the adsorbed molecules is called the surface diffusion. In this case the total flux is the summation of the pore flux and the surface flux. The pore flux is given by eq. (13.2-8) and the surface diffusion flux can be calculated from the following equation

$$N_{A,s} = -D_{\mu,\text{eff}} \frac{dC_\mu}{dx} \qquad (13.2\text{-}13)$$

where $D_{\mu,\text{eff}}$ is the effective surface diffusivity and C_μ is the adsorbed concentration. The integral of the above equation subject to constant boundary conditions is:

$$N_{A,s} = D_{\mu,\text{eff}} \frac{(C_{\mu,1} - C_{\mu,2})}{L} \qquad (13.2\text{-}14)$$

For simplicity we have assumed constant surface diffusivity although in general the surface diffusivity is a strong function of adsorbed concentration. Thus, the integral equation (13.2-14) is strictly applicable to the situation where the difference in concentration is small enough to ensure constant surface diffusivity or when the loading is low. For such a case, the total flux at steady state is:

$$N_{A,\text{eff}} = \frac{cD_{AB}}{q\,\sigma_{AB} L} \int_0^\infty \ln\left(\frac{1-\sigma_{AB} y_2 + D_{AB}/D_{K,A}(r)}{1-\sigma_{AB} y_1 + D_{AB}/D_{K,A}(r)}\right) f(r)\,dr + \frac{D_{\mu,\text{eff}}(C_{\mu,1}-C_{\mu,2})}{L} \qquad (13.2\text{-}15)$$

Usually the pore volume diffusion flux is characterized by the use of a non-adsorbing gas. When an adsorbing gas is used the contribution of the surface diffusion can be obtained by subtracting the pore diffusion flux from the total flux. Knowing the surface diffusion flux, the effective surface diffusivity then can be calculated.

13.3 Transient Diffusion Cell

The steady state diffusion cell method, despite its simplicity in operation and measurement, does not give us information about the dead-end pores. The information about the dead-end pores is important in two respects:
(a) Most of the adsorption capacity resides in the dead-end pores
(b) Transient operation of a solid containing dead-end pores might be controlled by the ability of the adsorbate molecules to enter the dead-end pores, usually micropores. This means that the time scale of the adsorption process is

controlled by the time it takes for the molecules to penetrate into the dead-end pores.

If these two aspects are needed to understand the given adsorption system, then the steady state diffusion cell is not the answer, but rather we must resort to the transient diffusion cell. Before addressing the utility of the transient mode in the determination of the dead-end pore characteristics, let us illustrate the method with a non-adsorbing gas. The difference between a non-adsorbing gas and an adsorbing gas is the formulation of the mass balance equation inside the particle. The mass balance equations for the two chambers are the same for both cases as there is no adsorption in the chambers.

In the transient operation of the diffusion cell, an inert gas is usually passed through the two chambers and their total pressures are adjusted to be the same. At time t = 0+ an impulse of tracer, either non-adsorbing gas or adsorbing gas, is injected into one chamber and its concentration is monitored at the other chamber. Perfect mixing is assumed in both chambers. This is usually satisfied with a proper design of the system, for example having the influent stream injected directly to the flat face of the particle.

13.3.1 Mass Balance around The Two Chambers

The mass balance equations describing the concentration change for the two chambers are:

$$V_1 \frac{dC_{b1}}{dt} = AD_a \frac{\partial C}{\partial x}\bigg|_{x=0} + F_1 C_{b1}^0(t) - F_1 C_{b1} \qquad (13.2\text{-}16a)$$

$$V_2 \frac{dC_{b2}}{dt} = -AD_a \frac{\partial C}{\partial x}\bigg|_{x=L} - F_2 C_{b2} \qquad (13.2\text{-}16b)$$

where V_1 and V_2 are volumes of the two chambers, A is the cross-sectional area of the pellet, F_1 and F_2 are volumetric flow rates into and out of the two chambers, and C_{b1}^0 is the time-varying input of the tracer into the chamber 1.

These two mass balance equations generally describe the concentration evolution in the two chambers. Since our purpose of using the diffusion cell to extract useful information about the diffusion, the cell design is usually such that the mass balance equations can be rendered to a simple form for subsequent simpler analysis. With this objective in mind, the chambers' volumes can be made small enough such that the gas residence times in the two chambers are much smaller than the diffusion time inside the particle. When this is satisfied experimentally, the mass balance equations (13.2-16) are reduced to:

$$C_{b1} = C_{b1}^0(t) \tag{13.2-17a}$$

$$F_2 C_{b2} = -AD_a \left.\frac{\partial C}{\partial x}\right|_{x=L} \tag{13.2-17b}$$

13.3.2 The Type of Perturbation

The transient operation of the diffusion cell depends on the shape of this input versus time. Usually the following three inputs in concentration are normally used:

Impulse
The impulse function takes the form

$$C_{b1}^0(t) = A \cdot \delta(t) \tag{13.2-18}$$

such that

$$\int_0^\infty F_1 \, C_{b1}^0(t) \, dt = AF_1 = Q \tag{13.2-19}$$

where Q is the amount of tracer injected into the system.

Square pulse
Injecting a tracer into the system as an impulse is usually not possible in practice as it takes a finite time to complete the injection. Let this injection time be t_0, the concentration of the tracer at the inlet of the chamber 1 is:

$$C_{b1}^0(t) = A \cdot [U(t) - U(t - t_0)] \tag{13.2-20}$$

such that

$$\int_0^\infty F_1 \, C_{b1}^0(t) \, dt = AF_1 t_0 = Q \tag{13.2-21}$$

The unit of A is mole/cc, while the unit of A in the case of impulse is mole-sec/cc as the delta function has an unit of sec^{-1}.

Step input
Instead of injecting the tracer as a pulse (impulse or square pulse), the tracer can be injected as a step injection. The step input is simply:

$$C_{b1}^0(t) = AU(t) \tag{13.2-22}$$

where A is the concentration of the step input (moles/cc).

Another type of input which is not normally used is the periodic input. With this input the standard frequency response method can be applied and the system parameters can be obtained from the amplitude and phase analysis.

13.3.3 Mass Balance in the Particle

As mentioned earlier the difference between the non-adsorbing gas and the adsorbing gas is the formulation of the mass balance in the particle.

13.3.3.1 Non-adsorbing Gases

For non-adsorbing gas, the mass balance in the particle is:

$$\varepsilon_T \frac{\partial C}{\partial t} = \varepsilon D_p \frac{\partial^2 C}{\partial x^2} \qquad (13.2\text{-}23a)$$

where ε is the porosity of the through pores, ε_T is the total porosity (that is porosity of the through pores plus the porosity of the dead-end pores) and D_p is the effective pore diffusivity in the through-pore network.

Since the total pressures of the two chambers are maintained the same, the viscous flow is either absent or negligible compared to the diffusive flow. Eq. (13.2-23a) only accounts for the diffusive flow.

The boundary conditions of the mass balance equation (13.2-23a) are:

$$x = 0; \qquad C = C_{b1} \qquad (13.2\text{-}23b)$$

$$x = L; \qquad C = C_{b2} \qquad (13.2\text{-}23c)$$

of which we have ignored the film resistances at the two flat surfaces of the particle. This is very reasonable as the system can be designed to minimize this resistance compared to the internal diffusional resistance.

13.3.3.2 Adsorbing Gases

When dealing with adsorbing gases, it is important to consider the topology of the adsorbed phase. We shall consider two topologies of the adsorbed phase. In the first topology, the adsorbed phase behaves like either a continuous surface connecting two ends of the pellet and hence the mobility of the adsorbed phase along the particle is accounted for (Figure 13.2-2). Activated carbon is an example of this topology. In the second topology, the adsorbed phase is a collection of transverse dead-end pores, and there is a resistance to molecular flow inside those pores. There are two sub-classes of this second topology. In the first sub-class the size of the dead end pores is of molecular dimension and these pores accommodate molecules only in the adsorbed state (Figure 13.2-3), while in the second sub-class the size of the dead-end pores is large enough to accommodate molecules in both free and adsorbed forms (Figure 13.2-4). Zeolite pellet is an example of the first sub-class, while alumina is a good example of the second sub-class. Let us now consider these two topologies.

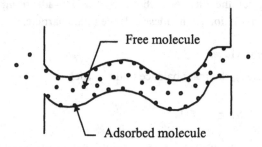

Figure 13.2-2: Porous solid with dead-end pores of molecular dimension

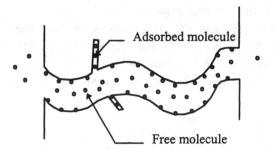

Figure 13.2-3: Porous solid with dead-end pores of molecular dimension

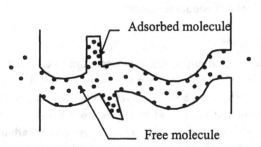

Figure 13.2-4: Porous solid with dead-end pores of dimension such that two forms of molecule exist

13.3.3.2.1 Topology 1: Continuous Surface joining The Two Ends of the Pellet

The mass balance equation describing the concentration change in the pellet is:

$$\varepsilon \frac{\partial C}{\partial t} + (1-\varepsilon)\frac{\partial C_\mu}{\partial t} = \varepsilon D_p \frac{\partial^2 C}{\partial x^2} + \frac{\partial}{\partial x}\left[(1-\varepsilon)D_\mu \frac{\partial C_\mu}{\partial x}\right] \quad (13.2\text{-}24)$$

where C_μ is the adsorbed concentration and D_μ is the diffusivity of the adsorbed species, which is generally a function of the adsorbed concentration. The pore diffusivity describes the diffusion of the free species in the through-pore network.

If the rate of mass exchange between the two phases is faster than the diffusion rate, we can invoke the local equilibrium between the two phases, that is:

$$C_\mu(x,t) = f[C(x,t)] \quad (13.2\text{-}25)$$

where f is the functional form for the adsorption isotherm.

For linear adsorption isotherm, $C_\mu(x,t) = K \cdot C(x,t)$, the mass balance equation (13.2-24) becomes:

$$\frac{\partial C}{\partial t} = D_{app} \frac{\partial^2 C}{\partial x^2} \quad (13.2\text{-}26a)$$

where the apparent diffusivity is given by:

$$D_{app} = \frac{\varepsilon D_p + (1-\varepsilon)K D_\mu}{\varepsilon + (1-\varepsilon)K} \quad (13.2\text{-}26b)$$

When K = 0 (that is non-adsorbing gas) the above equation reduces to:

$$\varepsilon \frac{\partial C}{\partial t} = \varepsilon D_p \frac{\partial^2 C}{\partial x^2} \quad (13.2\text{-}27)$$

13.3.3.2.2 Topology 2a: Dead-End Pores of Molecular Dimension

If the dead-end pores are of molecular dimension, adsorbate molecules in these pores are subject to the attraction force of the walls, therefore the only state in these pores is the adsorbed state. Assuming there is no adsorption on the wall of the through pores, the mass balance equation in the particle is:

$$\varepsilon \frac{\partial C}{\partial t} + (1-\varepsilon)\frac{\partial \overline{C_\mu}}{\partial t} = \varepsilon D_p \frac{\partial^2 C}{\partial x^2} \quad (13.2\text{-}28)$$

where ε is the porosity of the through-pore, and C_μ is the mean concentration in the dead-end pore and is defined as follows:

$$\overline{C}_\mu = \frac{1}{V_\mu} \int_V C_\mu \, dV_\mu \tag{13.2-29}$$

with V_μ being the volume of the solid containing the dead-end pores, that is the particle volume minus the through pore volume.

The concentration in the dead end pore is distributed along the pore, and its distribution is governed by the following mass balance equation in the dead-end pore:

$$\frac{\partial C_\mu}{\partial t} = \frac{1}{r^s} \frac{\partial}{\partial r}\left[r^s D_\mu(C_\mu) \frac{\partial C_\mu}{\partial r} \right] \tag{13.2-30}$$

where r is the distance along the dead end pore, and D_μ is the micropore diffusivity.

The boundary conditions of eq.(13.2-30) are:

$$r = 0; \quad \frac{\partial C_\mu}{\partial r} = 0 \tag{13.2-31a}$$

$$r = R; \quad C_\mu = f[C(x,t)] \tag{13.2-31b}$$

Eq. (13.2-31b) states that at the mouth of the dead-end pore, there is a local equilibrium between the molecules in the macropore and the molecules at the pore mouth of the dead-end pore.

For non-adsorbing gases, the mass balance equation (13.2-28) reduces to eq. (13.2-27).

13.3.3.2.3 Topology 2b: Dead-End Pores with Both Forms of Molecules

When the dead-end pores have a size such that they can accommodate both free and adsorbed molecules, the mass balance of the particle is:

$$\varepsilon \frac{\partial C}{\partial t} + (1-\varepsilon)\left[\varepsilon_\mu \frac{\partial \overline{C}_m}{\partial t} + (1-\varepsilon_\mu) \frac{\partial \overline{C}_\mu}{\partial t} \right] = \varepsilon D_p \frac{\partial^2 C}{\partial x^2} \tag{13.2-32}$$

where ε_μ is the porosity of the dead-end pore, defined as the ratio of the volume of dead-end pore to the volume of solid, C_m is the concentration of the free molecule in the dead-end pore, and C_μ is the concentration of the adsorbed molecule in the dead-end pore.

Carrying out the mass balance in the dead-end pore yields the following equation:

$$\varepsilon_\mu \frac{\partial C_m}{\partial t} + (1-\varepsilon_\mu)\frac{\partial C_\mu}{\partial t} = \varepsilon_\mu D_m \frac{1}{r^s}\frac{\partial}{\partial r}\left(r^s \frac{\partial C_m}{\partial r}\right) \quad (13.2\text{-}33)$$

Assuming a local equilibrium between the two phases inside the dead-end pore, we have the following relationship between the free molecule and the adsorbed molecule:

$$C_\mu(x,r,t) = f[C_m(x,r,t)] \quad (13.2\text{-}34)$$

The boundary conditions of eq. (13.2-33) are:

$$r = 0; \quad \frac{\partial C_m}{\partial r} = \frac{\partial C_\mu}{\partial r} = 0 \quad (13.2\text{-}35a)$$

$$r = R; \quad C_m = C(x,t) \quad (13.2\text{-}35b)$$

For non-adsorbing gases, the mass balance equation (13.2-32) reduces to eq. (13.2-23a) with $\varepsilon_T = \varepsilon + (1-\varepsilon)\varepsilon_\mu$.

13.3.4 The Moment Analysis

When the system equations are linear (that is when the adsorption isotherm is linear or the system is perturbed incrementally), we can apply the method of Laplace transform to solve the set of equations and obtain the inverse by either the method of residues or a numerical inversion scheme. For the two types of input, the impulse and the square input, the inversion is not necessary if we are interested in using the response to extract the system parameters. If this is our goal which is the case for the diffusion cell method, then the method of moment can be useful for this purpose.

Given a concentration response, the n-th moment is defined as follows:

$$m_n = \int_0^\infty t^n C(t)\, dt \quad (13.2\text{-}36)$$

and its normalized n-th moment is scaled against the zero-order moment:

$$\mu_n = \frac{m_n}{m_0} \quad (13.2\text{-}37)$$

The second central moment is of interest in the determination of parameter and is defined as:

$$\mu_2' = \frac{\int_0^\infty (t-\mu_1)^2 \, C(t) \, dt}{m_0} \qquad (13.2\text{-}38)$$

The method procedure is as follows.
(a) First a mathematical model is developed for a given system.
(b) Taking Laplace transform of the mathematical equations and obtain the solution for the concentration that we wish to utilize
(c) The zero-order, first-order and the second central moments are obtained from the solution obtained in the Laplace domain by using the following formula:

$$m_n = (-1)^n \lim_{s \to 0} \frac{d^n \overline{C}(s)}{ds^n} \qquad (13.2\text{-}39)$$

where $\overline{C}(s)$ is the Laplace transform of $C(t)$

$$\overline{C}(s) = \int_0^\infty e^{-st} \, C(t) \, dt \qquad (13.2\text{-}40)$$

(d) Matching the zero-order, first-order and second-order central moments of the theory with those obtained from the experimental response will yield equations to determine the necessary parameters.

We now first illustrate the method of moment by using the simplest case of non-adsorbing gas.

13.3.5 Moment Analysis of Non-Adsorbing Gas

For the case of non-adsorbing gas, the governing equations describing the concentration change in the particle and in the chambers are given in eqs. (13.2-23) and (13.2-17). Taking Laplace transform of those equations, we get:

$$s\overline{C} = D_p \frac{d^2 \overline{C}}{dx^2} \qquad (13.2\text{-}41)$$

$$x = 0; \qquad \overline{C} = \int_0^\infty e^{-st} C_{b1}^0(t) \, dt \qquad (13.2\text{-}42a)$$

$$x = L; \qquad F_2 \overline{C}_{b2} = -A\varepsilon D_p \frac{d\overline{C}}{dx}\bigg|_{x=L} \qquad (13.2\text{-}42b)$$

For the case of perfect impulse of tracer (eq. 13.2-18), eq. (13.2-42a) will become:

$$x = 0; \qquad \overline{C} = \int_0^\infty e^{-st} C_{b1}^0(t)\, dt = A \qquad (13.2\text{-}43)$$

The concentration of the tracer in the chamber 2, C_{b2}, is what we would like to monitor. Thus solving eqs. (13.2-41), (13.2-42b) and (13.2-43) for C_{b2}, we obtain the following normalized first-order moment:

$$\mu_1 = \frac{m_1}{m_0} = \frac{L^2}{6D_p} \frac{\left(\dfrac{3A\varepsilon D_p}{L} + F_2\right)}{\left(\dfrac{A\varepsilon D_p}{L} + F_2\right)} \qquad (13.2\text{-}44)$$

For very high flow rate of the carrier gas in the chamber 2, the normalized first-order moment will reduce to:

$$\lim_{F_2 \to \infty} \mu_1 = \frac{L^2}{6D_p} \qquad (13.2\text{-}45)$$

It is interesting but not surprising that the first-order moment given in the above equation is identical to the time lag dealt with in Chapter 12. They both represent the mean diffusion time of non-adsorbing molecule taken to diffuse from one end of the pellet to the other end.

The normalized first-order moment (eq. 13.2-44) can be matched with the experimental moment

$$(\mu_1)_{exp} = \frac{\int_0^\infty t\, C_{b2}(t)\, dt}{\int_0^\infty C_{b2}(t)\, dt} \qquad (13.2\text{-}46)$$

to obtain the pore diffusivity D_p. Thus the moment method allows us to extract the diffusivity without the need to obtain explicitly the temporal evolution of the tracer concentration in the chamber 2. Another advantage of the moment method is that there is no need to obtain a calibration curve relating the concentration to the signal of the detector as the proportionality constant cancels out in the numerator and the denominator on eq. (13.2-46).

Example 13.2-1: *Diffusion of helium in nitrogen through a Boehmite particle*

Dogu and Smith (1972) used pellets made by compressing porous Boehmite particles. They provided the following data about the system of non-adsorbing helium in nitrogen at 24 °C and 1 atm.

Pellet length, L	2.44 cm
Pellet diameter, D	1.35 cm
Macropore porosity, ε	0.480
Total porosity, ε	0.771
Mean macropore radius, r_M	2990 A
Chamber volume, $V_1 = V_2$	0.5 cc
Volumetric flow rate range, F_2	20 - 150 cc/min

By matching the experimental moment to the theoretical first-order moment, they obtain a value for the diffusivity:

$$\varepsilon D_p = 0.079 \text{ cm}^2/\text{sec}$$

This pore diffusivity is related to the combined diffusivity D_c as:

$$D_p = \frac{D_c}{q} \tag{13.2-47}$$

where

$$\frac{1}{D_c} = \frac{1-\sigma_{12}y_1}{D_{12}} + \frac{1}{D_{K,1}(r_M)} \tag{13.2-48}$$

Under the conditions carried out by Dogu and Smith, the mole fraction of helium is very low, and hence the combined diffusivity becomes:

$$\frac{1}{D_c} = \frac{1}{D_{12}} + \frac{1}{D_{K,1}(r_M)} \tag{13.2-49}$$

At 24 °C and 1 atm, the molecular diffusivity of helium in nitrogen is 0.713 cm²/sec and for a macropore radius of 2990 A, the Knudsen diffusivity is calculated as (Table 7.4-2):

$$D_{K,1} = (9700)(2990 \times 10^{-8} \text{ cm})\sqrt{\frac{297 \text{ K}}{4 \text{ g/gmole}}} = 2.5 \text{ cm}^2/\text{sec}$$

The combined diffusivity is:

$$D_c = \left[\frac{1}{D_{12}} + \frac{1}{D_{K,1}(r_M)}\right]^{-1} = \left(\frac{1}{0.713} + \frac{1}{2.5}\right)^{-1} = 0.555 \text{ cm}^2/\text{sec}$$

Using eq. (13.2-47), we obtain the following tortuosity factor:

$$q = \frac{\varepsilon D_c}{\varepsilon D_p} = \frac{(0.480)(0.555)}{0.079} = 3.37$$

13.3.6 Moment Analysis of Adsorbing Gas

The utility of the transient diffusion cell is not just useful for the study of a diffusion process involving a non-adsorbing gas, it can be used to study adsorption systems as well. Let us illustrate this with the case of pellet with dead-end pores of a size such that two forms of molecule (that is free and adsorbed molecules) are possible. In this case the mass balance equations describing the concentration distribution in the particle are given in eqs. (13.2-32) to (13.2-35). Usually when the diffusion cell method is used to extract parameters, conditions are chosen such that the adsorption isotherm is linear. Thus, the equilibrium relation of eq. (13.2-34) becomes:

$$C_\mu = K C_m \qquad (13.2\text{-}50)$$

Solving the mass balance equations for the case of linear isotherm subject to the boundary conditions (13.2-17) by the method of Laplace transform and from the solution we obtain the following moments when the input is an impulse (Dogu and Ercan, 1983):

$$\mu_1 = \frac{L^2\left[\varepsilon + (1-\varepsilon)\varepsilon_\mu + (1-\varepsilon)(1-\varepsilon_\mu)K\right]}{6\varepsilon D_p} \left[\frac{(3\varepsilon D_p A/L) + F}{(\varepsilon D_p A/L) + F}\right] \qquad (13.2\text{-}51)$$

$$\mu_2' = \frac{L^4\left[\varepsilon+(1-\varepsilon)\varepsilon_\mu+(1-\varepsilon)(1-\varepsilon_\mu)K\right]^2}{6\varepsilon^2 D_p^2} \frac{\left[(\varepsilon D_p A/L)^2+(2\varepsilon D_p AF/5L)+F^2/15\right]}{\left[(\varepsilon D_p A/L)+F\right]^2} +$$

$$\frac{L^2\left[(1-\varepsilon)\varepsilon_\mu+(1-\varepsilon)(1-\varepsilon_\mu)K\right]^2 R^2}{15\varepsilon D_p \varepsilon_\mu D_m(1-\varepsilon)} \frac{\left[(\varepsilon D_p A/L)^2+(4\varepsilon D_p AF/3L)+F^2/3\right]}{\left[(\varepsilon D_p A/L)+F\right]^2} \quad (13.2\text{-}52)$$

The first normalized moment contains the diffusion coefficient in the through-pore, while the second central moment contains the diffusion coefficients in both pores. Matching the first moment (eq. 13.2-51) with experimental moments will allow us to extract the macropore diffusivity, and matching the second moments we would obtain the micropore diffusivity. In matching the second moment, we require the high degree of accuracy of the experimental data as a drift in the tail of the response curve could give rise to the incorrect determination of the second moment.

Dogu and Ercan (1983) used this method to extract dynamic characteristics of a system of ethylene on α-alumnia. The following information is available for their system at 45°C and 1 atm

Micropartice radius, R	$= 20\ \mu$
Pellet length, L	$= 0.6$ cm
Cross-section area, A	$= 1.41$ cm^2
Through-pore porosity, ε	$= 0.36$
Micropore porosity, ε_μ	$= 0.578$
Henry constant, K	$= 40$

Matching the moments, they found

$$\varepsilon D_p = 0.0334 \text{ cm}^2/\text{sec}$$

$$\varepsilon_\mu D_m = 8.9 \times 10^{-6} \text{ cm}^2/\text{sec}$$

13.4 Conclusion

The methods presented in the last two sections can be applied to any other diffusion models in the pellet. Readers are encouraged to apply the method to their specific systems. Despite of the simplicity suggested by the method, the extraction of the micropore diffusivity (second order process) requires a very careful collection of experimental data. If micropore diffusion is dominating the dynamic uptake, the batch adsorber provides a better means to extract the micropore diffusivity, and this will be discussed in Chapter 15.

14

Adsorption and Diffusivity Measurement by a Chromatography Method

14.1 Introduction

Chapters 9 to 11 deal with the dynamic analysis of a single particle exposed to a constant bulk environment. The method of differential adsorption bed discussed in Chapter 11 is suitable for the application of the single particle analysis. A permeation method called the time lag method is useful for characterisation of diffusional flow, viscous flow and surface flow of pure gas through a single pellet (Chapter 12). The diffusion cell method either in steady state mode or transient mode is useful to characterize binary diffusional systems (Chapter 13). All these methods evolve around the analysis of a single particle and they complement each other in the characterization of diffusion and adsorption characteristics of a system. From the stand point of system set-up, the time lag and diffusion cell methods require a careful mounting of a particle or particles between two chambers and extreme care is exercised to avoid any gas by-passing the particle.

The single particle analysis is applicable to the differential adsorption bed or TGA method. Naturally, the single particle analysis is the simplest as it does not require the solution of any additional equation describing concentration variations outside the particle. The method presented in this chapter is the chromatography method and from the stand point of system set-up it is the easiest to set up. All it requires are the careful packing of particles into a cylindrical column, a means to inject some tracer into the column, and of course a means to monitor the exit concentration as a function of time. In the chromatography operation, the injection can be in almost any form provided the tracer concentration decreases to zero in finite time, for example a perfect impulse or a square input. The tracer will propagate down the column with a speed of which the magnitude depends on the affinity between the fluid and adsorbed phases. The spread of the exit concentration versus time depends on the speed of the molecules can diffuse in and out of the particle. Thus by measuring the exit concentration versus time, we can analyse for its mean retention, from which the affinity can be obtained, and its variance from which some information of the dynamic characteristics can be deduced. Usually the chromatography method is utilized to obtain the kinetic information within the particle, but the exit concentration response is affected not only by the diffusional resistance or any resistances within the particle but also by the axial dispersion or any nonideal behaviour of the axial flow along the column. Thus to extract the information on the resistances inside the particle, the axial dispersion contribution on the spread of the column response needs to be carefully isolated. It is this isolation process that makes the chromatography not so attractive in the reliable determination of the internal diffusional resistances. Table 14.1-1 shows the advantages and disadvantages of the various methods.

14.2 The methodology

The method of chromatography is very simple. Particles of uniform size or narrow particle size distribution are packed in a cylindrical column. An inert gas is introduced into the column until the column is stabilised in the sense that the detection at the column exit detects no variation in the signal. This is registered as the base line. After this has been achieved, a pulse of tracer, either in the form of an impulse or a square input, is introduced at the inlet of the column. If the tracer is nonadsorbing, it will exit the column at the mean retention of the system. However, if the tracer is an adsorbing solute, its movement down the column is retarded due to the affinity between the tracer and the particle, and this speed of propagation depends on the magnitude of this affinity. The stronger the affinity, the longer it takes for the tracer to exit the column. The exit concentration is monitored by an appropriate detection device, and it usually exhibits a bell shape response curve with the mean retention time being proportional to the affinity between the two phases. The spread of the curve is a complex function of all dispersion forces in the system. These dispersive forces are:

(a) axial dispersion
(b) film resistance
(c) all resistances within the particle
   (c1) pore diffusion resistance in macropore
   (c2) micropore diffusional resistance
   (c3) finite adsorption resistance

Table 14.1-1: Advantages and disadvantages of various methods

Method	Advantages	Disadvantages
Single particle (DAB) Chapter 9, 10, 11	• Reliable data • can deal with any type of system • isothermality is ensured. • can deal with any mixtures	• very time consuming
TGA Chapters 9, 10, 11	• quick collection of data	• flow rate is limited due to the instability of the balance at high flow rate • can not deal with more than on adsorbate unless coupled with other means such as mass spectrometry • nonisothermality if the adsorbate is strongly adsorbing
Time lag Chapter 12	• very easy to collect and analyse data	• difficult in mounting pellet • nonisothermal if high pressure is used • can deal with only pure component
Diffusion cell Chapter 13	• isothermal in steady state operation	• extraction of dead end pore requires the analysis of second moment • heat effect in transient operation of diffusion cell
Chromatography Chapter 14	• easy to set up and collection of data • quick if only affinity constant is required	• Data analysis is quite cumbersome • nonisothermal operation
Batch adsorber Chapter 15	• easy to set up and collection of data • useful for expensive adsorbates	• non-isothermal operation

If the goal of this method is to determine the internal resistances, the first two resistances (outside the particle) must be isolated either by known correlations or by way of experimental methods.

14.2.1 The General Formulation of Mass Balance Equation

The analysis of this method involves the formulation of mass balance equations describing the concentration distribution in two subsystems:
(a)   mass balance equation in the column
(b)   mass balance equations in the particle.

The mass balance equation describing the concentration distribution along the column accounts for the rate of hold up in the column, the axial dispersion, the convection term and the rate of mass transfer into the particle. This mass balance takes the following form:

$$D_{ax} \frac{\partial^2 C_b}{\partial z^2} - \frac{\partial}{\partial z}(uC_b) - a.J|_R = \varepsilon_b \frac{\partial C_b}{\partial t} \tag{14.2-1}$$

where C_b is the concentration of the tracer in the flowing fluid, D_{ax} is the axial dispersion coefficient, $J|_R$ is the mass transfer rate into the particle per unit interfacial surface area, a is the interfacial area per unit bed volume, ε_b is the bed porosity, and u is the superficial velocity.

If the particles are spherical of radius R, the interfacial area per unit bed volume is

$$a = \frac{3(1-\varepsilon_b)}{R} \tag{14.2-2}$$

The boundary conditions of the mass balance equation (14.2-1) generally take the form

$$z = 0; \quad C_b = C_0(t) \tag{14.2-3a}$$

$$z = L \quad \frac{\partial C_b}{\partial z} = 0 \tag{14.2-3b}$$

Strictly speaking, one should impose the more proper boundary condition at the entrance to allow for the axial dispersion at the inlet, that is

$$z = 0; \quad D_{ax} \frac{\partial C_b}{\partial z} = u(C_b - C_0(t)) \tag{14.2-4}$$

With the long column commonly used in the chromatography operation, the use of eq. (14.2-3a) is satisfactory.

The mass balance equation describing the concentration inside the particle can be written in the following general format:

$$\underline{F}(\underline{C}, x, t) = \underline{0} \qquad (14.2\text{-}5)$$

where the dimension of the vector $\underline{F}$ depends on the dimension of the concentration vector $\underline{C}$ which is defined as

$$\underline{C} = \begin{bmatrix} C_1 & C_2 & \cdots & C_n \end{bmatrix} \qquad (14.2\text{-}6)$$

where C_j represents the concentration of the j-th phase within the particle, with C_1 being the intra-particle concentration of fluid phase which is in contact with the bulk outside the particle. The other concentrations C_j (j = 2, 3, ..., n) represent those of different form in the adsorbed phase.

The boundary condition of eq. (14.2-5) is usually in the form of the film boundary condition, that is

$$x = R; \qquad J\big|_R = k_m \left(C_b - C_1 \big|_R \right) \qquad (14.2\text{-}7)$$

The specific form of eq. (14.2-5) depends on the assumption of the diffusion mechanism within the particle. We shall address this by applying to three specific cases in the next three sections.

14.2.2 The Initial Condition

The initial state of the column can be either in two following situations.
(a) the column is initially free of any tracer molecule
(b) the column is initially equilibrated with a tracer of a concentration C*.

In the first situation, a carrier fluid (which is usually an inert fluid but this is not necessary) is passed through the column, and once this is stabilised a tracer is injected into the column with a concentration of $C_0(t)$ at the inlet. The concentration is chosen such that the adsorption isotherm of this tracer towards the solid packing is linear. This results in a set of linear equations which permit the use of Laplace transform to obtain solution analytically. Knowing the solution in the Laplace domain, the solution in real time can be in principle obtained by some inversion procedure whether it be analytically or numerically. However, the moment method illustrated in Chapter 13 can be utilised to obtain moments from the Laplace solutions directly without the tedious process of inversion.

In the second situation, the column is initially equilibrated with a tracer having a concentration of C*, and once this is done the column is injected with a pulse of tracer having a concentration of C* + ΔC* where ΔC* << C*. Because of this small perturbation in the tracer concentration, the mass balance equations can be linearised around the concentration C*. The resulting linearised mass balance equations are then susceptible to linear analysis of Laplace transform. We shall illustrate more about this later.

The injection of tracer can be in either the following form. For the case of initially tracer-free column

$$z = 0; \quad C_b = A\,\delta(t) \tag{14.2-8}$$

or

$$z = 0; \quad C_b = C_0\left[U(t) - U(t - t_0)\right] = \begin{cases} C_0 & \text{for } 0 < t < t_0 \\ 0 & \text{for } t > t_0 \end{cases} \tag{14.2-9}$$

where $\delta(t)$ is the Dirac delta function and $U(t)$ is the step function.

For the case of column initially equilibrated with C^* concentration, the injection can be in either of the following form:

$$z = 0; \quad C_b = C^* + A\,\delta(t) \tag{14.2-10}$$

or

$$z = 0; \quad C_b = \begin{cases} C^* + \Delta C^* & \text{for } 0 < t < t_0 \\ C^* & \text{for } t > t_0 \end{cases} \tag{14.2-11}$$

14.2.3 The Moment Method

With either the impulse or square injection, the response of the exit concentration versus time exhibits a bell-shape curve, from which the moments can be obtained experimentally. The n-th moment is defined as follows:

$$m_n = \int_0^\infty t^n\, C_b(z,t)\, dt \tag{14.2-12}$$

and the normalised moment scaled against the zero order moment is defined as

$$\mu_n = \frac{m_n}{m_0} \tag{14.2-13}$$

The n-th central moment is defined as the moment relative to the centre of gravity of the chromatographic curve:

$$\mu_n' = \frac{1}{m_0} \int_0^\infty (t - \mu_1)^n\, C_b(z,t)\, dt \tag{14.2-14}$$

The set of governing linear equations are solved by the method of Laplace transform. The transform is defined as

$$\overline{C}_b = \int_0^\infty e^{-st} C_b(z,t)\, dt \qquad (14.2\text{-}15)$$

Thus if we know the solution in the Laplace domain, the n-th order moment can be readily obtained by taking the n-th derivative of $\overline{C}_b$ as shown below.

$$m_n = (-1)^n \left.\frac{d^n \overline{C}_b(s)}{ds^n}\right|_{s=0} \qquad (14.2\text{-}16)$$

By matching the theoretical moments given in eq. (14.2-16) with the experimental moments, we can obtain the various parameters of the system.

14.3 Pore Diffusion Model with Local Equilibrium

Let us first illustrate the chromatography method with the simplest diffusion mechanism within the particle: the pore volume diffusion. For this mechanism the mass balance equation (eq. 14.2-5) describing the concentration within the particle is (that is the function F of eq. 14.2-5 has the following form)

$$\underline{F}(\underline{C}, x, t) = \begin{cases} \varepsilon \dfrac{\partial C}{\partial t} + (1-\varepsilon)\dfrac{\partial C_\mu}{\partial t} - \dfrac{\varepsilon D_p}{r^2}\dfrac{\partial}{\partial r}\left(r^2 \dfrac{\partial C}{\partial r}\right) = 0 \\ \\ C_\mu - KC = 0 \end{cases} \qquad (14.3\text{-}1)$$

Here $\underline{C} = [C, C_\mu]$, where C is the pore fluid concentration, C_μ is the adsorbed concentration, K is the Henry constant and D_p is the pore diffusivity. Here we have assumed the particle is spherical.

For this model, the flux into the particle $J|_R$ (of eq. 14.2-1) takes the form:

$$J|_R = \varepsilon D_p \left.\frac{\partial C}{\partial r}\right|_R \qquad (14.3\text{-}2)$$

Thus, the set of governing equations for this case are

eq. (14.2-1): bed equation
eq. (14.2-3): bed boundary conditions
eq. (14.3-1): particle equations
eq. (14.2-7): particle boundary condition.

Solving these equations subject to the entrance condition (14.2-9) by the method of Laplace transform yields the solution for the exit concentration $C_b(L,s)$ in the Laplace domain. Making use of the formula (14.2-16), we obtain the following first normalised moment and the second central moment.

$$\mu_1 = \frac{L\varepsilon_b}{u}(1+\delta_0) + \frac{t_0}{2} \tag{14.3-3}$$

$$\mu_2' = \frac{2L\varepsilon_b}{u}(\delta_d + \delta_f + \delta_M) + \frac{t_0^2}{12} \tag{14.3-4}$$

where the parameter δ_0 is the capacity parameter defined as

$$\delta_0 = \frac{(1-\varepsilon_b)\varepsilon}{\varepsilon_b}\left[1 + \frac{(1-\varepsilon)K}{\varepsilon}\right] \tag{14.3-5}$$

The parameters δ_d, δ_f and δ_M are the parameters characterising the contribution of axial dispersion, film resistance and pore diffusion resistance towards the spread of the chromatographic response. They are given by:

$$\delta_d = \frac{D_{ax}}{\varepsilon_b}(1+\delta_0)^2\left(\frac{\varepsilon_b}{u}\right)^2 \tag{14.3-6a}$$

$$\delta_f = \frac{(1-\varepsilon_b)\varepsilon}{\varepsilon_b}\left[1 + \frac{(1-\varepsilon)K}{\varepsilon}\right]^2 \frac{R\varepsilon}{3k_m} \tag{14.3-6b}$$

$$\delta_M = \frac{(1-\varepsilon_b)\varepsilon}{\varepsilon_b}\left[1 + \frac{(1-\varepsilon)K}{\varepsilon}\right]^2 \frac{R^2}{15D_p} \tag{14.3-6c}$$

It is interesting to note that the contributions of these three dispersive processes are additive, and such additive property can be used to our advantage in the parameter determination.

14.3.1 Parameter Determination:

In principle, by matching the theoretical moments with the corresponding experimental moments, the adsorption and diffusion parameters can be extracted. To facilitate the reliability of the parameter extracted, we can make use of an inert tracer as well as an adsorbing tracer, and carry out experiments for both of them at different flow rates.

With the inert tracer (that is no adsorption, $K = 0$), we have the following first normalised moment.

$$(\mu_1)_{\text{INERT}} = \frac{L\varepsilon_b}{u}\left[1 + \frac{(1-\varepsilon_b)\varepsilon}{\varepsilon_b}\right] \tag{14.3-7}$$

At the same flow rate, the first moment of the adsorbing tracer is given by eq.(14.3-3). Taking the difference between the first moment for adsorbing tracer and that for the inert at the same flow rate, we get

$$\frac{\mu_1 - (\mu_1)_{\text{INERT}}}{\left[\dfrac{(1-\varepsilon_b)(1-\varepsilon)}{\varepsilon_b}\right]} = K\frac{\varepsilon_b L}{u} \tag{14.3-8}$$

which suggests that a plot of the LHS versus the group $(\varepsilon_b L/u)$ would yield a straight line whose slope is equal to the Henry constant. This shows the utility of using the inert gas and the variation of the flow rate in the experiment to obtain the Henry constant. The linearity of the isotherm can be experimentally confirmed by carrying out experimental runs at different concentrations of adsorbate in the pulse. If the same results are obtained as well as the chromatographic curve is symmetrical then the assumption of isotherm linearity is justified.

The dynamic parameters have to be determined from the second central moment as they only affect the spread of the response curve. Like the first normalised moment, we can also utilise the variation of the flow rate to systematically extract the dynamic parameters:
(a) axial dispersion coefficient
(b) film mass transfer coefficient
(c) pore diffusivity.

Rearranging eq. (14.3-4) as follows:

$$\frac{\left(\mu_2' - t_0^2/12\right)}{2\left(\dfrac{L}{u/\varepsilon_b}\right)} = \delta_f + \delta_M + \left(\frac{D_{ax}}{\varepsilon_b}\right)(1+\delta_0)^2\left(\frac{\varepsilon_b}{u}\right)^2 \tag{14.3-9}$$

where we see that the RHS is the summation of the three dispersion forces: pore diffusion, film diffusion and axial dispersion. They affect the spread of the chromatographic curve in an additive manner as we have discussed earlier. The diffusional resistance term (δ_M) is independent of velocity as it should be since flow variation outside the particle does not affect the internal resistance. This diffusional resistance is proportional to the square of the particle radius. The film resistance term (δ_f) is a linear function of particle radius and depends on the velocity though the film mass transfer coefficient.

If we plot the LHS of eq. (14.3-9) versus $(\varepsilon_b/u)^2$, we could obtain the axial dispersion from the slope and the intercept of such plot in $\delta_f + \delta_M$. We note that the intercept is corresponding to infinite velocity at which the contribution of the film mass transfer resistance is negligible, and hence the intercept is simply δ_M, from which the pore diffusivity can be extracted:

$$D_p = \frac{(1-\varepsilon_b)\varepsilon}{\varepsilon_b}\left[1+\frac{(1-\varepsilon)K}{\varepsilon}\right]^2 \frac{R^2}{15\,\delta_M} \qquad (14.3\text{-}10)$$

14.3.2 Quality of The Chromatographic Response

The chromatography response is characterised by the first normalised moment and the second central moment. The extent of spread of the chromatographic response can be measured as the ratio of the spread (second central moment) to the square of the first normalised moment:

$$\frac{\mu_2'}{(\mu_1)^2} = \frac{2}{L}\left[\frac{D_{ax}}{u} + \frac{(\delta_M+\delta_f)}{(1+\delta_0)^2}\left(\frac{u}{\varepsilon_b}\right)\right] \qquad (14.3\text{-}11)$$

For a given system, that is for given particle size, length, porosity, the degree of spread is a function of velocity. To obtain the explicit dependence of the degree of spread in terms of velocity, we need to express the axial dispersion in terms of velocity. Here we use the following expression for the axial dispersion (Ruthven, 1984).

$$D_{ax} = \varepsilon_b\left(\gamma_1 D_M + \gamma_2 2R\frac{u}{\varepsilon_b}\right) \qquad (14.3\text{-}12)$$

where D_M is the molecular diffusivity, and

$$\gamma_1 = 0.45 + 0.55\,\varepsilon \qquad (14.3\text{-}13a)$$

$$\gamma_2 \approx 0.5 \qquad (14.3\text{-}13b)$$

The parameter δ_f is related to the film mass transfer coefficient, which is a function of velocity. For a packed column, the film mass transfer coefficient can be calculated from the following correlation.

$$\frac{k_m(2R)}{D_M} = 2 + 1.1\left(\frac{u2R\rho}{\mu}\right)^{0.6}\left(\frac{\nu}{D_M}\right)^{1/3} \qquad (14.3\text{-}14)$$

As a first approximation, we take k_m as a constant. Substitution of eq. (14.3-12) into eq. (14.3-11) gives:

$$\frac{\mu_2'}{(\mu_1')^2} = \frac{2}{L}\left[2\gamma_2 R + \frac{\varepsilon_b \gamma_1 D_M}{u} + \frac{(\delta_M + \delta_f)}{(1+\delta_0)^2}\left(\frac{u}{\varepsilon_b}\right)\right] \qquad (14.3\text{-}15)$$

The contribution of the axial dispersion towards a smaller spread occurs when the velocity is increased. On the other hand, the contribution of the film and pore diffusional resistances is unfavourable with an increase in the velocity. The following figure (14.3-1) shows the two opposing effects on the degree of spread.

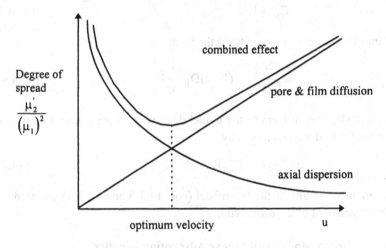

Figure 14.3-1: Plot of the Degree of Spread versus Velocity

Thus the chromatography operation is optimal when the function in the RHS of eq. (14.3-15) has a minimum. The optimal velocity is:

$$u \approx \frac{\varepsilon_b(1+\delta_0)\sqrt{\gamma_1 D_M}}{\sqrt{\delta_M + \delta_f}} \qquad (14.3\text{-}16)$$

An equivalent term commonly used in the chromatography literature is the height equivalent to a theoretical plate (HETP). It is related to the degree of spread as follows:

$$\text{HETP} = \frac{\mu_2' L}{(\mu_1')^2} \qquad (14.3\text{-}17)$$

14.4 Parallel Diffusion Model with Local Equilibrium

Very often that when dealing with adsorption of many gases and vapours in high surface area solids such as activated carbon and silica gel that surface diffusion can contribute significantly to the overall uptake. For this mass transfer mechanism, the mass balance equation describing the concentration distribution within the particle are:

$$\varepsilon \frac{\partial C}{\partial t} + (1-\varepsilon)\frac{\partial C_\mu}{\partial t} = \varepsilon D_p \frac{1}{r^2}\frac{\partial}{\partial r}\left(r^2 \frac{\partial C}{\partial r}\right) + (1-\varepsilon)D_\mu \frac{1}{r^2}\frac{\partial}{\partial r}\left(r^2 \frac{\partial C_\mu}{\partial r}\right) \quad (14.4\text{-}1a)$$

$$C_\mu = KC \quad (14.4\text{-}1b)$$

The flux into the particle $J|_R$ now takes the form:

$$J|_R = \varepsilon D_p \frac{\partial C}{\partial r}\bigg|_R + (1-\varepsilon)D_\mu \frac{\partial C_\mu}{\partial r}\bigg|_R \quad (14.4\text{-}2)$$

This model is identical in form to the model dealt with in the last section with the replacement of εD_p of the last model by

$$\varepsilon D_p + (1-\varepsilon)D_\mu K \quad (14.4\text{-}3)$$

Hence the moment results of the last model (eqs. 14.3-3 and 14.3-4) are used with the above replacement of the diffusivity.

14.5 Pore Diffusion Model with Linear Adsorption Kinetics

Section 14.3 dealt with the pore diffusion model and the rate of mass exchange between the two phases is much faster than the diffusion rate. In this section we shall consider the case where such mass exchange is comparable in rate to the diffusion, and this mass exchange can be described mathematically by the following equation:

$$\frac{\partial C_\mu}{\partial t} = k_a\left(C - \frac{C_\mu}{K}\right) \quad (14.5\text{-}1)$$

The flux into the particle for this model still takes the same form as eq. (14.3-2). The set of mass balance equations for this case are

eq. (14.2-1):	bed equation
eq. (14.2-3):	bed boundary condition

eq. (14.3-1a) and (14.5-1):   particle equation
eq. (14.2-7):   particle boundary condition.

The first normalised moment and the second central moment for this case of pore diffusion and finite mass exchange kinetics are:

$$\mu_1 = \frac{L\varepsilon_b}{u}(1+\delta_0) + \frac{t_0}{2} \tag{14.5-2}$$

$$\mu_2' = \frac{2L\varepsilon_b}{u}\left[\delta_1 + \frac{D_{ax}}{\varepsilon_b}(1+\delta_0)^2\left(\frac{\varepsilon_b}{u}\right)^2\right] + \frac{t_0^2}{12} \tag{14.5-3}$$

where

$$\delta_0 = \left[\frac{(1-\varepsilon_b)\varepsilon}{\varepsilon_b}\right]\left[1 + \frac{(1-\varepsilon)K}{\varepsilon}\right] \tag{14.5-4}$$

$$\delta_1 = \left[\frac{(1-\varepsilon_b)\varepsilon}{\varepsilon_b}\right]\left\{\frac{(1-\varepsilon)K^2}{\varepsilon k_a} + \frac{R^2\varepsilon}{15}\left[1 + \frac{(1-\varepsilon)K}{\varepsilon}\right]^2\left(\frac{1}{\varepsilon D_p} + \frac{5}{k_m R}\right)\right\} \tag{14.5-5}$$

We see in eq. (14.5-3) that the second central moment is the summation of four dispersion processes:

(a) finite adsorption kinetics
(b) pore diffusion resistance
(c) film diffusion resistance
(d) axial dispersion.

14.6 Bi-Dispersed Solids with Local Equilibrium

14.6.1 Uniform Grain Size

The utility of the chromatography method has been illustrated with a number of adsorption models:

(a) pore diffusion model with local equilibrium
(b) parallel diffusion model with local equilibrium
(c) pore diffusion model with finite adsorption kinetics

In these models the affinity constant K is obtained from the first normalised moment and the diffusion characteristics are obtained from the second central moment. The contribution of each resistance on the spread of the responses curve (that is the second central moment) is additive. This behaviour of the moment technique holds no matter what diffusion mechanisms are operating within the particle. We shall further illustrate this by applying the chromatography method to bi-dispersed solids, such as zeolites and molecular sieving carbon. In these solids, we have macropore diffusion in the void space between the grains and micropore diffusion in the channels within the grain. The exterior surface area of the grain is very small compared to the capacity volume within the grain and therefore the adsorption capacity on the grain exterior surface area is usually neglected.

Like the last three models dealt with so far, the difference between the models is the formulation of model equations within the particle. For this case, the mass balance equations within the pellet are:

$$\varepsilon \frac{\partial C}{\partial t} = \varepsilon D_p \frac{1}{r^2} \frac{\partial}{\partial r}\left(r^2 \frac{\partial C}{\partial r}\right) - \frac{3(1-\varepsilon)}{R_\mu} D_\mu \left.\frac{\partial C_\mu}{\partial r_\mu}\right|_{R_\mu} \quad (14.6\text{-}1a)$$

$$\frac{\partial C_\mu}{\partial t} = D_\mu \frac{1}{r_\mu^2} \frac{\partial}{\partial r_\mu}\left(r_\mu^2 \frac{\partial C_\mu}{\partial r_\mu}\right) \quad (14.6\text{-}1b)$$

$$r_\mu = R_\mu; \quad C_\mu = KC \quad (14.6\text{-}1c)$$

$$r = R; \quad \varepsilon D_p \frac{\partial C}{\partial r} = k_m(C_b - C) \quad (14.6\text{-}1d)$$

where ε is the macropore porosity, C is the concentration in the macropore, D_p is the macropore diffusivity, D_μ is the micropore diffusivity, C_μ is the concentration in the grain (mole/cc of the grain), R_μ is the grain radius, R is the particle radius, and K is the Henry constant. The micropore diffusivity is known to have a strong concentration dependence (Chapter 7). However, in the chromatography operation, the tracer concentration is usually very low, resulting in a linear isotherm (eq. 14.6-1c) and constant micropore diffusivity.

The set of governing equation is composed of
    eq. (14.2-1):   bed equation
    eq. (14.2-3):   bed boundary condition
    eqs. (14.6-1):   particle equations and boundary conditions.

These equations are linear and are susceptible to Laplace transform analysis, from which we can obtain the theoretical moments. The first normalised moment and the second central moment for this model are:

$$\mu_1 = \frac{L\varepsilon_b}{u}(1+\delta_0) + \frac{t_0}{2} \tag{14.6-2}$$

$$\mu_2' = \frac{2L\varepsilon_b}{u}(\delta_d + \delta_f + \delta_M + \delta_\mu) + \frac{t_0^2}{12} \tag{14.6-3}$$

where

$$\delta_0 = \left[\frac{(1-\varepsilon_b)\varepsilon}{\varepsilon_b}\right]\left[1 + \frac{(1-\varepsilon)K}{\varepsilon}\right] \tag{14.6-4}$$

$$\delta_d = \frac{D_{ax}}{\varepsilon_b}(1+\delta_0)^2\left(\frac{\varepsilon_b}{u}\right)^2 \tag{14.6-5}$$

$$\delta_f = \frac{(1-\varepsilon_b)}{\varepsilon_b}\frac{R\varepsilon^2}{3k_m}\left[1 + \frac{(1-\varepsilon)K}{\varepsilon}\right]^2 \tag{14.6-6}$$

$$\delta_M = \frac{(1-\varepsilon_b)}{\varepsilon_b}\frac{R^2\varepsilon}{15D_p}\left[1 + \frac{(1-\varepsilon)K}{\varepsilon}\right]^2 \tag{14.6-7}$$

$$\delta_\mu = \frac{(1-\varepsilon_b)}{\varepsilon_b}\frac{R_\mu^2\,K\,(1-\varepsilon)}{15\,D_\mu} \tag{14.6-8}$$

The Henry constant K is determined from the first normalised moment and all the kinetic parameters are determined from the second central moment. It is noted that the contributions of the axial dispersion and the film are the same for all models, and the contribution of the macropore diffusion is also the same in all models involving macropore diffusion. These contributions are additive, which makes the chromatography method simple in its analysis.

We summarise all the results in the following table, with the following notation to denote the various models.

Model A: Pore diffusion model with local equilibrium
Model B: Parallel diffusion model with local equilibrium
Model C: Pore diffusion with finite adsorption kinetics
Model D: Bi-dispersed model with local equilibrium

Table 14.6-1: Comparison of moments from various kinetic models

	A	B	C	D
First normalized moment	$\mu_1 = \dfrac{L\varepsilon_b}{u}(1+\delta_0) + \dfrac{t_0}{2}$			(14.6-9)
Second central moment	$\mu_2' = \dfrac{2L\varepsilon_b}{u}(\delta_d + \delta_f + \delta_M + \delta_a + \delta_\mu) + \dfrac{t_0^2}{12}$			(14.6-10)
δ_0	$\dfrac{(1-\varepsilon_b)\varepsilon}{\varepsilon_b}\left[1 + \dfrac{(1-\varepsilon)K}{\varepsilon}\right]$			(14.6-11)
δ_d	$\dfrac{D_{ax}}{\varepsilon_b}(1+\delta_0)^2\left(\dfrac{\varepsilon_b}{u}\right)^2$			(14.6-12)
δ_f	$\dfrac{(1-\varepsilon_b)}{\varepsilon_b}\dfrac{R\varepsilon^2}{3k_m}\left[1 + \dfrac{(1-\varepsilon)K}{\varepsilon}\right]^2$			(14.6-13)
δ_M	$\dfrac{(1-\varepsilon_b)}{\varepsilon_b}\dfrac{R^2\varepsilon^2}{15\varepsilon D}\left[1 + \dfrac{(1-\varepsilon)K}{\varepsilon}\right]^2$			(14.6-14)
εD	εD_p	$\varepsilon D_p + (1-\varepsilon)KD_s$	εD_p	εD_p
δ_a	0	0	δ_a	0
δ_μ	0	0	0	δ_μ

where

$$\delta_a = \dfrac{(1-\varepsilon_b)(1-\varepsilon)K^2}{\varepsilon_b k_a} \qquad (14.6\text{-}15)$$

$$\delta_\mu = \dfrac{(1-\varepsilon_b)}{\varepsilon_b}\dfrac{R_\mu^2 K(1-\varepsilon)}{15 D_\mu} \qquad (14.6\text{-}16)$$

14.6.2 Distribution of Grain Size:

The grains in bi-dispersed solids are usually not uniform in size due to our inability to tailor the grain size. Therefore to properly account for the diffusion characteristics inside the grain we must include the grain size distribution in the mathematical model. Let $f(R_\mu)$ be the grain distribution such that

$$f(R_\mu)d(\ln R_\mu) \qquad (14.6\text{-}17)$$

be the fraction of grain having radii between R_μ and $R_\mu + dR_\mu$.

The mass balances in the particle for this case are given by (14.6-1) with the exception that eq. (14.6-1a) is replaced by

$$\varepsilon \frac{\partial C}{\partial t} = \varepsilon D_p \frac{1}{r^2}\frac{\partial}{\partial r}\left(r^2 \frac{\partial C}{\partial r}\right) - \int_{-\infty}^{\infty} \frac{3(1-\varepsilon)}{R_\mu}\left[D_\mu \frac{\partial C_\mu}{\partial r_\mu}\bigg|_{R_\mu}\right]f(R_\mu)\,d(\ln R_\mu) \qquad (14.6\text{-}18)$$

Applying the moment analysis to the new set of equations (eqs. 14.2-1, 14.2-3, 14.6-1b to 14.6-1d, and 14.6-18), we obtain the first normalised moment and the second central moment given as in eqs. (14.6-2) to (14.6-7) and

$$\delta_\mu = \frac{(1-\varepsilon_b)}{\varepsilon_b}\frac{K(1-\varepsilon)}{15 D_\mu}\int_{-\infty}^{\infty} R_\mu^2\, f(R_\mu)\,d\ln R_\mu \qquad (14.6\text{-}19)$$

For a log normal distribution of the grain size

$$f(R_\mu) = \frac{1}{\sigma\sqrt{2\pi}}\exp\left[-\frac{\left(\ln R_\mu - \ln R_\mu^0\right)^2}{2\sigma^2}\right] \qquad (14.6\text{-}20)$$

the parameter δ_μ will take the form

$$\delta_\mu = \frac{(1-\varepsilon_b)}{\varepsilon_b}\frac{K(1-\varepsilon)}{15 D_\mu}\left(R_\mu^0\right)^2 \exp(\sigma^2) \qquad (14.6\text{-}21)$$

In the case of uniform grain size distribution (that is $\sigma = 0$), eq. (14.6-21) reduces to eq. (14.6-8) obtained in the earlier section for uniform grain size.

14.7 Bidispersed Solid (Alumina type) Chromatography

The bi-dispersed solid dealt with in Section 14.6 is zeolite type solid where the micropores within the grain are of molecular dimension, and hence they only accommodate one type of adsorbate within the micropore, namely the adsorbed species. In this section we shall consider another type of bi-dispersed solids where the micropores are large enough to accommodate adsorbate in both forms: free form as well as adsorbed form. For such cases, the mass balance equations inside the particle are:

$$\varepsilon_M \frac{\partial C}{\partial t} = \varepsilon_M D_p \frac{1}{r^2} \frac{\partial}{\partial r}\left(r^2 \frac{\partial C}{\partial r}\right) - \frac{3(1-\varepsilon_M)}{R_\mu} \varepsilon_\mu D_i \frac{\partial C_i}{\partial r_\mu}\bigg|_{R_\mu} \quad (14.7\text{-}1)$$

$$\varepsilon_\mu \frac{\partial C_i}{\partial t} + (1-\varepsilon_\mu)\frac{\partial C_\mu}{\partial t} = \varepsilon_\mu D_i \frac{1}{r_\mu^2} \frac{\partial}{\partial r_\mu}\left(r_\mu^2 \frac{\partial C_i}{\partial r_\mu}\right) \quad (14.7\text{-}2)$$

$$\frac{\partial C_\mu}{\partial t} = k_a\left(C_i - \frac{C_\mu}{K_i}\right) \quad (14.7\text{-}3)$$

where C is the concentration of the free species in the macropore, C_i is the concentration of the free species in the micropore, and C_μ is the adsorbed concentration in the micropore (mole/cc of the grain solid), ε_M is the macropore porosity (macropore volume/total particle volume), ε_μ is the micropore porosity (micropore column/grain volume), D_p is the macropore diffusivity, D_i is the micropore diffusivity of the free species, k_a is the adsorption rate constant and K_i is the equilibrium constant.

Applying the moment method, we obtain the following first normalised moment and the second central moment.

$$\mu_1 = \frac{L\varepsilon_b}{u}(1+\delta_0) + \frac{t_0}{2} \quad (14.7\text{-}4)$$

$$\mu_2' = \frac{2L\varepsilon_b}{u}(\delta_d + \delta_f + \delta_M + \delta_i) + \frac{t_0^2}{12} \quad (14.7\text{-}5)$$

where

$$\delta_0 = \frac{(1-\varepsilon_b)\varepsilon_M}{\varepsilon_b}\left[1 + \frac{(1-\varepsilon_M)\varepsilon_M}{\varepsilon_M} + \frac{(1-\varepsilon_M)(1-\varepsilon_\mu)K_i}{\varepsilon_M}\right] \quad (14.7\text{-}6)$$

$$\delta_d = \frac{D_{ax}}{\varepsilon_b}(1+\delta_0)^2 \left(\frac{\varepsilon_b}{u}\right)^2 \quad (14.7\text{-}7)$$

$$\delta_f = \frac{(1-\varepsilon_b)}{\varepsilon_b}\frac{R\varepsilon_M^2}{3k_m}\left[1 + \frac{(1-\varepsilon_M)\varepsilon_\mu}{\varepsilon_M} + \frac{(1-\varepsilon_M)(1-\varepsilon_\mu)K_i}{\varepsilon_M}\right]^2 \quad (14.7\text{-}8)$$

$$\delta_M = \frac{(1-\varepsilon_b)}{\varepsilon_b}\frac{R^2\varepsilon_M}{15D_p}\left[1 + \frac{(1-\varepsilon_M)\varepsilon_\mu}{\varepsilon_M} + \frac{(1-\varepsilon_M)(1-\varepsilon_\mu)K_i}{\varepsilon_M}\right]^2 \quad (14.7\text{-}9)$$

$$\delta_i = \frac{(1-\varepsilon_b)}{\varepsilon_b}\left\{\frac{K_i^2(1-\varepsilon_M)(1-\varepsilon_\mu)}{k_a} + \frac{\varepsilon_\mu R_\mu^2}{15D_i}\left[1 + \frac{(1-\varepsilon_M)\varepsilon_M}{\varepsilon_M}\right]^2\right\} \quad (14.7\text{-}10)$$

Note at ε_μ is the micropore porosity based on the grain volume. The micropore porosity based on the particle volume is

$$\varepsilon_m = (1-\varepsilon_M)\varepsilon_\mu \tag{14.7-11}$$

and the total porosity based on the particle volume is

$$\varepsilon_m + (1-\varepsilon_M)\varepsilon_\mu \tag{14.7-12}$$

14.8 Perturbation Chromatography

The chromatography analysis presented so far for a number of practical adsorption models illustrates its usefulness in determining the adsorption equilibria constant in the form of Henry constant and the various kinetics parameters. This technique usefulness is not limited to the very low concentration range where we extract the Henry constant, it can also be applied to any concentration and if applied appropriately we can obtain the slope of the adsorption isotherm at any concentration. The appropriate method is the perturbation chromatography and its operation is as follows. First the column is equilibrated with a concentration, say C*, until all void space within the column and particle have a solute concentration of C* and the adsorbed phase has a concentration of f(C*) where f is the functional form for the adsorption isotherm. After the column has been equilibrated with a flow of concentration C*, we inject into the column a pulse of adsorbate having a concentration of C* + ΔC* where ΔC* << C*. With this small perturbation in concentration, the responses of the concentration in the column and in the particle will take the following asymptotic form:

$$C_b \cong C^* + \Delta C_b \tag{14.8-1a}$$

$$C \approx C^* + \Delta C \tag{14.8-1b}$$

$$C_\mu \approx C^* + \Delta C_\mu \tag{14.8-1c}$$

We shall illustrate the principles of the perturbation chromatography on the adsorption model for pore diffusion with local equilibrium.

The mass balance equation describing the concentration distribution with the particle are:

$$\varepsilon \frac{\partial C}{\partial t} + (1-\varepsilon)\frac{\partial C_\mu}{\partial t} = \varepsilon D_p \frac{1}{r^2}\frac{\partial}{\partial r}\left(r^2 \frac{\partial C}{\partial r}\right) \tag{14.8-2a}$$

$$C_\mu = f(C) \tag{14.8-2b}$$

We substitute the expansions (eq. 14.8-1) into the balance equations for the bed (eq. 14.2-1) and the particle (eq. 14.8-2) and obtain the following necessary equations in terms of the perturbed variables.

$$\varepsilon_b \frac{\partial(\Delta C_b)}{\partial t} = D_{ax} \frac{\partial^2(\Delta C_b)}{\partial z^2} - \frac{\partial}{\partial z}(u\,\Delta C_b) - \frac{3(1-\varepsilon_b)}{R}\varepsilon D_p \frac{\partial(\Delta C)}{\partial r}\bigg|_R \quad (14.8\text{-}3a)$$

$$\varepsilon \frac{\partial(\Delta C)}{\partial t} + (1-\varepsilon)\frac{\partial(\Delta C_\mu)}{\partial t} = \varepsilon D_p \frac{1}{r^2}\frac{\partial}{\partial r}\left(r^2 \frac{\partial \Delta C}{\partial r}\right) \quad (14.8\text{-}3b)$$

$$\Delta C_\mu = \left[\frac{\partial f(C^*)}{\partial C}\right]\Delta C \quad (14.8\text{-}3c)$$

where $\partial f/\partial C|_{C^*}$ is the slope of the adsorption isotherm at the concentration C^*.

We see that the new set of equations in linear and is identical to that in Section 14.3 with the exception that the Henry constant in Section 14.3 is now replaced with the slope of the adsorption isotherm at C^*. Thus if the moment method is applied on the perturbed variable ΔC_b at the exit of the column, we will obtain the first normalised moment and the second central moment as given in eqs. (14.3-3) and (14.3-4), respectively with K replaced by

$$\frac{\partial f(C^*)}{\partial C} \quad (14.8\text{-}4)$$

This type of experiment can be repeated at different concentration C^* and the adsorption isotherm is readily obtained as

$$f(C) = \int_0^C \frac{\partial f(C)}{\partial C} dC \quad (14.8\text{-}5)$$

14.9 Concluding Remarks

We have illustrated the chromatography method and showed that the operation is straightforward, and the moment analysis is very simple in the task of parameter determination. Provided that some precautions are taken care of (such as careful packing, removal of heat release) the chromatography method is recommended as a quick way to learn about the affinity as well as the diffusion characteristics of the system.

15
Analysis of a Batch Adsorber

15.1 Introduction

The methods of time lag, diffusion cell and chromatography are useful to characterize the adsorption and diffusion properties of an adsorption system. Although in principle, those methods can be extended to cover situations such as multicomponent systems or nonlinear isotherm, but such extension is limited due to the limitations pointed out in Table 14.1-1. In this chapter we present another method, the batch adsorber method, as another tool for diffusion and adsorption characterisation. This method is easy to set up, and it can handle practically any complexities of an adsorption system. In practice, we can prepare the batch adsorber in a number of ways. For example, an amount of solid sample is put inside a closed reservoir and an amount of adsorbate is introduced into the reservoir and its content is stirred with some means of stirring. Another way is restraining the particle in a mesh basket attached to a stirrer and the stirrer is rotated, and an amount of adsorbate is then introduced into the reservoir. The third way is to simply pack an amount of solid in a small column connected to a closed loop, and then an amount of adsorbate is introduced into the loop and its content is circulated through the bed of solids at a speed such that the gas residence time in the loop is much less than the adsorption time. Among the three configurations, the last one is the most preferred choice as it is easy to pack the solid in a small column and the gas circulation rate through the column can be adjusted such that the heat released from the adsorption process can be effectively removed and hence maintaining the isothermality of the system. Common to these three configurations is the need for a means to monitor the adsorbate concentration in the gas phase. This is done by a number of means, such as the thermal conductivity detector (TCD) or FT-IR, and they allow the concentration to be monitored continuously.

15.2 The General Formulation of Mass Balance Equation

The mass balance equations for the batch adsorber are much simpler than those of the chromatography method. The mass balance involves the balance equation in the reservoir and the balance equations in the particle.

The mass balance equations in the particle depend on the diffusion and adsorption mechanisms inside the particle. Those equations have been discussed in great details in Chapters 9, 10 and 11. In this chapter we will address a number of simple diffusion and adsorption models of which analytical solutions are feasible in order to illustrate its application in parameter determination. Other complex models can also be used with the batch adsorber method but the parameter determination must be done numerically.

Common to all adsorption and diffusion mechanisms within the particle is the mass balance in the reservoir. Assuming perfect mixing, the mass balance equation in the reservoir is

$$V \frac{dC_b}{dt} = -A \cdot J\big|_R \tag{15.2-1}$$

where V is the volume of the reservoir, C_b is the concentration of the adsorbate in the reservoir, A is the total exterior surface area of all particles, and $J|_R$ the mass transfer into the particle per unit interfacial area.

If the particles are spherical in shape, the total exterior surface area is

$$A = \left(\frac{m_p}{\rho_p}\right) \frac{3}{R} \tag{15.2-2}$$

where m_p is the mass of the particle, ρ_p is the particle density (mass/volume of particle) and R is the particle radius. For particles of other shape, the total exterior surface area is

$$A = \left(\frac{m_p}{\rho_p}\right) \frac{(s+1)}{R} \tag{15.2-3}$$

where s = 0 for slab particle and R is its half-length, and s = 1 for very long cylinder and R is its radius.

As mentioned earlier, the mass balance equations inside the particle depends on the adsorption and diffusion mechanisms within the particle. We can write these mass balance equations for the particle as

$$\underline{F}(\underline{C}, x, t) = \underline{0} \tag{15.2-4}$$

where the dimension of the functional form $\underline{F}$ depends on the dimension of the concentration vector $\underline{C}$

$$\underline{C} = \begin{bmatrix} C_1 & C_2 & \cdots & C_n \end{bmatrix} \tag{15.2.5}$$

where C_j is the concentration of the adsorbate of the phase j within the particle. Let C_1 be the concentration of the free species and $C_2, C_3 \ldots C_N$ are concentrations of the adsorbed species in various forms.

15.2.1 The Initial Condition

The batch adsorber can be operated by either having the reservoir initially free of any adsorbate and at time $t = 0^+$ an amount of adsorbate is introduced into the reservoir. For this case the initial conditions are:

$$t = 0: \quad C_b = C_{bo}, \quad \underline{C} = \underline{0} \tag{15.2-6}$$

Another condition is that the adsorber is initially equilibrated with an adsorbate of concentration C_{b1} and at $t = 0+$, the concentration in the bulk is increased from C_{b1} to C_{b2}, that is

$$t = 0; \quad C_b = C_{b2} \quad \underline{C} = \begin{bmatrix} C_{b1}, f_2(C_{b1}), f_3(C_{b1}), \cdots, f_n(C_{b1}) \end{bmatrix} \tag{15.2-7}$$

15.2.2 The Overall Mass Balance Equation

Since the total mass is conserved within a closed reservoir, the mass loss from the bulk phase of the reservoir must be the same as the mass gain by the particle. This is achieved by simply integrating the mass balance of the particle (15.2-4) over the whole volume of the particle and adding the result to the mass balance of the reservoir to finally get

$$V\frac{dC_b}{dt} + \frac{d(N_T)}{dt} = 0 \tag{15.2-8}$$

where N_T is the total number of moles within the particle. This total amount is a function of the concentrations of adsorbate in the free and adsorbed forms, that is

$$N_T = N_T(<\underline{C}>) \tag{15.2-9a}$$

where

$$<\underline{C}> = \int_V \underline{C}\, dV \tag{15.2-9b}$$

798 Measurement Techniques

Integrating the overall differential mass balance equation (15.2-8) with respect to time from t=0 to t gives:

$$VC_b + N_T = VC_{b0} + N_{T0} \tag{15.2-10}$$

The LHS is the total amount of adsorbate at any time t and the RHS is that amount at t = 0. This is simply a statement of conservation of mass.

At steady state, the concentration of the free species within the particle is equal to the bulk concentration, that is

$$C_1 = C_b = C_\infty \tag{15.2-11}$$

and the adsorbed species concentrations are in equilibrium with the free species that is

$$C_{j,\infty} = f_j(C_\infty) \tag{15.2-12}$$

for j = 2, 3, , n, when f_j is the adsorption isotherm functional form for the adsorbed phase j. Then the conservation of mass at steady state will give:

$$VC_\infty + N_T(C_\infty, f_1(C_\infty), f_2(C_\infty), ..) = VC_{b0} + N_{T0} \tag{15.2-13}$$

This is a nonlinear algebraic equation in terms of C_∞, from which the steady state concentration can be obtained.

15.3 Pore Diffusion Model with Local Equilibrium

Let us first illustrate the batch adsorber analysis with a simple adsorption diffusion mechanism inside the particle: the pore volume diffusion with local equilibrium. For this mechanism, the mass balance equations describing the concentration distribution within the particle are:

$$\underline{F}(\underline{C}, x, t) = \begin{cases} \varepsilon \dfrac{\partial C}{\partial t} + (1-\varepsilon)\dfrac{\partial C_\mu}{\partial t} - \dfrac{\varepsilon D_p}{r^2}\dfrac{\partial}{\partial r}\left(r^2 \dfrac{\partial C}{\partial r}\right) = 0 \\ \\ C_\mu - f(C) = 0 \end{cases} \tag{15.3-1}$$

where $\underline{C} = [C, C_\mu]$ and f is the functional form for the adsorption isotherm. For this model, the flux $J|_R$ into the particle is:

$$J|_R = \varepsilon D_p \dfrac{\partial C}{\partial r}\bigg|_R \tag{15.3-2}$$

The boundary condition at the exterior surface of the particle is:

$$r = R; \quad \varepsilon D_p \frac{\partial C}{\partial r}\bigg|_R = k_m(C_b - C|_R) \qquad (15.3\text{-}3)$$

Eqs. (15.2-1) and (15.3-1 to 15.3-3) describe the evolution of the concentrations of all phases in the system. The evolution of these concentrations is constrained by the constant total number of moles in the system, that is the loss of mass in one sub-system is balanced by the gain of mass in another sub-system. This is achieved by multiplying eq. (15.3-1a) by $r^2 dr$ and integrating the result with respect to r from 0 to R, and finally adding the result to eq. (15.2-1) gives:

$$V\frac{dC_b}{dt} + \left\{\left(\frac{m_p}{\rho_p}\right)\left[\varepsilon\frac{d\overline{C}}{dt} + (1-\varepsilon)\frac{d\overline{C}_\mu}{dt}\right]\right\} = 0 \qquad (15.3\text{-}4a)$$

where

$$\overline{C} = \frac{3}{R^3}\int_0^R r^2\, C\, dr, \qquad \overline{C}_\mu = \frac{3}{R^3}\int_0^R r^2\, C_\mu\, dr \qquad (15.3\text{-}4b)$$

This equation simply states that the loss of mass in the reservoir VdC_b/dt is balanced by the gain of mass within the particle (the curly bracket term).

Integrating eq. (15.3-4a) with respect to time from 0 to t gives:

$$VC_b + \left(\frac{m_p}{\rho_p}\right)\left[\varepsilon\overline{C} + (1-\varepsilon)\overline{C}_\mu\right] = VC_{b0} + \left(\frac{m_p}{\rho_p}\right)\left[\varepsilon C_i + (1-\varepsilon)f(C_i)\right] \qquad (15.3\text{-}5)$$

where C_{b0} is the initial bulk concentration in the reservoir, and C_i is the initial concentration in the particle and $C_{\mu i} = f(C_i)$ is the adsorbed concentration which is in equilibrium with C_i. Eq. (15.3-5) is the statement of the conservation of mass.

At steady state, we must have

$$\lim_{t \to \infty} C_b = \lim_{t \to \infty} \overline{C} = C_\infty \qquad (15.3\text{-}6a)$$

$$\lim_{t \to \infty} \overline{C}_\mu = f(C_\infty) \qquad (15.3\text{-}6b)$$

Thus, the conservation of mass at steady state is:

$$VC_\infty + \left(\frac{m_p}{\rho_p}\right)[\varepsilon C_\infty + (1-\varepsilon)f(C_\infty)] = VC_{b0} + \left(\frac{m_p}{\rho_p}\right)[\varepsilon C_i + (1-\varepsilon)f(C_i)] \quad (15.3\text{-}7)$$

This is a nonlinear algebraic equation for C_∞, from which it can be solved to give the steady state concentration of the adsorbate in the reservoir. Explicit solution of eq. (15.3-7) depends on the specific form of the functional form of the adsorption isotherm f. Let us deal with the following three isotherms:
- (a) linear isotherm
- (b) rectangular isotherm
- (c) Langmuir isotherm

Linear isotherm
We have the following functional form for the linear isotherm

$$f(C) = KC \quad (15.3\text{-}8a)$$

The steady state concentration can be solved from eq. (15.3-7) and we get

$$C_\infty = \alpha C_{b0} + (1-\alpha) C_i \quad (15.3\text{-}8b)$$

where

$$\alpha = \frac{V}{V + \left(\dfrac{m_p}{\rho_p}\right)[\varepsilon + (1-\varepsilon)K]} < 1 \quad (15.3\text{-}8c)$$

We can see that when the reservoir is very large such that $\alpha \approx 1$, the steady state concentration C_∞ is approximately equal to the initial concentration of the reservoir, and when the reservoir is very small such that $\alpha \approx 0$ the steady state concentration is approximately equal to the initial concentration of the particle. Both of these extremes are physically expected.

Rectangular isotherm
The adsorption isotherm takes the form

$$f(C) = C_{\mu s} \quad (15.3\text{-}9)$$

The steady state concentration is:

$$C_\infty = \frac{VC_{b0} - \left(\dfrac{m_p}{\rho_p}\right)(1-\varepsilon)C_{\mu s}}{\left[V + \left(\dfrac{m_p}{\rho_p}\right)\varepsilon\right]} \tag{15.3-10}$$

In this case of irreversible isotherm, the initial concentration C_i inside the particle must be zero; otherwise for finite C_i (no matter how small it is) the irreversible adsorption isotherm states that the adsorbed phase ($f(C_i) = C_{\mu s}$) is completely saturated. Hence no adsorption kinetics will happen no matter what the initial bulk phase concentration is.

Eq. (15.3-10) also states that the steady state bulk phase concentration is positive when

$$VC_{b0} > \left(\frac{m_p}{\rho_p}\right)(1-\varepsilon)C_{\mu s}, \tag{15.3-11}$$

that is the initial amount in the reservoir (VC_{b0}) must be more than enough to saturate all the sites within the particle $(m_p/\rho_p)(1-\varepsilon)C_{\mu s}$. If the constraint (15.3-11) is not satisfied, that is the initial mass in the reservoir is not enough to saturate all adsorption sites inside the particle, then the steady state concentration is

$$C_\infty = 0, \tag{15.3-12}$$

and only a fraction of the particle is saturated with adsorbate; that is the particle is divided into two regions: the outer shell and the inner core. The outer shell is saturated with adsorbate while the inner core is free from any adsorbate. The position dividing the two regions at steady state is readily calculated from the mass balance that is

$$VC_{b0} = \left(\frac{m_p}{\rho_p}\right)\alpha . (1-\varepsilon) C_{\mu s} \tag{15.3-13}$$

where α is the volume fraction of the outer shell. If the particle is a sphere of radius R, then

$$\alpha = \frac{\dfrac{4}{3}\pi R^3 - \dfrac{4}{3}\pi R_{f,\infty}^3}{\dfrac{4}{3}\pi R^3} = 1 - \left(\frac{R_{f,\infty}}{R}\right)^3 \tag{15.3-14}$$

Here $R_{f,\infty}$ is the position demarcating the outer shell and the inner core at steady state. Combining eqs. (15.3-13) and (15.3-14), we obtain this position as given below:

$$R_{f,\infty} = R\left[1 - \frac{VC_{b0}}{\left(\dfrac{m_p}{\rho_p}\right)(1-\varepsilon)C_{\mu s}}\right]^{1/3} < R \qquad (15.3\text{-}15)$$

The existence of the adsorption front at steady state is only possible with the irreversible isotherm.

<u>Langmuir isotherm</u>
For Langmuir isotherm, the functional form is

$$C_\mu = f(C) = C_{\mu s}\frac{bC}{1+bC} \qquad (15.3\text{-}16)$$

Substituting this Langmuir equation into the conservation of mass equation at steady state (eq. 15.3-7) and solving for the solution, we obtain:

$$C_\infty = \left\{-A + \left[A^2 + 4Bb\left(V + \frac{m_p\varepsilon}{\rho_p}\right)\right]^{1/2}\right\}\left[2b\left(V + \frac{m_p\varepsilon}{\rho_p}\right)\right]^{-1} \qquad (15.3\text{-}17a)$$

where

$$A = V + \left(\frac{m_p}{\rho_p}\right)[\varepsilon + (1-\varepsilon)K] - B\cdot b\,; \qquad K = bC_{\mu s} \qquad (15.3\text{-}17b)$$

$$B = VC_{b0} + \left(\frac{m_p}{\rho_p}\right)\left[\varepsilon C_i + (1-\varepsilon)C_{\mu s}\frac{bC_i}{1+bC_i}\right] \qquad (15.3\text{-}17c)$$

Let us make an observation about the effect of nonlinearity on the steady state concentration. We choose the situation where the particle is initially free from adsorbate, that is $C_i = 0$. The steady state concentration for the three isotherms are tabulated below.

Isotherm	Steady state concentration	Eq #
Linear	$\dfrac{C_\infty}{C_{bo}} = \dfrac{V}{V + \left(\dfrac{m_p}{\rho_p}\right)\left[\varepsilon + (1-\varepsilon)K\right]}$	(15.3-18)
Langmuir	$\dfrac{C_\infty}{C_{bo}} = \dfrac{-A + \left[A^2 + 4Bb\left(V + \dfrac{m_p \varepsilon}{\rho_p}\right)\right]^{1/2}}{2bC_{bo}\left(V + \dfrac{m_p \varepsilon}{\rho_p}\right)}$; $B = VC_{bo}$	(15.3-19)
Irreversible	$\dfrac{C_{bo}}{C_{bo}} = \begin{cases} \dfrac{V - \left(\dfrac{m_p}{\rho_p}\right)\dfrac{(1-\varepsilon)C_{\mu s}}{C_{bo}}}{V + \left(\dfrac{m_p}{\rho_p}\right)\varepsilon} & \text{for } VC_{bo} > \dfrac{m_p}{\rho_p}(1-\varepsilon)C_{\mu s} \\ 0 & \text{otherwise} \end{cases}$	(15.3-20)

In the linear isotherm case, the extent of uptake C_∞/C_{bo} (given in eq. 15.3-18) is independent of the initial concentration used. This is not the case for the case of nonlinear isotherm. The following figure (15.3-1) shows a plot of this extent of uptake versus the initial bulk phase concentration

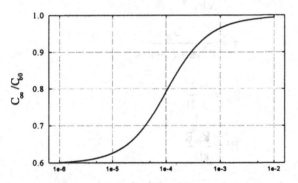

Figure 15.3-1: Plot of C_∞/C_{bo} versus C_{bo}

In generating the above figure, we have used for the following parameters, $V = 1000$ cm^3, $m_p = 10$g, $\rho_p = 1$g/cc, $\varepsilon = 0.33$, $b = 20\,000$ cc/mole, $C_{\mu s} = 0.005$ mole/cc.

15.3.1 Linear Isotherm

We have formulated the mass balance equation and considered the steady state concentration. Let us now address the dynamic behaviour for the case of linear isotherm.

$$f(C) = KC \tag{15.3-21}$$

Solving for the concentration in the reservoir by using the Laplace transform, the following solution is obtained for the bulk concentration

$$\frac{C_b}{C_{bo}} = \frac{B}{1+B} + \left(\frac{B}{3}\right)\sum_{n=1}^{\infty} a_n e^{-\lambda_n^2 \tau} \tag{15.3-22}$$

where

$$a_n = \frac{\left(1+\dfrac{\beta_n}{1-\beta_n}\right)^2}{\dfrac{1}{2}+\dfrac{B}{6}\left[\dfrac{1+(B/3)\lambda_n^2-\beta_n}{(1-\beta_n)^2}\right]+\dfrac{B}{3}\left(1+\dfrac{\beta_n}{1-\beta_n}\right)^2} \tag{15.3-23a}$$

$$B = \frac{V}{\left(\dfrac{m_p}{\rho_p}\right)[\varepsilon+(1-\varepsilon)K]} \tag{15.3-23b}$$

$$\tau = \frac{\varepsilon D_p t}{R^2[\varepsilon+(1-\varepsilon)K]} \tag{15.3-23c}$$

$$\beta_n = \frac{B}{3}\frac{\lambda_n^2}{Bi} \tag{15.3-23d}$$

$$Bi = \frac{k_m R}{\varepsilon D_p} \tag{15.3-23e}$$

The eigenvalues λ_n are positive roots of the following transcendental equation:

$$\lambda_n \cot\lambda_n - 1 = \frac{B}{3}\frac{\lambda_n^2}{1-\beta_n} \tag{15.3-24}$$

For infinite stirring (Bi → ∞), the coefficient a_n and the eigenvalues are obtained from:

$$a_n = \frac{18}{9 + 9B + B^2 \lambda_n^2} \quad (15.3\text{-}25a)$$

$$\lambda_n \cot \lambda_n - 1 = \frac{B}{3} \lambda_n^2 \quad (15.3\text{-}25b)$$

The first ten eigenvalues for two values of B (B = 1 and 10) are tabulated in the following table.

j	λ_j	
	B = 1	B = 10
1	3.7264	3.2316
2	6.6814	6.3302
3	9.7156	9.4564
4	12.7927	12.5901
5	15.8924	15.7270
6	19.0049	18.8654
7	22.1251	22.0048
8	25.2504	25.1447
9	28.3793	28.2849
10	31.5106	31.4255

The transient solution given in eq. (15.3-22) contains two terms. The first term is the steady state solution (which is simply eq. 15.3-18) and the second term is the transient contribution to the bulk concentration. This equation can be used to fit experimental data to extract the diffusion coefficient, in this case the pore diffusivity. This is usually done by a nonlinear optimization procedure, that is a value for the diffusivity is guessed and the theoretical curve is generated from eq. (15.3-22). The next step is to calculate the residual defined as

$$\text{Residual} = \sum_{j=1}^{n} \left[C_b(t_j) - C_b(t_j) \big|_{\exp} \right]^2 \quad (15.3\text{-}26)$$

where n is the number of data point. The optimisation procedure is then to minimize the above residual and when such minimum (usually local minimum) is found, the parameter obtained is the optimised parameter.

An alternative approach to the above numerical optimization involves only a simple linear plot. This procedure is as follows:
(1) For a known isotherm parameter, that is K, calculate B using eq. (15.3-23b) (or it can obtained from steady state concentration).

(2) Evaluate the eigenvalues from eq. (15.3-25b) and the coefficient a_n (eq. 15.3-25a). Then generate a plot of C_b/C_{bo} versus nondimensional time τ using eq. (15.3-22) as shown in Figure 15.3-2.
(3) From the experimental data $(t_j, C_b(t_j))$ obtain the nondimensional time τ_j corresponding to this data point. Repeat this for all experimental data points.
(4) Plot τ_j versus t_j and this should give a straight line (by the virtue of eq. 15.3-23c) with a slope of

$$\frac{\varepsilon D_p}{R^2[\varepsilon + (1-\varepsilon)K]} \qquad (15.3\text{-}27)$$

Knowing this slope the pore diffusivity can be calculated.

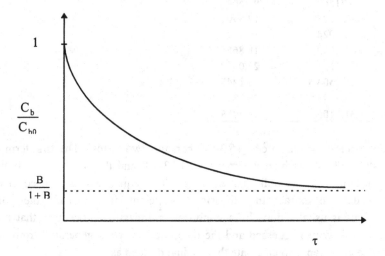

Figure 15.3-2: Plot of C_b/C versus τ (eq. 15.3-22)

15.3.2 Irreversible Adsorption Isotherm

When the adsorption isotherm is irreversible, the mass balance equations in the particle are (see Chapter 9 for more detail):

$$\frac{1}{r^2}\frac{\partial}{\partial r}\left(r^2 \frac{\partial C}{\partial r}\right) \cong 0 \qquad R_f < r < R \qquad (15.3\text{-}28a)$$

$$(1-\varepsilon)C_{\mu s}\frac{dR_f}{dt} = -\varepsilon D_p\frac{\partial C}{\partial r}\bigg|_{R_f} \tag{15.3-28b}$$

subject to:
$$r = R_f; \quad C = 0 \tag{15.3-29c}$$

$$r = R; \quad -\varepsilon D_p\frac{\partial C}{\partial r}\bigg|_R = k_m(C-C_b) \tag{15.3-29b}$$

The mass balance for the adsorbate in the reservoir is

$$V\frac{dC_b}{dt} = -\frac{3}{R}\left(\frac{m_p}{\rho_p}\right)\varepsilon D_p\frac{\partial C}{\partial r}\bigg|_R \tag{15.3-30}$$

Eqs. (15.3-28) simply state that the particle is divided into two regions separated by the position R_f. The process occurring in the saturated outer region is by pore diffusion and the rate of mass accumulation is that region is assumed negligible (that is $\partial C/\partial t \approx 0$). At the adsorption front ($R=R_f$), the adsorbate concentration is zero and the molar flux at the front is equal to the adsorption rate.

Solving eqs. (15.3-28) to (15.3-30) yields the following solution for the bulk concentration.

$$\frac{C_b}{C_{b0}} = 1 - \alpha(1-X^3) \tag{15.3-31}$$

where $X = R_f/R$ is a function of time, determined from the following equation:

$$\int_X^1 \frac{s^2\left(\frac{1}{s}-1+\frac{1}{Bi}\right)ds}{[1-\alpha(1-s^3)]} = \frac{\varepsilon D_p C_{b0}}{R^2(1-\varepsilon)C_{\mu s}}t \tag{15.3-32}$$

where

$$\alpha = \frac{\left(\frac{m_p}{\rho_p}\right)(1-\varepsilon)C_{\mu s}}{VC_{b0}} \tag{15.3-33a}$$

$$Bi = \frac{k_m R}{\varepsilon D_p} \tag{15.3-33b}$$

Eq. (15.3-31) can be used to match with the experimental data to extract the pore diffusivity. The procedure is done as follows:

(i) for a given data $(t_j, C_b(j))$, evaluate X_j from eq. (15.3-31). Knowing this X_j, evaluate the LHS of eq. (15.3-32) and let this value be H_j. Repeat this step for all data points.

(ii) Next, plot the LHS of eq. (15.3-32), that is H_j, versus t_j for all j. We get a straight line with a slope of

$$\frac{\varepsilon D_p C_{b0}}{R^2(1-\varepsilon)C_{\mu s}} \quad (15.3\text{-}34)$$

from which the pore diffusivity can be determined.

For the transient solution for the bulk concentration (eq. 15.3-31) to be positive, we must have:

$$\alpha < 1 \quad (15.3\text{-}35a)$$

that is

$$VC_{b0} > \left(\frac{m_p}{\rho_p}\right)(1-\varepsilon)C_{\mu s} \quad (15.3\text{-}35b)$$

the steady state concentration is achieved when the adsorption front R_f reaches the center of the particle (that is $R_f = 0$):

$$\frac{C_b}{C_{b0}} = 1 - \alpha = \frac{VC_{b0} - \left(\frac{m_p}{\rho_p}\right)(1-\varepsilon)C_{\mu s}}{VC_{b0}} \quad (15.3\text{-}36)$$

and the time it takes to reach this adsorption steady state concentration is

$$t^* = \frac{R^2(1-\varepsilon)C_{\mu s}}{\varepsilon D_p C_{b0}} \int_0^1 \frac{s^2\left(\frac{1}{s} - 1 + \frac{1}{Bi}\right)ds}{\left[1 - \alpha(1-s^3)\right]} \quad (15.3\text{-}37)$$

It takes finite time to reach the steady state and this is the property of the irreversible isotherm. In contrast, the time it takes to reach the steady state concentration in the case of linear isotherm is infinite and therefore to estimate the time scale of adsorption we have to use the time taken for the concentration to reach, say 95% of the steady state concentration.

If $\alpha > 1$, or

$$VC_{bo} < \left(\frac{m_p}{\rho_p}\right)(1-\varepsilon)C_{\mu s} \qquad (15.3\text{-}38)$$

That is the initial number of moles in the reservoir is not sufficient to saturate the particle, the steady state concentration will be zero and the final adsorption front is

$$X_\infty = \left(\frac{\alpha-1}{\alpha}\right)^{1/3} \qquad (15.3\text{-}39)$$

and the time it takes, for this case, to <u>empty</u> all adsorbate from the reservoir is:

$$t^0 = \frac{R^2(1-\varepsilon)C_{\mu s}}{\varepsilon D_p C_{bo}} \int_{X_\infty}^1 \frac{s^2\left(\frac{1}{s}-1+\frac{1}{Bi}\right)ds}{\left[1-\alpha(1-s^3)\right]} \qquad (15.3\text{-}40)$$

15.3.3 Nonlinear Adsorption Isotherm

When the adsorption isotherm is nonlinear, for example the Langmuir equation

$$C_\mu = C_{\mu s}\frac{bC}{1+bC} \qquad (15.3\text{-}41)$$

the mass balance equations (15.3-1a), (15.3-41) for the particle and the mass balance equation for the reservoir (15.3-4a) must be integrated numerically. The behaviour of this solution will be between those described in the last two sections for the case of linear and irreversible isotherms. For the parameter determination, the numerical solution can be matched with the experimental data by using a nonlinear optimisation to obtain the dynamic parameters.

15.4 Concluding Remarks

The batch adsorber method is a convenient method to obtain the dynamic parameter. The ease of such determination has been shown with systems following a linear isotherm or an irreversible isotherm. For systems following a nonlinear isotherm, the model equations must be solved numerically and such numerical solutions can be readily matched with the experimental data by using a nonlinear optimisation procedure. Software for nonlinear optimisation is readily available nowadays, and this will make the task of obtaining parameters easier. What we have shown so far in this chapter are situations under isothermal conditions.

Usually for the purpose of mass transfer parameter determination, care is always exercised to eliminate the heat transfer, such as the third configuration as mentioned in the introduction section (15.1).

Table of Contents of MatLab Programs

Disclaimer
This book contains a number of computer programs written in Matlab version 5.1. They are intended to use by the readers to understand the concepts of adsorption better, and they are for educational and instructional purposes only. Use of these programs for design purposes may be done at the owner's risk. Neither the author nor the publisher will bear any responsibility for the accuracy obtained from the enclosed programs

Program	Program description
ADSORB0	Adsorption kinetics of one species in a particle 1. Linear isotherm 2. Isothermal conditions
ADSORB1A	Adsorption kinetics of one species in a particle 1. Langmuir or Toth 2. Isothermal conditions
ADSORB1B	Adsorption kinetics of one species in a particle 1. Langmuir isotherm 2. Nonisothermal conditions
ADSORB1C	Adsorption kinetics of one species in a crystal 1. Langmuir isotherm 2. Isothermal conditions
ADSORB1D	Adsorption kinetics of one species in a crystal 1. Langmuir isotherm 2. Nonisothermal conditions

ADSORB1E	Adsorption kinetics of one species in a zeolite pellet composed of many crystals 1. Langmuir isotherm 2. Nonisothermal conditions 3. Zeolite pellet = many crystals
ADSORB3A	Adsorption kinetics of N species in a crystal under isothermal conditions 1. Extended Langmuir equation 2. Isothermal conditions 3. N species 4. Crystal
ADSORB5A	Adsorption kinetics of N species in a cylindrical pore under isothermal conditions 1. Extended Langmuir equation 2. Isothermal conditions 3. Cylindrical pore 4. N species 5. Viscous flow
ADSORB5B	Adsorption kinetics of N species in a particle under isothermal conditions 1. Extended Langmuir 2. Isothermal conditions 3. Porous particle 4. N species 5. Viscous flow and surface flow
ADSORB5C	Adsorption kinetics of N species in a particle under non-isothermal conditions 1. Extended Langmuir 2. Non-isothermal conditions 3. Particle 4. N species 5. Including viscous flow and surface flow
BIND	Calculation of binary D_{ij}
CAPILL2	Steady state diffusion flux in a capillary 1. Knudsen flow 2. Molecular diffusion flow
E_VS_D	Interaction energy versus half-width or pore radius
EIGENP	Eigenvalue in a single particle

FASTIAS	Fast IAS theory
FOWLER	Adsorption isotherm of Fowler-Guggenhein equation
HILL	Adsorption isotherm of Hill-deBoer equation
HK	Effective width of a slit pore as a function of reduced pressure using HK method
IAS	Ideal Adsorption Theory
ISO_FIT1	Pure component isotherm fitting
ISO_FIT2	Pure component fitting at multiple temperatures
MesoPSD1	Mesopore size distribution using double gamma PSD
P1043	Characteristics of 10_4_3 potential energy
PSD_MESO	Statistical thickness as a function of reduced pressure for mesopore
STEFTUBE	Stefan tube of a mixture of n species evaporating into an n-th species medium
UPTAKEP	Uptake calculation in a single particle

Most of these program can be run by typing the name of the program. For programs which require optimization routines, the user will be prompted for the guessing parameters and for approval whether the initial guess is acceptable. To change the parameters in the program, the user simply open the m files using either the NOTEPAD in Windows or the built-in m-editor of the MatLab version 5.1. The built-in m-editor is recommended because it can tell the user whether the syntax used is correct or not. In each program there is a section called "USER SUPPLY SECTION". There is a nomenclature in each of the program so that the user knows the meaning of each parameter. The user may change these parameters and save the file before executing the program within the MatLab worksheet.

For example, if the user wants to simulate the adsorption kinetics of an adsorbate in a particle under isothermal conditions and Langmuir isotherm, the user simply opens the ADSORB1A.M file, and will see the following "USER SUPPLY SECTION"

```
%--------------------------------------------------------------------------
% USER SUPPLY SECTION
s         = 0;              % particle shape factor
R         = 0.2;            % particle radius (cm)
porosity  = 0.33;           % particle porosity (-)
ci        = 0;              % initial gas concentration (mole/cc)
cb        = 1e-6;           % bulk gas concentration (mole/cc)
c0        = 1e-6;           % reference concentration (mole/cc)
b         = 1e6;            % adsorption affinity (cc/mole)
cmus      = 5e-3;           % maximum adsorption capacity (mole/cc)
biot      = 1000;           % biot number (-)
Dp        = 0.1;            % pore diffusivity (cm^2/sec)
Ds        = 1e-5;           % surface diffusivity (cm^2/sec)
%--------------------------------------------------------------------------
```

To change any of those parameters, for example the particle radius to 0.1, the user simply replace R = 0.2 in the "USER SUPPLY SECTION" to R = 0.1. Next save the file, and then return to the MatLab worksheet and execute the program by simply typing ADSORB1A.

Supporting programs

All the programs enclosed with this book require a number of supporting programs to run. Most of these programs are built-in programs within the MatLab version 5.1. However, some programs such as the Newton-Raphson program for solving nonlinear algebraic equations and collocation programs for solving boundary value problems are enclosed with the diskette. The collocation programs were adapted from algorithms developed by Villadsen and Michelsen (1978) (Chapter 3). Readers must refer to this excellent treatise for further exposition of the orthogonal collocation method.

Basic Instructions

Prior to running the programs enclosed in the diskette, the user should copy them into a sub-directory in a hard drive and make sure that the paths of these programs are added to the MatLab worksheet so that they can be executed.

Nomenclature

Symbol in text	Symbol in MatLab	Description	Units
a	a	amount adsorbed	mole/cc
a_n	a(n)	coefficient in expansion	-
a_m	am	projected area occupied by one molecule	nm^2/molecule
A	A	adsorption potential (=$R_g T \ln(P_0/P)$)	Joule/mole
b	b	Langmuir affinity constant	kPa^{-1}, Torr^{-1}
b_0	b0	Affinity constant at some reference temperature T_0	kPa^{-1}, Torr^{-1}
b_∞	b_infinity	Affinity constant at infinite temperature	kPa^{-1}, Torr^{-1}
B	B	structural parameter in Dubinin equation	(mole/Joule)^2
$\underline{B}$	B	inverse diffusion coefficient matrix	sec/m^2
B_0	B0	viscous flow parameter, = $r^2/8$ for straight capillary	m^2
Bi	Bi	Biot number for mass transfer	-
Bi_H	BiH	Biot number for heat transfer	-
c	c	free species concentration	mole/cc of gas
c_b	cb	bulk concentration	mole/cc of gas
c_i	ci	initial concentration	mole/cc of gas
c_0	c0	reference concentration	mole/cc of gas
C_{BET}	C_BET	constant in BET equation	-
C_p	Cp	heat capacity of solid	Joule/kg/K
C_f	Cf	heat capacity of fluid	Joule/kg/K
C_μ	cmu	adsorbed concentration	mole/cc solid
$C_{\mu b}$	cmub	adsorbed concentration in equilibrium with c_b	mole/cc solid
$C_{\mu i}$	cmui	adsorbed concentration in equilibrium with c_i	mole/cc solid
$C_{\mu s}$	cmus	saturation adsorbed concentration	mole/cc solid
$C_{\mu 0}$	cmu0	adsorbed concentration in equilibrium with c_0	mole/cc solid
d	d	pore half-width of slit shaped pore	nm
d_j		driving force in M-S formulation for component j	
D	D	diffusion coefficient	m^2/sec
D_{app}	Dapp	apparent diffusivity	m^2/sec
D_c	D_combined	combined diffusivity	m^2/sec
D_e	De	effective diffusivity	m^2/sec
D_p	Dp	pore diffusivity	m^2/sec
D_s	Ds	surface diffusivity	m^2/sec
D_μ	Dmu	diffusivity in adsorbed phase	m^2/sec

Nomenclature

Symbol in text	Symbol in MatLab	Description	Units
$D_{\mu T}$	DmuT	characteristic adsorbed diffusivity	m²/sec
$D_{\mu 0}$	Dmu0	diffusivity in adsorbed phase at reference T_0	m²/sec
D_μ^0	D0mu	corrected diffusivity	m²/sec
$D_{\mu 0}^0$	D0mu0	corrected diffusivity at reference temperature T_0	m²/sec
E	E	interaction energy between solid and adsorbing molecule	Joule/mole
E_d	Ed	activation energy for desorption	Joule/mole
E_L	EL	heat of liquefaction	Joule/mole
E_μ	Emu	activation energy for surface diffusion	Joule/mole
E_0	E0	characteristic energy used in Dubinin equation	Joule/mole
E_1	E1	interaction energy between solid and the first layer in BET	Joule/mole
f	f	fugacity	kPa, Torr
f_0	f0	fugacity at standard state	kPa, Torr
$f(C)$		functional form for the adsorption isotherm	mole/cc of solid
$f(r)$		pore size distribution	m⁻¹
$f(x)$		micropore size distribution	m⁻¹
F	fractional_uptake	fractional uptake	-
F		Hemholtz free energy	Joule/mole
$F(E)$		interaction energy distribution	mole/Joule
G		free energy	Joule/mole
h	h	heat transfer coefficient	Joule/m²/sec/K
H		enthalpy	Joule/mole
J	J	diffusive flux	mole/m²/sec
k_B	kB	Boltzmann constant, 1.38×10^{-23} Joule/molecule/K	J/molecule/K
k		parameter in Anderson's modified BET equation	-
k_a	k_ads	rate constant for adsorption	
k_d	k_des	rate constant for desorption	
k_e		effective thermal conductivity	Joule/m/sec/K
k_f		fluid thermal conductivity	Joule/m/sec/K
k_m	km	mass transfer coefficient	m/sec
K	K	Henry constant	-
K		Freundlich constant	
$K_n(x)$		eigenfunction	-
K_0	K0	Knudsen flow parameter, $= r/2$ for straight capillary	m
Kn		Knudsen number	-
L	L	length of porous medium	m
L	L	radius of a crystallite	m
L_c	Lc	length of a capillary	m
$LeBi$	LeBi	Lewis-Biot number for heat transfer	-

Symbol in text	Symbol in MatLab	Description	Units
m_p	mp	mass of particle	g
M	MW	molecular weight	g/mole
M(t)	M	amount adsorbed up to time t	mole
n	n	exponent in DA equation	-
n	n	number of sites per adsorbed molecule in Nitta equation	-
n	n	number of multilayer in multilayer BET equation	-
nc	ncomp	number of component in a mixture	-
N	N	molar flux	mole/m²/sec
N	N	number of interior collocation point	-
N_{AV}		Avogrado number	-
$N_{A,B}$		fluxes of component A and B	mole/m²/sec
p	p	partial pressure	kPa, Torr
p_b	pb	bulk pressure	kPa, Torr
p_i	pi	initial pressure	kPa, Torr
p_0	p0	reference pressure	kPa, Torr
P_{ads}		pressure at which condensation occurs upon adsorption	kPa, Torr
P_{des}		pressure at which condensation occurs upon desorption	kPa, Torr
P_0	P0	vapor pressure	kPa, Torr
p_T	pT	total pressure	kPa, Torr
q	q	tortuosity factor	-
Q	Q	heat of adsorption	Joule/mole
q_{iso}		isosteric heat of adsorption	Joule/mole
q_{net}		net heat of adsorption	Joule/mole
r	r	radial coordinate	m
r	r	pore radius	m
r	r	distance between the centers of two atoms or molecules	m
r_m		mean radius of curvature	m
r_0		distance between two atoms at which ε_{12} is zero	m
$r_{1,2}$		principal radii of a curved interface	m
R	R	particle radius	m
R_g	Rg	gas constant = 8.314 Joule/mole/K = 1.987 cal/mole/K = 82.05 cc-atm/mole/K = 8.314 × 10⁷ g-cm²-sec⁻²mole⁻¹ K⁻¹ = 8.314 × 10³ kg-m²sec⁻²kgmol⁻¹K⁻¹	
R_a		rate of adsorption	mole/sec
R_d		rate of desorption	mole/sec
R_μ	Rmu	radius of microparticle	m

Nomenclature

Symbol in text	Symbol in MatLab	Description	Units
s	s	heterogeneity parameter in Unilan equation	-
s	s	particle shape factor	-
S	S	entropy	Joule/mole/K
S_{ext}	S_ext	external surface area	m^2/g
S_g	Sg	specific surface area	m^2/g
t	t	time	sec
t	t	parameter in Toth isotherm	-
t	t	average thickness of the adsorbed layer	nm
t_{lag}	t_lag	time lag in the time lag analysis	sec
$t_{0.5}$	half_time	half time in fractional uptake	sec
T	T	temperature	K
T_b	Tb	bulk temperature	K
T_i	Ti	initial temperature	K
T_0	T0	reference temperature	K
u	u	superficial velocity	m/sec
u	u	pairwise adsorbate-adsorbate interaction energy	Joule/mole
v	v	interstitial velocity	m/sec
v_M	vm	liquid molar volume	cc/mole
$\bar{v}, v_T$		mean molecular speed	m/sec
V	V	volume	cc, m^3
V	V	volume of gas adsorbed	cc, m^3
V_m	Vm	volume of gas adsorbed to fill a monolayer	cc, m^3
V_p	Vp	specific pore volume	cc/g
w	w	adsorbate-adsorbate interaction energy	Joule/mole
W	W	specific amount adsorbed	mole/g, cc/g
W_0	W0	limiting amount adsorbed	mole/g, cc/g
x	x	mole fraction in the adsorbed phase	-
x	x	coordinate	m
x	x	micropore half width	nm
y	y	mole fraction in the gas phase	-
z	z	coordination number in adsorbate-adsorbate interaction	-
z	z	coordinate	m
z	z	reduced spreading pressure	mole/cc

Greek symbols

α	alpha	sticking coefficient	-
α	al	parameter for the Jacobi polynomial	-
α	alpha	exponent in the temperature dependence of D_p	-
β	beta	similarity coefficient	-
β	beta	heat transfer number = $QC_{\mu 0}/\rho C_p T_0$	-
β	be	parameter for the Jacobi polynomial in collocation method	-

Symbol in text	Symbol in MatLab	Description	Units
β		slip friction coefficient of viscous flow	
χ		parameter, function	-
δ	delta	thermal expansion coefficient of the maximum capacity	1/K
δ		distance between two adjacent adsorption sites	m
δ	delta	ratio of surface to pore volume diffusion	-
δ_0	delta0	ratio of surface to pore volume diffusion at T_0	-
ε	porosity	macropore and mesopore porosity	-
ε_μ	porosity_mu	micropore porosity	-
ε_b	porosity_b	bed porosity	-
ε_{12}		adsorption potential between two atoms or molecules	Joule
ε_{12}^*		depth of the adsorption potential minimum	Joule
ϕ		adsorption potential	Joule
γ	gamma	activity coefficient	-
γ	gammaE	activation number $= E_\mu/R_g T_0$	-
γ	gammaQ	heat of adsorption number $= Q/R_g T_0$	-
η_{nm}		eigenvalue	-
λ	lambda	mean free path	m
λ	lambda	Langmuir nondimensional parameter, $= bc, bp$	-
λ_b	lambdab	bC_b, bp_b	-
λ_i	lambdai	bc_i, bp_i	-
λ_0	lambda0	bc_0, bp_0	-
μ		chemical potential	Joule/mole
μ	viscosity	viscosity	kg/m/sec
η		nondimensional distance	-
ν		kinematic viscosity	m^2/sec
π		spreading pressure	Joule/m^3
θ	theta	fractional loading	-
θ	theta	non-dimensional temperature	-
θ_b	thetab	non-dimensional bulk temperature	-
θ_i	thetai	non-dimensional initial temperature	-
θ		angle of the liquid condensate at the solid	radian
θ		residence time	sec
ρ_p	rho_p	particle density	g/cc
ρ_μ	rho_mu	microparticle density	g/cc
ρC_p	rhoCp	volumetric heat capacity	Joule/cc/K
σ	sigma	surface tension	Joule/m^2
σ_{12}	sigma12	parameter in continuum diffusion, $= 1-\sqrt{M_1/M_2}$	-
σ_{12}	sigma12	average collision diameter	m
σ^2	sigma2	parameter in O'Brien and Myers equation	-
τ	tau	nondimensional time	-
τ		tortuosity ($=L/L_c$)	-
ω		frequency	1/sec

Symbol in text	Symbol in MatLab	Description	Units
ξ_n	zai(n)	n-th eigenvalue	-
Ψ		Functional form, parameter	-
ζ		eigenvalue	-
Δ		micropore half width variance	nm
Δh_i		enthalpy of immersion	Joule/mole
Γ		Gamma function	-
Λ		Function, defined in the Stefan-Maxwell equation	sec/m^2
		Upperscript	
α		phase α	
β		phase β	
σ		interface σ	
		Subscript	
0		reference temperature T_0 and pressure p_0	
b		bulk condition	
i		initial condition	
D		continuum diffusion	
K		Knudsen diffusion	
s		saturation	
s		surface	
T		total	
vis		viscous flow	
μ		adsorbed phase	

Constants & Units Conversion

SI prefixes

Prefix	Symbol	Multiplication factor
mega	M	10^6
kilo	k	10^3
deka	da	10
deci	d	10^{-1}
centi	c	10^{-2}
milli	m	10^{-3}
micro	µ	10^{-6}
nano	n	10^{-9}

Gas Constants R_g

1.987	cal gmole^{-1} K^{-1}
8.314	Joule gmole^{-1} K^{-1}
82.057	cm^3 atm gmole^{-1} K^{-1}
8.314×10^7	g-cm^2 sec^{-2} gmole^{-1} K^{-1}
8.314×10^3	kg m^2 sec^{-2} kgmole^{-1} K^{-1}
8314.34	Pa m^3 kgmole^{-1} K^{-1}

Length

	cm	m	in	ft
cm	1	10^{-2}	0.3937	0.0328
m	10^2	1	39.37	3.281
in	2.54	0.0254	1	0.08333
ft	30.48	0.3048	12	1

Mass

	g	kg	lb_m
g	1	10^{-3}	2.2046×10^{-3}
kg	10^3	1	2.2046
lb_m	453.59	0.45359	1

Volume

	cm^3	m^3	in^3	ft^3
cm^3	1	10^{-6}	0.061	3.5315×10^{-5}
m^3	10^6	1	61023.7	35.315
in^3	16.387	1.6387×10^{-5}	1	5.787×10^{-4}
ft^3	2.8317×10^4	0.028317	1728	1

Force

	g cm sec^{-2} (dynes)	kg m sec^{-2} (Newton)	lb_m ft sec^{-2}
g cm sec^{-2} (dynes)	1	10^{-5}	7.233×10^{-5}
kg m sec^{-2} (Newton)	10^5	1	7.233
lb_m ft sec^{-2}	1.3826×10^4	1.3826×10^{-1}	1

Pressure

	kg m^{-1} sec^{-2} (Newtons m^{-2})	lb_f in^{-2} (psia)^a	Atmospheres (atm)	mm Hg
kg m^{-1} sec^{-2}	1	1.4504×10^{-4}	9.8692×10^{-6}	7.5006×10^{-3}
lb_f in^{-2}	6.8947×10^3	1	6.8046×10^{-2}	5.1715×10^1
Atmospheres	1.0133×10^5	14.696	1	760
mm Hg	1.3332×10^2	1.9337×10^{-2}	1.3158×10^{-3}	1

Power

	g cm^2 sec^{-2} (ergs)	kg m^2 sec^{-2} (J)	cal	Btu
g cm^2 sec^{-2}	1	10^{-7}	2.3901×10^{-8}	9.4783×10^{-11}
kg m^2 sec^{-2}	10^7	1	2.3901×10^{-1}	9.4783×10^{-4}
cal	4.1840×10^7	4.1840	1	3.9657×10^{-3}
Btu	1.0550×10^{10}	1.0550×10^3	2.5216×10^2	1

Viscosity

	g cm^{-1} sec^{-1} (poises)	kg m^{-1} sec^{-1}	lb$_m$ ft^{-1} sec^{-1}
g cm^{-1} sec^{-1}	1	10^{-1}	6.7197 × 10^{-2}
kg m^{-1} sec^{-1}	10	1	6.7197 × 10^{-1}
lb$_m$ ft^{-1} sec^{-1}	1.4882 × 10^1	1.4882	1

Thermal conductivity

	kg m sec^{-3} K^{-1} (watts m^{-1} K^{-1})	lb$_m$ ft sec^{-3} F^{-1}	cal sec^{-1} cm^{-1} K^{-1}	Btu hr^{-1} ft^{-1} F^{-1}
kg m sec^{-3} K^{-1}	1	4.0183	2.3901 × 10^{-3}	5.7780 × 10^{-1}
lb$_m$ ft sec^{-3} F^{-1}	2.4886 × 10^{-1}	1	5.9479 × 10^{-4}	1.4379 × 10^{-1}
cal sec^{-1} cm^{-1} K^{-1}	4.1840 × 10^2	1.6813 × 10^3	1	2.4175 × 10^2
Btu hr^{-1} ft^{-1} F^{-1}	1.7307	6.9546	4.1365 × 10^{-3}	1

Diffusivity

	cm^2 sec^{-1}	m^2 sec^{-1}	ft^2 hr^{-1}
cm^2 sec^{-1}	1	10^{-4}	3.8750
m^2 sec^{-1}	10^4	1	3.8750 × 10^4
ft^2 hr^{-1}	2.5807 × 10^{-1}	2.5807 × 10^{-5}	1

Heat transfer coefficient

	kg sec^{-3} K^{-1} (watts m^{-2} K^{-1})	cal cm^{-2} sec^{-1} K^{-1}	Watts cm^{-2} K^{-1}	Btu ft^{-2} hr^{-1} F^{-1}
kg sec^{-3} K^{-1}	1	2.3901 × 10^{-5}	10^{-4}	1.7611 × 10^{-1}
cal cm^{-2} sec^{-1} K^{-1}	4.1840 × 10^4	1	4.1840	7.3686 × 10^3
Watts cm^{-2} K^{-1}	10^4	2.3901 × 10^{-1}	1	1.7611 × 10^3
Btu ft^{-2} hr^{-1} F^{-1}	5.6782	1.3571 × 10^{-4}	5.6782 × 10^{-4}	1

Mass transfer coefficient

	g cm^{-2} sec^{-1}	kg m^{-2} sec^{-1}	lb$_m$ ft^{-2} sec^{-1}	lb$_m$ ft^{-2} hr^{-1}
g cm^{-2} sec^{-1}	1	10^1	2.0482	7.3734 × 10^3
kg m^{-2} sec^{-1}	10^{-1}	1	2.0482 × 10^{-1}	7.3734 × 10^2
lb$_m$ ft^{-2} sec^{-1}	4.8824 × 10^{-1}	4.8824	1	3600
lb$_m$ ft^{-2} hr^{-1}	1.3562 × 10^{-4}	1.3562 × 10^{-3}	2.7778 × 10^{-4}	1

Appendices

Appendix 3.1: Isosteric Heat of the Sips Equation (3.2-18)

The Sips equation is (from eqs. 3.2-18):

$$\theta = \frac{(bP)^{1/n}}{1+(bP)^{1/n}} \tag{A3.1-1a}$$

where

$$b = b_\infty \exp\left(\frac{Q}{R_g T}\right); \quad \frac{1}{n} = \frac{1}{n_0} + \alpha\left(1 - \frac{T_0}{T}\right) \tag{A3.1-1b}$$

The isosteric heat is obtained from the van't Hoff equation:

$$\frac{(-\Delta H)}{R_g T^2} = \left(\frac{\partial \ln P}{\partial T}\right)_\theta \tag{A3.1-2}$$

We let $u = (bP)^{1/n}$, then the Sips equation written in terms of this new variable as

$$\theta = \frac{u}{1+u} \tag{A3.1-3}$$

At constant loading, we have $d\theta = 0$ and hence $du = 0$, that is:

$$du = n(bP) \ln(bP) \, d\left(\frac{1}{n}\right) + b \, dP + P \, db = 0 \tag{A3.1-4}$$

Using the temperature dependence of b and (1/n) given in eq. (A3.1-1b), we derive:

$$d\left(\frac{1}{n}\right) = \frac{\alpha T_0}{T^2} dT, \quad db = -b\frac{Q}{R_g T^2} dT \tag{A3.1-5}$$

Substitution eqs. (A3.1-5) into eq. (A3.1-4) and then into the van't Hoff equation yields:

$$(-\Delta H) = Q - n\alpha R_g T_0 \ln(bP) \tag{A3.1-6}$$

or in terms of the fractional loading

$$(-\Delta H) = Q - n^2\alpha R_g T_0 \ln\left(\frac{\theta}{1-\theta}\right) \tag{A3.1-7}$$

Appendix 3.2: Isosteric heat of the Toth equation (3.2-19)

The Toth equation is (eqs. 3.2-19)

$$\theta = \frac{bP}{\left[1+(bP)^t\right]^{1/t}} \tag{A3.2-1a}$$

where

$$b = b_\infty \exp\left(\frac{Q}{R_g T}\right); \quad t = t_0 + \alpha\left(1 - \frac{T_0}{T}\right) \tag{A3.2-1b}$$

Eqn. (A3.2-1a) can be written as:

$$\theta^t = \frac{(bP)^t}{1+(bP)^t} \tag{A3.2-2}$$

Taking the total differentiation of eq. (A3.2-2) at constant loading θ, we get:

$$\theta^t \ln\theta \, dt = \frac{\ln(bP)(bP)^t dt + t(bP)^{t-1} d(bP)}{(1+u)^2} \tag{A3.2-3}$$

where $u = (bP)^t$. Knowing the temperature dependence of t and b as in eq. (A3.2-1b), we evaluate the above equation and substitute into the van't Hoff equation (A3.1-2) to finally get:

$$(-\Delta H) = Q - \frac{1}{t}(\alpha R_g T_0)\left\{\ln(bP) - \left[1+(bP)^t\right]\ln\left[\frac{bP}{\left(1+(bP)^t\right)^{1/t}}\right]\right\} \tag{A3.2-4}$$

or in terms of the fractional loading

$$(-\Delta H) = Q - \frac{1}{t}(\alpha R_g T_0)\left\{\ln\left[\frac{\theta}{(1-\theta^t)^{1/t}}\right] - \frac{\ln\theta}{(1-\theta^t)}\right\} \tag{A3.2-5}$$

Appendix 3.3: Isosteric heat of the Unilan equation (3.2-23)

The Unilan equation (3.2-23) is:

$$\theta = \frac{1}{2s}\ln\left(\frac{1+be^{s}P}{1+be^{-s}P}\right) \qquad (A3.3\text{-}1a)$$

where

$$b = b_{\infty}\exp\left(\frac{\overline{E}}{R_gT}\right); \qquad s = \frac{\Delta E}{2R_gT} = \frac{E_{max}-E_{min}}{2R_gT} \qquad (A3.3\text{-}1b)$$

First we rewrite eq. (A3.3-1a) as follows:

$$e^{2s(\theta-1)} = \frac{e^{-s}+bP}{e^{s}+bP} \qquad (A3.3\text{-}2)$$

Taking the total differentiation of the above equation, we get

$$2(\theta-1)e^{2s(\theta-1)}ds + 2se^{2s(\theta-1)}d\theta = -\frac{(2+e^{s}bP+e^{-s}bP)}{(e^{s}+bP)^{2}}ds + \frac{e^{s}-e^{-s}}{(e^{s}+bP)^{2}}d(bP) \qquad (A3.3\text{-}3)$$

At constant fractional loading ($d\theta=0$), we have:

$$2(\theta-1)e^{2s(\theta-1)}ds + \frac{(2+e^{s}bP+e^{-s}bP)}{(e^{s}+bP)^{2}}ds = \frac{e^{s}-e^{-s}}{(e^{s}+bP)^{2}}d(bP) \qquad (A3.3\text{-}4)$$

From eq. (A3.3-1b), we obtain

$$ds = -\frac{\Delta E}{2R_gT^2}dT; \qquad d(bP) = bdP - bP\frac{\overline{E}}{R_gT^2}dT \qquad (A3.3\text{-}5)$$

Combining eqs. (A3.3-1a), (A3.3-4), and (A3.3-5), we obtain:

$$(-\Delta H) = R_gT^2\frac{\partial \ln P}{\partial T}$$

$$= \overline{E} + \frac{2(1-\theta)}{bP}\left[\frac{(e^{s}+bP)(e^{-s}+bP)}{(e^{s}-e^{-s})}\right]\left(\frac{\Delta E}{2}\right) - \left[\frac{2+e^{s}bP+e^{-s}bP}{e^{s}-e^{-s}}\right]\left(\frac{1}{bP}\right)\left(\frac{\Delta E}{2}\right) \qquad (A3.3\text{-}6)$$

Appendix 6.1: Energy potential between a species and surface atoms

The 12-6 potential energy between the molecule A and one surface atom as shown in the figure below is (eq. 6.10-1b):

$$\varphi_{12} = 4\varepsilon_{12}^* \left[\left(\frac{\sigma_{12}}{r}\right)^{12} - \left(\frac{\sigma_{12}}{r}\right)^{6} \right] \quad (A6.1\text{-}1)$$

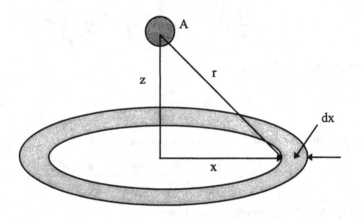

If the number of surface atom centers per unit area is n, the number of surface atom centers in the annulus as shown in the figure is $2n\pi x dx$. Therefore, the potential between the molecule and all the surface atom centers in that annulus is simply:

$$d\varphi_{1,SLP} = 2n\pi x dx \times 4\varepsilon_{12}^* \left[\left(\frac{\sigma_{12}}{r}\right)^{12} - \left(\frac{\sigma_{12}}{r}\right)^{6} \right] \quad (A6.1\text{-}2)$$

The potential between the molecule and the lattice plane of infinite extent is then simply the integral with respect to x from 0 to infinity:

$$\varphi_{1,SLP} = \int_0^\infty 2n\pi x dx \times 4\varepsilon_{12}^* \left[\left(\frac{\sigma_{12}}{r}\right)^{12} - \left(\frac{\sigma_{12}}{r}\right)^{6} \right] \quad (A6.1\text{-}3)$$

We note that $r^2 = x^2 + z^2$, and substitute this into the above integral, we finally obtain:

$$\varphi_{1,SLP} = 4n\pi\varepsilon_{12}^* \sigma_{12}^2 \left[\frac{1}{5}\left(\frac{\sigma_{12}}{z}\right)^{10} - \frac{1}{2}\left(\frac{\sigma_{12}}{z}\right)^{4} \right] \quad (A6.1\text{-}4)$$

Appendix 8.1: The momentum transfer of molecular collision

As a result of random motion of molecules distributed in space, molecule 1 collides with molecule 2. In such a collision, the momentum as well as the total energy of the system of two particles are conserved. There are two extremes of the collision. In one extreme, the collision is completely elastic, that is the kinetic energy is conserved. The sum of the kinetic energies of the two molecules before the collision is the same as that after the collision. In the other extreme, the two particles stick together after the collision, the collision is called completely inelastic.

As we have indicated above, even though the momentum and energy are conserved, the details of the motion after the collision are not determined, except for the two cases, the completely inelastic case and the elastic case in one dimension. Let us first investigate the case of completely inelastic, and then the other case of elastic collision.

Case 1: Completely Inelastic:

When the two bodies approach each other and stick together after the collision, the momentum and the total energy are conserved, that is:

$$m_1 u_1 + m_2 u_2 = (m_1 + m_2) u' \quad \text{(A8.1-1a)}$$

$$\frac{1}{2} m_1 u_1^2 + \frac{1}{2} m_2 u_2^2 = \frac{1}{2} (m_1 + m_2)(u')^2 + \text{Deformed energy} \quad \text{(A8.1-1b)}$$

where the prime denotes condition after the collision. Solving the conservation of momentum equation will give the following equation for the velocity after collision:

$$u' = \frac{m_1 u_1 + m_2 u_2}{m_1 + m_2} \quad \text{(A8.1-2)}$$

Substitute this final velocity into the total energy conservation equation to give the following expression for the deformed energy:

$$\text{Deformed Energy} = \frac{1}{2} \frac{m_1 m_2 (u_1 - u_2)^2}{m_1 + m_2} \quad \text{(A8.1-3)}$$

With the final velocity u' given in eq. (A8.1-2), the momentum transferred from the body 1 to the body 2 is (that is, the loss of momentum by the body 1):

$$m_1 (u_1 - u') = m_1 \left(u_1 - \frac{m_1 u_1 + m_2 u_2}{m_1 + m_2} \right) = \frac{m_1 m_2}{m_1 + m_2} (u_1 - u_2) \quad \text{(A8.1-4)}$$

We see that the momentum transferred from the body 1 to the body 2 is proportional to the velocities the collision. Similarly, the momentum transferred from the body 2 to the body 1 is:

$$m_2(u_2 - u') = m_2\left(u_2 - \frac{m_1 u_1 + m_2 u_2}{m_1 + m_2}\right) = \frac{m_1 m_2}{m_1 + m_2}(u_2 - u_1) \quad (A8.1-5)$$

which is the same as that of eq. (A8.1-4), except a difference in sign; this is the consequence of the momentum conservation.

Case 2: Elastic Collision

When two glassy bodies collide with each other in one dimensional, that is along the line connecting the two centers of the two bodies. The motion after this collision is also along the same line without rotation. The conservation of momentum and total energy will give:

$$m_1 u_1 + m_2 u_2 = m_1 u_1' + m_2 u_2' \quad (A8.1-6a)$$

$$\frac{1}{2} m_1 u_1^2 + \frac{1}{2} m_2 u_2^2 = \frac{1}{2} m_1 (u_1')^2 + m_2 (u_2')^2 \quad (A8.1-6b)$$

Solving these two equations for the two final velocities, we get:

$$u_1' = \left(\frac{m_1 - m_2}{m_1 + m_2}\right) u_1 + \left(\frac{2 m_2}{m_1 + m_2}\right) u_2 \quad (A8.1-7a)$$

$$u_2' = \left(\frac{m_2 - m_1}{m_1 + m_2}\right) u_2 + \left(\frac{2 m_1}{m_1 + m_2}\right) u_1 \quad (A8.1-7b)$$

The transfer of momentum from the body 1 to the body 2 is:

$$m_1(u_1 - u_1') = m_1\left[u_1 - \left(\frac{m_1 - m_2}{m_1 + m_2}\right) u_1 - \left(\frac{2 m_2}{m_1 + m_2}\right) u_2\right] = \frac{2 m_1 m_2}{m_1 + m_2}(u_1 - u_2) \quad (A8.1-8a)$$

Similarly, the transfer of momentum from the body 2 to the body 1 is:

$$m_2(u_2 - u_2') = m_2\left[u_2 - \left(\frac{m_2 - m_1}{m_1 + m_2}\right) u_2 - \left(\frac{2 m_1}{m_1 + m_2}\right) u_1\right] = \frac{2 m_1 m_2}{m_1 + m_2}(u_2 - u_1) \quad (A8.1-8b)$$

Thus, we see again that the transfer of momentum from one body to the other is again proportional to the difference of the two initial velocities, as we have observed for the case of completely inelastic.

After considering the two cases of completely elastic and elastic collision in one dimension, we see that the momentum transferred from the body 1 to body 2 is proportional to the difference of the two initial velocities. If we now assume that molecules behave like a rigid body, the above analysis is applicable to the case of collision between molecules. Thus, for the collision of two molecules of different type, the momentum transferred from the molecule 1 to the molecule 2 is proportional to their initial velocities before the collision. If the two molecules of the same type collide, there will be no net loss of the momentum of the type 1 molecule as the momentum is conserved for the type 1 family. Hence, the momentum **lost** by the type 1 depends only on the collision of the type 1 molecule and the type 2 molecule. This is a rather important result in the understanding of diffusion.

Appendix 8.2: Solving the Stefan-Maxwell equations (8.2-97 and 8.2-98)

The Stefan-Maxwell equation written in vector format can take the following form written in terms of fluxes (eq. 8.2-97):

$$c\frac{d\underline{y}}{dz} = -\underline{\underline{B}}(\underline{y})\underline{N} \quad \text{(A8.2-1)}$$

or it can be written in terms of mole fraction as follows (eq. 8.2-98):

$$\frac{d\underline{y}}{dz} = \underline{\underline{A}}(\underline{N})\underline{y} + \underline{\psi}(\underline{N}) \quad \text{(A8.2-2)}$$

The boundary conditions at two ends of the domain are:

$$z = 0; \quad \underline{y} = \underline{y}_0 \quad \text{(A8.2-3a)}$$

$$z = L; \quad \underline{y} = \underline{y}_L \quad \text{(A8.2-3b)}$$

Using the matrix-vector method, we can integrate eq. (A8.2-2) to obtain (Cole, 1968):

$$\underline{y} = \underline{y}_0 + \left\{\exp[z\underline{\underline{A}}(\underline{N})] - \underline{\underline{I}}\right\}\left\{\underline{y}_0 + [\underline{\underline{A}}(\underline{N})]^{-1}\underline{\psi}(\underline{N})\right\} \quad \text{(A8.2-4)}$$

where $\underline{\underline{I}}$ is the identity matrix. Evaluating this equation at $z = L$, we get:

$$\underline{y}_L = \underline{y}_0 + \left\{\exp[L\underline{\underline{A}}(\underline{N})] - \underline{\underline{I}}\right\}\left\{\underline{y}_0 + [\underline{\underline{A}}(\underline{N})]^{-1}\underline{\psi}(\underline{N})\right\} \quad \text{(A8.2-5)}$$

Combining these two equations yields:

$$\underline{y} - \underline{y}_0 = \left[\exp(z\underline{\underline{A}}) - \underline{\underline{I}}\right]\left[\exp(L\underline{\underline{A}}) - \underline{\underline{I}}\right]^{-1}(\underline{y}_L - \underline{y}_0) \quad \text{(A8.2-6)}$$

Differentiating the above equation with respect to z, we get the concentration gradient vector:

$$\frac{d\underline{y}}{dz} = \underline{\underline{A}}\exp(z\underline{\underline{A}})\left[\exp(L\underline{\underline{A}}) - \underline{\underline{I}}\right]^{-1}(\underline{y}_L - \underline{y}_0) \quad (A8.2\text{-}7)$$

Thus, the gradients at two end points of the domain are:

$$\left.\frac{d\underline{y}}{dz}\right|_{z=0} = \underline{\underline{A}}\left[\exp(L\underline{\underline{A}}) - \underline{\underline{I}}\right]^{-1}(\underline{y}_L - \underline{y}_0) \quad (A8.2\text{-}8)$$

$$\left.\frac{d\underline{y}}{dz}\right|_{z=L} = \underline{\underline{A}}\exp(L\underline{\underline{A}})\left[\exp(L\underline{\underline{A}}) - \underline{\underline{I}}\right]^{-1}(\underline{y}_L - \underline{y}_0) \quad (A8.2\text{-}9)$$

It is noted that the matrix $\underline{\underline{A}}$ in the above two equations is a function of the flux vector $\underline{N}$. To solve for this flux vector, we consider eq. (A8.2-1), rewritten in the form:

$$\underline{N} = -c\left[\underline{\underline{B}}(\underline{y})\right]^{-1}\frac{d\underline{y}}{dz} \quad (A8.2\text{-}10)$$

Evaluating eq. (A8.2-10) at two end points, we get the relations for the fluxes at the two end points:

$$\left.\underline{N}\right|_{z=0} = -c\left[\underline{\underline{B}}(\underline{y}_0)\right]^{-1}\left.\frac{d\underline{y}}{dz}\right|_{z=0} \quad (A8.2\text{-}11a)$$

$$\left.\underline{N}\right|_{z=L} = -c\left[\underline{\underline{B}}(\underline{y}_L)\right]^{-1}\left.\frac{d\underline{y}}{dz}\right|_{z=L} \quad (A8.2\text{-}11b)$$

At steady state, the fluxes at two ends must be the same. Thus, we can use either of the two equations to determine the flux through the medium. We take eq. (A8.2-11a) and substitute the gradient obtained in eq. (A8.2-8) to finally get:

$$\underline{N} = c\left[\underline{\underline{B}}(\underline{y}_0)\right]^{-1}\underline{\underline{A}}\left[\exp(L\underline{\underline{A}}) - \underline{\underline{I}}\right]^{-1}(\underline{y}_0 - \underline{y}_L) \quad (A8.2\text{-}12)$$

However, if we take (A8.2-11b) and substitute eq.(A8.2-9) into that equation, we get

$$\underline{N} = c\left[\underline{\underline{B}}(\underline{y}_L)\right]^{-1}\underline{\underline{A}}\exp(L\underline{\underline{A}})\left[\exp(L\underline{\underline{A}}) - \underline{\underline{I}}\right]^{-1}(\underline{y}_0 - \underline{y}_L) \quad (A8.2\text{-}13)$$

This equation is equivalent to eq.(A8.2-12).

Eqs. (A8.2-13) or (A8.2-12) represents n-1 nonlinear algebraic equations in terms of the flux vector $\underline{N}$. It can be solved quite readily with the Newton-Raphson method for the flux vector. The initial guess can be obtained from eq. (A8.2-10) by approximating the matrix B as that evaluated at the mean concentration

$$\underline{y}_m = (\underline{y}_0 + \underline{y}_L)/2 \qquad (A8.2\text{-}14)$$

Thus, the initial guess is:

$$\underline{N}^{(0)} = c\left[\underline{\underline{B}}(\underline{y}_m)\right]^{-1}(\underline{y}_0 - \underline{y}_L) \qquad (A8.2\text{-}15)$$

and the iteration formula is:

$$\underline{N}^{(k+1)} = \underline{N}^{(k)} - \left[\underline{\underline{J}}(\underline{N})^{(k)}\right]^{-1}\underline{f}^{(k)} \qquad (A8.2\text{-}16)$$

where J is the Jacobian of the vector f, defined as:

$$\underline{f} = \underline{N} - c\left[\underline{\underline{B}}(\underline{y}_0)\right]^{-1}\underline{\underline{A}}\left[\exp(\underline{\underline{L}}\underline{\underline{A}}) - \underline{\underline{I}}\right]^{-1}(\underline{y}_0 - \underline{y}_L) \qquad (A8.2\text{-}17a)$$

or

$$\underline{f} = \underline{N} - c\left[\underline{\underline{B}}(\underline{y}_L)\right]^{-1}\underline{\underline{A}}\exp(\underline{\underline{L}}\underline{\underline{A}})\left[\exp(\underline{\underline{L}}\underline{\underline{A}}) - \underline{\underline{I}}\right]^{-1}(\underline{y}_0 - \underline{y}_L) \quad (A8.2\text{-}17b)$$

For numerical computation, the Jacobian matrix is best obtained numerically. Usually, it takes about a few iterations to converge the above nonlinear equation.

Appendix 8.3: Collocation analysis of eqs. (8.3-16) and (8.3-17)

Equations (8.3-16) and (8.3-17) for the Loschmidt tube system are:

Section I	Section II
$\dfrac{\partial \underline{y}^I}{\partial t^*} = \dfrac{\partial}{\partial \eta_1}\left\{\left[\underline{\underline{B}}^*(\underline{y}^I)\right]^{-1}\dfrac{\partial \underline{y}^I}{\partial \eta_1}\right\}$ (A8.3-1a)	$\dfrac{\partial \underline{y}^{II}}{\partial t^*} = \left(\dfrac{L_1}{L_2}\right)^2 \dfrac{\partial}{\partial \eta_2}\left\{\left[\underline{\underline{B}}^*(\underline{y}^{II})\right]^{-1}\dfrac{\partial \underline{y}^{II}}{\partial \eta_2}\right\}$ (A8.3-2a)
$t^* = 0; \quad \underline{y}^I = \underline{y}^I(0)$ (A8.3-1b)	$t^* = 0; \quad \underline{y}^{II} = \underline{y}^{II}(0)$ (A8.3-2b)
$\eta_1 = 0; \quad \dfrac{\partial \underline{y}^I}{\partial \eta_1} = 0$ (A8.3-1c)	$\eta_2 = 0; \quad \dfrac{\partial \underline{y}^{II}}{\partial \eta_2} = 0$ (A8.3-2c)
$\eta_1 = 1$ & $\eta_2 = 1; \quad \underline{y}^I\big|_{\eta_1=1} = \underline{y}^{II}\big|_{\eta_2=1}$	(A8.3-1d)
$\eta_1 = 1$ & $\eta_2 = 1; \quad \left[\underline{\underline{B}}^*(\underline{y}^I)\right]^{-1}\dfrac{\partial \underline{y}^I}{\partial \eta_1}\bigg|_{\eta_1=1} = -\left(\dfrac{L_1}{L_2}\right)\left[\underline{\underline{B}}^*(\underline{y}^{II})\right]^{-1}\dfrac{\partial \underline{y}^{II}}{\partial \eta_2}\bigg|_{\eta_2=1}$	(A8.3-1e)

Expanding the RHS of eqs.(A8.3-1a) and (A8.3-2a), we have:

$$\frac{\partial y^I}{\partial t^*} = \left[\underline{\underline{B}}^*(\underline{y}^I)\right]^{-1} \frac{\partial^2 \underline{y}^I}{\partial \eta_1^2} + \frac{\partial\left\{\left[\underline{\underline{B}}^*(\underline{y}^I)\right]^{-1}\right\}}{\partial \eta_1} \frac{\partial \underline{y}^I}{\partial \eta_1} \quad \text{(A8.3-3)}$$

$$\left(\frac{L_2}{L_1}\right)^2 \frac{\partial \underline{y}^{II}}{\partial t^*} = \left[\underline{\underline{B}}^*(\underline{y}^{II})\right]^{-1} \frac{\partial^2 \underline{y}^{II}}{\partial \eta_2^2} + \frac{\partial\left\{\left[\underline{\underline{B}}^*(\underline{y}^{II})\right]^{-1}\right\}}{\partial \eta_2} \frac{\partial \underline{y}^{II}}{\partial \eta_2} \quad \text{(A8.3-4)}$$

Because of the symmetry at $\eta_1 = 0$ and $\eta_2 = 0$, we make the following transformation:

$$u_1 = \eta_1^2 \quad \text{(A8.3-5a)}$$

$$u_2 = \eta_2^2 \quad \text{(A8.3-5b)}$$

With these transformations, eqs.(A8.3-3) and (A8.3-4) become:

$$\frac{\partial \underline{y}^I}{\partial t^*} = \left[\underline{\underline{B}}^*(\underline{y}^I)\right]^{-1}\left[4u_1 \frac{\partial^2 \underline{y}^I}{\partial u_1^2} + 2\frac{\partial \underline{y}^I}{\partial u_1}\right] + 4u_1 \frac{\partial\left\{\left[\underline{\underline{B}}^*(\underline{y}^I)\right]^{-1}\right\}}{\partial u_1} \frac{\partial \underline{y}^I}{\partial u_1} \quad \text{(A8.3-6)}$$

and

$$\left(\frac{L_2}{L_1}\right)^2 \frac{\partial \underline{y}^{II}}{\partial t^*} = \left[\underline{\underline{B}}^*(\underline{y}^{II})\right]^{-1}\left[4u_2 \frac{\partial^2 \underline{y}^{II}}{\partial u_2^2} + 2\frac{\partial \underline{y}^{II}}{\partial u_2}\right] + 4u_2 \frac{\partial\left\{\left[\underline{\underline{B}}^*(\underline{y}^{II})\right]^{-1}\right\}}{\partial u_2} \frac{\partial \underline{y}^{II}}{\partial u_2} \quad \text{(A8.3-7)}$$

and the boundary conditions at the junction between the two sections (eqs. A8.3-1d and A8.3-1e) will be

$$u_1 = 1 \quad \& \quad u_2 = 1; \quad \underline{y}^I\big|_{u_1=1} = \underline{y}^{II}\big|_{u_2=1} \quad \text{(A8.3-8)}$$

$$u_1 = 1 \quad \& \quad u_2 = 1; \quad \left[\underline{\underline{B}}^*(\underline{y}^I)\right]^{-1}\frac{\partial \underline{y}^I}{\partial u_1}\bigg|_{u_1=1} = -\left(\frac{L_1}{L_2}\right)\left[\underline{\underline{B}}^*(\underline{y}^{II})\right]^{-1}\frac{\partial \underline{y}^{II}}{\partial u_2}\bigg|_{u_2=1} \quad \text{(A8.3-9)}$$

The collocation analysis is carried out in the u_1 - domain and u_2 - domain for Section 1 and Section 2, respectively. We choose N interior collocation points in Section 1, and for simplicity we also choose the same number of interior collocation points in Section 2.

Evaluating the mass balance equation in Section 1 (eq. A8.3-6) at the j-th interior collocation point, we have:

$$\frac{d\underline{y}_j^I}{dt^*} = \left[\underline{\underline{B}}^*(\underline{y}_j^I)\right]^{-1} \left[4u_{1,j} \frac{\partial^2 \underline{y}^I}{\partial u_1^2}\bigg|_{u_1=u_{1,j}} + 2\frac{\partial \underline{y}^I}{\partial u_1}\bigg|_{u_1=u_{1,j}} + 4u_{1,j} \frac{\partial \left\{\left[\underline{\underline{B}}^*(\underline{y}^I)\right]^{-1}\right\}}{\partial u_1}\bigg|_{u_1=u_{1,j}} \left(\frac{\partial \underline{y}^I}{\partial u_1}\bigg|_{u_1=u_{1,j}}\right)\right]$$

(A8.3-10)

for j = 1, 2, ..., N.

The derivatives at the collocation points are related to the functional values according to the following collocation formulas (Villadsen and Michelsen, 1978):

$$\frac{\partial^2 \underline{y}^I}{\partial u_1^2}\bigg|_{u_1=u_{1,j}} = \sum_{k=1}^{N+1} B_{jk} \underline{y}_k^I \qquad (A8.3\text{-}11a)$$

$$\frac{\partial \underline{y}^I}{\partial u_1}\bigg|_{u_1=u_{1,j}} = \sum_{k=1}^{N+1} A_{jk} \underline{y}_k^I \qquad (A8.3\text{-}11b)$$

$$\frac{\partial \left\{\left[\underline{\underline{B}}^*(\underline{y}^I)\right]^{-1}\right\}}{\partial u_1}\bigg|_{u_1=u_{1,j}} = \sum_{k=1}^{N+1} A_{jk} \left[\underline{\underline{B}}^*(\underline{y}_k^I)\right]^{-1} \qquad (A8.3\text{-}11c)$$

where $\underline{\underline{A}}$ and $\underline{\underline{B}}$ are first and second derivative matrices, 1978).

Substitution of eqs.(A8.3-11) into eq. (A8.3-10) yields:

$$\frac{d\underline{y}_j^I}{dt^*} = \left[\underline{\underline{B}}^*(\underline{y}_j^I)\right]^{-1} \sum_{k=1}^{N+1} C_{jk} \underline{y}_k^I + 4u_{1,j} \left\{\sum_{k=1}^{N+1} A_{jk} \left[\underline{\underline{B}}^*(\underline{y}_k^I)\right]^{-1}\right\} \left\{\sum_{k=1}^{N+1} A_{jk} \underline{y}_k^I\right\} \quad (A8.3\text{-}12a)$$

where

$$C_{jk} = 4u_{1,j} B_{jk} + 2A_{jk} \qquad (A8.3\text{-}12b)$$

We note that the summations in the above equation involve the boundary point at N+1. Splitting the last terms of the above summations, we get:

$$\frac{d\underline{y}_j^I}{dt^*} = \left[\underline{\underline{B}}^*(\underline{y}_j^I)\right]^{-1} \times \left[\sum_{k=1}^{N} C_{jk} \underline{y}_k^I + C_{j,N+1} \underline{y}_{N+1}^I\right] +$$

$$4u_{1,j} \left\{\sum_{k=1}^{N} A_{jk} \left[\underline{\underline{B}}^*(\underline{y}_k^I)\right]^{-1} + A_{j,N+1} \left[\underline{\underline{B}}^*(\underline{y}_{N+1}^I)\right]^{-1}\right\} \times \left\{\sum_{k=1}^{N} A_{jk} \underline{y}_k^I + A_{j,N+1} \underline{y}_{N+1}^I\right\}$$

(A8.3-13)

for j = 1, 2, ..., N.

Applying the similar collocation analysis to the mass balance equation of Section 2 (eq. A8.3-7), we obtain:

$$\left(\frac{L_2}{L_1}\right)^2 \frac{d\underline{y}_j^{II}}{dt^*} = \left[\underline{\underline{B}}^*(\underline{y}_j^{II})\right]^{-1} \times \left[\sum_{k=1}^{N} C_{jk} \underline{y}_k^{II} + C_{j,N+1} \underline{y}_{N+1}^{II}\right] +$$

$$4u_{2,j} \left\{ \sum_{k=1}^{N} A_{jk} \left[\underline{\underline{B}}^*(\underline{y}_k^{II})\right]^{-1} + A_{j,N+1} \left[\underline{\underline{B}}^*(\underline{y}_{N+1}^{II})\right]^{-1} \right\} \times \left\{ \sum_{k=1}^{N} A_{jk} \underline{y}_k^{II} + A_{j,N+1} \underline{y}_{N+1}^{II} \right\}$$

for j = 1, 2, ..., N (A8.3-14)

At the interface between the two sections, we have the continuity of concentration

$$\underline{y}_{N+1}^{I} = \underline{y}_{N+1}^{II} = \underline{\psi} \qquad (A8.3\text{-}15)$$

With this, eqs.(A8.3-13) and (A8.3-14) will become:

$$\frac{d\underline{y}_j^I}{dt^*} = \left[\underline{\underline{B}}^*(\underline{y}_j^I)\right]^{-1} \times \left[\sum_{k=1}^{N} C_{jk} \underline{y}_k^I + C_{j,N+1} \underline{\psi}\right] +$$

$$4u_{1,j} \left\{ \sum_{k=1}^{N} A_{jk} \left[\underline{\underline{B}}^*(\underline{y}_k^I)\right]^{-1} + A_{j,N+1} \left[\underline{\underline{B}}^*(\underline{\psi})\right]^{-1} \right\} \times \left\{ \sum_{k=1}^{N} A_{jk} \underline{y}_k^I + A_{j,N+1} \underline{\psi} \right\} \quad (A8.3\text{-}16)$$

for j = 1, 2, ..., N, and

$$\left(\frac{L_2}{L_1}\right)^2 \frac{d\underline{y}_j^{II}}{dt^*} = \left[\underline{\underline{B}}^*(\underline{y}_j^{II})\right]^{-1} \times \left[\sum_{k=1}^{N} C_{jk} \underline{y}_k^{II} + C_{j,N+1} \underline{\psi}\right] +$$

$$4u_{2,j} \left\{ \sum_{k=1}^{N} A_{jk} \left[\underline{\underline{B}}^*(\underline{y}_k^{II})\right]^{-1} + A_{j,N+1} \left[\underline{\underline{B}}^*(\underline{\psi})\right]^{-1} \right\} \times \left\{ \sum_{k=1}^{N} A_{jk} \underline{y}_k^{II} + A_{j,N+1} \underline{\psi} \right\} \quad (A8.3\text{-}17)$$

for j = 1, 2, ..., N.

The vector $\underline{\psi}$ is determined from the continuity of flux conditions (eq. A8.3-9). Written in terms of the collocation variables, we have:

$$\left[\underline{\underline{B}}^*(\underline{\psi})\right]^{-1} \left[\sum_{k=1}^{N} A_{N+1,k} \underline{y}_k^I + A_{N+1,N+1} \underline{\psi}\right] = -\left(\frac{L_1}{L_2}\right) \left[\underline{\underline{B}}^*(\underline{\psi})\right]^{-1} \left[\sum_{k=1}^{N} A_{N+1,k} \underline{y}_k^{II} + A_{N+1,N+1} \underline{\psi}\right]$$

(A8.3-18)

Cancelling $\left[\underline{\underline{B}}^*(\underline{\psi})\right]^{-1}$ and solving for $\underline{\psi}$ from the above equation, we get:

$$\underline{\psi} = -\frac{1}{\left(1+\frac{L_1}{L_2}\right)A_{N+1,N+1}} \sum_{k=1}^{N} A_{N+1,k}\left(\underline{y}_k^I + \frac{L_1}{L_2}\underline{y}_k^{II}\right) \qquad (8.3\text{-}19)$$

Solution Procedure:

Eqs.(A8.3-16) and (A8.3-17) represent 2N (n-1) ordinary differential equations with 2N (n-1) unknowns, where n is the number of components. They can be numerically integrated.

For the purpose of programming this problem in MATLAB language, we define the following concentration matrices for the two sections of the Loschmidt tube.

$$\underline{\underline{Y}}^I = \begin{bmatrix} y^I(1,1) & y^I(1,2) & y^I(1,3) & \cdots & y^I(1,N) \\ y^I(2,1) & y^I(2,2) & y^I(2,3) & \cdots & y^I(2,N) \\ \vdots & \vdots & \vdots & \ddots & \vdots \\ y^I(n-2,1) & y^I(n-2,2) & y^I(n-2,3) & \cdots & y^I(n-2,N) \\ y^I(n-1,1) & y^I(n-1,2) & y^I(n-1,3) & \cdots & y^I(n-1,N) \end{bmatrix} \qquad (A8.3\text{-}20)$$

$$\underline{\underline{Y}}^{II} = \begin{bmatrix} y^{II}(1,1) & y^{II}(1,2) & y^{II}(1,3) & \cdots & y^{II}(1,N) \\ y^{II}(2,1) & y^{II}(2,2) & y^{II}(2,3) & \cdots & y^{II}(2,N) \\ \vdots & \vdots & \vdots & \ddots & \vdots \\ y^{II}(n-2,1) & y^{II}(n-2,2) & y^{II}(n-2,3) & \cdots & y^{II}(n-2,N) \\ y^{II}(n-1,1) & y^{II}(n-1,2) & y^{II}(n-1,3) & \cdots & y^{II}(n-1,N) \end{bmatrix} \qquad (A8.3\text{-}21)$$

With these definitions, we have the following notation in MATLAB language

Y(:,j) ≡ mole fractions of all species at the j-th collocation point
Y(i,:) ≡ mole fractions of species "i" at all N interior collocation points

In terms of these MATLAB notations, eqs.(A8.3-16) and (A8.3-17) will become:

$$\frac{dY^I(:,j)}{dt^*} = \left[\underline{\underline{B}}^*(Y^I(:,j))\right]^{-1} \times \left[\sum_{k=1}^{N} C_{jk} Y^I(:,k) + C_{j,N+1}\psi(:)\right]$$

$$+ (4u_1)_j \left\{\sum_{k=1}^{N} A_{jk}\left[\underline{\underline{B}}^*(Y^I(:,k))\right]^{-1} + A_{j,N+1}\left[\underline{\underline{B}}^*(\psi(:))\right]^{-1}\right\} \times \left\{\sum_{k=1}^{N} A_{jk} Y^I(:,k) + A_{j,N+1}\psi(:)\right\}$$

$$(A8.3\text{-}22)$$

and

$$\left(\frac{L_2}{L_1}\right)^2 \frac{dY^{II}(:,j)}{dt^*} = \left[\underline{\underline{B}}^*(Y^{II}(:,j))\right]^{-1} \times \left[\sum_{k=1}^{N} C_{jk} Y^{II}(:,k) + C_{j,N+1}\psi(:)\right]$$

$$+ (4u_2)_j \left\{\sum_{k=1}^{N} A_{jk}\left[\underline{\underline{B}}^*(Y^{II}(:,k))\right]^{-1} + A_{j,N+1}\left[\underline{\underline{B}}^*(\psi(:))\right]^{-1}\right\} \times \left\{\sum_{k=1}^{N} A_{jk} Y^{II}(:,k) + A_{j,N+1}\psi(:)\right\}$$

(A8.3-23)

where:

$$\psi(:) = -\frac{1}{\left(1+\frac{L_1}{L_2}\right) A_{N+1,N+1}} \sum_{k=1}^{N} A_{N+1,k}\left[Y^{I}(:,k) + \frac{L_1}{L_2} Y^{II}(:,k)\right] \quad (A8.3\text{-}24)$$

For numerical integration by method such as the Runge-Kutta method, we have to put all the variables in the column format. This is done by defining a vector Ω as follows:

$$\Omega((i-1)N+j) = \underline{Y}^I(i,j) \text{ for } j = 1, 2, ..., N \text{ and } i = 1, 2, ... (n-1).$$

$$\Omega((n-1)N+(i-1)N+j) = \underline{Y}^{II}(i,j)$$

This column vector Ω has $2(n-1)N$ elements, with the first $(n-1)N$ elements for the section I and the last $(n-1)N$ elements for the section II. A programming code LOSCHMID.M based on this algorithm is available with this book.

Appendix 8.4: Collocation analysis of eqs. (8.4-13) to (8.4-15).

The governing equations (8.4-13 to 8.4-15) are rewritten here for clarity:

$$\frac{\partial \underline{y}}{\partial t^*} = \frac{\partial}{\partial \eta}\left\{\left[\underline{\underline{B}}^*(\underline{y})\right]^{-1}\frac{\partial \underline{y}}{\partial \eta}\right\} \quad (A8.4\text{-}1a)$$

$$\frac{d\underline{y}_0}{dt^*} = \alpha_0 \left[\underline{\underline{B}}^*(\underline{y})\right]^{-1} \frac{\partial \underline{y}}{\partial \eta}\bigg|_{\eta=0} \quad (A8.4\text{-}1b)$$

$$\frac{d\underline{y}_L}{dt^*} = -\alpha_L \left[\underline{\underline{B}}^*(\underline{y})\right]^{-1} \frac{\partial \underline{y}}{\partial \eta}\bigg|_{\eta=L} \quad (A8.4\text{-}1c)$$

$$\eta = 0; \quad \underline{y} = \underline{y}_0 \quad (A8.4\text{-}1d)$$

$$\eta = 1; \quad \underline{y} = \underline{y}_L \quad (A8.4\text{-}1e)$$

Expanding the RHS of eq.(A8.4-1a), we get:

$$\frac{\partial \underline{y}}{\partial t^*} = \left[\underline{\underline{B}}^*(\underline{y})\right]^{-1} \frac{\partial^2 \underline{y}}{\partial \eta^2} + \frac{\partial\left\{\left[\underline{\underline{B}}^*(\underline{y})\right]^{-1}\right\}}{\partial \eta} \frac{\partial \underline{y}}{\partial \eta} \quad (A8.4-2)$$

We now use N interior collocation points in the η domain, and together with the two boundary points ($\eta = 0$ and $\eta = 1$) we have N+2 interpolation points. The derivatives at the j-th interpolation point are given by:

$$\left.\frac{\partial \underline{y}}{\partial \eta}\right|_j = \sum_{k=1}^{N+2} A_{jk} \underline{y}_k \quad (A8.4-3a)$$

$$\left.\frac{\partial^2 \underline{y}}{\partial \eta^2}\right|_j = \sum_{k=1}^{N+2} A_{jk} \underline{y}_k \quad (A8.4-3b)$$

$$\left.\frac{\partial\left\{\left[\underline{\underline{B}}^*(\underline{y})\right]^{-1}\right\}}{\partial \eta}\right|_j = \sum_{k=1}^{N+2} A_{jk} \left[\underline{\underline{B}}^*(\underline{y}_k)\right]^{-1} \quad (A8.4-3c)$$

Evaluating eq.(A8.4-2) at the j-th interior collocation point, we get:

$$\frac{\partial \underline{y}_j}{\partial t^*} = \left[\underline{\underline{B}}^*(\underline{y}_j)\right]^{-1} \sum_{k=1}^{N+2} B_{jk} \underline{y}_k + \left\{\sum_{k=1}^{N+2} A_{jk}\left[\underline{\underline{B}}^*(\underline{y}_k)\right]^{-1}\right\} \times \left\{\sum_{k=1}^{N+2} A_{jk} \underline{y}_k\right\} \quad (A8.4-4)$$

for j = 2, 3, ..., N+1.

Eqs. (A8.4-1b) and (A8.4-1c), written in collocation variables, are:

$$\frac{d\underline{y}_1}{dt^*} = \alpha_0 \left[\underline{\underline{B}}^*(\underline{y}_1)\right]^{-1} \times \sum_{k=1}^{N+2} A_{1,k} \underline{y}_k \quad (A8.4-5)$$

$$\frac{d\underline{y}_{N+2}}{dt^*} = -\alpha_L \left[\underline{\underline{B}}^*(\underline{y}_{N+2})\right]^{-1} \times \sum_{k=1}^{N+2} A_{N+2,k} \underline{y}_k \quad (A8.4-6)$$

Eqs. (A8.4-4) to (A8.4-6) represent (n-1)(N+2) equations in terms of (n-1)(N+2) unknown variables. We now define the following concentration matrix.

$$\underline{\underline{Y}} = \begin{bmatrix} y(1,1) & y(1,2) & \cdots & y(1,N+1) & y(1,N+2) \\ y(2,1) & y(2,2) & \cdots & y(2,N+1) & y(2,N+2) \\ \vdots & \vdots & \ddots & \vdots & \vdots \\ y(n-1,1) & y(n-1,2) & \cdots & y(n-1,N+1) & y(n-1,N+2) \end{bmatrix} \quad (A8.4-7)$$

with

$Y(:,j) \equiv$ mole fractions of all species at the collocation point j
$Y(i,:) \equiv$ mole fractions of species i for all collocation points

Eqs. (A8.4-4) to (A8.4-6) will now become:

$$\frac{dY(:,1)}{dt^*} = \alpha_0 \left[\underline{\underline{B}}^*(Y(:,1))\right]^{-1} \times \sum_{k=1}^{N+2} A_{1,k} Y(:,k) \qquad (A8.4\text{-}8a)$$

$$\frac{dY(:,j)}{dt^*} = \left[\underline{\underline{B}}^*(Y(:,j))\right]^{-1} \times \sum_{k=1}^{N+2} B_{jk} Y(:,k)$$

$$+ \left\{\sum_{k=1}^{N+2} A_{jk} \left[\underline{\underline{B}}^*(Y(:,k))\right]^{-1}\right\} \times \left\{\sum_{k=1}^{N+2} A_{jk} Y(:,k)\right\} \qquad (A8.4\text{-}8b)$$

for j = 2, 3, ..., N+1

$$\frac{dY(:,N+2)}{dt^*} = -\alpha_L \left[\underline{\underline{B}}^*(Y(:,N+2))\right]^{-1} \times \sum_{k=1}^{N+2} A_{N+2,k} Y(:,k) \qquad (A8.4\text{-}8c)$$

The numerical integration of eqs. (A8.4-8) is done by the Runge-Kutta method, and the column vector for the integration is:

$$\Omega((i-1)(N+2)+j) = Y(i,j) \qquad \begin{matrix} j = 1,2,...,N+2 \\ i = 1,2,...,n-1 \end{matrix} \qquad (A8.4\text{-}9)$$

Appendix 8.5: The correct form of the Stefan-Maxwell equation

The use of the modified Stefan-Maxwell equation can be in three different forms, written as below.

In terms of mole fraction:

$$-\frac{dy_i}{dz} = \sum_{\substack{j=1 \\ j \neq i}}^{n} \frac{y_j N_i - y_i N_j}{cD_{ij}} + \frac{N_i}{cD_{K,i}} \qquad (A8.5\text{-}1a)$$

In terms of concentration:

$$-\frac{dc_i}{dz} = \sum_{\substack{j=1 \\ j \neq i}}^{n} \frac{y_j N_i - y_i N_j}{D_{ij}} + \frac{N_i}{D_{K,i}} \qquad (A8.5\text{-}1b)$$

In terms of partial pressure:

$$-\frac{1}{R_g T}\frac{dp_i}{dz} = \sum_{\substack{j=1\\j\neq i}}^{n}\frac{y_j N_i - y_i N_j}{D_{ij}} + \frac{N_i}{D_{K,i}} \qquad (A8.5\text{-}1c)$$

Haynes (1986) used the momentum argument to show that the correct form is eq.(A8.5-1c), which uses the partial pressure gradient as the driving force. When the system is isothermal, T is independent of z and hence eqs. (A8.5-1c) and (A8.5-1b) are equivalent. When the system is isobaric, eqs.(A8.5-1c) and (A8.5-1a) are equivalent. And finally these three equations are only equivalent when the system is isothermal and isobaric.

Using the momentum transfer method, the average collision number between two types of molecule is:

$$Z_{12} = \sigma_{12}^2 n_1 n_2 \sqrt{\frac{8\pi \kappa T}{m_{12}}} \qquad (A8.5\text{-}2)$$

where n_1 and n_2 are the molecular densities of the species 1 and species 2, respectively, m_{12} is defined as:

$$m_{12} = \frac{m_1 m_2}{m_1 + m_2} \qquad (A8.5\text{-}3)$$

The average collision between the molecule of type 1 and the surface of the pore is:

$$Z_{K,1} = \frac{1}{4} n_1 v_1 \qquad (A8.5\text{-}4)$$

where v_1 is the thermal velocity of the species 1, defined as follows:

$$v_1 = \sqrt{\frac{8\kappa T}{\pi m_1}} \qquad (A8.5\text{-}5)$$

When two molecules of different types collide with each other, each will move after the collision at a velocity, which is on average the same as the center of mass velocity. The center of mass velocity is:

$$v^* = \frac{m_1 v_1 + m_2 v_2}{m_1 + m_2} \qquad (A8.5\text{-}6)$$

Because of the change in the velocity, the change of the momentum of the species 1 by colliding with the molecules of the type 2 is:

$$m_1 v_1 - m_1 v^* = m_{12}(v_1 - v_2) \qquad (A8.5\text{-}7)$$

Assuming the collision of molecule of the type 1 to the wall will result in a net zero velocity in the direction along the pore (diffusive reflection), the transfer of momentum as a result of such collison is simply

$$m_1 v_1 \tag{A8.5-8}$$

The momentum balance along the pore coordinate z is:

$$-Z_{12}\left(\pi r^2 \Delta z\right)m_{12}(v_1 - v_2) - Z_{K,1}(2\pi r \Delta z)m_1 v_1 + \pi r^2 p_1\big|_z - \pi r^2 p_1\big|_{z+\Delta z} = 0 \tag{A8.5-9}$$

Simplifying the above equation and making appropriate substitution, we obtain the following equation:

$$-\frac{1}{R_g T}\frac{dp_1}{dz} = \frac{y_2 N_1 - y_1 N_2}{D_{12}} + \frac{N_1}{D_{K,1}} \tag{A8.5-10}$$

suggesting that the partial pressure is the correct description for the driving force.

Appendix 8.6: Equivalence of two matrix functions

Here we need to prove the equivalence of the LHS and RHS of the following equation:

$$\underline{\underline{C}}^{(1)}\left[\underline{\underline{C}}^{(1)} + \underline{\underline{C}}^{(2)}\right]^{-1}\underline{\underline{C}}^{(2)} \stackrel{??}{=} \underline{\underline{C}}^{(2)}\left[\underline{\underline{C}}^{(1)} + \underline{\underline{C}}^{(2)}\right]^{-1}\underline{\underline{C}}^{(1)} \tag{A8.6-1}$$

where $\underline{\underline{C}}$ is a square matrix.

Multiplying eq. (A8.6-1) by $\left[\underline{\underline{C}}^{(2)}\right]^{-1}$ from the left, we have:

$$\left[\underline{\underline{C}}^{(2)}\right]^{-1}\underline{\underline{C}}^{(1)}\left[\underline{\underline{C}}^{(1)} + \underline{\underline{C}}^{(2)}\right]^{-1}\underline{\underline{C}}^{(2)} \stackrel{?}{=} \left[\underline{\underline{C}}^{(1)} + \underline{\underline{C}}^{(2)}\right]^{-1}\underline{\underline{C}}^{(1)} \tag{A8.6-2}$$

Next, we left-multiply the above equation by $\left[\underline{\underline{C}}^{(1)} + \underline{\underline{C}}^{(2)}\right]$ and obtain:

$$\left[\underline{\underline{C}}^{(1)} + \underline{\underline{C}}^{(2)}\right]\left[\underline{\underline{C}}^{(2)}\right]^{-1}\underline{\underline{C}}^{(1)}\left[\underline{\underline{C}}^{(1)} + \underline{\underline{C}}^{(2)}\right]^{-1}\underline{\underline{C}}^{(2)} \stackrel{?}{=} \underline{\underline{C}} \tag{A8.6-3}$$

Now we multiply from the right with $\left[\underline{\underline{C}}^{(2)}\right]^{-1}$ and get:

$$\left[\underline{\underline{C}}^{(1)} + \underline{\underline{C}}^{(2)}\right]\left[\underline{\underline{C}}^{(2)}\right]^{-1}\underline{\underline{C}}^{(1)}\left[\underline{\underline{C}}^{(1)} + \underline{\underline{C}}^{(2)}\right]^{-1} \stackrel{?}{=} \underline{\underline{C}}^{(1)}\left[\underline{\underline{C}}^{(2)}\right]^{-1} \tag{A8.6-4}$$

We next multiply from the right with $\left[\underline{\underline{C}}^{(1)} + \underline{\underline{C}}^{(2)}\right]$:

$$\left[\underline{\underline{C}}^{(1)} + \underline{\underline{C}}^{(2)}\right]\left[\underline{\underline{C}}^{(2)}\right]^{-1}\underline{\underline{C}}^{(1)} \stackrel{?}{=} \underline{\underline{C}}^{(1)}\left[\underline{\underline{C}}^{(2)}\right]^{-1}\left[\underline{\underline{C}}^{(1)} + \underline{\underline{C}}^{(2)}\right] \tag{A8.6-5}$$

The LHS of the above equation is:

$$\text{LHS} = \underline{\underline{C}}^{(1)}\left[\underline{\underline{C}}^{(2)}\right]^{-1}\underline{\underline{C}}^{(1)} + \underline{\underline{C}}^{(1)} \qquad (A8.6\text{-}6)$$

while the RHS is:

$$\text{RHS} = \underline{\underline{C}}^{(1)}\left[\underline{\underline{C}}^{(2)}\right]^{-1}\underline{\underline{C}}^{(1)} + \underline{\underline{C}}^{(1)} \qquad (A8.6\text{-}7)$$

which is identical to the LHS. Thus

$$\underline{\underline{C}}^{(1)}\left[\underline{\underline{C}}^{(1)} + \underline{\underline{C}}^{(2)}\right]^{-1}\underline{\underline{C}}^{(2)} = \underline{\underline{C}}^{(2)}\left[\underline{\underline{C}}^{(1)} + \underline{\underline{C}}^{(2)}\right]^{-1}\underline{\underline{C}}^{(1)} \qquad (A8.6\text{-}8)$$

Appendix 8.7: Alternative Derivation of the Basic equation for Bulk-Knudsen and Viscous Flow

The diffusion flux equation written in vector form is:

$$\underline{N}^D = -\frac{1}{R_g T}\left[\underline{\underline{B}}(\underline{y})\right]^{-1}\frac{d\underline{p}}{dz} \qquad (A8.7\text{-}1a)$$

where the matrix $\underline{\underline{B}}$ is given in eq. (8.8-11a), rewritten here for clarity:

$$\underline{\underline{B}} = \begin{cases} \dfrac{1}{D_{K,i}} + \sum_{\substack{j=1 \\ j \neq i}}^{n}\dfrac{y_j}{D_{ij}} & \text{for } i = j \\ -\dfrac{y_i}{D_{ij}} & \text{for } i \neq j \end{cases} \qquad (A8.7\text{-}1b)$$

and similarly the viscous flux in vector form is:

$$\underline{N}^V = -\frac{1}{R_g T}\frac{B_0 P}{\mu}\frac{dP}{dz}\underline{y} \qquad (A8.7\text{-}2)$$

Therefore, the combined flux in vector form is the summation of the above two equations:

$$\underline{N} = -\frac{1}{R_g T}\left[\underline{\underline{B}}(\underline{y})\right]^{-1}\frac{d\underline{p}}{dz} - \frac{1}{R_g T}\frac{B_0 P}{\mu}\frac{dP}{dz}\underline{y} \qquad (A8.7\text{-}3)$$

Using the relation of partial pressure

$$\underline{p} = P\underline{y} \qquad (A8.7\text{-}4)$$

eq.(A8.7-3) will become:

$$\underline{N} = -\frac{P}{R_g T}\left[\underline{B}(\underline{y})\right]^{-1}\frac{d\underline{y}}{dz} - \frac{1}{R_g T}\frac{dP}{dz}\left\{\left[\underline{B}(\underline{y})\right]^{-1} + \frac{B_0 P}{\mu}\underline{\underline{I}}\right\}\underline{y} \qquad (A8.7-5)$$

where $\underline{\underline{I}}$ is the identity matrix.

Now we multiply both sides of eq.(A8.7-5) by the matrix $\underline{\underline{B}}$, we get:

$$\underline{\underline{B}}(\underline{y})\underline{N} = -\frac{P}{R_g T}\frac{d\underline{y}}{dz} - \frac{1}{R_g T}\frac{dP}{dz}\left\{\underline{\underline{I}} + \frac{B_0 P}{\mu}\underline{\underline{B}}(\underline{y})\right\}\underline{y} \qquad (A8.7-6)$$

We consider the vector $\underline{\underline{B}}(\underline{y})\underline{y}$. The element i of this vector is:

$$\sum_{j=1}^{n} B_{ij} y_j = \sum_{\substack{j=1 \\ j\neq i}}^{n} B_{ij} y_j + B_{ii} y_i \qquad (A8.7-7)$$

Making use of the definition of the matrix $\underline{\underline{B}}$ in eq.(A8.7-1b), we get:

$$\sum_{j=1}^{n} B_{ij} y_j = -\sum_{\substack{j=1 \\ j\neq i}}^{n} \frac{y_i y_j}{D_{ij}} + \frac{y_i}{D_{K,i}} + \sum_{\substack{j=1 \\ j\neq i}}^{n} \frac{y_i y_j}{D_{ij}} = \frac{y_i}{D_{K,i}} \qquad (A8.7-8)$$

This means that:

$$\underline{\underline{B}}(\underline{y})\underline{y} = \underline{\underline{\Lambda}}\underline{y} \qquad (A8.7-9)$$

where the matrix $\underline{\underline{\Lambda}}$ is the diagonal matrix with its i-th element being the inverse of the Knudsen diffusivity of the species i. Thus, eq. (A8.7-6) can now be written:

$$\underline{\underline{B}}(\underline{y})\underline{N} = -\frac{P}{R_g T}\frac{d\underline{y}}{dz} - \frac{1}{R_g T}\frac{dP}{dz}\left\{\underline{\underline{I}} + \frac{B_0 P}{\mu}\underline{\underline{\Lambda}}\right\}\underline{y} \qquad (A8.7-10)$$

which is identical to eq. (8.8-12).

Appendix 8.8: Derivation of Eq. (8.8-19a)

Starting from equation (8.8-6)

$$\sum_{\substack{j=1 \\ j\neq i}}^{n} \frac{y_j N_i - y_i N_j}{D_{ij}} + \frac{N_i}{D_{K,i}} = -\frac{P}{R_g T}\frac{dy_i}{dz} - \frac{y_i}{R_g T}\left(1 + \frac{B_0 P}{\mu D_{K,i}}\right)\frac{dP}{dz} \qquad (A8.8-1)$$

The diffusive flux is related to the flux and the total flux as:

$$N_i = J_i + y_i N_T \qquad (A8.8-2)$$

Substituting eq. (A8.8-2) into eq. (A8.8-1), we have:

$$\frac{J_i}{D_{K,i}} + y_i \frac{N_T}{D_{K,i}} + \sum_{\substack{j=1 \\ j \neq i}}^{n} \frac{y_j J_i - y_i J_j}{D_{ij}} = -\frac{P}{R_g T} \frac{dy_i}{dz} - \frac{y_i}{R_g T}\left(1 + \frac{B_0 P}{\mu D_{K,i}}\right)\frac{dP}{dz} \quad (A8.8\text{-}3)$$

Summing the above equation with respect to i from 1 to n, we get:

$$\sum_{i=1}^{n} \frac{J_i}{D_{K,i}} + N_T \sum_{i=1}^{n} \frac{y_i}{D_{K,i}} + \sum_{i=1}^{n}\sum_{\substack{j=1 \\ j \neq i}}^{n} \frac{y_j J_i - y_i J_j}{D_{ij}} = -\frac{P}{R_g T} \frac{d}{dz}\sum_{i=1}^{n} y_i - \frac{dP}{dz}\sum_{i=1}^{n}\frac{y_i}{R_g T}\left(1 + \frac{B_0 P}{\mu D_{K,i}}\right)$$

(A8.8.3)

The third term in the LHS and the first term in the RHS are zero. Hence, we get:

$$\sum_{i=1}^{n} \frac{J_i}{D_{K,i}} + N_T \sum_{i=1}^{n} \frac{y_i}{D_{K,i}} = -\frac{1}{R_g T}\frac{dP}{dz} - \frac{B_0 P}{\mu R_g T}\left(\sum_{i=1}^{n}\frac{y_i}{D_{K,i}}\right)\frac{dP}{dz} \quad (A8.8\text{-}4)$$

Solving for the total flux from eq.(A8.8-4), we have:

$$N_T = -\frac{\sum_{i=1}^{n} \frac{J_i}{D_{K,i}}}{\gamma} - \frac{1}{R_g T}\left(\frac{B_0 P}{\mu} + \frac{1}{\gamma}\right)\frac{dP}{dz} \quad (A8.8\text{-}5)$$

where

$$\gamma = \sum_{i=1}^{n} \frac{y_i}{D_{K,i}} \quad (A8.4\text{-}6)$$

Substituting the total flux of eq. (A8.8-5) into eq. (A8.8-3), we have:

$$\frac{J_i}{D_{K,i}} - \frac{y_i}{D_{K,i}}\sum_{j=1}^{n}\frac{J_j}{\gamma D_{K,j}} - \left(\frac{y_i}{D_{K,i}}\right)\left(\frac{1}{R_g T}\right)\left(\frac{B_0 P}{\mu} + \frac{1}{\gamma}\right)\frac{dP}{dz} + \sum_{\substack{j=1 \\ j \neq i}}^{n}\frac{y_j J_i - y_i J_j}{D_{ij}} =$$

$$-\frac{P}{R_g T}\frac{dy_i}{dz} - \frac{y_i}{R_g T}\left(1 + \frac{B_0 P}{\mu D_{K,i}}\right)\frac{dP}{dz}$$

Multiplying the first term in the LHS by γ, we have:

$$\frac{J_i}{\gamma D_{K,i}}\left(\sum_{j=1}^{n}\frac{y_j}{D_{K,j}}\right) - \frac{y_i}{D_{K,i}}\left(\sum_{j=1}^{n}\frac{J_j}{\gamma D_{K,j}}\right) - \left(\frac{y_i}{D_{K,i}}\right)\left(\frac{1}{R_g T}\right)\left(\frac{B_0 P}{\mu} + \frac{1}{\gamma}\right)\frac{dP}{dz} + \sum_{\substack{j=1 \\ j \neq i}}^{n}\frac{y_j J_i - y_i J_j}{D_{ij}} =$$

$$-\frac{P}{R_g T}\frac{dy_i}{dz} - \frac{y_i}{R_g T}\left(1 + \frac{B_0 P}{\mu D_{K,i}}\right)\frac{dP}{dz}$$

(A8.8-7)

Simplifying the terms, we get:

$$\sum_{j=1}^{n} \frac{y_j J_i - y_i J_j}{\gamma D_{K,i} D_{K,j}} + \sum_{\substack{j=1 \\ j \neq i}}^{n} \frac{y_j J_i - y_i J_j}{D_{ij}} = \left(\frac{y_i}{D_{K,i}}\right)\left(\frac{1}{R_g T}\right)\left(\frac{B_0 P}{\mu} + \frac{1}{\gamma}\right)\frac{dP}{dz} - \frac{P}{R_g T}\frac{dy_i}{dz} - \frac{y_i}{R_g T}\left(1 + \frac{B_0 P}{\mu D_{K,i}}\right)\frac{dP}{dz}$$

Simplifying the above equation further by combining the two terms in the LHS, we get:

$$\sum_{\substack{j=1 \\ j \neq i}}^{n} \frac{y_j J_i - y_i J_j}{\Delta_{ij}} = -\frac{P}{R_g T}\frac{dy_i}{dz} - \frac{y_i}{R_g T}\left(1 - \frac{1}{D_{K,i} \sum_{k=1}^{n}(y_k/D_{K,k})}\right)\frac{dP}{dz} \quad (A8.8\text{-}8)$$

where

$$\frac{1}{\Delta_{ij}} = \frac{1}{D_{ij}} + \frac{1}{D_{K,i} D_{K,j} \sum_{k=1}^{n}(y_k/D_{K,k})} \quad (A8.8\text{-}9)$$

Note that Δ_{ij} is symmetric, that is

$$\Delta_{ij} = \Delta_{ji} \quad (A8.8\text{-}10)$$

Summing eq. (A8.8-8) with respect to i from 1 to n would give zero as only (n-1) equations are independent. Thus (n-1) equations of the form (A8.8-8) and the following equation

$$\sum_{i=1}^{n} J_i = 0 \quad (A8.8\text{-}11)$$

will form a complete set of equations in terms of n variables J_i (i = 1, 2, ..., n). Knowing these diffusive fluxes, the total flux is then calculated from eq. (A8.8-5).

Appendix 8.9: Collocation Analysis of Model Equation (eq. 8.9-10)

The mass balance equation describing the pressure variation inside a cylindrical pore takes the following vector form (eq. 8.9-10):

$$\frac{\partial \underline{p}}{\partial \tau} = \underline{\underline{H}}(\underline{p}) \frac{\partial}{\partial \eta}\left\{[\underline{\underline{D}}_T \underline{\underline{B}}(\underline{p})]^{-1}\left[\frac{\partial \underline{p}}{\partial \eta} + \Phi \frac{\mu_0}{\mu_m(\underline{y})}\frac{1}{P_0}\frac{\partial P_T}{\partial \eta}(\underline{\underline{D}}_T \underline{\underline{\Lambda}}) \underline{p}\right]\right\} \quad (A8.9\text{-}1)$$

where D_T is the reference diffusivity, P_0 is the reference pressure and Φ is the non-dimensional viscous flow strength number which is defined as follows:

$$\Phi = \frac{B_0 P_0}{\mu_0 D_T} \tag{A8.9-2}$$

with B_0 being the viscous flow parameter ($B_0 = r^2/8$) and μ_0 is the reference viscosity. For pure diffusion inside the cylindrical pore, $\underline{\underline{H}}(\underline{p}) = \underline{\underline{I}}$ (identity matrix).

Expanding the RHS of eq. (A8.9-1), we get:

$$\frac{\partial \underline{p}}{\partial \tau} = \underline{\underline{H}}(\underline{p}) \left\{ \frac{\partial}{\partial \eta} \left[D_T \underline{\underline{B}}(\underline{p}) \right]^{-1} \left[\frac{\partial \underline{p}}{\partial \eta} + \Phi \frac{\mu_0}{\mu_m(\underline{y})} \frac{1}{P_0} \frac{\partial P_T}{\partial \eta} (D_T \underline{\underline{\Lambda}}) \underline{p} \right] \right.$$
$$\left. + \left[D_T \underline{\underline{B}}(\underline{p}) \right]^{-1} \left[\frac{\partial^2 \underline{p}}{\partial \eta^2} + \Phi \frac{\mu_0}{\mu_m(\underline{y})} (D_T \underline{\underline{\Lambda}}) \left(\frac{1}{P_0} \frac{\partial^2 P_T}{\partial \eta^2} \underline{p} + \frac{1}{P_0} \frac{\partial P_T}{\partial \eta} \frac{\partial \underline{p}}{\partial \eta} \right) \right] \right\} \tag{A8.9-3}$$

in which we have ignored the term

$$\Phi \frac{\partial}{\partial \eta} \left(\frac{\mu_0}{\mu_m(\underline{y})} \right) \frac{1}{P_0} \frac{\partial P_T}{\partial \eta} (D_T \underline{\underline{\Lambda}}) \underline{p} \tag{A8.9-4}$$

compared to the other terms in the second square bracket on the RHS of eq. (A8.9-3). This is reasonable if the viscosities of all components are not so much different.

The symmetry of the problem at $\eta = 0$ suggests the following transformation:

$$u = \eta^2 \tag{A8.9-5}$$

With this transformation, eq. (A8.9-3) can be rewritten as follows:

$$\frac{\partial \underline{p}}{\partial \tau} = \underline{\underline{H}}(\underline{p}) \left\{ 4u \frac{\partial}{\partial u} \left[D_T \underline{\underline{B}}(\underline{p}) \right]^{-1} \left[\frac{\partial \underline{p}}{\partial u} + \Phi \frac{\mu_0}{\mu_m(\underline{y})} \frac{1}{P_0} \frac{\partial P_T}{\partial u} (D_T \underline{\underline{\Lambda}}) \underline{p} \right] + \right.$$
$$\left. \left[D_T \underline{\underline{B}}(\underline{p}) \right]^{-1} \left[\left(4u \frac{\partial^2 \underline{p}}{\partial u^2} + 2 \frac{\partial \underline{p}}{\partial u} \right) + \Phi \frac{\mu_0}{\mu_m(\underline{y})} (D_T \underline{\underline{\Lambda}}) \left(\frac{1}{P_0} \left(4u \frac{\partial^2 P_T}{\partial u^2} + 2 \frac{\partial P_T}{\partial u} \right) \underline{p} + 4u \left(\frac{1}{P_0} \frac{\partial P_T}{\partial u} \right) \left(\frac{\partial \underline{p}}{\partial u} \right) \right) \right] \right\}$$

$$\tag{A8.9-6}$$

We now choose N interior interpolation points and together with the boundary point we have N+1 interpolation points. Evaluate eq. (A8.9-6) at the i-th interior collocation point, we get:

$$\frac{\partial \underline{p}_i}{\partial \tau} = \underline{\underline{H}}(\underline{p}_i)\left\{4u_i\left[\sum_{j=1}^{N+1}A_{ij}\left(D_T\underline{\underline{B}}(\underline{p}_j)\right)\right]^{-1}\left[\sum_{j=1}^{N+1}A_{ij}\underline{p}_j + \Phi\frac{\mu_0}{\mu_m(\underline{y}_i)}\frac{1}{P_0}\left(\sum_{j=1}^{N+1}A_{ij}P_{Tj}\right)(D_T\underline{\underline{A}})\underline{p}_i\right]\right.$$

$$\left.+\left[D_T\underline{\underline{B}}(\underline{p}_i)\right]^{-1}\left[\sum_{j=1}^{N+1}C_{ij}\underline{p}_j + \Phi\frac{\mu_0}{\mu_m(\underline{y}_i)}(D_T\underline{\underline{A}})\left(\frac{1}{P_0}\sum_{j=1}^{N+1}C_{ij}P_{Tj}\underline{p}_i + 4u_i\frac{1}{P_0}\left(\sum_{j=1}^{N+1}A_{ij}P_{Tj}\right)\left(\sum_{j=1}^{N+1}A_{ij}\underline{p}_j\right)\right)\right]\right\}$$

(A8.9-7)

for i =1, 2, …, N, where A_{ij}, B_{ij} are elements of the first and second derivative matrices, and C_{ij} is defined as follows:

$$C_{ij} = 4u_i B_{ij} + 2A_{ij} \qquad (A8.9\text{-}8)$$

The integration vector will contain the partial pressures of all components at N interior collocation points and is defined as follows:

$$\underline{\Omega} = \begin{bmatrix} \begin{Bmatrix} p(1,1) \\ p(1,2) \\ \vdots \\ p(1,N) \end{Bmatrix} & \text{partial pressure of the component 1 @ N interior collocation points} \\ \begin{Bmatrix} p(2,1) \\ p(2,2) \\ \vdots \\ p(2,N) \end{Bmatrix} & \text{partial pressure of the component 2 @ N interior collocation points} \\ \vdots & \\ \begin{Bmatrix} p(nc,1) \\ p(nc,2) \\ \vdots \\ p(nc,N) \end{Bmatrix} & \text{partial pressure of the nc - th component @ N interior collocation pts} \end{bmatrix}$$

(A8.9-9)

The length of this $\underline{\Omega}$ array is (nc × N), and the time derivative of this integration vector is given in eq. (A8.9-7). However to calculate the RHS of eq. (A8.9-7) we need the total pressure as well as the mole fractions of all components at N interior collocation points. This can be achieved with the following procedure which is suitable for programming.

Given a $\underline{\Omega}$ array, we generate a matrix $\underline{\underline{p}}$ with the i-th row and the j-th column for the i-th component and the j-th collocation point as follows:

$$\underline{\underline{p}} = \begin{bmatrix} p(1,1) & p(1,2) & p(1,3) & \cdots & p(1,N) \\ p(2,1) & p(2,2) & p(2,3) & \cdots & p(2,N) \\ \vdots & \vdots & \vdots & \ddots & \vdots \\ p(nc,1) & p(nc,2) & p(nc,3) & \cdots & p(nc,N) \end{bmatrix} \quad (A8.9\text{-}10)$$

Next we generate an augmented pressure matrix by adding the pressures at the bulk to the (N+1)-th column

$$\underline{\underline{p}} = \begin{bmatrix} \underline{\underline{p}} & \begin{matrix} p_b(1) \\ p_b(2) \\ \vdots \\ p_b(nc) \end{matrix} \end{bmatrix} \quad (A8.9\text{-}11)$$

Summing all elements of each column to give the total pressure and putting the total pressure as the last row of the matrix $\underline{\underline{p}}$ as follows:

$$\underline{\underline{p}} = \begin{bmatrix} p(1,1) & p(1,2) & \cdots & p(1,N) & p_b(1) \\ p(2,1) & p(2,2) & \cdots & p(2,N) & p_b(2) \\ \vdots & \vdots & \ddots & & \vdots \\ p(nc,1) & p(nc,2) & \cdots & p(nc,N) & p_b(nc) \\ p_T(1) & p_T(2) & \cdots & p_T(N) & p_T(N+1) \end{bmatrix} \quad (A8.9\text{-}12)$$

where

$$p_T(j) = \sum_{i=1}^{nc} p(i,j) \quad (A8.9\text{-}13)$$

for j = 1, 2, ..., N+1.

Knowing the above pressure matrix, we generate a mole fraction matrix:

$$\underline{\underline{Y}} = \begin{bmatrix} y(1,1) & y(1,2) & \cdots & y(1,N) & y_b(1) \\ y(2,1) & y(2,2) & \cdots & y(2,N) & y_b(2) \\ \vdots & \vdots & \ddots & \vdots & \\ y(nc,1) & y(nc,2) & \cdots & y(nc,N) & y_b(nc) \end{bmatrix} \quad (A8.9\text{-}14)$$

where

$$y(i,j) = \frac{p(i,j)}{p_T(j)} \quad (A8.9\text{-}15)$$

for i = 1, 2, ..., nc and j = 1, 2, ..., N+1.

The pressure matrix of eq. (A8.9-12) and the mole fraction matrix of eq. (A8.9-14) are used to calculate the RHS of eq. (A8.9-7). The RHS of eq. (A8.9-7) is a matrix of dimension (nc, N), which then can be integrated by the standard Runge-Kutta method with the integration vector as defined in eq. (A8.9-9). A programming code ADSORB5A based on this algorithm is available with this book.

Appendix 9.1: Collocation Analysis of a Diffusion Equation (9.2-3)

The linear Fickian diffusion equation (9.2-3) has the form (written in nondimensional form)

$$\frac{\partial y}{\partial \tau} = \frac{1}{x^s}\frac{\partial}{\partial x}\left(x^s \frac{\partial y}{\partial x}\right) \qquad (A9.1\text{-}1)$$

subject to the following initial and boundary conditions:

$$\tau = 0; \quad y = y_i \qquad (A9.1\text{-}2)$$

$$x = 0; \quad \frac{\partial y}{\partial x} = 0 \qquad (A9.1\text{-}3a)$$

$$x = 1; \quad y = y_b \qquad (A9.1\text{-}3b)$$

This problem has a symmetry at $x = 0$ so it is useful to utilize this by making the following transformation, $u = x^2$. With this transformation, the mass balance equation (A9.11) is:

$$\frac{\partial y}{\partial \tau} = 4u\frac{\partial^2 y}{\partial u^2} + 2(1+s)\frac{\partial y}{\partial u} \qquad (A9.1\text{-}4)$$

The domain $u \in (0,1)$ is now represented discretely by N interior collocation points. Taking the boundary point (u=1) as the (N+1)-th point, we have a total of N+1 interpolation points. According to the orthogonal collocation method, the first and second derivatives at these interpolation points are related to the functional values at all points as given below:

$$\left.\frac{\partial y}{\partial u}\right|_i = \sum_{j=1}^{N+1} A_{ij} y_j \qquad (A9.1\text{-}5)$$

$$\left.\frac{\partial^2 y}{\partial u^2}\right|_i = \sum_{j=1}^{N+1} B_{ij} y_j \qquad (A9.1\text{-}6)$$

for i = 1, 2,..., N+1. The matrices $\underline{\underline{A}}$ and $\underline{\underline{B}}$ are constant matrices once N+1 interpolation points have been chosen.

The mass balance equation (A9.1-4) is valid at any point within the u domain. Thus, evaluating that equation at the i-th interior collocation point we get:

$$\frac{\partial y_i}{\partial \tau} = \sum_{j=1}^{N+1} C_{ij} y_j \qquad (A9.1-7)$$

for i = 1, 2,..., N, where

$$C_{ij} = 4u_i B_{ij} + 2(1+s) A_{ij} \qquad (A9.1-8)$$

Since $y_{N+1} = y_b$, the above equation becomes:

$$\frac{\partial y_i}{\partial \tau} = \sum_{j=1}^{N} C_{ij} y_j + C_{i,N+1} y_b \qquad (A9.1-9)$$

or written in vector form:

$$\frac{\partial \underline{y}}{\partial \tau} = \underline{\underline{C}}\, \underline{y} + \underline{\psi} \qquad (A9.1-10)$$

where

$$\underline{y} = [y_1, y_2 \cdots y_N]^T \qquad (A9.1-11)$$

$$\underline{\psi} = y_b [C_{1,N+1}\ C_{2,N+1}\ \cdots\ C_{N,N+1}]^T \qquad (A9.1-12)$$

The linear vector equation (A9.1-10) can be readily integrated. A programming code ADSORB0 is provided to solve this set of equations. It gives at each value of time the solution for $\underline{y}$, and the mean concentration <y> is calculated from the Radau quadrature

$$<y> = \underline{w}(1{:}N) \cdot \underline{y} + w_{N+1} y_b \qquad (A9.1-13)$$

where $w_j (j = 1, 2, ..., N+1)$ are Radau quadrature weights.

Alternative solution

There is another way of solving the mass balance equation especially when the flux term

$$J = -\frac{\partial y}{\partial x} \quad \text{(A9.1-14)}$$

is taking a more complicated form. We write the mass balance as follows:

$$\frac{\partial y}{\partial \tau} = -\frac{1}{x^s}\frac{\partial}{\partial x}(x^s J), \quad \text{(A9.1-15)}$$

We make use of the transformation, $u = x^2$, and eq. (A9.0-15) can be rewritten as follows:

$$\frac{\partial y}{\partial \tau} = -2u^{(1-s)/2}\frac{\partial}{\partial u}\left[u^{s/2}\cdot J\right], \quad \text{(A9.1-16)}$$

where

$$J = -2\sqrt{u}\frac{\partial y}{\partial u} \quad \text{(A9.1-17)}$$

Combine the above two equations, we get:

$$\frac{\partial y}{\partial \tau} = 2u^{(1-s)/2}\frac{\partial}{\partial u}(F) \quad \text{(A9.1-18)}$$

where

$$F = 2u^{(1+s)/2}\cdot\frac{\partial y}{\partial u} \quad \text{(A9.1-19)}$$

Evaluating eq. (A9.1-18) at the i-th interior collocation point, we have:

$$\frac{\partial y_i}{\partial \tau} = 2u_i^{(1-s)/2}\sum_{j=1}^{N+1} A_{ij} F_j \quad \text{(A9.1-20)}$$

for $i = 1, 2, \ldots, N$. To evaluate the RHS of the above equation we need the value of F at all interpolation points. This is done by evaluating eq. (A9.1-19) at the j-th interpolation point:

$$F_j = 2u_j^{(1+s)/2}\sum_{k=1}^{N+1} A_{jk} y_k \quad \text{(A9.1-21)}$$

Knowing $y_{N+1} = y_b$, eq. (A9.1-20) now is a set of N equations involving N unknowns ($y_1, y_2, \ldots, y_N$). This set then can be integrated and is done in the programming code ADSORB0.

Appendix 9.2: The first ten eigenvalues for the three shapes of particle
Slab

Bi → ∞	Bi = 100	Bi = 10
1.5707 9633	1.5552 4513	1.4288 7001
4.7123 8898	4.6657 6514	4.3058 0141
7.8539 8163	7.7763 7408	7.2281 0977
10.9955 7429	10.8871 3010	10.2002 6259
14.1371 6694	13.9980 8974	13.2141 8568
17.2787 5959	17.1093 0726	16.2593 6123
20.4203 5225	20.2208 3419	19.3270 3429
23.5619 4490	23.3327 1880	22.4108 4833
26.7035 3756	26.4450 0575	25.5063 8299
29.8451 3021	29.5577 3581	28.6105 8194

Cylinder

Bi → ∞	Bi = 100	Bi = 10
2.4048 2556	2.3809 0166	2.1794 9660
5.5200 7811	5.4652 0700	5.0332 1198
8.6537 2791	8.5678 3165	7.9568 8342
11.7915 3444	11.6747 3543	10.9363 3020
14.9309 1771	14.7834 2086	13.9580 3045
18.0710 6397	17.8931 3665	17.0098 7821
21.2116 3663	21.0036 0098	20.0829 1063
24.3524 7153	24.1146 9932	23.1709 5712
27.4934 7913	27.2263 8726	26.2698 4148
30.6346 0647	30.3386 5250	29.3767 1749

Sphere

Bi → ∞	Bi = 100	Bi = 10
3.1415 92654	3.1101 8695	2.8363 0039
6.2831 8531	6.2204 3512	5.7172 4920
9.4247 7796	9.3308 0501	8.6587 0470
12.5663 7061	12.4413 5573	11.6532 0755
15.7079 6327	15.5521 4434	14.6869 3740
18.8495 5592	18.6632 2528	17.7480 6901
21.9911 4858	21.7746 4982	20.8282 2625
25.1327 4123	24.8864 6563	23.9217 9001
28.2743 3388	27.9987 1645	27.0250 1045
31.4159 2654	31.1114 4180	30.1353 5038

Appendix 9.3: Collocation analysis of eq. (9.2-47)

The mass balance equation for the case of non linear adsorption isotherm given in eqs. (9.2-47) is rewritten here for clarity

$$G(y)\frac{\partial y}{\partial t} = \frac{1}{\eta^s}\frac{\partial}{\partial \eta}\left[\eta^s H(y)\frac{\partial y}{\partial \eta}\right] \quad (A9.3\text{-}1)$$

The boundary conditions of eq. (A9.3-1) are:

$$\eta = 0; \quad \frac{\partial y}{\partial \eta} = 0 \quad (A9.3\text{-}2)$$

$$\eta = 1; \quad H(y)\frac{\partial y}{\partial \eta} = Bi(y_b - y) \quad (A9.3\text{-}3)$$

Expanding the RHS of eq. (A9.3-1), we get:

$$G(y)\frac{\partial y}{\partial t} = H(y)\left[\frac{1}{\eta^s}\frac{\partial}{\partial \eta}\left(\eta^s \frac{\partial y}{\partial \eta}\right)\right] + \frac{\partial H(y)}{\partial \eta}\frac{\partial y}{\partial \eta} \quad (A9.3\text{-}4)$$

The symmetry of the problem at $\eta = 0$ suggests the following transformation of the independent variable (Rice and Do, 1995), $u = \eta^2$. With this new variable, eq. (A9.3-4) will become:

$$G(y)\frac{\partial y}{\partial t} = H(y)\left[4u\frac{\partial^2 y}{\partial u^2} + 2(1+s)\frac{\partial y}{\partial u}\right] + 4u\frac{\partial H(y)}{\partial u}\frac{\partial y}{\partial u} \quad (A9.3\text{-}5)$$

Similarly, the boundary condition written in terms of the new independent variable u is:

$$u = 1; \quad H(y)\frac{\partial y}{\partial u} = \frac{Bi}{2}(y_b - y) \quad (A9.3\text{-}6)$$

To apply the orthogonal collocation method, we choose N interior collocation points in the spatial domain u, that is

$$0 < u_1, u_2, \cdots, u_N < 1$$

These N interior points together with the point at the boundary $u = 1$ will form N+1 interpolation points. Evaluating eq. (A9.3-5) at the j-th interior collocation point, we get:

$$G(y_j)\frac{\partial y_j}{\partial t} = H(y_j)\left[4u_j\frac{\partial^2 y}{\partial u^2}\bigg|_{u_j} + 2(1+s)\frac{\partial y}{\partial u}\bigg|_{u_j}\right] + 4u_j\frac{\partial H(y)}{\partial u}\bigg|_{u_j}\frac{\partial y}{\partial u}\bigg|_{u_j} \quad (A9.3\text{-}7)$$

for j = 1, 2, 3, ..., N. Here y_j is the value of y at the collocation point u_j, or; for short, collocation point j. The first and second derivatives at any interpolation points can be expressed in terms of the dependent variables $\underline{y}$, as given below:

$$\left.\frac{\partial y}{\partial u}\right|_{u_j} = \sum_{k=1}^{N+1} A_{jk} y_k \qquad (A9.3\text{-}8a)$$

$$\left.\frac{\partial^2 y}{\partial u^2}\right|_{u_j} = \sum_{k=1}^{N+1} B_{jk} y_k \qquad (A9.3\text{-}8b)$$

$$\left.\frac{\partial H(y)}{\partial u}\right|_{u_j} = \sum_{k=1}^{N+1} A_{jk} H(y_k) \qquad (A9.3\text{-}8c)$$

where $\underline{\underline{A}}$ and $\underline{\underline{B}}$ are known constant matrices for <u>a given set of N+1 interpolation points</u>. Readers are referred to Villadsen and Michelsen (1978) or Rice and Do (1995) for details.

Substitution of eqs. (A9.3-8) into eq. (A9.3-7) gives:

$$G(y_j)\frac{\partial y_j}{\partial t} = H(y_j)\sum_{k=1}^{N+1} C_{jk} y_k + 4u_j \left(\sum_{k=1}^{N+1} A_{jk} H(y_k)\right)\left(\sum_{k=1}^{N+1} A_{jk} y_k\right) \qquad (A9.3\text{-}9)$$

for j = 1, 2, 3, ..., N, where

$$C_{jk} = 4u_j B_{jk} + 2(1+s) A_{jk} \qquad (A9.3\text{-}10)$$

Eq. (A9.3-9) is valid for N interior collocation points. The equation for the (N+1)-th interpolation point is given in eq. (A9-3-6). Written that equation in terms of the collocation variables, we have:

$$H(y_{N+1})\sum_{k=1}^{N+1} A_{N+1,k} y_k = \left(\frac{Bi}{2}\right)(y_b - y_{N+1}) \qquad (A9.3\text{-}11)$$

from which we can solve for the concentration at the boundary (y_{N+1}) in terms of other dependent variables $y_1, y_2, ..., y_N$. The above equation is a nonlinear algebraic equation for y_{N+1} expressed in terms of N collocation values y_j (j=1, 2, ..., N). It can be solved by using the Newton-Raphson method, and the initial guess is:

$$(y_{N+1})_{guess} = \frac{y_b - \left(\frac{2}{Bi}\right) H(y_b) \sum_{k=1}^{N} A_{N+1,k} y_k}{1 + \frac{2}{Bi} A_{N+1,N+1}} \quad (A9.3\text{-}12)$$

Knowing this, eq. (A9.3-9) now represents a set of N coupled first order ordinary differential equations in terms of $y_1, y_2, ..., y_N$, which can be solved numerically by using integration techniques such as the Runge-Kutta method. The programming code for this problem is ADSORB1A.M.

Appendix 9.4: Collocation analysis of eqs. (9.3-19)

The mass and heat balance equations for a single component system with a non linear adsorption isotherm of eqs. (9.3-19) are rewritten here for clarity

$$G_1(y,\theta)\frac{\partial y}{\partial t} + G_2(y,\theta)\frac{\partial \theta}{\partial t} = \frac{1}{\eta^s}\frac{\partial}{\partial \eta}\left[\eta^s H(y,\theta)\frac{\partial y}{\partial \eta}\right] \quad (A9.4\text{-}1a)$$

$$\frac{d\theta}{dt} = \beta\left(\frac{C_0}{C_{\mu 0}}\right)(1+s)\left[H(y,\theta)\frac{\partial y}{\partial \eta}\right]_{\eta=1} - LeBi(\theta - \theta_b) \quad (A9.4\text{-}1b)$$

The boundary conditions of eq. (A9.4-1) are:

$$\eta = 0; \quad \frac{\partial y}{\partial \eta} = 0 \quad (A9.4\text{-}2a)$$

$$\eta = 1; \quad H(y,\theta)\frac{\partial y}{\partial \eta} = Bi(y_b - y) \quad (A9.4\text{-}2b)$$

Expanding the RHS of eq. (A9.4-1a), we get:

$$G_1(y,\theta)\frac{\partial y}{\partial t} + G_2(y,\theta)\frac{\partial \theta}{\partial t} = H(y,\theta)\left[\frac{1}{\eta^s}\frac{\partial}{\partial \eta}\left(\eta^s \frac{\partial y}{\partial \eta}\right)\right] + \frac{\partial H(y,\theta)}{\partial \eta}\frac{\partial y}{\partial \eta} \quad (A9.4\text{-}3)$$

The symmetry of the problem at $\eta = 0$ suggests the following transformation of the independent variable, $u = \eta^2$. With this new variable, eqs. (A9.4-3) and (A9.4-1b) will become:

$$G_1(y,\theta)\frac{\partial y}{\partial t} + G_2(y,\theta)\frac{\partial \theta}{\partial t} = H(y,\theta)\left[4u\frac{\partial^2 y}{\partial u^2} + 2(1+s)\frac{\partial y}{\partial u}\right] + 4u\frac{\partial H(y,\theta)}{\partial u}\frac{\partial y}{\partial u} \quad (A9.4\text{-}4a)$$

$$\frac{d\theta}{dt} = 2\beta\left(\frac{C_0}{C_{\mu 0}}\right)(1+s)\left[H(y,\theta)\frac{\partial y}{\partial u}\right]_{u=1} - \text{LeBi}(\theta - \theta_b) \qquad (A9.4\text{-}4b)$$

Similarly, the boundary condition written in terms of the new independent variable u is:

$$u = 1; \qquad H(y,\theta)\frac{\partial y}{\partial u} = \frac{\text{Bi}}{2}(y_b - y) \qquad (A9.4\text{-}5)$$

To apply the orthogonal collocation method, we choose N interior collocation points in the spatial domain u, $0 < u_1, u_2, \cdots, u_N < 1$. These N interior points together with the point at the boundary $u = 1$ will form N+1 interpolation points. Evaluating eq. (A9.4-4a) at the j-th interior collocation point and using formulas (A9.3-8), we get:

$$G_1(y_j,\theta)\frac{\partial y_j}{\partial t} + G_2(y_j,\theta)\frac{\partial \theta}{\partial t} = H(y_j,\theta)\sum_{k=1}^{N+1} C_{jk}y_k + 4u_j\left(\sum_{k=1}^{N+1} A_{jk}H(y_k,\theta)\right)\left(\sum_{k=1}^{N+1} A_{jk}y_k\right)$$

$$(A9.4\text{-}6)$$

for $j = 1, 2, 3, \ldots, N$, where

$$C_{jk} = 4u_j B_{jk} + 2(1+s)A_{jk} \qquad (A9.4\text{-}7)$$

Eq. (A9.4-6) is valid for N interior collocation points. The equation at the (N+1)-th interpolation point is given in eq. (A9.4-5). Written that equation in terms of the collocation variables, we have:

$$H(y_{N+1},\theta)\sum_{k=1}^{N+1} A_{N+1,k}y_k = \left(\frac{\text{Bi}}{2}\right)(y_b - y_{N+1}) \qquad (A9.4\text{-}8)$$

from which we can solve for the concentration at the boundary (y_{N+1}) in terms of other dependent variables $y_1, y_2, \ldots, y_N$, and θ. Eq. (A9.4-8) is a nonlinear algebraic equation in terms of y_{N+1} and it can be solved numerically by a Newton-Raphson method and the initial guess for that method is:

$$y_{N+1} = \frac{y_b - \left(\dfrac{2}{\text{Bi}}\right)H(y_b,\theta)\sum_{k=1}^{N} A_{N+1,k}y_k}{1 + \dfrac{2}{\text{Bi}}A_{N+1,N+1}} \qquad (A9.4\text{-}9)$$

Applying the collocation analysis to eq. (A9.4-4b), we get:

$$\frac{d\theta}{dt} = 2\beta\left(\frac{C_0}{C_{\mu 0}}\right)(1+s)H(y_{N+1},\theta)\sum_{j=1}^{N+1} A_{N+1,j}y_j - \text{LeBi}(\theta - \theta_b) \qquad (A9.4\text{-}10)$$

Eqs. (A9.4-6) and (A9.4-10) are N+1 coupled ordinary differential equations in terms of $y_1, y_2, ..., y_N$ and θ, which can be solved numerically by using any integration techniques. The programming code for this problem is ADSORB1B.M.

Appendix 9.5: Mass Exchange Kinetics Expressions

The following expression can describe the mass exchange kinetics between the fluid and adsorbed phases.

R_{ads}		Equilibrium isotherm	
$R_{ads} = k_a C \left(1 - \dfrac{C_\mu}{C_{\mu s}}\right)^n - k_d \left(\dfrac{C_\mu}{C_{\mu s}}\right)^n$	(A9.5-1a)	$C_\mu = C_{\mu s} \dfrac{(bC)^{1/n}}{1+(bC)^{1/n}}$	(Sips)
$R_{ads} = k_a C^{1/n} \left(1 - \dfrac{C_\mu}{C_{\mu s}}\right) - k_d \left(\dfrac{C_\mu}{C_{\mu s}}\right)$	(A9.5-1b)	$C_\mu = C_{\mu s} \dfrac{(bC)^{1/n}}{1+(bC)^{1/n}}$	(Sips)
$R_{ads} = k_a C \left[1 - \left(\dfrac{C_\mu}{C_{\mu s}}\right)^t\right]^{1/t} - k_d \left(\dfrac{C_\mu}{C_{\mu s}}\right)$	(A9.5-1c)	$C_\mu = C_{\mu s} \dfrac{bC}{\left[1+(bC)^t\right]^{1/t}}$	(Toth)

The corresponding equilibrium isotherm equations for these kinetic expressions are listed in the above table.

Appendix 9.6: Collocation Analysis of Model Equation (eq. 9.5-26)

The mass balance equation (9.5-26) takes the following vector form:

$$\frac{\partial \underline{p}}{\partial t} = \frac{\underline{\underline{H}}(\underline{p})}{\eta^s} \frac{\partial}{\partial \eta} \left\{ \eta^s \, \varepsilon \left[D_T \underline{\underline{B}}^{eff}(\underline{p}) \right]^{-1} \left[\frac{\partial \underline{p}}{\partial \eta} + \Phi \frac{\mu_0}{\mu_m(\underline{y})} \frac{1}{P_0} \frac{\partial P_T}{\partial \eta} (D_T \underline{\underline{A}}) \underline{p} \right] \right\}$$
$$+ \frac{\underline{\underline{H}}(\underline{p})}{\eta^s} \frac{\partial}{\partial \eta} \left\{ \eta^s (1-\varepsilon) \, \underline{\underline{G}}^*(\underline{p}) \frac{\partial \underline{p}}{\partial \eta} \right\}$$

(A9.6-1)

Expanding the RHS of the above equation, we get:

$$\frac{\partial \underline{p}}{\partial \tau} = \varepsilon \underline{H}(\underline{p}) \left\{ \left[D_T \underline{\underline{B}}^{eff}(\underline{p}) \right]^{-1} \left[\left(\frac{\partial^2 \underline{p}}{\partial \eta^2} + \frac{s}{\eta} \frac{\partial \underline{p}}{\partial \eta} \right) + \Phi \frac{\mu_0}{\mu_m(\underline{y})} (D_T \underline{\underline{\Lambda}}) \left(\frac{1}{P_0} \left(\frac{\partial^2 P_T}{\partial \eta^2} + \frac{s}{\eta} \frac{\partial P_T}{\partial \eta} \right) \underline{p} + \frac{1}{P_0} \frac{\partial P_T}{\partial \eta} \frac{\partial \underline{p}}{\partial \eta} \right) \right] $$
$$+ \frac{\partial}{\partial \eta} \left[D_T \underline{\underline{B}}^{eff}(\underline{p}) \right]^{-1} \left[\frac{\partial \underline{p}}{\partial \eta} + \Phi \frac{\mu_0}{\mu_m(\underline{y})} \frac{1}{P_0} \frac{\partial P_T}{\partial \eta} (D_T \underline{\underline{\Lambda}}) \underline{p} \right] \right\}$$
$$+ (1-\varepsilon) \underline{H}(\underline{p}) \left\{ \underline{\underline{G}}^*(\underline{p}) \left(\frac{\partial^2 \underline{p}}{\partial \eta^2} + \frac{s}{\eta} \frac{\partial \underline{p}}{\partial \eta} \right) + \frac{\partial}{\partial \eta} \left[\underline{\underline{G}}^*(\underline{p}) \right] \frac{\partial \underline{p}}{\partial \eta} \right\}$$

(A9.6-2)

The symmetry of the problem at $\eta = 0$ suggests the following transformation:
$$u = \eta^2 \tag{A9.6-3}$$

With this transformation, eq. (A9.6-2) becomes:

$$\frac{\partial \underline{p}}{\partial \tau} = \varepsilon \underline{H}(\underline{p}) \left\{ \left[D_T \underline{\underline{B}}^{eff}(\underline{p}) \right]^{-1} \left[\left(4u \frac{\partial^2 \underline{p}}{\partial u^2} + 2(1+s) \frac{\partial \underline{p}}{\partial u} \right) + \right.\right.$$

$$\left. \Phi \frac{\mu_0}{\mu_m(\underline{y})} (D_T \underline{\underline{\Lambda}}) \left(\frac{1}{P_0} \left(4u \frac{\partial^2 P_T}{\partial u^2} + 2(1+s) \frac{\partial P_T}{\partial u} \right) \underline{p} + 4u \left(\frac{1}{P_0} \frac{\partial P_T}{\partial u} \right) \left(\frac{\partial \underline{p}}{\partial u} \right) \right) \right]$$

$$+ 4u \frac{\partial}{\partial u} \left[D_T \underline{\underline{B}}^{eff}(\underline{p}) \right]^{-1} \left[\frac{\partial \underline{p}}{\partial u} + \Phi \frac{\mu_0}{\mu_m(\underline{y})} \frac{1}{P_0} \frac{\partial P_T}{\partial u} (D_T \underline{\underline{\Lambda}}) \underline{p} \right] \right\}$$

$$+ (1-\varepsilon) \underline{H}(\underline{p}) \left\{ \underline{\underline{G}}^*(\underline{p}) \left(4u \frac{\partial^2 \underline{p}}{\partial u^2} + 2(1+s) \frac{\partial \underline{p}}{\partial u} \right) + 4u \frac{\partial}{\partial u} \left[\underline{\underline{G}}^*(\underline{p}) \right] \frac{\partial \underline{p}}{\partial u} \right\}$$

(A9.6-4)

We now choose N interior collocation points and together with the boundary point at $u = 1$ we have N+1 interpolation points. Evaluate eq. (A9.6-4) at the i-th interior collocation point, we get:

$$\frac{\partial \underline{p}_i}{\partial \tau} = \varepsilon \underline{H}(\underline{p}_i) \left\{ \left[D_T \underline{\underline{B}}^{eff}(\underline{p}_i) \right]^{-1} \left[\sum_{j=1}^{N+1} C_{ij} \underline{p}_j + \Phi \frac{\mu_0}{\mu_m(\underline{y}_i)} (D_T \underline{\underline{\Lambda}}) \left(\frac{1}{P_0} \sum_{j=1}^{N+1} C_{ij} P_{Tj} \underline{p}_i + 4u_i \frac{1}{P_0} \sum_{j=1}^{N+1} A_{ij} P_{Tj} \sum_{j=1}^{N+1} A_{ij} \underline{p}_j \right) \right] \right.$$

$$+ 4u_i \left[\sum_{j=1}^{N+1} A_{ij} \left(D_T \underline{\underline{B}}^{eff}(\underline{p}_j) \right) \right]^{-1} \left[\sum_{j=1}^{N+1} A_{ij} \underline{p}_j + \Phi \frac{\mu_0}{\mu_m(\underline{y}_i)} \frac{1}{P_0} \left(\frac{1}{P_0} \sum_{j=1}^{N+1} A_{ij} P_{Tj} \right) (D_T \underline{\underline{\Lambda}}) \underline{p}_i \right] \right\}$$

$$+ (1-\varepsilon) \underline{H}(\underline{p}_i) \left\{ \underline{\underline{G}}^*(\underline{p}_i) \sum_{j=1}^{N+1} C_{ij} \underline{p}_j + 4u_i \left[\sum_{j=1}^{N+1} A_{ij} \underline{\underline{G}}^*(\underline{p}_j) \right] \left(\sum_{j=1}^{N+1} A_{ij} \underline{p}_j \right) \right\}$$

(A9.6-5a)

for i = 1, 2, ..., N, where
$$C_{ij} = 4u_i B_{ij} + 2(1+s) A_{ij} \tag{A9.6-5b}$$

The integration vector will contain the partial pressures of all components at N interior collocation points and is defined as follows:

$$\underline{\Omega} = \begin{bmatrix} \left.\begin{array}{l} p(1,1) \\ p(1,2) \\ \vdots \\ p(1,N) \end{array}\right\} \text{partial pressures of the component 1 @ N interior collocation points} \\ \left.\begin{array}{l} p(2,1) \\ p(2,2) \\ \vdots \\ p(2,N) \end{array}\right\} \text{partial pressures of the component 2 @ N interior collocation points} \\ \vdots \\ \left.\begin{array}{l} p(nc,1) \\ p(nc,2) \\ \vdots \\ p(nc,N) \end{array}\right\} \text{partial pressures of the nc - th component @ N interior collocation pts} \end{bmatrix}$$

(A9.6-6)

The length of this $\underline{\Omega}$ array is (nc × N), and the time derivative of this integration vector is given in eq. (A9.6-5a). However to calculate the RHS of eq. (A9.6-5a) we need the total pressure as well as the mole fractions of all components at N interior collocation points. This can be achieved with the procedure suitable for programming as described in Appendix 8.9 (eqs. A8.9-10 to A8.9-15). A programming code ADSORB5B for this problem is provided with this book.

Appendix 9.7: Collocation Analysis of Eq. (9.6-24)

Expanding the mass balance equation (9.6-24a) gives:

$$\frac{\partial \underline{p}}{\partial \tau} = \frac{d\theta}{d\tau} \underline{\underline{H}}(\underline{p},\theta) \left[\frac{\varepsilon}{(1+\theta)} \underline{p} - (1-\varepsilon)(1+\theta) R_g T_0^2 \frac{\partial \underline{f}(\underline{p},T)}{\partial T} \right]$$

$$+ \varepsilon \underline{\underline{H}}(\underline{p},\theta) \left\{ \left[D_T \underline{\underline{B}}(\underline{y},P_T,T) \right]^{-1} \left[\left(\frac{\partial^2 \underline{p}}{\partial \eta^2} + \frac{s}{\eta} \frac{\partial \underline{p}}{\partial \eta} \right) + \Phi\left(\frac{\mu_0}{\mu_m} \right) (D_T \underline{\underline{\Lambda}}) \left(\frac{1}{P_{T0}} \left(\frac{\partial^2 P_T}{\partial \eta^2} + \frac{s}{\eta} \frac{\partial P_T}{\partial \eta} \right) \underline{p} + \frac{1}{P_{T0}} \frac{\partial P_T}{\partial \eta} \frac{\partial \underline{p}}{\partial \eta} \right) \right] \right.$$

$$+ \frac{\partial \left[D_T \underline{\underline{B}}(\underline{y},P_T,T) \right]^{-1}}{\partial \eta} \left[\frac{\partial \underline{p}}{\partial \eta} + \Phi\left(\frac{\mu_0}{\mu_m} \right) \left(\frac{1}{P_{T0}} \frac{\partial P_T}{\partial \eta} \right) (D_T \underline{\underline{\Lambda}}) \underline{p} \right] \right\}$$

$$+ (1-\varepsilon) \underline{\underline{H}}(\underline{p},\theta) \left[\underline{\underline{G}}^*(\underline{p},\theta) \left(\frac{\partial^2 \underline{p}}{\partial \eta^2} + \frac{s}{\eta} \frac{\partial \underline{p}}{\partial \eta} \right) + \frac{\partial \underline{\underline{G}}^*(\underline{p},\theta)}{\partial \eta} \frac{\partial \underline{p}}{\partial \eta} \right] \quad \text{(A9.7-1)}$$

The problem has a symmetry at $\eta=0$. To make use of this, we make the following transformation

$$u = \eta^2 \qquad (A9.7\text{-}2)$$

In terms of this new independent variable, the mass balance equation (A9.7-1) becomes:

$$\frac{\partial \underline{p}}{\partial \tau} = \frac{d\theta}{d\tau} \underline{\underline{H}}(\underline{p},\theta) \left[\frac{\varepsilon}{(1+\theta)} \underline{p} - (1-\varepsilon)(1+\theta) R_g T_0^2 \frac{\partial \underline{f}(\underline{p},T)}{\partial T} \right]$$

$$+ \varepsilon \underline{\underline{H}}(\underline{p},\theta) \left\{ \left[D_T \underline{\underline{B}}(y,P_T,T) \right]^{-1} \left[4u \frac{\partial^2 \underline{p}}{\partial u^2} + 2(1+s) \frac{\partial \underline{p}}{\partial u} \right] \right.$$

$$+ \Phi\left(\frac{\mu_0}{\mu_m}\right) (D_T \underline{\underline{\Lambda}}) \left(\frac{1}{P_{T0}} \left(4u \frac{\partial^2 P_T}{\partial u^2} + 2(1+s) \frac{\partial P_T}{\partial u} \right) \underline{p} + 4u \left(\frac{1}{P_{T0}} \frac{\partial P_T}{\partial u} \right) \left(\frac{\partial \underline{p}}{\partial u} \right) \right)$$

$$+ 4u \frac{\partial \left[D_T \underline{\underline{B}}(y,P_T,T) \right]^{-1}}{\partial u} \left[\frac{\partial \underline{p}}{\partial u} + \Phi\left(\frac{\mu_0}{\mu_m}\right) \left(\frac{1}{P_{T0}} \frac{\partial P_T}{\partial u} \right) (D_T \underline{\underline{\Lambda}}) \underline{p} \right] \right\}$$

$$+ (1-\varepsilon) \underline{\underline{H}}(\underline{p},\theta) \left[\underline{\underline{G}}^*(\underline{p},\theta) \left(4u \frac{\partial^2 \underline{p}}{\partial u^2} + 2(1+s) \frac{\partial \underline{p}}{\partial u} \right) + 4u \frac{\partial \underline{\underline{G}}^*(\underline{p},\theta)}{\partial u} \frac{\partial \underline{p}}{\partial u} \right] \qquad (A9.7\text{-}3)$$

Also in terms of the new independent variable, the heat balance equation (9.6-24b) becomes:

$$\frac{d\theta}{d\tau} = 2(1+s) \frac{\beta\varepsilon}{(1+\theta)} \left(\frac{Q}{Q_0} \right) \left\{ \left[D_T \underline{\underline{B}}(y,P_T,T) \right]^{-1} \left[\frac{1}{P_{T0}} \cdot \frac{\partial \underline{p}}{\partial u} + \Phi\left(\frac{\mu_0}{\mu_m} \right) \left(\frac{1}{P_{T0}} \frac{\partial P_T}{\partial u} \right) (D_T \underline{\underline{\Lambda}}) \left(\frac{1}{P_{T0}} \underline{p} \right) \right] \right\}_{u=1}$$

$$+ 2(1+s) \frac{\beta(1-\varepsilon)}{(1+\theta)} \left(\frac{Q}{Q_0} \right) \left[\underline{\underline{G}}^*(\underline{p},\theta) \left(\frac{1}{P_{T0}} \frac{\partial \underline{p}}{\partial u} \right) \right]_{u=1} - \text{Le Bi}(\theta - \theta_b) \qquad (A9.7\text{-}4)$$

In the orthogonal collocation method, the first and second derivatives at specific interpolation points within the domain are given by:

$$\left. \frac{\partial \underline{p}}{\partial u} \right|_{u_i} = \sum_{j=1}^{N+1} A_{ij} \underline{p}_j \qquad (A9.7\text{-}5a)$$

$$\left.\frac{\partial^2 \underline{p}}{\partial u^2}\right|_{u_i} = \sum_{j=1}^{N+1} B_{ij}\underline{p}_j \tag{A9.7-5b}$$

Written in terms of the collocation variables, the mass and heat balance equations are:

$$\frac{\partial \underline{p}_i}{\partial \tau} = \frac{d\theta}{d\tau}\underline{\underline{H}}(\underline{p}_i,\theta)\left[\frac{\varepsilon}{(1+\theta)}\underline{p}_i - (1-\varepsilon)(1+\theta)R_g T_0^2 \frac{\partial \underline{f}(\underline{p}_i,T)}{\partial T}\right]$$

$$+ \varepsilon \underline{\underline{H}}(\underline{p}_i,\theta)\left\{\left[D_T \underline{\underline{B}}(\underline{y}_i, P_{T_i}, T)\right]^{-1}\left[\sum_{j=1}^{N+1}C_{ij}\underline{p}_j + \Phi\left(\frac{\mu_o}{\mu_m(\underline{y}_i)}\right)(D_T\underline{\underline{\Lambda}}(\theta))\left(\frac{1}{P_{TO}}\sum_{j=1}^{N+1}C_{ij}P_{T_j}\underline{p}_i\right.\right.\right.$$

$$+ 4u_i\left(\frac{1}{P_{TO}}\sum_{j=1}^{N+1}A_{ij}P_{T_j}\right)\left(\sum_{j=1}^{N+1}A_{ij}\underline{p}_j\right)\right]$$

$$+ 4u_i\sum_{j=1}^{N+1}A_{ij}\left[D_T\underline{\underline{B}}(\underline{y}_j, P_{T_j}, T)\right]^{-1}\left[\sum_{j=1}^{N+1}A_{ij}\underline{p}_j + \Phi\left(\frac{\mu_o}{\mu_m}\right)\left(\frac{1}{P_{TO}}\sum_{j=1}^{N+1}A_{ij}P_{T_j}\right)(D_T\underline{\underline{\Lambda}})\underline{p}_i\right]\right\}$$

$$+ (1-\varepsilon)\underline{\underline{H}}(\underline{p}_i,\theta)\left[\underline{\underline{G}}^*(\underline{p}_i,\theta)\sum_{j=1}^{N+1}C_{ij}\underline{p}_j + \left(4u_i\sum_{j=1}^{N+1}A_{ij}\underline{\underline{G}}^*(\underline{p}_j,\theta)\right)\left(\sum_{j=1}^{N+1}A_{ij}\underline{p}_j\right)\right] \tag{A9.7-6}$$

Note that

$$\underline{p}_{N+1} = \underline{p}_b \tag{A9.7-7a}$$

$$P_{T_{N+1}} = \langle \underline{p}_b \rangle \tag{A9.7-7b}$$

where $\langle \rangle$ denotes the summation of all the elements of the vector.

Similarly applying the transformation of eq. (A9.7-2), the heat balance equation (A9.7-4) can be written in terms of collocation variables as:

$$\frac{d\theta}{d\tau} = 2(1+s)\frac{\beta\varepsilon}{(1+\theta)}\left(\frac{Q}{Q_0}\right)\bullet\left\{\left[D_T\underline{\underline{B}}(\underline{y}_b, P_{T_b}, T)\right]^{-1}\left[\frac{1}{P_{TO}}\sum_{j=1}^{N+1}A_{N+1,j}\underline{p}_j\right.\right.$$

$$+ \Phi\left(\frac{\mu_o}{\mu_m}\right)\left(\frac{1}{P_{TO}}\sum_{j=1}^{N+1}A_{N+1,j}P_{T_j}\right)(D_T\underline{\underline{\Lambda}})\left(\frac{1}{P_{TO}}\underline{p}_b\right)\right]\right\}$$

$$+ 2(1+s)\frac{\beta(1-\varepsilon)}{(1+\theta)}\left(\frac{Q}{Q_0}\right)\bullet\left[\underline{\underline{G}}^*(\underline{p}_b,\theta)\left(\frac{1}{P_{TO}}\sum_{j=1}^{N+1}A_{N+1,j}\underline{p}_j\right)\right] - \text{Le Bi}(\theta-\theta_b). \tag{A9.7-8}$$

Eqs. (A9.7-6) and (A9.7-8) represent $nc \times N + 1$ equations for $nc \times N + 1$ unknowns of $\underline{p}_j$ (j = 1, 2, ..., N) and θ. They can be readily integrated using any integration routines. The procedure for programming these equations is illustrated in Appendices A8.9 and A9.6. The integration vector for this problem is shown in the following equation:

$$\begin{bmatrix} \left.\begin{array}{l} p(1,1) \\ p(1,2) \\ \vdots \\ p(1,N) \end{array}\right\} \text{partial pressures of the first component at N collocation points} \\ \left.\begin{array}{l} p(2,1) \\ p(2,2) \\ \vdots \\ p(2,N) \end{array}\right\} \text{partial pressures of the second component at N collocation points} \\ \vdots \\ \left.\begin{array}{l} p(nc,1) \\ p(nc,2) \\ \vdots \\ p(nc,N) \end{array}\right\} \text{partial pressures of the nc - th component at N collocation points} \\ \theta \quad\quad \text{Nondimensional temperature} \end{bmatrix}$$

A programming code ADSORB5C.M is provided with this book.

Appendix 10.1: Orthogonal Collocation Analysis of Eqs. (10.2-38) to (10.2-40)

The mass balance equations (10.2-38 to 10.2-40) are rewritten here for convenience:

$$\frac{\partial x}{\partial \tau} = \frac{1}{\eta^s} \frac{\partial}{\partial \eta} \left[\eta^s H(x) \frac{\partial x}{\partial \eta} \right] \quad \text{(A10.1-1a)}$$

The boundary conditions and initial condition are:

$$\eta = 0; \quad \frac{\partial x}{\partial \eta} = 0 \quad \text{(A10.1-1b)}$$

$$\eta = 1; \quad x = 1 \quad \text{(A10.1-1c)}$$

$$\tau = 0; \quad x = 0 \quad \text{(A10.1-1d)}$$

Due to the symmetry of the problem (eq. A10.1-1b), we introduce the following transformation $u = \eta^2$, and with this transformation, the mass balance equation will become:

$$\frac{\partial x}{\partial \tau} = H(x)\left[4u\frac{\partial^2 x}{\partial u^2} + 2(1+s)\frac{\partial x}{\partial u}\right] + 4u\frac{\partial H(x)}{\partial u}\frac{\partial x}{\partial u} \quad \text{(A10.1-2a)}$$

$$u = 1; \quad x = 1 \quad \text{(A10.1-2b)}$$

We choose N interior collocation points and the boundary point (u = 1) to form a set of N+1 interpolation points. Evaluating eq.(A10.1-2a) at the interior collocation point j as the equation is not valid at the boundary point, we get:

$$\frac{\partial x_j}{\partial \tau} = H(x_j)\left(\sum_{k=1}^{N} C_{jk} x_k + C_{j,N+1} x_{N+1}\right) +$$

$$4u_j\left[\sum_{k=1}^{N} A_{jk} H(x_k) + A_{j,N+1} H(x_{N+1})\right]\left[\sum_{k=1}^{N} A_{jk} x_k + A_{j,N+1} x_{N+1}\right] \quad \text{(A10.1-3)}$$

for $j = 1, 2, 3, ..., N$, with $x_{N+1} = 1$ by virtue of eq. (A10.1-2b). In obtaining eq. (A10.1-3) we have used the collocation formulas (A9.3-8). Knowing the concentration values x at N+1 interpolation points, the average concentration <x> is calculated from a Radau formula:

$$\langle x \rangle = \sum_{k=1}^{N+1} w_k x_k \quad \text{(A10.1-4)}$$

where w_k is the Radau quadrature weights. A programming code ADSORB1C for this problem is provided with this book.

Appendix 10.2: Orthogonal Collocation Analysis of Eqs. (10.3-8) to (10.3-10)

The mass and heat balance equations (10.3-8 to 10.3-10) for a nonisothermal single crystal are:

$$\frac{\partial x}{\partial \tau} = \frac{1}{\eta^s}\frac{\partial}{\partial \eta}\left[\eta^s \varphi(\theta) H(x) \frac{\partial x}{\partial \eta}\right] \quad \text{(A10.2-1a)}$$

$$\frac{d\theta}{d\tau} = \beta\frac{d\langle x \rangle}{d\tau} - \text{LeBi} \cdot \theta \quad \text{(A10.2-1b)}$$

The boundary conditions are:

$$\eta = 0; \quad \frac{\partial x}{\partial \eta} = 0 \tag{A10.2-1c}$$

$$\eta = 1; \quad x\big|_1 = \frac{\dfrac{\lambda_\infty \phi(\theta)}{1+\lambda_\infty \phi(\theta)} - \dfrac{\lambda_0}{1+\lambda_0}}{\dfrac{\lambda_\infty}{1+\lambda_\infty} - \dfrac{\lambda_0}{1+\lambda_0}} \tag{A10.2-1d}$$

The initial condition is:

$$\tau = 0; \quad x = 0; \quad \theta = \frac{T_0 - T_b}{T_b} \tag{A10.2-1e}$$

Due to the symmetry of the problem, we introduce the following transformation $u = \eta^2$, and with this transformation, the mass balance equation (A10.2-1a) will become:

$$\frac{\partial x}{\partial \tau} = \varphi(\theta)H(x)\left[4u\frac{\partial^2 x}{\partial u^2} + 2(1+s)\frac{\partial x}{\partial u}\right] + 4u \cdot \varphi(\theta)\frac{\partial H(x)}{\partial u}\frac{\partial x}{\partial u} \tag{A10.2-2a}$$

$$u = 1; \quad x\big|_1 = \frac{\dfrac{\lambda_\infty \phi(\theta)}{1+\lambda_\infty \phi(\theta)} - \dfrac{\lambda_0}{1+\lambda_0}}{\dfrac{\lambda_\infty}{1+\lambda_\infty} - \dfrac{\lambda_0}{1+\lambda_0}} \tag{A10.2-2b}$$

We choose N interior collocation points and the boundary point (u=1) to form a set of N+1 interpolation points. Evaluating eq.(A10.2-2a) at the collocation point j, we get:

$$\frac{dx_j}{d\tau} = \varphi(\theta)H(x_j)\left(\sum_{k=1}^{N} C_{jk}x_k + C_{j,N+1}x_{N+1}\right) +$$
$$4u_j\varphi(\theta)\left[\sum_{k=1}^{N} A_{jk}H(x_k) + A_{j,N+1}H(x_{N+1})\right]\left(\sum_{k=1}^{N} A_{jk}x_k + A_{j,N+1}x_{N+1}\right) \tag{A10.2-3a}$$

where the boundary point x_{N+1} is a function of temperature, given by:

$$x_{N+1} = \frac{\dfrac{\lambda_\infty \phi(\theta)}{1+\lambda_\infty \phi(\theta)} - \dfrac{\lambda_0}{1+\lambda_0}}{\dfrac{\lambda_\infty}{1+\lambda_\infty} - \dfrac{\lambda_0}{1+\lambda_0}} \tag{A10.2-3b}$$

The time derivative of the average concentration $<x>$ in the heat balance equation can be obtained from the mass balance equation (A10.2-1a) as:

$$\frac{d\langle x \rangle}{d\tau} = (1+s)\varphi(\theta) \cdot H(x_{N+1})\frac{\partial x}{\partial \eta}\bigg|_1 = 2(1+s)\varphi(\theta) \cdot H(x_{N+1})\frac{\partial x}{\partial u}\bigg|_1 \quad \text{(A10.2-4)}$$

Evaluating this equation using the collocation variables, we get:

$$\frac{d\langle x \rangle}{d\tau} = 2(1+s)\varphi(\theta) \cdot H(x_{N+1})\left(\sum_{k=1}^{N} A_{N+1,k} x_k + A_{N+1,N+1} x_{N+1}\right) \quad \text{(A10.2-5)}$$

Substituting eq.(A10.2-5) into the heat balance equation (A10.2.-1b), we get:

$$\frac{d\theta}{d\tau} = 2(1+s)\beta \cdot \varphi(\theta) \cdot H(x_{N+1})\left(\sum_{k=1}^{N} A_{N+1,k} x_k + A_{N+1,N+1} x_{N+1}\right) - LeBi \cdot \theta \quad \text{(A10.2-6)}$$

Eqs.(A10.2-3a) and (A10.2-6) form the complete set of coupled ordinary differential equations, from which they can be integrated to obtain the concentration profile as well as temperature. The programming code ADSORB1D.M was written in MatLab language to solve this problem. The integration vector contains the first N elements being the concentration at N interior collocation points and the last element is the temperature:

$$\underline{\Omega} = \begin{bmatrix} x_1 & x_2 & \cdots & x_N & \theta \end{bmatrix}^T \quad \text{(A10.2-7)}$$

The rate of change of this integration vector with respect to time is:

$$\frac{d\underline{\Omega}}{d\tau} = \begin{bmatrix} \frac{dx_1}{d\tau} & \frac{dx_2}{d\tau} & \cdots & \frac{dx_N}{d\tau} & \frac{d\theta}{d\tau} \end{bmatrix}^T \quad \text{(A10.2-8)}$$

The first N elements of this derivative vector are given in eq.(A10.2-3a), and the last element is given in eq.(A10.2-6).

Appendix 10.3: Order of Magnitude of Heat Transfer Parameters

Some typical values used in the simulation are given in the following table:

$\dfrac{(1-\varepsilon_M)C_{\mu 0}}{\rho_p}$	0.004 mmole/g
E_a	15,000 Joule/mole
Q	30,000 Joule/mole
C_p	1 Joule/g/K
T_0	300 K
h	10×10^{-4} Joule / cm^2 / sec/ K

R	0.05 cm
$\varepsilon_M D_{p0}$	0.01 cm²/sec
C_0'	1×10^{-6} mole / cc
$\gamma = \dfrac{Q}{R_g T_0}$	12
$\alpha_1 = \dfrac{E}{R_g T_0}$	6
$\beta_1 = \dfrac{(1-\varepsilon_M) C_{\mu 0}}{\rho_p} \dfrac{(-\Delta H)}{C_p T_0}$	0.4
$\beta_2 = \dfrac{hR\left[\varepsilon_M + (1-\varepsilon_M) C_{\mu 0}/C_0\right]}{\rho_p C_p \varepsilon_M D_{p0}}$	20

The following table gives the thermal properties of some systems

System	k_e J/s/cm/K	k_f J/s/cm/K	C_p J/g/K	C_f J/g/K	Q kJ/mole	h J/cm²/s/K
benzene/silica gel	$4 - 7 \times 10^{-4}$	8.5×10^{-5}	0.75		42	$3 - 100 \times 10^{-4}$
water/silica gel		2×10^{-3}	1.05	1.88		
water/zeolite	3×10^{-3}	1.3×10^{-3}	0.8	1.88		30×10^{-4}
CO$_2$/zeolite 10A		1.3×10^{-3}	0.8	0.85		
benzene/charcoal			1.2	2.2	43.5	
CCl$_2$H$_2$/carbon					18.7	$2 - 7 \times 10^{-4}$
heptane/zeolite 5A			0.963		41.57	

References:
Brunovska, Ilavsky and Kukurukova, Coll Czech Chem Commun, 50, 1341 (1985)
Haul and Stremming, J. Coll Inter Sci., 97, 348 (1984)
Haul and Stremming, in Characterization of porous solids, edited by Under (1988)
Ilavsky, Brunovska, and Klavacek, Chem Eng Sci., 35, 2475 (1980)
James and Phillips, J. Chem Soc 1066 (1954)
Kanoldt and Mersmann, in Fundamentals of adsorption (1986)
Meunier and Sun, J. Chem Soc Farad Trans I, 84, 1973 (1988)
Sun and Meunier, in Fundamentals of Adsorption, 1986.

Appendix 10.4: Collocation Analysis of Eqs. (10.4-45)

The non-dimensional mass and heat balance equations (10.4-45) are:

$$\frac{\partial x}{\partial \tau} = \gamma \frac{1}{\zeta^{s_\mu}} \frac{\partial}{\partial \zeta}\left[\zeta^{s_\mu} \, \varphi(\theta) \, H(x,\theta) \frac{\partial x}{\partial \zeta}\right] \qquad (A10.4\text{-}1a)$$

$$\zeta = 1; \quad x = F(y,\theta) \qquad (A10.4\text{-}1b)$$

$$\sigma \frac{\partial y}{\partial \tau} + (1-\sigma)\frac{\partial \bar{x}}{\partial \tau} = (1+\theta)^\alpha \frac{1}{\eta^s}\frac{\partial}{\partial \eta}\left(\eta^s \frac{\partial y}{\partial \eta}\right) \qquad (A10.4\text{-}1c)$$

$$\eta = 1 \; ; \; -(1+\theta)^\alpha \frac{\partial y}{\partial \eta} = Bi(y - y_b) \qquad (A10.4\text{-}1d)$$

$$\frac{d\theta}{d\tau} = \beta \frac{d\bar{\bar{x}}}{d\tau} - LeBi\,(\theta - \theta_b) \qquad (A10.4\text{-}1e)$$

The volumetric average concentrations are defined as follows:

$$\bar{x}(\eta,\tau) = (1+s_\mu)\int_0^1 \zeta^{s_\mu} \, x(\zeta,\eta,\tau) \, d\zeta \qquad (A10.4\text{-}2a)$$

$$\bar{\bar{x}}(\tau) = (1+s)\int_0^1 \eta^s \, \bar{x}(\eta,\tau) \, d\eta \qquad (A10.4\text{-}2b)$$

The change in the average concentration in eq.(A10.4-1c) can be replaced by the flux into the micro-particle. This is done as follows. Multiplying eq.(A10.4-1a) by ζ^{s_μ} and integrating the result with respect to ζ, we get the following expression for the mean adsorbed concentration along the pellet coordinate:

$$\frac{\partial \bar{x}}{\partial \tau} = \gamma \, \varphi(\theta) \, (1+s_\mu)\left[H(x,\theta)\frac{\partial x}{\partial \zeta}\right]_1$$

Combining eqs. (A10.4-1c) and the above equation, we have:

$$\sigma \frac{\partial y}{\partial \tau} + (1-\sigma)\gamma \, \varphi(\theta) \, (1+s_\mu)\left[H(x,\theta)\frac{\partial x}{\partial \zeta}\right]_1 = (1+\theta)^\alpha \frac{1}{\eta^s}\frac{\partial}{\partial \eta}\left(\eta^s \frac{\partial y}{\partial \eta}\right) \quad (A10.4\text{-}3)$$

Now we apply the orthogonal collocation method. Noting the symmetry at the centers of the micro-particle and pellet, we make the following transformation:

$$v = \zeta^2; \qquad u = \eta^2 \qquad (A10.4\text{-}4)$$

With this transformation, eqs. (A10.4-1a) and (A10.4-3) will become:

$$\frac{\partial x}{\partial \tau} = \gamma\, \varphi(\theta)\left\{H(x,\theta)\left[4v\frac{\partial^2 x}{\partial v^2} + 2(1+s_\mu)\frac{\partial x}{\partial v}\right] + 4v\frac{\partial H(x,\theta)}{\partial v}\frac{\partial x}{\partial v}\right\} \quad \text{(A10.4-5)}$$

$$\sigma\frac{\partial y}{\partial \tau} + 2(1-\sigma)\gamma\,\varphi(\theta)(1+s_\mu)\left[H(x,\theta)\frac{\partial x}{\partial v}\right]_1 = (1+\theta)^\alpha\left[4u\frac{\partial^2 y}{\partial u^2} + 2(1+s)\frac{\partial y}{\partial u}\right]$$
$$\text{(A10.4-6)}$$

The boundary conditions for the above equations are:
$$v = 1; \qquad x = F(y,\theta) \quad \text{(A10.4-7)}$$

$$u = 1; \quad -(1+\theta)^\alpha \frac{\partial y}{\partial u} = \frac{Bi}{2}(y - y_b) \quad \text{(A10.4-8)}$$

We now choose M interior collocation points in the micro-particle, and evaluate eq.(A10.4-5) at the interior collocation point k to get:

$$\frac{\partial x_k}{\partial \tau} = \gamma\,\varphi(\theta)\left\{H(x_k,\theta)\sum_{\ell=1}^{M+1} C_{k\ell}^\mu x_\ell + 4v_k\left(\sum_{\ell=1}^{M+1} A_{k\ell}^\mu H(x_\ell,\theta)\right)\left(\sum_{\ell=1}^{M+1} A_{k\ell}^\mu x_\ell\right)\right\} \quad \text{(A10.4-9)}$$

for k = 1, 2, ..., M, where

$$C_{k\ell}^\mu = 4v_k B_{k\ell}^\mu + 2(1+s_\mu)A_{k\ell}^\mu \quad \text{(A10.4-10)}$$

The adsorbed concentration at the exterior surface of the micro-particle is in equilibrium with the concentration in the macropore, that is:

$$x_{M+1} = F(y,\theta) \quad \text{(A10.4-11)}$$

Next we evaluate eq.(A10.4-9) at the interior collocation point i along the pellet coordinate and get:

$$\frac{d x_{i,k}}{d\tau} = \gamma\,\varphi(\theta)\left\{H(x_{i,k},\theta)\sum_{\ell=1}^{M+1} C_{k\ell}^\mu x_{i,\ell} + 4v_k\left(\sum_{\ell=1}^{M+1} A_{k\ell}^\mu H(x_{i,\ell},\theta)\right)\left(\sum_{\ell=1}^{M+1} A_{k\ell}^\mu x_{i,\ell}\right)\right\}$$
$$\text{(A10.4-12)}$$

for k = 1, 2, ..., M and i = 1, 2, ..., N+1. Here $x_{i,k}$ is the adsorbed concentration at the k point along the micro-particle co-ordinate and the point i along the pellet co-ordinate. Note that this equation is valid up to the point N+1 in the pellet co-ordinate. The adsorbed concentration at the exterior surface of the micro-particle at the collocation i, $x_{i,M+1}$, can be calculated from eq. (A10.4-11), that is

$$x_{i,M+1} = F(y_i, \theta) \quad \text{(A10.4-13)}$$

Next, we evaluate the mass balance equation along the pellet co-ordinate (A10.4-6) at the i-th collocation point::

$$\sigma \frac{\partial y_i}{\partial \tau} + 2(1-\sigma)\gamma \, \varphi(\theta)(1+s_\mu) \, H(x_{i,M+1},\theta) \sum_{l=1}^{M+1} A^\mu_{M+1,l} \, x_{i,l} = (1+\theta)^\alpha \sum_{j=1}^{N+1} C_{ij} \, y_j \quad (A10.4\text{-}14)$$

for i = 1, 2, ..., N, where

$$C_{ij} = 4u_i B_{ij} + 2(1+s) A_{ij} \quad (A10.4\text{-}15)$$

The concentration at the exterior surface of the pellet is obtained from eq.(A10.4-8), that is:

$$y_{N+1} = \frac{y_b - \dfrac{2}{Bi}(1+\theta)^\alpha \sum_{j=1}^{N} A_{N+1,j} \, y_j}{1 + \dfrac{2}{Bi}(1+\theta)^\alpha A_{N+1,N+1}} \quad (A10.4\text{-}16)$$

The temperature equation written in terms of discrete variables is:

$$\frac{d\theta}{d\tau} = \beta \sum_{i=1}^{N+1} w_i \frac{d\overline{x}_i}{d\tau} - LeBi(\theta - \theta_b) \quad (A10.4\text{-}17)$$

or

$$\frac{d\theta}{d\tau} = 2\beta \, \gamma \, \varphi(\theta)(1+s_\mu) \sum_{i=1}^{N+1} w_i \, H(x_{i,M+1},\theta) \sum_{l=1}^{M+1} A^\mu_{M+1,l} \, x_{i,l} - LeBi(\theta - \theta_b) \quad (A10.4\text{-}18)$$

Eqs. (A10.4-12), (A10.4-14) and (A10.4-18) form a set of (N+1)(M+1) equations in terms of (N+1)(M+1) unknowns:
1. $x_{i,k}$ (i=1,2,...,N+1; k=1,2,...,M)
2. y_i (i=1,2,...,N)
3. θ

Appendix 10.5: Orthogonal Collocation Analysis of Eq. (10.5-22)

The governing equation for the case of multicomponent diffusion in a zeolite crystal (eq. 10.5-22) is:

$$\frac{\partial \underline{C}_\mu}{\partial t} = \frac{1}{\eta^s} \frac{\partial}{\partial z} \left\{ \eta^s \left[\underline{\chi}(\underline{C}_\mu) \right] \frac{\partial \underline{C}_\mu}{\partial \eta} \right\} \quad (A10.5\text{-}1a)$$

$$\eta = 0; \quad \frac{\partial \underline{C}_\mu}{\partial \eta} = \underline{0} \qquad (A10.5\text{-}1b)$$

$$\eta = 1; \quad \underline{C}_\mu = \underline{C}_{\mu b} \qquad (A10.5\text{-}1c)$$

$$\tau = 0; \quad \underline{C}_\mu = \underline{C}_{\mu i} \qquad (A10.5\text{-}1d)$$

We will solve this problem by using the method of orthogonal collocation (Rice and Do, 1995). We first expand the RHS of eq. (A10.5-1a) to get:

$$\frac{\partial \underline{C}_\mu}{\partial \tau} = \left[\underline{\underline{\chi}}(\underline{C}_\mu)\right] \frac{1}{\eta^s} \frac{\partial}{\partial \eta}\left(\eta^s \frac{\partial \underline{C}_\mu}{\partial \eta}\right) + \frac{\partial\left[\underline{\underline{\chi}}(\underline{C}_\mu)\right]}{\partial \eta} \frac{\partial \underline{C}_\mu}{\partial \eta} \qquad (A10.5\text{-}2)$$

We note that the problem is symmetrical at $\eta = 0$, it is then logical to define a new variable to account for this symmetry, $u = \eta^2$. With this transformation, eq. (A10.5-2) becomes:

$$\frac{\partial \underline{C}_\mu}{\partial \tau} = \left[\underline{\underline{\chi}}(\underline{C}_\mu)\right]\left[4u\frac{\partial^2 \underline{C}_\mu}{\partial u^2} + 2(1+s)\frac{\partial \underline{C}_\mu}{\partial u}\right] + 4u\frac{\partial\left[\underline{\underline{\chi}}(\underline{C}_\mu)\right]}{\partial u}\frac{\partial \underline{C}_\mu}{\partial u} \qquad (A10.5\text{-}3)$$

The boundary condition for this equation is:

$$\eta = 1; \quad \underline{C}_\mu = \underline{C}_{\mu b} \qquad (A10.5\text{-}4)$$

The domain $\{u \in [0, 1]\}$ of the particle is discretized with N interior collocation points, and we use the point at the surface (u=1) as the additional point; thus, there is a total of N+1 interpolation points. Using the Lagrangian interpolation polynomial passing these N+1 points, we obtain the following relations for the first and second derivatives, written in terms of the functional values:

$$\left.\frac{\partial \underline{C}_\mu}{\partial u}\right|_j = \sum_{k=1}^{N+1} A_{jk}\,\underline{C}_{\mu,k}\,; \quad \left.\frac{\partial \underline{\underline{\chi}}(\underline{C}_\mu)}{\partial u}\right|_j = \sum_{k=1}^{N+1} A_{jk}\,\underline{\underline{\chi}}(\underline{C}_{\mu,k})\,; \quad \left.\frac{\partial^2 \underline{C}_\mu}{\partial u^2}\right|_j = \sum_{k=1}^{N+1} B_{jk}\,\underline{C}_{\mu,k}$$

$$(A10.5\text{-}5)$$

where $\underline{\underline{A}}$ and $\underline{\underline{B}}$ matrices are known matrices after the number of interpolation points are chosen (Rice & Do, 1995).

Evaluating eq. (A10.5-3) at the interior collocation points (j = 1, 2, ..., N), and then substituting eqs. (A10.5-5) into the resulting equation, we obtain the following:

$$\frac{\partial \underline{C}_{\mu,j}}{\partial \tau} = \left[\underline{\underline{\chi}}(\underline{C}_{\mu,j})\right]\left[\sum_{k=1}^{N+1} C_{jk}\,\underline{C}_{\mu,k}\right] + 4u_j\left\{\sum_{k=1}^{N+1} A_{jk}\left[\underline{\underline{\chi}}(\underline{C}_{\mu,k})\right]\right\}\left(\sum_{k=1}^{N+1} A_{jk}\,\underline{C}_{\mu,k}\right) \quad (A10.5\text{-}6)$$

where

$$C_{jk} = 4u_j B_{jk} + 2(1+s)A_{jk} \quad (A10.5\text{-}7)$$

The concentration vector at the interpolation point (N+1) is simply the concentration at the boundary, that is:

$$\underline{C}_{\mu,N+1} = \underline{C}_{\mu b} \quad (A10.5\text{-}8)$$

Substitute this equation into eq. (A10.5-6), we finally obtain (for j = 1, 2, ..., N):

$$\frac{\partial \underline{C}_{\mu,j}}{\partial t} = \left[\underline{\underline{\chi}}(\underline{C}_{\mu,j})\right]\left[\sum_{k=1}^{N} C_{jk}\underline{C}_{\mu,k} + C_{j,N+1}\underline{C}_{\mu b}\right] +$$

$$4u_j \left\{\sum_{k=1}^{N} A_{jk}\left[\underline{\underline{\chi}}(\underline{C}_{\mu,k})\right] + A_{j,N+1}\left[\underline{\underline{\chi}}(\underline{C}_{\mu b})\right]\right\}\left(\sum_{k=1}^{N} A_{jk}\underline{C}_{\mu,k} + A_{j,N+1}\underline{C}_{\mu b}\right) \quad (A10.5\text{-}9)$$

For the purpose of programming, we define a concentration matrix as follows:

$$\underline{\underline{\Omega}} = \begin{bmatrix} C_\mu(1,1) & C_\mu(1,2) & \cdots & C_\mu(1,N) \\ C_\mu(2,1) & C_\mu(2,2) & \cdots & C_\mu(2,N) \\ \vdots & \vdots & \ddots & \vdots \\ C_\mu(nc,1) & C_\mu(nc,2) & \cdots & C_\mu(nc,n) \end{bmatrix} \quad (A10.5\text{-}10)$$

where the i-th row represents the i-th species, and the j-th column represents the j-th collocation point. Here nc is the number of components. We use the notation of MATLAB to represent the concentration vector. There are two vectors to deal with:

$$\Omega(:,j) = \text{concentrations of all species at the j-th collocation point} \quad (A10.5\text{-}11)$$

$$\Omega(i,:) = \text{concentrations of component i at all collocation points} \quad (A10.5\text{-}12)$$

With the notation of eqs. (A10.5-11) and (A10.5-12), the discretized mass balance equation of eq. (A10.5-9) can be rewritten as:

$$\frac{\partial \Omega(:,j)}{\partial t} = \left[\underline{\underline{\chi}}(\Omega(:,j))\right]\left[\sum_{k=1}^{N} C_{jk}\Omega(:,k) + C_{j,N+1}\underline{C}_{\mu b}\right] +$$

$$4u_j\left[\sum_{k=1}^{N} A_{jk}\left[\underline{\underline{\chi}}(\Omega(:,k))\right] + A_{j,N+1}\left[\underline{\underline{\chi}}(\underline{C}_{\mu b})\right]\right] \times \left(\sum_{k=1}^{N} A_{jk}\Omega(:,k) + A_{j,N+1}\underline{C}_{\mu b}\right) \quad (A10.5\text{-}13)$$

Eqs. (A10.5-13) is coded in a program ADSORB3.M.

Appendix 10.6: Orthogonal collocation analysis of eqs. (10.6-25)

The mass balance equations for the case of nonisothermal crystal is:

$$\frac{\partial C_\mu}{\partial \tau} = \frac{1}{\eta^s} \frac{\partial}{\partial \eta}\left[\eta^s \underline{\underline{\chi}}(\underline{C}_\mu, \theta) \frac{\partial \underline{C}_\mu}{\partial \eta}\right] \quad \text{(A10.6-1)}$$

$$\eta = 0; \quad \frac{\partial \underline{C}_\mu}{\partial \eta} = \underline{0} \quad \text{(A10.6-2a)}$$

$$\eta = 1; \quad \underline{C}_\mu = \underline{f}(\underline{p}_b, \theta) \quad \text{(A10.6-2b)}$$

It is now noted that the boundary condition at the surface is no longer constant because of the variation of the isotherm f with respect to temperature.

The heat balance equation is:

$$\frac{d\theta}{d\tau} = \underline{\beta}' \bullet \frac{d\langle \underline{C}_\mu \rangle}{d\tau} - \text{LeBi}\,(\theta - \theta_b) \quad \text{(A10.6-3)}$$

where $\langle \underline{C}_\mu \rangle$ is the volumetric average concentration, defined as:

$$\langle \underline{C}_\mu \rangle = (1+s) \int_0^1 \eta^s \underline{C}_\mu(\eta, \tau) d\eta \quad \text{(A10.6-4)}$$

Using the mass balance equation (eq. A10.6-1), we evaluate the rate of change of the mean concentration and substitute it into the heat balance equation (A10.6-3) to give:

$$\frac{d\theta}{d\tau} = (s+1)\underline{\beta}' \bullet \left[\underline{\underline{\chi}}(\underline{C}_\mu, \theta) \frac{\partial \underline{C}_\mu}{\partial \eta}\right]_{\eta=1} - \text{LeBi}\,(\theta - \theta_b) \quad \text{(A10.6-5)}$$

Using the transformation $u = \eta^2$ because of symmetry, the mass and heat balance equations will become:

$$\frac{\partial \underline{C}_\mu}{\partial \tau} = \underline{\underline{\chi}}(\underline{C}_\mu, \theta)\left[4u\frac{\partial^2 \underline{C}_\mu}{\partial u^2} + 2(1+s)\frac{\partial \underline{C}_\mu}{\partial u}\right] + 4u\frac{\partial \underline{\underline{\chi}}(\underline{C}_\mu, \theta)}{\partial u}\frac{\partial \underline{C}_\mu}{\partial u} \quad \text{(A10.6-6a)}$$

$$\frac{d\theta}{d\tau} = 2(s+1)\underline{\beta}' \bullet \left[\underline{\underline{\chi}}(\underline{C}_\mu, \theta)\frac{\partial \underline{C}_\mu}{\partial u}\bigg|_{u=1}\right] - \text{LeBi}\,(\theta - \theta_b) \quad \text{(A10.6-6b)}$$

The interpolation points are chosen with N interior collocation points and the point at the boundary (u=1). Evaluating the mass balance equation (eq. A10.6-6a) at the interior point j, we get:

$$\frac{d\underline{C}_{\mu,j}}{d\tau} = \underline{\underline{\chi}}(\underline{C}_{\mu,j},\theta)\left[\sum_{k=1}^{N}C_{jk}\underline{C}_{\mu,k} + C_{j,N+1}\underline{C}_{\mu,N+1}\right] +$$

$$4u\left[\sum_{k=1}^{N}A_{jk}\underline{\underline{\chi}}(\underline{C}_{\mu,k},\theta) + A_{j,N+1}\underline{\underline{\chi}}(\underline{C}_{\mu,N+1},\theta)\right] \times \left(\sum_{k=1}^{N}A_{jk}\underline{C}_{\mu,k} + A_{j,N+1}\underline{C}_{\mu,N+1}\right)$$

(A10.6-7)

The concentration vector at the boundary u=1 is given in eq. (A10.6-2b), that is

$$\underline{C}_{\mu,N+1} = \underline{f}(\underline{p}_b,\theta)$$

We have completed the collocation analysis of the mass balance equation, now we turn to the heat balance equation:

$$\frac{d\theta}{d\tau} = 2(s+1)\left[\underline{\beta}' \bullet \left(\underline{\underline{\chi}}(\underline{C}_{\mu,N+1},\theta)\left(\sum_{k=1}^{N}A_{N+1,k}\underline{C}_{\mu,k} + A_{N+1,N+1}\underline{C}_{\mu,N+1}\right)\right)\right] - LeBi(\theta - \theta_b)$$

(A10.6-8)

Eqs. (A10.6-8) and (A10.6-9) represent nc.N + 1 equations in terms of nc.N+1 unknowns $\underline{C}_{\mu,j}$ ($j = 1,2,\cdots,N$) and θ. To help with the coding, we introduce the following matrix:

$$\underline{\underline{\Omega}} = \begin{bmatrix} C_\mu(1,1) & C_\mu(1,2) & \cdots & C_\mu(1,N) \\ C_\mu(2,1) & C_\mu(2,2) & \cdots & C_\mu(2,N) \\ \vdots & \vdots & \ddots & \vdots \\ C_\mu(nc,1) & C_\mu(nc,2) & \cdots & C_\mu(nc,N) \end{bmatrix} \quad (A10.6-9)$$

eqs. (A10.6-7) and (A10.6-8) become:

$$\frac{d\Omega(:,j)}{d\tau} = \underline{\underline{\chi}}(\Omega(:,j),\theta)\left[\sum_{k=1}^{N}C_{jk}\Omega(:,k) + C_{j,N+1}\underline{C}_{\mu,N+1}\right]$$

$$+ 4u_j\left[\sum_{k=1}^{N}A_{jk}\underline{\underline{\chi}}(\Omega(:,k),\theta) + A_{j,N+1}\underline{\underline{\chi}}(\underline{C}_{\mu,N+1},\theta)\right]\left[\sum_{k=1}^{N}A_{jk}\Omega(:,k) + A_{j,N+1}\underline{C}_{\mu,N+1}\right]$$

(A10.6-10)

$$\frac{d\theta}{d\tau} = 2(1+s)\left[\underline{\beta}^1 \bullet \underline{\underline{\chi}}(\underline{C}_{\mu,N+1},\theta)\left(\sum_{k=1}^{N}A_{N+1,k}\Omega(:,k) + A_{N+1,N+1}\underline{C}_{\mu,N+1}\right)\right] - LeBi(\theta - \theta_b)$$

(A10.6-11)

where $\Omega(:,j)$ is the j-th column vector of the matrix $\underline{\underline{\Omega}}$ (eq. A10.6-9).

Appendix 12.1: Laplace Transform for the Finite Kinetic Case

The mass balance equations for the case of finite kinetics and the final equilibrium isotherm is linear are (eqs. 12.5-2 and 12.5-8):

$$\varepsilon \frac{\partial C}{\partial t} + (1-\varepsilon)\frac{\partial C_\mu}{\partial t} \approx \varepsilon D_p \frac{\partial^2 C}{\partial x^2} \qquad (A12.1\text{-}1a)$$

$$\frac{dC_\mu}{dt} = k_a\left(C - \frac{C_\mu}{K}\right) \qquad (A12.1\text{-}1b)$$

where k_a is the rate of adsorption and K is the Henry constant.

The initial and boundary conditions for the case of adsorption and diffusion in an initially clean porous medium are:

$$\begin{aligned} t &= 0; \quad C = C_\mu = 0 \\ x &= 0; \quad C = C_0 \\ x &= L; \quad C \approx 0 \end{aligned} \qquad (A12.1\text{-}1c)$$

Taking the Laplace transform of the mass balance eq. (A12.1-1b), we get:

$$s\overline{C}_\mu = k_a\left(\overline{C} - \overline{C}_\mu / K\right) \qquad (A12.1\text{-}2)$$

from which we obtain the adsorbed concentration in terms of the gas phase concentration as follows:

$$\overline{C}_\mu = \frac{k_a \overline{C}}{s + k_a / K} \qquad (A12.1\text{-}3)$$

Next taking the Laplace transform of eq. (A12.1-1a) and making use of eq. (A12.1-3), we obtain the following second order ODE in terms of the gas phase concentration:

$$\frac{d^2 \overline{C}}{dx^2} - \alpha^2(s) \cdot \overline{C} = 0 \qquad (A12.1\text{-}4)$$

where $\alpha(s)$ is defined as follows:

$$\alpha^2(s) = \frac{\varepsilon s + \dfrac{(1-\varepsilon)sk_a}{s + k_a / K}}{\varepsilon D_p} \qquad (A12.1\text{-}5)$$

The boundary conditions for eq. (A12.1-4) in the Laplace domain are:

$$x = 0; \quad \overline{C} = \frac{C_0}{s}$$
$$x = L; \quad \overline{C} \approx 0 \qquad (A12.1\text{-}6)$$

Solving the ODE of eq. (A12.1-4) subject to boundary conditions (A12.1-6), we obtain the following solution for the gas phase concentration written in terms of the hyperbolic functions:

$$\overline{C} = \frac{C_0}{s} \frac{\sinh[\alpha(s)(L-x)]}{\sinh[\alpha(s)L]} \qquad (A12.1\text{-}7)$$

One can obtain the inverse for the gas phase concentration by the method of residues, but if we are mainly interested in the amount collected at the outgoing surface of the porous medium, we just simply obtain the amount collected in the Laplace domain and then find its inverse by using the method of residues.

The flux at the exit and the amount collected at the exit are given by:

$$J_L = -\varepsilon D_p \left.\frac{\partial C}{\partial x}\right|_L$$

$$Q_L = A \int_0^t J_L \, dt \qquad (A12.1\text{-}8)$$

respectively. Taking the Laplace transform of the amount collected (eq. A12.1-8), we get:

$$\overline{Q}_L = -A(\varepsilon D_p)\frac{1}{s}\left.\frac{d\overline{C}}{dx}\right|_L \qquad (A12.1\text{-}9)$$

Taking the derivative of eq. (A12.1-7) with respect to x and putting the result into eq. (A12.1-9), we get the following solution for the amount collected at the outgoing surface in the Laplace transform domain:

$$\overline{Q}_L = A(\varepsilon D_p)C_0 \frac{1}{s^2} \frac{\alpha(s)}{\sinh[\alpha(s)L]} \qquad (A12.1\text{-}10)$$

The inverse of this function can be found by the method of residues. The poles are zero, and

$$\alpha(s_n)L = jn\pi \qquad (A12.1\text{-}11)$$

To find the residue at the pole zero, we simply find the behaviour of the solution when s approaches zero. Using the Taylor series, we get:

$$\alpha(s) \approx \left[\varepsilon + (1-\varepsilon)K\right]^{1/2} \frac{\sqrt{s}}{\sqrt{\varepsilon D_p}} \left\{ 1 - \frac{1}{2} \frac{(1-\varepsilon)K^2}{k_a\left[\varepsilon + (1-\varepsilon)K\right]} s + \cdots \right\}$$

$$\sinh[\alpha(s)L] \approx L\left[\varepsilon + (1-\varepsilon)K\right]^{1/2} \frac{\sqrt{s}}{\sqrt{\varepsilon D_p}} \left\{ 1 - \frac{1}{2} \frac{(1-\varepsilon)K^2}{k_a\left[\varepsilon + (1-\varepsilon)K\right]} s + \frac{L^2\left[\varepsilon + (1-\varepsilon)K\right]}{6} \frac{s}{\varepsilon D_p} + \cdots \right\}$$

Thus:

$$\frac{\alpha(s)}{s^2 \cdot \sinh[\alpha(s)L]} \approx \frac{1}{Ls^2} \left\{ 1 - \frac{L^2}{6} \frac{\left[\varepsilon + (1-\varepsilon)K\right]}{\varepsilon D_p} s + \cdots \right\}$$

Therefore, the residue corresponding to the zero pole is:

$$\mathrm{Residue}(s=0) = \frac{A(\varepsilon D_p)C_0}{L} \left\{ t - \frac{L^2}{6} \frac{\left[\varepsilon + (1-\varepsilon)K\right]}{\varepsilon D_p} \right\}$$

This is basically the long time bahaviour of the amount collected in the outgoing reservoir.

References

Abbasi, M.H. and J.W. Evans, *AIChEJ*, **29** (1983) 617.
Adamson, A.W., *Physical Chemistry of Surfaces*, InterScience Publishers, New York (1984).
Adzumi, H., *Bull. Chem. Soc. Jap.*, **12** (1937a) 199.
Adzumi, H., *Bull. Chem. Soc. Jap.*, **12** (1937b) 285.
Adzumi, H., *Bull. Chem. Soc. Jap.*, **12** (1937c) 292.
Adzumi, H., *Bull. Chem. Soc. Jap.*, **12** (1937d) 304.
Adzumi, H., *Bull. Chem. Soc. Jap.*, **14** (1939) 343.
Aharoni, Ch. and M.J.B. Evans, *Fundamentals of Adsorption*, Proc. IVth Int. Conf., Kyoto, (1992) 17.
Akanni, K.A. and J.W. Evans, *Chem. Eng. Sci.*, **42** (1987) 1945.
Amankwah, K.A.G. and J.A. Schwarz, *Carbon*, **33** (1995) 1313.
Anderson, R.B., *J. Am. Chem. Soc.*, **68** (1946) 686.
Anderson, J.R. and K.C. Pratt, *Introduction to Characterization and Testing of Catalysts*, Academic Press, Sydney, 1985.
Aranovich, G.L., *Russ. J. Phys. Chem.*, **62** (1988) 1561.
Aranovich, G.L., *Russ. J. Phys. Chem.*, **63** (1989) 845.
Aranovich, G.L., *Russ. J. Phys. Chem.*, **64** (1990) 83.
Aranovich, G.L., *Langmuir*, **8** (1992) 736.
Arnold, K.R. and H.L. Toor, *AIChEJ*, **13** (1967) 909.
Asaeda, M., M. Nakano and R. Toei, *J. Chem. Eng. Jap.*, **7** (1974) 173.
Asaeda, M., S. Yoneda and R. Toei, *J. Chem. Eng. Jap.*, **7** (1974) 93.
Ash, R., R.W. Baker and R.M. Barrer, *Proc. Roy. Soc.*, **A299** (1967) 434.
Ash, R., R.W. Baker and R.M. Barrer, *Proc. Roy. Soc.*, **304A** (1968) 407.
Ash, R., R.M. Barrer and J.H. Petropoulos, *Brit. J. Appl. Phys.*, **14** (1963) 854.
Ash, R., R.M. Barrer, F.R.S. and C.G. Pope, *Proc. Roy. Soc.*, **A271** (1963) 1.
Ash, R., R.M. Barrer, F.R.S. and C.G. Pope, *Proc. Roy. Soc.*, **A271** (1963) 19.
Ash, R. and D.M. Grove, *Trans. Farad. Soc.*, **56** (1960) 1357.
Aylmore, L.A.G. and R.M. Barrer, *Proc. Roy. Soc.*, **A290** (1966) 477.
Barrer, R.M., *Diffusion in and Through Solids*, Cambridge Univ. Press, Cambridge (1941).

Barrer, R.M., *Phil. Mag.*, **35** (1944) 802.
Barrer, R.M., *J. Phys. Chem.*, **57** (1953) 35.
Barrer, R.M., in *Zeolites and Clay Minerals as Sorbents and Molecular Sieves*, Academic Press, London (1978).
Barrer, R.M. and E. Strachan, *Proc. Roy. Soc.*, **A231** (1955) 52.
Barrett, E.P., L.G. Joyner and P.P. Halenda, *J. Phys. Chem.*, **73** (1951) 373.
Barrie, J.A., H.G. Spencer and A. Quig, *J. Chem. Soc. Farad. Trans.* I, **71** (1975) 2459.
Bering, B. P. and V.V. Serpinsky, *Dokl. Akad. Nauk. SSSR*, **148** (1963) 1331.
Bering, B.P., A.L. Myers and V.V. Serpinsky, *Dokl. Akad. Nauk. SSR*, **193** (1970) 119.
Bering, B.P., M.M. Dubinin and V.V. Serpinsky, *J. Colloid Interface Sci.*, **21** (1966) 378.
Bering, B.P., M.M. Dubinin and V.V. Serpinsky, *J. Colloid Interface Sci.*, **38** (1972) 185.
Bhatia, S.K., *Chem. Eng. Sci.*, **41** (1986) 1311.
Bhatia, S.K., *AIChEJ*, **34** (1988) 1094.
Bird, R. B., W. E. Stewart and E.N. Lightfoot, *Transport Phenomena*, John Wiley & Sons, New York (1960).
Biswas, J., D.D. Do, P.F. Greenfield and J.M. Smith, *Applied Catalysis*, **32** (1987) 235.
Blake, F.C. *Trans. AIChE*. **14** (1922) 15.
Bloomquist and Clark, *Ind. Eng. Chem. Anal. Ed.*, **12** (1940) 61.
Brecher, L.E., D.C. Frantz and J.A. Kostecki, *Chem. Eng. Prog. Symp. Ser.*, **63** (1967) 25.
Brecher, L.E., J.A. Kostecki and D.T. Camp, *Chem. Eng. Prog. Symp. Ser.*, **63** (1967) 18.
Broekhoff, J.C.P. and J.H. de Boer, *J. Catal.*, **9** (1967) 14.
Broekhoff, J.C.P. and J.H. de Boer, *J. Catal.*, **9** (1967) 15.
Brown, L.F. and B.J. Travis, *Chem. Eng. Sci.*, **38** (1983) 843.
Brunauer, S., L.S. Deming, E. Deming and E. Teller, *J. Am. Chem. Soc.*, **62** (1940) 1723.
Brunauer, S., P.H. Emmett and E. Teller, *J. Am. Chem. Soc.*, **60** (1938) 309.
Brunauer, S., K.S. Love and R.G. Kennan, *J. Am. Chem. Soc.*, **64** (1942) 751.
Brunauer, S., R.Sh. Mikhail and E.E. Boder *J. Colloid Interface Science*, **25** (1967) 353.
Burganos, V.N. and S.V. Sotirchos, *AIChEJ*, **33** (1967) 1678.
Burgess, C.G.V., D.H. Everett and S. Nuttall, *Langmuir*, **6** (1990) 1734.
Burghardt, A., *Chem. Eng. Process*, **21** (1986) 229.
Carman, P.C., *Dis. Farad. Soc.*, **3** (1948) 72.
Carman, P.C., *Proc. Roy. Soc.*, **A211** (1952) 526.
Carman, P.C., *Flow of Gases through Porous Media*, Academic Press, New York (1956).
Carman, P.C. and F.A. Raal, *Proc. Roy. Soc. London*, **A209** (1951) 38.
Carty, R. and T. Schrodt, *Ind. Eng. Chem. Fundam.*, **14** (1975) 276.
Cen, P.L. and R.T. Yang, *AIChEJ*, **32** (1986) 1635.
Cerro, R.L. and J.M. Smith, *AIChEJ*, **16** (1970) 1034.
Chang, H-C., *AIChEJ*, **29** (1983) 846.
Chen, Y.D., J.A. Ritter and R.T. Yang, *Chem. Eng. Sci.*, **45** (1990) 2877.
Chen, J.S. and F. Rosenberg, *Chem. Eng. Commun.*, **99** (1991) 77.
Chen, S.G. and R.T. Yang, *Langmuir*, **10** (1994) 4244.

Chen, Y.D. and R.T. Yang, *AIChEJ*, **37** (1991) 1579.
Chiang, A.S., A.G. Dixon and Y.H. Ma, *Chem. Eng. Sci.*, **39** (1984) 1461.
Clausing, Von P., *Annal der Physik*, **5** (1932) 961.
Cochran, T.W. and R.L. Kabel and R.P. Danner, *AIChEJ*, **31** (1985) 268.
Cochran, T.W., R.L. Kabel and R.P. Danner, *AIChEJ*, **31** (1985) 2075.
Cohan, L.H., *J. Am. Chem. Soc.*, **60** (1938) 433.
Cohan, L.H., *J. Am. Chem. Soc.* **66** (1940) 98.
Costa, E., J.L. Sotelo, G. Calleja and C. Marron, *AIChEJ*, **27** (1981) 5.
Costa, E., G. Calleja and F. Domingo, *AIChEJ*, **31** (1985) 982.
Costa, C. and A. Rodrigues, in *Adsorption at the Gas-Solid and Liquid-Solid Interface*, edited by J. Rouquerol and K. J.W. Sing, Elsevier, Amsterdam, (1985) 125.
Crank, J., *The Mathematics of Diffusion*, Clarendon Press, Oxford (1975).
Cranston, R.W. and F.A. Inkley, *Advances in Catalysis*, **9** (1957) 143.
Cunningham, R.S. and C.J. Geankoplis, *Ind. Eng. Chem. Fund.*, **7** (1968) 429.
Cunningham, R.S. and C.J. Geankoplis, *Ind. Eng. Chem. Fund.*, **7** (1968) 535.
Cunningham, R.E. and R.J.J. William, *Diffusion in Gases amd Porous Media*, Plenum Press, New York (1980).
Damkohler, G., *Z. Phys. Chem.*, **A174** (1935) 222.
Darken, L.S., *Trans. AIME*, **175** (1948) 184.
Daynes, H.A., *Proc. Roy. Soc.*, **97A** (1920) 286.
De Boer, J.H., *The Dynamical Character of Adsorption*, Clarendon Press, Oxford (1968).
Deepak, P.D. and S.K. Bhatia, *Chem. Eng. Sci.*, **49** (1994) 245.
Dedrick, R.L. and R.B. Breckmann, *Chem. Eng. Prog. Symp. Sre.*, **63** (1967) 68.
Defay and Prigogine, *Surface tension and adsorption*, Longman, London (1966).
Derjaguin, B.C., *Compt. Rend. Acad Sci URSS*, **53** (1946) 623.
Do, D.D., *Chem. Eng. Commun.* **23** (1983) 27.
Do, D.D., *Ind. Eng. Chem. Fund.*, **25** (1986) 321.
Do, D.D., *Chem. Eng. Sci.*, **44** (1989) 1707.
Do, D.D., *Chem. Eng. Sci.*, **45** (1990) 1373.
Do, D.D., in *Equilibria and Dynamics of Gas Adsorption on Heterogeneous Solid Surfaces*, edited by W. Rudzinski, W.A. Steele and G. Zgrablich, Elsevier, (1997) 777.
Do, H.D. and D.D. Do, *Chem. Eng. Sci.* (1998) in press.
Do, D.D. and R.G. Rice, *Chem. Eng. Commun.*, **107** (1991) 151.
Dogu, T. and G. Dogu, *Chem. Eng. Comm.*, **103** (1991) 1.
Dogu, G. and C. Ercan, *Canadian Journal of Chem. Eng.*, **61** (1983) 660.
Dogu, G. and J.M. Smith, *AIChEJ*, **21** (1975) 58.
Doong, S.J. and R.T. Yang, *Ind.Eng. Chem. Res.*, **27** (1988) 630.
Dubinin, M.M., *Chemistry and Physics of Carbon*, **2** (1966) 51.
Dubinin, M.M., *J. Colloid Interface Science*, **23** (1967) 487.
Dubinin, M.M., in *Adsorption - Desorption Phenomena*, edited by Ricca, Academic Press, London (1972) 3.

Dubinin, M.M., *Progress in Surface and Membrane Science*, **9** (1975) 1.
Dubinin, M.M., *Carbon*, **17** (1979) 505.
Dubinin, M.M., I.T. Erashko, O. Kadlec, V.I. Ulin, A.M. Voloshchuk and P.P. Zolotarev, *Carbon*, **13** (1975) 193.
Dubinin, M.M. and L.V. Radushkevich, *Dokl. Akad. Nauk. SSSR*, **55** (1947) 327.
Dubinin, M.M. and H.F. Stoeckli, *J. Colloid Interface Science*, **75** (1980) 34.
Dubinin, M.M. and Timofeev, *Proc. Acad. Sci. USSR*, **54** (1946) 701.
Dubinin, M.M. and E.D. Zaverina, *Zh. Fiz. Khimii*, **23** (1949) 1129.
Dubinin, M.M., Zaverina, E.D. and Serpinsky, V.V., *J. Chem. Soc.* (1955) 1760.
Dubinin, M.M., Zaverina, E.D. and Serpinsky, V.V., *Carbon*, **19** (1981) 402.
Dullien, F.A.L., *Porous Media: Fluid Transport and Pore Structure*, Academic Press, New York, (1979).
Duncan, J.B. and H.L. Toor, *AIChEJ.*, **8** (1962) 38.
Dunne, J. and A.L. Myers, *Chem. Eng. Sci.*, **49** (1994) 2941.
Earnshaw, J.W. and J.P. Hobson, *J. Vacuum Sci. Technol.*, **5** (1968) 19.
Edeskuty, F. and N.R. Amundson, *Ind. Eng. Chem.*, **44** (1952) 1698.
Edeskuty, F.J. and N.R. Amundson, *J. Phys. Chem.*, **56** (1952) 148.
Eiden and Schlunder, *Chem. Eng. Process.*, **28** (1990) 13.
Epstein, N., *Chem. Eng. Sci.*, **44** (1989) 777.
Evans, R.B. III, G.M. Watson and E.A. Mason, *J. Chem. Phys.* **35** (1961) 2076.
Everett, D.H., *Langmuir*, **6** (1990) 1729.
Everett, D.H., *Journal Colloid Interface Science*, **38** (1975) 125.
Everett, D.H. and J.M. Haynes, *Journal Colloid Interface Sciences*, **38** (1972) 125.
Everett, D.H. and J.C. Powl, *J. Chem. Soc. Farad. Trans. I.*, **72** (1976) 619.
Foster, A.G., *Trans. Farad. Soc.*, **28** (1932) 645.
Foster, A.G., *Proc. Roy. Soc.*, **A146** (1934) 129.
Frenkel, J., *Kinetic Theory of Liquid*, Dover, New York (1946).
Freundlich, H., *Trans. Farad. Soc.*, **28** (1932) 195.
Frisch, H.L., *J. Phys. Chem.*, **60** (1956) 1177
Frisch, H.L., *J. Phys. Chem.*, **61** (1957) 93.
Frisch, H.L., *J. Phys. Chem.*, **62** (1958) 401.
Frisch, H.L., *J. Phys. Chem.*, **63** (1959) 1249.
Furusawa,T. and J.M. Smith, *AIChEJ.*, **19** (1973) 401.
Gavalas, G.R. and S. Kim, *Chem. Eng. Sci.*, **36** (1981) 1111.
Gilliland, E.R., R. Baddour, G.P. Perkinson and K.J. Sladek, *Ind. Eng. Chem. Fund.*, **13** (1974) 95.
Gilliland, E.R., R.F. Baddour and J.L. Russell, *AIChE Journal*, **4** (1958) 90.
Glessner, A.J. and A.L. Myers, *Chem.Eng. Prog. Symp. Ser.*, **65** (1969) 73.
Goodknight, R.C. and I. Fatt, *J. Phys. Chem.*, **65** (1961) 1709.
Goodknight, R.C., W.A. Klikoff, Jr., and I. Fatt, *J. Phys. Chem.*, **64** (1960) 1162.
Graham, T. *Phil. Trans. Roy. Soc. London*, **4** (1846) 573.

Graham, T., *J. Quart. Sci.* **2** (1829) 74. Reprinted in *Chemical and Physical Researches*, Edinburgh Univ. Press, Edinburgh (1876) 28.
Graham, T., *Phil. Mag.* (1833) 2, 175, 269, 351. Reprinted in *Chemical and Physical Researches*, Edinburgh Univ. Press, Edinburgh (1876) 44.
Grant, R.J. and M. Manes, *Ind. Eng. Chem. Fundam.*, **5** (1966) 490.
Gregg, S.J. and K.S.W. Sing, *Adsorption, Surface Area and Porosity*, Academic Press, New York (1982).
Gurvitch, L., *J. Phys. Chem. Soc. Russ.*, **47** (1915) 805.
Gusev, V., J.A. O'Brien, C.R.C. Jensen and N.A. Seaton, *Fundamentals of Adsorption*, edited by M.D. LeVan, Kluwer publishers, Boston (1996) 337.
Hacskaylo, J.J. and M.D. LeVan, *Langmuir*, **1** (1985) 97.
Halsey, G., *J. Chem. Phys.*, **16** (1948) 931.
Haq, N. and D.M. Ruthven, *Journal of Colloid and Interface Science*, **112** (1986) 155.
Haq, N. and D.M. Ruthven, *Journal of Colloid and Interface Science*, **112** (1986) 164.
Harkins, W.D. and G. Jura, *J. Chem. Phys.*, **11** (1943) 431.
Haynes, H.W. Jnr., *Chem. Eng. Education*, Winter edition (1986) 22.
Haynes, H.W. and P.N. Sharma, *AIChEJ*, **19** (1973) 1043.
Hazlitt, J.D., C.C. Hsu and B.W. Wojciechowski, *J. Chem. Soc. Farad Trans.*, **75** (1979) 602.
Helfand, E., H.L. Frisch and J.L. Lebowitz, *J. Chem. Phys.*, **34** (1961) 1037.
Helfferich, F.G., *Chem. Eng. Edu.*, Winter Edition, **23** (1992).
Higashi, K., H. Ito and J. Oishi, *J. Atomic Energy Society of Japan*, **5** (1963) 24.
Hill, T.L., *Advances in Catalysis*, **4** (1952) 211.
Hobson, J.P., *Can. J. Phys.*, **43** (1965) 1934.
Hobson, J.P., *Can. J. Phys.*, **43** (1965) 1941.
Hobson, J.P., *J. Phys. Chem.*, **73** (1969) 2720.
Hobson, J.P., *Adv. Coll. Interface Science*, **4** (1974) 79.
Hobson, J.P. and R.A. Armstrong, *J. Phys. Chem.*, **67** (1963) 2000.
Hobson, J.P. and J.W. Earnshaw, *J. Vacuum Sci. Technol.*, **43** (1967) 257.
Honig, J.M. and C.R. Mueller, *J. Phys. Chem.*, **66** (1962) 1305.
Hoogschagen, J., *Ind. Eng. Chem.*, **47** (1955) 906.
Hoogschagen, J., *J. Chem. Phys.*, **21** (1953) 2096.
Horvath, G. and K. Kawazoe, *J. Chem. Eng. Japan*, **16** (1983) 470.
Hu, X. and D.D. Do, *Langmuir*, **10** (1994) 3296.
Huang, T.C. and L.T. Cho, *Chem. Eng. Commun.*, **75** (1989) 181.
Jackson, R., *Transport in Porous Catalysts*, Elsevier, New York (1977).
Jaeger, J.C., *Quart. Appl. Math*, **8** (1950) 187.
Jaeger, J.C., *Trans. Farad. Soc.*, **42** (1946) 615.
Jagiello, J. and J.A. Schwartz, *J. Colloid Inter. Sci.*, **154** (1992) 225.
Jaroniec, M., *Langmuir*, **3** (1987) 795.
Jaroniec, M., X.Lu and R. Madey, *Chemica Script*, **28** (1988) 369.

Jaroniec, M. and R. Madey, *Physical Adsorption on Heterogeneous Solids*, Elsevier, Amsterdam (1988).
Jaroniec, M. and R. Madey, *J. Phys. Chem.* **92** (1988) 3986.
Jolley, L. B. W., *Summation of Series*, Dover, New York (1961).
Kapoor, A. and R.T. Yang, *Gas Separation & Purification*, **3** (1989) 187.
Kapoor, A. and R.T. Yang, *Chem. Eng. Science*, **45** (1990) 3261.
Kapoor, A., J.A. Ritter and R.T. Yang, *Langmuir*, **5** (1989) 1118.
Karavias, F. and A.L. Myers, *Chem. Eng. Sci.* **47** (1991) 1441.
Karger, J. and D.M. Ruthven, *Diffusion in Zeolites and other Microporous Solids*, John Wiley & Sons, New York (1992).
Kaviany, M., *Principles of Heat Transfer in Porous Media*, Springer-Verlag, New York, (1991).
Kawazoe, K. and Y. Takeuchi, *J. Chem. Eng. Japan*, **7** (1974) 431.
Keller, J.U., in *Fundamentals of Adsorption V*, edited by M.D. LeVan, Kluwer Academic Publishers, Boston, Massachusetts (1996) 865.
Kiesling, R.A., J.J. Sullivan and D.J. Santeler, *J. Vac. Sci. Technol.*, **15** (1978) 771.
Kirchheim, R., *Acta Metall.*, **35** (1987) 271.
Knaff, G. and E.U. Schlunder, *Chem. Eng. Process*, **19** (1985) 167.
Knudsen, M., *Annal. der Physil (Leipzig)*, **28** (1909) 75.
Komiyama, H. and J.M. Smith, *AIChEJ.*, **20** (1974) 728.
Koricik, M. and A. Zikanova, *Z. Phys. Chemie, Leibzig*, **250** (1972) 250.
Kozeny, J., S.B. Akad. Wiss. Wien, *Abt. IIa*, **136** (1927) 271.
Kramers, H.A. and J. Kistemaker, *Physica* **10** (1943) 699.
Kraus, G. and J.W. Ross, *J. Phys. Chem.*, **57** (1953) 334.
Kraus, G., J.W. Ross and L.A. Girifalco, *J. Phys. Chem.*, **57** (1953) 330.
Krishna, R., *Chem. Eng. Science*, **45** (1990) 1779.
Kundt, A. and E. Warburg, *Ann. Physik* **155** (1875) 337, 525, *Phil. Mag.* **50** (1875) 53.
Langmuir, I., *J. Am. Chem. Soc.*, **40** (1918) 1361.
Lavanchy, A., M. Stockli, C. Wirz and F. Stoeckli, *Ads. Sci. Tech.* (1996) 537.
Lee, L.K., *AIChEJ*, **24** (1978) 531.
Lee, C.V. and C. Wicke, *Ind. Eng. Chem.*, **46** (1954) 2381.
LeVan, M.D. and T. Vermeulen, *J. Phys. Chem.*, **85** (1981) 3247.
Levenspiel, O., *Engineering Flow and Heat Exchange*, Plenum Press, New York (1984).
Leypoldt and Gough, *J. Phys. Chem.*, **84** (1980) 1058.
Lopatkin, A.A., *Russ. J. Phys. Chem.*, **61** (1987) 431.
Ludolph, R.A., W.R. Vieth and H.L. Frisch, *J. Phys. Chem.*, **83** (1979) 2793.
Ma, Y.H. and C. Mancel, *Adv. Chem. Ser.*, **121** (1973) 392.
Markham, E.D. and A.F. Benton, *J. Am. Chem. Soc.*, **53** (1931) 497.
Masamune, S. and J.M. Smith, *AIChEJ*, **11** (1965) 41.
Masamune, S. and J.M. Smith, *AIChEJ.*, **10** (1964) 246.

Mason, E.A. and A. P. Malinauskas, *Gas Transport in Porous Media: The Dusty Gas Model,* Elsevier, Amsterdam (1983).
Mayfield, P.L. and D.D. Do, *Ind. Eng. Chem. Res.*, **30** (1991) 1262.
McBain, J.W., *Trans. Farad. Soc.*, **14** (1919) 202.
Mehta, S.D. and R.P. Dannes, *Ind. Eng. Chem. Fundam.*, **24** (1985) 325.
Millikan, R.A., *Physical Review*, **21** (1923) 224.
Misra, D.N., *J. Chem. Phys.*, **52** (1970) 5499.
Mohanty, K.K., J.M. Ottino and H.T. Davis, *Chem. Eng. Sci.*, **37** (1982) 905.
Mulder, M., *Basic Principles of Membrane Technology*, Kluwer Academic Publishers, Dordrecht (1991).
Myers, A.L., *Fundamentals of Adsorption I*, edited by A.L. Myers and G. Belfort (1983) 365.
Myers, A.L. and J.M. Prausnitz, *AIChEJ.*, **11** (1965) 121.
Myers, A.L. and D. Valenzuela, *J.Chem. Eng. Jap.*, **19** (1986) 392.
Nakahara, T. M. Hirata and S. Komatsu, *J. Chem. Eng. Data*, **26** (1981) 161.
Nemeth, E.J. and E.B. Stuart, *AIChEJ*, **16** (1970) 999.
Neogi, P. and E. Ruckenstein, *AIChEJ*, **26** (1980) 787.
Neretnieks, I., *Chem. Eng. Sci.*, **31** (1976) 107.
Neufeld, *J. Chem. Phys.*, **57** (1972) 1100.
Nguyen, X.Q., Z. Broz and P. Uchytil, *J. Chem. Soc. Faraday Trans.*, **88** (1992) 3553.
Nitta, T., T. Shigetomi, M. Kuro-Oka and T. Katayama, *J. Chem. Eng. Jap..*, **17** (1984) 39.
Nitta, T. A. Yamaguchi, N. Tokunaga and T. Katayama, *J. Chem. Eng. Jap.*, **24** (1991) 312.
Nitta, T. and A. Yamaguchi, *J. Chem. Eng. Jap.*, **25** (1993) 420.
Nitta, T. and A. Yamaguchi, *Langmuir*, **9** (1993) 2618.
O'Brien, J.A. and A.L. Myers, *J. Chem. Soc. Faraday Trans.*, **80** (1984) 1467.
O'Brien, J.A. and A.L. Myers, *Ind. Eng. Chem. Process Des. Dev.*, **24** (1985) 1188.
O'Brien, J.A. and A.L. Myers, *Ind. Eng. Chem. Res.*, **27** (1988) 2085.
Ochoa-Tapia, A.J., Del Rio P., J.A. and S. Whitaker, *Chem. Eng. Science*, **48** (1993) 2061.
Okazaki, M., H. Tamon and R. Toei, *AIChEJ.*, **27** (1981) 262.
Ozawa, S., S. Kusumi and Y. Ogino, *J. Colloid Interface Science*, **56** (1976) 83.
Patel, P.V. and J.B. Butt, *Chem. Eng. Science*, **27** (1972) 2175.
Pismen, L.M., *Chem. Eng. Sci.*, **29** (1974) 1227.
Pollack, H.O. and H.L. Frisch, *J. Phys. Chem.*, **63** (1959) 1022.
Polte, W. and A. Mersmann, *Fundamentals of Adsorption*, edited by A. Liapis, Santa Barbara, California, May 4-9 (1986).
Raghavan, N.S. and D.M. Ruthven, *Chem. Eng. Sci.*, **40** (1985) 699.
Rasmunson, A., *Chem. Eng. Science*, **37** (1982) 787.
Rayleigh, L. *Proc. Roy. Soc.*, **156A** (1936) 350.
Redhead, P.A., *Langmuir*, **12** (1995) 763.
Reed, E.M. Jr. and J.B. Butt, *J. Phys. Chem.*, **75** (1971) 133.
Reich, R., W. Ziegler and K. Roger, *Ind. Eng. Chem. Proc. Des. Dev.*, **19** (1980) 336.

Reid, R. C., J.M. Praustnitz and B.E. Poling, *The Properties of Gases and Liquids*, McGraw Hill, New York (1983).
Remick, R.R. and C.J. Geankoplis, *Ind. Eng. Chem. Fundam.*, **9** (1970) 206.
Remick, R.R. and C.J. Geankoplis, *Ind. Eng. Chem. Fund.*, **12** (1973) 214.
Remick, R.R. and C.J. Geankoplis, *Chem. Eng. Sci.* **29** (1974) 1447.
Reyes, S. and K.F. Jensen, *Chem. Eng. Science*, **40** (1985) 1723.
Reyes, S. and K.F. Jensen, *Chem. Eng. Science*, **41** (1986) 333.
Reyes, S. and K.F. Jensen, *Chem. Eng. Science*, **41** (1986) 345.
Rice, R.G. and D. D. Do, *Applied Mathematics and Modeling for Chemical Engineers*, John Wiley & Sons, New York (1995).
Richter, E., W. Schutz and A.L. Myers, *Chem. Eng. Sci.*, **44** (1989) 1609.
Rivarola, J.B. and J.M. Smith, *Ind. Eng. Chem. Fund.*, **3**, (1964) 308.
Rogers, W.A., R.S. Buritz and D. Alpert, *Journal of Applied Physics*, **25** (1954) 868.
Ross, S. and J.P. Olivier, *On Physical Adsorption*, InterScience Publishers, New York (1964).
Rothfeld, L., *AIChEJ*, **9** (1963) 19.
Rozwadowski, M. and R. Wojsz, *Carbon*, **22** (1984) 363.
Ruckenstein, E., A.S. Baidyanathan and G.R. Youngquist, *Chem. Eng. Sci.*, **26** (1971) 1305.
Rutherford, S.W. and D.D. Do, *Adsorption*, **3** (1997) 283.
Ruthven, D.M., *Principles of Adsorption and Adsorption Processes*, John Wiley & Sons, New York (1984).
Ruthven, D.M. and M. Goddard, *Zeolites*, **6** (1986) 275.
Rudzinski, W., K. Nieszporek, H. Moon and H.K. Rhee, *Chem. Eng. Sci.*, **50** (1995) 2641.
Rudzinski, W. and D.H. Everett, *Adsorption of Gases on Heterogeneous Surfaces*, Academic Press, San Diego (1992).
Saito, A. and H.C. Foley, *AIChEJ.*, **37** (1991) 429.
Sams, J.R., G. Constabaris and G. Halsey, *J. Phys. Chem.* **64** (1960) 1689.
Samsonov, *Handbook of the Physical and Chemical Properties*, Plenum, New York (1968).
Satterfield, C.N. and H. Ino, *Ind. Eng. Chem. Fundam.*, **7** (1968) 214.
Schneider, P. and J.M. Smith, *AIChEJ*, **14** (1968) 762.
Schneider, P. and J.M. Smith, *AIChEJ*, **14** (1968) 886.
Schull, C.G., *J. Am. Chem. Soc.*, **70** (1948) 1405.
Scott, D.S. and F.A.L. Dullien, *AIChEJ.* **8** (1962) 113.
Scott, D.S. and F.A.L. Dullien, *AIChEJ.* **8** (1962) 293.
Sing, K.S.W., D.H. Everett, R.A.W. Haul, L. Moscon, R.A. Pierotti, J. Rouguerol and T. Siemieniewska, *Pure Appl. Chem.*, **57** (1985) 603.
Sips, R., *J. Chem. Phys.*, **16** (1948) 490-495.
Sircar, S., *Surface Science*, **164** (1985) 393.
Sircar, S., *Carbon*, **25** (1987) 39.
Sircar, S., *Ind. Eng. Chem. Res.*, **30** (1991) 1032.
Sircar, S. and A.L. Myers, *Chem. Eng. Sci.*, **28** (1973) 489.
Sircar, S. and A.L. Myers, *AIChE Symp. Ser.*, **80**, No. 233 (1984) 55.

Slygin, A. and A. Frumkin, *Acta Physicochim. USSR,* **3** (1935) 791.
Smith, D.M., *Ind. Eng. Chem. Fund.* **23** (1984) 265.
Smith, J.M., *Chemical Engineering Kinetics,* McGraw-Hill, New York, (1956).
Smith, J.M., *Chemical Engineering Kinetics,* McGraw-Hill, New York, (1970).
Sotirchos, S.V. and V.N. Burganos, *AIChEJ,* **34** (1988) 1106.
Staudt, R., F. Dreisbach and J.U. Keller, *Fundamentals of Adsorption,* edited by M.D. LeVan, Kluwer (1995).
Steele, W.A., *The Interaction of Gases with Solid Surfaces,* Pergamon Press, Oxford (1974).
Stoeckli, H.F., *J. Colloid Interface Science,* **59** (1977) 184.
Stoeckli, H.F., L. Ballerini and S. De Bernadini, *Carbon,* **27** (1989) 501.
Stoeckli, F. and D. Huguenin, *J. Chem. Soc. Faraday Trans.,* **88** (1992) 737.
Stoeckli, F. and F. Kraehenbuehl, *Carbon,* **19** (1981) 353.
Stoeckli, F. and F. Kraehenbuehl, *Carbon,* **22** (1989) 297.
Stoeckli, F., F. Kraehenbuehl and D. Morel, *Carbon,* **21** (1993) 589.
Stoeckli, F., F. Kraehenbuehl, C. Quellet, B. Schnitter, *J. Chem. Soc. Farad. Trans. 1,* **82** (1986) 3439.
Stoeckli, F., T. Jakubov and A. Lavanchy, *J. Chem. Soc. Farad. Trans.,* **90** (1994) 783.
Sun, L.M. and F. Meunier, *Chem. Eng. Sci.,* **42** (1987) 1585.
Sun, L.M. and F. Meunier, *Chem. Eng. Sci.,* **42** (1987) 2899.
Suwanayuen, S. and R.P. Danner, *AIChEJ.,* **26** (1980) 68.
Suzuki, M., *Adsorption Engineering,* Kodansha, Tokyo (1990).
Suzuki, M. and K. Kawazoe, *Journal of Chemical Engineering of Japan,* **7** (1974) 346.
Szepesy, L. and V. Illes, *Acta Chim. Hung.* **35** (1963) 37, 53, 245.
Talu, O. and R.L. Kabel, *AIChEJ,* **33** (1987) 510.
Talu, O. and A.L. Myers, *AIChEJ.,* **34** (1988) 1887.
Talu, O. and I. Zwiebel, *AIChEJ.,* **32** (1986) 1263.
Taylor, R. and R. Krishna, *Multicomponent Mass Transfer,* John Wiley & Sons, New York (1993).
Testin, R. and E.B. Stuart, *Chem. Eng. Prog. Symp. Ser.,* **63** (1966) 10.
Tomadakis, M.M. and S.V. Sotirchos, *Chem. Eng. Sci.,* **48** (1993) 3323.
Ugrozov, V.V. and P.P. Zolotarev, *Russian Journal of Physical Chemistry,* **56** (1982) 10.
Urano, K., Y. Koichi and J. Nakazawa, *J. Colloid and Interface Science,* **79** (1981) 136.
Valenzuela, D.P. and A.L. Myers, *Adsorption Equilibrium Data Handbook,* Prentice Hall, New Jersey (1989).
Van Amerongen, G.J., *Journal of Applied Physics,* **17** (1946) 972.
Vignes, A., *Ind. Eng. Chem. Fundam.,* **5** (1966) 189.
Villadsen, J. and M.L. Michelsen, *Solution of Differential Equation Models by Polynomial Approximation,* Prentice Hall, New Jersey (1978).
Wakao, N. and S. Kaguei, *Heat and Mass Transfer in Packed Beds,* Gordon and Breach Science Publishers, New York (1982).
Wakao, N. and J.M. Smith, *Chem. Eng. Science,* **17** (1962) 825.

Walker, P.L.Jr., L.G. Austin and S.P. Nandi, *Chemistry and Physics of Carbon*, **2** (1966) 257.
Wang, C-T. and J.M. Smith, *AIChEJ*, **29** (1983) 132.
Weber, S., *Kgl. Danske Videnskab. Selskab. Mat. Fys. Medd.*, **28** (1954).
Wheeler, A., *Catalysis Symposia*, Gibson island conferences, June (1945).
Whitaker, S., *Ind. Eng. Chem. Res.*, **30** (1991) 978.
White, L., *J. Phys. Chem.*, **51** (1947) 644.
Wicke, E., *Kolloid Z.*, **93** (1940) 129.
Wicke, E. and P. Hugo, *Z. Phys. Chem.*, **28** (1961) 401.
Wicke, E. and R. Kallenbach, *Kolloid Z.*, **97** (1941) 135.
Wilson, J. *Am. Chem. Soc.*, **86** (1964) 127.
Yang, R.T., *Gas Separation by Adsorption Processes*, Butterworth, New York, (1987).
Yang, R.T., *Gas Separation by Adsorption Processes*, Imperial College Press, London, (1997).
Yang, R.T., J.B. Fenn and G.L. Haller, *AIChEJ*, **19** (1973) 1052.
Yun, J., H. Park and H. Moon, *Korean J. Chem. Eng.*, **13** (1996) 246.
Zhou, C., F. Hall, K. A. M. Gasem and R.L. Robinson, *Ind. Eng. Chem. Res.*, **33** (1994) 1280.
Zeldowitsch, J.B., *Acta Physicochim. USSR*, **1** (1935) 961.
Zolotarev, P.P. and V.V. Ugrozov, *Russ. J. Phys. Chem.*, **56** (1982) 510.
Zolotarev, P.P., V.V. Ugrozov and I.A. Yabko, *Russ. J. Phys. Chem.*, **58** (1984) 584.
Zolotarev, P.P., V.V. Ugrozov and I.A. Yabko, *Russ. J. Phys. Chem.*, **58** (1984) 1010.

Index

A

Adsorbed film thickness 126
Adsorbent
    Activated carbon 4
    Alumina 3
    Silica gel 3
    Zeolite 6
Adsorption
    Potential 154, 156
    Processes 7
alpha-method 147
Aranovich isotherm 101

B

BDDT classification 94
BET isotherm 84
    Characteristics 89
    Surface area determination 91
BET n-layers 96
Biot number 534, 536
BLK approach 268

C

Capillary 391
    Converging & diverging 359
    parallel capillaries 362, 372, 487
Characteristic energy 155
Chromatography 775
    response quality 784
Cohan equation 117
Cranston-Inkley method 136

D

Darcy equation 374
Darken relation 412
de Boer method 140
Differential adsorption bed 689
    Procedure 690
Diffusion
    Bimodal 634
    Heterogeneous 679
    Knudsen 348
    Molecular 387
    Parallel diffusion 521
    Surface 399
Diffusion cell 755
Diffusivity
    Apparent 523
    Corrected 606
    Combined 525
    Knudsen 354

Pore	522
Surface	354
Dispersive force	151
Distribution function	
Exponential	264
Gamma	264
Gaussian	265
Log-normal	265
Rayleigh	265
Shifted Gamma	265
Uniform	264
Weibull	156
Dubinin-Astakhov	159
Heterogeneity parameter n	161
Theoretical basis	171
Water adsorption	163
Dubinin-Radushkevich	77, 156
Isosteric heat	80, 168
Super-critical adsorbates	162
Dubinin-Serpinski	164
Dubinin-Stoeckli	185

E

Energy distribution	257, 264
Enthalpy of immersion	168
Extended Langmuir equation	191
Equilibrium approach	195
Kinetics approach	191

F

Fast IAS theory	222
Algorithm	231
FHH isotherm	107
Fractional uptake	537, 609
Freundlich	
Isotherm	50
Isosteric heat	57
Frisch method	718

G

Graham law of diffusion	367, 482
Graphite slit pore	
Number density per unit area	286
Number density per unit volume	289
Gurvitch rule	157

H

Harkins-Jura isotherm	31, 103
Heat transfer coeficient	565, 570
Heterogeneous adsorption isotherm	
BLK approach	268
Hobson approach	270
Isosteric heat	265
Langmuir	252
Horvath & Kawazoe method	315
Hypothetical pure component pressure	203
Hysteresis loop	112, 142

I

Ideal Adsorption Solution Theory	198
Algorithm	208
Basic theory	198
Interaction energy	282
one lattice layer	284
one lattice layer with sub-layers	309
one slab	287
two lattice layers	290
two lattice layers with sublayers	310
two slabs	296
Irreversible isotherm	551
Isotherms	
Aranovich	101
BET	84
BET - n layers	96
Dubinin-Astakhov	159
Dubinin-Radushkevich	77

Gibbs	21
Fowler-Guggenheim	26
Freundlich	50
Harkins-Jura	31, 103
Hill-de Boer	24
Jovanovich	82
KST	76
Langmuir	13
Nitta	35
O'Brien-Myers	222
Sips	57
Temkin	82
Toth	64
Unilan	70
Volmer	22
VSM-Wilson	43
VSM-Flory-Huggin	44
Isotherm types	94

K

Kelvin equation	113
Generalized equation	115

L

Langmuir	
Isotherm	13
Isosteric heat	17
LeVan-Vermeulen approach	234
Lewis relationship	205
Loschmidt	449

M

Maxwell-Stefan approach	
Capillary	475
Loschmidt tube	449
Molecular diffusion	415
Molecular-Knudsen diffusion	470
Molecular-Knudsen-viscous flow	495
Non-ideal fluids	462
Stefan tube	431
Two bulbs method	457
Media	
Consolidated	367
Unconsolidated	365, 376
Micropore	
Volume filling	150
Mixture isotherm	
Extended Langmuir	191
IAS theory	198
Potential theory	246
RAS theory	240

P

Pore	
Closed pore	758
Dead end pore	758
Through pore	758
Pore size	
IUPAC classification	2
Distribution of mesopore	119
Distribution of micropore	177, 183

R

Real Adsorption Solution theory	240
Redhead isotherm	108

S

Separation	
Equilibrium	1
Kinetics	1
Steric	1
Sips isotherm	57
Isosteric heat	63
temperature dependence	61

Multicomponent systems	215
Statistical film thickness	126, 137
Stefan tube	343, 431
Surface diffusion	399
Diffusivity	403
Temperature dependence	404
Models	406

T

t-method	143
Thermal velocity	349
Thermodynamics	
Correction factor	606
Surface phase	18
Time lag	701
Topography	
Patchwise	257
Random	257
Tortuosity	338
Tortuosity factor	364
Toth isotherm	64
Isosteric heat	67
Temperature dependence	66

U

Unilan isotherm	70
Isosteric heat	73
Temperature dependence	72

V

Viscous flow	369
Viscous flow parameter	369
Volmer equation	22
Volume filling	150

W

Wheeler-Schull's method	130
Wicke-Kallanbach diffusion cell	344, 755

Z

Zeolite	
Crystal	634
Pellet	634